Advances in the
Science and Technology of
Titanium Alloy Processing

The cover art shows an optical micrograph and an orientation image for a Ti-6Al-2Sn-4Zr-2Mo forging. The micrograph reveals a microstructure of fine equiaxed alpha grains in a matrix of transformed beta. The orientation image, which covers an area 100 times larger than that of the micrograph, reveals the occurrence of microtexture not obvious in the micrograph. Photos courtesy of Dr. Andrew Woodfield, GE Aircraft Engines, Evendale, Ohio.

Advances in the
Science and Technology of
Titanium Alloy Processing

Proceedings of an international symposium
sponsored by the TMS Titanium and Shaping and Forming committees
and held at the 125th TMS Annual Meeting and Exhibition
in Anaheim, California,
February 5-8, 1996

Edited by

I. Weiss

R. Srinivasan

P.J. Bania

D. Eylon

and

S.L. Semiatin

A Publication of

TMS
Minerals • Metals • Materials

A Publication of The Minerals, Metals & Materials Society
420 Commonwealth Drive
Warrendale, Pennsylvania 15086
(412) 776-9000

The Minerals, Metals & Materials Society is not responsible for statements or opinions and is absolved of liability due to misuse of information contained in this publication.

Printed in the United States of America
Library of Congress Catalog Number 97-70338
ISBN Number 0-87339-324-4

Authorization to photocopy items for internal or personal use, or the internal or personal use of specific clients, is granted by The Minerals, Metals & Materials Society for users registered with the Copyright Clearance Center (CCC) Transactional Reporting Service, provided that the base fee of $3.00 per copy is paid directly to Copyright Clearance Center, 27 Congress Street, Salem, Massachusetts 01970. For those organizations that have been granted a photocopy license by Copyright Clearance Center, a separate system of payment has been arranged.

© 1996

If you are interested in purchasing a copy of this book, or if you would like to receive the latest TMS publications catalog, please telephone 1-800-759-4867 (U.S. only), or 412-776-9000, Ext. 270.

PREFACE

This book contains the proceedings of the symposium on "Advances in the Science and Technology of Titanium Alloy Processing," held at the AIME Annual Meeting in Anaheim, California during February 5-8, 1996. The symposium was sponsored by the Titanium and the Shaping and Forming committees of TMS-AIME, and was a part of the regular programming of TMS. Both invited and contributed papers were presented in eight sessions over four days.

Programming for the symposium had great breadth and depth. The subjects ranged from hot and cold working to heat treatment, casting, joining, and machining of conventional titanium alloys, as well as titanium aluminide alloys. In the hot working sessions, processing fundamentals and applications were reviewed for a wide range of conventional titanium alloys and titanium aluminide alloys. The state of the art with respect to constitutive behavior, microstructure evolution, and failure were highlighted. Challenges in making large plan area parts via hot forging were examined by several authors. In the sessions on cold working, new developments, particularly for beta titanium alloys, were reviewed. Among other topics, recent work on cold extrusion, rod drawing, and sheet forming was discussed with respect to both the fundamentals of such processes as well as technological applications which were both aerospace and non-aerospace in nature. As in the metalworking sessions, papers in the heat treatment and joining sessions dealt with a wide range of alloy classes including alpha/beta and beta alloys and gamma and orthorhombic titanium aluminide alloys. New developments in the machining of titanium alloys, notably those based on cryogenic techniques, were also presented and discussed as a means of remedying the longstanding challenges in this area.

The book is divided into four sections, namely, Hot Working, Cold Working, Heat Treatment, and Joining and Casting, representing the four days of programming. All of the papers were reviewed for publication by the session organizers, I. Weiss and R. Srinivasan, (Wright State University), and the co-editors of this volume, D. Eylon (University of Dayton), P.J. Bania (Timet Corporation), and S.L. Semiatin (AF Materials Directorate, Wright Laboratory).

I. Weiss
R. Srinivasan
P.J. Bania
D. Eylon
S.L. Semiatin
Dayton, OH
October, 1996

TABLE OF CONTENTS

PREFACE v

HOT WORKING

HOT WORKING OF TITANIUM ALLOYS - AN OVERVIEW 3
    S.L. Semiatin, V. Seetharaman, I. Weiss

MICROSTRUCTURAL DEVELOPMENT IN A TITANIUM ALLOY 75
    G. Shen, D. Furrer, J. Rollins

REDUCING THE DIMENSIONAL VARIABILITY OF Ti-6Al-4V EXTRUDATES BY
PROCESS MODELING AND DIE DESIGN 83
    D. Damodaran, R. Shivpuri, T. Esposito, K. Galyon

INFLUENCE OF STARTING MICROSTRUCTURE ON THE HIGH TEMPERATURE
PROCESSING OF CP TITANIUM 93
    S. Guillard, M. Thirukkonda, P.K. Chaudhury

MICROSTRUCTURAL EVOLUTION IN Ti-6Al-4V DURING HOT
DEFORMATION 101
    M.B. Gartside

SOLUTE SOFTENING OF ALPHA TITANIUM - HYDROGEN ALLOYS 109
    O.N. Senkov, J.J. Jonas

SOLUTE STRENGTHENING IN BETA TITANIUM - HYDROGEN ALLOYS 117
    O.N. Senkov, J.J. Jonas

MEETING THE CHALLENGE OF HOT WORKING TITANIUM ALLOYS INTO
COST EFFICIENT FINISHED COMPONENTS 125
    G.W. Kuhlman

THE DESIGN, PRODUCTION, AND METALLURGY OF ADVANCED, VERY
LARGE, TITANIUM AEROSPACE FORGINGS 143
    T.E. Howson, R. G. Broadwell

A MECHANISTIC STUDY OF THE HOT WORKING OF CAST GAMMA Ti-Al 153
    D.A. Hardwick, P.L. Martin

EFFECT OF MICROSTRUCTURE ON CAVITATION AND FAILURE BEHAVIOR
DURING SUPERPLASTIC DEFORMATION OF A NEAR-GAMMA TITANIUM
ALUMINIDE ALLOY 161
    C.M. Lombard, A.K. Ghosh, S.L. Semiatin

LAMELLAR TO EQUIAXED GRAIN STRUCTURE BY TORSIONAL
DEFORMATION OF Ti-6Al-2Sn-4Zr-2Mo ALLOY 169
    G. Welsch, I. Weiss, D. Eylon, F.H. Froes

INFLUENCE OF THERMOMECHANICAL PROCESSING ON THE MATERIAL
PROPERTIES OF Ti-6%Al-4%V/TiC COMPOSITES 185
    P. Wanjara, R.A.L. Drew, S. Yue

PRINCIPLES OF TITANIUM ALLOYS STRUCTURE CONTROL WITH THE
PURPOSE OF INCREASING THEIR MECHANICAL PROPERTIES 193
    M. Brun, G. Shachanova

HOT WORKING OF CP TITANIUM 201
    K. Ray, W.J. Poole, A. Mitchell, E.R. Hawbolt

COLD WORKING

COLD FORMING OF TITANIUM ROUNDS AND FLATS 211
    C. Pepka

Ti-22Al-23Nb ALLOY FOILS COLD ROLLED FROM STRIP CAST BY THE
PLASMA MELT OVERFLOW PROCESS 219
    T.A. Gaspar, I.M. Sukonnik

ADIABATIC SHEAR BANDS FORMED BY PUNCHING IN Ti-6Al-4V 225
    Y. Kosaka, J.E. Kosin

THE EFFECT OF RESIDUAL WORK ON THE AGING RESPONSE OF BETA
ALLOYS 233
    P.G. Allen, P.J. Bania

COLD AND WARM WORKING OF LCB TITANIUM ALLOY 241
    I. Weiss, R. Srinivasan, M. Saqib, N. Stefansson, A. Jackson, S.R. LeClair

COLD FORMABILITY OF TIMETAL® 21S SHEET MATERIAL 259
    J. Reshad, I. Weiss, R. Srinivasan, T.F. Broderick, S.L.Semiatin

COLD EXTRUSION OF TITANIUM ALLOYS AND TITANIUM-MMCs 271
    H.W. Wagener, J. Wolf

DEVELOPMENT OF PROCESSING METHODS FOR Ti-13Nb-13Zr 283
    E.W. Robare, C.M. Bugle, J.A. Davidson, K.P. Daigle

TITANIUM ALUMINIDE FOIL PROCESSING 293
    C.C. Wojcik, R. Roessler, R. Zordan

ABRASIVE WATERJET CUTTING OF TITANIUM VENT SCREENS FOR THE
F-22, NEXT-GENERATION AIR SUPERIORITY FIGHTER 301
    H. R. Phelps

TEMPERATURE DISTRIBUTION IN CRYOGENIC MACHINING OF TITANIUM
BY FINITE ELEMENT ANALYSIS 309
    S.Y. Hong, Y. Ding

MANUFACTURE OF FASTENERS AND OTHER ITEMS IN TITANIUM
ALLOYS 319
    V.A. Volodin, I.A. Vorobiov

HYDROGEN TECHNOLOGY AS NEW PERSPECTIVE TYPE OF TITANIUM
ALLOY PROCESSING 331
    B.A. Kolachev, A.A. Ilyin, V.K. Nosov

HYDROGEN INFLUENCE ON MACHINING OF TITANIUM ALLOYS 339
    B.A. Kolachev, Y.B. Egorova, V.D. Talaev

HEAT TREATMENT

HEAT TREATMENT OF TITANIUM ALLOYS: OVERVIEW 349
    R.R. Boyer, G. Lütjering

SELECTION OF HEAT TREATMENT OPTIMUM TECHNOLOGIES FOR
INTRICATELY SHAPED TITANIUM ALLOY ARTICLES 369
    A.N. Lozhko, G.Z. Malkin

PROCESSING, MICROSTRUCTURE, AND PROPERTIES of β-CEZ 379
    J.O. Peters, G. Lütjering, M. Koren, H. Puschnik, R.R. Boyer

PROCESSING OF NANOSTRUCTURED GAMMA TiAl BY MECHANICAL
ALLOYING AND HOT ISOSTATIC PRESSING 387
    F.H. Froes, C. Suryanarayana, N. Srisukhumbowornchai, X. Chen,
    D.K. Mukhopadhyay, M.L. Överçoglu, K. Brand, J. Hebeisen

EFFECTS OF HEAT TREATMENT ON MATRIX MICROSTRUCTURE,
INTERFACIAL REACTIONS AND FATIGUE CRACK GROWTH RESISTANCE
OF TITANIUM METAL MATRIX COMPOSITES 397
    S.V. Sweby, P.Bowen

FLOW BEHAVIOR OF A MECHANICALLY ALLOYED AND HIPPED
NANOCRYSTALLINE γ-TiAl 405
    R.S. Mishra, A.K. Mukherjee, D.K. Mukhopadhyay, C. Suryanarayana,
    F.H. Froes

EFFECT OF COOLING RATE FROM MILL ANNEALING TEMPERATURE ON
STRESS CORROSION THRESHOLD OF TITANIUM 6Al-4V ELI BETA
ANNEALED 413
    R. Briggs

HEAT TREATMENT OF TITANIUM ALLOYS.. 421
    J R. Wood, P. A. Russo

ROLE OF OXYGEN ON TRANSFORMATION KINETICS IN TIMETAL-21S
TITANIUM ALLOY 435
    M.A. Imam, C.R. Feng

PROCESSING - MICROSTRUCTURE- PROPERTY RELATIONSHIPS IN Ni
MODIFIED CORONA 5 451
    P.L. Martin, J. C. Fanning

A STUDY OF GRAIN BOUNDARY NUCLEATED α PRECIPITATES IN AN $\alpha+\beta$
TITANIUM ALLOY 459
    K. Ameyama, H. Fujiwara, H. Kawakami, A. Mitchell

TITANIUM-TANTALUM ALLOY DEVELOPMENT 467
    J.D. Cotton, J.F. Bingert, P.S. Dunn, D.P. Butt, R.W. Margevicius

EFFECTS OF HEAT TREATMENT ON MICROSTRUCTURE, STRENGTH, AND
TOUGHNESS OF TIMETAL-21S 473
    R.K. Bird, T.A. Wallace, W.D. Brewer

ISOTHERMAL DECOMPOSITION OF TIMETAL 21S 481
    Q. Li, R. Crooks

EFFECTS OF MICROSTRUCTURE AND OXYGEN CONTENT ON THE
FRACTURE BEHAVIOUR OF THE $\alpha + \beta$ TITANIUM ALLOY
Ti-4Al-4Mo-2Sn-0.5Si wt.% (IMI 550) 489
    L.J. Hunter, M. Strangwood

PHASE STABILITY IN A Ti-45.5Al-2Nb-2Cr ALLOY 497
    V. Seetharaman, C.M. Lombard, S.L. Semiatin

MICROSTRUCTURE EVOLUTION IN AN ORTHORHOMBIC TITANIUM
ALUMINIDE AS A FUNCTION OF TEMPERATURE-STRAIN-TIME 507
    R.W. Hayes, C.G. Rhodes

HYDROGEN TECHNOLOGY OF SEMIPRODUCTS AND FINISHED GOODS
PRODUCTION FROM HIGH-STRENGTH TITANIUM ALLOYS 517
    A.A. Ilyin, V.K. Nosov, M.Y. Kollerov, A.A. Krastilevsky, S.V. Scvortsova,
    A.V. Ovchinnikov

JOINING AND CASTING

A STUDY OF DIFFUSION-BONDING OF DISSIMILAR TITANIUM ALLOYS 527
    G. Das, C. Barone

EXPLOSION WELDS BETWEEN TITANIUM AND DISSIMILAR METALS 539
    J.G. Banker, V.D. Linse

TITANIUM/STEEL EXPLOSION BONDED CLAD FOR AUTOCLAVES AND
REACTORS 549
    J.G. Banker, A.L. Forrest

LASER WELDING OF Ti-6.8Mo-4.5Fe-1.5Al (TIMETAL® LCB) ALLOY 561
    I. Weiss, N. Stefansson, W.A. Baeslack III, J. Harley, P.G. Allen, P.J. Bania

COATING OF TITANIUM WITH CHROMIUM TO ENABLE PORCELAIN-
TITANIUM BONDING FOR DENTAL RESTORATION 585
    R.R. Wang, G.E. Welsch, O.R. Monteiro, I.G. Brown

THE CORRELATION OF PRIMARY CREEP RESISTANCE TO THE HEAT
TREATED MICROSTRUCTURE IN INVESTMENT CAST Ti-Al GAMMA
ALLOYS 603
    D.Y. Seo, T.R. Bieler, D. E. Larsen

SIMULATION OF MULTICOMPONENT EVAPORATION IN ELECTRON BEAM
MELTING AND REFINING 623
    A. Powell, J. Van Den Avyle, B. Damkroger, J. Szekely

CASTING TECHNOLOGY FOR GAMMA TITANIUM ALUMINIDE VALVES 631
    M.M. Keller, W.J. Porter III, P.E. Jones, D. Eylon

THERMOHYDROGEN TREATMENT OF SHAPE CASTED TITANIUM
ALLOYS 639
    A.A. Ilyin, A.M. Mamonov, Y.N. Kusakina

COMPUTERIZED SPECIFICATION SYSTEM FOR TITANIUM PROCESSING 647
    D. Z. Sokol

SUBJECT INDEX 653

AUTHOR INDEX 659

HOT WORKING

HOT WORKING OF TITANIUM ALLOYS - AN OVERVIEW

S.L. Semiatin,[*] V. Seetharaman,[‡] and I. Weiss[§]

[*] Materials Directorate, Wright Laboratory, WL/MLLN,
Wright-Patterson Air Force Base, OH 45433-7817
[‡] UES, Inc., 4401 Dayton-Xenia Rd., Dayton, OH 45432
[§] Wright State University, Dayton OH 45435

Abstract

The thermomechanical processing of ingot metallurgy titanium alloys is summarized with special emphasis on microstructure evolution and workability considerations for alpha/beta, beta, alpha-two titanium aluminide, orthorhombic titanium aluminide, and gamma titanium aluminide alloys. The conversion of ingot structures to fine equiaxed wrought structures is addressed. In this regard, the breakdown of lamellar microstructures, the occurrence of cavitation/wedge cracking, and the development of crystallographic texture are described. Special methods to breakdown the difficult-to-work titanium aluminide alloys are also summarized. These methods include canned hot extrusion and canned conventional forging. Secondary processes such as bare and pack sheet rolling, superplastic forming of sheet, and closed-die forging are also reviewed. The emergence of non-standard methods for microstructure control, e.g., forging of metastable microstructures, is summarized as well.

Introduction

Titanium alloys are often considered to be among the most difficult-to-process metals. This belief is a result of the necessity to hot work these materials at relatively low temperatures, typically 0.65 to 0.75 of the solidus temperature, in order to control microstructure and to obtain the properties that make them attractive for aerospace and other applications. At such low temperatures, the flow stress of most titanium alloys is fairly high and increases sharply with small decreases in temperature during conventional, nonisothermal deformation processes. Moreover, the hot workability of many titanium alloys decreases rapidly with falling temperature. Thus, the processing window may be quite narrow compared to other engineering materials such as aluminum alloys and steels. For these reasons, net- or near-net-shape processing is usually not possible unless isothermal- or near-isothermal metalworking techniques are utilized. With the development of titanium alloys based on intermetallic phases, such as alpha-two, orthorhombic, and gamma, even greater challenges have had to be met with regard to the design of hot-working processes.

The principal thrust of the present review is to summarize a number of the scientific aspects of the hot working of ingot metallurgy titanium and titanium aluminide alloys. In particular, attention is focused on microstructure and texture evolution, hot workability, and novel thermomechanical processes. The discussion is organized by major alloy subclass, i.e., alpha/beta, beta, alpha-two, orthorhombic, and gamma alloys. Unifying concepts applicable to several alloy subclasses are also discussed.

Hot Working of Alpha/Beta Alloys

In this section, the hot working of alpha/beta alloys is reviewed. At room temperature, commercial alpha/beta alloys typically contain 10 to 20 percent beta phase with the balance being alpha phase (1). So-called near-alpha alloys usually contain less than 10 percent beta. However, because the methods used to control microstructure are similar for both near-alpha and alpha/beta alloys, the hot-working approach is similar for the two subclasses.

The typical ingot metallurgy approach for processing alpha/beta alloys comprises (i) ingot production via vacuum arc melting, (ii) primary ingot breakdown in the beta phase field, often finishing at temperatures below the beta transus (temperature at which alpha+beta $\rightarrow$ beta), (iii) recrystallization annealing in the beta phase field, and (iv) secondary processing in the alpha/beta phase field. Microstructure and texture evolution are a function of the process parameters used for each of these operations and are discussed below.

Vacuum Arc Melting and As-Cast Microstructure

Vacuum arc melting (VAM) is the principal method for producing titanium alloy ingots ranging in size from 75 mm to 1.5 m diameter. The process comprises the melting (and often remelting for refinement purposes) of a consumable electrode by means of a dc arc under vacuum (or a low partial pressure of argon). The molten metal thus produced solidifies in a water cooled copper crucible. The melting and solidification rates are a function of the melting power (i.e., melting current and voltage), the thermophysical properties of the ingot material, and the heat transfer coefficient characterizing the ingot-crucible interface. Among the principal advantages of the process are the removal of dissolved gases such as hydrogen and nitrogen, the minimization of undesirable trace elements having high vapor pressures, and the minimization of macrosegregation (2,3). Grain structures that are produced are strongly dependent on the rate of cooling during solidification of the molten alloy. In general, higher cooling rates result in finer grain structures and more homogeneous as-cast microstructures.

The macrostructure of as-cast alpha/beta titanium ingots melted by VAM usually consists of three zones: a surface region or chilled layer of fine, equiaxed grains produced by the rapid cooling of liquid metal contacting the side and bottom surfaces of the water-cooled copper crucible, a region of large columnar grains (each grain often being several to many thousand microns in dimension) growing inwardly from the chilled layer, and another zone of equiaxed grains lying along the ingot axis (4). This type of solidification pattern often gives rise to a form of macrosegregation, known as beta flecks, in alloys with beta stabilizing elements such as chromium, copper, iron, and manganese (3,4). The ratio of the solute in the solid to that in the liquid, or the partition coefficient, is substantially less than unity for these elements. Hence, these elements partition to the liquid phase during solidification, lower its melting point, and segregate to the center of the ingot. When ingots with such macrosegregation are worked or heat treated just below the nominal beta transus temperature of the alloy, regions of single-phase beta, or beta flecks, with a somewhat lower transus temperature are thus formed. Because of the large scale over which such chemical inhomogeneities are found, their elimination during thermomechanical processing can be quite difficult. On the other hand, Hayakawa, *et al.* (4) have shown that the segregation that causes beta fleck problems can be alleviated by using faster solidification rates and tapered electrodes during VAM processing.

Another structural inhomogeneity often developed in the vacuum arc melting of alpha/beta alloys is the defect known as tree rings which are concentric patterns noticeable in macroetched ingot cross sections (3). The rings outline liquid/solid interfaces of different composition which existed during melting. Small variations in heat transfer, erratic arc behavior, and stirring influence the rate at which the liquid/solid interface advances and hence the tendency to form such composition variations.

Microstructure Evolution During Ingot Breakdown

The coarse and inhomogeneous structure developed during arc melting of alpha/beta titanium alloy ingots is typically broken down by preheating and forging in the single phase beta field. For this purpose, ingots are preheated at a temperature approximately 100°C to 150°C above the beta transus temperature (Table I). Forging itself is usually conducted in hydraulic presses whose deformation rate provides a compromise between very low rates that would lead to excessive die chill (and thus increased flow stress and reduced workability at lower temperatures) and very high rates (and thus increased flow stresses and reduced workability due to strain rate effects). Two types of breakdown forging operations are common for conventional titanium alloys - cogging and upsetting. In cogging, the round ingot is sidepressed between flat dies incrementally along its length. After each bite, the ingot is rotated and indexed forward. By this means, a longer, smaller diameter billet is produced. In some cogging operations, the tooling consists of one vee-die and one flat die. This die design decreases the level of secondary tensile stresses generated during cogging of the ingot and is beneficial for alloys/starting microstructures with limited workability. The second common breakdown operation, upsetting, consists of compressing the ingot along its axis between flat dies.

Table I Beta Transus Temperatures and Recommended Hot-Working Temperatures for Near-Alpha and Alpha/Beta Titanium Alloys (5)

Alloy	Beta Transus Temperature (°C)	Hot Working Temperature Range (°C) Ingot Breakdown	Hot Working Temperature Range (°C) Secondary Working
Ti-5Al-2.5Sn	1040	1120-1175	900-1010
Ti-8Al-1Mo-1V	1040	1120-1175	930-1010
Ti-6Al-4V	995	1095-1150	860-980
Ti-6Al-2Sn-4Zr-2Mo	990	1095-1150	920-975
Ti-6Al-2Sn-2Zr-2Mo-2Cr	980	1070-1130	870-955
Ti-6Al-6V-2Sn	945	1035-1095	845-915
Ti-6Al-2Sn-4Zr-6Mo	940	1030-1090	850-910
Ti-4.5Al-5Mo-1.5Cr	925	1015-1070	850-910
Ti-17 (Ti-5Al-2Sn-2Zr-4Cr-4Mo)	885	990-1050	800-865

Upsetting is often used as a precursor to cogging to impart additional deformation that may be required to obtain a well wrought structure in a prespecified billet size.

During breakdown processing, the ingot loses heat from two main sources-chilling of the hot metal in contact with the cooler dies and radiation to the surroundings. Not surprisingly, the temperature losses are greater for the surface layers of the ingot than those at the center; the temperature at the center may even increase because of deformation heating effects. Breakdown forging is usually controlled so that the final reduction is conducted at temperatures below the beta transus in order to impart a reasonable level of hot work to the ingot for subsequent heat treatment but without exceeding the workability limits of typical alpha/beta titanium alloys.

Because of the temperature variation generated across the ingot section during the breakdown operation, the forged microstructure may also be expected to be somewhat nonuniform with varying amounts of equiaxed (primary) alpha and transformed (Widmanstatten) microstructure across the section. To minimize this variation, breakdown forging is usually followed by a heat treatment at a temperature 50 to 75°C above the beta transus. The stored work in the ingot drives the recrystallization process to completion. Heating time is kept to a minimum to prevent excessive grain growth. Even with careful controls, however, as-heat treated grain sizes are of the order of 500 μm. Following heat treatment, ingots are cooled to room temperature, during which process alpha phase forms on the prior beta grain boundaries and a transformed structure of platelet/basketweave alpha is formed within the beta grains via a Widmanstatten transformation (Figure 1a). Fast cooling rates from the beta field are preferred to minimize the thickness of grain boundary alpha phase, which is deleterious to subsequent subtransus hot workability, and to obtain as fine a transformed structure within the prior beta grains as possible.

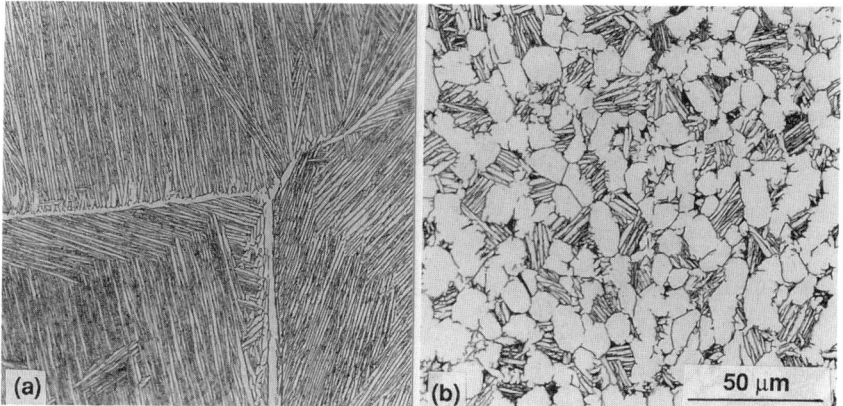

Figure 1 - Wrought microstructures developed in Ti-6Al-2Sn-4Zr-2Mo after hot-working in (a) the beta or (b) the alpha + beta phase field. (6)

Globularization of Widmanstatten Alpha

Following recrystallization heat treatment in the beta field, billet material of alpha/beta alloys is often hot worked below the beta transus to obtain a microstructure of fine, equiaxed (primary) alpha (grain size ≈ 5 to 10 μm) in a matrix of transformed beta (Figure 1b). Hot working is conducted via techniques such as press forging/cogging, radial forging, and extrusion. The kinetics and mechanisms that control the globularization of the platelet, or Widmanstatten, alpha morphology have been investigated by a number of researchers (6-17).

Globularization Kinetics. Using isothermal, hot compression testing of Ti-6Al-2Sn-4Zr-2Mo* with a prior-beta grain size of 400 μm, Semiatin et al. (6) demonstrated that two distinct temperature-strain rate regimes exist with regard to microstructure evolution during subtransus deformation of the Widmanstatten alpha microstructure. For deformation at a strain rate of 10^{-2} s^{-1}, the transition between these is at T_β-60°C (where $T_\beta \equiv$ the beta transus temperature); at a strain rate of 10 s^{-1}, the transition temperature is approximately T_β-20°C. For the lower temperature regime, deformation is highly nonuniform resulting in regions of intense localized shear in which the Widmanstatten colonies breakup and begin to spheroidize after a 50 percent reduction (Figures 2a, b). During subsequent subtransus heat treatment at T ≈ T_β-40°C, the regions of high deformation statically globularize to an equiaxed alpha morphology (Figures 2c, d). For hot working in the high temperature regime, the nonuniform deformation features are not observed after a 50 percent reduction although a variety of coarsened alpha morphologies are produced (Figure 3). Post deformation heat treatment of these latter structures increases the amount of equiaxed alpha.

---

* In this section and later, the compositions of alpha/beta and beta alloys are expressed in *weight* percent, and those of the titanium aluminide alloys are expressed in *atomic* percent.

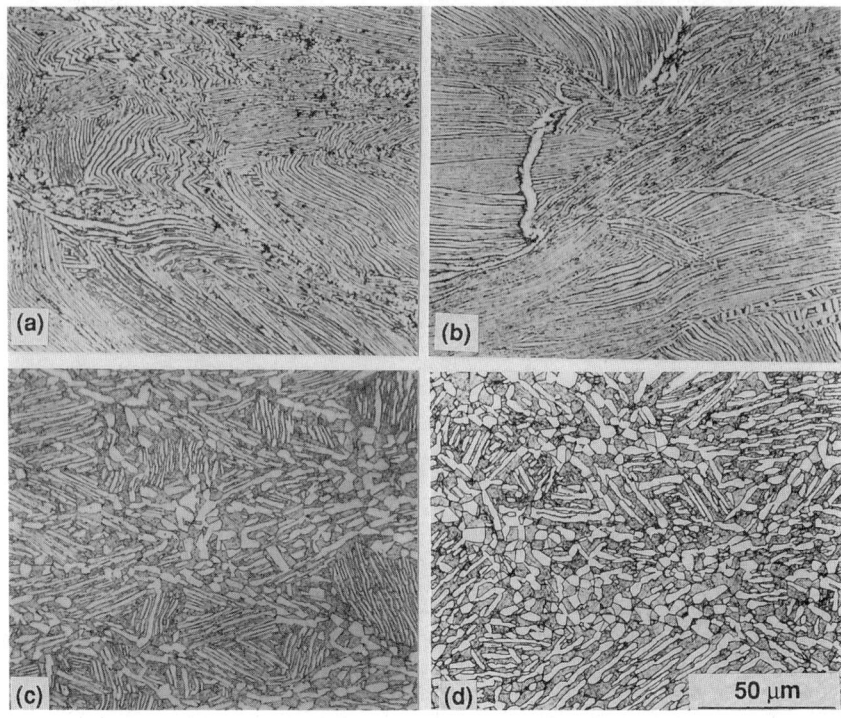

Figure 2 - Microstructures developed in Ti-6Al-2Sn-4Zr-2Mo with a Widmanstatten alpha preform structure after hot compression to a 50 percent reduction at 900°C and a strain rate of (a,c) 10^{-2} s^{-1} or (b,d) 10 s^{-1}. The samples in (c,d) were given a post-deformation heat treatment of 955°C for 2 hours. The compression axis is vertical in all micrographs. (6)

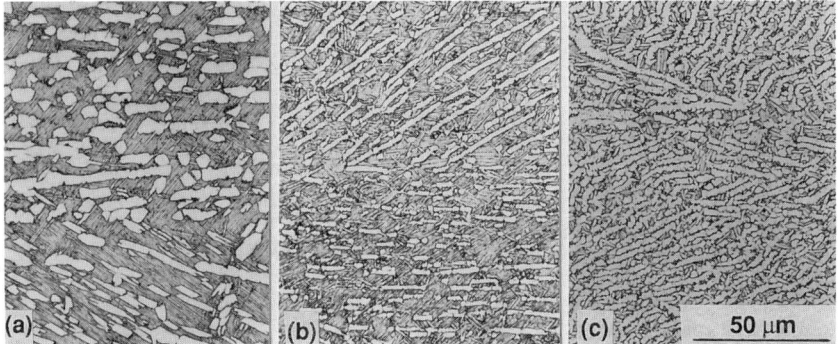

Figure 3 - Microstructures developed in Ti-6Al-2Sn-4Zr-2Mo with a Widmanstatten alpha preform structure after hot compression to a 50 percent reduction at 980°C and strain rates of (a) 10^{-3} s^{-1}, (b) 10^{-1} s^{-1}, or (c) 10 s^{-1}. The compression axis is vertical in all micrographs. (6)

Several investigations have elucidated the effect of strain level and mode of deformation on globularization behavior during hot deformation. For example, Semiatin, et al. (6) showed that after an upset reduction of approximately 65 percent (or an effective strain $\bar{\varepsilon} \approx 1.0$) at $T = T_\beta - 90°C$ and a strain rate of 2×10^{-3} s^{-1}, the Ti-6Al-2Sn-4Zr-2Mo alloy contained an appreciable fraction of essentially undeformed Widmanstatten colonies. A similar result was found by Malcor, et al. (7) for the forging of Ti-6Al-4V at $T_\beta - 30°C$ and a strain rate of 1.4 s^{-1}. In this study, only partial spheroidization was observed after $\bar{\varepsilon} = 1.02$, as in the work of Semiatin, et al., but essentially full globularization had been achieved after $\bar{\varepsilon} = 1.43$.

The work of Rauch, et al. (8) and Korshunov, et al. (9) revealed a strong effect of mode of deformation on globularization. Rauch, et al. torsion tested the same Ti-6Al-2Sn-4Zr-2Mo material used by Semiatin, et al. (6). Hot torsion tests were conducted at $T_\beta - 80°C$ and an effective strain rate of 10^{-2} s^{-1}. In this deformation mode, quite large effective strains, of the order of $\bar{\varepsilon} \approx 5.0$, were required to complete the breakup and dynamic spheroidization of the Widmanstatten microstructure.

In a more wide-ranging effort, Korshunov and his coworkers determined the globularization kinetics of the Russian alloy VT9 (Ti-6.6Al-3.5Mo-1.7Zr-0.27Si) which had been beta annealed and air cooled. Samples were isothermally deformed various amounts at 960°C and an effective strain rate of 10^{-3} s^{-1} via tension, torsion, concurrent tension and torsion, reversed torsion, alternating tension and torsion, and sequential open-die forging along three orthogonal directions. The rate of globularization with strain was found to be greatest for tension and concurrent tension and torsion (Figure 4). Monotonic torsion deformation yielded comparable

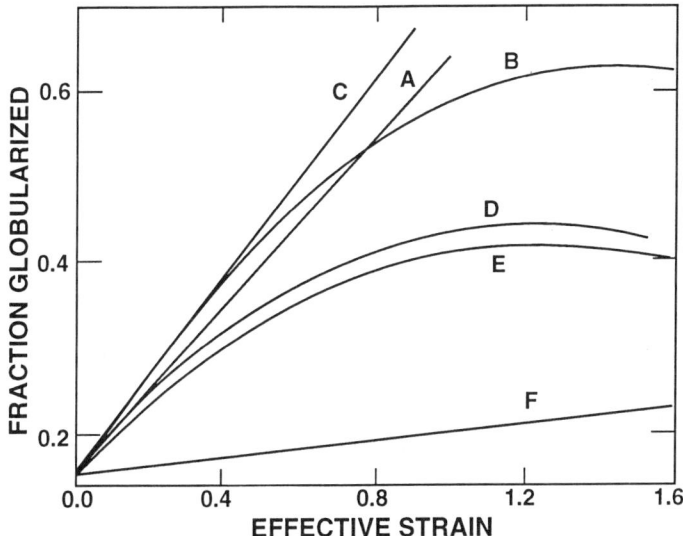

Figure 4 - Fraction globularized as a function of effective strain during hot working of Ti-6.6Al-3.5Mo-1.7Zr-0.27Si (with a Widmanstatten alpha preform structure) at a temperature of 960°C and a strain rate of 10^{-3} s^{-1} via A: tension, B: torsion, C: concurrent tension and torsion, D: reversed torsion, E: alternating tension and torsion, and F: sequential open-die forging along three orthogonal directions. (9)

kinetics at low strains, but the globularization rate decayed noticeably at effective strains of the order of unity. This finding is in qualitative agreement with the results of Rauch, et al. (8). The most surprising result of the Korshunov work was the relatively low rates of globularization for the three non-monotonic loading paths. In particular, the rate was especially low for the sequential, or so-called 'abc', forging process. In fact, the structure was only 50 percent globularized after a total effective strain of approximately 10 in this deformation mode. Although not discussed by Korshunov, this behavior can be rationalized in terms of the possible absence of kinking of the lamellar plates, a phenomenon which the investigators mentioned above have correlated with high local strains that may drive the globularization process. Aggregate theory modeling of the platelet kinking process, such as has been performed by Kad, Asaro, and their coworkers (10,11), may provide insight into these observations.

The recent work of Semiatin and Seetharaman on Ti-6Al-4V (12) confirmed several of the above trends and shed light on the effect of prior-beta grain size on globularization kinetics as well. This research comprised hot compression tests at temperatures between T_β-175°C and T_β-35°C and strain rates between 10^{-3} and $1 s^{-1}$ on samples thermomechanically processed to yield Widmanstatten microstructures with grain sizes of approximately 100 μm or 500 μm. With regard to globularization kinetics, deformation of a 500 μm grain size sample at T_β-35°C and $10^{-3} s^{-1}$, for example, yielded a structure which was approximately 10 percent globularized after a 40 percent reduction ($\bar{\varepsilon}$ = 0.51) and 40 percent globularized after a 65 percent reduction ($\bar{\varepsilon}$ = 1.05) (Figures 5a,b), largely in agreement with the work of Semiatin, et al. (6) and Malcor, et al. (7). By contrast, deformation of a 100 μm grain size sample under identical conditions gave rise to a structure which was approximately 10 percent globularized after 40 percent reduction but 90 percent globularized after 65 percent reduction (Figure 5c, d). Thus, the finer grain size substantially enhanced the kinetics. It is also important to note the relatively small strain increment over which the globularization occurred for the 100 μm samples, a result in broad agreement with the findings of previous workers using coarser grain size samples.

Globularization Mechanisms. The work on Ti-6Al-4V by Margolin, Weiss, and their colleagues (13-17) provides insight into the mechanisms of globularization of Widmanstatten alpha during hot working or during heat treatment following hot working. Margolin and Cohen (13,14) proposed models for the globularization of both the lamellar plates within the prior-beta grains as well as grain-boundary alpha. The mechanism by which the former process proceeds is shown in Figure 6. Initially, recrystallized alpha grains are formed within the alpha plate (Figure 6b). Surface tension requirements do not permit a 180° dihedral angle to exist between the alpha/alpha boundary and the beta plate boundary, however. Thus, a driving force is provided for the movement of some beta phase into the alpha/beta boundaries and a simultaneous rotation of the alpha/beta boundaries toward one another (Figure 6c). Such a rotation enlarges the recrystallized alpha grain to a size larger than the original plate thickness bringing it into contact with the adjacent alpha plate, which may or may not have recrystallized (Figure 6d). The Margolin and Cohen model for globularization of grain-boundary alpha is illustrated in Figure 7. Again, the mechanism is based on recrystallization of alpha within the grain-boundary layer and the surface-tension driven redistribution of beta phase accompanied by pinchoff of the recrystallized-alpha grains and Widmanstatten-alpha side plates.

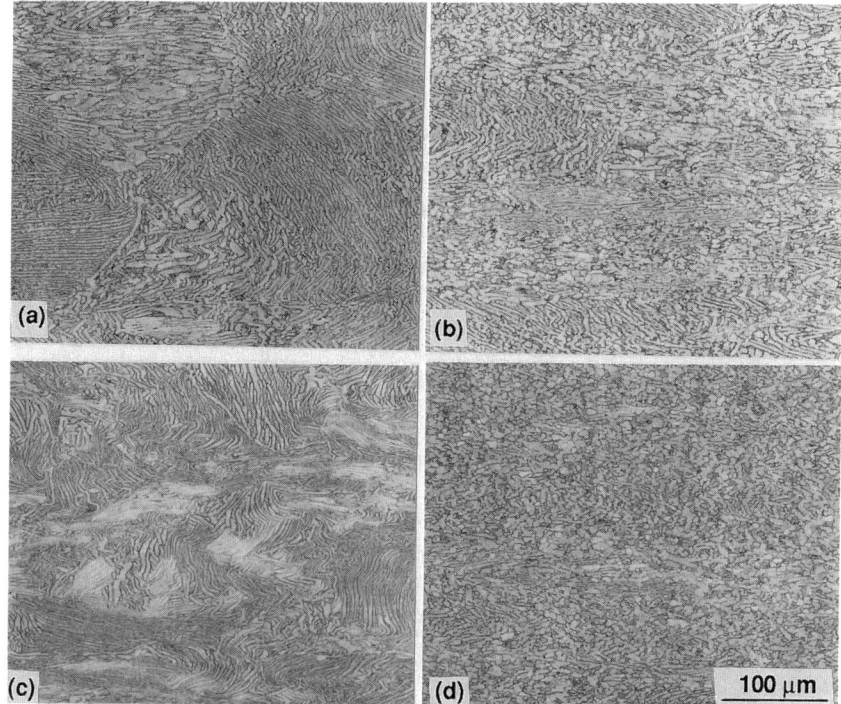

Figure 5 - Microstructures developed in Ti-6Al-4V beta annealed to yield a Widmanstatten alpha microstructure with prior-beta grain sizes of (a,b) 500 μm or (c,d) 100 μm and then hot compressed at 955°C, 10^{-3} s^{-1} to height reductions of (a,c) 40 percent or (b,d) 65 percent. The compression axis is vertical in all micrographs. (12)

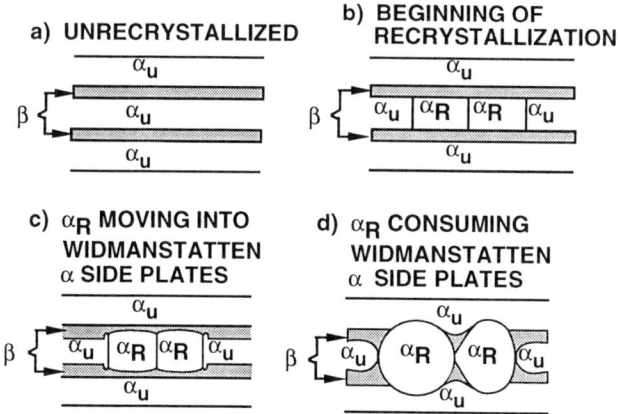

Figure 6 - Schematic illustration of the mechanism of globularization of Widmanstatten alpha in alpha/beta titanium alloys. (13)

a) UNRECRYSTALLIZED

b) BEGINNING OF RECRYSTALLIZATION

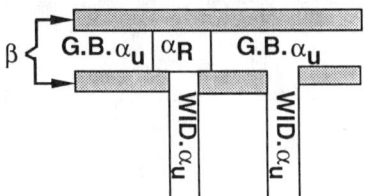

c) RECRYSTALLIZATON PROGRESSING

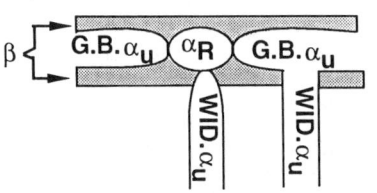

d) α_R REMAINS IN G.B. α STRUCTURE; WID. α IS PINCHED OFF

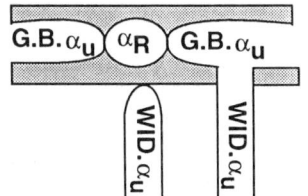

Figure 7 - Schematic illustration of the mechanism of globularization of grain-boundary alpha in alpha/beta titanium alloys. (13)

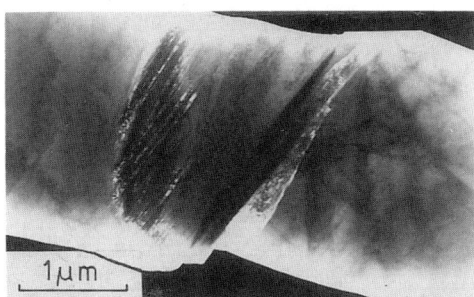

Figure 8 - TEM micrograph showing localized shear deformation of an alpha plate in a subtransus hot-worked sample of Ti-6Al-4V with a Widmanstatten alpha starting microstructure. (15)

The investigations of Weiss, et al. (15-17) expanded upon the work of Margolin and Cohen. From a phenomenological standpoint, Weiss, et al. found that a certain critical strain (dependent on deformation mode) was required to reduce the aspect ratio of the alpha platelets. In contrast to Margolin and Cohen, Weiss, et al. did not invoke recrystallized-alpha grains as the source of alpha pinchoff. Rather, the main driving force was hypothesized to be the formation of intense shear bands within the alpha lamellae (Figure 8), such as may be a result of the kinking-type deformation discussed by Semiatin, et al. (6), or the formation of subgrains spaced at periodic intervals along the length of the lamellae. In either case, the development of such alpha/alpha interfaces in contact with beta platelets gives rise to surface tension driven penetration of the alpha plates by beta phase. The depth of penetration by the beta phase depends on the balance of the interfacial energies of alpha/alpha versus alpha/beta boundaries. For thin alpha plates (3 to 6 μm in thickness), the beta phase was found to penetrate completely. On the other hand, the beta penetration depth was found to be a small fraction of the plate thickness for coarse alpha plates (30 to 40 μm).

The work of Margolin and Weiss suggests that the equiaxed alpha grain size can be refined by hot working of microstructures with thin alpha plates. This approach may be limited to thin section products, however, for which sufficiently fast cooling rates to obtain fine Widmanstatten structures can be imposed.

Microstructure Evolution During Secondary Processing

Following ingot breakdown, wrought alpha/beta titanium alloys are often further hot worked via processes such as open or closed-die conventional forging, extrusion, and plate and sheet rolling (5). The microstructures that are developed depend greatly on the processing temperature relative to the beta transus as well as final heat treatment conditions. Even with careful process controls, however, microstructure variations can be developed within hot worked materials. As with ingot breakdown, the source of such variations can often be related to temperature nonuniformities due to die chilling, radiation heat losses prior to and following hot working operations, and nonuniform cooling following heat treatment. Several cases in which temperature nonuniformities may be particularly important are sheet rolling (for which pack rolling is a frequently used alternative for titanium alloys) and the production of heavy-section forgings (18,19). For example, in heavy section forgings processed below the beta transus, the percentage and size of primary, equiaxed alpha may show noticeable variation from the interior to the outer portions of the parts (19).

The most common type of secondary hot working is conducted in the alpha+beta phase field (20). After such hot working, the structure is characterized by deformed or equiaxed primary alpha in a transformed beta matrix. Optimal control of the microstructure developed during subtransus secondary hot working is obtained for alpha/beta alloys which exhibit a weak-to-moderate variation of alpha phase with temperature. With this goal in mind, so-called beta approach curves have been determined for a range of titanium alloys. The beta approach rate has been found to be a function of both macro- and micro- alloy content. For example, controlled additions of small amounts of the interstitial elements carbon and oxygen have been used to control the beta approach curves for several alpha/beta alloys.

Following alpha+beta hot working, a number of different heat treatments are used to develop various final properties (20-22). These heat treatments include duplex annealing (DA), solution heat treatment and aging (STA), beta annealing (βA), beta quenching (βQ), recrystallization annealing (RA), and mill annealing (MA) (Table II). Properties that can be enhanced by alpha+beta working followed by various heat treatments are summarized in Table III (20).

A less common type of secondary hot working of alpha/beta titanium alloys is beta processing. As its name implies, most, if not all, of the hot work is conducted in the beta-phase field (20). The Widmanstatten or acicular alpha microstructures that are developed by beta working are used to obtain enhanced fracture properties such as toughness, fatigue crack propagation resistance, and creep resistance with only moderate losses in strength and ductility relative to those that characterize equiaxed alpha microstructures (23). This attractive blend of properties is obtained

Table II Final Heat Treatments for Alpha/Beta Titanium Alloys (21)

Heat Treatment Designation	Heat Treatment Cycle	Microstructure
Duplex anneal (DA)	Solution treat at 50-75°C below T_β, air cool, and age for 2-8 h at 540-675°C	Primary α plus Widmanstatten $\alpha + \beta$ regions
Solution treat and age (STA)	Solution treat at ~40°C below T_β, water quench*, and age for 2-8 h at 535-675°C	Primary α plus tempered α' or a $\beta + \alpha$ mixture
Beta anneal (βA)	Solution treat at ~15°C above T_β, air cool, and stabilize at 650-760°C for 2 h	Widmanstatten $\alpha + \beta$ colony microstructure
Beta quench (βQ)	Solution treat at ~15°C above T_β, water quench, and temper at 650-760°C for 2 h	Tempered α'
Recrystallization anneal (RA)	925°C for 4 h, cool at 50°C/h to 760°C, air cool	Equiaxed α with β at grain-boundary triple points
Mill anneal (MA)	$\alpha + \beta$ hot work + anneal at 705°C for 30 min to several hours and air cool	Incompletely recrystallized α with a small volume fraction of small β particles

* In more heavily β-stabilized alloys such as Ti-6Al-2Sn-4Zr-6Mo or Ti-6Al-6V-2Sn, solution treatment is followed by air cooling. Subsequent aging causes precipitation of α phase to form an $\alpha + \beta$ mixture.

only if large reductions in the beta field are utilized to obtain partially unrecrystallized, elongated beta grain macrostructures (23,24).

Details of microstructure evolution during beta forging have been discussed by Uginet (25) and Malcor, et al. (7). For Ti-6Al-2Sn-4Zr-2Mo, Uginet found that the microstructure after a fifty percent reduction $(\bar{\epsilon} = 0.69)$ at $T_\beta+40°C$ consisted of elongated beta grains whose boundaries were decorated by fine recrystallized grains. A similar microstructure was observed by Malcor, et al. in torsion samples of Ti-6Al-4V deformed to $\bar{\epsilon} = 1.2$. Malcor, et al. also noted such a microstructure in hammer forged Ti-6Al-4V pancakes preheated at temperatures no higher than $T_\beta+100°C$. Higher preheat temperatures resulted in structures comprising elongated beta grains alone. These results can be rationalized by the findings of Sheppard and Norley (24) who concluded that alpha/beta alloys dynamically recover during hot working in the beta field, but may subsequently *statically* recrystallize if sufficient hot work has been imposed. A similar conclusion regarding the dynamic softening mechanism during beta working was reached by Liu and Baker (26).

A variant of beta forging is beta extrusion. Okada, et al. (27) produced Ti-6Al-6V-2Sn extrusions with an excellent combination of strength, ductility, fatigue resistance, and toughness by beta extrusion followed by water quenching and heat treatment. The microstructure thus produced comprised fine alpha and beta particles without any acicular alpha phase.

The benefits of low working pressure and good die fill during beta forging can also be realized through a process known as Hydrovac (28). As applied to alpha/beta alloys, the process consists of hydrogenating the alloy, hot forging, and, finally, vacuum annealing to remove the hydrogen. Being a beta stabilizing element,

Table III Property Enhancements via Thermomechanical Processing of Alpha/Beta Titanium Alloys with Standard Compositions (20)

Alloy	Forge Process*	H.T. Process‡	Mechanical Property Effects§				
			Str	K1c	Sm.Fat.	Not.Fat.	FCGR	LCF
Ti-6Al-4V	α+β	MA	----------Baseline----------				
Ti-6Al-4V	α+β	RA	=	+	=	=	=	=
Ti-6Al-4V	β1	MA	-	+	-	≥	+	-
Ti-6Al-4V	β2	MA	-	+	--	+	+	-
Ti-6Al-4V	βHDF	MA	=	+	-	+	+	-
Ti-6Al-4V	α+β	βA	-	+	-	+	++	-
Ti-6Al-4V	β2	Dupl. STAN	=	+	≤	+	+	-
Ti-6Al-4V	α+β	Dupl. STA	+	-	+	=	=	+
Ti-6Al-6V-2Sn	α+β	MA	----------Baseline----------				
Ti-6Al-6V-2Sn	α+β	RA	=	+	=	=	=	=
Ti-6Al-6V-2Sn	β1	MA	-	+	-	≥	+	-
Ti-6Al-6V-2Sn	β2	MA	-	+	--	+	+	-
Ti-6Al-2Sn-4Zr-6Mo	α+β	STA	----------Baseline----------				
Ti-6Al-2Sn-4Zr-6Mo	α+β	Dupl. STA	=	≥	=	=	≥	+
Ti-6Al-2Sn-4Zr-6Mo	α+β	Dupl. βA	=	+	≤	+	+	-
Ti-6Al-2Sn-4Zr-6Mo	β2	STA	=	+	≤	+	+	-
Ti-6Al-2Sn-4Zr-6Mo	α+β	HSTA	+	=	=	=	≥	+
Ti-6Al-2Sn-4Zr-2Mo-0.2Si	α+β	STA	----------Baseline----------				
Ti-6Al-2Sn-4Zr-2Mo-0.2Si	β2	STA	≤	+	-	+	+	-
Ti-6Al-2Sn-4Zr-2Mo-0.2Si	β3	STA	≤	+	-	+	+	-
Ti-6Al-2Sn-4Zr-2Mo-0.2Si	α+β	βA	-	++	-	+	+	-
Ti-6Al-2Sn-4Zr-2Mo-0.2Si	α+β	Dupl. βA	=	++	-	+	++	-

* Forging process codes: α+β - alpha+beta working; β1 - beta preforming only; β2- beta preform/block; β3 - beta finish forging; βHDF - beta hot die forging

‡ Heat treatment process codes: MA - mill anneal; RA - recrystallization anneal; βA- beta anneal; STA/STAN - solution treat and age/anneal; Dupl - duplex heat treatment; HSTA - high solution treatment temperature and age

§ Mechanical property codes: Str - strength; K1c - toughness; Sm. Fat. - smooth high cycle fatigue; Not. Fat. - notched high cycle fatigue; FCGR - fatigue crack growth resistance; LCF - low cycle fatigue

hydrogen lowers the beta transus, thus enabling the use of lower forging temperatures to obtain flow stresses comparable to those of unhydrogenated material. It has also been found that the eutectoid decomposition of beta (beta → alpha + hydride) offers a valuable method to refine the wrought microstructure. A major drawback of the process, however, relates to the basic need to hydrogenate the alloy and thus concomitant section size limitations.

Crystallographic Texture

Because of its large effect on a variety of mechanical properties (29,30), crystallographic textures developed during hot working of alpha/beta titanium alloys have been well documented. These textures are a strong function of deformation mode (e.g., rolling, extrusion), deformation temperature, and percent reduction. For the alpha, or hexagonal, phase component, textures are often represented by (0002) pole figures which depict the spatial distribution of the normals (poles) to the basal planes on stereographic projections. Because of the uncertainty of the crystallite orientation with respect to rotations about the basal pole, a second pole figure (e.g., a pole figure for the poles of the ($10\bar{1}0$) prism planes) is sometimes measured. These latter pole figures are often reported in conjunction with the (0002) pole figures or used to determine ideal orientations (in terms of preferred crystallographic planes and directions) that are plotted directly on the (0002) pole figures.

The two most common texture components for the alpha phase in flat rolled alpha/beta titanium plate and sheet products are the basal and transverse textures (Figure 9). In the basal texture, the basal poles are aligned with the sheet normal; for the transverse texture, the basal poles are parallel to the transverse direction of the sheet. Williams and Stark (21) have summarized the textures developed during unidirectional hot rolling of Ti-6Al-4V (Figure 10). As shown in this figure, the hot-working temperature relative to the beta-transus temperature ($\approx 990°C$ for Ti-6Al-4V) has a strong influence on the deformation textures that are developed. The most noticeable changes occur as the working temperature is gradually increased through the regime in which the proportions of the alpha and beta phases are approximately equal (i.e., $T \approx 900$ to $925°C$). In this manner, the texture can be changed from a combination of basal and transverse to one that is relatively weak to one which is essentially transverse at temperatures just below the beta transus. Hot rolling above the beta transus gives rise to a substantially different texture which is related to the combined effects of the beta phase deformation texture, ($\bar{1}11$) [110], and the decomposition of the beta phase (32). It has also been found that extensive cross-rolling below 900°C gives rise to a simple basal texture. The degree of property anisotropy resulting from the various textures is illustrated in Table IV.

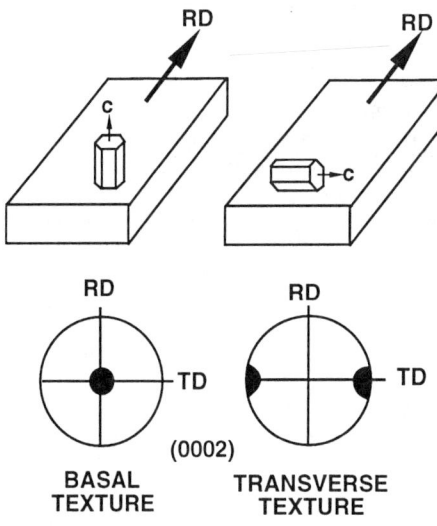

Figure 9 - Schematic illustration of crystallite orientations and the corresponding basal plane ((0002)) pole figures for alpha titanium with (a) a basal texture or (b) a transverse texture. (31)

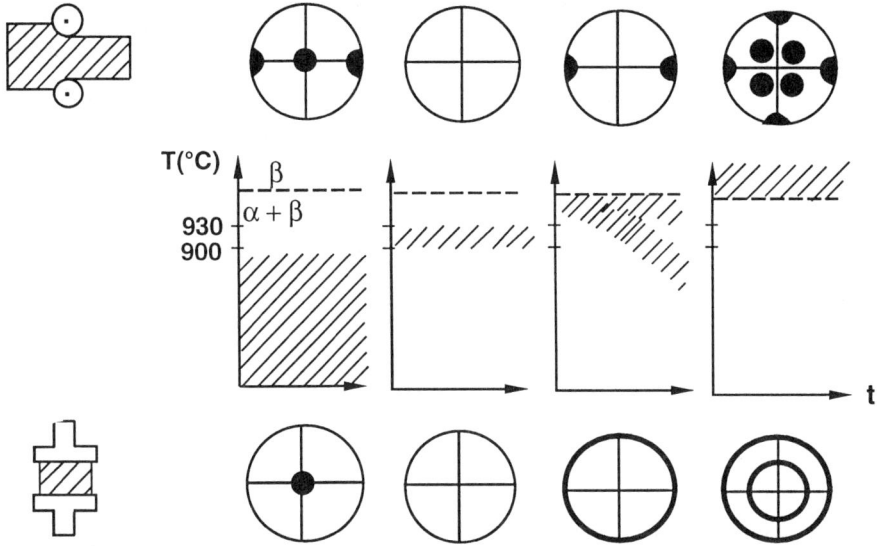

Figure 10 - Schematic illustration of the relationship between working method (flat rolling at top, axisymmetric compression at bottom), working temperature, and alpha-phase texture (in terms of (0002) pole figures) for Ti-6Al-4V. (21)

Table IV Tensile Properties of Ti-6Al-4V Sheet Samples Hot Worked to Develop Different Texture Components (31)

Texture Type	Tension Axis	Young's Modulus (GPa)	0.2 Pct. Yield Strength (MPa)	UTS (MPa)	Fracture Strain
Basal/transverse	RD	107	1120	1650	0.62
Basal/transverse	45° to RD	113	1055	1560	0.76
Basal/transverse	TD	123	1170	1515	0.55
Transverse	RD	113	1105	1540	0.57
Transverse	45° to RD	120	1085	1610	0.76
Transverse	TD	126	1170	1665	0.70
Basal	RD, 45°, TD	109	1120	1505	0.70

Peters and Luetjering (31) have provided a quantitative view of texture development for Ti-6Al-4V plate which was beta annealed, water quenched, and then hot rolled to a seventy-five percent reduction over a wide range of temperatures. Beta annealing and water quenching prior to rolling produced an essentially texture-free material whose microstructure consisted of fine lamellar alpha+beta. Texture measurements after rolling revealed the variation of basal and transverse texture components for either unidirectional rolling (Figure 11a) or cross-rolling (Figure 11b). For unidirectional rolling, deformation at low subtransus temperatures gave rise to strong basal and transverse texture components in the alpha, or predominant, phase. The intensity of both components decreased as a rolling temperature of 900°C was approached. Above this temperature, the development of the transverse texture component in the alpha phase was correlated to the development of a strong beta phase texture, the absence of dynamic recrystallization of beta, and the subsequent decomposition of the deformed beta. When cross-rolling was conducted, the transverse texture component was almost totally suppressed at low temperatures, giving rise to a symmetric basal texture (Figure 11b). For rolling at temperatures above 900°C, cross-rolling produced a transverse texture similar to that from unidirectional rolling.

Based on the above results, Peters and Luetjering (31) were perhaps the first to propose a method for the independent control of texture and microstructure in hot rolled alpha/beta alloys. The type of texture is primarily determined by the type of rolling (unidirectional vs. cross rolling) and rolling temperature. As long as the level of residual hot work retained after rolling is high enough, a variety of equiaxed alpha and bimodal (approximately equal proportions of alpha and transformed beta) microstructures can then be obtained by varying the subsequent subtransus annealing time and temperature. For example, for Ti-6Al-4V, short annealing times at 800°C produce a fine equiaxed alpha structure, long annealing times at 800°C yield a coarse equiaxed structure, and short annealing times at 955°C followed by water quenching and annealing at 800°C produce a bimodal microstructure. As

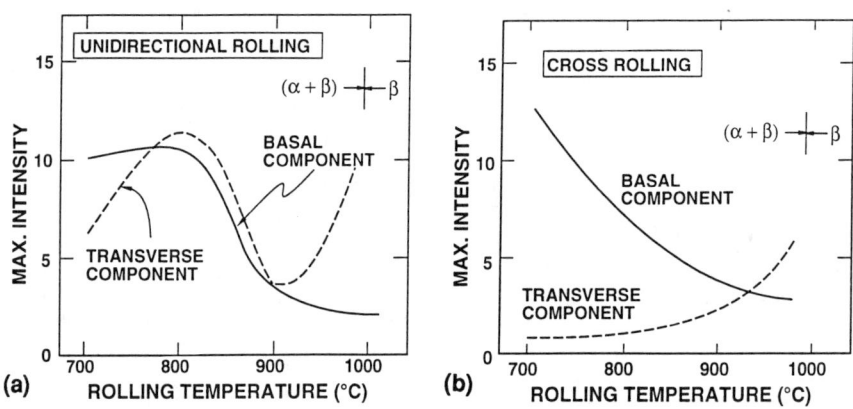

Figure 11 - Maximum intensities of the basal and transverse texture components developed in the alpha-phase during hot-rolling to a 4:1 reduction of prior-beta-annealed-and-water-quenched plate of Ti-6Al-4V. Deformation consisted of either (a) unidirectional rolling or (b) cross rolling. (31)

mentioned earlier, annealing above the beta transus tends to give a random texture (31,33) and a transformed microstructure whose morphology is determined by cooling rate from the annealing temperature and any subsequent subtransus heat treatments.

Deformation modes other than rolling produce their own distinctive textures as well. For example, textures that are rotationally symmetric (i.e., "wire" textures) are produced by axisymmetric hot-working methods such as pancake forging and round-to-round extrusion. In these cases, pole figures typically consist of circularly symmetric distributions of basal poles, such as are shown in the bottom portion of Figure 10 for upsetting types of deformations (21). As for rolling, the specific texture components are very dependent on hot-working temperature.

Although pole figures provide a global, or 'average', measure of crystallographic texture, more microscopic measures of preferred orientations, or so-called microtextures, are now being sought (34,35). For alpha/beta titanium alloys, these microtextures are a result of the Burgers relations between the beta and alpha phases. Thus, a given beta grain would provide a certain orientation of its alpha and beta plates during subtransus decomposition. In the absence of recrystallization, the breakdown and globularization of these plates would then result in a similar orientation of all of the primary alpha grains within a given prior-beta grain. Hence, depending on the prior-beta grain size, a wrought product might contain large regions of a specific orientation (Figure 12). It is now thought that certain microtexture components in wrought alpha/beta titanium alloys might provide easy fracture initiation or propagation sites and therefore deleteriously affect properties such as fatigue performance.

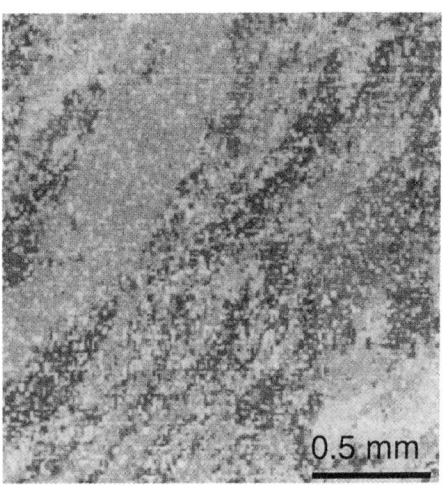

Figure 12 - Representation of microtexture developed in a thermomechanically processed sample of Ti-6Al-2Sn-4Zr-2Mo. (34)

Isothermal Forging and Superplastic Sheet Forming

Because of their large commercial significance, isothermal forging (IF) and superplastic sheet forming (SPF) of alpha/beta titanium alloys deserve special mention in this review. Both processes rely on the superplastic nature of the fine, equiaxed alpha microstructures developed during subtransus deformation processing. These microstructures provide low flow stresses ($\bar{\sigma}$) and high values of the strain rate sensitivity index $\left(m = \partial \log \bar{\sigma} / \partial \log \dot{\bar{\varepsilon}}\right)$. High values of m provide excellent die-filling capability (IF) and resistance to strain localization (IF and SPF). Because of these attributes, complex parts can be made to net or near-net shape thus reducing final machining/material costs and a large number of assembly operations. The discussion below is divided into two subsections - plastic flow phenomena during IF and SPF and novel thermomechanical processes developed for IF and SPF.

Plastic Flow of Superplastic Alpha/Beta Titanium Alloys.

The superplastic properties of alpha/beta alloys are a result of the fine, two-phase microstructure of equiaxed alpha and beta grains. At appropriate strain rates, these microstructures deform by diffusion-controlled, grain boundary sliding thus giving rise to low flow stresses ($\bar{\sigma}$) and high values of the strain rate sensitivity index (m) as well as deformed microstructures which retain much of their original equiaxed nature (36). The presence of two phases helps to limit the growth of either to a large extent, thereby enabling the retention of the plastic properties to large strains.

The data in Figure 13 illustrate the flow behavior of various wrought titanium alloys (with equiaxed microstructures) and the benefits that can be accrued by working at low strain rates, such as are feasible under IF conditions. As an example, if Ti-6Al-4V were conventionally hot forged at 940°C, chilling of the workpiece surface to a temperature of only 870°C could lead to severe deformation inhomogeneities because the flow stress at this lower temperature is about three times that at the nominal forging temperature. This effect is eliminated by using isothermal forging. Figure 13 also shows that a decrease in strain rate by a factor of a hundred or a thousand, which is possible in IF, leads to a substantial reduction in forging pressures for Ti-6Al-4V. Hence, for a given press capacity, parts can be forged isothermally at low strain rates and at workpiece preheat temperatures lower than those used in higher strain rate conventional forging even if die chilling could be avoided. Of course, due to the economics associated with production rate, there is a limit to how low the IF forging rate can be.

Spurred by the desire to maximize the processing rate, a number of investigators have studied in detail the flow phenomenology of alpha/beta titanium alloys with equiaxed microstructures. Figure 14 shows the flow stress dependence on strain rate for Ti-6Al-4V as measured by Lee and Backofen (38); these data were obtained for sheet material, but the trends are similar for forging bar stock (39). The sigmoidal nature of the plots is most marked at 900 to 950°C. For temperatures in this range, the $\log \bar{\sigma}$ - $\log \dot{\bar{\varepsilon}}$ plots are steepest at $\dot{\bar{\varepsilon}} \approx 10^{-4}$ s^{-1}. This observation is confirmed by the m value data in Figure 15. Thus, the most favorable conditions for IF or SPF of Ti-6Al-4V are T = 925°C, $\dot{\bar{\varepsilon}} = 10^{-4}$ s^{-1}.

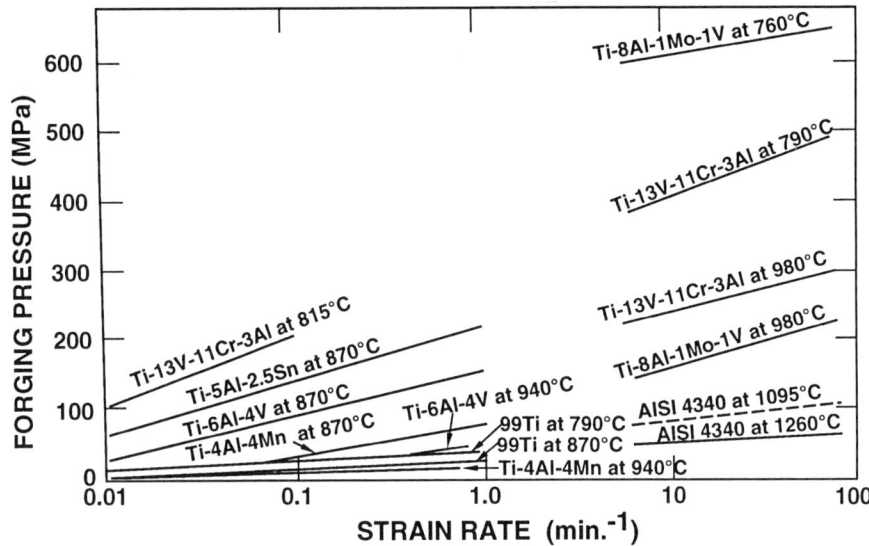

Figure 13 - Effect of strain rate on forging pressures at various forging temperatures for several titanium alloys and AISI 4340 steel. (37)

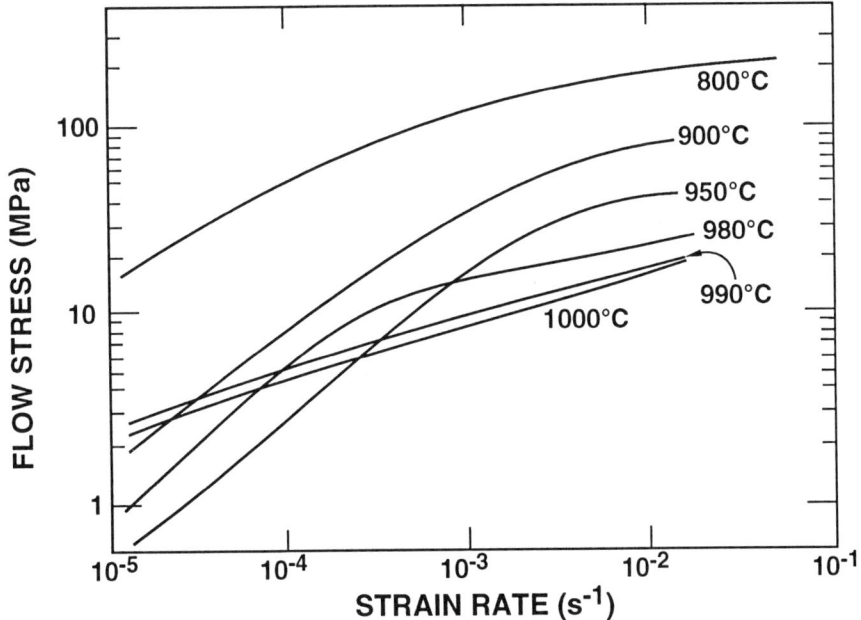

Figure 14 - Flow stress dependence on strain rate and temperature for Ti-6Al-4V sheet. (38)

The flow stress dependence on temperature and strain rate for several titanium alloys has also been explored by Sastry and his coworkers (40). Besides measuring similar flow stress data for Ti-6Al-4V, these researchers confirmed and expanded upon the findings of Lee and Backofen (38) pertaining to the temperature at which maximum m is obtained. As shown in Figure 16, this temperature was found to be 915°C, independent of strain rate. This temperature corresponds to that at which the alpha and beta phase proportions are approximately equal. In view of this observation, Sastry, et al. hypothesized that a similar dependence of flow stress on temperature and strain rate, and hence m values, would be obtained in other alpha/beta alloys when the volume fractions of the two phases were equal. To a first order, these proportions are a function of T/T_β, where T denotes the test temperature, and T_β is the beta transus temperature. Thus, it should be possible to normalize flow stress data versus T/T_β. This is indeed what was found for Ti-8Al-1Mo-1V, Ti-3Al-2.5V, and Ti-6Al-4V (Figure 17). Because m for Ti-6Al-4V is a maximum at $T = 0.93\ T_\beta$, maximum m's are expected for other alpha/beta alloys at an equal fraction of their beta transus temperatures.

Sastry, et al. (40) also verified the effect of alpha grain size on the flow stress and m value dependence on strain rate in alpha/beta alloys such as Ti-6Al-4V. At strain rates of the order of 10^{-4} s^{-1}, a difference in alpha grain size of a factor of 2.5 led to differences in flow stress of about a factor of 2 and a difference in m values of about a factor of fifty percent (Figure 18). At higher strain rates, at which grain boundary sliding and grain boundary diffusion phenomena are less pronounced, the flow stress and m values were found to be almost independent of alpha grain size.

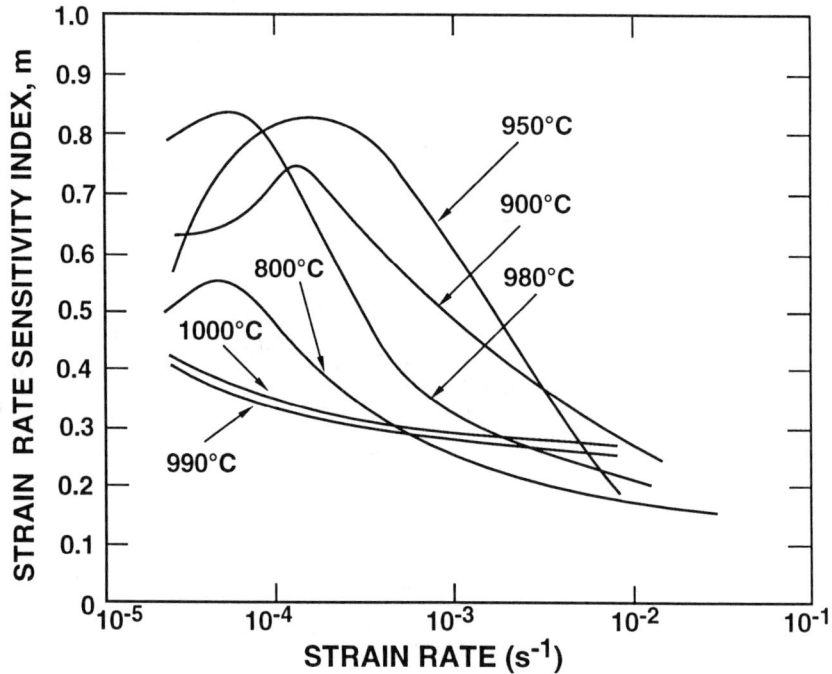

Figure 15 - Dependence of the strain-rate-sensitivity index ('m value') of Ti-6Al-4V sheet on strain rate and temperature. (38)

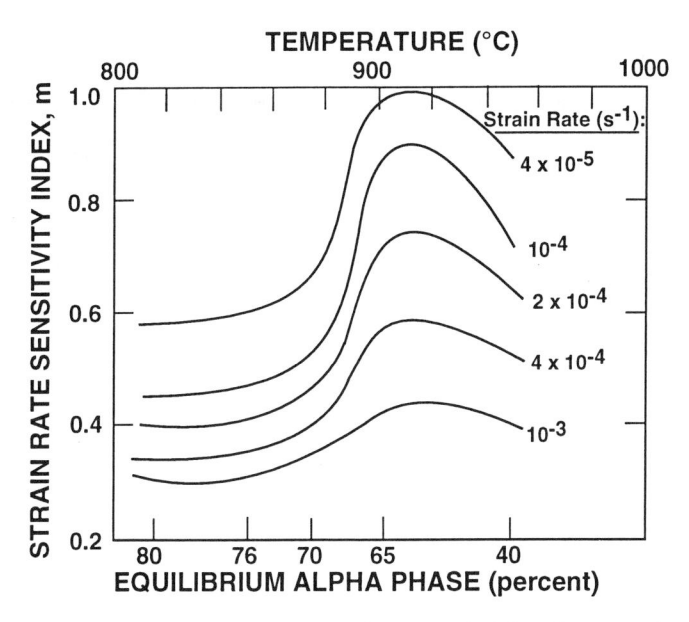

Figure 16 - Effect of temperature and volume fraction of alpha phase on the strain-rate-sensitivity index of Ti-6Al-4V. (40)

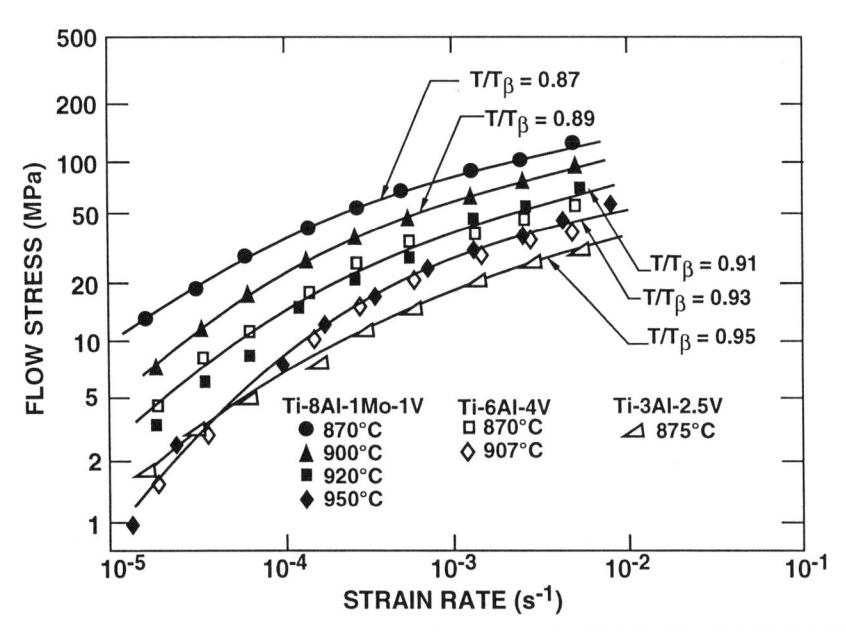

Figure 17 - Strain rate dependence of the flow stress for Ti-8Al-1Mo-1V, Ti-6Al-4V, and Ti-3Al-2.5V, showing the unique relationship between flow stress at constant strain rate and T/T_β. (40)

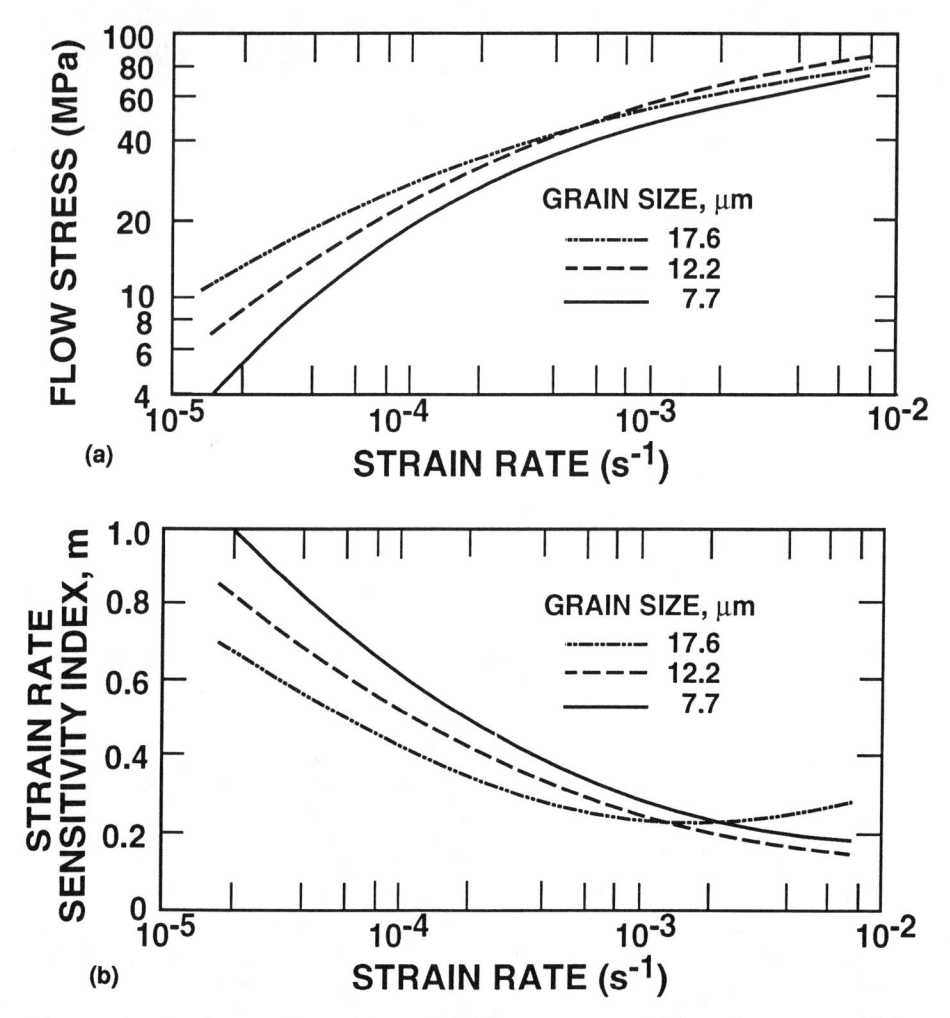

Figure 18 - Strain rate dependence of (a) flow stress and (b) strain-rate-sensitivity index for Ti-6Al-4V with various initial alpha grain sizes tested at 850°C (volume percent primary, equiaxed alpha = 88). (40)

Because of the very strong dependence of m values on grain size, static and dynamic grain size changes during processing can significantly affect ductility during SPF and, to a lesser extent, die filling capability during IF. Rosenblum, Furushiro, Hamilton, and their coworkers (41-44) have documented such effects. The most frequent occurrence is dynamic grain growth, as demonstrated by Hamilton and Ghosh (44) (Figure 19); however, under certain conditions, grain refinement and m increasing with strain have been documented (42,43).

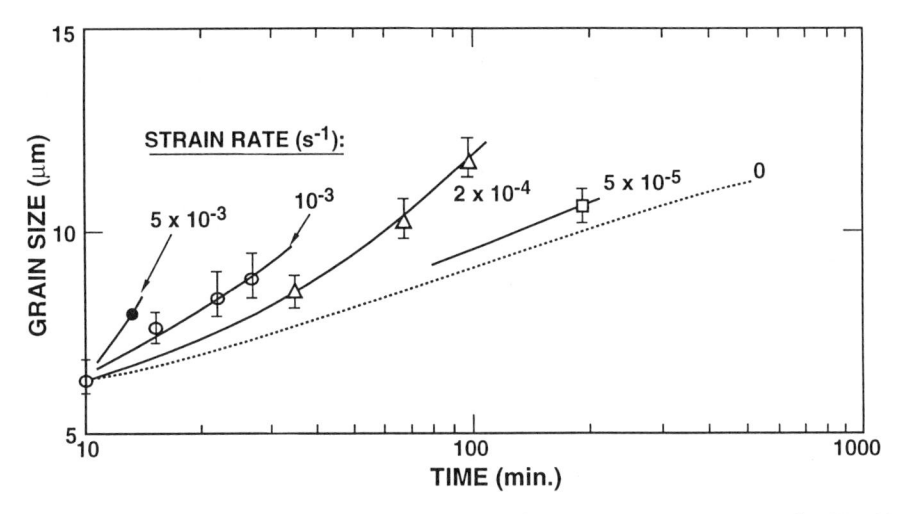

Figure 19 - Comparison of dynamic and static alpha grain growth kinetics for Ti-6Al-4V with an initial grain size of 6.4 μm tension tested at 927°C. (44)

Novel Thermomechanical Processes. A number of novel thermomechanical processes have been developed to enhance or utilize the superplastic properties of alpha/beta titanium alloys. Guided by the work of Weiss, et al. (15-17) on globularization of lamellar microstructures, Inagaki (45) developed a warm working method to refine the alpha grain size in alloys such as Ti-6Al-4V, thereby reducing flow stress and substantially increasing elongation. In brief, plate samples of Ti-6Al-4V, Ti-6Al-2Sn-4Zr-2Mo, and Ti-6Al-2Sn-4Zr-6Mo were beta annealed at 1050°C and then water quenched to produce an acicular α' martensite (Figure 20a). The plates were then warm rolled at 750°C to a reduction of 95 percent. The resulting structures contained very fine ($\leq$ 2 μm) alpha grains (Figure 20b). This microstructure yielded tensile elongations of greater than 2000 percent and flow stresses of approximately 35 MPa at 850°C and a strain rate 10^{-2} s^{-1}. Elongations were lower, however, when tension testing was conducted at a strain rate of 10^{-3} s^{-1} or when test samples were given a heat treatment at a temperature between 750°C and 1000°C prior to testing; both of these effects can be attributed to coarsening of the microstructure, either dynamically or statically.

To determine the effect of starting microstructure on warm-rolled structure and subsequent plastic properties, Inagaki also rolled as-received, commercial Ti-6Al-4V plate to a 95 reduction at 750°C. In this material, the starting microstructure comprised elongated alpha, 15 to 25 μm in size, and some transformed beta (Figure 21a). After rolling at 750°C, the alpha grains were thinned and elongated in the rolling direction; however, the deformation was very nonuniform with heavy shear banding being evident (Figure 21b). This considerably coarser, nonuniform structure had a flow stress of 43 MPa and total elongation of only 825 percent when tested at 850°C and a strain rate of 10^{-2} s^{-1}. These observations clearly indicate the

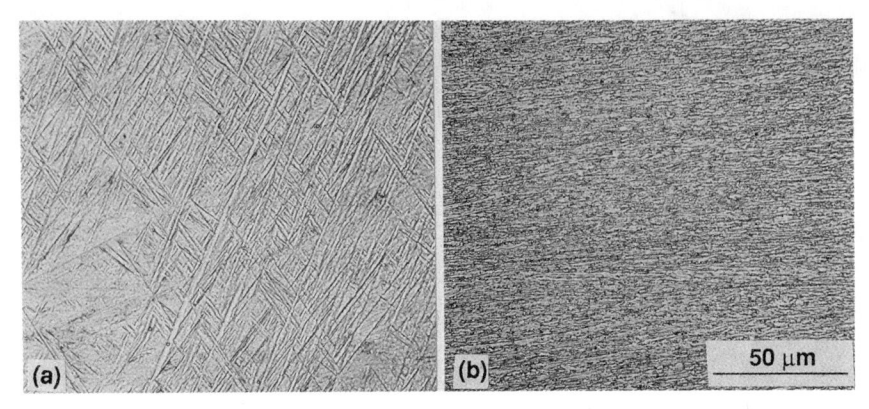

Figure 20 - Microstructures in Ti-6Al-4V developed via (a) beta annealing and water quenching of hot-rolled plate, or (b) beta annealing, water quenching, and warm rolling at 750°C to a 95 percent thickness reduction. (45)

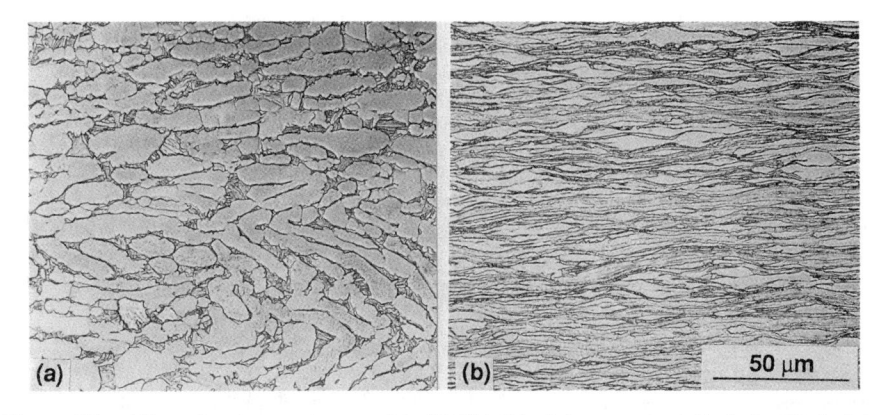

Figure 21 - Microstructures observed in Ti-6Al-4V: (a) commercially hot-rolled plate or (b) commercially hot-rolled plate subsequently warm rolled at 750°C to a 95 percent thickness reduction. (45)

effect of initial lamellar microstructure on the grain size of the broken down / globularized structure and can be rationalized in the context of the work of Weiss, et al. (15-17).

In work similar to that of Inagaki, Peters and Luetjering (31) demonstrated that a fine globularized microstructure (alpha grain size ≈ 2 μm) could be obtained in plate and sheet material of Ti-6Al-4V by a process consisting of beta annealing, water quenching, rolling at 800°C to $\bar{\varepsilon} = 0.9$, and final annealing at 800°C for one hour. Higher rolling temperatures or longer annealing times gave considerably coarser (~ 12 μm) alpha grain sizes.

In related work, Brantingham and Thomas (46) investigated the effect of concurrent dynamic globularization on superplastic characteristics. Tension tests were conducted at 925°C on Ti-6Al-4V sheet samples in the as-hot-rolled condition or after annealing treatments at temperatures between 785°C and 900°C. In the as-hot rolled condition, the microstructure contained a highly elongated alpha phase (approximately 1 µm thick x 10 to 20 µm long). Samples of this material had noticeably higher m values than those samples tested after heat treating (Figure 22). The lower m's in the annealed samples were attributed to globularization and alpha grain growth during annealing or preheating prior to tension testing. Such microstructural changes were limited during the testing of the as-hot-rolled material condition. In addition, dynamic globularization in this condition may have provided a source of microstructure refinement and therefore resulted in higher m values.

Several methods have been utilized to reduce the processing time for superplastic sheet forming operations on alpha/beta titanium alloys. These include thermomechanical processing such as that developed by Inagaki (45) to obtain very fine grain sizes which give rise to high m's and high elongations at higher than typical strain rates. Another approach is that developed by Hamilton and his coworkers (47) in which forming pressures and strain rates are continuously adjusted to minimize overall processing time. In its simplest form, the technique involves initially high strain rates at which the strain hardening rate is high, thereby providing flow stability. At high deformation levels at which strain hardening has decayed to a low value, the deformation rate is decreased resulting in higher m values and sufficient flow stability. By this means, the overall processing time is decreased relative to that required if the process had been conducted entirely at the lower strain rate.

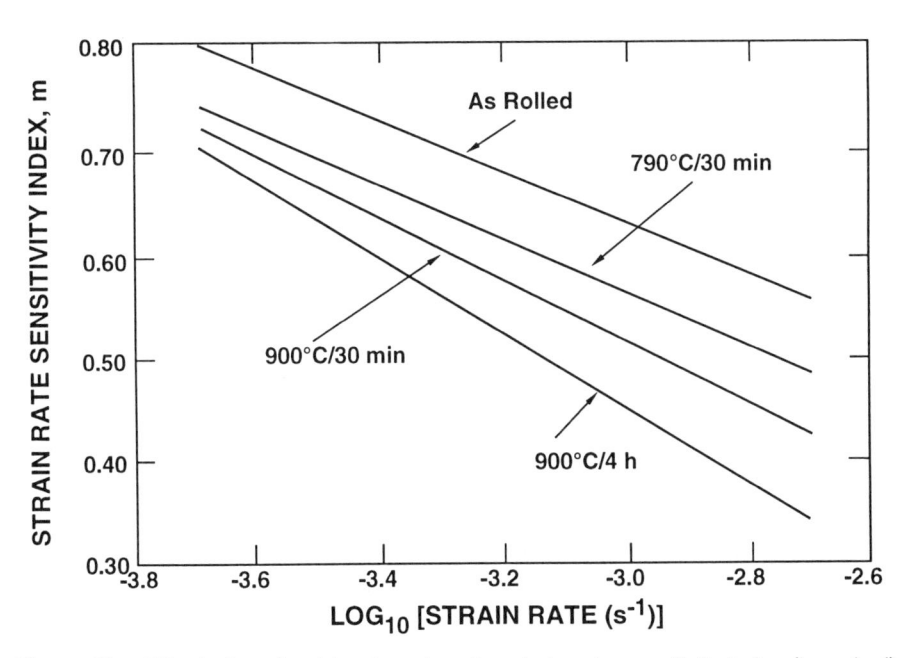

Figure 22 - Effect of pre-heat treatment on the strain rate sensitivity index ('m value') of hot rolled Ti-6Al-4V sheet. (46)

The final thermomechanical processing approach to be discussed in this subsection concerns an isothermal forging technique to produce parts with a graded microstructure and thus graded properties in alpha/beta titanium alloys (48). The preform for the method consists of a beta annealed shape contoured such that a variable deformation is imposed within it during the final isothermal forging step. Although the beta annealed microstructure with a coarse prior-beta grain size is not superplastic, it still exhibits high enough m values, ~ 0.25, (49) to provide adequate die fill for near-net shape processing. In the as-forged condition or forged-and-subtransus-annealed condition, workpiece regions which have undergone large strains develop a globularized structure while those which have suffered limited deformation retain the lamellar preform microstructure. Overall design of processes such as this requires a firm understanding of the effect of process variables on globularization as well as the utilization of computer simulations to establish proper preform and die geometries to produce the desired deformation field.

Workability Considerations

Control of hot-working parameters to ensure adequate workability is equally as important as the control of microstructure and texture for alpha/beta titanium alloys. During primary breakdown operations in the beta phase field, most alloys have excellent failure resistance. Typically, the only problems encountered are superficial chill cracks associated with the formation of a brittle, oxygen-enriched surface layer (alpha case). These flaws are readily removed by billet conditioning (grinding, turning, etc.) prior to subtransus hot working. During subtransus hot working, workability becomes a more major consideration. Defects during these operations may be classified into three major categories: fracture-related, shear-localization-controlled, and gross metal flow irregularities.

Fracture-Related Defects. The principal fracture-related defects developed during subtransus hot working of alpha/beta titanium alloys are wedge cracking and cavitation (Figure 23). Also known in the metalworking industry as "strain-induced porosity", these defects are usually initiated during the early stages of the working of the Widmanstatten microstructure previously developed during primary ingot breakdown. Wedge cracking is microcracking that initiates at beta-grain-boundary triple points; cavitation refers to microvoid formation at beta-grain edges. These

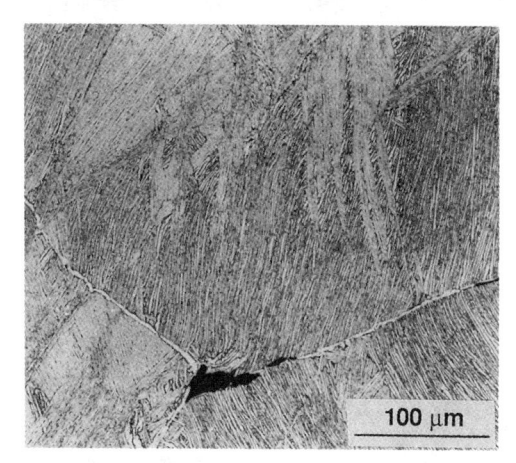

Figure 23 - Wedge cracking in hot worked Ti-6Al-2Sn-4Zr-2Mo with a Widmanstatten alpha preform microstructure. (50)

defects are produced by microscopically inhomogeneous deformation which leads to large stress concentrations at grain interfaces (10,11). At high strain rates, the stresses cannot be relieved by diffusion or plastic flow processes, and voids are opened. At low strain rates, on the other hand, such processes serve to relax the stress concentrations and accommodate the imposed strain prior to microvoid formation. In regions of the workpiece subjected to secondary tension stresses (e.g., bulged free surfaces), the voids may grow to produce macroscopic fracture. Alternatively, voids which have been initiated and not grown substantially may be closed during subsequent hot working provided large enough levels of compressive deformation are imposed (51).

The work of Raj (52), Matsumoto, *et al.* (53), Semiatin and Seetharaman (54), and Suzuki, *et al.* (55) provides insight into the phenomenology and mechanism of wedge cracking and cavitation. In Raj's work, plane-strain tension tests were conducted on beta processed Ti-6Al-2Sn-4Zr-2Mo samples having a basketweave microstructure with grain-boundary alpha. The strain rate and test temperature had a marked influence on workability. Specifically, a sharp transition from very ductile behavior at low strain rates and high temperatures to semi-brittle behavior due to intergranular failure at high strain rates and lower temperatures was observed (Figure 24).

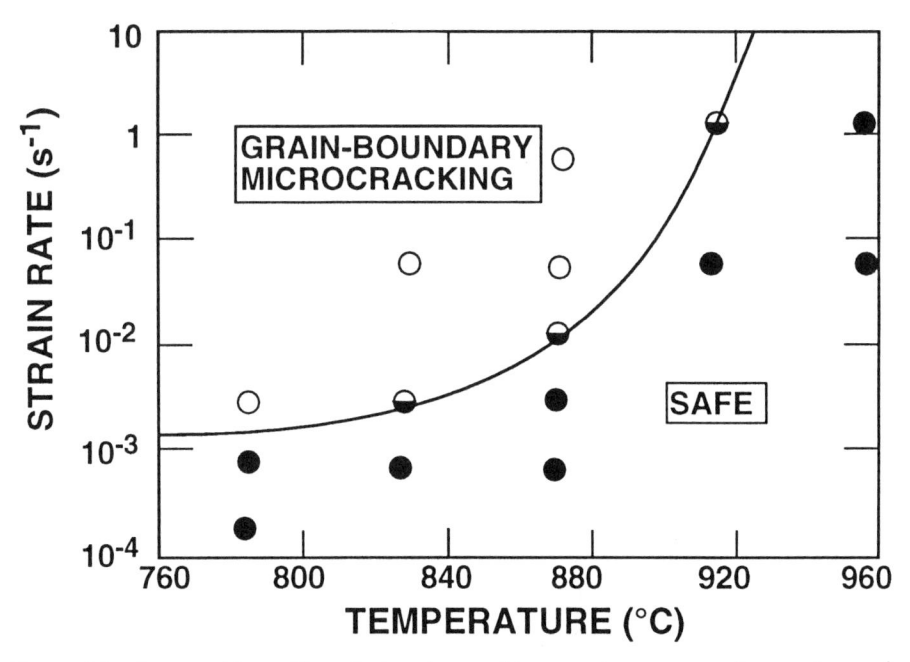

Figure 24 - Locus of transition strain rates and temperatures separating regimes of brittle intergranular failure from ductile behavior during plane-strain tension testing of Ti-6Al-2Sn-4Zr-2Mo with a Widmanstatten alpha microstructure. (52)

Matsumoto, Semiatin, Suzuki, and their coworkers provided further and more quantitative data on intergranular failure by using hot tension tests. Matsumoto, *et al.* (53) performed such tests at $\dot{\varepsilon} = 1$ s^{-1} on Ti-6Al-4V in four microstructural conditions: (1) equiaxed alpha, (2) deformed, partially-globularized Widmanstatten alpha, (3) beta-annealed-and-quenched acicular alpha, and (4) beta-annealed-and-air-cooled Widmanstatten alpha with grain-boundary alpha. The workability of the four microstructural variants was expressed in terms of the reduction in area (R.A.) as a function of test temperature (Figure 25). All of the microstructural conditions except that consisting of a transformed structure with grain-boundary alpha exhibited reasonably high ductility (R.A. between 75 and 100 percent) over the entire temperature range. By contrast, the transformed structure with grain-boundary alpha had excellent ductility at temperatures greater than 900°C, but showed a sharp drop in R.A. to values of the order of 50 percent at temperatures below 900°C. These low ductility failures were ascribed to the embrittling effect of the grain-boundary alpha phase which gave rise to intergranular failures. With these workability results, Matsumoto was able to predict the occurrence of intergranular failures during the subtransus hot rolling of Ti-6Al-4V bar with a Widmanstatten microstructure.

The effect of prior-beta grain size and strain rate on the hot ductility of Ti-6Al-4V samples containing a transformed structure with grain-boundary alpha was established by Semiatin and Seetharaman (54). To this end, tensile bars with a fine equiaxed alpha microstructure were subjected to short-time beta annealing followed by controlled cooling to produce microstructures with prior-beta grain sizes of 100 μm or 500 μm. Tension tests were conducted between 650 and 950°C at nominal strain rates of 0.01, 0.1, and 3 s^{-1}. For a given grain size, the R.A.-versus-temperature behavior was relatively insensitive to strain rate. Furthermore, similar to the results of Matsumoto, *et al.*, the ductility values were very high for test temperatures near the beta transus but showed a noticeable drop at lower temperatures due to the occurrence of wedge cracking and intergranular failure (Figure 26). The principal effect of grain size was evidenced by the rate of ductility loss with decreasing temperature. Unlike the results Matsumoto, *et al.*, the loss of ductility occurred over a broad temperature range. For the 100 μm-grain-size samples, the R.A. decreased from 100 percent to approximately 50 percent when the temperature was lowered from 950°C to 750°C. For the 500 μm-grain-size samples, the R.A. decreased from 100 percent to approximately 40 percent over temperatures of 950°C to 800°C, respectively.

In yet another investigation of the hot ductility of Ti-6Al-4V, Suzuki and Eylon (55) tested as-cast samples of the alloy which had a prior-beta grain size of 2000 μm. These samples exhibited a decrease in R.A. from 95 percent at and above the transus temperature to approximately 30 percent at 900°C. Thus, based on these observations and those of Semiatin and Seetharaman, it may be concluded that as the grain size increases the transition temperatures over which the ductility drops increase and the ductility trough is lower. Suzuki, *et al.* also noted that the effect of strain rate in the range of 0.005 to 5 s^{-1} on ductility was almost within experimental scatter.

<u>Shear-Localization Defects</u>. Alpha/beta titanium alloys are particularly susceptible to shear-localization defects such as shear bands and shear cracks. During conventional hot forging, for example, heat losses from the workpiece into the dies as well as frictional effects give rise to dead metal/chill zones in contact with the dies. Simultaneously, the interior portions of the workpiece may be undergoing

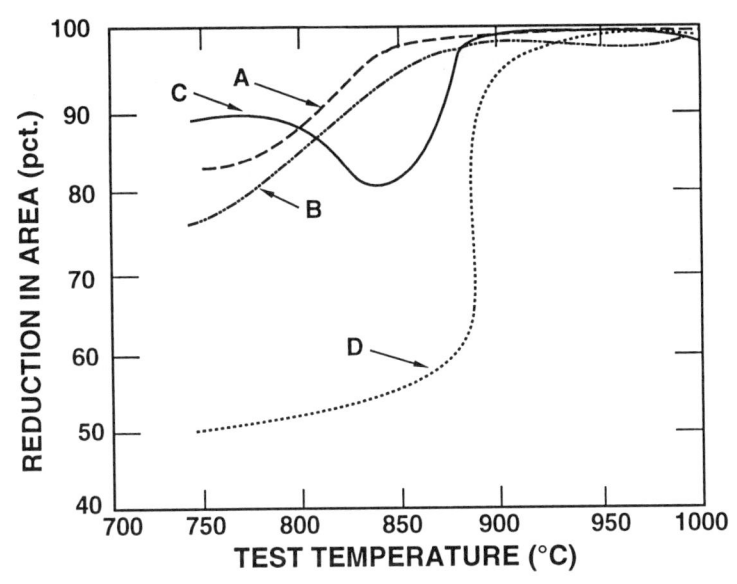

Figure 25 - Temperature dependence of tension-test reduction in area for Ti-6Al-4V samples with a microstructure of A: equiaxed alpha, B: deformed, partially-globularized Widmanstatten alpha, C: beta-annealed-and-water-quenched acicular alpha, or D: beta-annealed-and-air-cooled Widmanstatten alpha with extensive grain-boundary alpha. (53)

Figure 26 - Failure morphologies in Ti-6Al-4V tension samples beta annealed to produce a Widmanstatten alpha microstructure with a prior-beta grain size of 500 µm and pulled to fracture at (a) a low or (b) a high subtransus temperature. (54)

large deformations. Because of this nonuniformity in deformation, shear bands and shear cracks develop between the regions of low and high deformation. Although such defects may develop during axisymmetric processes such as pancake forging, the tendency for shear localization is greatest in plane-strain modes of flow such as the forging of airfoils.

The conditions under which shear bands and shear cracks may form during conventional (nonisothermal) forging of Ti-6Al-2Sn-4Zr-2Mo (Ti-6242) with either an equiaxed or Widmanstatten microstructure were determined by Semiatin and Lahoti (56). Round bars of this material were sidepressed between flat dies in either a mechanical press $(\dot{\varepsilon} \approx 25 \text{ s}^{-1})$ or a hydraulic press $(\dot{\varepsilon} \approx 1 \text{ s}^{-1})$ to reductions between 15 and 80 percent. Sample preheat temperatures were either 915°C (T_β-75°C) or 980°C (T_β-10°C), and die temperatures were 190°C or 345°C. At the higher sample preheat temperature, the dependence of flow stress on temperature for Ti-6242 is weak, whereas, at the lower temperature, this dependence is very large. From this research, it was established that working speed, through its effect on the overall time of deformation and heat transfer, is the most important process variable affecting shear band severity during conventional hot forging. Slow working, for example in a hydraulic press, leads to larger thermal gradients and more severe shear bands than rapid working such as occurs in mechanical presses. It was also found that the large amounts of heat transfer occurring under slow-working-rate conditions may lower the workpiece temperature into a regime of low workability and thus cause cracking along the shear bands. This effect was used to explain the occurrence of cracking in Ti-6242 samples nonisothermally sidepressed in a hydraulic press (Figure 27). For the processing conditions studied by Semiatin and Lahoti, shear localization could only be avoided by relatively fast deformation in a mechanical press using samples preheated to a temperature at which the flow stress is not sensitive to temperature (i.e., 980°C).

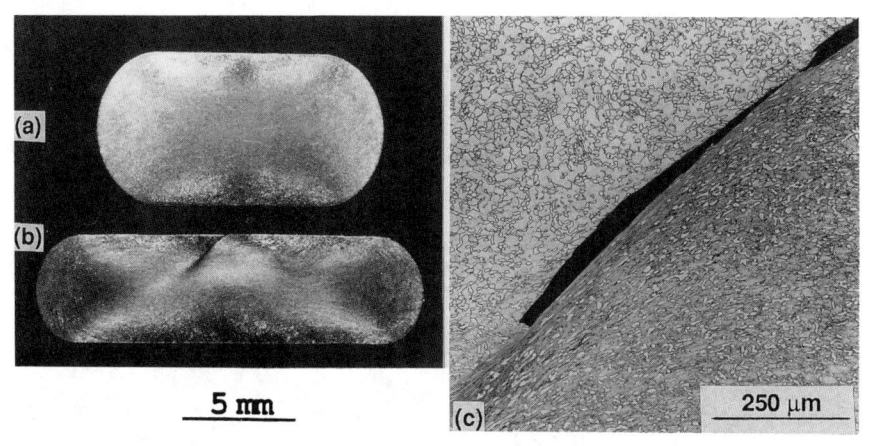

Figure 27 - Transverse metallographic sections (a,b) and micrograph (c) of the region with a shear band and a shear crack from the section shown in (b) of round bars of Ti-6Al-2Sn-4Zr-2Mo with an equiaxed alpha microstructure which were nonisothermally sidepressed in a hydraulic press ($\dot{\varepsilon} \approx 1 \text{ s}^{-1}$). Sample preheat temperature was 980°C (T_β - 10°C), die temperature was 345°C. Reductions were (a) 33 percent and (b) 52 percent. (56)

Even in the absence of heat transfer effects, as in isothermal hot forging, shear localization may occur. In these instances, the conditions that promote shear band formation are plane strain deformation conditions and material flow stress properties characterized by a high rate of flow softening and low strain rate sensitivity index (m) (49,57). High rates of flow softening in alpha/beta titanium alloys have several main sources: (1) forging at low temperatures and high strain rates at which the flow stress, and thus deformation-heating-induced softening, are high or (2) microstructure-based softening such as occurs during the breakdown of coarse-grain lamellar microstructures.

Metal Flow Defects. Gross metal flow defects are most commonly found in conventional, closed-die hot forging of alpha/beta titanium alloys and other difficult-to-work materials. These defects may or may not involve the formation of fractures or regions of localized flow. Often they result from factors such as improper choice of preform shape, poor die design, or poor choice of lubricant or process variables. All of these may contribute to the formation of flaws such as laps, flow-through defects, extrusion defects, etc. Laps are defects that form when metal folds over itself during forging (58,59). This may occur in rib-web forgings at a variety of locations. One such location is the web in a forging in which the preform web is too thin. During finish forging, such a web may buckle, causing a lap to form. Another location is a rib in which metal is made to flow nonuniformly (Figures 28 and 29). Most often, a lap at this location is a result of an excessively sharp fillet radius in the forging die.

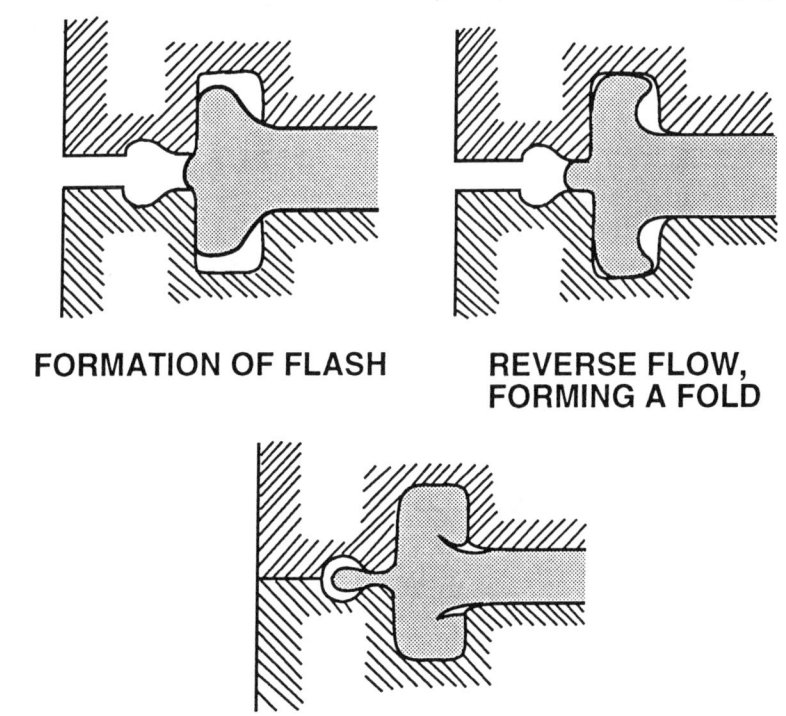

FORMATION OF FLASH

REVERSE FLOW, FORMING A FOLD

DEFECT IN FINISHED FORGING

Figure 28 - Schematic illustration of lap formation in the rib of a rib-web part due to improper preform geometry. (60)

Flow-through defects are flaws that form when metal is forced to flow past a recess after it has filled or has ceased to deform because of chilling (58). Similar to laps in appearance, flow through defects may be shallow but are indicative of an undesirable grain flow pattern or shear band that extends much deeper into the forging. Flow-through defects may also occur when trapped lubricant forces metal to flow past an impression.

Extrusion-type defects are formed when centrally located ribs formed by extrusion-type flow draw too much metal from the main body or web of the forging. A defect similar to a pipe cavity is thus formed (Figure 30) (58). Means of minimizing the occurrence of these defects include increasing the thickness of the web or designing the forging with a small rib opposite the larger rib.

Laps and flow-through defects may also involve the entrapment of oxides or lubricants. When this occurs, the metal is incapable of rewelding itself back together under the high working stresses, resulting in what is termed a cold shut.

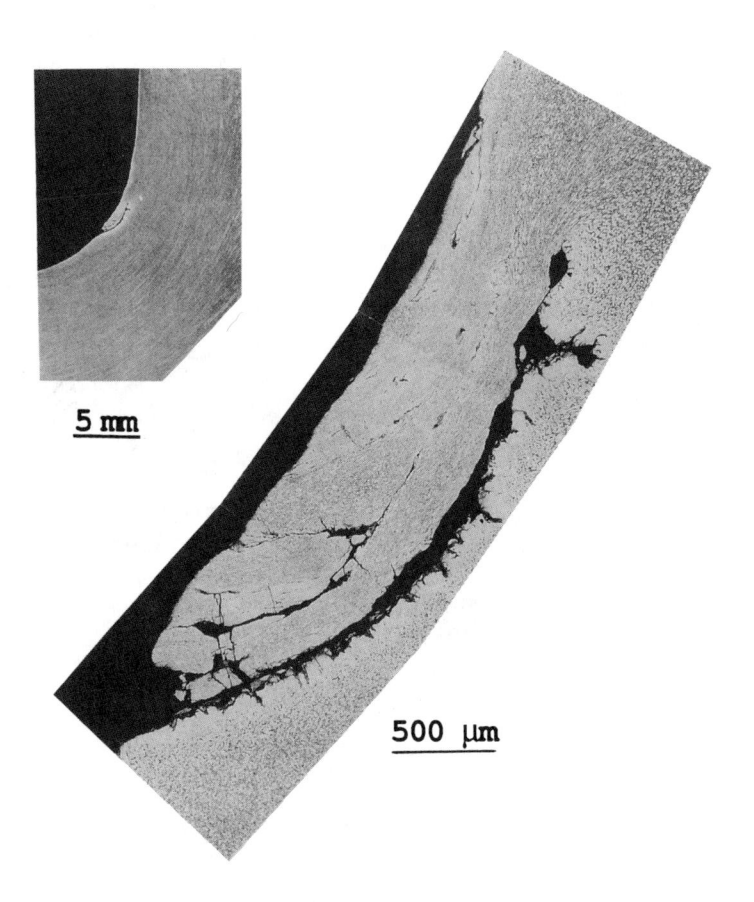

Figure 29 - Macrograph and micrograph of a lap defect in a Ti-6Al-4V bulkhead forging. (61)

Hot Working of Beta Titanium Alloys

Beta titanium alloys are materials which have sufficient additions of iron, chromium, vanadium, or molybdenum to suppress phase transformations when quenched from the high temperature beta field. This class of alloys may be broken into two subclasses, near-beta and metastable beta alloys. Compositions of several common alloys are given in Table V. Near-beta alloys, such as Ti-10V-2Fe-3Al, are hot worked in both the beta as well as alpha+beta phase fields to develop a primary alpha phase within the matrix of beta grains. The metastable beta alloys such as Ti-15-3-3-3, Beta C, and Beta III, being more highly beta stabilized, have lower beta transus temperatures than the near-beta alloys. The high flow stresses of metastable beta alloys at hot working temperatures limits hot working to supertransus temperatures; precipitation of second phases for property control is therefore done during static heat treatments.

EXTRUSION DEFECT POSSIBLE IF w>t

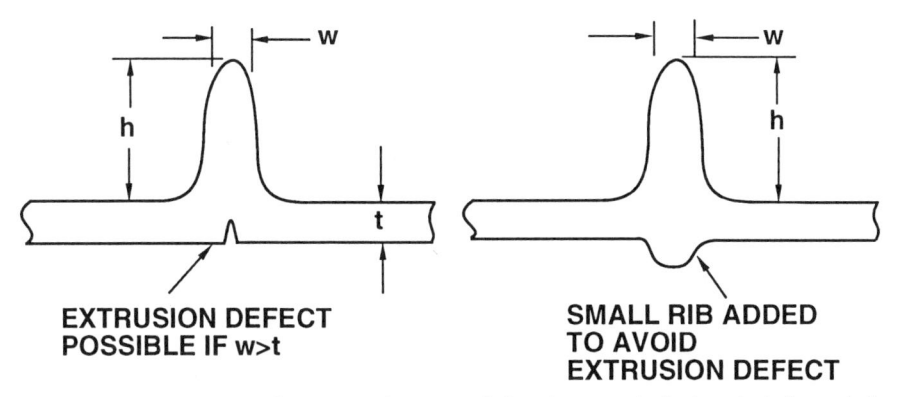

SMALL RIB ADDED TO AVOID EXTRUSION DEFECT

Figure 30 - Illustration of an extrusion-type defect in a centrally located rib and die-design modification used to avoid such defects. (58)

Table V Composition and Beta Transus Temperatures of Several Commercially Important Beta Titanium Alloys (20)

Alloy	V	Al	Fe	Si	Mo	Cr	Sn	Zr	Nb	Tβ (°C)
Ti-10-2-3	10.0	3.0	2.0	---	---	---	---	---	---	805
Ti-15-3-3-3	15.0	3.0	0.30	---	---	3.0	3.0	---	---	770
Beta C	8.0	3.5	0.30	---	4.0	6.0	---	4.0	---	795
Beta III	---	---	0.35	---	11.5	----	4.5	6.0	---	745
Timetal®21S	---	3.0	---	0.20	15.0	---	---	---	2.8	807

Because of the high level of alloying, control of composition uniformity in arc melted ingots of beta alloys can be difficult. Problems such as beta flecks, discussed above with regard to alpha/beta alloys, are among the most serious and require careful control of melting parameters to avoid. Similarly, the grain structure in cast beta alloy ingots can be quite coarse (upwards of 25 mm) and vary from columnar to equiaxed across the section. Because the microstructural objectives of hot working are different for near-beta and metastable beta alloys, the processing of the two subclasses is discussed separately.

Near-Beta Alloys

Near-beta alloys are typically hot worked at supertransus and subtransus temperatures to produce a beta grain structure with a controlled volume fraction and morphology of alpha-phase precipitate. Initial ingot breakdown is conducted via hot working and annealing at supertransus temperatures. For an alloy such as Ti-10V-2Fe-3Al with a beta transus T_β of approximately 800°C, the hot working step is usually performed at $T \approx 900$ to 1000°C followed by beta annealing at comparable or slightly higher temperatures. The exact choice of breakdown parameters may be aided by the utilization of processing maps such as that developed by Weiss and Froes (62). One such map, shown in Figure 31, indicates those combinations of

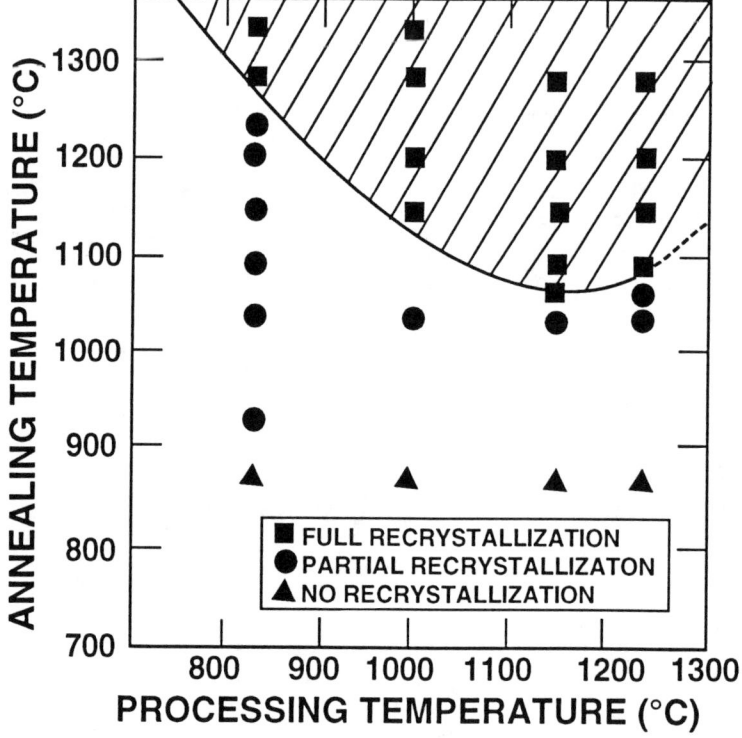

Figure 31 - Processing map indicating the effect of deformation and post-deformation annealing temperatures on the microstructures developed in cast samples of Ti-10V-2Fe-3Al upset to 65 percent height reduction. (62)

forging and heat treatment parameters (assuming a 65 percent reduction during forging and a one hour heat treatment time) that yield a uniform, equiaxed beta grain structure (2 to 3 mm in size) from ingot material with columnar grains as large as 50 mm. After the initial breakdown passes, additional primary working and heat treatment in the beta field may be conducted to yield beta grain sizes of the order of 250 μm in commercial ingots. Schroder and Duerig (63) even report on the processing of Ti-10V-2Fe-3Al to produce a grain size of 10 μm. However, the flow stresses at conventional strain rates for the very-fine grained product were found to be similar to those of the coarser-grain counterparts.

Secondary processing of wrought near-beta alloys such as Ti-10V-2Fe-3Al typically involves an initial working step in the beta field and working through the transus or finish hot working in the alpha+beta field. The amount of hot work imposed in the alpha+beta field is usually controlled to be approximately 5 to 20 percent. Such working results in an optimal size and volume fraction of primary alpha phase particles and thus a good balance of strength, ductility, and fracture toughness (64). Working solely in the beta field followed by air cooling would produce grain boundary alpha phase and some acicular/platelet alpha within the grains, a microstructure that would provide good toughness but low ductility. A small amount of alpha+beta work partially breaks up the grain boundary alpha and reduces the aspect ratio of the acicular/platelet alpha. Chen, et al. (64) have shown however that a novel final heat treatment of the alpha+beta worked alloy, comprising a double solution heat treatment and age sequence, provides the best combination of alpha phase morphology and thus properties. For Ti-10V-2Fe-3Al, this heat treatment consists of 780°C/2 hours/air cool + 774°C/2 hours/water quench + 510°C/8 hours. The first solution treatment takes the product of the alpha+beta worked material and normalizes the microstructure by an exposure close to, but below, the beta transus. The short duration of this first solution treatment is necessary to ensure that the primary alpha retains a high degree of its acicular morphology. During subsequent air cooling, the alpha grows slightly but retains its basic shape. The second solution treatment at a temperature close to, but below, that of the first allows some globular alpha to form. The water quench after this treatment preserves the metastable beta for subsequent hardening during the final aging heat treatment.

The low hot-working temperatures for near-beta alloys such as Ti-10V-2Fe-3Al make them excellent candidates for hot-die and isothermal forging. Although the near-beta alloys are not superplastic at typical hot-working temperatures, the improved die fill, elimination or reduction of die chill, and ability to use inexpensive die materials can often be used to justify the utilization of such techniques over conventional forging practices.

Metastable Beta Alloys

The objective of hot working of metastable beta alloys is simply to control grain size and grain shape. Thus, working processes are often conducted at temperatures 30 to 120°C above the beta transus with multiple working and intermediate reheating/heat treatment steps (20). Although supertransus working of beta alloys involves higher forging pressures than supertransus hot working of alpha/beta alloys, the beta alloys possess equally good workability during such working processes. Service properties in these alloys are developed by a variety of direct-age or solution-treat-and-age heat treatments.

Hot Working of Alpha-Two Titanium Aluminide Alloys

Alpha-two base alloys were the first titanium aluminide intermetallic materials to be hot worked. Most compositions are based on the Ti$_3$Al compound with an ordered hexagonal DO$_{19}$ crystal structure. In addition to the binary composition Ti-25Al, the vast majority of hot working of alpha-two materials has been conducted on two alloy compositions now commonly referred to as simply "alpha-two" (Ti-24Al-11Nb) and "super alpha-two" (Ti-25Al-10Nb-3V-1Mo).

The hot working of alpha-two and super alpha-two is very analogous to that for conventional alpha/beta titanium alloys because of the similarity between the two classes with regard to equilibrium phases and microstructural morphologies. Both classes have a high-temperature beta phase whose decomposition into alpha or alpha-two plus beta* mixtures with various morphologies is central to the control of properties. The major differences lie in the higher difficulty with which ingots are manufactured and broken down and limitations with regard to the extent of the processing window for secondary hot working of the alpha-two alloys. These and other aspects are addressed in the sections below.

Ingot Production and Breakdown

As for alpha/beta alloys, alpha-two ingots are usually made by triple vacuum arc remelting. Following casting, ingots are usually either hot charged directly for conversion or cooled to room temperature for surface conditioning followed by reheating and conversion. In the latter practice, reheating is often done in a gradual, or stepped, manner to minimize temperature gradients and thus thermal stresses; direct charging of large alpha-two ingots (~ 750 mm diameter) into hot furnaces may lead to thermal stresses large enough to cause thermal cracking of the low ductility intermetallic alpha-two alloys (65).

Initial breakdown passes for alpha-two ingots are conducted via cogging in the beta phase field, typically at temperatures between T_β and T_β + 125°C. Final breakdown passes may involve preheating and working at somewhat lower temperatures, i.e., $T \geq T_\beta$ - 150°C to impart some work in the alpha-two + beta phase field. The wrought product thus produced is sometimes then given a beta-field recrystallization heat treatment prior to secondary hot working (66), much like the approach used for conventional alpha/beta titanium alloys.

The precise extent of the processing window for breakdown of ingots of alpha-two, Ti-24Al-11Nb, was determined by the examination of the flow stress behavior and microstructure evolution by Semiatin, et al. (67). Specifically, flow stress data from isothermal hot compression tests on cast material (Figure 32) were plotted in terms of $\log_{10} \bar{\sigma}(\bar{\varepsilon} = 0.05)$ versus 1/T. The plots for various strain rates exhibited a trilinear dependence (Figure 33). The break in each plot at 1000/T(K) $\approx$ 0.740 corresponds to T = 1078°C or a value near the measured beta transus for this alloy (i.e., 1075°C). The other breaks in the $\log_{10} \bar{\sigma}$ versus 1/T plots observed at lower temperatures (i.e., higher values of 1/T), on the other hand, could be ascribed

---

* Depending on specific alloy compositions, the disordered beta phase in alpha-two alloys undergoes an ordering reaction at a temperature below the beta transus. This ordered phase is referred to as β_o or B2. In the discussion to follow, no distinction is made between high temperature beta and B2.

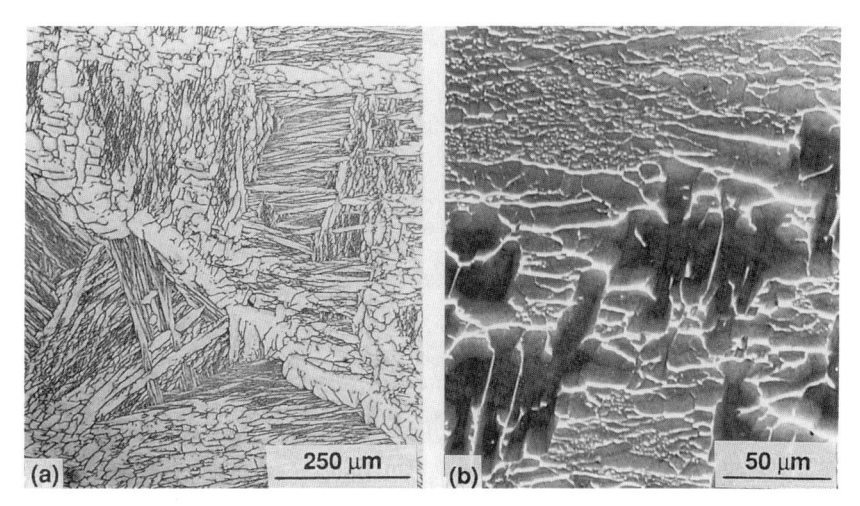

Figure 32 - Typical microstructure in a cast alpha-two titanium aluminide alloy (Ti-24Al-11Nb): (a) polarized light micrograph and (b) backscattered micrograph. (67)

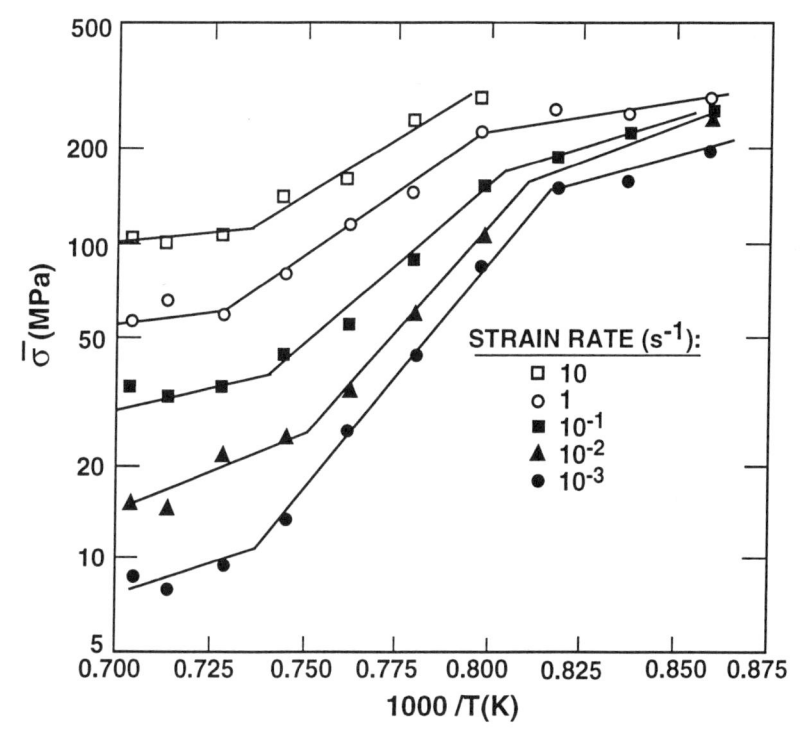

Figure 33 - Log $\bar{\sigma}(\bar{\varepsilon} = 0.05)$ versus 1/T (K) behavior determined via hot compression tests on the cast alpha-two titanium aluminide alloy Ti-24Al-11Nb. (67)

neither to phase transformation nor to the regime in which the volume percent of the beta phase is rapidly increasing.

The accompanying microstructure observations (Figure 34) revealed that the behavior associated with these latter transitions was due to a change in deformation conditions from those typical of hot working to ones representative of warm working. The microstructures developed in the warm-working regime (T < T_β - 100°C) were indicative of a worked material; there was no evidence of globularization, and a number of wedge cracks and cavities were observed (Figures 34a,b). By contrast, material deformed at T_β - 100°C was partially globularized with very few wedge cracks or cavities (Figure 34c). At higher subtransus temperatures, a change from essentially a primary alpha-two structure to one indicative of the presence of a substantial amount of beta phase at the deformation temperature was found (Figure 34d). The transformed beta matrix in this instance retained no evidence of deformation, but the primary alpha-two was not fully globularized, at least not for the deformation imposed for these experiments. The specific temperatures at which the transition from warm to hot working occurred for the various strain rates were found to be related by a fixed value of the Zener-Hollomon parameter, $Z = \dot{\varepsilon} \exp(Q/RT)$, where Q and R denote an activation energy and the gas constant, respectively.

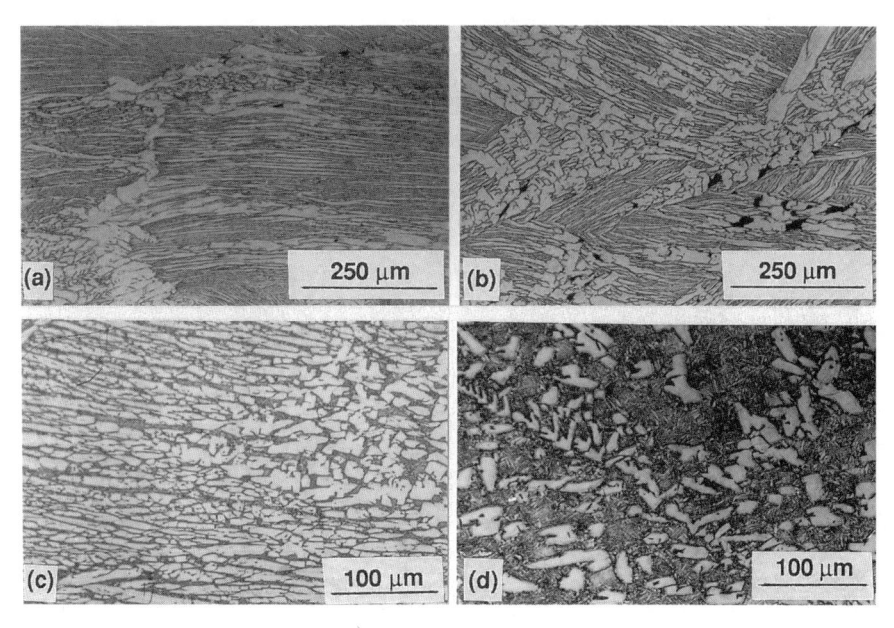

Figure 34 - Microstructures developed in a cast alpha-two titanium aluminide alloy (Ti-24Al-11Nb) compression tested to a 50 percent reduction in height at $(\dot{\varepsilon} = 1\ s^{-1})$ and test temperatures of (a,b) 890°C, (c) 980°C, or (d) 1040°C. Micrographs (a,c,d) were shot near the center of the test samples, and micrograph (b) was taken near the lateral free surface. The compression axis is vertical in all micrographs. (67)

Secondary Hot Working

A number of secondary hot-working processes for alpha-two alloys have been investigated. These include isothermal and hot-die forging, pack rolling of sheet, and superplastic forming.

Ward, et al. (68) investigated various thermomechanical processes to control the microstructure and properties of super alpha-two, Ti-25Al-10Nb-3V-1Mo. The processes comprised hot-die pancake forging of wrought billets at various temperatures below the beta transus in order to vary the volume fraction of globular primary alpha-two phase. Post forging heat treatments were also devised to develop two distinct scales of beta matrix decomposition (secondary alpha-two) products. These treatments consisted of either a direct age by quenching into salt after hot forging or an oil quench after forging followed by an alpha-two + beta solution treatment and subsequent salt bath age. The direct aging process yielded a transformation product of very fine Widmanstatten alpha-two plates in a beta matrix. The fineness was a result of the large number of heterogeneous nucleation sites present in the heavily deformed beta matrix. On the other hand, coarser secondary alpha-two plates were formed when the forged pancakes were oil quenched and then reheated for solution treatment and aging. The coarser plates in this case were attributed to recovery or recrystallization of the beta matrix during the solution treatment, thereby leaving a smaller number of nucleation sites in the beta phase and transformation conditions dominated by growth rather than nucleation. Moreover, the size of the prior-beta grains in both microstructures was controlled by the size and volume fraction of the primary alpha-two phase which pinned the beta grain boundaries. The beta grains were decorated as well by alpha-two in the solution-treated-and-aged microstructure.

Extensive work on hot pack rolling of alpha-two alloys was conducted by Bassi and his coworkers (69-71). Most of the efforts dealt with rolling of 0.15 mm-thick super alpha-two foil from 3-mm thick preforms. Bassi and Peters (71) determined the effect of three different thermomechanical processing sequences on microstructure and properties. The sequences consisted of (1) hot rolling high in the alpha-two + beta field, followed by rapid cooling to room temperature, and, finally, alpha-two + beta annealing, (2) beta annealing followed by direct hot rolling lower in the alpha-two + beta field, and (3) beta annealing followed by hot rolling high in the alpha-two + beta field to a 93 percent reduction, and final rolling lower in the alpha-two + beta field. The microstructures (Table VI) contained various amounts and morphologies of alpha-two, ordered beta (B2), orthorhombic (O) phase, and the metastable ω phase. The bend ductilities of the foils were found to be heavily dependent on microstructure (Table VI). The best ductility was obtained for the sequence involving rolling low in the alpha-two + beta regime (sequence 2) which led to a microstructure with a large volume fraction of globular and elongated primary alpha-two.

Superplastic forming of alpha-two and super alpha-two was studied by Dutta and Banerjee (72) and Yang, et al. (73), respectively. The material used by Dutta and Banerjee had been broken down by beta forging and then hot-rolled low in the alpha-two + beta phase field to obtain a fine alpha-two+beta structure with a primary alpha-two size of approximately 4 µm. Guided by results for optimum superplastic behavior in alpha/beta titanium alloys, Dutta and Banerjee conducted tension tests at $\dot{\varepsilon} \approx$ 3 to 6 x 10^{-4} s^{-1} and T = 960, 980, or 1020°C, i.e., temperatures spanning those at which the alpha-two and beta phases comprise equal volume fractions.

Table VI Properties and Microstructures of
Hot Rolled Super Alpha-Two Foils (71)

TMP Sequence	Bend Ductility (pct.) L	Bend Ductility (pct.) T	Vickers Hardness L	Vickers Hardness T	Microstructure
1	4.4	3.1	389	429	B2 + ω matrix; 5-10 pct. globular and elongated alpha-two; grain-boundary alpha-two or O phase
2	16.0	3.3	404	404	B2 + ω matrix; 30-35 pct. globular and elongated alpha-two
3	2.9	0	499	475	Fully transformed structure of fine alpha-two (or O) laths in prior B2 grains

Surprisingly, the highest m value ($\approx$ 0.6) and elongation (520 percent) were obtained for a microstructure containing 75 percent alpha-two (T = 980°C = T_β - 120°C). Furthermore, the best elongation for the alpha-two alloy was only about one-half that typically found for alpha/beta titanium alloys. Three possibilities were postulated to explain the inferior behavior of alpha-two compared to the conventional titanium alloys: (1) low diffusivity in the beta phase and hence poor ability to accommodate strain, (2) the different nature of the interphase interfaces compared to the disordered alloys and (3) grain growth effects.

Yang, et al. determined the superplastic behavior of super alpha-two sheet with an elongated primary alpha-two morphology. As for the alpha-two material, maximum values of m were approximately 0.6, and peak elongation was only 570 percent. This elongation was achieved at $\dot{\varepsilon}$ = 1.5 x 10^{-4} s^{-1} and T = 980°C.

Crystallographic Texture

Similarities between alpha-two and alpha/beta titanium alloys in microstructure development are also mirrored in crystallographic texture. In this regard, Verma and Ghosh (74) measured textures on hot rolled foils of Ti-24Al-11Nb. The processing sequence comprised pancake forging of wrought stock in either the beta field (T = 1130°C = T_β + 5°C) or high in the alpha-two + beta regime (T = 1060°C = T_β - 65°C). Sections of 4 mm thickness were cut from the pancakes and then hot pack rolled at 985°C (T_β - 140°C) in successive campaigns, first to 1-mm thick sheet and then to 0.2 mm-thick foil via either unidirectional rolling or cross rolling. The texture of the alpha-two phase was then determined (Table VII). As for the alpha/beta alloys rolled at temperatures considerably below the beta transus temperature (Figure 10), the textures contained both basal and transverse components. For the unidirectionally hot rolled alpha-two material, the transverse component of the alpha-two texture was considerably stronger than the basal component irrespective of the initial forging temperature. In contrast, for the alpha/beta alloy Ti-6Al-4V rolled at a similar decrement below its beta transus, the transverse texture component of the alpha phase is only slightly stronger than the basal component, i.e., 8.5 versus 7.0 times random (Figure 11a). Verma and Ghosh (74) also noted a basal/transverse type

Table VII Texture Components Developed in Alpha-Two Phase of Ti-24Al-11Nb Foils Hot Rolled at 985°C (T$_\beta$-140°C) (74)

Rolling Preform Microstructure	Rolling Method	(0002) Texture Component (x random) Basal	Transverse
Widmanstatten alpha-two	unidirectional	3.0	12.0
Equiaxed alpha-two	unidirectional	3.2	9.6
Equiaxed alpha-two	cross	3.4	2.6

Table VIII Knoop Hardness Data for Ti-24Al-11Nb Foils Hot Rolled at 985°C (T$_\beta$-140°C) (74)

Rolling Preform Microstructure	Rolling Method	Knoop Hardness (kg/mm^2)* K1	K2	K3
Widmanstatten alpha-two	unidirectional	279	343	389
Equiaxed alpha-two	unidirectional	280	293	320
Equiaxed alpha-two	cross	301	295	440

* K1 taken on transverse direction cross section, long direction of indenter parallel to transverse direction; K2 taken on rolling direction cross section, long direction of indenter parallel to rolling direction; K3 taken on an in-plane sample, long direction of indenter parallel to rolling direction.

texture from cross rolling with the basal component being stronger in this instance. A similar *qualitative* trend can be found for the alpha texture in Ti-6Al-4V cross-rolled at a comparable subtransus temperature (Figure 11b), although again there are quantitative differences.

As for textured alpha/beta titanium sheet products, hot-rolled alpha-two sheet and foil products may be expected to exhibit a noticeable anisotropy in room-temperature mechanical properties, as was found by Verma and Ghosh (Table VIII).

Hot Working of Orthorhombic Titanium Aluminide Alloys

In an attempt to obtain a better blend of specific strength, toughness, and creep properties compared to alpha-two alloys, the orthorhombic titanium aluminide alloys have been developed (75,76). These alloys are based on the Ti$_2$AlNb phase with orthorhombic crystal geometry, a geometry related to the hexagonal symmetry of alpha-two (Ti$_3$Al) through specified distortions in the closed-packed planes (75-77).

While still under a great deal of development, orthorhombic compositions based on Ti-22Al-23Nb, Ti-22Al-27Nb, and quaternary variants of these alloys have received the greatest attention. Like the alpha-two alloys, the high-temperature phase for orthorhombic alloys is beta which becomes ordered B2 at lower temperatures depending on composition. The equilibrium phases below the beta transus are orthorhombic and beta/B2 in Ti-22Al-27Nb and low oxygen content (≲ 250 wppm) Ti-22Al-23Nb. Higher oxygen levels (> 250 wppm) in Ti-22Al-23Nb give rise to a three-phase mixture of orthorhombic, beta/B2, and alpha-two (78).

The melting and working of orthorhombic titanium aluminide alloys has evolved from processing knowledge developed for conventional alpha/beta alloys and alpha-two titanium aluminide alloys. Early melting was conducted via conventional vacuum arc melting. More recently, plasma, cold hearth melting has been utilized to improve the compositional uniformity of the high niobium content of these alloys in ingot form. Primary ingot breakdown usually consists of hot working (e.g., cogging, upsetting) and intermediate reheating/annealing in the single phase beta field followed by subtransus hot working (via rolling, radial forging, etc.) at temperatures as low as approximately 100 to 150°C below the transus. Such working produces a primary globular orthorhombic or alpha-two phase in a transformed B2 matrix. As shown by Semiatin and Smith (79), the rapid precipitation of second phases on the beta-grain boundaries at temperatures below the transus effectively pins these boundaries. Thus, large levels of hot work below the transus typically result in highly elongated prior-beta grain structures (Figure 35) and a noticeable crystallographic texture. Aging of the beta matrix at temperatures below the hot working temperature can be used effectively to precipitate additional (secondary) orthorhombic or alpha-two, usually in an acicular form (78).

Rhodes, et al. (78) investigated crystallographic texture development during hot and cold rolling of low oxygen content Ti-22Al-23Nb sheet and foil. After subtransus hot rolling at 1000°C (= T_β - 20°C for this alloy), a two-phase (orthorhombic + B2) microstructure was produced. The orthorhombic phase exhibited a strong, cube-like (100) [022] texture, whereas the ordered beta phase had a moderately strong cube-type texture, (100) <011>. The latter texture is typical for deformation textures in cold worked bcc metals (80), and its occurrence is expected in the absence of recrystallization of beta phase hot worked below the transus temperature.

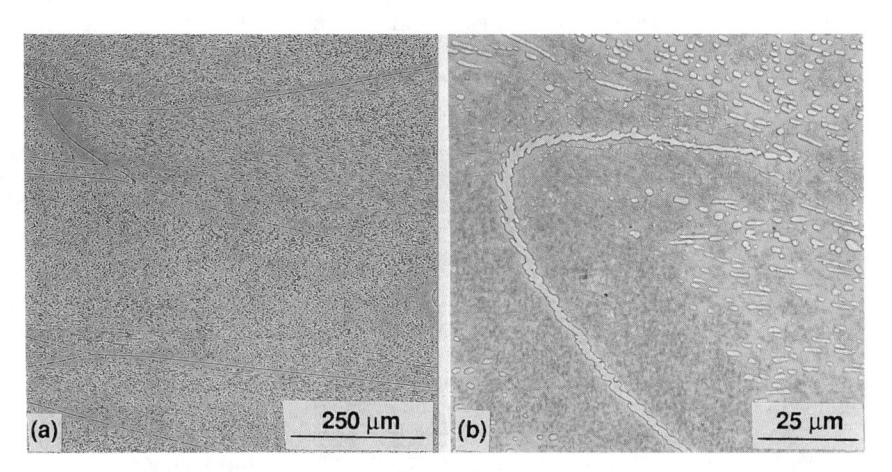

Figure 35 - Microstructure (at two levels of magnification) of subtransus hot rolled sheet of Ti-22Al-23Nb, showing precipitate morphology and elongated prior-beta grain structure. The rolling direction is horizontal, and the thickness direction is vertical in both micrographs. (79)

Hot Working of Gamma Titanium Aluminide Alloys

The titanium aluminide alloys based on the gamma, or TiAl, phase present the greatest challenge in terms of hot working. As for Ti$_3$Al and Ti$_2$AlNb, TiAl is an ordered intermetallic phase; it has a face-centered-tetragonal (fct) crystal structure. The near-gamma (consisting of gamma + alpha-two and sometimes a small amount of beta) and single-phase gamma alloys typically contain between 45 and 48 or between 50 and 52 atomic percent aluminum, respectively, as well as 0.3 to 2.5 atomic percent of secondary alloying elements such as niobium, chromium, manganese, vanadium, tantalum, and tungsten. Because of the improved ability to control microstructure in two-phase alloys as well as their more attractive properties (81), most current attention is being focused on near-gamma alloys.

In broad terms, the hot-working of near-gamma titanium aluminide alloys bears a number of similarities to the processing of alpha/beta alloys. A high temperature second phase (alpha) is used as a structure-control phase for the near-gamma alloys in much the same manner as the beta phase for alpha/beta alloys. Hence, the choice of working or heat treatment temperature relative to the alpha transus T_α (temperature at which alpha + gamma $\rightarrow$ alpha) for near-gamma alloys plays an equally important role vis-à-vis processing relative to the beta transus T_β for alpha/beta alloys. For most near-gamma alloys, however, the alpha transus temperature is only 100 to 150°C below the solidus temperature. Hence, alpha-phase-field hot working is done to a much lesser degree than beta working in alpha/beta alloys because of rapid grain growth and the accompanying loss in microstructure control and workability.

The typical ingot metallurgy approach for processing near-gamma and single-phase gamma titanium aluminide alloys usually comprises (i) ingot production, (ii) ingot breakdown with or without intermediate and final heat treatment, and (iii) secondary processing.

Ingot Production and Ingot Structure

Three principal methods have been used successfully to melt near-gamma and single-phase gamma titanium aluminide ingots; these are induction skull melting, vacuum arc melting, and plasma (cold hearth) melting. The first of these, induction skull melting, has been used primarily to produce small diameter (approx. 75 to 125 mm) ingots for laboratory research. In this method, the charge is induction melted in a specially-designed, water-cooled copper crucible and then poured into a graphite mold. Large ingots with diameters usually between 200 and 350 mm are made by the other two techniques. Thermal stresses developed by nonuniform temperature fields during arc and plasma melting may be quite large, especially for larger diameter ingots, and thus give rise to cracking of the low-ductility gamma titanium aluminide alloys (82).

For a given alloy composition, the cast structure is at least qualitatively similar irrespective of melting method and ingot size. The broad features of cast structure development are most easily understood with reference to the binary titanium-aluminum phase diagram in the region of the equiatomic composition (Figure 36). The most important feature of the phase diagram with regard to solidification structure is the occurrence of a double cascading peritectic reaction. The reaction gives rise to dendritic regions of alpha-two and gamma lamellae (which have evolved from the high temperature beta and alpha phases) and interdendritic regions of nominally single- phase gamma which are last to solidify from the melt

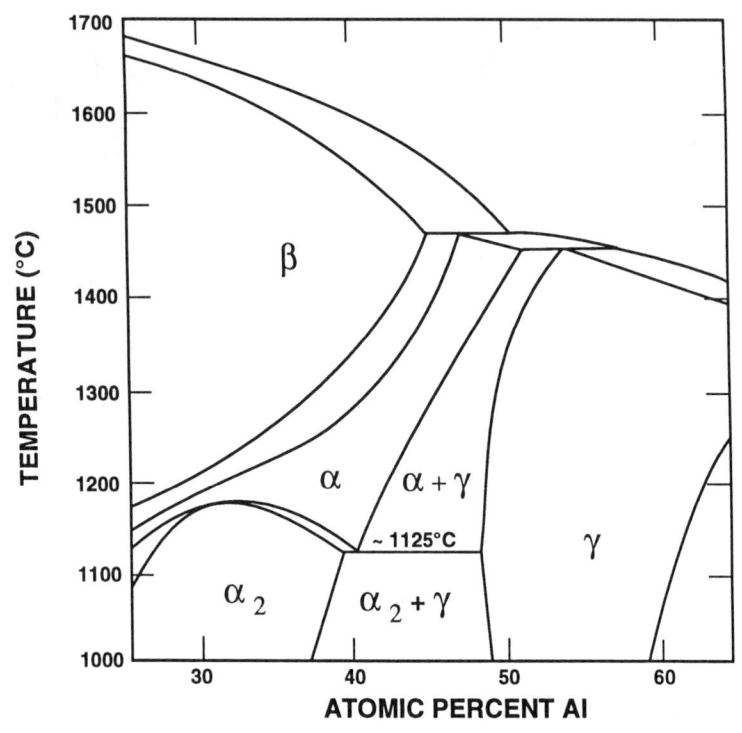

Figure 36 - Portion of the binary titanium-aluminum phase diagram of interest in the processing of near-gamma and single-phase gamma titanium aluminide alloys. (83)

Figure 37 - Polarized light optical micrographs of microsegregation in (a) cast plus HIP'ed Ti-47.3Al-2.0Nb-1.7Mn and (b) cast plus HIP'ed plus isothermally forged Ti-48Al-2.5Nb-0.3Ta. (84)

(Figure 37). This *microsegregation* is unavoidable in both the near-gamma alloys containing approximately 46 to 48 atomic percent aluminum as well as the nominally single-phase gamma alloys with 48 to 55 atomic percent aluminum. Although this segregation has been most thoroughly documented for binary titanium-aluminum alloys (83), related research on a wide range of multicomponent materials (85) has shown that the form and extent of segregation are similar in these alloys as well. The only major exception to these trends has been observed in alloys containing large amounts ($\geq$ 0.5 atomic percent) of tantalum or tungsten which tend to stabilize the beta phase as well as hinder homogenization by thermal treatment alone (86). Small amounts of beta phase are also stabilized by alloying additions of chromium in the range of 1 to 3 atomic percent (87).

The size of the lamellar grains in as-cast near-gamma alloys with 46 to 48 atomic percent aluminum is typically 100 to 300 μm. A much wider range of grain sizes may be found in near-gamma alloys with less than 46 atomic percent aluminum, or materials that freeze without an interdendritic gamma phase which can pin the alpha grain boundaries. In these alloys, it appears as though cooling rate below the solidus temperature plays an important role. For example, a 75-mm diameter induction skull melted ingot of Ti-45.5Al-2Cr-2Nb exhibited an alpha grain size of 150 μm, whereas a 200-mm diameter, vacuum arc remelted ingot of the same alloy had a grain size of approximately 700 μm (88).

Homogenization Heat Treatment

Extensive research has been conducted on the near-gamma alloys to develop homogenization treatments to eliminate the microsegregation characterized predominantly by the interdendritic phase. Not surprisingly, the majority of this work has shown that processes conducted solely in the alpha plus gamma phase field are insufficient to dissolve the interdendritic gamma fully to produce a completely homogeneous structure (Figure 37b); these processes have included various heat treatments, hot isostatic pressing (e.g. 1260°C/175 MPa/4h), and metalworking operations. Thus, attention has been focused on homogenization heat treatments in the single-phase alpha field, (84,89) or, more recently, in the alpha plus beta phase field (90).

Homogenization in the single phase alpha field has been conducted on a variety of cast plus HIP'ed and prior-worked near-gamma titanium aluminide alloys. All of the results reveal a very sharp increase in homogenization kinetics with relatively small increases in temperature above the alpha transus. As might be expected, however, homogenization in the single-phase alpha field can give rise to grain growth which rapidly accelerates as the pinning influence of the dissolving gamma phase is lost. Hence, a short exposure time in this phase field is preferable in order to minimize grain growth and thereby retain a measure of hot workability for subsequent deformation processing. Semiatin, *et al.* (84) have shown that a homogenization heat treatment of 20 minutes at a temperature approximately 60°C above the alpha transus is optimal with regard to full dissolution while minimizing high temperature exposure and alpha grain growth. Such a treatment also tends to lend itself to the homogenization of finite size ingots in which the center-to-surface temperature lag during furnace heat treatment is an important consideration in obtaining a uniform macrostructure.

An alternate and very useful method for homogenizing near-gamma alloys with aluminum content less than approximately 46 atomic percent has recently been developed by Martin, et al. (90). In their approach, the interdendritic gamma phase is dissolved by heat treatment in the alpha plus beta field. The presence of two phases minimizes grain growth and hence concomitant losses in workability.

Ingot Breakdown Techniques

Three production-scaleable processes have been successfully utilized to breakdown the cast structure of gamma titanium aluminide alloys - isothermal forging, conventional (canned) forging, and conventional (canned) extrusion. For each approach, the ingot is usually HIP'ed or given a homogenization heat treatment prior to working. HIP'ing of both near-gamma and single-phase-gamma alloys is usually done at 1260°C and a pressure of approximately 175 MPa; near-gamma alloys with an aluminum content less than approximately 46 atomic percent may be HIP'ed at a slightly lower temperature to avoid an incursion into the single-phase alpha field and a large amount of alpha grain growth during the long thermal exposure typical of the process.

Isothermal Forging. Isothermal forging to breakdown the coarse ingot structure typically consists of pancaking cylindrical preforms to reductions between 4:1 and 6:1 at temperatures between 1065 and 1175°C and nominal strain rates between 10^{-3} and 10^{-2} s^{-1}. Under these conditions, the ductility is usually fairly high, and sufficient hot work is imparted to globularize the lamellar structure at least partially in near-gamma alloys (percent spheroidization ≈ 50) or fully recrystallize single-phase gamma alloys.

The kinetics of globularization of the lamellar microstructure have been investigated by Seetharaman and Semiatin (91). For this purpose, fully lamellar microstructures with prior alpha grain sizes between 80 and 900 μm were developed in Ti-45.5Al-2Cr-2Nb using a special forging and heat treatment process (92). Isothermal hot compression tests were conducted at 1093°C and strain rates of 0.001, 0.1, and 1 s^{-1} on samples with the various grain sizes. The flow curves from these tests showed a very strong dependence of peak flow stress and flow softening rate on grain size; the larger grain-size samples exhibited significantly higher peak stresses and flow softening rates (Figure 38). Examination of the as-compressed microstructures revealed that dynamic globularization initiated at and proceeded inward from the prior-alpha grain boundaries (Figure 39). The grain interiors showed evidence of moderate-to-extensive kinking of the lamellae, depending on the orientation of the lamellae relative to the applied load. However, there was no evidence of the kinking giving rise to globularization unlike the behavior commonly observed for the deformation of alpha/beta titanium alloys with Widmanstatten alpha microstructures. With regard to the rate of breakdown of the lamellar microstructure, globularization kinetics were found to increase as the strain rate decreased, for a given alpha grain size, and to decrease with increasing alpha grain size for a given strain rate (Figure 40). In most cases, the dependence of percent globularization on strain followed an Avrami (sigmoidal) behavior, at least approximately.

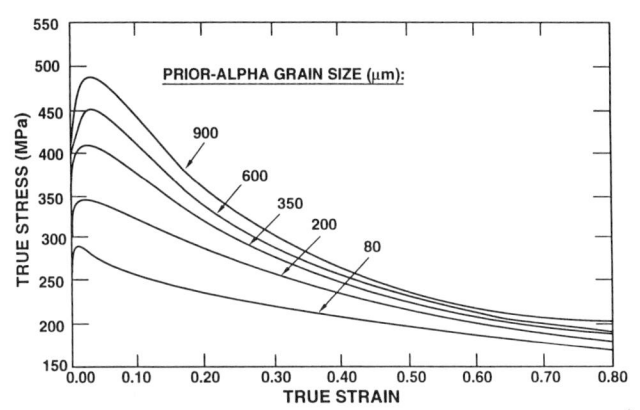

Figure 38 - Flow curves from isothermal hot compression tests at 1093°C and $\dot{\varepsilon}$ = 0.1 s^{-1} on Ti-45.5Al-2Cr-2Nb thermomechanically processed to yield lamellar microstructures with prior-alpha grain sizes between 80 and 900 μm. (91)

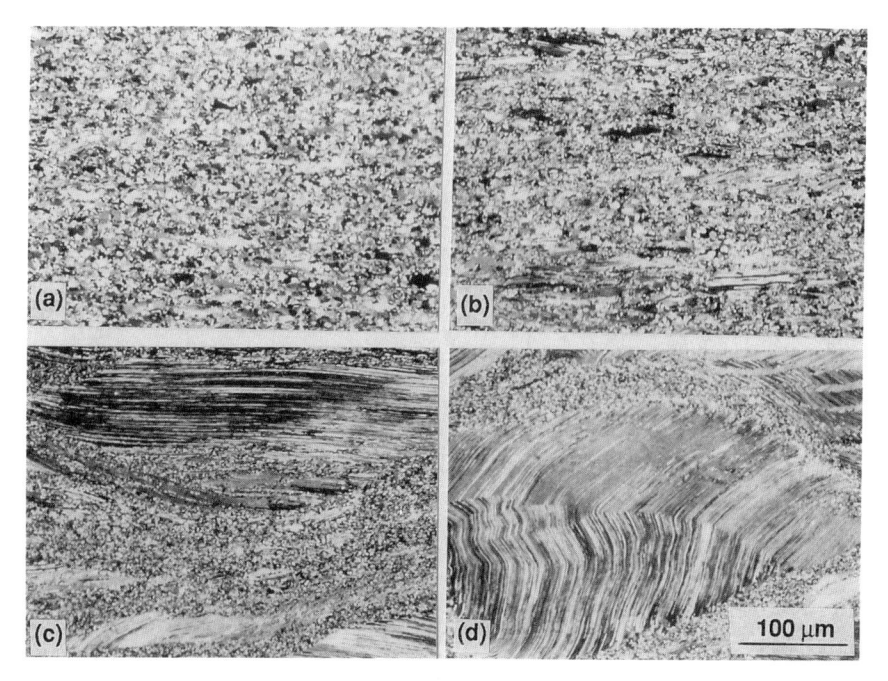

Figure 39 - Polarized light optical microstructures developed in Ti-45.5Al-2Cr-2Nb samples isothermally upset to a 75 percent reduction at 1093°C and $\dot{\varepsilon}$ = 0.1 s^{-1}. Prior to compression testing, the samples had been processed to yield lamellar microstructures with prior-alpha grain sizes of (a) 80 μm, (b) 200 μm, (c) 600 μm, or (d) 900 μm. The compression axis is vertical in all micrographs. (91)

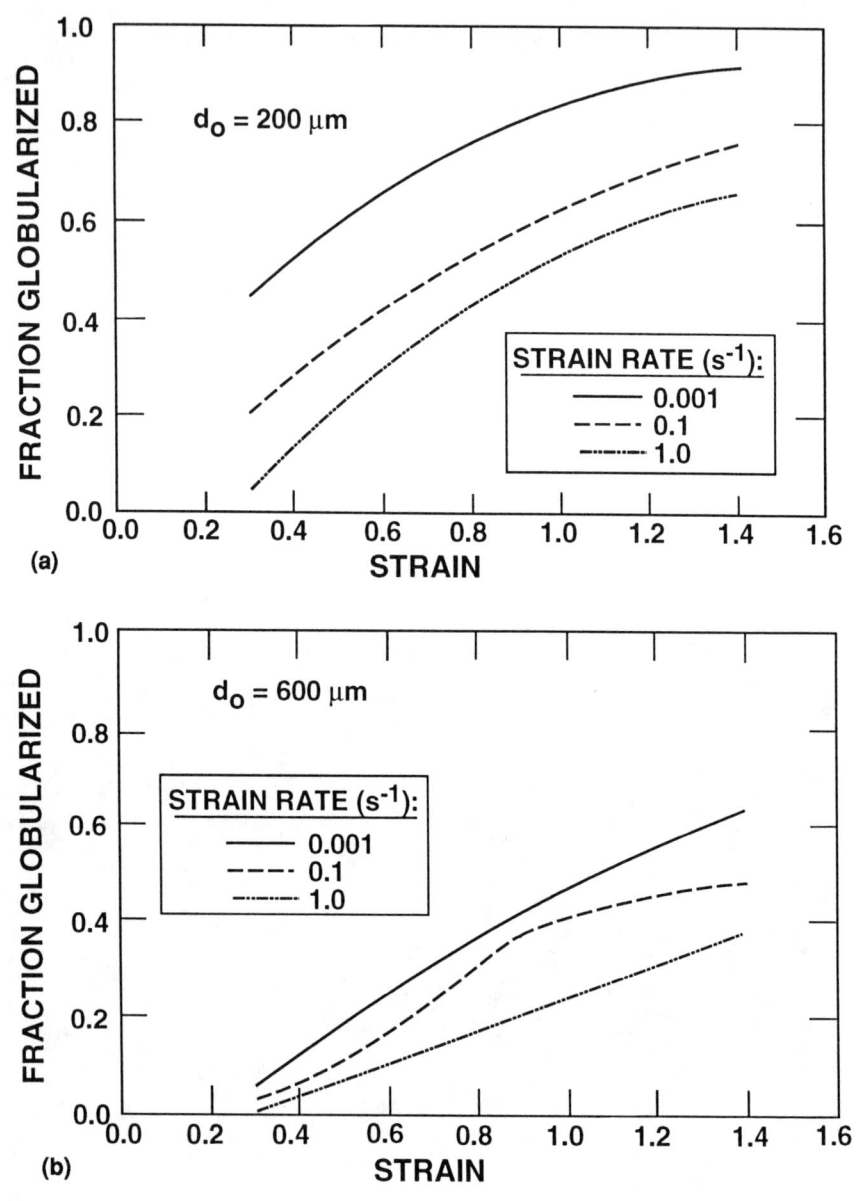

Figure 40 - Fraction globularized microstructure as a function of height strain for samples of Ti-45.5Al-2Cr-2Nb isothermally upset at 1093°C and various constant strain rates. Prior to compression testing, the samples had been processed to yield lamellar microstructures with prior-alpha grain sizes of either (a) 200 µm or (b) 600 µm. (91)

The recrystallization behavior of single-phase gamma was determined by Seetharaman and his coworkers via isothermal hot compression tests on Ti-49.5Al-2.5Nb-1.1Mn (93,94). Again, the percent recrystallized structure varied in a sigmoidal fashion with strain (Figure 41). The logarithm of the recrystallized grain size exhibited a linear dependence on inverse temperature and inverse dependence on strain rate for deformation in the single-phase field (Figure 42). Deformation at higher temperatures resulted in hot-working in the two-phase (alpha + gamma) field and thus decreasing grain size with increasing temperature due to the pinning effect of the second phase.

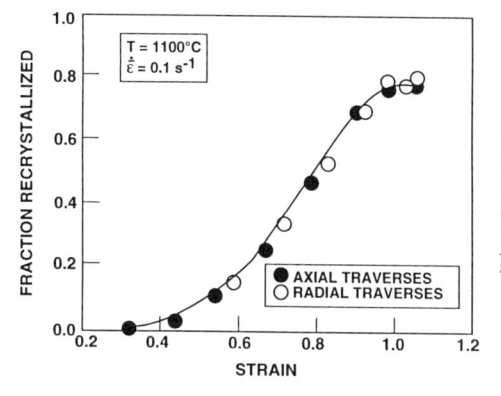

Figure 41 - Fraction recrystallized structure as a function of height strain for cast samples of Ti-49.5Al-2.5Nb-1.1Mn isothermally upset at 1100°C and a constant strain rate of $\dot{\varepsilon} = 0.1$ s^{-1}. (93)

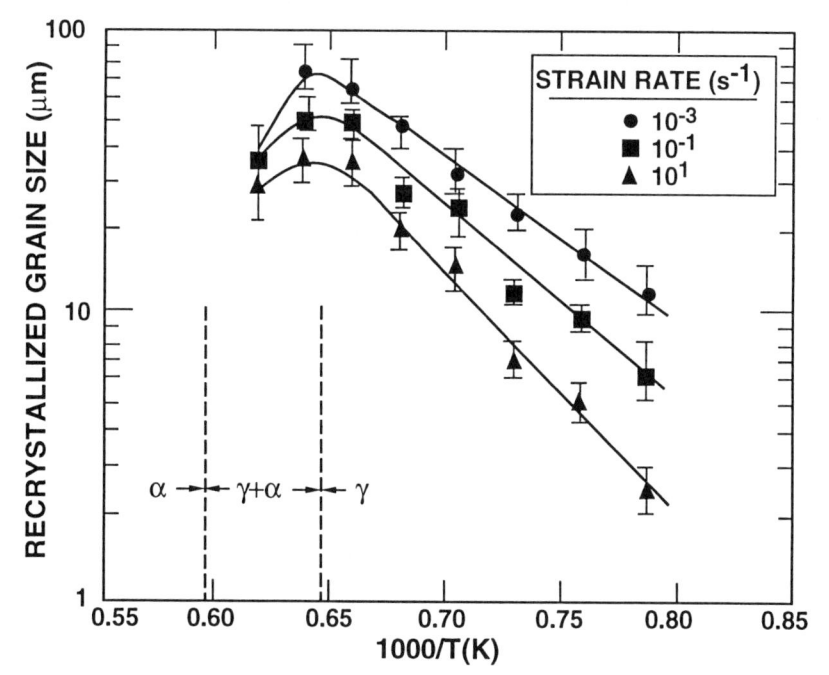

Figure 42 - Temperature dependence of dynamically recrystallized grain size in cast samples of Ti-49.5Al-2.5Nb-1.1Mn isothermally upset at various constant strain rates. (94)

Several novel isothermal forging practices have been developed to enhance the rate of globularization or recrystallization. For the lamellar, near-gamma alloys these include the utilization of a short dwell period (~15 minutes) on the dies midway through the forging stroke in order to effect an increment of static globularization (Figures 43a,b) as well as the use of a two step forging process with an off-line, intermediate furnace heat treatment in the alpha plus gamma phase field (86,95). Although more expensive from a production standpoint, the latter of these two modified practices permits higher overall reductions through the ability to relubricate after the intermediate heat treatment. Another novel practice for near-gamma alloys, known as alpha forging, has evolved from thermomechanical processing principles developed originally for enhancing the properties of high-strength steels. To be specific, alpha forging is analogous to ausforming of steels in which refined microstructures and higher strengths are obtained by deformation of a metastable (high temperature) austenite phase. The corresponding practice for breakdown of near-gamma titanium aluminide ingots comprises billet preheating high in the alpha plus gamma phase field, cooling as rapidly as possible to an isothermal forging temperature substantially lower in this two-phase field, and then forging immediately (95). Success of the process depends, of course, on the ability to retain a large

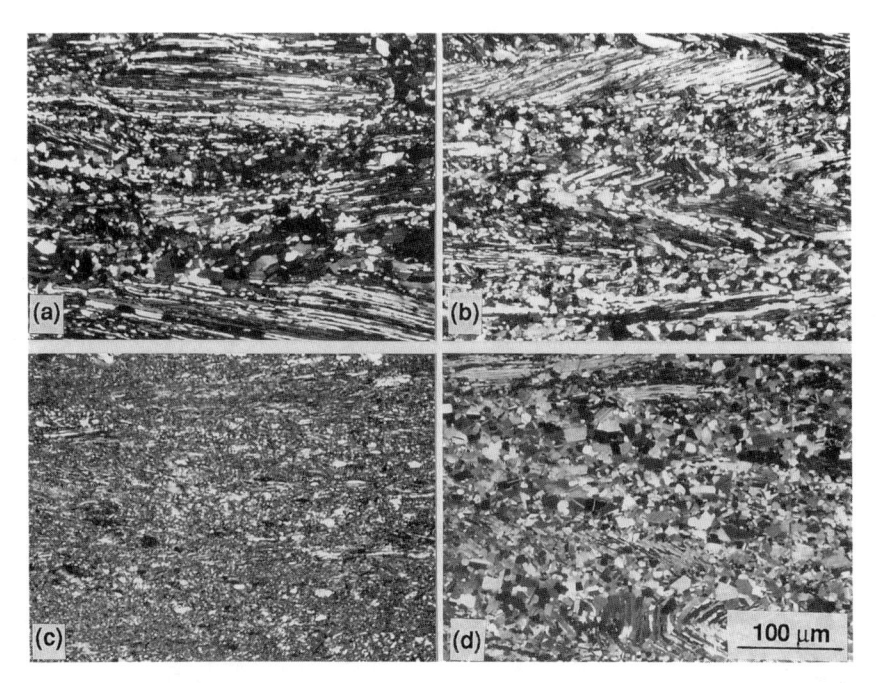

Figure 43 - Polarized light optical microstructure developed in Ti-45.5Al-2Cr-2Nb pancakes upset at 1150°C to a 6:1 reduction using (a) "standard isothermal forging practice, (b) isothermal forging with a 15 minute dwell after the first 2:1 reduction, (c) conventional forging ($\dot{\varepsilon}$ ~ 1 s^{-1}), and (d) alpha forging. (95)

percentage of the metastable alpha phase during cooling. Therefore, the technique is most well-suited for small ingot mults which can be cooled rapidly. A demonstration of the process for the breakdown of cast plus HIP'ed Ti-45.5Al-2Cr-2Nb was described in reference 95. In this example, the preform was preheated at 1260°C (40°C below the alpha transus), cooled to 1150°C within 60s, and then isothermally forged to a 6:1 reduction using a standard, constant ram speed corresponding to a nominal strain rate of 0.0015 s^{-1}. The as-forged pancake in this case exhibited an almost totally globularized microstructure (Figure 43d), which contrasts sharply to the partially broken down microstructure obtained in the same material via standard isothermal forging practice (Figure 43a).

Another variant of standard isothermal forging practice, known as "smart forging", was developed to refine the grain size of single-phase gamma alloys (96). The process relies on three behaviors exhibited by gamma alloys and other difficult-to-work materials: (1) an effective strain of approximately 0.7 to 1.0 is required at typical isothermal forging temperatures and strain rates to achieve full (or nearly full) dynamic recrystallization, (2) the dynamically recrystallized grain size is reduced as the strain rate is increased (Figure 42), and (3) hot workability increases as the grain size is reduced. These trends are applied in smart forging through the utilization of a ram speed which is increased substantially during the forging stroke. As in standard isothermal forging practice, a rather low initial ram speed is used to a reduction between 2:1 and 3:1. At this reduction, a steady state, recrystallized grain size has been achieved, and workability is much improved relative to that of the coarse ingot structure. Therefore, at this point, the ram speed can be increased, typically by one or two orders of magnitude, without workpiece fracture. With this increase in ram speed, the material undergoes another cycle of dynamic recrystallization to a yet finer grain size. For example, Semiatin, *et al.* (96) have obtained a gamma grain size of 30 µm for Ti-51Al-2Mn by this means, compared to the ~150 µm grain size using standard (constant ram speed) isothermal forging. Another advantage of smart forging is the reduction in total forging time as compared to that for the standard practice.

<u>Conventional (Canned) Hot Forging.</u> Conventional hot pancake forging of cast plus HIP'ed near-gamma titanium aluminide alloys such as Ti-45.5Al-2Cr-2Nb and Ti-48Al-2Cr has been successfully demonstrated (95, 97-99). In this process, the dies are usually at ambient or slightly higher (~200°C) temperatures. To minimize die chilling and thus the tendency for fracture, strain rates typical of conventional hot working processes (i.e., ~ 1s^{-1}) are used. However, even with these strain rates, the workpiece must be canned to produce a sound forging. Because of can-workpiece flow stress differences and heat transfer effects, uniform flow of typical can materials (e.g., type 304 stainless steel) and gamma titanium aluminide preforms can be difficult to achieve. To remedy this problem, Jain, *et al.* (97) applied FEM techniques to design cans and select process variables. It was shown that moderately uniform gamma pancake thicknesses can be achieved through a judicious choice of can geometry and can-workpiece insulation. It was demonstrated that the optimal can geometry includes end caps which are chamfered and flush with the outer (lateral) wall of the can. The optimal end cap thickness was found to be a complex function of process variables and material properties. For example, for the forging of a 60-mm diameter x 86-mm length gamma billet in a type 304 stainless steel can, the optimal end cap thickness was approximately 6 mm. The FEM simulation results of Jain, *et al.* also showed that an optimal ram speed (to provide the most uniform flow) can be defined. This speed is a function of the complex deformation and heat transfer processes involved in canned forging.

Because of the higher strain rates involved in conventional hot forging, as compared to those in isothermal forging, more hot work is imparted by the conventional process conducted at the same nominal workpiece temperature and to the same level of reduction. Thus, the as-forged microstructure from conventional hot forging of near-gamma alloys is typically finer, more uniform, and contains very little if any remnant lamellar colonies (Figure 43c). In addition, with optimal can design and insulation, temperature and deformation nonuniformities within the gamma preform can be minimized during conventional forging, and relatively uniform macrostructure and microstructure throughout wrought pancakes are obtained (95).

Conventional (Canned) Hot Extrusion. Considerable effort has been expended to develop conventional (canned) hot extrusion techniques for the breakdown of a variety of near-gamma and single-phase gamma ingot materials. As with conventional hot forging, the selection of can materials and geometry, can-workpiece insulation, and process variables is extremely important with regard to obtaining sound wrought products with attractive microstructures (94,100).

Typical process variables for conventional hot extrusion to breakdown the cast structure of gamma titanium aluminide alloys include ram speeds of 15 to 50 mm/s, reductions between 4:1 and 12:1, and preheat temperatures ranging from 1050°C to 1450°C. Dies with streamline or conical geometry have been used with equal success in round-to-round extrusion. Streamline dies have also been employed in producing round-to-rectangle extrusions to make sheet bar having a width-to-thickness ratio as large as 6:1.

Can materials for conventional hot extrusion are usually type 304 stainless steel (for preheat temperatures of 1250°C or lower) or either Ti-6Al-4V or commercial purity titanium (for preheat temperatures higher than 1250°C). Even with canning, however, substantial temperature (and hence microstructural) nonuniformities may develop during extrusion due to the complex interaction of heat transfer and deformation heating effects. The temperature nonuniformities are most marked for the extrusion of billets of small diameter, i.e., O(75 mm) (Figure 44). These temperature nonuniformities can be decreased, but not eliminated, by the use of insulation between the billet and can. One of the best materials for reducing heat losses has been found to be woven silica fabric (101), although other materials such as various foil alloys are also effective. Nevertheless, even with such measures, the temperature gradients are sufficiently large to produce noticeable radial microstructure variations in the extrudate. For example, Seetharaman, et al. (94) found that the gamma grain size varied from 6 μm to 14 μm from the surface to the center of a Ti-49.5Al-2.5Nb-1.1Mn workpiece extruded at 1050°C to a 6:1 reduction. A similar effect is seen in the "TMP extrusion" (102,103) of near-gamma alloys to obtain fully-lamellar microstructures. This extrusion technique involves billet preheating at or just below the alpha transus temperature. Deformation heating raises the workpiece temperature well into the alpha phase field, thereby promoting recrystallization of single-phase alpha which then transforms to the lamellar structure during cooldown. A typical variation in alpha grain size from the surface to the center of a Ti-45Al-2Cr-2Nb extrusion hot worked by the "TMP extrusion" technique is shown in Figure 45 (104).

Two novel conventional (canned) hot extrusion processes have been developed and applied to breakdown gamma titanium aluminide alloys. These are referred to as "controlled-dwell" extrusion and "equal channel angular extrusion". The controlled-dwell extrusion process is a technique aimed at overcoming difficulties associated with the extrusion of a harder workpiece material such as a gamma titanium aluminide alloy in a softer, inexpensive can material such as a stainless steel. At a given temperature, the flow stress mismatch may be so great that the relative flow of the workpiece and can during extrusion may become nonuniform, sometimes leading to can thinning and failure and then gross fracture of the workpiece when it contacts the cold tooling. This problem of the flow stress mismatch is overcome to a large extent by preheating the canned workpiece in a furnace (or induction heater), removing the assembly from the furnace, and allowing it to air cool for a prespecified, or controlled, dwell period prior to extrusion. The purpose of the controlled dwell is to set up a temperature differential between the can and the workpiece in order to make their respective flow stresses more nearly equal and thus to enhance the uniformity of metal flow during the deformation process. Such a practice contrasts sharply with standard techniques in which efforts are usually made to *minimize* the transfer time. The controlled-dwell technique has been applied for the extrusion of a number of near-gamma as well as single-phase gamma alloys (105).

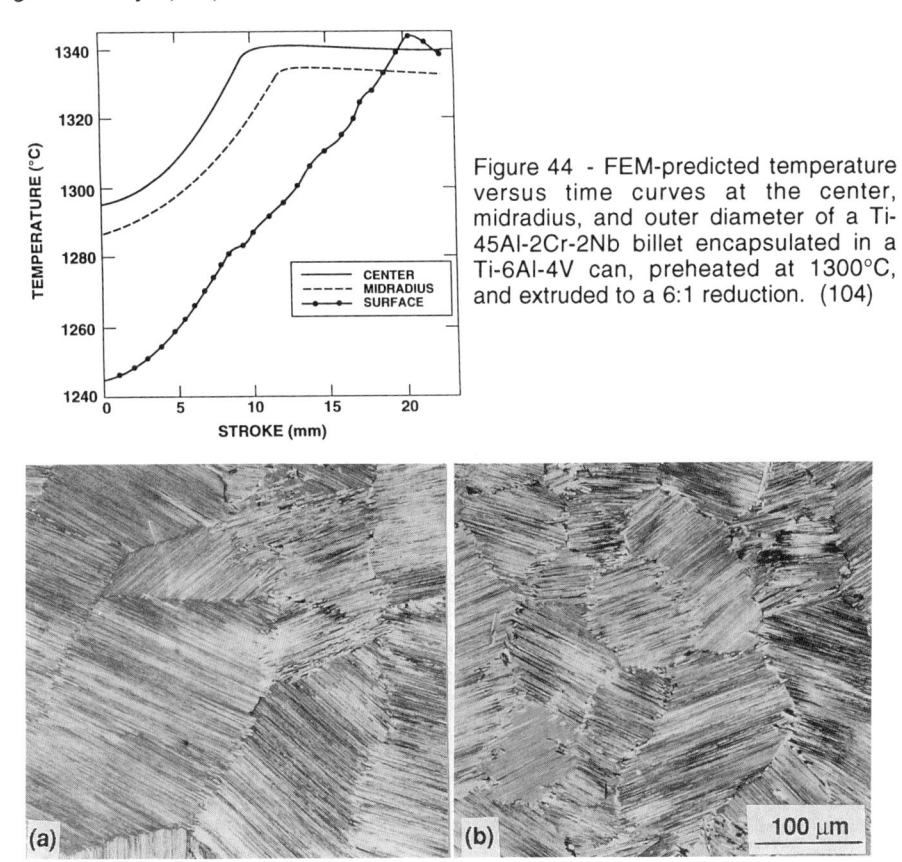

Figure 44 - FEM-predicted temperature versus time curves at the center, midradius, and outer diameter of a Ti-45Al-2Cr-2Nb billet encapsulated in a Ti-6Al-4V can, preheated at 1300°C, and extruded to a 6:1 reduction. (104)

Figure 45 - Polarized light optical microstructures developed in a canned Ti-45Al-2Cr-2Nb billet preheated at the alpha transus temperature and extruded to a 6:1 reduction: (a) center of extrudate, (b) outer diameter of extrudate. (104)

Developed in the former Soviet Union, the equal channel angular extrusion (ECAE) process (106) offers an attractive alternative for the primary breakdown of conventional and advanced alloys. In the process, an ingot (or prior-worked billet) is extruded through a channel consisting of two continuous sections situated at an angle 2ϕ to each other (Figure 46). The imposed strain per pass is a function of the channel angle. Much larger strains can also be imposed in a given set of tooling through the use of multi-pass extrusion which is possible because the cross-sectional area of the workpiece is not changed. The ability to impart large deformations without a change in cross-section permits smaller ingots to be melted to obtain a given size of semi-finished product. Hence, the process is especially useful for materials prone to macrosegregation during the casting of large ingots. Other advantages of ECAE are moderate working pressures (compared to extrusion through converging or shear dies) and the ability to control crystallographic and mechanical texture during multi-pass ECAE by judicious rotation of the workpiece between passes. The broad feasibility of using ECAE for breakdown of canned gamma titanium aluminide ingot materials was demonstrated recently (107).

Secondary Processing

The development of uniform, fine microstructures during breakdown of ingot-cast gamma titanium aluminide alloys leads to improved workability with regard to both fracture resistance and reduced flow stresses. These improvements are useful in secondary processes such as sheet rolling, superplastic forming of sheet, and isothermal, closed-die forging.

Two major techniques for rolling of sheet have evolved from the early work on near-isothermal, hot pack rolling conducted by Hoffmanner, et al. (108). These methods are conventional hot pack rolling and bare isothermal rolling. With regard to the former process, pack design (e.g., cover thickness, use of parting agents, etc.) and the selection of rolling parameters have been aided by the development of models which quantify the temperature transients and hence the stresses during rolling (109,110) as well as information on the hot workability of gamma titanium aluminide alloys (111). For the near-gamma titanium aluminide alloys, rolling is usually most easily conducted in the alpha plus gamma phase field at temperatures approximately 40 to 150°C below the alpha transus using reductions per pass of 10 to 15 percent and rolling speeds that produce effective strain rates of the order of 1 s^{-1}. Using these parameters, sheets as large as 400 mm x 700 mm and ranging in thickness from 0.2 to 2.0 mm have been produced (112). In addition, a variety of microstructures have been developed in rolled sheet products (for example, those in Figure 47 for Ti-45.5Al-2Cr-2Nb), some with gamma grain sizes as small as 2 µm (110). Sheets rolled and then "direct" heat treated have exhibited an even wider range of microstructures (113). The conventional hot pack rolling of nominally single-phase gamma alloys has been conducted with far less success. Alloys of this class which have been rolled successfully are generally those with lower aluminum content (~48 to 52 atomic percent) which can be heated into the alpha plus gamma phase field. Presumably, the presence of two phases in such instances retards grain growth during preheating and reheating and thus the accompanying sharp losses in hot workability.

An alternate sheet fabrication technique, involving rolling of *uncanned* gamma titanium aluminide preforms under isothermal conditions, has been developed and demonstrated (on a laboratory scale) by Kobe Steel, Ltd. (114,115). The rolling equipment includes a mill with 300-mm wide, 60-mm diameter, ceramic work rolls and 150-mm diameter TZM molybdenum backup rolls. The rolls and gamma workpiece are enclosed in a vacuum chamber and heated under an argon atmosphere. To date, Ti-46Al and Ti-51Al binary alloys have been rolled to 0.75 to 1.0 mm-thick, 150 mm-wide sheet from 3.0 mm-thick preforms using this equipment. Typical processing parameters include a preform/roll temperature between 1000 and 1100°C, rolling speed between 2 and 6 mm/minute, and reduction per pass between 5 and 15 percent. For the reduction per pass and rolling speed utilized, the effective strain rate of the preform as it is rolled is approximately 10^{-3} s^{-1}, or a rate at which the workability of gamma alloys is good in both the as-cast and wrought forms. Unfortunately, these very low strain rates lead to relatively long processing times. However, the microstructures produced by isothermal rolling (115) are similar to those produced by conventional, hot pack rolling conducted under higher temperature/higher strain rate conditions.

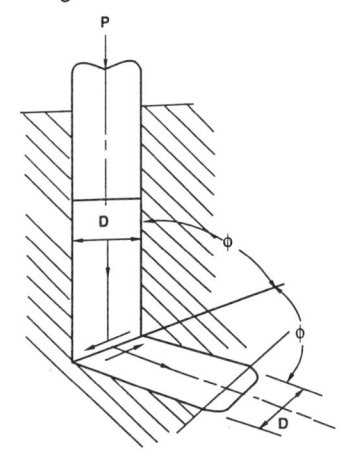

Figure 46 - Schematic illustration of the equal channel angular extrusion process. (106)

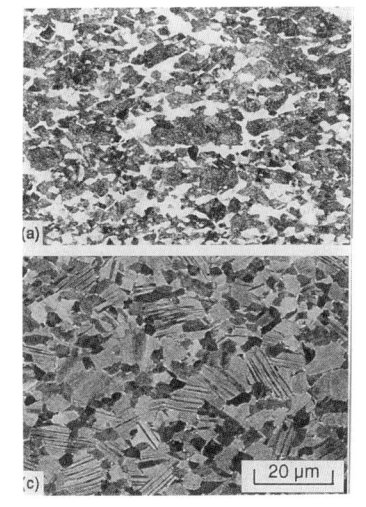

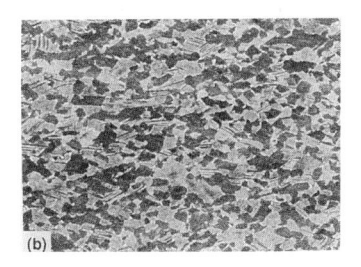

Figure 47 - Backscattered micrographs of the microstructure developed in Ti-45.5Al-2Cr-2Nb sheets after pack rolling at nominal (furnace) temperatures of (a) 1175°C, (b) 1225°C, or (c) 1260°C. (110)

The fact that fine-grained microstructures can be developed in two-phase (alpha-two + gamma) or three-phase (alpha - two + gamma + beta) near-gamma titanium aluminide alloys during ingot breakdown and/or rolling suggests that these materials might be prime candidates for superplastic forming. With this possibility in mind, Lombard (116) reviewed a number of the phenomenological observations for these materials in the literature (Table IX). The alloys investigated had a wide range of aluminum contents, microstructures, and degrees of microstructural refinement. In addition, the materials were tested over a wide range of temperatures and strain rates. For most of the test conditions, the m values generally ranged from 0.4 to 0.8, or conditions under which tensile ductilities in the range of approximately 800 to 8000 percent might be expected (125). In the vast majority of the cases, however, the observed tensile ductilities were much less (i.e., ~200 to 500 percent), thus suggesting the influence of fracture processes such as cavitation in controlling formability. A noteworthy exception to this general trend was the achievement of an elongation of approximately 1000 percent at T = 1200°C and a nominal strain rate of 10^{-3} s^{-1} in rolled, fine-gamma grain samples of Ti-45.5Al-2Cr-2Nb (124). Under these test conditions, the m value was estimated to be between 0.4 and 0.5. From a mechanistic standpoint, the high m values obtained in gamma titanium aluminide alloys have been explained using models developed for conventional metallic materials. These models include those based on either grain-boundary sliding (accommodated by diffusional flow) or grain-boundary deformation (core-mantle model). In particular, Hashimoto,et al.(126) invoked the core-mantle model to explain the higher elongations in chromium-bearing gamma aluminide alloys. According to their hypothesis, the beta phase (stabilized by the chromium addition) which surrounds the gamma grains serves as the mantle providing a source of geometrically necessary dislocations.

In isothermal, closed-die forging, the high m values developed in fine-grained gamma titanium aluminide alloys deformed at low strain rates have been utilized to make parts such as jet-engine blades (103). Forging results obtained thus far have shown excellent die-filling capability even in complex areas such as blade root sections. The enhanced metal flow of the fine-grained titanium aluminide alloys has also spurred efforts to develop higher rate, conventional forging processes (using

Table IX Hot Tension Test Data for Near-Gamma Titanium Aluminide Alloys (116)

Alloy	Test Temperature (°C)	Strain Rate (s^{-1})	m	Elongation (pct.)	Reference
Ti-49.5Al	947	10^{-4}	0.40	>200	117
Ti-48.1Al-0.8Mo	1000	5×10^{-5}	0.76	230	118
Ti-49.8Al	850	8.3×10^{-4}	0.46	260	119
Ti-43Al	1050	2.5×10^{-4}	---	275	120
Ti-47Al	1050	2.8×10^{-4}	0.70	398	121
Ti-46.1Al-3.1Cr	1200	5.4×10^{-4}	0.57	450	122
Ti-47.3Al-1.6Cr-1.9Nb-0.5Si-0.4Mn	1280	8×10^{-5}	0.65	470	123
Ti-45.5Al-2Cr-2Nb	1200	10^{-3}	0.45	980	124

unheated tooling) for parts such as automotive engine valves (103). In this regard, insight into microstructure changes during the nonisothermal forging of wrought near-gamma titanium aluminide alloys has been provided in the work of Seetharaman and Semiatin (127). Specifically, the lamellar decomposition of the metastable, high temperature alpha phase and its subsequent globularization under transient temperature conditions were elucidated.

Crystallographic Texture

Because of the tetragonality of the fct crystal structure of TiAl and the possible variation in critical resolved shear stresses for the various deformation modes (via ordinary dislocations, superlattice dislocations, or twinning), the development of crystallographic texture in gamma and near-gamma titanium aluminide alloys may have significant influence on final properties. Fukutomi, Hartig, Mecking, and their coworkers (128-130) have pioneered the investigation of texture for these alloys.

Fukutomi, et al. (128) measured textures in arc melted samples of nearly single-phase Ti-50Al subjected to a 6:1 reduction via hot compression under low strain rate, high temperature conditions (e.g., $\dot{\varepsilon} = 10^{-4}$ s^{-1}, T = 1200°C) or higher strain rate, low temperature conditions (e.g., $\dot{\varepsilon} = 10^{-3}$ s^{-1}, T = 1000°C). In both instances, dynamic recrystallization was observed. However, the recrystallization mechanism and the type of texture differed. For the low strain rate, high temperature deformation regime, recrystallization was noted to occur via a mechanism comprising grain boundary bulging; the crystallites *behind* the bulged boundaries continued to deform during recrystallization of those in front of the boundaries. By this means, very strong fiber textures (8 x random) formed; these consisted of (101) planes parallel to the compression plane. (Note that in the TiAl crystal structure of alternating 002 planes of titanium and aluminum atoms, planes and directions which would be in the same family for a disordered fcc crystal, are *not* identical. Thus texture specifications for TiAl must connote specific planes and directions.) Later, more refined texture analysis by Mecking and Hartig indicated that these sharp textures were closer to (302). In contrast to these low strain rate, high temperature textures, very weak (101) textures (2 x random) were found to develop during hot compression of Ti-50Al under higher strain rate, low temperature test condition. In these cases, recrystallization occurred by nucleation and growth of many finer grains at the initial gamma grain boundaries, thus randomizing the texture.

Hartig, et al. (129) measured deformation textures similar to that observed by Fukutomi, et al. (128) in samples of Ti-50Al upset at $\dot{\varepsilon} = 10^{-6}$ s^{-1} and T = 450°C. Although the material did not recrystallize under these conditions, the texture was relatively weak, presumably because of the light reduction (less than 2:1). In parallel work, Hartig, et al. also measured textures in sheets of Ti-50Al pack rolled in multiple passes to a 4:1 reduction at T = 1050°C and $\dot{\varepsilon} \approx 0.25$ s^{-1}. After this deformation, a moderate (3 x random) cube-like texture component (100) [010] ((100) ‖ rolling plane, [010] ‖ rolling direction), which is similar to the common recrystallization or cube texture in fcc sheets, was measured. In addition, an attempt was made to interpret the marked anisotropy in room temperature yield strength measured for the rolled sheet. This was done using the measured texture and classical aggregate theory modified to account for possible critical resolved shear stress anisotropies between ordinary dislocation and superdislocation slip systems. Although in qualitative agreement with observations, the aggregate theory calculations predicted a property anisotropy much less than that measured, thereby suggesting the importance of mechanical texturing of the alpha-two phase as an important factor in controlling properties.

Mecking and Hartig (130) also found a moderate cube-like texture component in the gamma phase after the hot-rolling of the near-gamma titanium aluminide alloy Ti-48Al-2Cr. In addition, a weak crystallographic texture (2 x random) was measured after extrusion of the near-gamma alloy Ti-46Al-1Cr-0.2Si; however, a very noticeable mechanical texture in terms of the orientation of the lamellar alpha-two and gamma plates was found to develop as well.

Workability Considerations

The near-gamma and single-phase gamma titanium aluminide alloys have a somewhat lower hot workability compared to the conventional alpha/beta and beta titanium alloys. This behavior, in conjunction with the substantially higher temperatures required in general for hot-working of the intermetallic alloys, has spurred research to quantify and understand the failure modes of the gamma alloys under hot working conditions. As for conventional titanium alloys, failures are frequently of two types: fracture-controlled and flow-localization-controlled.

Fracture-Controlled Failure. As for alpha/beta titanium alloys, the phenomenon of wedge crack and cavity formation appears to play an important role with respect to fracture. Research now suggests that two major regimes can be defined with regard to overall fracture behavior. One is a low-temperature, high-strain rate regime in which wedge crack initiation occurs at very low deformation levels and leads to very brittle, intergranular type failures. Such fracture behavior is of utmost importance in the design of conventional, high-rate working operations such as forging, rolling, and extrusion. The other regime comprises higher-temperature, low-strain rate deformation in which cavity formation, growth, and coalescence occurs and is a gradual process, thereby giving rise to moderate-to-high hot ductility. An understanding of this type of fracture response is important with regard to the design of processes such as isothermal forging.

The work of Seetharaman, *et al.* (131) and Semiatin, *et al.* (111) has sought to provide a first step toward quantifying the factors that control the brittle, intergranular mode of fracture during hot working of gamma titanium aluminide alloys. In the former work, the hot tension behavior of Ti-49.5-2.5Nb-1.1Mn in both the cast plus HIP'ed condition and the wrought condition was established. Both material conditions exhibited a complex relationship between reduction in area, test temperature, and strain rate (Figure 48). However, in broad terms, the data revealed transitions from brittle behavior (with extensive wedge cracking) to ductile behavior over a rather narrow temperature. For each material condition, the brittle-to-ductile transition temperature increased with increasing strain rate (Figure 49). Furthermore, the transition temperatures for a given strain rate were higher for the coarse-grained, cast material than for the finer-grain, wrought material. An Arrhenius-type of analysis of the transition-temperature data yielded values of activation energy comparable to those that describe the dynamic recrystallization of gamma titanium aluminide alloys during hot compression testing (132). From this analysis, it was thus concluded that the onset of dynamic recrystallization was the mechanism by which brittle fracture was suppressed.

The observations of wedge crack formation in isothermal, hot-compression testing of wrought Ti-48Al-2.5Nb-0.3Ta also showed a strong effect of strain rate on fracture behavior (111). For temperatures below the alpha transus of this alloy

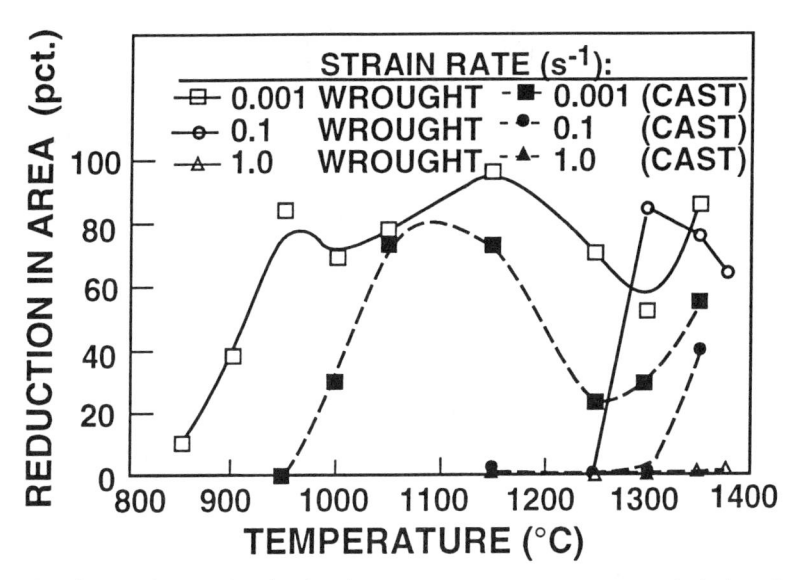

Figure 48 - Dependence of reduction in area on test temperature and strain rate for hot tension tests on cast or wrought (extruded + recrystallization heat treated) samples of Ti-49.5Al-2.5Nb-1.1Mn. (131)

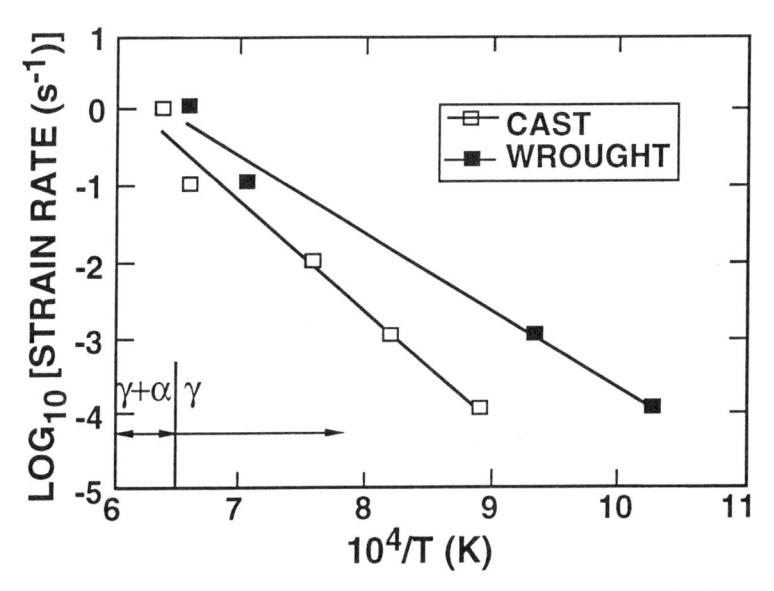

Figure 49 - Relationship between strain rate and inverse of the ductile-to-brittle transition temperature for the hot tension behavior of cast and wrought (extruded + recrystallization heat treated) Ti-49.5Al-2.5Nb-1.1Mn. (131)

(1360°C), the severity of wedge cracking increased with decreasing temperature and increasing strain rate (Table X); as in the work of Seetharaman, *et al.* (131), these conditions correspond to those requiring high applied stresses for hot working. In addition, a second deformation regime, consisting of high strain rates in the single phase alpha field (e.g., T = 1370°C), was also found to give rise to wedge cracks (Figure 50). In these instances, the flow stresses are much lower, but supertransus preheating gives rise to considerable coarsening of the microstructure. The results of Seetharaman, *et al.* and Semiatin, *et al.* thus demonstrate the importance of both applied stress and the scale of microstructural features in controlling wedge cracking. Preliminary analysis suggests that a fracture criterion in terms of the product of the applied stress and square root of the grain size, much like the Griffith criterion, may be useful in predicting brittle failures due to wedge cracking.

Table X Occurrence of Wedge Cracks in Wrought
Ti-48Al-2.5Nb-0.3Ta Hot Compression
Samples Deformed to 50 Pct. Height Reduction (111)

Temperature (°C)	$\dot{\varepsilon} = 0.1\ s^{-1}$	$\dot{\varepsilon} = 1\ s^{-1}$	$\dot{\varepsilon} = 10\ s^{-1}$
1150	X	XX	XXX
1260	X	XX	XXX
1315	X	X	X
1370	X	XX	XXX

X - no wedge cracks; XX - some wedge cracks; XXX - many wedge cracks

Figure 50 - Optical micrographs from polished-and-etched sections of wrought Ti-48Al-2.5Nb-0.3Ta samples isothermally upset to a 2:1 height reduction at $\dot{\varepsilon}$ = 10 s^{-1} and test temperatures of (a) 1260°C (T$_\alpha$-100°C) or (b) 1370°C (T$_\alpha$+10°C), indicating presence of wedge cracks. The compression axis is vertical in both micrographs. (111)

The work of Semiatin, et al. (111) also revealed that the occurrence of wedge cracking during hot compression testing could be used to predict gross fracture during pack rolling of gamma titanium aluminide alloys. In this operation, secondary (rolling direction) tensile stresses are generated when a relatively high flow stress material (the titanium aluminide) is packed and rolled within a material with a lower flow stress (133). These tensile stresses cause microscopic wedge cracks to propagate giving rise to fractures that lie transverse to the rolling direction.

Secondary tensile stresses also play an important role in the more ductile failures that result from cavity formation, growth, and coalescence which occur under low strain rate deformation conditions. One example of such failures are the cracks that develop at the bulged free surfaces of gamma titanium aluminide alloys during open-die forging processes. The kinetics of this type of failure were analyzed by Seetharaman, et al. (134) who investigated free surface cracking of Ti-49.5Al-2.5Nb-1.1Mn during pancake forging of cylindrical mults. The degree of bulge during pancake forging, and thus the level of the tensile stresses, is a function of the instantaneous aspect ratio of the workpiece and the die-workpiece interface friction conditions. For the specific geometry and interface friction involved in the experiments of Seetharaman, et al., it was found that the free-surface fracture could be predicted using the maximum tensile work criterion first proposed by Cockcroft and Latham (135) (for ductile fracture under cold-working conditions). To this end, the "critical" tensile work to fracture, C*, was measured in uniaxial tension tests and compared to the work C done through the maximum tensile stress at the equator of the bulge developed during isothermal pancake forging, as estimated using a finite element modeling (FEM) technique. This comparison is illustrated in Figure 51. The broken line corresponds to the critical value C* for strain rates $\dot{\varepsilon}$ between 10^{-3} and 10^{-2} s^{-1} at a temperature of 1150°C. The shaded bands are the FEM - predicted values for C as a function of height strain ε for friction factors between 0.2 and 0.4, or the typical range of measured friction factors. For $\dot{\varepsilon} = 10^{-3}$ s^{-1}, the values of C ($\bar{\varepsilon}$)

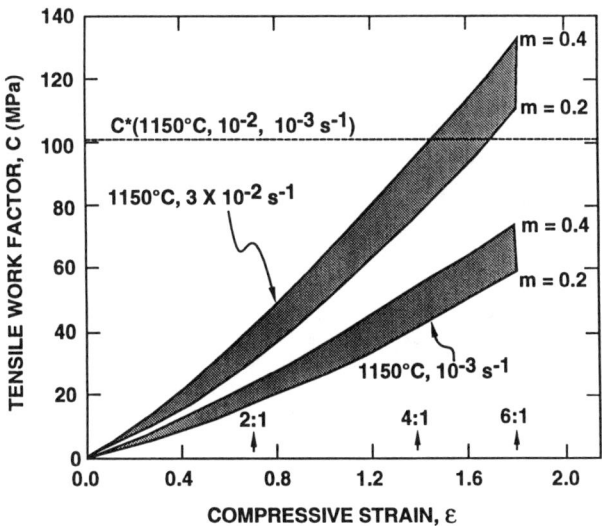

Figure 51 - Comparison of the tensile work factor C during isothermal forging of cast + HIP'ed Ti-49.5Al-2.5Nb-1.1Mn pancakes and the critical tensile work for fracture, C*. (134)

are well below the critical damage factor C*; thus free surface fracture would not be expected. Experimentally, it was found that for reductions as high as 6:1 at this strain rate and a temperature of 1150°C, no free surface fractures were observed, thus validating the predictions. On the other hand, for $\dot{\varepsilon} = 3 \times 10^{-2}$ s^{-1}, C is below C* for a 2:1 reduction, but becomes comparable to C* at a 4:1 reduction. Thus, a tensile type of fracture is predicted at deformations near the latter reduction, as indeed was observed.

Limits on the hot workability of gamma titanium aluminide alloys due to the generation of secondary tensile stresses and thus cavitation can be extended by superimposing high levels of hydrostatic pressure (136). The resultant improvements in ductility are analogous to the well-documented effects of pressure on the ductility of a number of conventional metallic alloys.

Shear-Localization-Controlled Failure. As for conventional titanium alloys, shear localization and shear fracture during hot working of gamma titanium aluminide alloys are most common in deformation modes that are plane strain or simple shear in nature. For example, the workability of the near-gamma alloy Ti-45.5Al-2Cr-2Nb during equal channel angular extrusion (ECAE) was shown to be limited by shear localization (107). A relatively sound product was produced by extrusion at 1250°C of cast plus HIP'ed material of this composition canned in type 304 stainless steel, but shear bands and gross shear cracks were developed when extrusion was attempted at 1150°C (Figure 52). These observations were explained in terms of the effect of extrusion temperature on the magnitude of the flow localization, or 'alpha', parameter in simple shear. The alpha parameter is defined as the ratio of the normalized flow softening rate (from stress-strain curves not corrected for deformation heating effects) to the strain rate sensitivity exponent (i.e., m value); a large amount of research on a variety of metals has established that marked flow localization may be expected for values of the alpha parameter of about 4 to 5 or greater (125). The cast plus HIP'ed Ti-45.5Al-2Cr-2Nb alloy exhibits a sharp increase in the alpha parameter and thus susceptibility to shear failure with decreasing temperature (Figure 53). In particular, for extrusion at 1150°C or 1250°C, this parameter is approximately 6 or 3, respectively. Thus, based on an alpha = 5 criterion, noticeably nonuniform flow would be expected for the lower temperature extrusion, but not the higher, as was observed. On the other hand, values of the flow localization parameter for Ti-45.5Al-2Cr-2Nb in the *wrought* condition (Figure 53) are somewhat lower, thus suggesting the design of multipass ECAE sequences involving an initial high temperature pass for ingot breakdown followed by lower temperature passes for microstructure refinement.

Future Outlook

The hot-working of titanium alloys is a fairly mature technology. To a large extent, engineering breakthroughs have paced the introduction of new process technology. However, the development of a number of concepts regarding microstructure and texture evolution and workability have led to improvements in process design and control as well as improvements in product quality, yield, and performance. This fundamental understanding has been and will continue to be instrumental in the development and commercialization of intermetallic alloys such as the titanium aluminides. Because of the narrow processing window and very high working temperatures associated with these materials, such a science-based approach, coupled with the development and utilization of advanced computer models, will be critical to their future development as engineering materials.

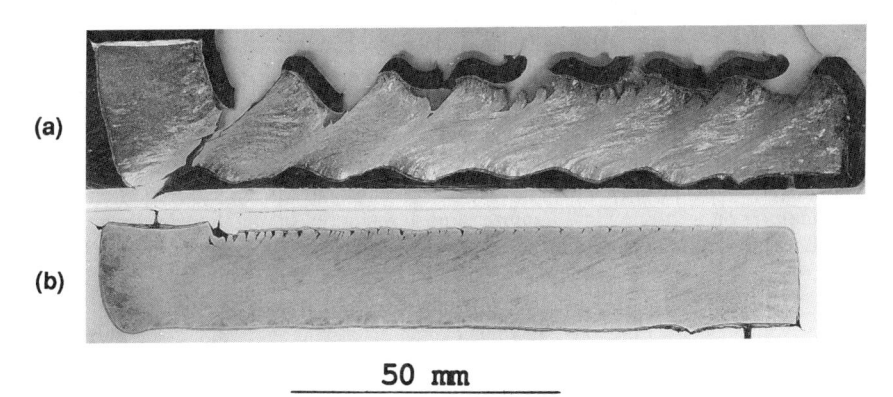

Figure 52 - Micrographs of sections of canned samples of cast + HIP'ed Ti-45.5Al-2Cr-2Nb deformed via equal channel angular extrusion at (a) 1150°C or (b) 1250°C. (107)

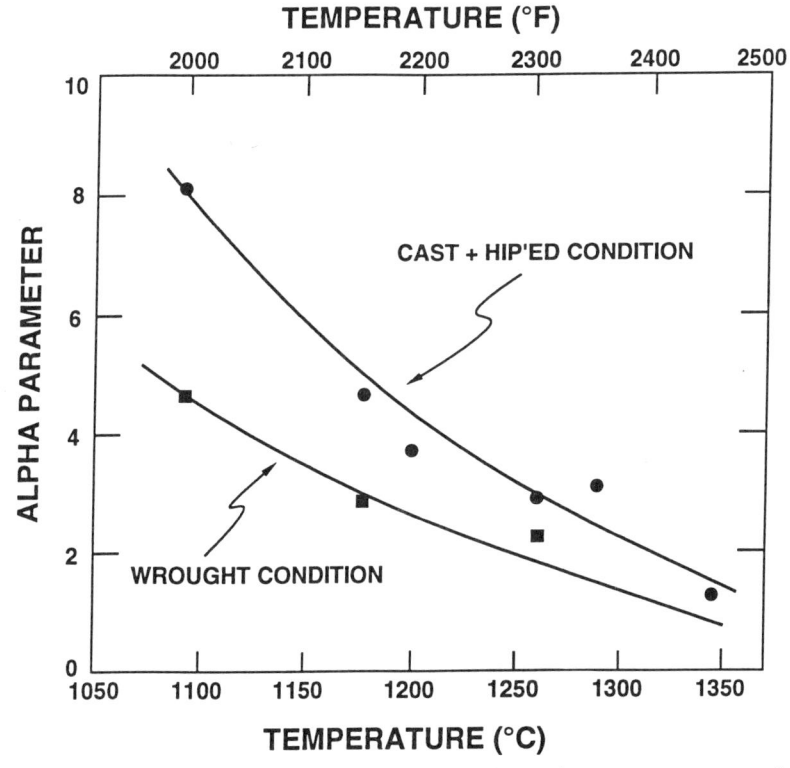

Figure 53 - Dependence of the flow localization (alpha) parameter on preheat temperature and material condition for Ti-45.5Al-2Cr-2Nb deformed in simple shear at an effective strain rate of 1 s^{-1}. (107)

Acknowledgments

Much of the work summarized here has been supported over many years by the basic research and exploratory development programs of the US Air Force. In particular, the enthusiastic support and encouragement of the management of the Air Force Office of Scientific Research and the Wright Laboratory Materials Directorate have been vital to these efforts. Technical discussions with many colleagues over a number of years have helped greatly in the formulation of much of the work contained in this review. In this regard, a special thanks is extended to M.K. Alam, D.D. Bhatt, J.D. Bryant, H.M. Burte, P. Dadras, D.M. Dimiduk, R.L. Goetz, A.L. Hoffmanner, S.K. Jain, V.K. Jain, J.J. Jonas, W.R. Kerr, Y-W. Kim, G.D. Lahoti, P.L. Martin, P.A. McQuay, S.I. Oh, A.R. Rosenstein, J.F. Thomas, Jr., C.H. Ward, and T.L. Wardlaw. The assistance of L. Farmer and J. Paine in preparing the manuscript is gratefully acknowledged. Drs. A.P. Woodfield and H. Inagaki graciously supplied several of the figures used in this paper.

References

(In the references listed below, citations to the various International Titanium Conferences have been abbreviated as follows: ITC 3 - J.C. Williams and A.F. Belov, eds., *Titanium and Titanium Alloys: Scientific and Technological Aspects* (New York, NY: Plenum Press, 1982); ITC 4 - H. Kimura and O. Izumi, eds., *Titanium '80: Science and Technology* (Warrendale, PA: TMS, 1980); ITC 5 - G. Luetjering, U. Zwicker, and W. Bunk, eds., *Titanium: Science and Technology* (Oberursel, Germany: Deutsche Gesellschaft für Metallkunde e.V., 1985); ITC 6 - P. Lacombe, R. Tricot, and G. Béranger, eds., *Sixth World Conference on Titanium* (Les Ulis Cedex, France: Societé Française de Metallurgie, 1988); ITC 7 - F.H. Froes and I.L. Caplan, eds., *Titanium '92: Science and Technology* (Warrendale, PA: TMS, 1993).)

1. E.W. Collings, *The Physical Metallurgy of Titanium Alloys* (Materials Park, OH: ASM International, 1984).
2. A. Choudhury and E. Weingartner, "Vacuum Arc Remelting", in *Metals Handbook, Volume 15: Casting* (Materials Park, OH: ASM International, 1988), 406-408.
3. F.H. Froes, D. Eylon, and H.B. Bomberger, "Melting, Casting, and Powder Metallurgy", *Lesson 8, Titanium and Its Alloys* (Materials Park, OH: Materials Engineering Institute, ASM International, 1994).
4. H. Hayakawa, N. Fukada, M. Koizumi, and H.G. Suzuki, "A New Method to Produce Segregation-Free Ingot", in ITC 7, 2317-2324.
5. J.S. Myers, "Primary Working", *Lesson 9, Titanium and Its Alloys* (Materials Park, OH: Materials Engineering Institute, ASM International, 1994).
6. S.L. Semiatin, J.F. Thomas, Jr., and P. Dadras, "Processing-Microstructure Relationships for Ti-6Al-2Sn-4Zr-2Mo-0.1Si", *Metall. Trans. A*, 14A (1983), 2363-2374.
7. J.G. Malcor, F. Montheillet, and B. Champin, "Mechanical and Microstructural Behavior of Ti-6Al-4V Alloy in the Hot Working Range", in ITC 5, 1495-1502.
8. E. Rauch, G.R. Canova, and J.J. Jonas, unpublished research, McGill University, Montreal, Canada, 1982.
9. A.A. Korshunov, F.U. Enikeev, M.I. Mazurskii, G.A. Salishchev, A.V. Muravlev, P.V. Chistyakov, and O.O. Dimitriev, "Effect of Method of High Temperature Loading on Transformation of Lamellar Structure in VT9 Titanium Alloy", *Russ. Metall.*, 3 (May-June 1994), 103-108.

10. B.J. Lee, S. Ahzi, B.K. Kad, and R.J. Asaro, "On the Deformation Mechanisms in Lamellar Ti-Al Alloys", *Scripta Metall. et Mater.*, 29 (1993), 823-828.
11. M. Dao, B.K. Kad, and R.J. Asaro, "Deformation and Fracture Under Compressive Loading in Lamellar TiAl Microstructures", *Phil. Mag. A*, 72 (1995), in press.
12. S.L. Semiatin and V. Seetharaman, unpublished research, Wright Laboratory Materials Directorate, Wright-Patterson AFB, OH, 1995.
13. H. Margolin and P. Cohen, "Evolution of the Equiaxed Morphology of Phases in Ti-6Al-4V", in ITC 4, 1555-1561.
14. H. Margolin and P. Cohen, "Kinetics of Recrystallization of Alpha in Ti-6Al-4V", in ITC 4, 2991-2997.
15. I. Weiss, G.E. Welsch, F.H. Froes, and D. Eylon, "Mechanisms of Microstructure Refinement in Ti-6Al-4V", in ITC 5, 1503-1510.
16. I. Weiss, F.H. Froes, D. Eylon, and G.E. Welsch, "Modification of Alpha Morphology in Ti-6Al-4V by Thermomechanical Processing", *Metall. Trans. A*, 17A (1986), 1935-1947.
17. G. Welsch, I. Weiss, D. Eylon, and F.H. Froes, "Shear Deformation and Breakup of Lamellar Morphology in Ti-6Al-2Sn-4Zr-2Mo Alloy", in ITC 6, 1289-1293.
18. S.L. Semiatin, J.C. Soper, and R. Shivpuri, "A Simple Model for Conventional Hot Rolling of Sheet Materials", *Metall. and Mater. Trans. A*, 25A (1994), 1681-1692.
19. R.B. Sparks and J.E. Coyne, "Microstructural Control of Large Section Titanium Forgings", in ITC 3, 413-422.
20. G.W. Kuhlman, "A Critical Appraisal of Thermomechanical Processing of Structural Titanium Alloys", in Y-W. Kim and R.R. Boyer, eds., *Microstructure/Property Relationships in Titanium Aluminides and Alloys* (Warrendale, PA: TMS, 1991), 465-491.
21. J.C. Williams and E.A. Starke, Jr., "The Role of Thermomechanical Processing in Tailoring the Properties of Aluminum and Titanium Alloys", in G. Krauss, ed., *Deformation, Processing, and Structure* (Materials Park, OH: ASM International, 1984), 279-354.
22. R.I. Jaffee, G. Luetjering, and T.M. Rust, "Production and Properties of Bimodal Ti-6Al-4V Blades for Steam Turbine Application", in ITC 5, 1081-1088.
23. J.E. Coyne, C.J. Scholl, and D.E. Batzer, "Secondary Working of Bar and Billet", *Lesson 10, Titanium and Its Alloys* (Materials Park, OH: Materials Engineering Institute, ASM International, 1994).
24. T. Sheppard and J. Norley, "Deformation Characteristics of Ti-6Al-4V", *Materials Science and Technology*, 4 (1988), 903-908.
25. J.F. Uginet, "Thermomechanical Treatment of Various Beta Processed Titanium Alloys", in ITC 7, 463-471.
26. Y. Liu and T.N. Baker, "Deformation Characteristics of IMI 685 Titanium Alloy Under Beta Isothermal Forging Characteristics", *Materials Science and Engineering*, A197 (1995), 125-131.
27. M. Okada, Y. Sato, M. Monkawa, and Y. Hayamizu, "High Strength Extruded Shapes of Ti-6Al-6V-2Sn Alloy for Airframe Materials", *The Sumitomo Search*, 46 (April 1991), 38-41.
28. W.R. Kerr, P.R. Smith, M.E. Rosenblum, F.J. Gurney, Y.R. Mahajan, and L.R. Bidwell, "Hydrogen as an Alloying Element in Titanium (Hydrovac)", in ITC 4, 2477-2486.
29. F. Larson and A. Zarkades, *Properties of Textured Titanium Alloys*, Report MCIC-74-20 (Columbus, OH: Metals and Ceramics Information Center, Battelle's Columbus Laboratories, 1974).
30. G. Luetjering and A. Gysler, "Critical Review-Fatigue", in ITC 5, 2065-2083.

31. M. Peters and G. Luetjering, "Control of Microstructure and Texture in Ti-6Al-4V", in ITC 4, 925-935.
32. A. Tanabe, T. Nishimura, M. Fukuda, K. Yoshida, and J. Kihara, "The Formation of Hot Rolled Texture in Commercially Pure Titanium and Ti-6Al-4V Alloy Sheets", in ITC 4, 937-945.
33. A. Sommer, M. Creager, S. Fumishiro, and D. Eylon, "Texture Development in Alpha + Beta Titanium Alloys", in ITC 3, 1863-1874.
34. A.P. Woodfield, M.D. Gorman, R.R. Corderman, J.A. Sutliff, and B. Yamron, "Effect of Microstructure on Dwell Fatigue Behavior of Ti-6242," in Eighth World Titanium Conference, Birmingham, England, October, 1995.
35. B.L. Adams, S.I. Wright, and K. Kunze, "Orientation Imaging: The Emergence of a New Microscopy", *Metall. Trans. A*, 24A (1993), 819-831.
36. C.H. Hamilton and A.K. Ghosh, "Superplastic Sheet Forming", in *Metals Handbook, Volume 14: Forming and Forging* (Materials Park, OH: ASM International, 1988), 852-873.
37. T. Altan, F.W. Boulger, J.R. Becker, N. Akgerman, and H.J. Henning, *Forging Equipment, Materials, and Practices*, Report MCIC HB-03 (Columbus, OH: Metals and Ceramics Information Center, Battelle's Columbus Laboratories, 1973).
38. D. Lee and W.A. Backofen, "Superplasticity in Some Titanium and Zirconium Alloys", *Trans. TMS-AIME*, 239 (1967), 1034-1040.
39. C.C. Chen and J.E. Coyne, "Deformation Characteristics of Ti-6Al-4V Under Isothermal Forging Conditions", *Metall. Trans. A*, 7A (1976), 1931-1941.
40. S.M.L. Sastry, R.J. Lederich, T.L. Mackay, and W.R. Kerr, "Generalized Relations Between Metallurgical and Process Parameters for Superplastic Forming of Titanium Alloys", *J. Metals*, 35 (January 1983), 48-53.
41. M.E. Rosenblum, P.R. Smith, and F.H. Froes, "Microstructural Aspects of Superplastic Forming of Titanium Alloys", in ITC 4, 1013-1024.
42. N. Furushiro, H. Ishibashi, S. Shimoyama, and S. Hori, "Factors Influencing the Ductility of Superplastic Ti-6Al-4V Alloy", in ITC 4, 993-1000.
43. N. Furushiro, P.X. Ping, and S. Hori, "Microstructural Change During High Temperature Deformation of a Ti-6Al-2Sn-4Zr-6Mo Alloy", in ITC 6, 1157-1160.
44. A.K. Ghosh and C.H. Hamilton, "Mechanical Behavior and Hardening Characteristics of a Superplastic Ti-6Al-4V Alloy", *Metall. Trans. A*, 10A (1979), 699-706.
45. H. Inagaki, "Enhanced Superplasticity in High Strength Ti Alloys", *Z. für Metallkunde*, 86 (1995), 643-650.
46. J.L. Brantingham and D.E. Thomas, "Effect of Recrystallization on the Superplastic Behavior of Ti-6Al-4V", in *Titanium 1986: Products and Applications* (Dayton, OH: Titanium Development Assoc., 1987), 1073-1086.
47. X.D. Ding, H.M. Zbib, C.H. Hamilton, and A.E. Bayoumi, "On the Optimization of Superplastic Blow-Forming Processes", *J. Mat. Eng. Perf.*, 4 (1995), 474-485.
48. G.D. Lahoti, S.L. Semiatin, S.I. Oh, T. Altan, and H.L. Gegel, "Development of Process Models to Produce a Dual-Property Titanium Alloy Compressor Disk", in D.F. Hasson and C.H. Hamilton, eds., *Advanced Processing Methods for Titanium* (Warrendale, PA: TMS, 1982), 23-39.
49. S.L. Semiatin and G.D. Lahoti, "Deformation and Unstable Flow in Hot Forging of Ti-6Al-2Sn-4Zr-2Mo-0.1Si", *Metall. Trans. A*, 12A (1981), 1705-1717.
50. S.L. Semiatin and G.D. Lahoti, "The Forging of Metals", *Scientific American*, 245 (August, 1981), 98-106.
51. J.C. Malas, H.L. Gegel, S.I. Oh, and G.D. Lahoti, "Metallurgical Validation of a Finite-Element Program for the Modeling of Isothermal Forging Processes", in *Experimental Verification of Process Models* (Materials Park, OH: ASM International 1983), 358-372.

52. R. Raj, unpublished research, Cornell University, Ithaca, NY, 1982.
53. T. Matsumoto, M. Nishigaki, M. Fukuda, and T. Nishimura, "Effect of Heat Treatment on the Surface Defect and Mechanical Properties of Ti-6Al-4V Rolled Bar", in ITC 5, 617-623.
54. S.L. Semiatin and V. Seetharaman, unpublished research, Wright Laboratory Materials Directorate, Wright-Patterson AFB, OH, 1995.
55. H.G. Suzuki and D. Eylon, "Hot Ductility of Titanium Alloys - A Comparison with Carbon Steels", *ISIJ Inter.*, 33 (1993), 1270-1274.
56. S.L. Semiatin and G.D. Lahoti, "The Occurrence of Shear Bands in Nonisothermal, Hot Forging of Ti-6Al-2Sn-4Zr-2Mo-0.1Si", *Metall. Trans. A*, 14A (1983), 105-115.
57. S.L. Semiatin and G.D. Lahoti, "The Occurrence of Shear Bands in Isothermal, Hot Forging", *Metall. Trans. A*, 13A (1982), 275-288.
58. A.M. Sabroff, F.W. Boulger, and H.J. Henning, *Forging Materials and Practices* (New York, NY: Rheinhold Book Company, 1968).
59. T.L. Subramanian, N. Akgerman, and T. Altan, "Application of CAD/CAM to Precision Isothermal Forging of Titanium Alloys", Technical Report AFML-TR-77-108 (Columbus, OH: Battelle's Columbus Laboratories, July 1977).
60. A. Chamouard, *Closed-Die Forging, Part I* (Paris, France: Dunod, 1964).
61. F.N. Lake and D.J. Moracz, "Comparison of Major Forging Systems", Technical Report AFML-TR-71-112 (Cleveland, OH: TRW, Inc., May, 1971).
62. I. Weiss and F.H. Froes, "The Processing Window for the Near Beta Ti-10V-2Fe-3Al Alloy", in ITC 5, 499-506.
63. G. Schröder and T.W. Duerig, "Forgeability of Beta-Titanium Alloys Under Isothermal Forging Conditions", in ITC 5, 585-593.
64. C.C. Chen, J.A. Hall, and R.R. Boyer, "High Strength Beta Titanium Alloy Forgings for Aircraft Structural Applications", in ITC 4, 457-466.
65. D.A. Wagner and E.W. Johansen, unpublished research, General Electric Aircraft Engines, Cincinnati, OH, 1990.
66. J.A. Hall, unpublished research, Allied Signal Propulsion Engines, Phoenix, AZ, 1993.
67. S.L. Semiatin, K.A. Lark, D.R. Barker, V. Seetharaman, and B. Marquardt, "Plastic Flow Behavior and Microstructural Development in a Cast Alpha-Two Titanium Aluminide", *Metall. Trans. A*, 23A (1992), 295-305.
68. C.H. Ward, J.C. Williams, A.W. Thompson, D.G. Rosenthal, and F.H. Froes, "Fracture Mechanisms in Titanium Aluminide Intermetallics", *Mem. et Etudes Scien. Rev. Metall.*, 86 (1989), 647-653.
69. C. Bassi, J.A. Peters, and J. Wittenauer, "Processing Titanium Aluminide Foils", *JOM*, 41 (September, 1989), 18-20.
70. J. Wittenauer, C. Bassi, and B. Walser, "Hot Deformation Characteristics of Nb-Modified Ti$_3$Al", *Scripta Metall.*, 23 (1989), 1381-1386.
71. C. Bassi and J.A. Peters, "Effect of Thermomechanical Processing on Microstructures and Properties of Titanium Aluminide Foils", *Scripta Metall. et Mater.*, 24 (1990), 1363-1368.
72. A. Dutta and D. Banerjee, "Superplastic Behavior in a Ti$_3$Al-Nb Alloy", *Scripta Metall. et Mater.*, 24 (1990), 1319-1322.
73. H.S. Yang, P. Jin, E. Dalder, and A.K. Mukherjee, "Superplasticity in a Ti$_3$Al-Base Alloy Stabilized by Nb, V, and Mo", *Scripta Metall. et Mater.*, 25 (1991), 1223-1228.
74. R. Verma and A.K. Ghosh, "Microstructural and Textural Changes During Rolling of Alpha-Two Titanium Aluminide Foils", in ITC 7, 995-1002.
75. R.G. Rowe, "The Mechanical Properties of Titanium Aluminides Near Ti-25Al-25Nb", in Y-W. Kim and R.R. Boyer, eds., *Microstructure/Property Relationships in Titanium Aluminides and Alloys* (Warrendale, PA: TMS, 1991), 387-398.

76. R.G. Rowe, D.G. Konitzer, A.P. Woodfield, and J.C. Chesnutt, "Tensile and Creep Behavior of Ordered Orthorhombic Ti$_2$AlNb-Based Alloys" in L.A. Johnson, D.P. Pope, and J.O. Stiegler, eds., *High-Temperature Ordered Intermetallic Alloys IV* (Pittsburgh, PA: MRS, 1991), 703-708.
77. L.A. Bendersky, W.J. Boettinger, and A. Roytburd, "Coherent Precipitates in the BCC/Orthorhombic Two-Phase Field of the Ti-Al-Nb System", *Acta Metall. et Mater.*, 39 (1991), 1959-1969.
78. C.G. Rhodes, J.A. Graves, P.R. Smith, and M.R. James, "Characterization of Orthorhombic Titanium Aluminide Alloys", in R. Darolia, et al., eds., *Structural Intermetallics* (Warrendale, PA: TMS, 1993), 45-52.
79. S.L. Semiatin and P.R. Smith, "Microstructural Evolution During Rolling of Ti-22Al-23Nb Sheet", *Mat. Sci Eng.*, A202 (1995), 26-35.
80. G. Wasserman and J. Grewen, *Textures of Metallic Materials* (Berlin, Germany: Springer-Verlag, 1962).
81. Y-W. Kim, "Intermetallic Alloys Based on Gamma Titanium Aluminide", *J. Metals*, 41 (July, 1989), 24-30.
82. M.K. Alam, S.L. Semiatin, and Z. Ali, "Thermal Stress Development During Vacuum Arc Remelting and Permanent Mold Casting", submitted to *Trans. ASME, J. Eng. Ind.*, 1995.
83. C. McCullough, J.J. Valencia, C.G. Levi, and R. Mehrabian, "Phase Equilibria and Solidification in Ti-Al Alloys", *Acta Metall.*, 37 (1989), 1321-1336.
84. S.L. Semiatin, R. Nekkanti, M.K. Alam, and P.A. McQuay, "Homogenization of Near-Gamma Titanium Aluminides: Analysis of Kinetics and Process Scaleup Feasibility", *Metall. Trans. A*, 24A (1993), 1295-1306.
85. J.D. Bryant and S.L. Semiatin, "Segregation in Multicomponent Ingot Metallurgy Gamma Titanium Aluminides", *Scripta Metall. et Mater.*, 25 (1991), 449-453.
86. P.L. Martin, C.G. Rhodes, and P.A. McQuay, "Thermomechanical Processing Effects on Microstructure in Alloys Based on γ-TiAl", in R. Darolia, et al., eds., *Structural Intermetallics* (Warrendale, PA: TMS, 1993), 177-186.
87. D.S. Shih, S-C. Huang, G.K. Scarr, H. Jang, and J.C. Chesnutt, "The Microstructural Dependence of Mechanical Properties of Ti-48Al-2Cr-2Nb", in Y-W. Kim and R.R. Boyer, eds., *Microstructure/Property Relationships in Titanium Aluminides and Alloys* (Warrendale, PA: TMS, 1991), 135-148.
88. S.L. Semiatin and V. Seetharaman, unpublished research, Wright Laboratory Materials Directorate, Wright-Patterson AFB, OH, 1992.
89. S.L. Semiatin and P.A. McQuay, "Segregation and Homogenization of a Near-Gamma Titanium Aluminide", *Metall. Trans. A*, 23A (1992), 149-161.
90. P.L. Martin, unpublished research, Rockwell Science Center, Thousand Oaks, CA, 1992.
91. V. Seetharaman and S.L. Semiatin, unpublished research, Wright Laboratory Materials Directorate, Wright-Patterson AFB, OH, 1995.
92. S.L. Semiatin, D.S. Lee, and D.M. Dimiduk, Air Force Invention 21,465, Wright Laboratory Materials Directorate, Wright-Patterson AFB, OH, 1994.
93. V. Seetharaman, C.M. Lombard, R.L. Goetz, and J.C. Malas, unpublished research, Wright Laboratory Materials Directorate, Wright-Patterson AFB, OH, 1991.
94. V. Seetharaman, J.C. Malas, and C.M. Lombard, "Hot Extrusion of a Ti-Al-Nb-Mn Alloy", in L.A. Johnson, et al., eds., *High-Temperature Ordered Intermetallic Alloys IV* (Pittsburgh, PA: MRS, 1991), 889-894.
95. S.L. Semiatin, V. Seetharaman, and V.K. Jain, "Microstructure Development During Conventional and Isothermal Hot Forging of a Near-Gamma Titanium Aluminide", *Metall. and Mater. Trans. A*, 25A (1994), 2753-2768.

96. S.L. Semiatin, P.A. McQuay, and V. Seetharaman, "A Novel Process for Breakdown Forging of Coarse-Grain Intermetallic Alloys", *Scripta Metall. et Mater.*, 29 (1993), 1235-1240.
97. V.K. Jain, R.L Goetz, and S.L. Semiatin, "Can Design for Nonisothermal Pancake Forging of Gamma Titanium Aluminide Alloys", *Trans. ASME, J. Eng. Ind.*, 118 (1996), in press.
98. K. Wurzwallner, H. Clemens, P. Schretter, A. Bartels, and C. Koeppe, "Forming of γ-TiAl Alloys", in I. Baker, et al., eds., *High-Temperature Ordered Intermetallic Alloys V* (Pittsburgh, PA: MRS, 1993), 867-872.
99. H. Clemens, P. Schretter, K. Wurzwallner, A. Bartels, and C. Koeppe, "Forging and Rolling of Ti-48Al-2Cr on Industrial Scale", in R. Darolia, et al., eds., *Structural Intermetallics* (Warrendale, PA: TMS, 1993), 205-214.
100. V. Seetharaman, L. Dewasurendra, A.B. Chaudhary, J.T. Morgan, and J.C. Malas, "Modeling of Hot Extrusion of a Gamma Titanium Aluminide Alloy", in N. Chandra and J.N. Reddy, eds., *Advances in Finite Deformation Problems in Materials Processing and Structures* (New York, NY: ASME, 1991), 97-109.
101. R.L. Goetz, V.K. Jain, and C.M. Lombard, "Effect of Core Insulation on the Quality of the Extrudate in Canned Extrusions of γ-Titanium Aluminide", *J. Mater. Proc. Techn.*, 35 (1992), 37-60.
102. Y-W. Kim and D.M. Dimiduk, "Method to Produce Gamma Titanium Aluminide Articles Having Improved Properties", U.S. Patent 5,226,985, July 13, 1993.
103. Y-W. Kim, "Ordered Intermetallic Alloys, Part III: Gamma Titanium Aluminides", *JOM*, 46 (7) (1994), 7, 30-40.
104. R.L. Goetz, S.L. Semiatin, and S-C. Huang, unpublished research, Wright Laboratory Materials Directorate, Wright-Patterson AFB, OH, 1994.
105. S.L. Semiatin, V. Seetharaman, R.L. Goetz, and V.K. Jain, "Controlled Dwell Extrusion of Difficult-to-Work Alloys", U.S. Patent 5,361,477, November 8, 1994.
106. V.M. Segal, V.I. Reznikov, A.E. Drobyshevskiy, and V.I. Kopylov, "Plastic Working of Metals by Simple Shear", *Russian Metallurgy*, 1 (1981), 99-105.
107. S.L. Semiatin, V.M. Segal, R.L. Goetz, R.E. Goforth, and T. Hartwig, "Workability of a Gamma Titanium Aluminide Alloy During Equal Channel Angular Extrusion", *Scripta Metall. et Mater.*, 33 (1995), 535-540.
108. A.L. Hoffmanner and D.D. Bhatt, unpublished research, Battelle Memorial Institute, Columbus, OH, 1977.
109. S.L. Semiatin, M. Ohls, and W.R. Kerr, "Temperature Transients During Hot Pack Rolling of High Temperature Alloys", *Scripta Metall. et Mater.*, 25 (1991), 1851-1856.
110. S.L. Semiatin and V. Seetharaman, "Deformation and Microstructure Development During Hot-Pack Rolling of a Near-Gamma Titanium Aluminide Alloy", *Metall. and Mater. Trans. A*, 26A (1995), 371-381.
111. S.L. Semiatin, D.C. Vollmer, S. El-Soudani, and C. Su, "Understanding the Failure of Near Gamma Titanium Aluminides During Rolling", *Scripta Metall. et Mater.*, 24 (1990), 1409-1413.
112. S.L. Semiatin, N. Frey, C.R. Thompson, and D.C. Volmer, unpublished research, Battelle Memorial Institute, Columbus, OH, 1989.
113. V. Seetharaman and S.L. Semiatin, "Microstructures and Mechanical Properties of Rolled Sheets of a Gamma Titanium Aluminide Alloy", in Y-W. Kim, R. Wagner, and M. Yamaguchi, eds., *Gamma Titanium Aluminides* (Warrendale, PA: TMS, 1995), 753-760.
114. N. Fujitsuna, Y. Miyamoto, and Y. Ashida, "Microstructural Change of Ti-46 a/o Al by Multi-Step Working", in R. Darolia, et al., eds., *Structural Intermetallics*, (Warrendale, PA: TMS, 1993), 187-194.

115. A. Morita, N. Fujitsuna, and H. Shigeo, "Isothermal Rolling of TiAl Sheet", *Symp. Proc. for Basic Technologies for Future Industries High-Performance Materials for Severe Environments 4th Meeting* (Tokyo, Japan: Japan Industrial Technology Association, 1993), 215-223.
116. C.M. Lombard, unpublished research, Wright Laboratory Materials Directorate, Wright-Patterson AFB, OH, 1995.
117. M. Nobuki and T. Tsujimoto, "Superplasticity of TiAl Intermetallics", in Y. Han, ed., *Advanced Structural Materials* (Amsterdam: Elsevier, 1991), 791-796.
118. T. Maeda, M. Okada, and Y. Shida, "Superplasticity in Ti-Rich TiAl", in S. Hori, M. Tokizane, and N. Furushiro, eds., *Superplasticity in Advanced Materials* (Osaka, Japan: Japan Society for Research on Superplasticity, 1991), 311-316.
119. R.M. Imayev, O.A. Kaibyshev, and G.A. Salishev, "Mechanical Behavior of Fine Grained TiAl Intermetallic Compound-I. Superplasticity", *Acta Metall. et Mater.*, 1992 (40), 581-587.
120. S.C. Cheng, J. Wolfenstine, and O.D. Sherby, "Superplastic Behavior of Two-Phase Titanium Aluminide", *Metall. Trans. A*, 23A (1992), 1509-1513.
121. T. Wajata, K. Isonishi, D. Ameyama, and M. Tokizane, "Superplasticity in TiAl and Ti$_3$Al Two Phase Material Made from PREP Powder", *ISIJ Inter.*, 33 (1993), 884-888.
122. N. Masahashi, Y. Mizuhara, M. Matsuo, T. Hanamura, M. Kimura, and K. Hashimoto, "High Temperature Deformation Behavior of Titanium - Aluminide Based Gamma Plus Beta Microduplex Alloy", *ISIJ Inter.*, 31 (1991), 728-737.
123. W.B. Lee, H.S. Yang, Y-W. Kim, and A.K. Mukherjee, "Superplastic Behavior in a Two-Phase TiAl Alloy", *Scripta Metall. et Mater.*, 29 (1993), 1403-1408.
124. C.M. Lombard, A.K. Ghosh, and S.L. Semiatin, "Superplastic Deformation of a Rolled Gamma Titanium Aluminide Alloy", in Y-W. Kim, R. Wagner, and M. Yamaguchi, eds., *Gamma Titanium Aluminides* (Warrendale, PA: TMS, 1995), 579-586.
125. S.L. Semiatin and J.J. Jonas, *Formability and Workability of Metals* (Materials Park, OH: ASM International, 1984).
126. K. Hashimoto, N. Masahashi, Y. Mizuhara, H. Fujii, and M. Matsuo, "Superplasticity of Thermomechanically Processed Gamma Titanium Aluminides", in Y-W. Kim and R.R. Boyer, eds., *Microstructure/Property Relationships in Titanium Aluminides and Alloys* (Warrendale, PA: TMS, 1991), 253-262.
127. V. Seetharaman and S.L. Semiatin, "Influence of Temperature Transients on the Hot Workability of a Two-Phase Gamma Titanium Aluminide Alloy", *Metall. and Mater. Trans. A*, 27A (1996), in press.
128. H. Fukutomi, S. Takagi, K. Aoki, M. Nobuki, H. Mecking, and T. Kamijo, "Effect of Deformation Conditions on Texture Formation During Dynamic Recrystallization of the Intermetallic Compound TiAl", *Scripta Metall. et Mater.*, 25 (1991), 1681-1684.
129. Ch. Hartig, X.F. Fang, H. Mecking, and M. Dahms, "Textures and Plastic Anisotropy in TiAl", *Acta Metall. et Mater.*, 40 (1992), 1883-1894.
130. H. Mecking and Ch. Hartig, "Effects of Processing on Texture, Microstructure, and Related Properties of TiAl Alloys", in Y-W. Kim, R. Wagner, and M. Yamaguchi, eds., *Gamma Titanium Aluminides* (Warrendale, PA: TMS, 1995), 525-538.
131. V. Seetharaman, S.L. Semiatin, C.M. Lombard, and N.D. Frey, "Deformation and Fracture Characteristics of a Gamma Titanium Aluminide at High Temperatures", in I. Baker, et al., eds., *High-Temperature Ordered Intermetallic Alloys V* (Pittsburgh, PA: MRS, 1993), 513-518.

132. V. Seetharaman and C.M. Lombard, "Plastic Flow Behavior of a Ti-Al-Nb-Mn Alloy at High Temperatures", in Y-W. Kim and R.R. Boyer, eds., *Microstructure/Property Relationships in Titanium Aluminides and Alloys* (Warrendale, PA: TMS, 1991), 237-251.
133. S.L. Semiatin and H.R. Piehler, "Formability of Sandwich Sheet Materials in Plane Strain Compression and Rolling", *Metall. Trans. A*, 10A (1979), 97-107.
134. V. Seetharaman, R.L. Goetz, and S.L. Semiatin, "Tensile Fracture Behavior of a Cast Gamma-Titanium Aluminide", in L.A. Johnson, *et al.*, eds., *High-Temperature Ordered Intermetallic Alloys IV* (Pittsburgh, PA: MRS, 1991), 895-900.
135. M.G. Cockcroft and D.J. Latham, "A Simple Criterion of Fracture for Ductile Metals", N.E.L. Report No. 240, National Engineering Laboratory, East Kilbride, Glasgow, Scotland, 1966.
136. D. Watkins, H.R. Piehler, V. Seetharaman, C.M. Lombard, and S.L. Semiatin, "Effect of Hydrostatic Pressure on the Hot-Working Behavior of a Gamma Titanium Aluminide", *Metall. Trans. A*, 23A (1992), 2669-2672.

MICROSTRUCTURAL DEVELOPMENT IN A TITANIUM ALLOY

G. Shen, D. Furrer, and J. Rollins

Ladish Co., Inc., P.O. Box 8902, Cudahy, WI 53110-8902

Abstract

Methods for the control and manipulation of titanium alloy microstructures have been traditionally developed by iterative trials. Variations in titanium heating, deformation and cooling practice result in greatly varied microstructural morphologies and volume fractions for the constituent alloy phases.

Efforts are being undertaken to develop an analytical tool to predict the development of titanium alloy microstructures under heating, hot working, and cooling conditions. Results of this tool have been successfully correlated to actual forge shop applications. Further utilization of this technology will allow more rapid process development for robust forged titanium components with ability to predict resultant microstructures, and subsequent mechanical properties.

Introduction

In the very competitive world of manufacturing, tools of all types are being developed and utilized to speed-up design and optimization of manufacturing processes. In the forging industry, as in many industries, computer process modeling is being used to help study and predict processing results electronically before any hardware or prototypes are initiated on the manufacturing shop floor.

The forging industry has used finite element method (FEM) codes for many years to study bulk deformation processing by many production manufacturing techniques [1,2]. FEM analysis of metal flow has allowed greater forge geometry contour refinements, reduced input material weight, eliminated lap formations, reduced initial engineering cycle times, nearly eliminated all shop floor trials prior to production, and allowed optimum processing parameters to be integrated into manufacturing methods for materials with very tight processing windows. Additionally, computer modeling of thermal processing (ie. pre-heating, cooling, and heat treatment) has also provided great insight and benefit into manufacturing process design [3].

Metallurgists and process engineers are being placed under greater pressure to design manufacturing processes and methods which will result in right-the-first-time components. The use of well established computer models has allowed process parameter sensitivity analysis, which is vital for process optimization. Current computer process modeling tools are being further refined and extended to new areas such as microstructural development modeling to further assist metallurgists and process engineers.

The goal of the current effort is to develop an FEM based tool that will be practical and will allow an accurate prediction of titanium alloy microstructures during secondary processing operations. Ti-6-2-4-2 has been selected as the alloy for this effort, and all material related parameters developed for the model equations will be for this material.

Microstructural models have been previously developed for a number of materials and processes such as rolled steel alloys [4,5], forged nickel-base superalloys [6,7], and for titanium ingot conversion [8,9], and secondary titanium forging. Much of the previous work which has allowed accurate microstructure predictions has been based on fundamental physical metallurgy principles, as is the current effort.

Experimental Procedure

The current program is focused on secondary processing of Ti-6-2-4-2, and will not include ingot conversion, or primary processing. For the purpose of a more general model with application to the broad Ti-6-2-4-2 alloy range, separate heats from two mills have been characterized in the current study. The ß transus for the material obtained from mill #1 was 996°C and for mill #2, 1004°C.

The procedures or steps involved in the development of the microstructure model in this program were:

(1) Preheating Tests:

Heating experiments were conducted to investigate the as-preheated microstructure under different preheat temperatures and hold times. The temperature used in preheating tests included:

ß-167°C; ß-111°C; ß-56°C; ß-42°C; ß-28°C; ß-14°C; ß-8°C; ß+14°C; ß+28°C; ß+42°C

The hold times used in preheating tests were 1 minute, 10 minutes, 1 hour, 4 hours, and 8 hours.

It is critical to understand the microstructural changes which are taking place during heat-up of billet material for secondary forging operations. Without an accurate understanding of the grain size and microstructure (% of each constituent phase), an accurate starting point for the simulation of the microstructural evolution during the deformation process is impossible.

(2) Compression Tests:

Right circular cylinder upset tests were used to simulate a broad range of forging parameters. The variables involved in compression tests were: as-preheated microstructure, temperature, strain, and strain rate. The as-preheated microstructure variables used in α-ß compression tests were percentage primary α, and α grain size. For ß compression testing, the starting microstructural variable investigated was ß grain size. Temperatures of ß-42°C, ß-28°C, ß-14°C, ß+14°C, ß+28°C, ß+42°C were used in the compression tests, along with three true strains (0.5, 0.8 and 1.2) and three strain rates (0.002/s, 0.02/s, 2/s). Compression tests were run on both an MTS compression test stand for isothermal conditions and a Gleeble testing unit for non-isothermal conditions.

(3) Solution Heat Treatment and Cooling Tests:

Heat treatment experiments were carried out at three different temperatures below and above the ß transus to investigate transformation kinetics of the ß-phase during cooling at different rates. For these tests, samples were soaked at different temperatures and cooled at controlled rates to specific intermediate temperatures. At the various intermediate temperatures during the controlled cooling cycle, the cooling was interrupted by rapid quenching. Each specimen was metallographically evaluated to develop an understanding of the transformation and growth kinetics of the transformation products.

(4) Generic Forging and Heat Treatment Tests:

Generic forging trials were conducted with both low strain rate (hydraulic press) and high strain rate (hammer) processing routes for use in initial model evaluations. Post forging heat treatments were also conducted on the generic components to study and compare the resultant microstructure to model predictions.

(5) Finite Element Method (FEM) Simulations:

FEM computer analyses were performed for all of the pre-heating test samples, compression test samples, heat treat samples, and generic forgings to define processing histories for each test specimen for use in subsequent evaluation work. These computer process model evaluations were conducted to allow determination of quantitative values of strain, strain rate, temperature, and process history at specific locations for correlation with metallographic analyses of these specific locations within the experimental samples.

Regression analysis was used to fit the experimental data to equations to determine alloy specific constants. Equations for the prediction of microstructures were fully defined based on the experimental results and were implemented into an environment provided by a commercial computer software DEFORM for the prediction of microstructural development during preheating, forging, and heat treatment for production parameters and boundary conditions that are accurately known.

Results and Discussion

The goal of the current effort is the development of accurate and practical mathematical models for prediction of microstructural evolution during preheating, deformation, and heat treatment of Ti-6-2-4-2 components.

(1) Modeling Microstructural Development During Preheating:

The model for preheating must include alloy phase relationships as functions of temperature and time, as well as grain growth kinetics. The model developed includes both a representation of the alloy's ß approach curve and the ß grain growth rate information obtained from preheating experiments. The ß approach curves define the primary α content for preheating forging stock in the α-ß regime. The ß approach curve for one of the heats investigated is shown in Figure 1.

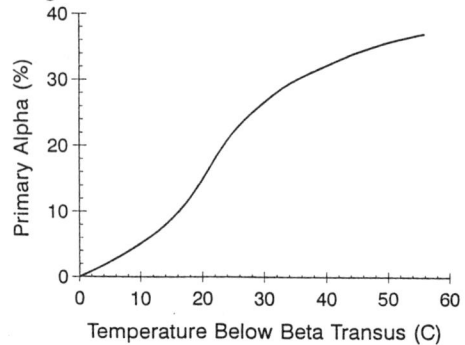

Figure 1. Beta approach curve developed experimentally for one of the heats of Ti-6-2-4-2 investigated in this program.

The curve shows the equilibrium percentage of primary α versus preheating temperature. As seen from Figure 1, the change in the primary α content with temperature is more pronounced for preheating temperatures from ß-28°C to ß-14°C.

The general equation for ß grain growth during ß temperature range preheating, which describes the final grain size with different preheat temperatures and hold times, is in the form:

$$d^n - d_0^n = A\, t\, \exp(-Q/[RT]) \qquad (1)$$

Where d= final grain size (μm), d_0= starting grain size (μm) t= time (s), Q= activation energy for grain growth (J/mol), R= gas constant (J/(mol*K)), T=temperature (K), and A and n are material specific constants.

Using the FEM predicted temperature-time history, the model can predict the as-preheated beta grain size for the preheating process. A prediction of the beta grain size for preheating of Ti-6-2-4-2 forging mult is shown in Figure 2. This figure shows normalized beta grain size contours for an as-preheated forging mult.

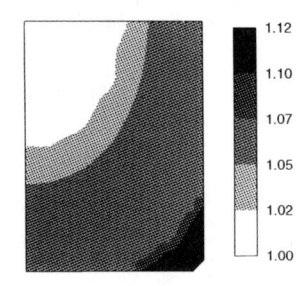

Figure 2. Plot of predicted as-preheated ß grain size contours (normalized) for a billet prior to forging.

(2) Modeling Microstructural Development During Deformation:

The local microstructure obtained from the compression specimens was correlated with the FEM prediction of the strain, average temperature and strain rate, as well as the as preheated structure for the compression tests. Empirical equations for the prediction of microstructural evolution during both α-ß forging and ß forging processes were developed.

The procedure for the prediction of the microstructure development in α-ß forging is as follows: The local percentage primary α is determined based on the local temperature history in the deformation process and the ß approach curve.

The procedure for the prediction of the beta grain size resulting from beta forging is divided into two steps:

(a) prediction of the as-preheated grain size using the procedure noted in section (1). The as-preheated grain size is used as the starting grain size.

(b) prediction of the as-forged beta grain size which depends on the starting as-preheated grain size, and the strain, temperature and strain rate experienced during the deformation process. (Equation (2)).

$$d = F(d_0, e, Z) \tag{2}$$

where d=beta grain size, d_0=as-preheated beta grain size, e = effective strain, Z= Zener-Hollomon parameter (the temperature compensated strain rate).

An example of an output of the model for the prediction of beta grain size distribution within a beta forged component is shown in Figure 3. Contours of normalized beta grain size and actual experimental data are shown in this figure. It can be seen that the model prediction agrees well with the experimental data.

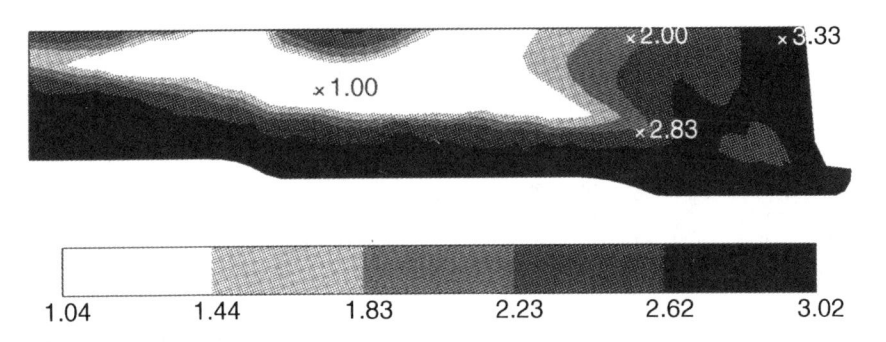

Figure 3. Predicted and experimental ß grain size contours and data points (normalized) for a section of a forged component.

(3) Modeling Microstructural Development During Heat Treatment:

Transformation kinetics of titanium alloys play a significant role in final microstructural results. Time-Temperature-Transformation (TTT) curves and relationships provide essential information regarding what type and quantity of various transformation products will be developed by various post thermal process cooling cycles.

The development of TTT information can be practically accomplished by running a series of isothermal transformation, or continuous cooling transformation tests. It is known that during the normal cooling of forged titanium components from the beta region, several transformation products are possible; grain boundary alpha can nucleate and grow, and acicular alpha can nucleate and grow in the beta phase. During the normal cooling of forged titanium components from the alpha-beta phase field, equiaxed alpha can form through growth onto already existing primary alpha, grain boundary alpha can nucleate and grow, and acicular alpha can nucleate and grow in the beta phase.

Previous titanium microstructual modeling efforts [8] incorporated transformation kinetic information to allow prediction of grain boundary alpha width for partially converted ingot material which was cooled from a beta processing step. This information was in-turn incorporated into a model to predict the amount of alpha-beta work required for full recrystallization of the grain boundary alpha into fine equiaxed particles.

The heat treatment of forged titanium parts is a necessary procedure for the control of the microstructure. The data obtained from the heat treatment and cooling tests was used for the development of the empirical microstructural relationships for the heat treatment process. Two empirical equations were used for the prediction of the microstructure in the heat treatment of α-ß forged Ti-6-2-4-2 parts. The first is the equation for the prediction of the percentage equiaxed α during cooling. The other is the equation for the prediction of the acicular α platelet width.

The microstructural data obtained from the heating and cooling experiments showed that equiaxed α structures can be modified during cooling. Slow cooling results in growth of equiaxed α. An equation was developed to describe the transformation process, and it is in the form:

$$F^n - F_0^n = A\, t \exp(Q/[RT]) \qquad (3)$$

Where the F = percentage of equiaxed α, F_0 = percentage of equiaxed α at the starting point, t = time (s), Q = activation energy for transformation, T = temeprature (K), and A and n are material specific constants.

The equation developed for the prediction of the acicular alpha width is also in a similar form:

$$w^n - w_0^n = A\, t \exp(Q/[RT]) \qquad (4)$$

where w_0 and w are the width of the acicular alpha platelets, in microns, at the beginning and at any instantaneous stage respectively, and the definitions of the other notations are the same as noted above for equation (3).

An example of the application of the equation for the prediction of the microstructure in the heat treatment of a forging in the alpha-beta phase field is shown in Figures 4 (a) & (b). Figure 4 (a) is the prediction for the percentage of equiaxed alpha and Figure 4 (b) for width of acicular alpha. The experimental results from metallographic analysis of the actual forgings is shown for specific locations. Figure 5 (a) and (b) show photomicrographs of Ti-6-2-4-2 material which was cooled from the same alpha-beta solution heat treatment at different rates. The variation in equiaxed alpha content and acicular alpha platelet width can be readily seen.

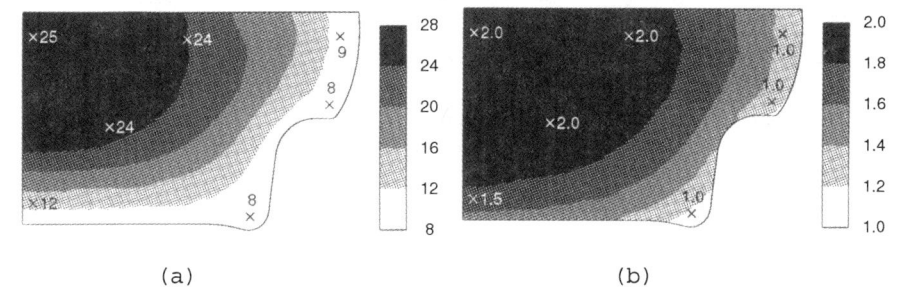

Figure 4. Predictions of (a) percentage of equiaxed α and (b) the acicular alpha platelet width for a part section after heat treatment.

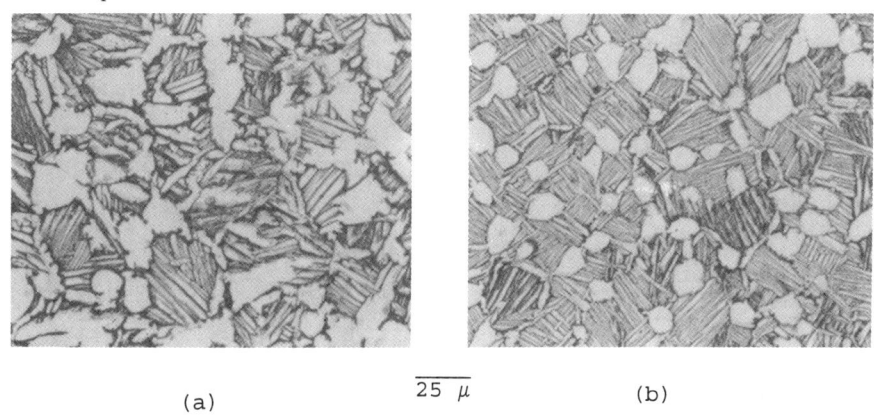

Figure 5. Photomicrographs of Ti-6-2-4-2 material which was cooled from the same solution heat treatment temperature at (a) a slow and (b) a fast rates.

Conclusions

The use of modern forging design tools such as FEM computer modeling, has greatly increased the speed in which manufacturing methods and shapes are being designed and implemented. Development and utilization of tools to predict and analyze microstructural development during deformation processing and heat treatment has increased the ability of forging companies to respond to the needs of aerospace and general industrial customers for reduced component cycle time from design to delivery.

Previous work and current efforts have shown that microstructural modeling can be successfully accomplished for titanium alloys. The current efforts have resulted in numerous achievements including:

1. A microstructural model has been developed and implemented for secondary forging of alloy Ti-6-2-4-2.

2. Model predictions of beta grain size development during forging has shown excellent correlation to shop trials.

3. Model predictions of types and percentages of various alloy phases resulting form post-thermal process cooling cycles has shown excellent correlation to shop trials.

Acknowledgements

The authors would like to thank Dr. L. Semiatin for discussions regarding this program and processing experiments, and Dr. R. Shivpuri and the Ohio State University for discussions on this program and assistance with laboratory experiments. The authors would also like to thank Ladish Co., Inc. for support of this work.

References

1. W.T. Wu, and S.I. Oh, "ALPIDT - A General Purpose FEM Code for Simualtion of Nonisothermal Forging Processes", Proceedings of North American Manufacturing Research Conference (NAMRC) XIII, Berkley, CA, p.449.

2. S. Kobayashi, S.I. Oh, and T. Altan: in *Metal Forming and the Finite Element Method*, Oxford University Press, New York, NY, 1989, PP. 222-2243.

3. P. Tarin and D. Kitchen, SteCal, (Materials Park, OH; ASM, 1992).

4. C.M. Sellars: in *Hot Working and Forming Processes*, C.M Sellars and G.J. Davies, eds., TMS, London, 1979, PP. 3-15.

5. H. Yada: Proc. Int. Symp. for Accelerated Cooling of Rolled Steel, G.E. Ruddle and A.F. Crawley, eds., Pergamon Press, Canada, 1987, pp. 105-120.

6. G. Shen, S.L. Semiatin, and R.Shivpuri, "Modeling Microstructural Developemtn during the Forging of Waspaloy", Met. Trans. A, 26A, (1995), 1795-1802.

7. Balaji, T. Furman, J. Rollins, and R. Shankar, "Prediction of Microstructures in Forged Waspaloy Components using Computer Process Modeling", Proceedings of the North American Forging Technology Conference, ASM/FIA, Dallas, 1994.

8. S.P. Fox and D.F. Neal, "The Role of Computer Modelling in the Development of Large Scale Primary Forging of Titanium Alloys", to be published in the proceedings of the Eighth World Conference on Titianium, Birmingham, UK, 1995.

9. G. Dumas, et al., "Control of Large Diameter Ti-6-4 Billets Microstructures Through Modelling of the Forging Process", to be published in the proceedings of the Eighth World Conference on Titianium, Birmingham, UK, 1995.

REDUCING THE DIMENSIONAL VARIABILITY OF TI64 EXTRUDATES BY PROCESS MODELING AND DIE DESIGN

Dinesh Damodaran*, Rajiv Shivpuri*, Tony Esposito[+], Keith Galyon[+]

* Department of Industrial, Welding and Systems Engineering,
Ohio State University, Columbus, OH
[+] Plymouth Tube Company, Extrusion Mill, Hopkinsville, KY

Abstract

Dimensional variability, or variability in the cross sectional shape of the extrudate continues to be a problem during the hot extrusion of titanium alloys. Many tryouts are required to determine the operating conditions, for optimum production, due to the complex relationships among the process variables. In this study, the entire hot extrusion process is numerically modeled including the induction heating, billet transfer, glass lubrication and metal flow during extrusion. The process model is then used to investigate the interactions between the major process variables that effect metal flow. Control of these variables and die design parameters used to reduce the flow imbalance and consequently the dimensional variability. Production trials have verified the efficacy of this approach.

Introduction

Very few studies exist on dimensional variation in extrusion. A previous study on dimensional variability by Pugliano and Misiolek [1], concluded that billet temperature gradients, equipment misalignment, billet preparation, and improper lubrication can have significant effects on dimensional variability of stainless steel tubing. A design of experiments procedure conducted at Plymouth Tube Company concluded that the top causes of dimensional variability were the extrusion pressure, the breakout pressure and the speed of extrusion. Other causes of dimensional variability were the type of heater used, the initial temperature of the heater, the die design and the billet outer diameter for a given shape. In the hot extrusion of aluminum, Tashiro et al [2] concluded that the parameters that influence dimensional variability the most were the initial billet temperature, the initial die temperature and the extrusion speed.

The Production System for Ti-64 extrudates

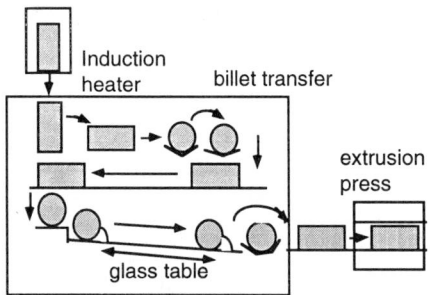

Figure 1 - Schematic of operations involved during hot extrusion of Ti-64 billet

The production system for Ti64 extrudates (Figure 1), consists of the following operations at the Plymouth Tube Company. Billets of diameter 203.2 mm (8") are cut into lengths of 698.5 mm (27.5 ") A nose radius of 10 mm (0.39") is then machined on the front edge of the billet to enable stable flow of the glass lubricant. The billets are heated in vertical induction furnaces by placing them nose up in the coil chamber. The heating schedule is monitored by an infrared probe though a glass window of the furnace. A KW-sec counter measures the amount of energy supplied to the heater. The Sejournet method is used for lubrication. In this process, the heated billet is rolled over a table sprinkled with powdered glass. Prior to insertion of the billet into the hot extrusion container two scrap glass cookies and a fiber pad are placed in front of the die. The prelubricated billet is quickly inserted into the container followed by the dummy block and the extrusion cycle is started.

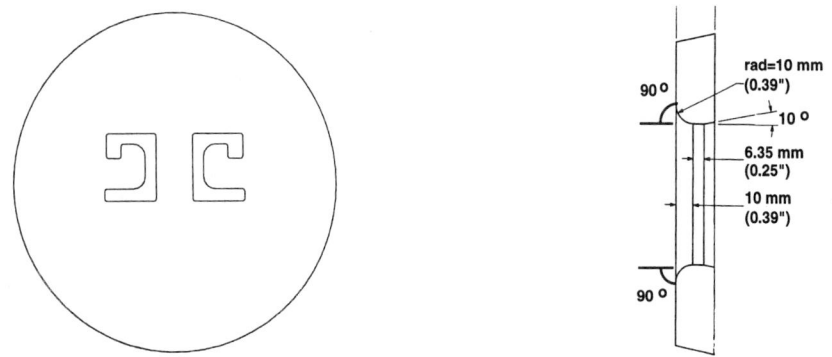

Figure 2a - Port shape used during extrusion of Ti-64   Figure 2b - Flat faced radiused die

The dies are cast from H-13 tool steel. In some cases, selected areas of the die opening which could experience increased heating during extrusion are coated with molybdenum followed by zirconia. The extrusions are carried out on a 17.8 MN Farell hydraulic press.

The container of the press has an internal diameter of around 216 mm (8.5") and is preheated and maintained at a temperature of around 371 C (700 F). The cross sectional shape selected for the analysis is shown in figure 2. This extrudate tends to show distortions and variations in cross sectional shape. It was decided to analyze the whole extrusion process, in order to understand the effect of various process parameters on metal flow.

Induction Heating Model

In induction heating, heat is generated within the billet, due to eddy currents which dissipate energy and heat the billet. The mathematical analysis of induction heating processes can be quite complex for all but the simplest of workpiece geometries. The induction heating model is based on an analytical solution given by Baker [3] for a one dimensional heat flow problem. The assumptions used are that the cylinder is supposed to be infinitely long, so that end effects are neglected and hence temperature at any point in the billet is supposed to be a function only of the radius "r" and the time "t". If the billet is heated by a constant power density then the rise in the center temperature of the billet (ΔT_c) can be given by the following equation.

$$\Delta T_c = \left(\frac{E\eta}{t(2\pi R_o L_o)}\right)\frac{R_o}{k_b}\left(2\left(\frac{k_b}{\rho_b c_b}\right)\left(\frac{t}{R_o^2}\right) - \frac{1}{4}\right) \qquad (1)$$

where E represents the energy in KW-sec that is supplied to the billet and η is the efficiency of induction heating, R_o is the radius of the billet, L_0 is the length of the billet, k_b is the thermal conductivity of the billet, ρ_b is the density of the billet, C_b is the specific heat of the billet and t is the time of induction heating. The surface temperature (T_s) can be computed using the following equation.

$$T_s = T_c + \left(\frac{E\eta}{t(2\pi R_o L_o)}\right)\left(\frac{R_o}{k_b}\right)F_1 \qquad (2)$$

where F_1 is a factor which accounts for the depth of current penetration of the induction heating [3], which depends among other factors, on the frequency of the induction heating. Lower frequencies enhance temperature uniformity by providing large penetration depths. The disadvantage is that they lead to low heating efficiencies.

Thermal properties of the Ti-6Al-4V billet were obtained from literature [4]. The flow stress curves of Ti-6Al-4V were obtained from literature and are given in figure 3 [5]. The heat transfer coefficient was taken as 7.5 N/sec/mm/C (0.002548 BTU/sec/in^2/F). Besides this, the pressure stroke curve of the extrusion process was almost constant, hence it was decided that the friction factor "m" should be 0.01. This is supported by numerous observations of smooth almost frictionless metal flow with glass lubrication [6].

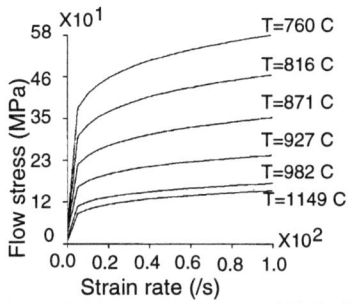

Figure 3 - Flow stress curves of Ti-64 [5]

The amount of energy used for induction heating the billet was around 152350 KW-sec. The time of heating was around 846 seconds. The radius of the billet, after it was upset to the container radius, was 108 mm (4.25"). The heating of the billet was followed by a soak period of 90 seconds. The soak period consisted of short periods, during which the heat was turned on and off. In order to simplify the model, it was decided to model the entire soak period in two stages, first a stage where the heat was turned on, followed by the second stage where the heat was turned off. Based on the heater soak setting, it was decided to model the soak period of 90 seconds as a stage where the heat was turned on for 47 seconds, followed by a period where the heat was turned off for 43 seconds. The process reaches steady state within a very short time and thus it was decided to model only an 203.2 mm (8") length of the extrusion process in order to simplify the model. After the induction heating and soaking process, the hot billet is transferred over a glass table, where it is coated with glass and then transferred to the container of the press. This transfer process takes 45 seconds. The temperature profile of the billet after the transfer process is given in figure 4a. The billet dwells in the container for approximately 5 seconds before the extrusion cycle is started.

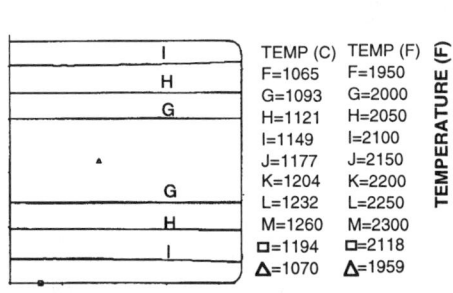

Figure 4a - Temperature profile after billet transfer

Figure 4b - Time temperature profile of points on the center and surface of the billet

The temperature-time profile of two points, one on the surface and another in the center were tracked during the induction heating, soaking and transfer phases, and this is plotted in figure 3b. The temperature difference at these two locations was around 222 C (400 F) during the heating, phase but reduced to 129 C (230 F) after soaking. Both predicted temperature 1183 C (2162 F) and measured temperature 1185 C (2165 F) at surface after soak were close enough to provide confidence in the model.

Extrusion model

In order to find out the effects of die design and other parameters on the metal flow, it was decided to model the extrusion process using the finite element method. In the past, studies using the upper bound method to analyze three dimensional material flow in the extrusion of shapes have been conducted by many researchers. Udagawa used two dimensional FEM to investigate titanium tube extrusion [4]. Shivpuri and Momin [7] also used two dimensional FEM to simulate twisting and surface defects in the extrusion of 304 Stainless Steel. The actual metal flow during extrusion of sections is three dimensional in nature. However when three dimensional finite element codes are applied to the simulation of hot extrusion using glass lubrication, they require enormous amounts of computational time. In hot extrusion, the typical reduction in area is around 95 % and die angles involved during hot extrusion using glass lubrication are 90 degrees. The elements thus become severely deformed during the deformation process. Three dimensional simulations have been conducted at Ohio State University using DEFORM-3D software. However, it took around two weeks to complete an isothermal simulation on an IBM RISC/6000 workstation. Besides this, the metal flow was very distorted (due to remeshing problems) and did not correspond well with experimental observations.

Thus it was decided to use a two dimensional finite element code, DEFORM, to model the extrusion process. This meant that slices had to be taken at various locations and each slice had to be treated independently of the other (plane strain). This approach has been used earlier by Shivpuri [7], and the results were quite encouraging. A similar approach has also been used by Tashiro [2]. They performed a number of plane strain extrusion experiments to investigate causes of twisting in three dimensional extrusions of sections.

Figure 5a - Schematic diagram of the 3-D extrudate

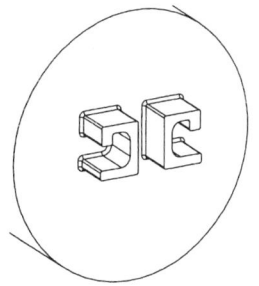

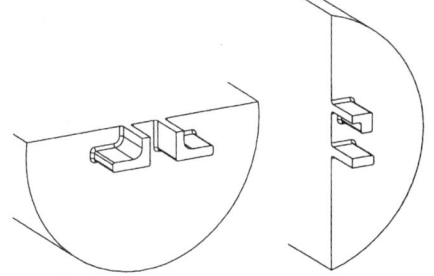

Figure 5b - Section AA

Figure 5c - Section BB

Figure 5a shows the schematic diagram of the Ti64 3-D extrudate. Figures 5b, and 5c show the procedure of taking slices of the section shape in order to perform two dimensional plane strain extrusion simulations. Two sections AA, and BB were selected for performing plane strain simulations. The sections were selected such that the width of the section was greater than the thickness of the section. This should increase the accuracy of the plane strain assumption at these sections. The procedure of selection of the sections is similar to that used by Tashiro [2] and Shivpuri [7]. It is preferable to select sections such that they pass through the centroid of the extrudate, thus they will pass through the extrudate.

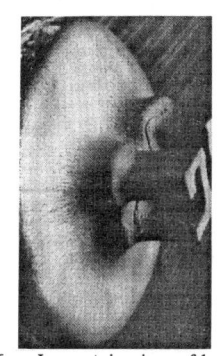

Figure 6a - Isometric view of butt end

Figure 6b - Side view of butt end

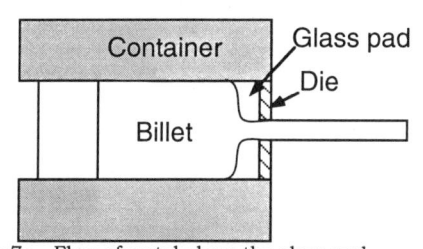

Figure 7a - Flow of metal along the glass pad

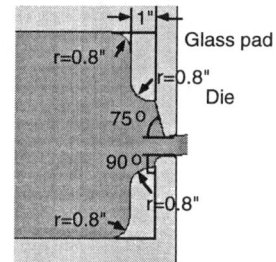

Figure 7b - The glass pad billet interface

The isometric and side views of the butt end of the extrudate are shown in figures 6a and 6b. It can be seen that the glass pad interface is at a 90 degree angle. In the isometric view, the path of the metal flow on the front face of the butt end can also be seen. It is seen that the material flow is relatively smooth. Once the sections were selected, it was necessary to define the die shape along each section. In the actual case, the billet does not flow along the die all the way. In fact, it actually flows along the glass pad which is placed in front of the die (see fig. 7a). Close to the aperture of the die, where the glass pad has already broken through, the billet flows along the die. It is quite difficult to predict the shape of the glass pad/billet/die interface. A Coordinate Measuring Machine (CMM) was used to take measurements of the extrudate butt end. Details of the glass pad geometry are shown in figure 7b.

Distortion of the extrudate

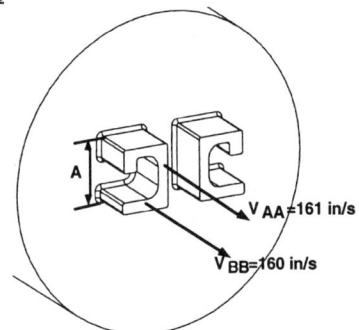

Figure 8 - Calculated velocity vectors at the exit for different sections

In order to understand the reason for the distortion of the extrudate, exit velocities were calculated for the different sections AA, BB and CC. The ram velocity was 277 mm/s (10.9 in/s). Now for each section, the extrusion ratio (ratio of initial thickness to final thickness) is known. Thus the exit velocity for any section can be computed by the simple mass conservation formula.

$$A^o V^o = A^f_{sec} V^f_{sec} \qquad (3)$$

where A^o and V^o are the initial area and ram velocity respectively, and A^f_{sec} and V^f_{sec} are the final area and exit velocity at the section sec, respectively. The computed exit velocity vectors are as shown in figure 8. It can be seen from figure 8, that the velocity of the section BB is much slower than that of section AA. This can cause the two horizontal legs of each C section of the extrudate to move closer to each other, resulting in an undersizing of dimension A. This trend has been observed experimentally. It is desired to offset this trend and thus it is necessary to investigate the effect of die design on metal flow.

The effect of die angles on metal flow

In today's industrial practice, the design of extrusion dies is still an art rather than a science. Invariably the die orifice is corrected to achieve the required control of cross sectional dimensions, straightness and surface quality. However this aspect of titanium extrusion has received far less attention than for aluminum extrusion where detailed design criteria have been established to ensure close dimensional control and minimum distortion of the extruded product. Angled 'lead-ins' to re-entrant and thin portions of the die are used to improve lubrication and metal flow in these difficult areas. Shivpuri [7] had investigated the effect of symmetrical die angles on either side on metal flow. However, no work has been reported on the effect of unsymmetrical die angles on metal flow. In order to investigate the effect of these die angles on metal flow, different die designs were investigated. The standard design for hot extrusion uses die angles of 90 degrees on either side of the aperture. Figures 9a, 9b and 9c show the metal flow (section AA) for different die designs. The trend observed in the metal flow, is that the metal tends to curve in the direction where there is a sharper corner radius. It has been observed, that when angles are machined on the die, the corner radius is around 2.5 mm (0.1"). The standard corner radius is 10 mm (0.39"). The larger corner radius seems to allow a larger local velocity, thus bending the extrudate strand towards the side with the smaller corner radius. For example, with the 75 deg. slope, (figure 9a), which has the smaller corner radius, the metal curves towards the 75 degree side. Therefore in figure 9a, the metal twists outwards and in figure 9b, it twists inwards. When both angles are symmetrical, or if both angles are 90 degrees, (figure 9c), such as in the standard design, the metal seems to come out relatively straight, and a small curvature is due to eccentricity of the port (center of port and die do not coincide). It also seems that the metal flow is more sensitive to the angle of the tongue as compared to the angle of the die. The effect of modifying die angles on the runout angle for section BB is shown in figures 10a and 10b.

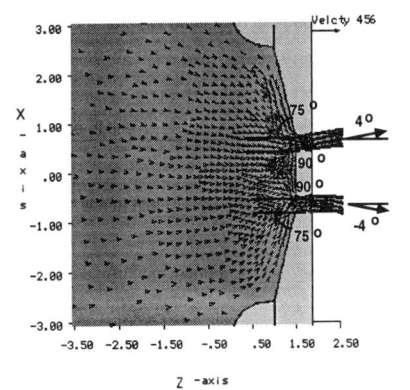

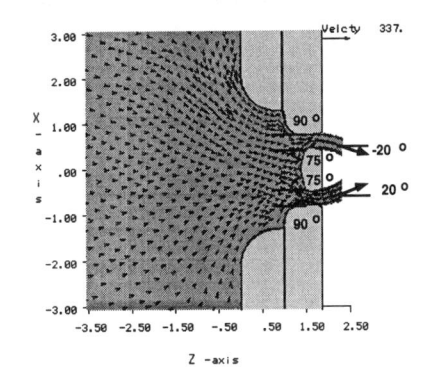

Figure 9a - 75 deg. upper die, 90 deg. upper and lower tongues, 75 deg. lower die (Sec. AA)

Figure 9b - 90 deg. upper die, 75 deg. upper and lower tongue, 90 deg. lower die (Sec. AA)

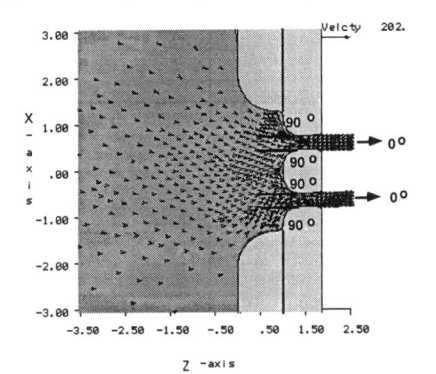

Figure 9c - Standard die design (90 degrees all over) (Sec. AA)

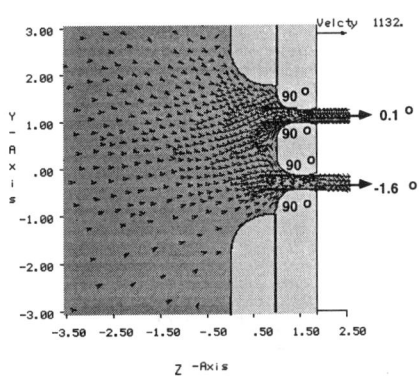

Figure 10a - Standard die design (90 degrees all over (Sec. BB)

Figure 10b - 75 deg. upper die, 90 deg. upper and lower tongue, 75 deg. lower die (Sec. BB)

The trend is consistent with earlier observations. The two legs move farther apart if angles of 75 deg. are provided. Since the simulation is performed with the assumption of plane strain, the runout angles are much larger than those observed during the actual experiment. This is because in the three dimensional metal flow case, there is a mechanical connection between the different sections, and this reduces the curling of the sections.

The effect of die land on the metal flow

In order to find out the effect of die land on the metal flow, the die land was varied from 2.5 mm (0.1") to 10 mm (0.4"). Simulations were performed for the case with an upper die angle of 75 deg. and a lower die angle of 90 deg. Increasing the die land reduces the runout angle, as shown in figure 11, however this effect reduces beyond a threshold.

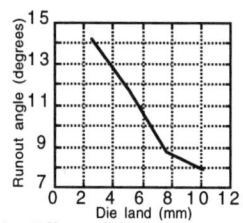

Figure 11 - Effect of die land on runout angle

The effect of heating temperature and soak times on runout angle

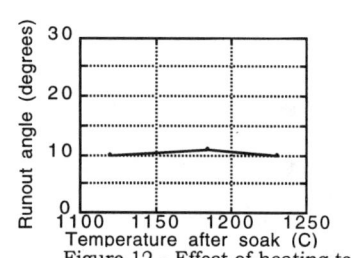

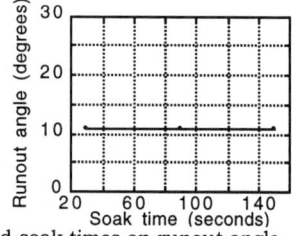

Figure 12 - Effect of heating temperatures and soak times on runout angle

The effect of heating temperature and soaking times on the metal flow were investigated by carrying out simulations with different heating temperatures and soak times. The runout angle with a design having an upper die angle of 75 degrees and a lower die angle of 90 degrees was measured. As can be seen, (figure 12), there is not much effect of the temperature and soak times on the runout angle.

Glass lubrication model

The temperature of three points on the billet were tracked and the following temperatures were noted. The temperature of the billet falls rapidly during extrusion, since the hot billet is in contact with the cooler container. The time temperature plots are given below.

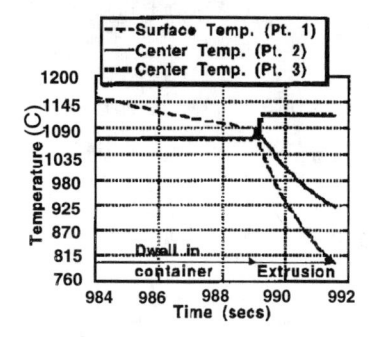

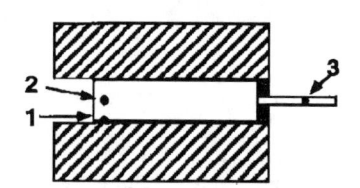

Figure 13a - Time temp. plots of pt.'s 1, 2, 3   Figure 13b - Pt.'s 1,2,3 selected for tracking

For the temperature tracking, the whole 27.5 in. billet was modeled, since after reaching steady state, the temperature variation in the extrusion billet can be modeled essentially as a heat transfer process of a hot cylinder inside a container. Now from figure 13a, it can be seen that the temperature of points 1 and 2 drop drastically after the start of extrusion at 989 sec. since they are in contact with the colder container for the longest time. On the other hand, point 3 is extruded out quickly, thus after a small temperature increase due to deformation heating, it cools in air slowly, hence its temperature remains almost constant compared to that of points 1 and 2. Thus the highest temperature that the extrudate has is around 1121 C (2050 F) and the minimum temperature is 778 C (1433 F). Now, the extrusion was modeled at ram speeds of 127 mm/s (5 in/s), 277 mm/s (10.9 in/s) and 432 mm/s (17 in/s). The maximum and minimum surface temperatures for these cases is as given in Table 1 below. Thus it can be seen that depending upon the speed, the temperature ranges involved are different. Thus the glass lubricant will perform differently depending upon the ram speed. For the above ram speeds, the theoretically calculated maximum and minimum glass thicknesses, based on Baques's model [6] are also given in the table. It can be seen that at higher ram speeds the lubricant thickness becomes thinner. On the other hand, at low ram speeds, the thickness can become too thick (causing a bad surface finish) and also the temperature may reduce so much that melting of the glass can stop. Thus there is an optimum range of speed for good performance of the lubricant.

Table 1: Temperature ranges and theoretical thicknesses of lubricant film at different speeds

Ram Speed (mm/s)	Max. temp (C)	Min. Temp (C)	t_{glass}^{max}	$t_{glass}^{min.}$
127	1121	692	62.0 microns	48.4 microns
277	1121	778	39.6 microns	32.9 microns
432	1121	925	29.0 microns	26.6 microns

Application of models to reduce distortion

The die design that were used initially to produce this shape had die angles of 90 degrees everywhere, and no slopes (fig. 14). It was observed that with the standard design, the most difficult dimension to maintain was the dimension A. It was observed that usually, this proved to be undersized. In order to increase this dimension and make it come closer to the desired (as extruded) dimension of 41.4 mm (1.63"), it was decided to change the die design so as to enable metal in the upper and lower horizontal legs to flow away from each other. The simulations of section BB showed that if the die angles were changed to 75 degrees on the upper and lower dies and 90 degrees on the tongue, the legs tended to move apart causing an increase in dimension A. Hence it was decided to use slopes of 75 degrees on the upper and lower horizontal legs and slopes of 90 degrees on the upper and lower tongue regions (figure 15). Experiments were conducted and the dimensions of A obtained by the standard design were compared with those obtained with the proposed design. It can be seen that the die design with slopes causes the dimension to open up as expected (figure 16). This resulted in significant improvement in the quality of the extrudate and also helped it to be within dimensional tolerance.

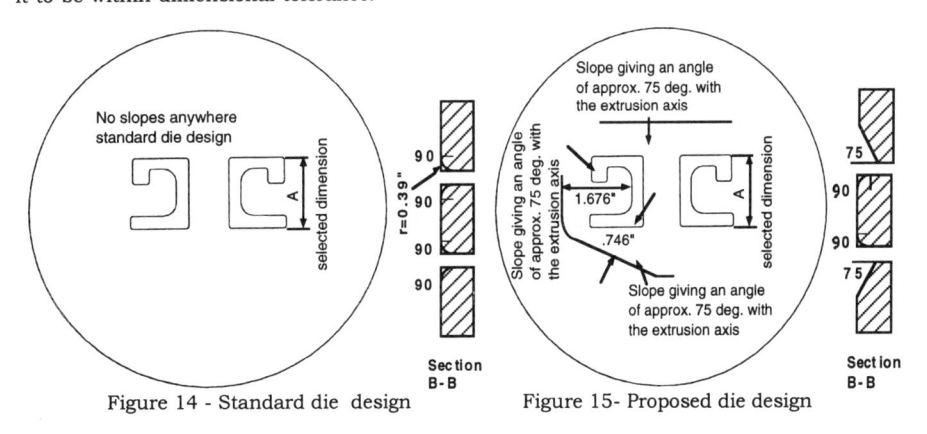

Figure 14 - Standard die design

Figure 15 - Proposed die design

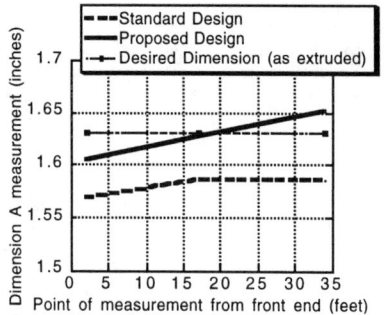

Figure 16 - Comparison of as extruded dimension A with standard and proposed designs

Conclusions

Effects of various extrusion conditions on the metal flow characteristics were investigated. It can be concluded that die design has a strong influence on the distortion of the extrudate. When die designs using unsymmetrical die angles were used, it was observed that the metal tends to flow in the direction with the smaller die angle. No strong correlation was observed between the soaking temperature, soak times and metal flow. There is an optimum range of speed for which the lubricant performs well. It is possible to reduce the distortion of the extrudate by judicious die design. An existing die design was modeled and shape variability calculated. Based on the results of the process model a new die design was selected. This die design indeed resulted in significant improvements in dimensional variability in actual production.

References

1. V. Pugliano and W. Misiolek, "Dimensional Control in Extrusion of Stainless Steel Tubing", (Proceedings of ASM's 15'th Conference on Emerging Methods for the production of tubes, bars and shapes, Clearwater, Florida, 2-3 May, 1994), 7-10.

2. Y. Tashiro, H. Yamasaki and N. Ohneda, "Extrusion Conditions and Metal Flow to Minimize Both Distortion and variance of Cross Sectional Shape", Showa Aluminum Corporation Report, Japan, (1992).

3. R. M. Baker, "Classical heat flow problems applied to induction billet heating", AIEE Transactions, 77, (1958), 106-112.

4. T. Udagawa, E. Kropp and T. Altan, "Simulation of Titanium Tube extrusion by FEM with automated remeshing capability", (Report No. ERC/NSM-B-89-27, The Ohio State University, Columbus, 1989).

5. G. Shen, "A temperature compensation technique for the correction of flow stress data in non-isothermal forging", (MS Thesis, Ohio State University, 1990).

6. P. Baque, J. Pantin and G. Jacob, "Theoretical and Experimental Study of the glass lubricated extrusion process", Journal of lubrication technology, Transactions of ASME, (1975), 18-24.

7. Shivpuri, R., Momin, S., "Computer Aided Design of Dies to control Dimensional quality of extruded shapes", Annals of CIRP, 41, (1), (1992), 275-279.

INFLUENCE OF STARTING MICROSTRUCTURE ON THE HIGH TEMPERATURE PROCESSING OF CP TITANIUM ALLOY

S. Guillard, M. Thirukkonda and P.K. Chaudhury

Concurrent Technologies Corporation
1450 Scalp Avenue, Johnstown, PA 15904

Abstract

Commercially pure (CP) titanium grades 2 and 3 have been tested in compression at temperatures and strain rates corresponding to those encountered in industrial hot working operations, i.e., temperatures from 625 to 850 C, and strain rates from 0.001 to 20 s^{-1}. Most deformation conditions led to either inhomogeneous flow, shear banding, or cracking of the grade 2 material, whereas the grade 3 material deformed homogeneously under most conditions. In addition, the flow stresses of the grade 2 material were higher than those of the grade 3 material by approximately 200 to 700% for the peak flow stresses and 150 to 250% for the steady-state flow stresses, although grades 2 and 3 materials had very similar chemical compositions. These differences were traced to the difference in starting microstructures, grade 2 exhibiting a Widmanstätten structure in contrast to grade 3 exhibiting slightly elongated α grains. Samples of the grade 3 material were heat treated to duplicate the microstructure of the grade 2 material, and mechanically tested under the same processing conditions. The results, including the flow localization characteristics and flow stresses, are compared to study the influence of starting microstructure on hot forming of CP titanium. It is suggested that to insure successful manufacturing of CP titanium alloys, formation of fine Widmanstätten structure should be prevented during prior processing.

Introduction

A recent study [1] of grade 2 Commercially Pure (CP) titanium showed that deformation under most conditions in the 500 - 750°C temperature range, and at strain rates from 0.001 s^{-1} to 5 s^{-1}, led to severe flow instability. Examination of the samples following compression testing revealed that the mode of flow localization evolved from localized bulging at low strain rates (Figures 1-a) to adiabatic shear banding at high strain rates (Figures 1-b), in good agreement with the evolution of the flow localization parameter, α [2]. Such severe flow instability has also been observed in other titanium alloys such as Ti-6Al-2Sn-4Zr-2Mo-0.1Si and the widely used Ti-6Al-4V [3,4], and is clearly undesirable for forming operations. Because the as-received grade 2 CP titanium exhibited a fine Widmanstätten structure, Figure 2, it was suggested that the flow localization was due to the combined effect of the break-up of this structure during deformation and the differences in temperature and friction conditions between the two specimen-die interfaces.

The intent of the present investigation was to confirm this suggestion by examining the deformation behavior of a similar titanium alloy not displaying the Widmanstätten structure.

(a) Inhomogeneous flow : T = 625°C, $\dot{\varepsilon}$ = 0.001 s^{-1}

(b) Shear banding + Fracture : T = 625°C, $\dot{\varepsilon}$ = 5.0 s^{-1}

Figure 1. Photomacrographs of the macroetched longitudinal section of the deformed specimen of grade 2 CP titanium at a magnification of 4X for two deformation modes.

Experimental Procedure

The material considered was commercially available grade 3 CP titanium which was received in the form of a bar, rolled at a temperature just below the β-transus and, therefore, exhibiting elongated α grains, with no evidence of Widmanstätten structure, Figure 3. Cylindrical test specimens with a diameter of 12.7 mm and a height of 15.9 mm were machined for compression testing. Isothermal, constant strain rate compression tests were carried out in vacuum on a 244-kN servo-hydraulic MTS machine, the test temperatures being varied from 500 to 850°C and the strain rates being varied from 0.001 to 5 s^{-1}. The platens were coated with boron nitride to minimize the frictional effects. Prior to compression, the specimen was held at the test temperature for about 10 minutes to avoid thermal gradients within the specimen. The load-stroke data were acquired by the data acquisition system during the test and later converted to true stress - true strain data. Photomacrographs of the cylindrical surface of the deformed specimens were obtained. Longitudinal sections of the specimens were examined both macroscopically and microscopically, after etching with a modified Kroll's reagent containing 10% HF, 5% HNO_3, and balance H_2O.

Figure 2. As-received microstructure of grade 2 CP titanium.

Figure 3. As-received microstructure of grade 3 CP titanium.

Results

Flow behavior of grade 3 CP titanium

Typical flow curves of grade 3 CP titanium at selected strain rates of 0.01 s^{-1} and 1 s^{-1} are shown in Figure 4 for various test temperatures. The flow curves have been corrected for adiabatic heating using the procedure outlined by Thomas and Srinivasan [5]. At all strain rates and temperatures, the curves exhibit hardening before eventually reaching a steady-state stress. Visual inspection of the deformed specimens revealed homogeneous flow, and the specimens were free from surface cracking. Some barreling due to friction at the specimen-platen interface was observed. Details of the flow characteristics as a function of the initial microstructure are discussed in the next section.

(a)

(b)

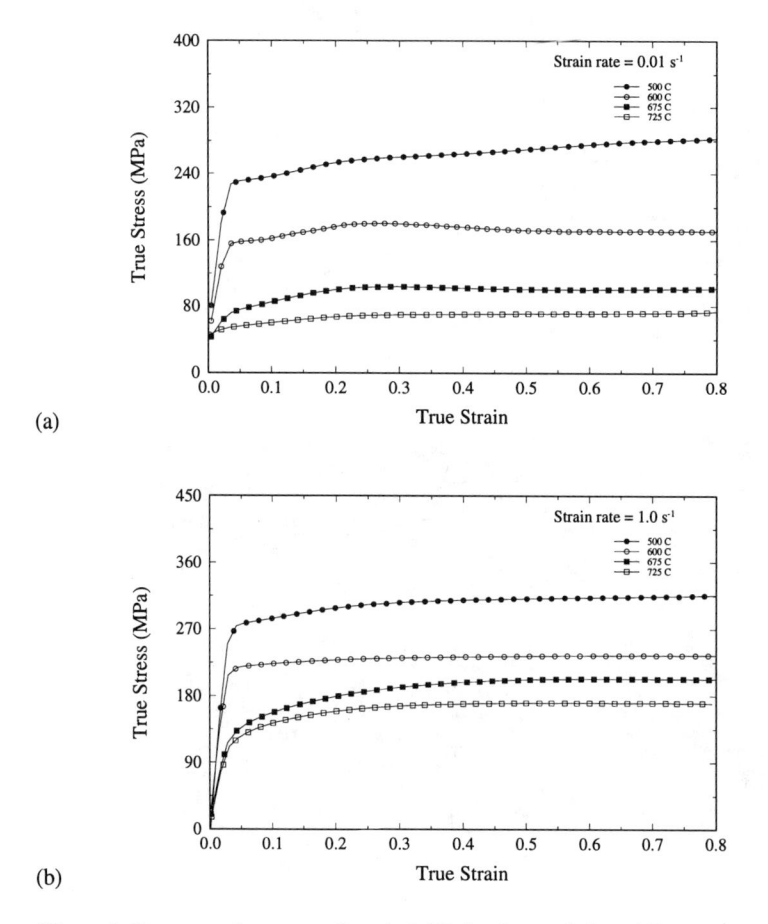

Figure 4. Stress - strain curves of grade 3 CP titanium at Selected Temperatures and at (a) 0.01 s^{-1} and (b) 1 s^{-1}.

Discussion

Comparison of flow characteristics of grade 2 and grade 3 CP titaniums

The flow behavior of grade 3 titanium differs widely from that of grade 2. For example, Figure 5 shows the stress - strain curves of both grades under the same deformation conditions of 650°C and 1 s^{-1}. Grade 2 titanium first reaches a peak before experiencing softening, probably due to the breakdown of the Widmanstätten structure. By contrast, grade 3 titanium shows hardening before reaching steady state. However, the most striking difference is the significantly higher flow stress exhibited by grade 2, 425 MPa, compared to grade 3, 210 MPa. The same is true of the other deformation conditions considered in this study. Table I lists the ratio of the steady-state flow stress of grade 2 titanium to that of grade 3 titanium.

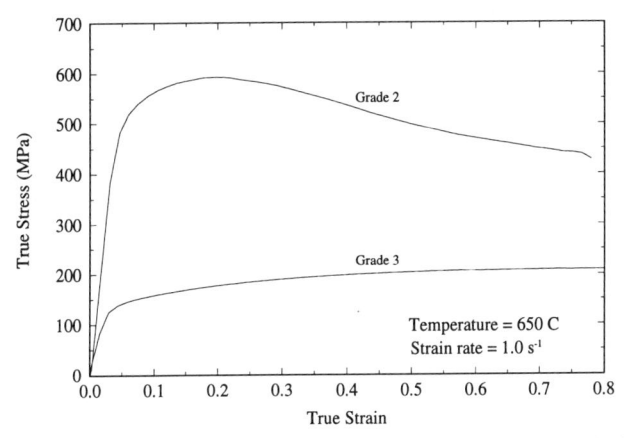

Figure 5. Stress - strain curves of grade 2 and grade 3 CP titaniums at 650°C and 1 s^{-1}.

Table I. Ratio of the Steady-State Stress of CP Titanium Grade 2 to that of CP Titanium Grade 3 under Various Deformation Conditions

	0.001 s^{-1}	0.01 s^{-1}	0.1 s^{-1}	1 s^{-1}
600°C	3.7	3.2	2.7	N/A
650°C	3.1	3.1	2.2	2.1
725°C	3.1	3.1	2.7	1.9

Moreover, grade 3 CP titanium displayed homogeneous flow under most deformation conditions, in sharp contrast to grade 2 CP titanium. Figure 6 examplifies this by showing samples of both grade 2 and grade 3 titaniums which have both been deformed at 850°C and 1 s^{-1}. It clearly shows that grade 2 titanium displays shear banding and a rough external surface with cracks, Figure 6-a, whereas grade 3 titanium has deformed homogeneously and shows a smooth external surface, Figure 6-b. Three factors can be considered as possible reasons for the difference in flow behavior: the friction and temperature conditions, the chemical compositions, and the initial microstructures. Friction and temperature effects were the same in both materials as the exact same experimental procedure was followed in both cases. Chemistry differences, namely the amount of interstitials and iron, are known to significantly affect the mechanical properties of titanium alloys at room temperature; when the nominal compositions are scrupulously inspected, the interstitial + iron content differential between grade 2 and grade 3 is 0.12 wt.%, grade 3 having the higher amount. In that case, the flow stress at room temperature varied from 343 MPa for grade 2 to 440 MPa for grade 3 [6]. At elevated temperatures, however, this difference in flow stress is expected to disappear. As shown in Table II, grade 3 has indeed the higher amount of interstitial + iron and, therefore, chemistry differences cannot account for its dramatically low flow stress.

(a)

(b)

Figure 6. Photomicrographs of the cylindrical surface (left), and the macroetched longitudinal section (right) of (a) grade 2 CP titanium and (b) grade 3 CP titanium (Magnification 4X; deformation at 850°C and 1s^{-1}).

Table II. Chemical Compositions of Grade 2 and Grade 3 Titanium Alloys

	Elements (wt.%)						
Material	C	H	O	N	Fe	Interstitials + Fe	Ti
Grade 2	.10	.004	.125	.007	.03	.266	99.73
Grade 3	.032	.004	.28	.006	.20	.522	99.48

The third factor must, therefore, be responsible for most of the observed differences. In order to examine this suggestion, a heat treatment was developed to produce a Widmanstätten structure in grade 3 CP titanium.

Development of Widmanstätten structure in grade 3

The heat treatment to produce a Widmanstätten structure in grade 3 CP titanium consisted of heating the samples to 950°C for 30 minutes, followed by rapid cooling. This treatment resulted in a microstructure consisting of colonies having the same size as those of the grade 2 titanium, 160 μm. However, the maximum controlled cooling rate which could be obtained was of the order of 230°C per minute, and led to a fairly coarse needle structure, Figure 7.
The heat treated grade 3 samples were tested in compression under various conditions of temperature and strain rate. Figure 8 shows a comparison of the flow curves of grade 2, grade 3, and heat treated grade 3 titanium, tested at 850°C and 1 s^{-1}. It reveals that the flow stress of the heat treated grade 3 titanium is higher than that of its as received counterpart, but it remains

well below that of grade 2 titanium. Because of the similar nature of the microstructures of grade 2 and heat treated grade 3 titanium, the remaining difference can only be attributed to the fineness of the Widmanstätten structure. This aspect has already been studied in alpha zirconium where the same effect of microstructure on flow stress, although to a much more limited extent, has been observed and was further related to internal stress contributions from the transformation substructures [7].

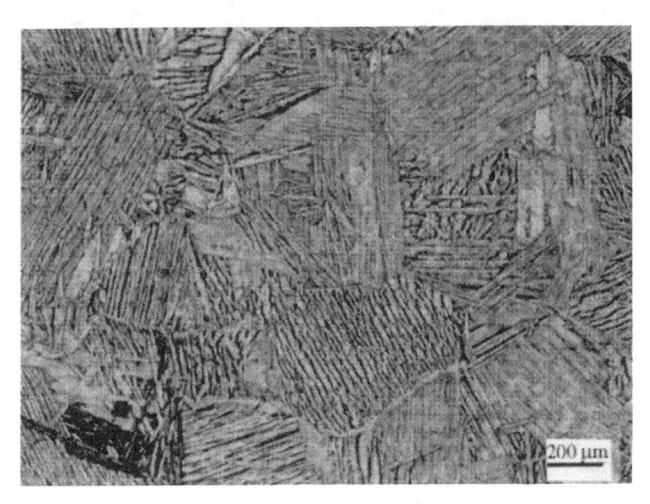

Figure 7. Microstructure of grade 3 CP titanium following heat treatment.

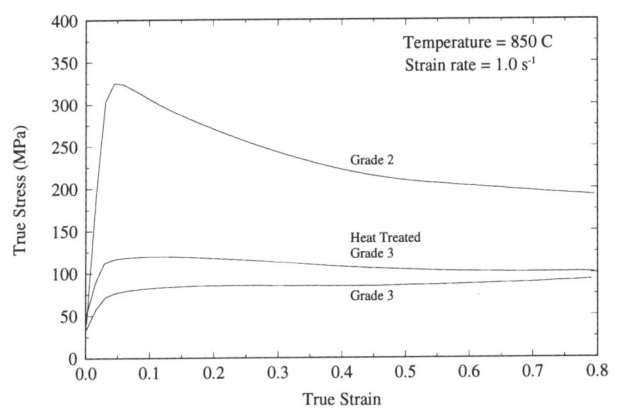

Figure 8. Stress - strain curves of grade 2, grade 3, and heat treated grade 3 CP titaniums at 850°C and 1 s^{-1}.

Figure 9 shows a macrograph and micrograph of a heat treated grade 3 titanium sample following compression at 850°C and 1 s^{-1}. It can be compared to Figure 6 to reveal that as the starting microstructure was changed from elongated grains to coarse Widmanstätten, the degree of flow non-uniformity was slightly increased; compare Figure 6-b to Figure 9. However, the change is much more dramatic when going from a coarse Widmanstätten structure to a fine one;

compare Figure 9 to Figure 6-a. This seems to indicate that an equiaxed starting microstructure should be preferred for forming CP titanium, but that a Widmanstätten structure can be acceptable as it is not the Widmanstätten per se which is unfavorable for forming, but rather its fineness.

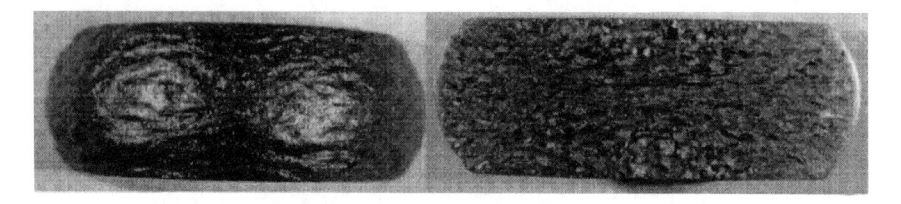

Figure 9. Photomicrographs of the cylindrical surface, and the macroetched longitudinal section of heat treated grade 3 CP titanium at a magnification of 4X following deformation at 850°C and 1 s^{-1}.

Summary

1. The high temperature compressive flow behavior of grade 3 CP titanium has been studied.
2. The very significant differences in flow behavior of grade 3 CP titanium and a previously tested grade 2 CP titanium are due to the differences in initial microstructure.
3. It has been shown that although an equiaxed starting microstructure should be preferred for forming CP titanium, a Widmanstätten microstructure is acceptable, provided its needles are not too fine.

References

1. M. Thirukkonda, P.K. Chaudhury, D. Zhao and J.J. Valencia, "Flow Instability During High Temperature Deformation of Grade 2 CP Titanium", Microstructure/Property Relationships of Titanium Alloys, S. Ankem and J.A. Hall, eds., TMS (1994) 17-26.
2. J.J. Jonas, R.A. Holt and G.E. Coleman, "Plastic Instability and Flow Localization in Ti-6242", Res Mechanica Letters, 1 (1981) 97-103.
3. S.L. Semiatin and G.D. Lahoti, "Deformation and Unstable Flow in Hot Forging of Ti-6Al-2Sn-4Zr-2Mo-0.1Si", Met.Trans., 12A (1981) 1705-1717.
4. A. Wang and H.J. Rack, "High Temperature Flow Localization Coarse Grain β-Processed Ti-6Al-4V", Seventh World Titanium Conference, San Diego, CA (1992).
5. J.F. Thomas, Jr. and R. Srinivasan, Computer Simulation in Materials Science, R.J. Arsenault, J.R. Beeler, Jr. and D.M. Esterling, eds., ASM, Metals Park, OH (1986) 269-290.
6. Properties and Selection: Nonferrous Alloys and Special-Purpose Materials, ASM Handbook (formerly 10th edition), (1990) 594.
7. D.J. Abson and J.J. Jonas, "Hot Compression Behavior of Thermo-Mechanically Processed Alpha Zirconium", Metals Technology (October 1977) 462.

MICROSTRUCTURAL EVOLUTION IN Ti-6Al-4V

DURING HOT DEFORMATION

M.B.Gartside

IRC in Materials for High Performance Applications
The University of Birmingham
Edgbaston
Birmingham B15 2TT
United Kingdom.

Abstract

Evolution of microstructure and its effect on mechanical behaviour has been investigated in Ti-6Al-4V for different initial α and β morphologies. Hot tensile tests were performed in the temperature range 720-920°C and at strain rates of 10^{-4}-10^{-2}s^{-1}. The stress strain behaviour exhibited a maximum stress followed by a decrease in flow stress which was most pronounced at low temperatures, high strain rates and for the transformed β microstructures. Optical microscopy of the transformed β microstructures showed shearing and fragmentation of the α needles, particularly at high temperatures, leading to a softer microstructure of spherical α particles. In the as-received microstructure some rearrangement of the α and β grains was observed. These results indicate that rearrangement of the microstructure is responsible, to some extent, for the development in flow stress seen in this alloy at high temperatures.

Introduction

Ti-6Al-4V is a two-phase α+β titanium alloy which, in the mill annealed condition, has a microstructure consisting of fine equiaxed α grains and transformed β grains with some residual β. At hot working temperatures the microstructure consists of α and β grains, the volume fraction of β increasing with temperature.

The stress-strain behaviour of two-phase alloys, such as the alloy used here, αβ brass and αγ stainless steel, is complex and does not necessarily obey a simple rule of mixtures law such as that used by Ankem et al. [1]. Roberts and Ottenberg [2] investigated the stress-strain behaviour of 60-40 brass at different temperatures and hence different β volume fractions. They found that at high volume fractions of α or β, deformation was concentrated in the major constituent, but that at approximately equal volume fractions of the two phases, deformation tended to occur in the softer β phase.

When Ti-6Al-4V is deformed at elevated temperatures the stress-strain behaviour exhibits a maximum in flow stress at a small strain followed by a decrease in flow stress towards a steady state value. This behaviour has been seen in tensile tests [3], torsion tests [4,5], compression tests [6] and isothermal forging [7]. The decrease in flow stress is more pronounced for the acicular transformed β microstructures but is also seen in material in the mill-annealed condition. The flow stress development in this alloy has been attributed to a number of different causes. Recrystallisation of the α phase, deformation heating, the formation of a softer texture and the rearrangement of microstructural components have been proposed.

The objective of the present investigation was to determine how the microstructure of Ti-6Al-4V evolves with deformation and to relate these changes to the mechanical behaviour of the alloy.

Experimental Procedure

Mill-annealed Ti-6Al-4V plate with a thickness of 25mm was used in this study. Samples of the as-received plate were heat treated in the β phase field at 1015°C for 1 hour followed by air cooling to room temperature to produce a transformed β microstructure. Half of these samples were then subjected to a temper at 950°C for 24 hours followed by air cooling to room temperature to form thicker transformed β needles.

Testpieces machined from the as-received and heat treated material were tested in tension at temperatures of 720, 820 and 920°C and at strain rates of the order of 10^{-4}, 10^{-3} and $10^{-2} s^{-1}$. The tests were carried out using a perturbed strain rate, varied by ±10% of the nominal value, in order to determine the strain rate sensitivity, m. Within each strain rate step a constant strain was used, the cross-head speed calculated by microcomputer with the assumption that the specimen gauge remained parallel. Microstructural observations were carried out on the relatively undeformed material from the head of the specimen and on the deformed material from the specimen gauge.

Results and Discussion

Stress-Strain Behaviour

Fig. 1 shows stress-strain curves for material with the globular α+β microstructure tested at a temperature of 820°C and strain rates of the order of 10^{-4}, 10^{-3} and $10^{-2} s^{-1}$. In most cases the stress rises to a maximum and then decreases levelling out, at the lower strain rates, to a steady-state value. The effects of necking can be seen in the curves for material tested at 10^{-3} and $10^{-2} s^{-1}$ where the gradients of the curves change. The decrease in flow stress is most pronounced for the highest strain rate and least pronounced at the lowest strain rate. The m values for the curves are marked on the graph, they are all quite high but the m value at the lowest strain rate is 0.5. This indicates that at temperatures greater than 820°C and at a strain rate of $10^{-4} s^{-1}$ the material exhibits superplasticity. At a temperature of 920°C and a strain rate of $10^{-4} s^{-1}$ an m value of 0.54 is seen, again indicating superplasticity. This was the only stress-strain curve not to exhibit a decrease in flow stress.

The stress-strain curves shown in Fig. 2 are for material with the globular α+β microstructure tested at a strain rate of $10^{-2}s^{-1}$ with temperatures of 720, 820 and 920°C. The maximum flow stress and the extent of the flow stress drop are greatest for the material tested at the lowest temperature.

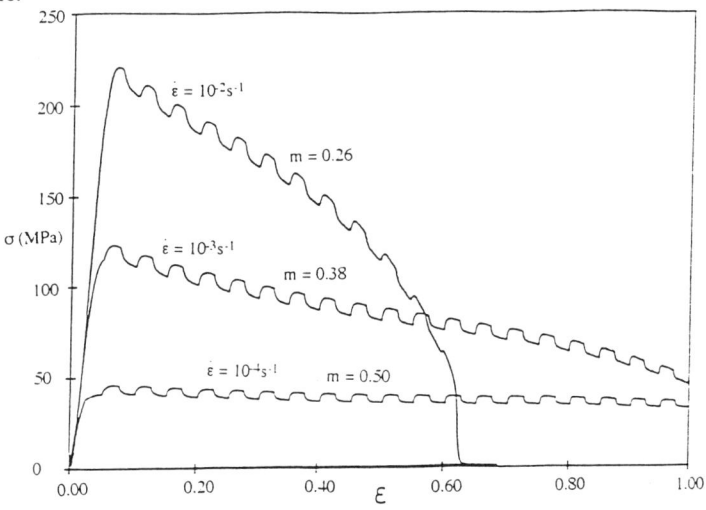

Fig. 1. True stress-true strain graphs for the globular α+β material tested in tension at a temperature of 820°C and strain rates of 10^{-4}, 10^{-3} and $10^{-2}s^{-1}$.

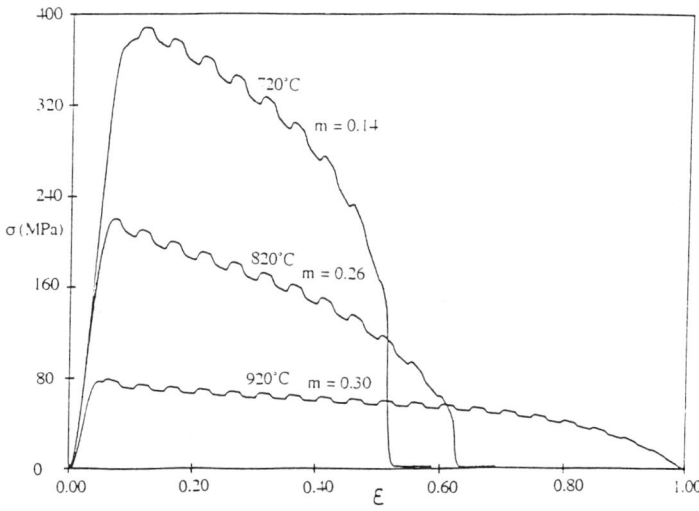

Fig. 2. True stress-true strain graphs for the globular α+β material tested at a strain rate of $10^{-2}s^{-1}$ and temperatures of 720, 820 and 920°C.

Fig. 3 shows stress-strain curves for material with each of the three microstructures tested at a temperature of 820°C and at a strain rate of $10^{-3}s^{-1}$. The globular α+β material underwent a larger amount of tensile strain to failure than the heat treated material. The material with the tempered transformed β microstructure exhibited the highest maximum flow stress and the least ductility. The heat treated material showed quite high m values, but these microstructures do not give such high strain rate sensitivities as the superplastic globular α+β material.

103

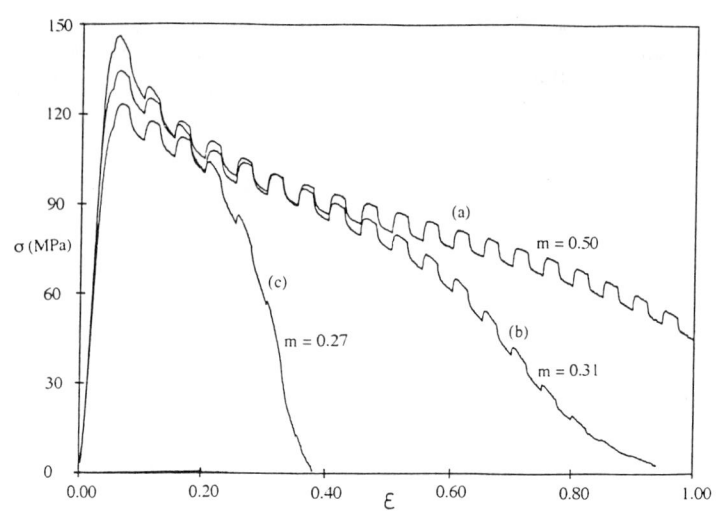

Fig. 3. Stress-strain graphs for material tested at a temperature of 820°C and a strain rate of 10^{-3} s^{-1}. Curve (a) represents the globular α+β material, curve (b) the β heat treated material and curve (c) the β heat treated and tempered material.

Microstructural Observations

Transformed β Microstructures

Figure 4 shows typical undeformed microstructures for material with the transformed β microstructure (Fig. 4a) and for material with the tempered transformed β microstructure (Fig. 4b). The morphology of the α phase differs, the tempered samples having shorter thicker α needles and a greater amount of α precipitated on prior β grain boundaries.

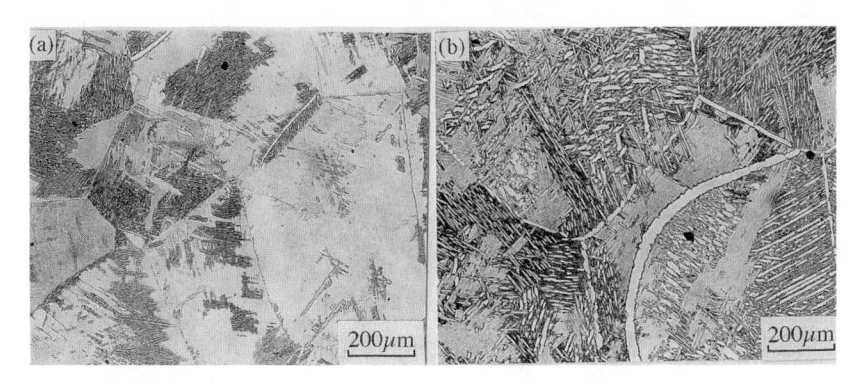

Fig. 4. Optical micrographs of undeformed material from the heads of samples with, (a) the β heat treated microstructure and (b) the β heat treated and tempered microstructure.

Fig. 5 shows deformed microstructures for both heat treatments deformed at different temperatures and strain rates. In each case the α needles have altered their morphology during the test. At high strain rates and low temperatures deformation is largely confined to the neck region. Fig. 5a shows the shearing and break-up of the α needles in material close to the fracture surface of a tempered sample. The thick layers of α phase precipitated at the prior β grain boundaries have nucleated several large voids. The fracture surfaces of these samples have a slightly faceted appearance, suggesting that the coalescence of these voids was the cause of failure.

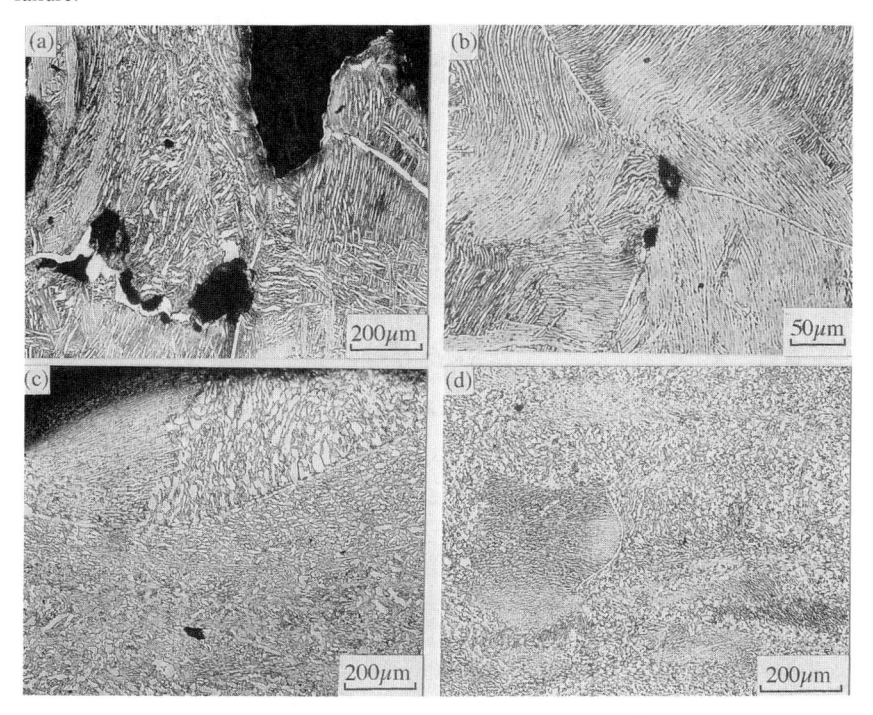

Fig. 5. Optical micrographs of deformed material tested at temperatures, strain rates and tensile strains of, (a) 820°C, $10^{-3}s^{-1}$ and 0.34, (b) 920°C, $10^{-2}s^{-1}$ and 0.9, (c) 920°C, $10^{-4}s^{-1}$ and 1.61 and (d) 920°C, $10^{-4}s^{-1}$ and 1.68. Initial microstructures were tempered transformed β (a, c) and transformed β (b, d).

Fig. 5b shows extensive shearing of α needles in a cross-section of the gauge of a β heat treated sample. This shearing is associated with a large number of cracks at prior β grain boundary triple points. It is likely that the deformation by shearing of constrained material is accommodated by the initiation and growth of a grain boundary triple point crack, particularly where there is α precipitated at the triple point.

At low strain rates and high temperatures deformation is less localised and the α needles are more broken up. Fig. 5c shows a typical microstructure from the neck of a tempered sample. α needles parallel to the tensile axis appear to break up by a process of "pinching-off", whereby the α needles are converted during deformation to α grains still aligned with the tensile axis. Weiss et al. [8] suggested that the mechanism for this phenomenon is penetration of β along high angle boundaries in the α needles. During deformation dynamic recovery occurs in the α needles forming sub-grain boundaries and grain boundaries across the needle. Penetration of β occurs along high angle boundaries causing globular α grains to be pinched off. Those α needles perpendicular to the tensile axis are broken up by a combination of shearing and penetration of β through heavily sheared regions. The resulting microstructure consists of irregular α grains interconnected along the original α needle.

The same mechanisms also operate in the β heat treated material; this can be seen in Fig. 5d. The thinner α needles allow easier penetration by the β phase producing α grains that are more equiaxed.

Under all test conditions the start microstructure was broken up by the combination of deformation and temperature to produce a more globular microstructure. Deformation is an important factor in this break-up since, as can be seen in Fig. 4, there was no break-up of the microstructure in the heads of the samples, which were subjected to temperature alone. In specimens that form necks during testing there is a strain distribution ranging from a maximum strain at the lowest cross-sectional area to a minimum strain at the highest cross-sectional area. Microstructural observation of areas that have been subjected to different levels of strain shows that the amount of break-up increases with increasing strain.

The globular microstructure produced by deformation consisted of hard α grains in a softer β matrix. This microstructure flows more easily than the transformed β microstructure that was present at the start of the test. As the samples were deformed, the volume fraction of globular α increased, lowering the flow stress to produce the type of stress-strain behaviour observed in these samples.

The break-up of the start microstructure was most complete in the samples subjected to the highest temperature and the lowest strain rate. The penetration of the α needles by the β phase, either along high angle boundaries or heavily sheared areas, occurs by a process of diffusion. As diffusion is most rapid at higher temperatures, a low strain rate to a given strain allows more time for diffusion to occur.

Globular α+β Microstructure

Fig. 6 shows a typical microstructure for relatively undeformed material from the head of a sample tested at 820°C. At each of the test temperatures the initial morphology of the α and β grains was similar to this, but the volume fractions of α and β varied with test temperature. There is some directionality in the microstructure, in many areas clumps of α grains are elongated and aligned with other clumps.

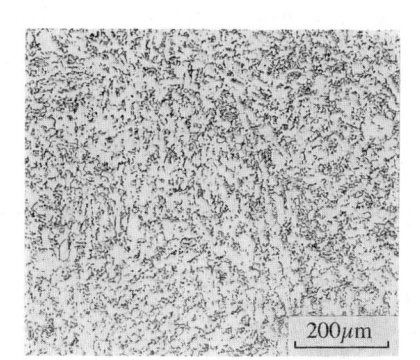

Fig. 6 Optical micrograph of undeformed material with the globular α + β microstructure.

Fig. 7a shows the microstructure close to the fracture surface of a sample deformed at the lowest temperature and the highest strain rate. There is extensive cavitation in the region immediate to the fracture, which was probably the cause of failure. The α grains adjacent to the fracture are more equiaxed than those that are further away which have a slightly elongated shape, like that of the α grains in the undeformed material. This indicates that at lower temperatures and higher strain rates deformation is largely concentrated in the necked region.

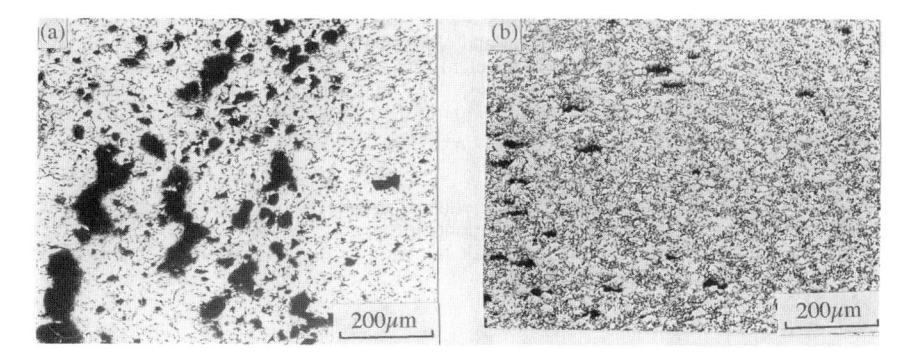

Fig. 7. Optical micrographs of deformed material with the globular α+β microstructure tested at (a) 720°C, $10^{-2}s^{-1}$ and (b) 820°C, $10^{-3}s^{-1}$. The samples were tested to tensile strains of (a) 0.6 and (b) 1.7.

The microstructure of material from the neck of a sample tested at an intermediate temperature and strain rate is shown in Fig. 7b. This sample also exhibits cavitation but the pores are neither as large nor as widespread as those in Fig. 7a. The α and β grains in this sample are much more equiaxed than those in the undeformed material shown in Fig. 6. The most equiaxed grains are at the centre of the necked region but the deformation is not as localised as that of the samples tested at 720°C.

Fig. 8 shows sections from the necked region of samples tested at the highest temperature and the lowest strain rate to tensile strains of 0.2, 0.6 and 1. As the tensile strain increases the elongated clumps of α grains are broken down. This process appears to occur by the "pinching-off" mechanism seen in the transformed β microstructures. The β phase penetrates the clumps of α along α/α grain boundaries thus pinching off discrete, more equiaxed, α grains surrounded by β. There may be some rearrangement of β grains, but it is difficult to verify this since on cooling the majority of the β phase transforms to martensitic α' or Widmanstatten α needles.

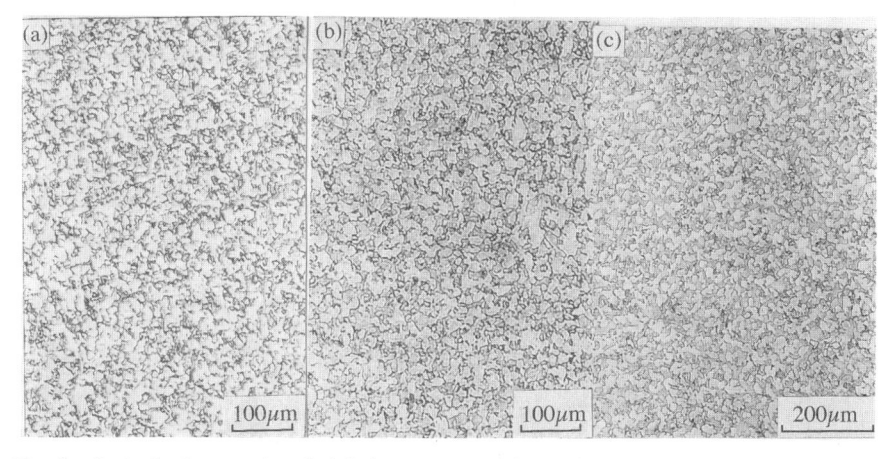

Fig. 8. Optical micrographs of globular α+β material tested at a temperature of 920°C and a strain rate of $10^{-4}s^{-1}$ to tensile strains of (a) 0.2, (b) 0.6 and (c) 1.

The mechanism for the flow stress development in material with the globular microstructure appears to be spheroidization of the irregular and elongated α grains. As the α grains become more equiaxed they become separated from the clumps of α grains of which they were originally a part and are surrounded by the softer β phase. This microstructure is similar to the softer globular microstructure formed during deformation of the transformed β microstructures. The more equiaxed microstructure flows more easily than the starting microstructure and seems to account for the flow stress development.

The change in the extent of the flow stress drop with microstructure can be explained by optical examination of the start microstructures. The globular α+β microstructure is very close to the softer equiaxed microstructure so only a small degree of softening is possible. The tempered transformed β microstructure is the hardest of the three and so has the greatest potential for softening.

Conclusions

When Ti-6Al-4V is subjected to tensile tests at temperatures in the temperature range 720-920°C and the strain rate range 10^{-4}-$10^{-2}s^{-1}$ the stress-strain behaviour is characterised by a maximum flow stress followed by a decrease in flow stress, except when the material exhibits superplasticity. The decrease in flow stress is most pronounced at the lowest temperature, the highest strain rate and for the transformed β material.

In the transformed β material the decrease in flow stress is caused by the break-up of the α needles to form a softer more globular microstructure, and in the globular α+β material the decrease in flow stress appears to be caused by the spheroidizing of elongated, irregular clumps of α grains. Changes in α morphology seem to notionally explain the development of flow stress in all cases.

References

1. S. Ankem *et al.*, "The Effect of Volume Per Cent of Phases on the High Temperature Tensile Deformation of Two-phase Ti-Mn Alloys", Mater. Sci. Eng., A111 (1989), 51-61.

2. W. Roberts, in Deformation, Processing and Structure (Metals Park,OH: ASM, 1982), 109-184.

3. Y. Ito and A. Hasegawa, in Proc. 4th Int. Conf. on Titanium (Kyoto, Japan: Japan Institute of Metals, 1980), 983-992

4. T. Sheppard and J. Norley, "Deformation Characteristics of Ti-6Al-4V", Mater. Sci. Tech.,4 (1988), 903-908.

5. K. Takahashi *et al.*, "Microstructural Evolutions of Different Ti-6Al-4V Titanium Alloy Morphologies During and After Hot Working Simulated by Torsion Testing", Rev. Met.-C.I.T., 90 (1993), 599-610.

6. S.M.L. Sastry, P.S. Pao and K.K.Sankaran, in Proc. 4th Int. Conf. on Titanium (Kyoto, Japan: Japan Institute of Metals, 1980), 873-886.

7. C.C. Chen, in Titanium and Titanium Alloys ed. J.C. Williams and A.F. Belov, (New York, NY: Plenum Press, 1982), 397-411.

8. I.Weiss *et al.*, "Modification of Alpha Morphology in Ti-6Al-4V by Thermomechanical Processing", Metall. Trans., 17A (1986), 1935-1947.

SOLUTE SOFTENING OF ALPHA TITANIUM - HYDROGEN ALLOYS

Oleg N. Senkov[*] and John J. Jonas

Department of Metallurgical Engineering, McGill University
3450 University Street, Montreal, Quebec, Canada H3A 2A7
[*]On leave from the Institute of Solid State Physics, Russian Academy of Sciences
Chernogolovka, Moscow District, Russia 142432

Abstract

Compression tests were carried out within the alpha phase field on a series of titanium - hydrogen alloys. The dependence of the flow stress on temperature, strain rate and hydrogen content was studied. Alloying with hydrogen leads to the noticeable softening of alpha titanium over the temperature range 500 to 800 °C; dynamic strain aging due to the interaction of dislocations with iron, carbon and oxygen atoms is also observed over this range. The softening is attributed to the weakening of these interactions brought about by the addition of hydrogen.

Introduction

The addition of interstitial elements to titanium generally results in appreciable hardening [1]. Hydrogen is an exception as it can produce noticeable softening in both alpha and alpha+beta alloys at elevated temperatures. Because hydrogen is a beta stabilizer, this softening was initially attributed [2,3] to the increased proportion of the softer bcc beta phase. According to a recent investigation [4] of the effect of phase composition and hydrogen level on the deformation behavior of titanium-hydrogen alloys, this explanation does indeed apply to the alpha+beta alloys. However, it is now clear that hydrogen produces marked softening in the single hcp alpha phase as well, although its addition leads to noticeable hardening of the bcc beta phase.

The present paper deals with certain peculiarities of the hydrogen-induced softening of alpha titanium. The flow stress, strain rate sensitivity, work hardening rate and strain anisotropy are presented as functions of strain rate, temperature and hydrogen concentration, and the possible mechanisms of softening are discussed.

Experimental Materials and Procedures

The base material in this study was a titanium of technical purity. The chemical composition of this material is given in Table 1. Cylindrical specimens 8 mm in diameter and 12 mm in height were prepared for compression testing by machining. The specimens were then alloyed with increasing amounts of hydrogen by a method described elsewhere [4]. Specimens with hydrogen concentrations of 0.05, 2.7, 3.8, 4.7 and 5 atomic percent were studied. All the specimens were annealed at 800 °C prior to testing. The mechanical tests were carried out on an MTS machine at strain rates of 10^{-4} to 1 s^{-1} and over the temperature range 500 to 800 °C. The hydrogen content of each specimen was remeasured after deformation. More detailed information about the experimental method is given in the paper cited above [4].

Table 1. Chemical composition of base specimens (at.%).

Al	V	Fe	H	C	O	N
0.426	0.255	0.058	0.047	0.067	0.011	0.001

Results and Discussion

Flow curves and work hardening rate.

Some typical flow curves determined on the unalloyed and alloyed materials are presented in Figures 1a and 1b. All the specimens exhibited conventional strain hardening when they were deformed at strain rates above 10^{-3} s^{-1}. At lower strain rates, some flow softening was detected, followed by a steady state of straining. Serrations were observed on some of the curves. Alloying with 2.7 at. % hydrogen decreased the yield and flow stresses at all stages of deformation, although the addition of more than 2.7 % hydrogen (up to 5 %) did not have much further effect on the curves.

The dependence of the work hardening rate on strain and hydrogen content is depicted in Figure 2 for two different strain rates and three different temperatures. At 1 s^{-1} and 500 °C (Figure 2a), the work hardening rate is inversely proportional to the strain initially and then decreases linearly with strain beyond $\varepsilon = 0.1$. By contrast, at 1 s^{-1} and 700 °C, the work hardening rate is inversely proportional to the strain right up to $\varepsilon = 0.5$. When the strain rate was reduced to 10^{-3} s^{-1} (Figure 2b), the work hardening rate decreased rapidly to zero at the beginning of deformation and remained at a low level on further straining. Alloying with hydrogen decreased the work hardening rate under all conditions (Figures 2a and 2b).

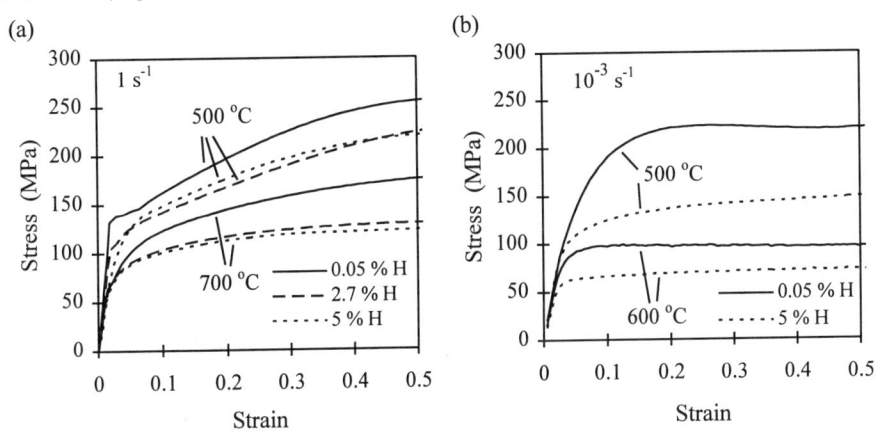

Figure 1 - Effect of hydrogen content and temperature on alpha titanium flow curves. (a) 1 s^{-1} and (b) 10^{-3} s^{-1}.

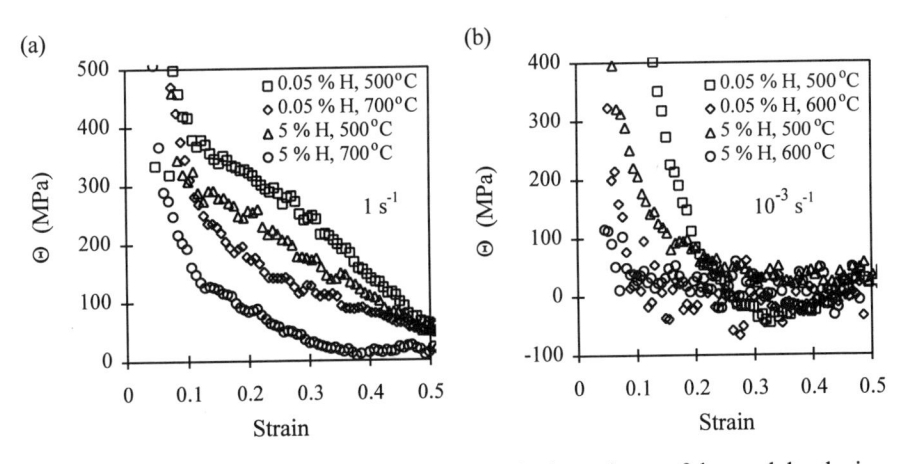

Figure 2 - Effect of hydrogen addition on the strain dependence of the work hardening rate in alpha titanium tested at (a) 1 s^{-1} and (b) 10^{-3} s^{-1}.

Effect of strain rate and temperature on the flow stress.

The strain rate dependence of the flow stress of unalloyed titanium is compared with that of the hydrogen-enriched material in Figure 3a. Three types of behavior were observed, each with a different rate dependence. In the first, the flow stress increased with strain rate; in the second, it decreased; and in the third, it increased once again. Similar anomalies were detected in the temperature dependence of the flow stress, Figure 3b. A plateau or bulge was observed over the temperature range 600 to 650 °C when the unalloyed specimens were deformed at 1 s^{-1}; when the strain rate was decreased to 10^{-3} s^{-1}, the temperature at which the flow stress was unexpectedly high decreased to 500 °C.

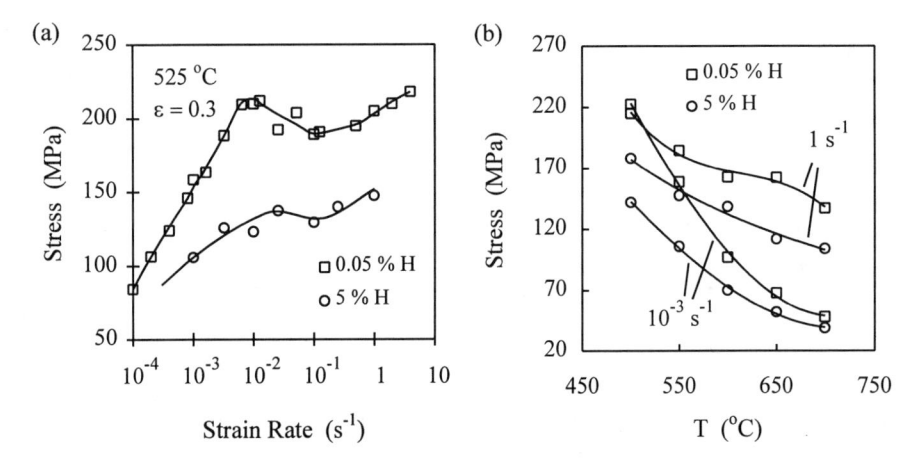

Figure 3 - Dependence of the flow stress on (a) strain rate and (b) temperature in titanium containing 0.05 and 5 at.% H.

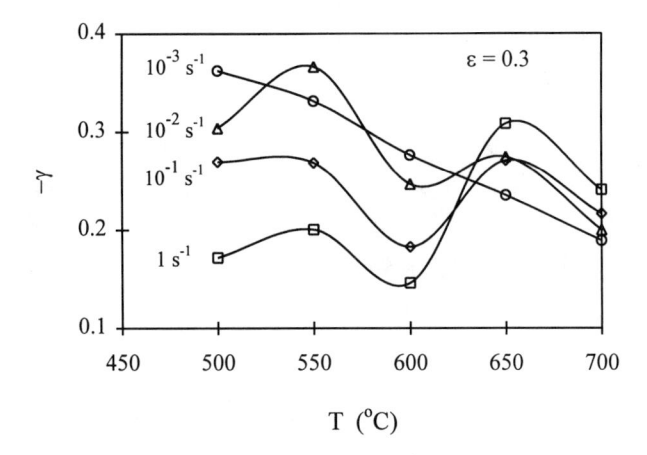

Figure 4 - Effect of strain rate and temperature on the relative softening attributable to the addition of 5 at.% hydrogen.

It can be seen from Figures 3a and 3b that the addition of hydrogen decreases the flow stress and smoothes out these anomalies as well. The relative softening, defined as γ = (σ_a-σ_u)/σ_u, where σ_a and σ_u are the flow stresses of the alloyed and unalloyed specimens, respectively, is plotted in Figure 4 as a function of temperature. Several softening peaks are evident; these are located at the same temperatures and strain rates as ones identified above with the abnormal behavior of the unalloyed titanium.

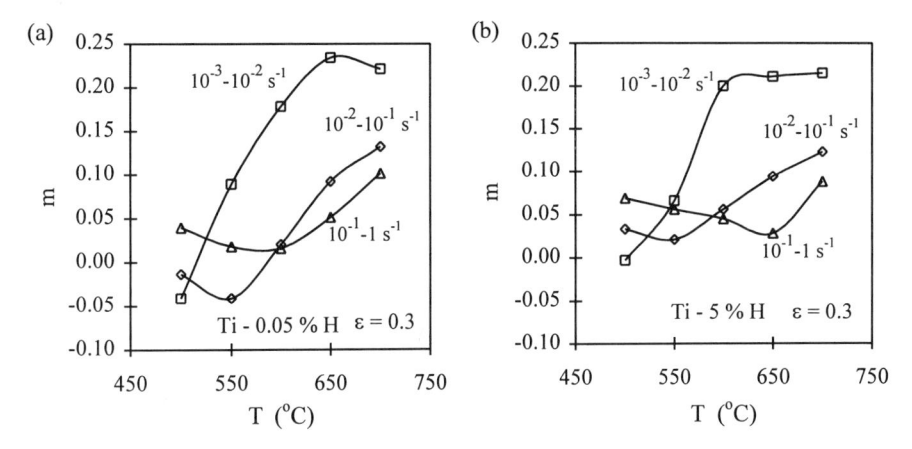

Figure 5 - Temperature dependence of the rate sensitivity m of alpha titanium containing (a) 0.05 and (b) 5 at.% H.

The temperature dependence of the rate sensitivity $m = (\partial \ln\sigma/\partial \ln\dot\varepsilon)_T$ is displayed in Figures 5a and 5b for the unalloyed and hydrogen-enriched specimens, respectively. Values of m were calculated over three strain rate intervals. At low strain rates, 10^{-3} to 10^{-2} s^{-1}, both series of specimens displayed negative values of m at 500 °C. When the temperature was raised, m increased continuously, attaining its normal value of about 0.22 above 600 °C. In the rate interval 10^{-2} to 10^{-1} s^{-1}, the rate sensitivity displayed lower values, and the minimum was observed at about 550 °C. The rate sensitivity decreased still further in the highest strain rate range, and the temperature associated with the minimum was now raised to 600 °C (0.05 % H) and 650 °C (5 % H). Alloying with hydrogen led to an m decrease in the high temperature range, but to an increase in the low temperature range. The values of m were positive as a rule in the hydrogen-enriched specimens, while m was negative over a wide temperature range in the unalloyed specimens.

Mechanical Anisotropy.

The addition of hydrogen affects the cross-sectional shape of deformed specimens. The strain anisotropy due to the presence of texture can be described in terms of the R coefficient:

$$R = ln(d_1/d_0)/ln(d_2/d_0), \qquad (1)$$

where d_0 is the initial diameter of a specimen, and d_1 and d_2 are the long and short axis values for the deformed specimen, respectively. The results obtained in the present samples are illustrated in Figure 6. Here it can be seen that the anisotropy is at a maximum (=2.6) when the unalloyed material is deformed at the lowest temperature. Increasing the hydrogen concentration promotes mechanical isotropy, as does increasing the temperature. The main decrease in the R coefficient (from 2.6 to 1.4) takes place in the alpha phase field. Such a hydrogen-induced decrease in the strain anisotropy supports the view [6,7] that additional slip systems are activated when the hydrogen level is increased. On increasing the temperature still further, formation of the isotropic bcc beta phase leads to still more uniform deformation in the alpha+beta and beta phase fields.

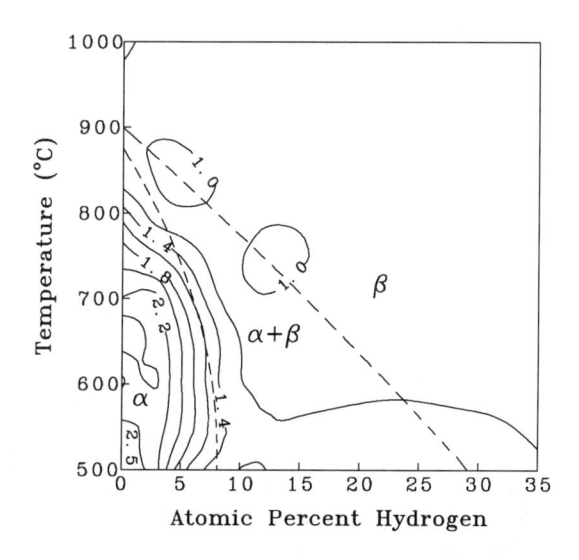

Figure 6 - Dependence on temperature of deformation and hydrogen concentration of the strain anisotropy coefficient R at a strain of 0.6 [4].

Dynamic strain aging and hydrogen-induced softening.

The anomalies in the strain rate and temperature dependences of the flow stress described above are manifestations of the dynamic strain aging (DSA) that takes place under experimental conditions of testing. The alloying elements responsible for DSA in the present temperature range are iron, carbon and oxygen [5]. These elements are all present in the base material (see Table 1). Alloying with hydrogen has been shown to decrease the extent of DSA [5]. This occurs when the presence of hydrogen in solid solution prevents other solute atoms from segregating to mobile dislocations, a tendency facilitated by its high mobility.

The hydrogen atmosphere not only prevents other atoms from settling on dislocations, it also decreases the strength of the latter's interactions with various obstacles [6]. As it occupies octahedral interstices in the lattice, hydrogen also appears to decrease the diffusivity frequency factor of other interstitial atoms; this effect even extends to substitutional elements such as iron, whose diffusion behavior also involves these interstices. By decreasing the effectiveness of DSA, dissolved hydrogen decreases the work hardening rate and flow stress in this range of behavior. The fact that the maximum softening attributable to the addition of hydrogen is observed at temperatures and strain rates where work hardening rate peaks occur in unalloyed titanium supports this point of view. For this reason, hydrogen-induced softening is expected to be peculiar to impure titanium and therefore to disappear in high purity titanium.

Conclusions

1. Alloying with 2.7 to 5 at.% hydrogen leads to noticeable softening in alpha titanium of technical purity. Both the flow stress and the work hardening rate decrease when hydrogen is added.
2. The maximum amounts of hydrogen-induced softening are observed at the temperatures and strain rates associated with dynamic strain aging peaks in unalloyed titanium. These strain aging peaks are in turn associated with the presence of impurities such as iron, carbon and oxygen.

Acknowledgments

This work was supported by the Natural Sciences and Engineering Research Council of Canada.

References

1. P. Lacombe, "Influence of Impurities on the Mechanical Properties of Titanium and Titanium Alloys," Titanium: Science and Technology, vol. 4, ed. G. Lütjering, U. Zwicker, and W. Bunk (Münich: Deutsche Gesellschaft für Metallkunde E.V., 1984), 2705-2721

2. V.K. Nosov and B.A. Kolachev: Hydrogen-Induced Plasticization of Titanium Alloys at Hot Deformation (Moscow: Metallurgia, 1986).

3. W.R. Kerr, R.R. Smith, M.E. Rosenblum, F.J. Gurney, Y.R. Mahajan and L.R. Bidwell: "Hydrogen as an Alloying Element in Titanium (Hydrovac)," Titanium 80: Science and Technology, vol. 4, ed. H. Kimura and O. Izumi (Warrendale, PA: TMS-AIME, 1980), 2477-2486.

4. O.N. Senkov and J.J. Jonas, "Effect of Phase Composition and Hydrogen Level on the Deformation Behavior of Titanium-Hydrogen Alloys," Metall. & Mater. Trans. A (1996, in press).

5. O.N. Senkov and J.J. Jonas, "Dynamic Strain Aging and Hydrogen-Induced Softening in Alpha Titanium," Metall. & Mater. Trans. A (1996, in press).

6. H.K. Birnbaum and P. Sofronis, "Hydrogen-Enhanced Localized Plasticity - a Mechanism for Hydrogen-Related Fracture," Mater. Sci. Eng., A176 (1994), 191-202.

7. O.N. Senkov, "Microstructure Evolution in a Hydrogen-Alloyed Titanium Alloy During Deformation at Elevated Temperatures," Strength of Materials, ed. H. Oikawa et al. (Sendai: The Japan Institute of Metals, 1994), 635-638.

SOLUTE STRENGTHENING IN BETA TITANIUM - HYDROGEN ALLOYS

Oleg N. Senkov* and John J. Jonas

Department of Metallurgical Engineering, McGill University
3450 University Street, Montreal, Quebec, Canada H3A 2A7
*On leave from the Institute of Solid State Physics, Russian Academy of Sciences
Chernogolovka, Moscow District, Russia 142432

Abstract

Compression tests were carried out on a series of titanium - hydrogen alloys containing up to 31.5 atomic % of dissolved hydrogen. The tests were performed within the beta phase field. The dependence of the flow stress on temperature, strain rate and hydrogen content was studied. The addition of hydrogen to beta titanium leads to noticeable hardening, the amount of which increases with hydrogen concentration according to a quadratic law. Analysis of the activation parameters of plastic flow shows that they depend weakly on hydrogen concentration while the pre-exponential term decreases exponentially when the hydrogen level is raised.

Introduction

Modern beta titanium alloys are candidates for many applications due to their excellent strength, toughness, formability, and corrosion resistance. Furthermore, because of the high solubility limit for hydrogen, they are less sensitive to hydrogen embrittlement than the alpha alloys [1]. When held in a hydrogen atmosphere at elevated temperatures, beta alloys can absorb up to forty atomic percent of this element. Although it is known that interstitial impurities can have significant effects on the mechanical properties of beta titanium [2], little information is available regarding the influence of dissolved hydrogen. Nevertheless, temporary alloying with hydrogen is now recognized as a useful method for improving the hot workability, microstructure and final properties of titanium alloys [3]. The former is considered to be a consequence of the increased volume fraction of the beta phase attributable to hydrogen addition.

In the present investigation, the flow stresses of beta titanium containing increasing levels of hydrogen were determined as a function of temperature and strain rate. Constitutive equations and activation parameters for the steady state flow of these Ti-H alloys were deduced as well.

Experimental

Hot rolled rods of titanium of technical purity were used to prepare cylindrical specimens 8 mm in diameter and 12 mm in height for compression testing. The base material contained the following impurities in atomic %: Al - 0.426, V - 0.255, Fe - 0.058, O - 0.011, C - 0.067, H - 0.047. The specimens were alloyed with up to 31.5 at.% of hydrogen by holding at 800°C in pure hydrogen at various pressures for 2.5 hours. Specimens with hydrogen concentrations of 0.05, 2.7, 5, 9, 12, 15, 23.5 and 31.5 atomic % were prepared in this way.

The $(\alpha+\beta)/\beta$ phase transformation temperature was determined for each of the alloys using the method described in Ref. [4]. Their values are plotted in Figure 1.

Specimens were heated to temperature at 1 °C/s and held for 10 minutes before deformation. The tests were carried out on an MTS hydraulic compression machine over the temperature range 500 to 1030 °C and using constant true strain rates of 10^{-3} to 1 s^{-1}. More detailed information about the experimental method is given in the above publication [4].

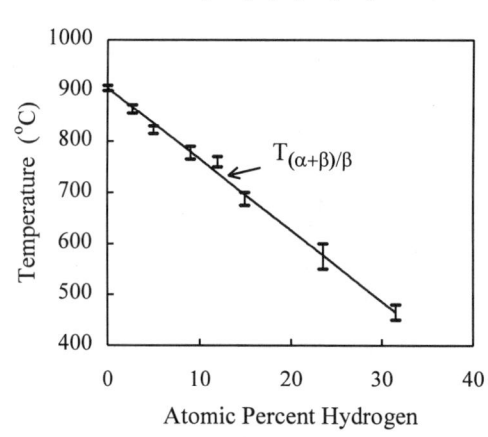

Figure 1 - $(\alpha+\beta)/\beta$ transformation temperatures of the present Ti - H alloys.

Results and Discussion

Examples of the flow curves obtained in the beta phase field are shown in Figure 2. A steady state of flow prevails once the strain has exceeded 0.1-0.2. The flow stress associated with steady state flow decreases with temperature and increases with strain rate and hydrogen concentration.

Hydrogen-Induced Hardening

Hydrogen produces unusually strong hardening in beta Ti - H alloys, as illustrated in Figure 3. The steady state flow stress increases with hydrogen concentration, as described by the relation:

$$\Delta\sigma = \sigma - \sigma(0) = \phi C_H^2 \quad (1)$$

Here $\sigma(0)$ and σ are the steady state flow stresses of the unalloyed and alloyed materials, and ϕ is a parameter that depends on temperature and strain rate. At temperatures below 900 °C, $\sigma(0)$ was obtained by extrapolation of the experimental σ vs. C_H dependence to $C_H = 0$. The corresponding values of ϕ are indicated next to each curve.

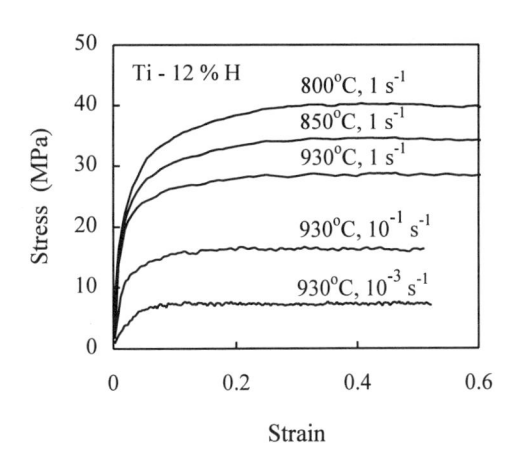

Figure 2 - Flow curves of the Ti - 12 % H alloy at three temperatures and three strain rates.

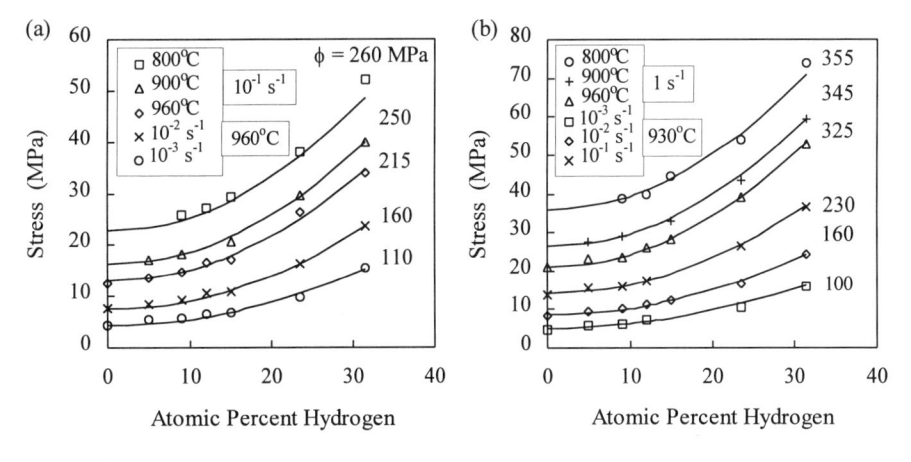

Figure 3 - Dependence of the steady-state flow stress on hydrogen concentration at a series of temperatures and strain rates.

Common theories of solute hardening generally lead to a power law expression for the dependence of the flow stress on solute concentration, $\sigma \propto C^p$, with values of p = 1/2 or 3/4 [5]. Such relations cannot account for the strong hydrogen-induced hardening observed in beta titanium. Nevertheless, p values as high as 1 to 2 have been observed to apply to solute hardening in several bcc alloys [5-9], as well as in some close packed alloys with high solute concentrations [8-10].

Particular mechanisms have been proposed to explain such behavior [5-11]. Some of these models are based on solute drag and may not apply to beta Ti - H alloys tested at elevated temperatures. This is because of the high mobility of hydrogen, which enables it to migrate more quickly than the dislocations by a factor 10^4 to 10^5 over the temperature range 500 to 1000 °C. Instead, it appears likely that short range ordering takes place at hydrogen concentrations above 10 % [12]. Although dislocation glide is expected to destroy such ordering, in this case it may be restored continuously during deformation because of the high mobility of hydrogen. Such a mechanism could lead to a quadratic dependence of the flow stress on solute concentration [11] and to a value of $\phi \approx 52v^2/a^3kT$, where a is the Ti lattice parameter and v is the interaction energy between Ti and H. Taking ϕ = 300 MPa, a = 3.3·10⁻¹⁰ m, and T = 1100 K, the reasonable value $v \approx -0.01$ eV is obtained. The activation energy of plastic flow increases with hydrogen concentration in this case.

Effect of Hydrogen on the Activation Parameters of Steady-State Flow

Examples of the rate dependence of the steady state flow stress measured at a series of temperatures are shown in Figures 4a and 4b for two titanium-hydrogen alloys. Data analysis shows that such behavior can be described by the following general relation [13]:

$$\dot{\varepsilon} = \dot{\varepsilon}_o(\sigma,T) \cdot \exp(-\Delta G(\sigma,T)/kT) \tag{2}$$

Here the pre-exponential term $\dot{\varepsilon}_o$ is proportional to the product of the rate at which dislocations attempt to overcome the obstacles to flow and the strain produced by a successful fluctuation, ΔG is the Gibbs free energy of activation required to surmount the barrier, and k is Boltzmann's constant.

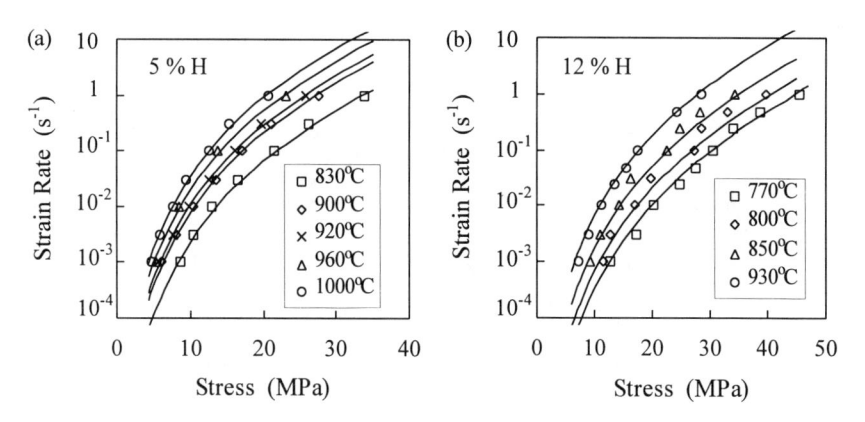

Figure 4 - Stress dependence of the strain rate for (a) Ti-5%H and (b) Ti-12%H.

The stress dependences of the pre-exponential term and of the Gibbs free energy were estimated for *each* of the present alloys using the technique described in Ref. 14. First, ΔG was calculated from the experimentally determined quantities:

$$\Delta G = kT^2 \frac{\left(\partial \ln\dot{\varepsilon}/\partial T\right)_\sigma + \left(\partial \ln\dot{\varepsilon}/\partial \ln\sigma\right)_T \left(d\ln\mu/dT\right)}{1 - T\left(d\ln\mu/dT\right)} \qquad (3)$$

where μ is the shear modulus. Then, having determined ΔG, $\dot{\varepsilon}_o$ was calculated from Equation (2). The values of ΔG and $\dot{\varepsilon}_o$ derived in this way are presented in Figures 5a to 5d for two of the alloys. It can be seen that ΔG is a function of both flow stress and temperature, and that $\dot{\varepsilon}_o$ is a power function of the modulus-reduced flow stress. These dependences can be expressed as:

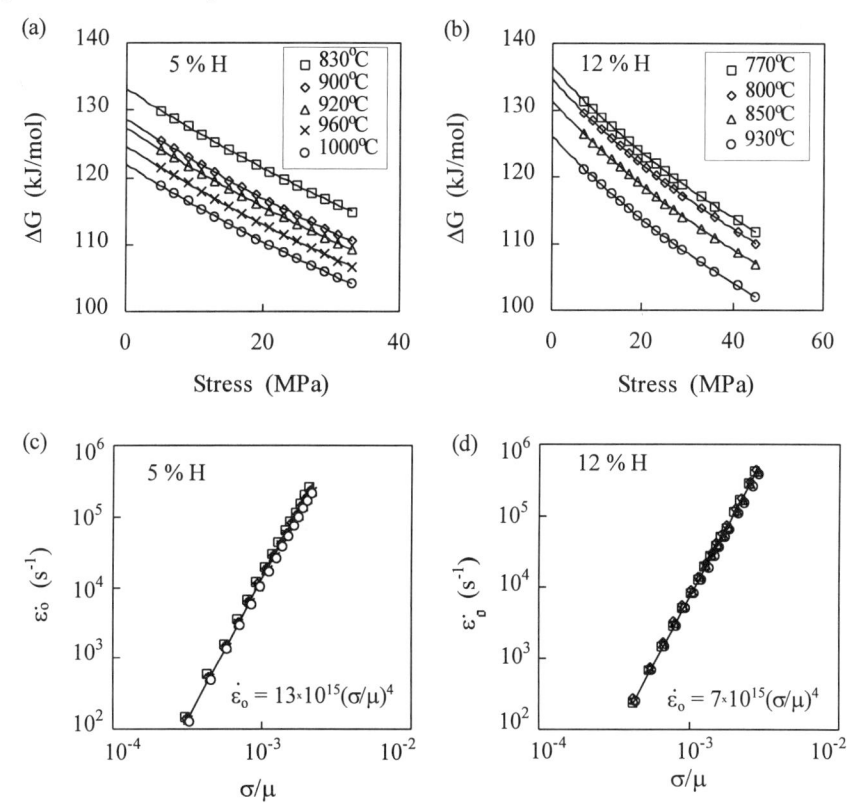

Figure 5 - Stress dependence of (a, b) the Gibbs free energy, and (c, d) the pre-exponential term.

$$\Delta G = \Delta G_o(T) - v\sigma/M, \quad \text{and} \quad \dot{\varepsilon}_o = A_1(\sigma/\mu)^{n_o} \qquad (4)$$

where $\Delta G_o(T) = \Delta H_o - T\Delta S_o$ is the Gibbs free energy at zero stress, $\Delta H_o = \alpha\mu_o b^3$ and $\Delta S_o = \alpha b^3 (d\mu/dT)$ are the zero stress enthalpy and entropy of activation, respectively, v is the activation volume, M=2 is the Taylor factor, n_o is about 4.0±0.2 for each of the alloys, A_1 is a composition-dependent parameter, b is the Burgers vector, μ_o is the shear modulus at 0 K, and α is a parameter that characterizes the activation mechanism [13]. Values of μ for pure titanium were used in these calculations because the dependence of μ on hydrogen concentration has not yet been established. It was however found that μ affects markedly the parameters ΔG and A_1, but not n_o.

The activation volume was therefore determined from the experimental data using the expression which does not include the shear modulus [14]:

$$v = kTM[(\partial \ln \dot{\varepsilon} / \partial \sigma)_T - n_o/\sigma] \quad (5)$$

The activation volume depends very weakly on hydrogen concentration and on the stress above 20 MPa, but increases rapidly when the stress is reduced to zero (Fig. 6). The approximate value $v = 1.6 \pm 0.2 \cdot 10^{-27}$ m³ (=68b³) applies to all the alloys above 20 MPa.

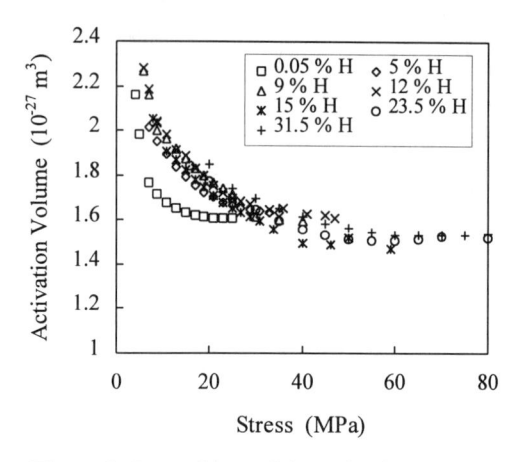

Figure 6 - Dependence of the activation volume on the steady state flow stress.

Thus Equations (2) and (4) can be combined to give:

$$\dot{\varepsilon} = A_2 \sigma^4 \exp\left(-\frac{\Delta H_o - v\sigma/M}{kT}\right) \quad (6)$$

where $A_2 = A_1 \mu^{-4} \exp[\alpha b^3 (d\mu/dT)/k]$ contains all the shear-modulus-dependent terms. The parameters A_2 and ΔH_o could not be calculated directly for the alloys with the present technique [14] because the shear moduli are not known for the hydrogen-enriched materials. These quantities were therefore determined separately for each alloy by fitting Equation (6) to the experimental data.

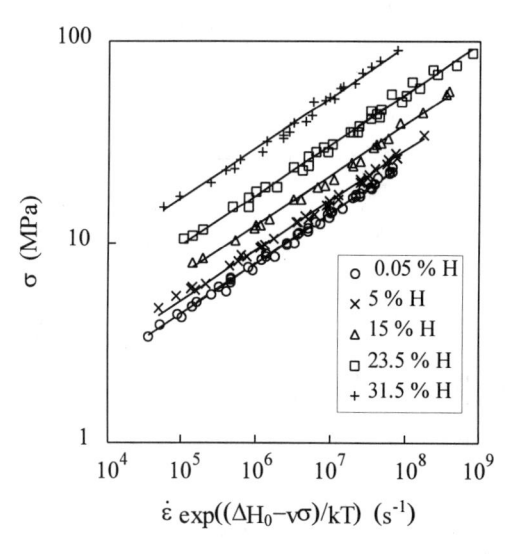

Figure 7 - Dependence of the steady-state flow stress on the temperature-compensated strain rate calculated from Equation 6 (solid lines). These fits can be compared with the experimental data obtained on five of the present materials.

The enthalpy $\Delta H_o = 1.97 \pm 0.06$ eV (190±6 kJ/mol) was found to be nearly identical for all seven materials, which gives $\alpha \approx 0.6$. However, A_2 decreases sharply with hydrogen concentration, according to the expression:

$$A_2 = A_3 \exp(-30 C_H^{1.5}) \qquad (7)$$

where $A_3 = 260$ s^{-1}MPa^{-4}. The dependence of the flow stress on the temperature compensated strain rate calculated from Equation (6) is plotted in Figure 7 in the form of solid lines for the present series of Ti-H alloys. These lines can be compared with the experimental data and it is apparent that good agreement is observed.

These results indicate that dissolved hydrogen does not noticeably change the activation enthalpy or activation volume of plastic flow but has a strong influence on the magnitude of the pre-exponential term. The experimental ΔH_o values are in good agreement with literature data for both the self-diffusion and high temperature deformation of β titanium [15,16]. Such values of ΔH_o, together with $\alpha = 0.6$ and the activation volume of $v = 68$ b^3 suggests that steady state flow may be controlled by the motion of jogged screw dislocations [13].

The sharp decrease in A_2 can be attributed to two alternative sources. On the one hand, the two order of magnitude decrease in the pre-exponential can be associated with a decrease in the density of thermally activatable sites [17]. Alternatively, it could perhaps be caused by a decrease in the temperature dependence of the shear modulus, $d\mu/dT$. In the latter case, $d\mu/dT$ must decrease by a factor of 3 on the addition of 31 at. % H. However, no data are available regarding the effect of hydrogen addition on the modulus.

Conclusions

1. Dissolved hydrogen produces appreciable hardening in beta titanium. The dependence of the flow stress on hydrogen concentration can be expressed by a quadratic law.

2. The activation volume and zero stress activation enthalpy of steady state flow are nearly independent of hydrogen concentration. However, the pre-exponential term decreases significantly on the addition of hydrogen. Such a behavior can be interpreted in terms of the decrease in the density of thermally activatable sites brought about by hydrogen addition.

Acknowledgments

This work was supported by the Natural Sciences and Engineering Research Council of Canada.

References

1. H.G. Nelson, "Hydrogen and Titanium Base Materials," Critical Issues in the Development of High Temperature Structural Materials, ed. N.S. Stoloff, D.J. Duquette and A.F. Giamei (Warrendale, PA: The Metallurgical Society, 1993), 455-464.

2. W.R. Tyson, "Interstitial Strengthening of Titanium Alloys," The Science, Technology and Application of Titanium, ed. R.I. Jaffee and N.E. Promisel (Oxford: Pergamon, 1970), 479-487.

3. F.H. Froes, D. Eylon, and C. Suryanarayana, "Thermochemical Processing of Titanium Alloys," JOM, 42 (3) (1990), 26-29.

4. O.N. Senkov and J.J. Jonas, "Effect of Phase Composition and Hydrogen Level on the Deformation Behavior of Titanium - Hydrogen Alloys," Metall. & Mater. Trans. A (1996, in press).

5. M.Z. Butt and P. Feltham, "Solid-Solution Hardening. Review," J. Mater. Sci., 28 (10) (1993), 2557-2576.

6. E.W. Collings and H.G. Gegel, "Physical Principles of Solid Solution Strengthening in Alloys," Physics of Solid Solution Strengthening, ed. E.W. Collings and H.L. Gegel, (New York: Plenum Press, 1975), 147-182.

7. H. Suzuki, "Solid Solution Hardening," Strength of Metals and Alloys, vol. 3, ed. P. Haasen, V. Gerold and G. Kostorz (Toronto: Pergamon Press, 1979), 1595-1614.

8. W.R. Tyson, "Solution Hardening by Interstitials in Closed-Packed Materials," Physics of Solid Solution Strengthening, ed. E.W. Collings and H.L. Gegel (New York: Plenum Press, 1975), 47-77.

9. J.M. Calligan and D. Goldman, "Metal Interstitial Solid Solution Strengthening," Strength of Metals and Alloys, vol. 2, ed. P. Haasen, V. Gerold and G. Kostorz (Toronto: Pergamon Press, 1979), 983-988.

10. M.J. Blackburn and J.C. Williams, "Strength, Deformation Modes and Fracture in Titanium - Aluminium Alloys," Trans. ASM, 62 (1969), 398-409.

11. P.A. Flinn, "Solute Hardening of Close-Packed Solid Solutions," Acta Metall., 6 (10) (1958), 631-635.

12. S.L. Ames and A.D. McQuillan, "The Resistivity-Temperature-Concentration Relationships in Beta-Phase Titanium-Hydrogen Alloys," Acta Metall., 4 (1956), 602-610.

13. U.F. Kocks, A.S. Argon, and M.F. Ashby, "Thermodynamics and Kinetics of Slip," Progress in Materials Science, vol. 19 (Oxford: Pergamon Press, 1975).

14. T. Surek, L.G. Kuon, M.J. Luton and J.J. Jonas, "Linear Elastic Obstacles: Analysis of Experimental Results in the Case of Stress Dependent Pre-Exponentials," Rate Processes in Plastic Deformation of Materials, ed. J.C.M. Li and A.K. Mukherjee (Metals Park, OH: American Society for Metals, 1975), 629-655.

15. H. Nakajima and M. Koiwa, "Diffusion in Titanium," ISIJ International, 31 (8) (1991), 757-766.

16. H.J. Frost and M.F. Ashby, Deformation - Mechanism Maps (Oxford: Pergamon Press, 1982).

17. O.N. Senkov and J.J. Jonas, "Effect of Strain Rate and Temperature on the Flow Stress of Beta-Phase Titanium - Hydrogen Alloys," Metall. & Mater. Trans. A, (1996 in press).

MEETING THE CHALLENGE OF HOT WORKING TITANIUM ALLOYS INTO COST EFFICIENT FINISHED COMPONENTS

G. W. Kuhlman

Alcoa Forged Products, Cleveland, Ohio USA

Abstract

Titanium alloys are among the most difficult to fabricate engineering materials. Management of titanium thermomechanical processing (TMP) in hot working processes (forging, rolling, extrusion) has become a skillfully executed science, utilizing sophisticated process engineering, process modeling, computer design and computer control technologies, rather than relying solely upon art or the experience of a designer or an operator. At the heart of overcoming the challenges of cost-efficient titanium component fabrication is the growing body of science and knowledge available on the interactions between TMP and hot working processes for optimal part fabrication and mechanical properties and capture of TMP and deformation science in component fabrication through highly refined engineering analysis, design and operating tools. Utilizing concurrent engineering, sophisticated deformation simulators, artificial intelligence and expert systems, FEM and other modeling technologies, and state-of-the-art computer controlled equipment, the titanium forger, roller or extruder has unparalleled power at his disposal in terms of enhanced product and tooling design, depth of deformation understanding, accurate process models and equipment capabilities. With these tools, fabrication of high value-added titanium alloy components is more efficiently accomplished. Reviewed in this paper are the critical elements and state-of-the-art in the key technologies essential to cost efficient titanium component manufacture: Concurrent Engineering; Computer Aided Engineering; Thermomechanical Processing; Deformation Mechanics and Process Modeling; Hot Working Process Design: and Linked Material-Thermal-Process-Structure-Mechanical Properties Models.

Background:

Titanium alloys are among the most difficult to fabricate engineering materials. Due to high raw material and fabrication costs, titanium components frequently have substantial potential for capture of superior value to the final customer through optimization of the manufacturing stream elements from raw material to finished part. To achieve the most cost efficient and highest value-added finished titanium components via forging, extrusion or rolling hot working processes, regardless of the part's end market use, the manufacturer generally operates at the limit of one or more of the elements of the components manufacturing stream which spans raw material through hot working to final machining and finishing.

Major process elements of Ti alloy manufacturing streams typically include raw material form, alloy, dies and tooling, fabrication, hot working and thermal equipment, hot working processes, thermomechanical processes, machining process and equipment and finishing processes and equipment. To successfully and consistently operate at the component's manufacturing stream's limits, Ti alloy fabricator(s) must have full command of all "software" and "hardware" elements of his processes, including finished part design, raw material product form design, die/tooling design and manufacture and finally, component/raw material product form fabrication process design, optimization and execution on the factory floor.

The computer and computer aided design, manufacturing, engineering, control and modeling technologies have effected a quantum leap change in engineering and manufacturing process approaches to component manufacture from difficult to fabricate, costly titanium alloys. In fact, customer demand upon titanium alloy component fabricators has led the drive in both aerospace and other markets to extend the frontiers of the science of computer aided design and manufacture because of the extremely high economic benefits realized from design and manufacturing stream optimization of hot worked titanium raw materials and raw material products and the finished components produced from them.

For example, consider the manufacturing stream of a large, closed-die forged titanium alloy component for an airframe application. There are multiple, interactive and iterative steps from conception of the part detail/component through the user's formalized design, structures, producibility, materials and processes, procurement and manufacturing groups' reviews leading to final raw material product design (in this example a closed-die forging) for subsequent fabrication by a supplier, who in turn is likely to have raw material and tool and die suppliers, as well other subcontractors in the part's fabrication stream. This multi-step design stream from concept to raw product design has, here-to-fore, been a "process" characterized by long flow times, e.g. 12 to 14 months or more, to complete the "sequential" nature of the process in the past.

To the final detail's and necessary raw material product form's design lead time then must be added the lead and flow times to produce the specified Ti raw material product form and finally, the lead and flow times in the user's final fabrication, metal removal and finishing processes. Ti raw material product form lead and flow times may range from 3 months (for simple shapes such as bar or plate) to 12 months or more depending upon the product selected and resource utilization levels. User's lead and flow times vary with part criteria but are typically at least 6 months to as much as 18 months.

Thus, the total time before an airframe customer may begin the final manufacturing processes to convert a new titanium raw material product into a finished part then may be 18 to 24 months or more. Total time to complete the processing of detail and make it ready for use in an airframe may range from 24 to 40 months, requiring a very costly inventory to keep the "pipeline full." Such long total flow times on new aerospace components are typical when the critical, multiple elements of the part and raw material product design (and fabrication) process are executed by individuals/groups within the end user's organization and then the supporting supplier tiers (e.g. forger, metal producer, etc.) act upon the component sequentially or separately.

Concurrent Engineering, along with other technologies discussed below, is an important new and emerging technology that is key to realizing major reductions in aerospace part lead and flow times. Computer aided engineering, design, manufacture, modeling and control technologies discussed herein when combined with Concurrent Engineering, hot working process design, TMP and other technologies reviewed below have demonstrated reduction of the overall flow time from concept to a finished Ti alloy component ready for use by 50% or more (e.g. from ≥ 24 months to ≤ 12 months). In a similar fashion, these computer aided and concurrent engineering technologies materially enhance the design and manufacturing process's reliability and capability to achieve "first round hits," e.g. successful part manufacture the first time, to more than 95%.

Discussed below are the Key Titanium Alloy Hot Working Science and Technologies for Ti part manufacture being exploited today and summarized is the critical framework that has developed and is being fine-tuned to exploit success -- as is measured in terms of cost and time savings. Each of these critical technologies is discussed in terms of the current state-of-the-art and challenges, problems and/or opportunities for new science.

Key Titanium Hot Working Science and Technologies for Part Manufacture:

The challenge of successful quantum changes in titanium alloy part manufacture by collapse of manufacturing lead/flow time and cost is being met head-on through exploitation of several key technologies reviewed below. Taken overall, all of these technologies have one objective: **To make the Ti alloy raw material product form, especially forgings, easier for the customer to acquire and to use in his manufacturing facilities.**

(1) <u>Concurrent Engineering</u>: A process wherein the raw material manufacturer (and his expertise) is brought into and impacts the design and product development processing stream (e.g. through Design to Build Teams, Integrated Product Design Teams, et. al.) early in the final component and raw material product form design cycle. The raw material producer is afforded direct input into critical decision making on relationships between the raw material and the finished component.

(2) <u>Computer Aided Engineering (CAE), Artificial Intelligence (A.I.) and Knowledge Intensive Design (KID)</u>: CAE, A.I. and KID are computer-based technologies that facilitate part and tooling design, process modeling and optimization, product and tooling process engineering, part and tooling process optimization, etc. These technologies include Expert and Knowledge-based Systems for part and tooling design and manufacture, thermomechanical processing design, manufacturing process selection, equipment control, quality assurance, microstructure and mechanical properties optimization or prediction, non-destructive evaluation (NDE), and testing, inspection and quality assurance processes.

(3) <u>Thermomechanical Processing (TMP)</u>: TMP, the combination of hot working and thermal or heat treatment processes to achieve microstructural and therefrom mechanical property objectives, is at the heart of all Ti alloy raw material product form fabrication, especially forgings and plate. TMP of all Ti alloys is a highly

refined metallurgical engineering process that is now managed and exploited through A.I. and computer-based modeling technologies.

(4) <u>Deformation Mechanics and Process Modeling</u>: Deformation Mechanics is defined by powerful, efficient computer-based models (e.g. DEFORM, MARC, ANSYS, ABAQUS, et. al.) that are routinely available to raw material product manufacturers (forgers, extruders, etc.). Deformation/Process modeling manages in 2D and/or 3D engineering problems both for the component and its tooling, permitting simulation of engineering solutions quickly and without costly trial and error. All of these models utilize Constitutive Law (Hyperbolic Sine or other curves) of state variables including flow stresses, etc. where experimental data are developed on sophisticated deformation simulators and then translated to universally usable Law. Outputs include both thermal and deformation analyses important to efficient product manufacture.

(5) <u>Hot Working Process Design</u>: Hot working process design for Ti raw product forms, especially closed-die Ti alloy forgings, requires creative combinations of process engineering, tooling design, equipment utilization and thermomechanical and deformation processes to achieve cost efficient hot worked shapes with desired structures and properties. Several recent very large Ti alloy die forging applications have challenged the creative forging engineering process and metallurgical engineering techniques to achieve critical properties in Ti alloy closed-die forging components that have advanced the state-of-the-art.

(6) <u>Linked Material - Thermal - Process - Structure - Properties Models</u>: Manufacturing streams for titanium raw materials and finished components are multiple step, interdependent processes leading to the desired end point. Thus, with advent of powerful 2D/3D modeling codes and computer systems, it is becoming possible to link the upstream and downstream process elements, e.g. raw material plus hot working plus thermal treatment for example, to track and then optimize structure, properties, and other product attributes (including cost) throughout the manufacturing stream. The science of linked models is in its infancy and rate of change of this technology is dramatic.

Following is a discussion for each of the above critical technology elements, all of which are exploited and contribute to the objective of more cost efficient Ti alloy component manufacture, the following issues:

- Critical Elements of the Technology.
- State-of-the-Art of the Technology.

<u>CONCURRENT ENGINEERING</u>:

Achieving the Goal of Making It Easier for the Customer to Use Our Product.

Concurrent (or Simultaneous) Engineering techniques in customer-supplier interactions is not new; it is, however, now being extensively exploited within the Aerospace and Automotive industries with a fervor and in a manner that is changing, re-inventing the paradigm of customer-supplier relationships. Through Concurrent Engineering the raw material supplier (or fabricator) becomes an integral part of the user's Producibility/Manufacturing Team and lends his expertise to key decisions on the raw material - finished component relationships. Concurrent Engineering is being exploited extensively, for this technology is clearly an extremely important enabler in the quest for quantum leap changes in part manufacturing time and costs. Fundamental to successful Concurrent Engineering activities are the following Critical (or Enabling) Elements of this technology.

(1) <u>Customer-Supplier Partnering</u>: It is a challenge to break down the traditional, frequently adversarial, relationships between customer and supplier and to reach a true "partnership" arrangement. Such partnerships are not just verbal but are based upon sound communications and trust that are not easily achieved and may require unique commercial arrangements.

(2) <u>Pre-Competitive Raw Material Supplier Selection</u>: For open communications and fullest participation by both parties in the Concurrent Engineering process, it is necessary to establish, in a pre-competitive fashion, commercial arrangements that retain the best financial interests of both parties. Pre-competitive supplier selection is a challenge in Aerospace/Automotive markets where multiple bids, lowest price supplier selection is typically the method of operation of procurement groups. However, rewards for re-engineering the competitive bidding process to before the design process is complete, rather than after are great.

(3) <u>Open Systems Communications</u>: Under the umbrella of Items 1 and 2 above, it now becomes possible and desirable, through electronic and other techniques, for communications between supplier and customer to exploit fully open systems, where each participates, with appropriate confidentiality safeguards, in the Wide Area Network (WAN) and/or Local Area Networks (LAN) of the other.

(4) <u>"Best Available Practice" Philosophy For Design, Manufacturing, Processes and Product</u>: Capture of "Best Available Practices" is an essential part of the necessary collapse of Ti part manufacturing costs and flow/lead times. Through concurrent engineering, supplier and customer alike bring to the collaboration ready access to and capture of each other's Best Practices. In this process, communications through partnering arrangements are structured to protect each participant's technologies and thereby encourage full disclosure.

(5) <u>Integrated Product Definition or Design (IPD) or Design to Manufacture (DTM) Teams</u>: Ti alloy and other materials component users in several markets, but particularly Aerospace, now utilize internal IPD or DTM teams to improve the direction, control and management of component and raw material product form design and acquisition process. A major objective of these teams is to enhance the "manufacturability" of components by raw material product form design and selection.

Frequently, IPD/DTM teams may be driven by program objectives to select product forms which may compromise raw or end product performance in order to realize cost or time savings. IPD/DTM teams, therefore, represent a major opportunity for Concurrent Engineering interaction with the Ti alloy raw material supplier (e.g. forger, extruder, etc.) to maximize benefits to both the user and supplier tiers.

Figure 1 illustrates a Flow Diagram for the Aerospace Component Design and Procurement Process - a "Past State" for Integrated Product Definition Teams. The diagram illustrates the inputs to the IPD that establish the detail or component design (and in this instance will lead to the design of a closed-die Ti forging), the typical user elements in an IPD and interaction between them, and outputs of the IPD to the user's supply base. Note that in the "Past State" case, the raw material form supplier (in this instance a forger) is outside the IPD and is near the end of the process, such that interactions with the forger are through Procurement and any inputs the forger may have to improve the raw forging design or product are difficult to communicate to the IPD. For example, alternate plans of manufacture created by the supplier may not be considered due to the press of time. With the process in Figure 1, the typical flow time from detail design to raw material product design to placing purchase

FIGURE 1

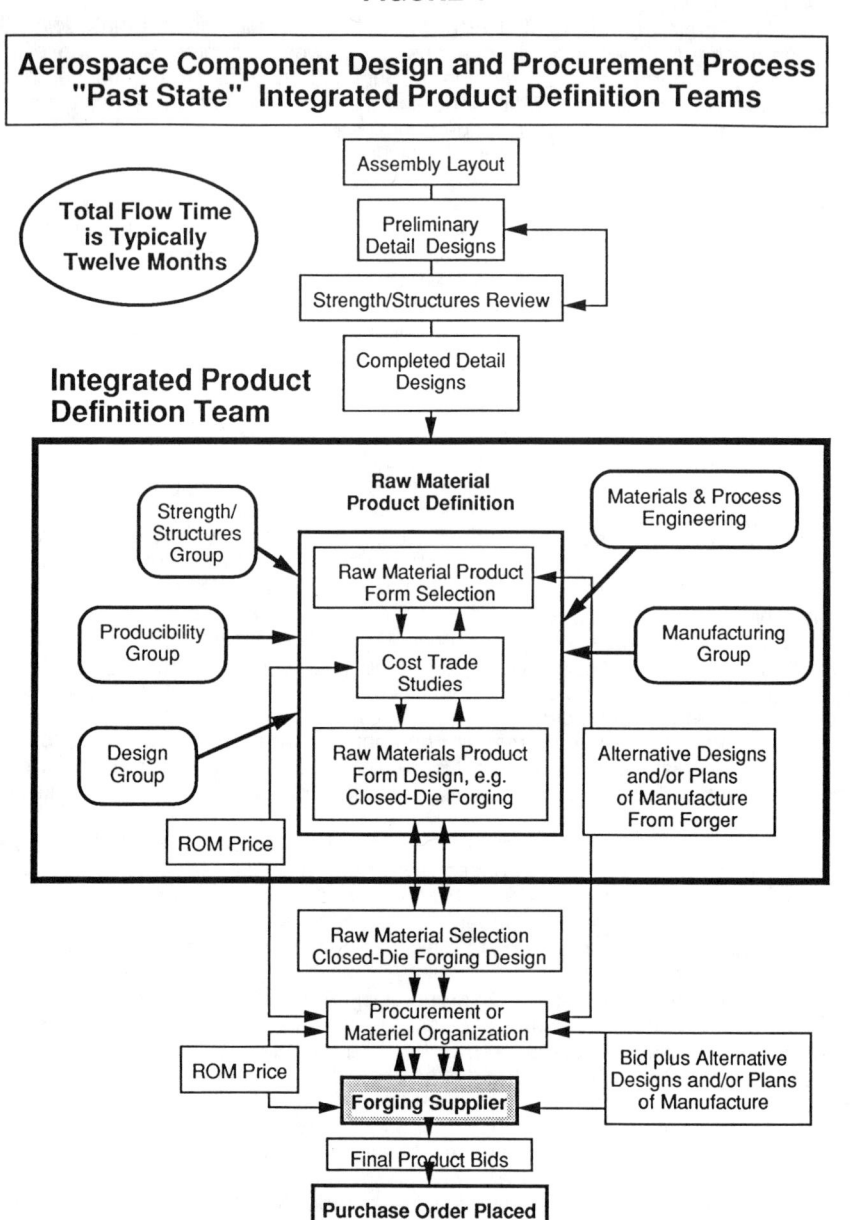

Aerospace Component Design and Procurement Process "Past State" Integrated Product Definition Teams

- Assembly Layout
- Preliminary Detail Designs
- Strength/Structures Review
- Completed Detail Designs

Total Flow Time is Typically Twelve Months

Integrated Product Definition Team

Raw Material Product Definition
- Strength/Structures Group
- Producibility Group
- Design Group
- Raw Material Product Form Selection
- Cost Trade Studies
- Raw Materials Product Form Design, e.g. Closed-Die Forging
- Materials & Process Engineering
- Manufacturing Group
- Alternative Designs and/or Plans of Manufacture From Forger
- ROM Price

- Raw Material Selection Closed-Die Forging Design
- Procurement or Materiel Organization
- ROM Price
- Bid plus Alternative Designs and/or Plans of Manufacture
- Forging Supplier
- Final Product Bids
- **Purchase Order Placed**

orders is typically twelve (12) months. Further, the IPD/DTM teams do not have the benefit of inputs or interactions with their supplier base that may well have measurable and critical impact on product cost and flow times.

On the other hand, Figure 2 describes the "Current State" where the Concurrent Engineering principles and technology elements described above are being fully captured and exploited to enhance the raw material design and procurement process. Note that the Forging supplier is a full participant in the IPD/DTM team, where electronic, computer-based communication systems make that possible. By integrating the forger into the IPD, he is now available to contribute the key information described in the lower right hand box of Figure 2. All of these inputs are key to a more efficient raw material design and procurement process and cost/time savings.

As is illustrated, the forger does not, however, participate in the customer's upstream detail or component design processes, because these may be sensitive and proprietary to the user, particularly in Aerospace. He does however become a full participant in the IPD process of raw material product design. Here the forger lends his expertise to assist in and expedite the design process. With the Current State, Concurrent Engineering process the typical flow time for raw material (die forging typically) design is reduced to 8 - 10 months, a 20 to 40% reduction from the Past Process in Figure 1.

Further, because of the enhanced communication and product data exchange processes between user and supplier, including electronic model and design data interchange, through the Concurrent Engineering process, the supplier's processes of forging part and die/tooling design, forging process design, raw material procurement and final part optimization and fabrication are collapsed significantly in lead and flow times, by as much as 40 to 60%. With the process described in Figure 2, closed-die Ti forgings (including very large, state-of-the-art parts) have been designed, tools manufactured, parts fabricated and parts inspected and approved without ever producing a paper drawing -- all documents and records were handled, managed, stored and retrieved electronically.

With capture of the combined Concurrent Engineering and IPD/DTM Team Processes described in Figure 2, the following are State-of-the Art for large, closed-die Titanium alloy forgings:
• Die/Tooling Lead-times are reduced by 40% or more, (e.g. from 30 to 36 weeks to 18 to 20 weeks).
• "Tryout" Articles and/or Multiple "Die Tryout" Iterations to develop the process are eliminated, leading to immediate limited scale forged part production.
• Overall lead and flow times for Titanium alloy forgings have been reduced by 40 to 60%.
• "First Round Hits" in successful fabrication of new parts has been increased to >95%.

COMPUTER AIDED ENGINEERING:

The Backbone of Concurrent Engineering Processes and Customer - Supplier communications.

The elements and potency of Computer Aide Engineering (CAE) have been extensively discussed and the rate of change of the body of engineering experience and capabilities of CAE is extremely rapid and is focused upon both internal user and suppliers interactions and external, between customer and raw material suppliers, communication processes.

FIGURE 2

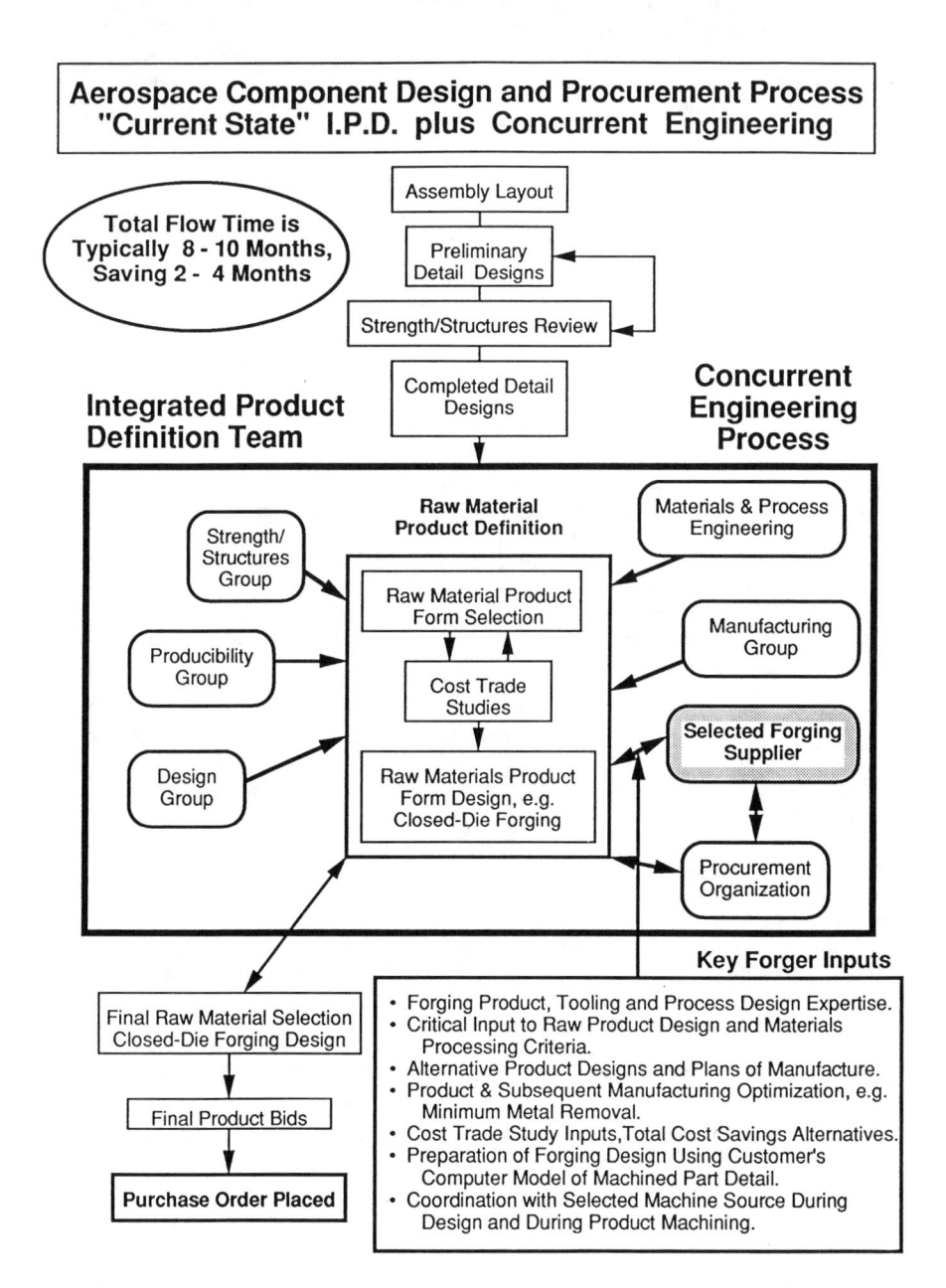

Aerospace Component Design and Procurement Process "Current State" I.P.D. plus Concurrent Engineering

Total Flow Time is Typically 8 - 10 Months, Saving 2 - 4 Months

- Assembly Layout
- Preliminary Detail Designs
- Strength/Structures Review
- Completed Detail Designs

Integrated Product Definition Team

Concurrent Engineering Process

Raw Material Product Definition

- Strength/Structures Group
- Producibility Group
- Design Group
- Materials & Process Engineering
- Manufacturing Group
- Selected Forging Supplier
- Procurement Organization

- Raw Material Product Form Selection
- Cost Trade Studies
- Raw Materials Product Form Design, e.g. Closed-Die Forging

- Final Raw Material Selection Closed-Die Forging Design
- Final Product Bids
- **Purchase Order Placed**

Key Forger Inputs
- Forging Product, Tooling and Process Design Expertise.
- Critical Input to Raw Product Design and Materials Processing Criteria.
- Alternative Product Designs and Plans of Manufacture.
- Product & Subsequent Manufacturing Optimization, e.g. Minimum Metal Removal.
- Cost Trade Study Inputs, Total Cost Savings Alternatives.
- Preparation of Forging Design Using Customer's Computer Model of Machined Part Detail.
- Coordination with Selected Machine Source During Design and During Product Machining.

(1) Essential sub-technologies in CAE that are captured and exploited for Ti alloy component and raw material (forging in particular) process design and optimization and manufacturing process control for reduction of time and costs and linkages between the customer, forger and other supplier tiers are illustrated in Figure 3 and discussed below.

- Computer Aided Design (CAD) and Manufacturing (CAM) for raw material part (e.g. closed-die forging) and tooling design. The CAD design is post-processed via CAM for tool and die manufacturing. This data transfer, from the Forger to a Die Shop for example, is frequently paperless. Figure 3 refers to Alcoa Forged Products CAD System (Computervision Cadds 5). There are other excellent CAD design systems available from CATIA, UNIGRAPHICS and many others. Communications between systems is usually managed through IGES techniques that are now widely available.

- Computer Numerically Controlled (CNC) Part and Die Manufacturing: Via CAM, forged part and forging tooling is manufactured on CNC controlled milling, turning or die sinking machines and now also includes using CAM in some instances to program forging presses for the manufacture a variety of open die forgings.

- Stereolithography (SLA) and Laminated Objective Modeling (LOM) along with other types of "Desktop" manufacturing systems are utilized for manufacture of prototypes, die sinking models, and in some cases disposable/low cost tooling. Prototypes serve many purposes from Sales/Marketing tools to part geometry verification and visualization, to Quality Assurance tools.

- Coordinate Measuring Machines (CMM): For part dimensional verification. CMM's typically employ CAD databases and present the opportunity for enhanced, more efficient Quality Control activities.

- Finite Element Analysis (FEA): FEA is a potent technology employed both for structural analysis in finished part/component design (e.g. codes include ANSYS, MECHNICA, MARC and others) and Forging/Extrusion Process Modeling (predominate codes are DEFORM, MARC, NIKE, ABAQUS, and others). FEA in process analysis has been proven to be a very effective tool for high temperature materials such as Ti alloy and Ni/Co-base superalloys because these materials are very difficult to fabricate and usually are processed at or close to equipment, material or process limits, especially hot working.

(2) The second critical technology element, arising from broad scale deployment of CAD/CAM throughout key markets for Ti (e.g. Aerospace and Automotive), is being driven critical need for successful cost reduction of Ti components. This technology is customer development of 3D Finished Part Models (or suggested raw forging or other product form geometry's) that are suitable for transmission from customer to supplier via the Internet, World Wide Web and/or IGES systems. In parallel, some more sophisticated customers have begun to "automate" the design of some raw materials (e.g. Boeing Wichita's Automated Forging Design [AFD] and similar process by McAir and others). Variants of AFD are utilized to develop first concepts of other raw material product forms including castings and extrusions.

(3) As an adjunct to the above cited critical CAE technologies is the rapid changing area of Artificial Intelligence (A.I.) and/or Knowledge Intensive Design/Modeling (KID, KIM). Successful capture of expert knowledge and experience in design and processing of difficult to fabricate materials like Ti is potent. AI/KID/KIM are exploited to design parts or raw material product forms like forgings, develop optimal processes for manufacture of Ti raw material products, design and manufacture tools and dies and create/guide critical deformation and thermal processes (TMP).

FIGURE 3

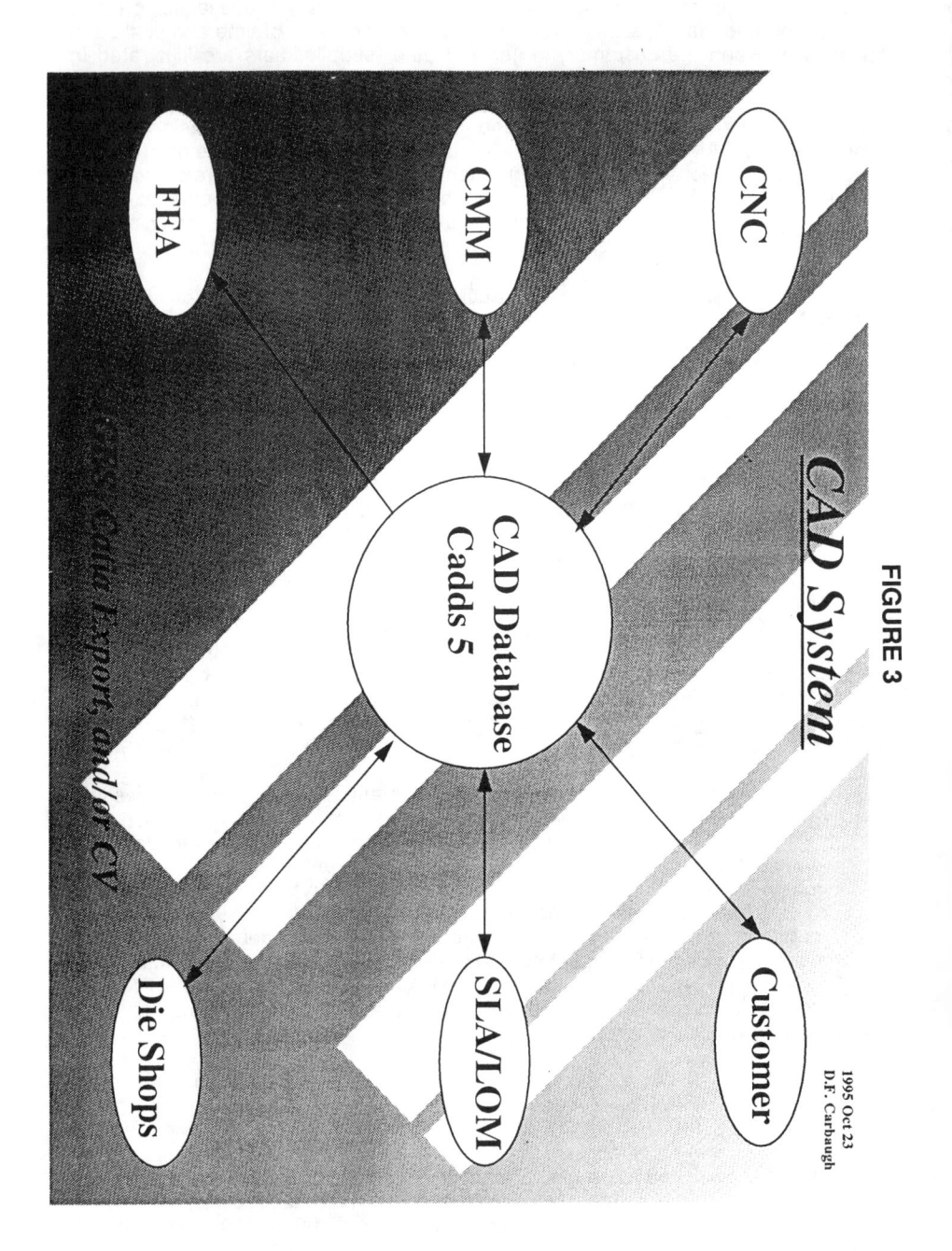

CAD System

The above universe of CAE technologies is exploited to greater or lesser extents in the fabrication of every Ti alloy raw material product form. Because forgings are so difficult to produce and costly, this product tends to employ more of these tools than other forms. State-of-the-Art in CAE as exemplified by forged shapes is as follows:

- Virtually 100% of Titanium alloy Closed-die Forgings, Castings and Extrusions are designed via CAE technologies.
- Through AFD, KID and KIM programs and design and processing systems, "Paperless" forging design, tooling design, tooling manufacture, forging manufacture and forging part verification has been achieved and results transmitted to the customer through a secure Internet Transmission sequence.
- For Titanium parts described by an axis of rotation, 3D CAD design is the norm. For web-rib Titanium forged shapes, 3D design techniques are utilized on 40% or more of the parts being produced.
- AFD, KID, KIM and other CAE technologies have reduced total part and raw product form design times by 60% or more.

THERMOMECHANICAL PROCESSING:

Process-Structure-Properties Are Exploited for Optimal End Part Performance.

Most titanium alloys, but especially alpha-beta and metastable beta alloys, represent a quintessential opportunity to develop and capture in even the largest, most challenging Ti alloy component and on a commercial scale that covers the full range of Ti products, including very large forgings, sophisticated TMP techniques. TMP is, therefore, an essential part of cost efficient Ti component manufacture and several critical technology elements prevail as important to satisfactory exploitation of this technology.

(1) Ti alloys represent a unique capability for capture of even the most sophisticated TMP techniques. As is illustrated in Figure 4, virtually all important aspects of Ti alloys' microstructure may be successfully manipulated or controlled by TMP. Microstructure-process interactions in most Ti alloys are well understood and available to the user and supplier alike in the part and process design of critical Ti components.

(2) Successful reduction to commercial practice and exploitation of Ti alloy TMP is based upon extensive study of TMP-structure-properties relationships since the inception of commercial scale Ti alloy fabrication and has recently been materially assisted by computer-aided design, analysis and process control technologies, including process design.

(3) TMP design and execution is a critical element in Ti alloy Concurrent Engineering processes so that the most rational, cost efficient process-properties-product form interactions are selected and utilized.

(4) The final role of TMP is to enhance Ti alloy hot working characteristics and processes in order to assist in shape making capabilities and Ti alloy part sophistication. Some of the very large Ti alloy closed-die forging shapes produced today and establish the state of the art are facilitated in their fabrication via well executed TMP.

TMP process maps, as illustrated in Figure 5, are generally the output of TMP development. State-of-the-Art in Titanium Alloy Thermomechanical Processing development and design includes:

FIGURE 4

Critical Microstructural Features of Titanium Alloys Controllable by Thermomechanical Processing

Feature	Description
I. Alpha morphology	Equiaxed primary alpha
	Elongated primary alpha
	Widmanstatten primary alpha
	Bi-modal alpha
	Colony alpha
	Secondary alpha
II. Prior beta grain morphology/size	Grain shape (aspect ratio)
	Coarse grain size (>0.05 in)
	Fine grain size (<0.05 in)
	Mixed mode grain size
III. Grain boundary/ Sub-grain boundary constituents	Alpha films
	Grain boundary alpha
	Interface phase
IV. Nature of matrix:	Low strength
	Medium strength
	High Strength

FIGURE 5

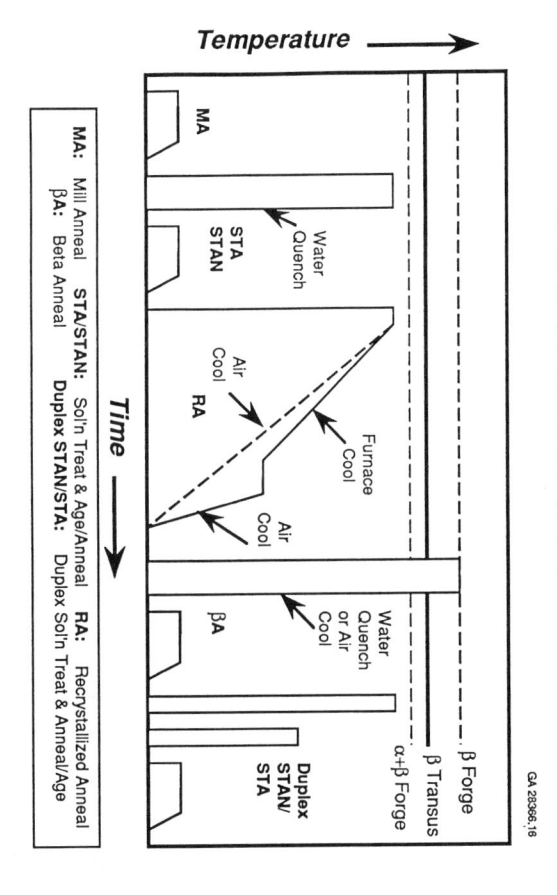

Forging Temperature and Thermal Treatment Cycles For Ti-6Al-4V and Ti-6Al-6V-2Sn

- TMP in Titanium Alloys, even TMP with multiple steps, is routinely captured in repeatable, well-executed production-scale Hot Working processes including in fabrication of the largest, most complex titanium alloy forgings and other components currently produced.
- Titanium alloy TMP principles, even for emerging alloys, are highly evolved, well documented and committed to 2D and in some instance 3D process models.
- TMP in Titanium Alloys is readily matched to raw material product form, finished product and/or raw material process and mechanical properties criteria.
- TMP has to date been a "Metallurgist-based" process but has evolved into a FEA/FEM-based structure-process-properties prediction system that utilizes KIM and AI for computer-based modeling.

DEFORMATION MECHANICS AND PROCESS MODELING:

For Successful Operation at the Limit of Titanium Raw Materials, Processes and Hot Working and Thermal Equipment.

Deformation mechanics and process modeling combines the disciplines and technologies of CAE and TMP into successful resolution of forging and process engineering problems encountered in the commercial fabrication of difficult to produce Titanium alloy components. Deformation Mechanics input into Process Models permits the exploration, examination and visualization of the deformation and thermal processes (e.g. forging, heat treatment, quenching, aging, etc.) and important process-material-product form dependent outputs such as temperature, strain, strain rate, etc. The latter are generally dependent variables the forging designer, engineer or metallurgist wishes to control through manipulation of the thermal and deformation sequencing and working conditions.

The Critical Technology Elements that make-up Deformation Mechanics and Process Modeling technologies include the following and are particularly pertinent to Titanium alloy part fabrication.

(1) Hot working and thermomechanical processing of titanium alloy components is usually at the limits of deformation process, tooling, raw material and equipment capabilities. Thus, Process Modeling is a high payback of analysis costs.

(2) FEA/FEM process modeling is predicated upon highly evolved Deformation Mechanics and Computer Processing Technologies. Advent of PC-based codes has expanded utilization; however, complex problems still require extensive computer and engineering "clock" time to achieve solutions.

(3) FEA/FEM is an excellent tool for development, evaluation and reduction to practice of "Best Available Techniques" for study and optimization of Least-Cost, Quality-Enhanced and a Time Efficient process and product therefrom.

(4) FEA/FEM while incrementally costly, when combined with CAE, has been found to be of particularly high value-added in difficult to fabricate materials such as titanium and has materially assisted in extending the state-of-the-art in Titanium alloy closed-die forging size.

(5) TMP capture and input the Titanium Alloy FEA/FEM and Deformation Mechanics techniques and problem solving is now highly evolved and time efficient. It is cost-effective to perform FEA/FEM analysis on most Titanium alloy forging shapes.

A key enabler in the growth of Deformation Mechanics and Process Modeling has been the development and capture of user friendly and computer efficient codes such as SFTC's DEFORM, available in both 2D and 3D. DEFORM will conduct both axisymmetric and plane strain models and employs a rigid viscoplastic formulation in solving both isothermal and non-isothermal deformation problems. Necessary inputs to DEFORM are

    A. A CAD Geometry Model of the Part.
    B. Material Constitutive Law (from Deformation Simulator).
    C. Thermal Properties of the workpiece and tooling.
    D. A velocity or strain rate profile.
    E. A friction model.

Outputs from DEFORM (or other FEM/FEA Deformation/Process Models) are easily interpreted charts, graphs or data compilations which illustrate features of the worked part that range from the final geometry, temperature profiles of workpiece and dies, strain profiles of workpiece and dies, strain rate profiles, state variable profiles (e.g. hardness, microstructure, etc.), defect generation (e.g. folds, lap, flow throughs, aberrant grain flow, etc.) and defect location, tooling and workpiece interactions, workpiece stresses, tooling stresses and profiles, metal (grain) flow, etc. All of these outputs serve as important inputs to Engineering analysis of the particular problem at hand and assist in rapid resolution.

For commercial exploitation of Deformation Mechanics and Process Modeling technologies, following are State-of-the-Art for Titanium alloy components:
    • FEA/FEM Process Modeling of Titanium Alloy parts in routinely used for the following:

    A. Finished Part Design.
    B. Raw Materials Product and Tooling Design.
    C. Thermal and Deformation Process Selection.
    D. Process Control Criteria and Strategies.
    E. Process Efficiency and Cost Collapse.
    F. Failure Analyses.
    G. Product Optimization, both Design and Processing.
    H. Extending Asset Utilization.

    • Key FEA/FEM Models including DEFORM, MARC, DYNA, etc. are now routinely used to solve 3D problems.
    • Execution of 2D FEA/FEM models on Personal Computers is now widely available.
    • Several Aerospace customers now require suppliers to conduct 2D/3D FEA models of Critical Process Elements as integral parts of Product/Process Qualification packages.

HOT WORKING PROCESS DESIGN FOR TITANIUM ALLOYS:

Creative Process Engineering to Make the Challenging, State-of-the-Art Ti Part Possible and a Production Routine.

Clearly, the computer-based technologies of CAE, TMP, Deformation Mechanics and FEA/FEM modeling have had immeasurable impact on successful engineering and fabrication of even the most challenging Ti component. Other, frequently creative, engineering technologies, under the heading of Hot Working Process Design, have also been necessary to realize the current state-of-the-art in large Ti alloy component fabrication success.

Critical technology elements in Hot Working Process Design include:
    (1) Asset utilization (management) techniques for enhanced equipment productivity and value and to assist in optimal decision making on competing issues in equipment maintenance, control and operation.

(2) Creative Hot Working (Forging) product design, Process Engineering and Utilization of Deformation Process Design Tools. Here both computer-based and human engineering skills are combined synergistically.

(3) Thermal and Deformation Processes and process equipment that is in control and capable.

(4) Microstructural manipulation of the alloy for enhanced producibility, shape sophistication, collapse of the manufacturing stream and cost reduction.

(5) Creative engineering solutions that extend the process and/or product limits and re-establish the state-of-the-art.

The Forging Process is understood to have twenty-one (21) process variables (listed below in descending order of importance) that to greater or lesser extents influence the process's outcomes and success. The Forging Engineer and Process Engineer address each in the ultimate design of the process stream for each part. This part of the undertaking remains as much art as is the science of CAE, TMP, and FEA/FEM modeling.

- (1) Part Temperature
- (2) Die Temperature
- (3) Strain
- (4) Strain Rate, Head Speed
- (5) Lubricant composition
- (6) Lubricant application
- (7) Lubricant temperature
- (8) Material microstructure
- (9) Material flow stress, flow behavior
- (10) Dwell time
- (11) Die surface condition, coating, surface treatment
- (12) Die material and behavior under load
- (13) Part surface condition
- (14) Part coating
- (15) Press pressure and/or Pressure application profile (strain rate)
- (16) Energy application or consumption
- (17) Number of die sets, die design, die geometry, flash geometry, etc.
- (18) Forging or dies process sequencing (e.g. preform, blocker, finisher)
- (19) Press and dimensional control strategy, e.g. pressure vs. position control
- (20) Press head approach and/or stroke sequencing
- (21) Transfer times

Specification and/or management of each of these forging (hot working) process variables is an essential element of Hot Working Process Design. The State-of-the-Art of this technology for Ti alloy components is the size of Ti alloy closed-die forging now producible. For the U.S. Navy F-18E/F, single piece, web-rib Ti-6Al-4V Wing Carry Through Bulkheads are 5,500 sq.in. PVA. U.S. Air Force F-22 single forward fuselage bulkhead in Ti-6Al-4V is 8,000 sq.in. in PVA. The Ti alloy forgings in these two aircraft clearly re-establish the State-of-the-Art and exploit all of the technologies described above.

LINKED MULTI-PROCESS-THERMAL-STRUCTURE-PROPERTIES-MODELS:

Reaching in the Ultimate Objective: Computer-based prediction of product characteristics through multiple process steps.

Currently, while the computer-based technologies reviewed above are having a potent effect on Ti alloy component manufacturing cost and time, CAE, TMP, FEA/FEM Models, etc. are essentially "unit processes" managing or analyzing single process steps. Moving these models into 3D has resulted in a major increase in computer power required. Moving to the next level of modeling the full manufacturing stream is today only beginning to enter research and development.

It is anticipated that a Quantum Leap in process modeling is required for continued success in satisfying the ever increasing market and customer focus on Part Manufacturing Time and Cost Reduction especially in Aerospace and Automotive markets. Linked models represent that Quantum Leap. Model linkage exploits highly evolved modeling systems, enhanced understanding of fundamental material behaviors (at the atomistic level) and even further growth in computing power requirements.

Linked design to multiple step deformation and thermal to microstructure and properties models appear to be the next paradigm shift required in today's re-engineering environment for Titanium alloy component users, metal producers and part fabricators. State-of-the-Art in linked model for titanium alloys is that currently process-microstructure enhancement models have been proposed and demonstrated. In other material systems, linked 2D deformation, thermal, quench and recrystallization and strength models have been demonstrated in 2X24 Aluminum alloy and in a high strength Ni-base superalloy (Udimet 720).

CRITICAL CHALLENGES:

Titanium alloy Hot Working Processes and all of its ancillary and supporting technologies have advanced and continue to advance to meet the need of users and producers alike. In addition to the continuing challenge to refine and improve Ti Hot Working Process and products it manufactures, there are two other major technical issues at play in large Ti alloy fabricated components:

• The Clean Metal Initiative: Expanded capture of hearth melting processes and enhanced and automated control of VAR melting -- both necessary to achieve the level of quality that is essential for the critical Ti alloy components being used.

• Strain Induced Porosity (SIP): The incidence of SIP has increased with the growth in demand for and the State-of-the-Art in very large, heavy section parts, including closed-die forgings used on new Military aircraft, the F-18E/F and F-22. SIP is characterized as porosity at grain boundaries or grain triple points, usually found in ultrasonic inspection. SIP is difficult to heal with hot deformation but can be resolved with HIP processing; however, at considerable cost. Fundamental deformation process-structure-SIP understanding has been investigated but is not fully resolved. Solutions to the incidence of SIP are generally empirical process modifications and are not always effective. Tools discussed above under Deformation Mechanics, Structures-Process Interactions, CAE, FEA/FEM are all potentially usable to resolve this problem. Needed is research that treats SIP as a state variable that can be reduced to Constitutive Law, then modeled and thermal and hot working processes optimized on the computer before committing costly Ti ingots.

CONCLUSIONS:

(1) Airframe and Engine titanium component customers' priority on manufacturing cost reduction has had a profound effect on the supply base:
- Changing the definition of "value".
- Driving focus on control of time (lead and flow times).
- Driving raw material product cost reduction.
- Driving part consolidation.

(2) Titanium alloy hot working and cost effective part manufacture has entered a new phase, with critical new technologies making major contributions to successful part manufacture:
- Concurrent Engineering.
- CAE, A.I., KID and KIM.
- TMP and Deformation Process Modeling.
- Linked Material-Processing-Product Models.

(3) Currently the outstanding titanium hot working metallurgy and technical issues affecting large Ti alloy product forms are
- Capture of the Clean Metal Initiative.
- Resolution of the Strain Induced Porosity Problem.

THE DESIGN, PRODUCTION, AND METALLURGY OF ADVANCED, VERY LARGE, TITANIUM AEROSPACE FORGINGS

T. E. Howson and R. G. Broadwell
Wyman-Gordon Forgings
North Grafton, MA 01536 USA

Abstract

This paper will describe the overall systems required by the forging vendor, working with the airframe customers and final machining houses, for producing large closed-die titanium airframe components. Forging design, including process modeling and CAD, die manufacture, forging and heat treat procedures, non-destructive evaluation, and metallurgical and mechanical properties will be described. The review will cover four different forgings later listed in the text.

Background

Considerations of lower overall cost and higher reliability and performance in advanced military and commercial aircraft have fostered the technological advancement of very large, one piece titanium forgings, for both airframes and engines. Large one-piece forgings reduce the problem of joining structural components together in the structure, reduce weight, reduce cost and usually enhance performance. Examples of such parts for structural application are bulkheads, longerons, and wing carry through members, as well as landing gear and landing gear components. Gas turbine engines for inflight vehicles are likewise increasing greatly in size, requiring very large rotating disks of one piece design, incorporating stub shafts or other coupling segments. Wyman-Gordon Forgings has been actively engaged for the past several years in the concurrent design, and manufacture of these forgings from such alloys as Ti-6Al-4V, Ti-6Al-4V ELI, Ti-10V-2Fe-3Al, Ti-6Al-2Sn-2Zr-2Mo-2Cr (Si), and Ti-17. This review will be concerned with the airframe sector primarily.

The advantages of titanium alloys over other metallic materials of construction for aerospace applications are well known. The principal reason behind the selection of titanium alloys for airframe (and engine) application is weight savings. Titanium alloys possess higher strengths than aluminum and strength similar to many steels. The higher strength material would be selected for areas where volume constraints are a consideration. The section thickness required for a landing gear, made from aluminum or lower strength steel, for example, might be too large to fit into the space available in the surrounding structure.

Titanium is virtually immune to corrosion in the environments experienced by commercial, and for that matter, military aircraft. In this same vein, there is a large galvanic potential between aluminum and graphite found in carbon-fiber reinforced polymers. Titanium exhibits excellent compatibility with such composites. These intrinsic advantages of titanium are multiplied several times as the plan area of the part becomes larger. The removal of the need for bolts, nuts, rivits, splices and welds makes for a lighter, more efficient, and in the long run, lower cost airframe structure.

These days, ingots weighing up to 20,000 lbs. (9,074 kg) and typically 36 inches (91 cm) in diameter are available in a variety of alloys. Increased chemical homogeneity and freedom from harmful defects are hallmarks of today's ingot technology. This makes possible large diameter, heavy weight billet multiples for conversion to the largest plan view area aircraft forgings.

The Alloys Involved

It is important now to discuss briefly the four titanium alloys that are used for the parts described. They are listed in order of when the alloy was developed through pilot plant and first came on the scene. The chemical composition of the four alloys is shown in Table 1.

Table I
Titanium Alloys Used

Ti-6Al-4V (STD) - 0.20% Oxygen (max)   Ti-10V-2Fe-3Al - 0.13% Oxygen(max)
Ti-6Al-4V (ELI) - 0.13% Oxygen (max)   Ti-6Al-2Sn-2Zr-2Mo-2Cr (Si) -
                                           0.13% Oxygen (max)

There is not too much more that can be said regarding this virtues of standard formulation Ti-6Al-4V. It was born out of U.S. Government contracts from Watertown Arsenal, for Ordnance application; it came to being in the mid to late 1950's. Since then, Ti-6Al-4V has become, and remains the "work-horse" alloy for the titanium universe, logging many pounds, miles, hours, cycles, etc. in a variety of airframe, engine, and ordnance applications over the past 35 years.

The need for Ti-6Al-4V (ELI) extra low interstitial came initially from the desire to use titanium alloys for their strength to weight ratio at cryogenic temperatures. Pressure vessels immersed in liquid oxygen for missile applications are a good example. Ti-6Al-4V, with reduced aluminum, iron, and oxygen (0.13% max.) contents was shown to have improved ductility and fracture characteristics over standard Ti-6Al-4V at temperatures as low as -423°F (-253°C). Later into the 1970's, the science of fracture mechanics became more exacting for airframe, and even ordnance armor applications, at room and slightly elevated temperatures. The strength to weight ratio appeal of Ti-6Al-4V ELI continues to meet higher KIC requirements under those conditions as well.

Ti-10V-2Fe-3Al emerged from alloy development efforts at Timet in the mid 1970's. The first aircraft producer exhaustive evaluation was done in consideration of certain critical, high strength-high toughness parts on the B-1 bomber. Later evaluations by other airframe companies have resulted in significant production applications for several critical airframe components. Ti-10V-2Fe-3Al is a high strength-high static (KIC) toughness alloy that is "producible" in most all respects. Metallurgically it is a near-beta alloy. It offers the opportunity to use near-net shape forging techniques and develops both high strength and excellent toughness in heavy sections. To the forger, it can mean lower temperature, isothermal-type forging, using economical die materials.

Ti-6-22-22S was developed in the mid 1970's by the RMI Titanium Co., as a deep hardenable, elevated temperature alpha/beta alloy for thick section forgings - initially for turbine engine application. For a variety of non-technical reasons, however, the alloy has not been significantly exploited in commercial applications until recently. Early development work had shown Ti-6-22-22S to have excellent strength at room and elevated temperatures, and good fracture toughness, fatigue crack growth resistance and fatigue properties under a variety of conditions, using conventional alpha/beta thermomechanical processing techniques. Another plus is that Ti-6-22-22S displays an elastic modulus about 10% higher than competitive alpha/beta alloys. For Ti-6-22-22S applications in forgings requiring the highest resistance to fatigue crack growth, beta processing procedures were developed, involving either beta forging and sub-transus heat treatment, or alpha/beta forging followed by beta-type heat treatment, in an effort to achieve the best balance of strength, ductility, toughness, and fatigue crack growth resistance.

General Considerations

Wyman-Gordon and other major forgers have been working together with the prime airframe and engine manufacturers for many years, overcoming design and potential production problems, on very large and somewhat smaller parts, realizing long term cost reduction and lighter, more efficient structures in the process. The prime customers usually supply the part drawings, and specifications for the type of material, heat treat condition, non destructive tests, and mechanical test plans.

At the outset there are several major decision points regarding the raw material that will go into the forging:
- The starting stock dimensions
- The Wyman-Gordon approval and designated material supplier(s)
- Any special Wyman-Gordon requirements over and above the customer's that will provide greater assurance for a good part, e.g., chemistry restrictions, special grain size needs, etc.

The manufacturing methods considerations involve the following factors (which are particularly sensitive for very large forgings):
- The alloy to be used and the maximum heat treat section size in which the required properties can be developed
- The thermomechanical processing (TMP) requirements that will provide the appropriate microstructures and properties (process modeling is a key tool)
- Cost considerations, involving the number of dies to be used, what type of press to use, and whether "right and left" hand forgings can be made in the same sequence, what processing steps are necessary, etc.
- The design of the tooling which includes an analysis of forging loads based in part upon material flow considerations, and die design standards.

Process design at the forger includes determination of the following:
- Number and type of forge operations
- Part complexity and size
- Type and size of forge equipment for each operation
- Forge temperature, time at temperature, cooling rates, etc.
- The determination of design cover, tolerance build up
- In process cleaning, machining and inspection
- Type of heat treatment
- Physical testing requirements, non-destructive testing (NDT), inspection, etc.

Therefore, the finished forging is really designed by both the prime customer and the forger. The forging can be machined all over or have some as-forged surfaces, or be precision forged.

Design And Manufacturing Processes

Several of the very large, closed-die titanium forgings developed over the past few years and produced in Wyman-Gordon's 35,000 ton or 50,000 ton hydraulic forging presses are listed in Table 2.

Table 2
Current Large, Closed-Die Titanium Forgings at Wyman-Gordon

Component	Alloy	Plan View Area	Ship Weight
Lockheed F-22 Bulkheads	Ti-6Al-4V ELI Ti-6-22-22S	6,300 - 8,800 in^2 (40,645 - 56,775 cm^2)	1,700 - 3,500 lbs. (771 - 1,588 kg.)
Boeing 747 Main Landing Gear Beams	Ti-6Al-4V	6,200 in^2 (40,000 cm^2)	2,850 lbs. (1,293 kg.)
McDonnell Douglas C-17 Pylon Panel	Ti-6Al-4V	4,000 in^2 (25,806 cm^2)	1,730 lbs. (785 kg.)
Boeing 777 Landing Gear Truck Beam	Ti-10V-2Fe-3Al	1,900 in^2 (12,258 cm^2)	3,175 lbs. (1,440 kg.)

The more recent achievements include the Ti-6Al-4V ELI and Ti-6-22-22S F-22 bulkheads, which have increased the plan view area of closed-die titanium forgings produced at Wyman-Gordon to 8,800 sq. in. (56,775 sq. cm.), and the C-17 Pylon Panel. The largest Ti-10-2-3 forging is now the Boeing 777 landing gear truck beam, at 3,175 lbs. ship weight.

Design

The 747 main landing gear Ti-6Al-4V forgings were initially designed and developed over 20 years ago. In recent years the forgings have been aggressively redesigned to remove weight. The redesigns were produced after CAD/CAM tools came on-line, and at Wyman-Gordon the system used has been ANVIL 5000 produced by MCS. Conventional 2-D prints are produced.

For the F-22 both Ti-6-22-22S and Ti-6Al-4V ELI bulkhead forgings have been produced. The Ti-6-22-22S bulkhead was the first produced, as a proof of concept to demonstrate the feasibility of producing the bulkheads as one piece forgings. The design was somewhat generic; it incorporated the bulkhead features that existed at that time, but since the design was going to change the design was not as tight as it could be. Subsequent F-22 bulkheads were redesigned and were made using Ti-6Al-4V ELI after the bill of materials was changed. The Ti-6Al-4V ELI bulkhead shown in Table 7 is an early redesign. Like the proof of concept forging, it was produced to a generous design to protect design changes and also meet schedules. It was produced in a blocker die only and called a "slabber" forging. The initial Ti-6-22-22S bulkhead was designed and drawn using 2-D CAD drawings. The subsequent Ti-6Al-4V ELI F-22 bulkhead forging designs have been created using CATIA, produced by Dassault, which was brought on-line at Wyman-Gordon two years ago. The Lockheed machined part models, created in CATIA, were read directly into a Wyman-Gordon CATIA workstation, and a 3-D finish forging design was built around the customer-supplied part. Blocker dies were produced on the ANVIL 5000 CAD system as 2-D drawings. The C-17 pylon panel was a customer designed forging, as was the 777 landing gear truck beam. The information on the final forge configuration was exchanged in print form.

Process Modeling

Computer simulation of forging and heat treatment has been integrated into the design process for axisymmetric forgings for a number of years. Modeling has successfully predicted die fill, formation of forging defects, strain, strain rate, temperature, die stresses, and many other features. Modeling of 2-D cross sections and some 3-D metal flow was carried out for several of the parts in Table 2. For the bulkheads, metal flow in a number of 2-D cross sections was simulated to help determine the best location of the preformed material in the blocker forging dies to achieve die fill. For the 777 truck beam, 2-D models of metal flow were run to help determine the preform die design that, together with the customer-designed finish forge geometry, would yield the required minimum amount of strain (for property development) in the finish forge step. This modeling helped to identify die designs that achieved properties above minimum requirements, but due to the limitations of the 2-D modeling of the 3-D process, an optimum design solution was not achieved. Now that 3-D models can be run, the finish forge step is being further simulated to attempt to achieve the best balance of tensile strength, ductility, and fracture toughness.

The difficulties of 3-D process modeling include the need to have fully defined 3-D models of the work pieces and dies, the size of models and the times needed for solving them, and accurately predicting the boundary conditions of friction, temperature, etc. In addition, recent work on process modeling has been focused on understanding the relationships between initial microstructures, thermomechanical history, and final microstructures and properties, and building that understanding into the material models so that the process simulation can predict microstructures and properties. These material models, along with 3-D modeling capability, are going to change the design of structural forgings from an experience-based art to more of a science.

Die Manufacture

Dies for the forgings listed in Table 2 have been machined using three different approaches. The two more traditional methods are manual die sinking working off of the 2-D print, and copy milling using models and masters. In copy milling a wood model of the forging is produced, a Fiberglas impression (master) of the model is taken, and after inspections to insure that the model master is correct, the die block is machined by driving the machine tools with probes that trace the Fiberglas impression. The third approach is to directly sink dies using

CNC. At Wyman-Gordon the software used to define the tooling paths and to create or refine the surface model of the forging has been SABRE 5000 produced by Gerber. This software tool is now being changed over to CATIA.

The 747 MLG beam dies and the 777 landing gear truck beam dies were produced using models and masters. The C-17 pylon panel dies were manually machined using 2-D prints outside of Wyman-Gordon. The F-22 bulkhead dies were produced using a combination of models and masters and CNC die sinking, in order to utilize all existing capability at Wyman-Gordon and meet customer schedules. In the near future, die sinking will be entirely CNC. The geometry of the forging will be defined as a 3-D model in CATIA, and the model will drive computer analysis, die sinking, and inspection.

Forging, Heat Treatment, And Processing

Many of the details about how the structural titanium forgings of large plan view area are produced on our 50,000 ton press are proprietary. In general, the forging sequence after ingot-to-billet conversion is an initial shaping operation on an open die 2,000 ton press followed by blocking and finish forging operations on the 50,000 ton press. Existing furnace and handling capabilities in the forge shop have been adequate for the large new bulkheads.

Heat treatment of all of the parts described consists of solutioning and annealing in carbottom or box furnaces, utilizing specialized fixtures, if needed, to maintain straightness. An elaborate fan cooling set-up is utilized to achieve the cooling rates after the solution heat treatment required to meet properties.

Ultrasonic And Surface Inspection Of Large Titanium Airframe Forgings

Ultrasonic inspection of titanium product can be accomplished intermediate during the forging process, in the starting bar or preform stage, or near the end of the process as a more complex shape. The primary types of internal defects that ultrasonics searches for are melt related. During inspection of bar, however, it is not uncommon to detect surface cracks as well. Ultrasonics is also used to locate pipe for cropping off the ends of the bar. Relative to the subject parts the inspection of all four was carried out on the starting bar or plate only. The surface of the material is prepared by surface "billet" grinders smooth enough to permit the required inspection sensitivities.

The F-22 Ti-6Al-4V ELI or Ti-6-22-22S bulkhead is inspected as plate to a Class A (#3 flat-bottomed hole) sensitivity. Scanning is performed from one side using the longitudinal mode and then a 4-direction refracted longitudinal mode for detection of defects not lying perpendicular to a straight sound wave. Both the Boeing 747 MLG beam and the 777 landing gear truck beam are inspected as rectangular bar to either MIL-STD-2154 Class A (#3 flat-bottomed hole) or MIL-STD-2154 Class B (#5 flat-bottomed hole) respectively. The typical ultrasonic equipment used is a Krautkramer-Branson USIP-12, with a 5 MHz, .75" O, flat-faced transducer. As forgings, no surface inspection other than visual is used on any of the three parts discussed in this paper.

RTA

The ready-to-assemble concept is a relatively new venture that is gaining popularity with major airframe producers. It involves the forging manufacturer supplying to the prime manufacturer a fully machined, non-destructively tested and painted part, produced from a forged product. This concept involves new procedures and philosophies, because a purchase order from a prime puts the responsibility for all manufacturing details on the forging house. But the process illustrates the steps that are required from part design to installation on the aircraft.

The process starts off with the manufacturing of tooling, programming and a tape try out (TTO). Upon completion of TTO the forger is responsible to do a first article inspection (FAI) which shall release the tooling, programming, and techniques for production parts. Furthermore, the first production part is also subject to FAI It also includes NDT and final surface conditioning. This is to be done in conjunction with the prime contractor. Note that parts can have up to approximately 4000 blueprint characteristics (inclusive of SPC-keys) requiring witnessing, reviewing documentation, and/or over inspection. The dynamics of this business requires the forging supplier to have personnel available to perform these previously mentioned inspections as well as to troubleshoot quality problems and incorporate engineering changes that constantly arise during redesigning or developing the base program.

Process Control And Continuous Improvement

At Wyman-Gordon, the model for process control and continuous improvement is based upon:

- The Boeing D1-9000 specification Advanced Quality System (AQS) and,
- The Lockheed Appendix Y Process Control requirements

The process involves establishing the key characteristics through risk analysis, determining the process steps and identifying when to measure for the key characteristics, documenting the AQS control plan, charting the measurements and determining if the key characteristics are in

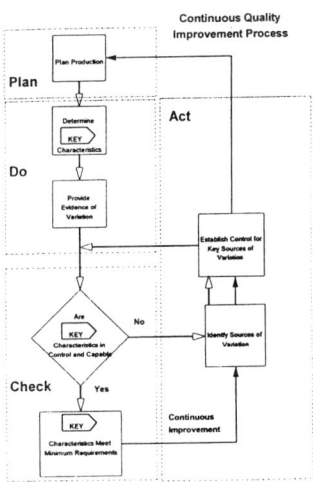

Figure 1. Continuous Quality Improvement Process

statistical control, and calculating the process capability ratio (Cpk). If the key characteristics do not meet the minimum required capability, potential sources of process variation must be identified and controlled.

The flowchart for the continuous quality improvement process is shown in Figure 1. The "Plan-Do-Check-Act" sequence is adequately pictured in the slide. It is a fact that the process can involve not only the forging manufactures, but also the prime customer and forgers sub-tiers, as well, if it is to be totally effective.

Mechanical Properties

The manufacturing operations and typical cut-up mechanical properties for the four forgings are shown in Tables 3-7.

Table 3
Boeing 747 Main Landing Gear Beam Data

Part Forging Procedure:

Operations	Dies	Forge Temp. (°F)	Equipment
Draw to Print	Open	1750	Cogging Press
Preform	Closed	1750	50K Ton Press
Block	Closed	1750	50K Ton Press
Finish	Closed	1750	50K Ton Press

Billet Starting Size: 18" RCS
Part Heat Treat Procedure:
Anneal: 1300°F, 2 hrs. furnace cool to 1000°F, hold 2 hrs. with furnace shut off, then air cool.

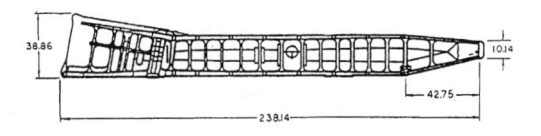

Mechanical Property Ranges:
0.2% y.s.	(KSI)	120-134
UTS	(KSI)	130-145
Elong.	(%)	10-17
RA	(%)	25-41

Part Specification: MIL-T-9047G, Am. #2

Table 4
McDonnell Douglas C-17 Pylon Panel Data

Part Forging Procedure:

Operations	Dies	Forge Temp. (°F)	Equipment
Draw to Print	Open	1750	Cogging Press
Finish	Closed	1750	50K Ton Press

Table 4 (continued)

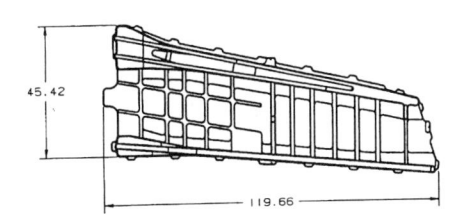

Billet Starting Size: 32" x 10"
Part Heat Treat Procedure:
Anneal: 1300°F, 2 hrs., air cool
Mechanical Property Ranges:
0.2% y.s.	(KSI)	125-140 (< 4 in. thick),	124-129 (> 4 in. thick)
UTS	(KSI)	135-142 (< 4 in. thick),	137-150 (> 4 in. thick)
Elong.	(%)	15 (< 4 in. thick),	11-16 (> 4 in. thick)
RA	(%)	31-37 (< 4 in. thick),	20-39 (> 4 in. thick)

Part Specification (Material): DMS 1583

Table 5
Boeing 777 Landing Gear Truck Beam Data

Part Forging Procedure:
Operations	Dies	Forge Temp. (°F)	Equipment
Draw to Print	Open	1400	Cogging Press
Block	Closed	1400	50K Ton Press
Finish	Closed	1400	50K Ton Press

Part Heat Treat Procedure:
Solution: 1400°F, 2 hrs. water quench
Age: 940°F, 10 hrs. air cool
Mechanical Property Ranges:
0.2% y.s.	(KSI)	155-168
UTS	(KSI)	166-182
Elong.	(%)	5-13
RA	(%)	11-47
Typical Fracture Toughness (KSI √in)		42

Part Specification: BMS 7-260G

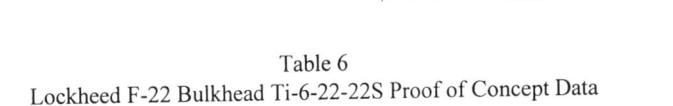

Table 6
Lockheed F-22 Bulkhead Ti-6-22-22S Proof of Concept Data

Part Forging Procedure:
Operations	Dies	Forge Temp. (°F)	Equipment
Preform	Open	1690	Cogging Press
Finish	Closed	1690	50K Ton Press

Table 6 (continued)

Billet Starting Size: 7" x 48" (plate)
Part Heat Treat Procedure:
Beta Solution: Beta plus 50°F, 1/2 hr. (min.), fan cool to 900°F, air cool
Alpha Beta Solution: Beta minus 50°F, 1 hr. (min.), fan cool to 900°F, air cool
Age: 975°F, 8 hrs. air cool
Mechanical Property Ranges:

0.2% y.s.	(KSI)	134-142
UTS	(KSI)	152-163
Elong.	(%)	7-10
RA	(%)	14-18

Typical Fracture Toughness (KSI $\sqrt{in}$)   84
Part Specification: 5PTM5T01

Table 7
Lockheed F-22 Slabber Bulkhead, Ti-6Al-4V ELI

Part Forging Procedure:

Operations	Dies	Forge Temp. (°F)	Equipment
Block Die Only	Closed	1725	50K Ton Press

Part Heat Treat Procedure:
Beta Solution: Beta plus 50-75°F, 1/2 hrs. (min.), fan cool
Anneal: 1375°F, 3 hrs., air cool
Mechanical Property Ranges:

0.2% y.s.	(KSI)	109-116
UTS	(KSI)	124-129
Elong.	(%)	7-10
RA	(%)	18-25

Typical Fracture Toughness (KSI $\sqrt{in}$)   112
Part Specification: 5PTM5T04

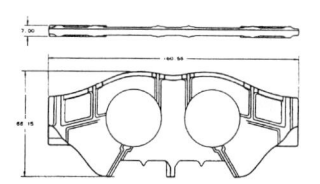

Summary

The direction for the future in large, critical airframe parts is toward single piece manufacture wherever possible, in order to gain the advantages of improved structure reliability and overall lower cost. A result of this trend will be ever increasing concurrent engineering and cooperation between the airframe prime contractor, the forger, and the finish machine and preparation vendors.

This paper has described the planning and process flow involved in the concurrent engineering, the material selection, the design, the production, and the properties of four large titanium forgings. The lessons still to be learned and the advances in process modeling, processing, and resulting forgings will be of considerable value to the integrity of airframe structure of the future.

A MECHANISTIC STUDY OF THE HOT WORKING OF CAST GAMMA Ti-Al

D. A. Hardwick and P. L. Martin

Rockwell Science Center,
1049 Camino Dos Rios,
Thousand Oaks, CA 91360

Abstract

Compression workability testing, coupled with extensive metallographic characterization, has been used to study the evolution of microstructural refinement and chemical homogeneity in two near-γ TiAl alloys: Ti-46Al-0.2B and Ti-48Al-0.2B. Homogenization of the casting segregation present in these materials was complicated by their peritectic solidification path. Due to its solidification pathway, the low Al alloy was inherently more homogenous but heat treatment temperatures of ≈1350°C were required to obtain complete homogenization. For the high Al alloy a temperature of ≈1400°C was necessary to establish chemical and microstructural homogeneity. Both alloys exhibited a fully lamellar $\gamma+\alpha_2$ microstructure, with a large colony size, following homogenization. Hot working of the cast materials at temperatures up to 1150°C retained the chemical segregation. This was most obvious in the hyper-peritectic alloy in which the segregated regions were merely compressed and elongated perpendicular to the deformation direction. In both alloys, this microstructural inhomogeneity was retained through a subsequent recrystallization heat treatment at 1200°C. Hot working of materials that had undergone an homogenization heat treatment resulted in the breakup of the initial fully lamellar microstructures into isolated islands of lamellae surrounded by fine bands of dynamically recrystallized grains. In the high Al alloy, subsequent heat treatment at 1200°C resulted in the breakup of a large proportion of the lamellae into equiaxed $\gamma+\alpha_2$.

This work was supported by the US Air Force Office of Scientific Research.

Introduction

A consensus has begun to develop in the aerospace industry that the γ-TiAl-base alloys could replace heavier nickel-base alloys in some applications (1). The Al content of gamma alloys of most interest for future applications fall in the range 45-48 at% Al. As shown in the binary phase diagram, Figure 1, this composition range straddles both the peritectic composition of ≈46.5 at% and the eutectoid reaction. The chemical inhomogeneities and microstructures generated by these phase transitions must be reckoned with in cast materials. Furthermore, the development of TiAl alloys has reached the stage where the addition of refractory elements, at the 1-4at% level, is quite common (1). Such additions exacerbate the problems of segregation as they diffuse more slowly than either Ti or Al (2). The elimination of segregation in these alloys through hot working has been, and continues to be, a subject receiving considerable attention (3-5).

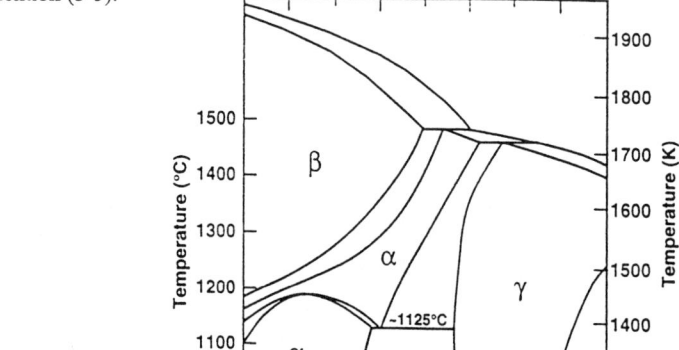

Figure 1 - TiAl equilibrium phase diagram

In this study, compression workability testing has been coupled with metallographic characterization to investigate the influence of primary ingot breakdown on microstructural refinement and chemical homogeneity. The necessity for an initial homogenization step prior to ingot breakdown was addressed by the testing of as cast and cast + homogenized material. The alloy compositions chosen for this study straddled the peritectic composition: Ti-45at%Al, a hypo-peritectic composition, and Ti-47at%Al, a hyper-peritectic alloy. Boron was incorporated into the alloys, as such additions can retard elevated temperature grain growth (6).

Experimental

Alloy Composition and Heat Treatment

A 100mm diameter ingot of each composition was produced by Duriron, Dayton, OH. The ingots were prepared using an induction skull melting technique and the melts were poured into graphite molds. The aim and actual compositions are given in Table I.

A series of heat treatments was performed to determine the microstructural response to homogenization. If there can be said to be an industry standard for the heat treatment of cast γ ingots, it is a HIP cycle of 1260°C for 4 hours. Therefore, heat treatments of 4 hours were carried out at 1250°C and the temperature was increased by 50°C increments until homogenization was achieved. Specimens for heat treatment were wrapped in Ta foil and encapsulated in quartz tubes backfilled with a partial pressure of argon. All heat treatments were finished with an air cool during which, the samples remained in their glass capsules.

Table I - Composition of Ingots (at%)

Aim Composition	Al	B	C	O (appm)	O (wppm)
Ti-47Al-0.2B	48.0	0.18	0.03	1533	650
Ti-45Al-0.2B	45.7	0.16	0.03	1528	640

Hot Workability

Specimens for compression testing were electro-discharge-machined (EDM) from 12.5mm thick slices taken from the top of the Duriron ingots. The compression cylinders were ≈ 7.6mm diameter and were removed from a circle having a radius of approximately three quarters of the total ingot radius. The sequential cutting of the cylinders around the radius of the ingot left a raised vertical line of material down the side of each cylinder; with this line as a reference mark, the orientation of each sample with respect to the radial direction in the ingot could be easily determined. The ends of these right circular cylinders were polished flat and perpendicular to the cylinder axis. Hot compression testing was carried out in a Centorr furnace mounted on an Instron test machine. All testing was conducted in a vacuum of 5×10^{-3} Pa or better. The compression platens were silicon nitride backed with W rams and the specimens were lubricated with boron nitride to reduce frictional effects. The silicon nitride platens were not deformed by the testing procedure. Extensometry, using LVDT sensors attached to the W rams, was used to record the displacement. Both the load and the extension were recorded digitally. Temperature was recorded by a thermocouple placed near the compression specimen.

Microstructural Characterization

The principal microstructural characterization method used was backscatter image SEM on unetched electropolished surfaces. This method provides both chemical and microstructural information as the image is formed through a combination of atomic number (Z) contrast and orientation due to electron channeling.

Results and Discussion

Cast and Homogenized Microstructures

Ti-45.7Al-0.2B The microstructure of the hypo-peritectic alloy in the cast condition and following homogenization heat treatment is shown in Figure 2. In the as-cast condition, the alloy is fully lamellar, as shown in Figure 2a, and backscatter contrast in the SEM reveals little compositional segregation. The Ti-Al phase diagram, shown in Figure 1, indicates that during solidification, the first material to solidify would be a low Al β (≈40at%Al) but that solidification to single phase α of the peritectic composition of 46.5at%Al will intervene before significant coring of the dendrites occurs. The maximum composition difference between dendrite core and the interdendritic region is thus ≈6at%; this small gradient in composition would be largely smoothed out during the residence time in the single phase α region as the ingot cools. On cooling below 1325°C, the single phase α will transform to lamellae of α + γ. Heat treatment at 1250°C and 1300°C had little effect on either the microstructure or the segregation pattern revealed by the backscatter contrast. However, heat treatment at 1350°C removes all trace of the casting segregation as shown in Figure 2b. At this temperature, the alloy was in the single phase α region as revealed by the finer lamellar spacing compared with the cast material, a result of air cooling of a small specimen. The locations of former interdendritic regions are identified by the stable boride particles; this is particularly easy to see in the large grain on the right-hand side of Figure 2b.

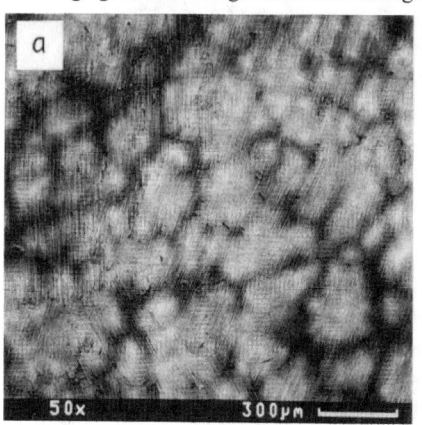

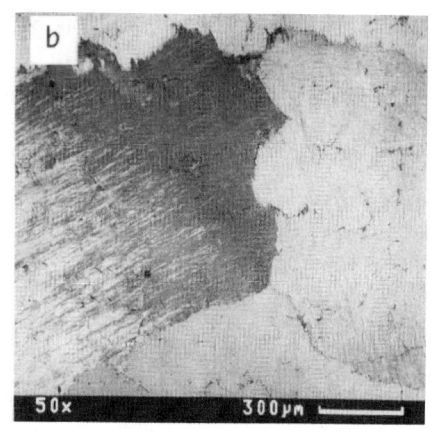

Figure 2 - Microstructure of Ti-45.7Al-0.2B:
(a) as cast and (b) after an homogenization heat treatment at 1350°C for 4 hours.

Ti-48.0Al-0.2B The microstructure of the hyper-peritectic alloy in the cast condition and following homogenization heat treatment is shown in Figure 3. During casting of this alloy, the first material to solidify will be β at 42at%Al. The liquid will continue to be enriched with Al until the last liquid to solidify will contain ≥50 at%Al. The maximum difference in composition between dendrite core and the interdendritic region is greater than for the hypo-peritectic alloy and, in addition, the material must cool through a two phase region. The dendrite cores will transform to α, but the high Al interdendritic regions transform to single phase γ which will impede compositional rearrangements. Diffusion will have to take place through ordered γ rather than through disordered α. The cast microstructure, Figure 3a, reflects this solidification sequence; the Ti-rich regions are fully lamellar while the interdendritic Al-rich areas are equiaxed γ. Following heat treatment at 1250°C, Figure 3b, the microstructure contains considerably more equiaxed γ. The Ti-rich dendrite cores remain fully lamellar and appear to decrease in volume fraction after this heat treatment. After heat treatment at 1300°C, the lamellar regions expand while the equiaxed interdendritic regions shrink. At temperature, the lamellar regions will be single phase α, facilitating diffusional rearrangement within these regions, while the interdendritic regions will remain as ordered γ phase. The major change affecting the γ phase will be dissolution and transition to α phase at the α/γ boundaries. As a result of this rearrangement, the microstructure after heat treatment at 1300°C appears virtually identical to that in the as-cast condition. Following heat treatment at 1350°C, the contraction of the equiaxed γ phase has almost reached its conclusion and the microstructure appears more segregated than it was in the cast condition. In the temperature range between 1350°C and 1400°C, the alloy becomes single phase α. During heat treatment at 1400°C, segregation dissipates and the air cooled microstructure is almost fully lamellar but with some isolated regions of equiaxed γ phase. Once again, locations of former interdendritic regions are revealed by the stable boride particles.

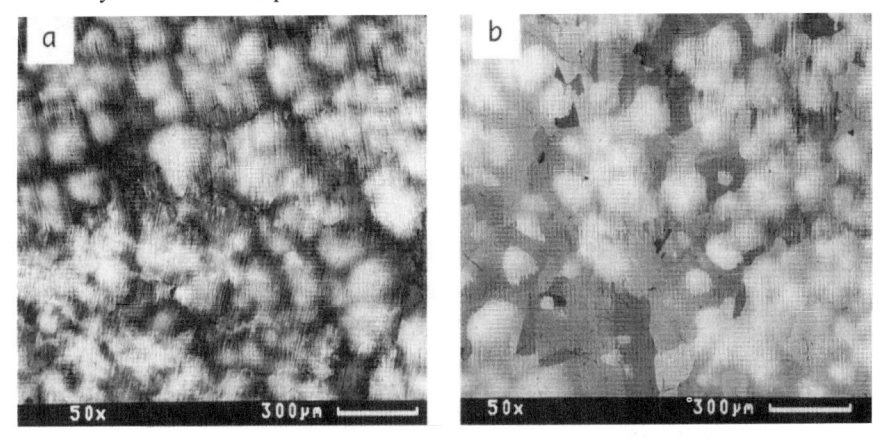

Figure 3 - Microstructure of Ti-48.0Al-0.2B: (a) as cast and (b) following heat treatment for four hours at 1250°C.

Hot Workability

The initial workability test parameters were selected to maximize the data relevant for the forging of preforms from the 4" inch diameter ingots. The test temperatures were 1050°C, 1100°C and 1150°C; all tests were conducted at an initial strain rate of $10^{-3}s^{-1}$. Specimens were tested in each of two microstructural conditions: as-cast and cast+homogenized. The homogenization temperature was that which ameliorated the compositional segregation; i.e. 1350°C for the hypo-peritectic alloy and 1400°C for the hyper-peritectic alloy.

Table II - Summary of Results from Hot Compression Testing; $\dot{\varepsilon} = 10^{-3}s^{-1}$

	Sample	Test Temp. (°C)	Peak Stress (MPa)	Total Strain (%)
as cast	Ti-45.7Al-0.2B	1150	210	50.1
		1100	270	49.9
		1050	415	49.4
	Ti-48.0Al-0.2B	1150	214	50.4
		1100	248	50.0
		1050	314	50.7
homogenized	Ti-45.7Al-0.2B	1150	183	50.1
		1100	278	49.9
		1050	434	49.4
	Ti-48.0Al-0.2B	1150	194	50.4
		1100	323	50.0
		1050	357	50.7

Table II presents the data obtained during the hot compression tests. In general, the as-cast material was slightly stronger than the homogenized material at 1150°C but slightly softer at the lower testing temperatures of 1100° and 1050°C. Of the alloy compositions, the low Al alloy was the stronger material. In all cases, the stress-strain curve was characterized by a peak followed by a gradual drop of 40-50% in flow stress to a steady state value.

Hot-worked Microstructures

The hot compression data provide some basic information on the stresses necessary to work the material at a given temperature but the more important information is gained by examining the response of the microstructure to the deformation strain. Each alloy exhibited a similar response to deformation at all three temperatures. The following section gives a detailed description and analysis of the results obtained at 1150°C.

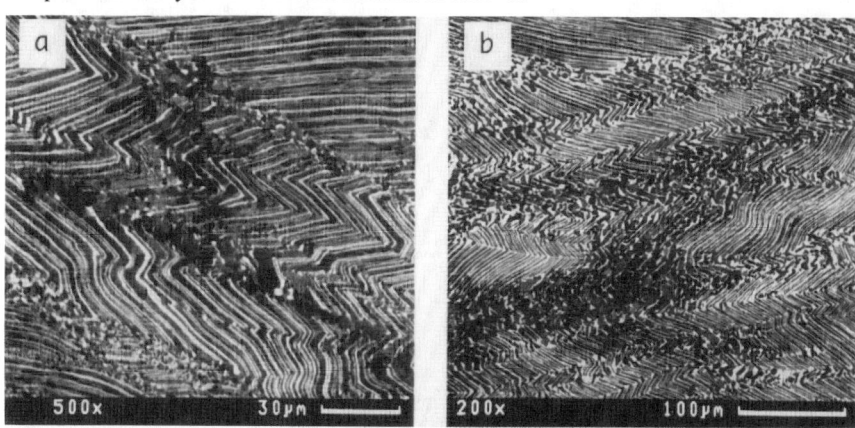

Figure 4 - Microstructure of cast Ti-45.7Al-0.2B: (a) after 50% strain at 1150°C and (b) following a post deformation heat treatment at 1200°C for 2hours.

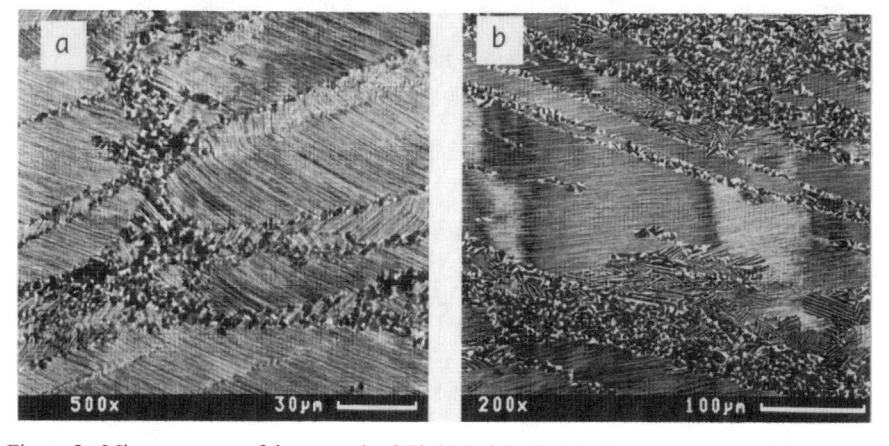

Figure 5 - Microstructure of homogenized Ti-45.7Al-0.2B: (a) after 50% strain at 1150°C and (b) following a post deformation heat treatment at 1200°C for 2 hours.

Ti-45.7Al-0.2B After compression at 1150°C, the cast lamellar microstructure is traversed by intense shear bands. Figure 4a shows that thin bands of very fine equiaxed grains have formed along these shear bands. Heat treatment at 1200°C for 2 hours yields the microstructure shown in Figure 4b. The bent lamellae caused by passage of the shear bands during compressive deformation are retained in the heat treated microstructure.

Compressive deformation of the homogenized material resulted in a similar microstructure, but deformation of the fine lamellae results in more closely-spaced shear bands, and thicker bands of fine equiaxed grains, Figure 5a. The equiaxed grains were most probably formed by dynamic recrystallization of the heavily worked material in the shear bands. When this material is heat treated at 1200°C the lamellar regions are still retained but the influence of the deformation strain on the lamellar colonies is highlighted by the variation in backscatter contrast within individual lamellar colonies. This variation in contrast is not due to a change in Z, but to a change in orientation, most probably due to deformation-induced lattice rotation of the lamellar colony. The equiaxed γ grains formed by dynamic recrystallization have grown and the α phase has precipitated from the γ phase in the form of plates parallel to the 4 close-packed 111 planes of the γ phase.

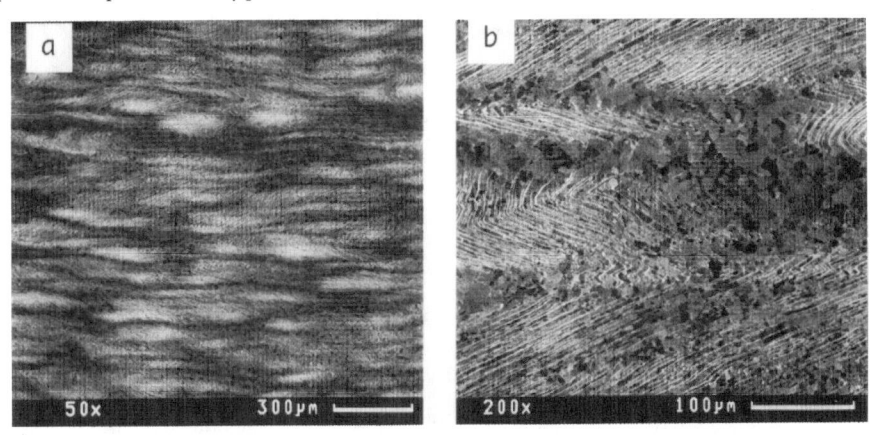

Figure 6 - Deformed microstructure of cast Ti-48.0Al-0.2B.

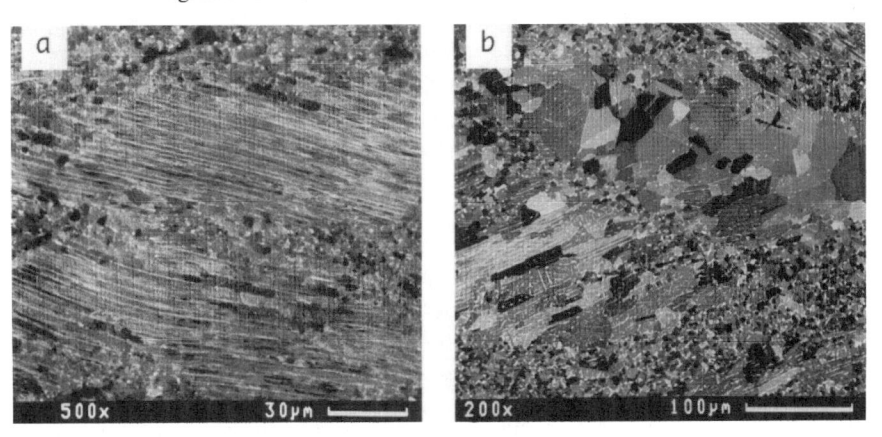

Figure 7 - Deformed microstructure of homogenized Ti-48.0Al-0.2B:
(a) as deformed and (b) after 2 hours at 1200°C

Ti-48.0Al-0.2B As would be expected from the heat treatment study, deformation of the highly segregated hyperperitectic alloy at 1150°C results in little diffusional rearrangement. The high Z regions that were the dendritic cores have merely been elongated perpendicular to the deformation direction, as shown in Figure 6a; the compression axis was vertical. At higher magnifications, the "banded" nature of the deformed microstructure can be easily seen, Figure 6b. Heat treatment at 1200°C leads to grain growth in the equiaxed γ regions but the microstructure remains banded.

Deformation of the fully lamellar, homogenized material yields the microstructure shown in Figure 7a. The microstructure consists of islands of lamellar colonies surrounded by regions of fine, equiaxed, dynamically recrystallized grains. Heat treatment at 1200°C for 2 hours yields the microstructure shown in Figure 7b. The α_2/γ lamellar regions have begun to break-up and spherodize while grain growth has occurred in regions of α_2-free, equiaxed γ grains.

Conclusions

As a result of this work, the following conclusions can be drawn:

i) The optimal homogenization heat treatment for the hypo-peritectic alloy, based on dispersion of both chemical and microstructural inhomogeneities, was 1350°C, while the hyper-peritectic alloy required 1400°C.

ii) An homogenization heat treatment prior to hot working was beneficial for obtaining a more uniform hot worked microstructure.

iii) Heat treatment after the hot working operation was benficial in obtaining a more uniform structure when the starting material was in the homogenized condition.

References

1. Y-W. Kim, "Ordered Intermetallic Alloys Part III: Gamma Titanium Aluminides", Journal of Metals, 46 (7) (1994), 30-39.

2. D.A. Hardwick and P.L. Martin: "Mechanistic Study of Hot Working of a Cast Multi-Phase Near-γ TiAl Alloy", to be published in Proceedings of the 35th Annual Conference of Metallurgists, (Candian Institute of Metallurgy, 1996).

3. P.L. Martin, C.G. Rhodes and P.A. McQuay, "Thermomechanical Processing Effects on Microstructure in Alloys Based on Gamma TiAl", in Structural Intermetallics, edited by R. Darolia et al. (TMS, 1993) 177-186.

4. S.L. Semiatin,"Wrought Processing of Ingot-Metallurgy Gamma TiAl Alloys", in Gamma Titanium Aluminides, edited by Y-W. Kim, R. Wagner and M. Yamaguchi (TMS, 1995) 509-524.

5. P.L. Martin, S.K. Jain and M.A. Stucke,"Microstructure and Mechanical Properties of Wrought Near-γ TiAl Alloys", in Gamma Titanium Aluminides, edited by Y-W. Kim, R. Wagner and M. Yamaguchi, (TMS, 1995) 727-736.

6. P.L. Martin and D.A. Hardwick, "Microstructure and Properties of Near-γ Alloys Based on Ti-47Al-2Nb-2Cr-1Mo", to be published in Proceedings of the 8th World Conference on Titanium, edited by M. Loretto et al. (Institute of Materials, London, 1996).

EFFECT OF MICROSTRUCTURE ON CAVITATION AND FAILURE BEHAVIOR DURING SUPERPLASTIC DEFORMATION OF A NEAR-GAMMA TITANIUM ALUMINIDE ALLOY

C. M. Lombard*, A. K. Ghosh** and S. L. Semiatin*

*WL/MLLN
Wright-Patterson AFB, OH 45433-7817

**Dept. of Materials Science and Engineering, University of Michigan,
Ann Arbor, MI 48109-2136

Abstract

The uniaxial hot tension behavior of sheet of a near-γ titanium aluminide alloy (Ti-45.5Al-2Cr-2Nb) was determined in the as-rolled condition (initial grain size $\approx$ 3 to 5 µm) and a rolled-and-heat treated (1175°C/4 hours) condition (initial grain size $\approx$ 10 to 12 µm). Microstructure evolution, cavitation rates, and failure modes were established via constant strain rate tests at 10^{-3} sec^{-1} and test temperatures between 900 and 1200°C. After extensive deformation, all specimens exhibited grain refinement at the lower test temperatures; at higher test temperatures, the as-rolled specimens revealed grain growth and the heat treated samples showed grain refinement. For both initial microstructural conditions, the failure mode was established as predominantly cavitation/fracture controlled. Cavity growth rates were greatest at lower temperatures and in the heat treated specimens with the coarser initial grain size; the higher cavitation rates in the heat treated specimens led to total elongations only approximately one-half those of the as-rolled material. For both initial microstructural conditions, the optimum forming temperature was 1200°C. At this temperature, the maximum elongations, minimum cavitation rates, and lowest flow stresses were observed.

Introduction

Near-γ titanium aluminide alloys have received considerable interest with regard to their superplastic characteristics and the potential use of superplastically formed parts in high temperature aerospace applications. In one recent study [1], fine grained, as-rolled sheet specimens of the near-γ alloy Ti-45.5Al-2Cr-2Nb (at. pct.) were pulled in tension at a nominal strain rate of $10^{-3} \sec^{-1}$ and temperatures between 900° and 1200°C. Microstructural examination after testing revealed cavitation in all specimens and grain refinement at 900-1000°C, a stable grain size at 1100°C, and grain growth at 1200°C. The optimum forming temperature was 1200°C, at which the maximum elongation (980%) and minimum cavitation were found.

In the present work, a more in-depth investigation of microstructure evolution and mechanical behavior during hot tension testing was undertaken on the same material used in Reference 1. This characterization was done, however, on material in the as-rolled as well as with material in a coarser-grain, rolled-and-heat treated (1177°C/4 hours) condition. To obtain insight into the micromechanisms controlling plastic flow under nominally superplastic conditions, post-formed microstructures, cavitation rates, and fracture modes were documented and analyzed.

Experimental Procedure

A Ti-45.5Al-2Cr-2Nb ingot was triple arc melted at TIMET Corp., (Henderson, NV) in the form of a 190 mm diameter ingot with 635 mm length. The ingot was isostatically hot pressed to seal casting porosity and then forged to a 5:1 reduction at 1177°C and a nominal strain rate of $10^{-3} \sec^{-1}$. Subsequent processing was done via pack rolling. For this purpose, preforms of 10.44 mm thickness were machined from the billet and welded inside Ti-6Al-4V cans. After preheating at 927°C for 20 min., the packs were transferred to a furnace at 1254°C, held for 30 min., removed, and then rolled; intermediate reheats were done between passes. The packs were rolled using a rolling speed of 7.62 m/sec and a nominal 12.5 pct. reduction per pass with the total reduction being on the order of 5:1. After the final pass, the packs were slow cooled in vermiculite. The packs were decanned; some of the rolled sheets were glass coated and heat treated at 1177°C for four hours followed by slow cooling in vermiculite.

The sheet material was ground to 1.524 mm thickness prior to machining tensile samples by electro-discharge machining. The bulk of the testing was performed using 12.7 mm gage length samples with 3.175 mm width. For higher elongation tests, 6.35 mm gage length, 1.588 mm width samples were employed. Tension testing was done in argon with the test chamber being evacuated and backfilled three times. Heating was done with a tungsten wire mesh furnace. Prior to pulling, each specimen was allowed to equilibrate at the test temperature for 30 min. All tests were conducted to failure at a constant true axial strain rate ($\dot{\epsilon}$) of $10^{-3} \sec^{-1}$ by continually increasing the crosshead speed (v) based on the equation $v = \dot{\epsilon} L_0 \exp(\dot{\epsilon} t)$ in which L_0 is the initial gage length. This relation assumes a nominally uniform deformation along the gage length, which was indeed confirmed to a first order by post-tension test measurements.

After testing, axial through-thickness metallographic sections were prepared using standard techniques and examined using polarized light optical microscopy and backscattered or secondary electron imaging (EI) in a field emission scanning electron microscope. To quantify cavitation behavior, 500X optical micrographs were taken at various locations along the length of the metallographic samples, and a point counting technique was then utilized to estimate cavity volume fraction. These cavity volume fractions were associated with local axial strains, ϵ_L, determined from measurements of the local thickness strain, ϵ_T, and the local width strain, ϵ_W ($\epsilon_L \approx -\epsilon_W - \epsilon_T$, neglecting the effects of cavitation on the change in volume).

Results and Discussion

Microstructure Evolution

<u>Starting Microstructures</u> The microstructures of the as-rolled and rolled-and-heat treated Ti-45.5Al-2Cr-2Nb alloy showed marked differences. In the as-rolled (AR) condition, the

Fig. 1. Polarized light, optical microstructure of AR Ti-45.5Al-2Cr-2Nb.

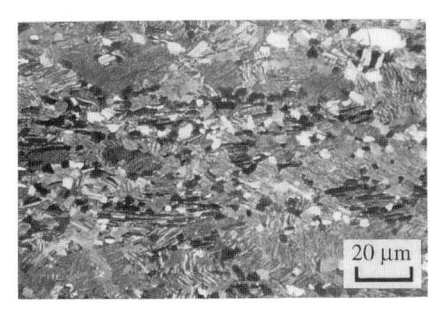

Fig. 2. Polarized light, optical microstructure of HT Ti-45.5Al-2Cr-2Nb.

microstructure was two-thirds recrystallized equiaxed γ and α_2 grains of 3 to 5 μm diameter (Fig. 1). Regions of untransformed γ + α_2 lamellae constituted the remaining one-third of the structure. (The phases were identified by backscattered EI: γ (dark), α_2 (white), and β (white); the composition of the phases was confirmed previously be electron microprobe analysis [1].) After heat treatment (HT) at 1177°C/4 hours, an equiaxed γ structure, with an average grain size between 10 and 12 μm was developed (Fig. 2). γ phase represented approximately 75 pct. of the microstructure with the remainder being 20 pct. α_2 and 5 pct. β. The α_2 was found at the γ grain boundaries or in rather blocky α_2 + γ lamellar colonies of 5 μm size. The straight lamellae in these colonies indicated that they were formed during heat treatment rather than being remnant from the rolling operation. The β grains were typically adjacent to α_2 grains.

Microstructure Evolution During Preheating Prior to Tension Testing A study of the heat treatment of the AR material in the 900-1200°C temperature range with a thirty minute hold was done previously to establish the effect of hold time prior to tension testing on microstructural changes [1]. It showed that the volume fraction of equiaxed γ grains remained essentially unchanged from the AR state after heat treatment at 900°C, but it increased with temperature to 95 pct. recrystallized γ grains at 1200°C. While the equiaxed volume fraction increased with increasing temperature, the recrystallized grain size remained narrowly distributed in the 3 to 5 μm range for all heat treatment temperatures. Similar heat treatment studies were conducted for this work using the material given the 1177°C/4 hours heat treatment. The results for this latter material showed that the γ grain size did not change significantly at any temperature. This was expected inasmuch as the initial treatment temperature (1177°C) was greater than three of the preheat temperatures and close to the fourth.

Microstructure Evolution During Tension Testing (Grip) Grip sections of tested specimens, in which the effective strain is essentially zero, were examined to determine the effects of preheating + test time at temperature on microstructure evolution; for the strain rate used and elongations of the program material, test times were of the order of thirty minutes or less. Examination of AR specimens showed that γ grain size increased slightly with increasing temperature, whereas it remained essentially unchanged in material given the 1177°C/4 hour pretreatment.

Microstructure Evolution During Tension Testing (Near the Failure Site) Microstructural observations for the region near the failure site are summarized in Table I and Figs 3, 4, and 5. These results indicated several similarities and a number of differences for specimens tested in the AR and HT conditions. For example, the γ grain size near the failure site was the same for both initial microstructures after testing at the same temperature. This observation can be ascribed to the occurrence of dynamic recrystallization which gives a steady state grain size dependent only on deformation strain rate and temperature [2]. The remaining observations are best summarized with regard to the specific test temperature. After large deformation at 900°C, the AR specimen exhibited untransformed (remnant) lamellar colonies, recrystallized γ grains, numerous finer grains (≈ 0.5 μm in size), and extensive cavitation (Fig. 3a). The finer grains

Table I. Microstructural Data (Near the Failure Site)

Temperature (°C)	Condition	γ Grain Size (μm)	α_2 Grain Size (μm)	$\varepsilon_{Thickness}$	ε_{Width}
900*	AR	1-2	0.5	-1.09	-0.82
1000*	AR	3	1.5	-1.63	-1.34
1100*	AR	5	5	-1.30	-1.20
1200**	AR	7 ##	10 ##	-2.07	-1.93
900*	HT	1-2	5 #	-0.53	-0.36
1000*	HT	4	1.5	-1.14	-0.84
1100*	HT	5	5	-1.20	-1.03
1200**	HT	7 ##	8 ##	-1.42	-1.25

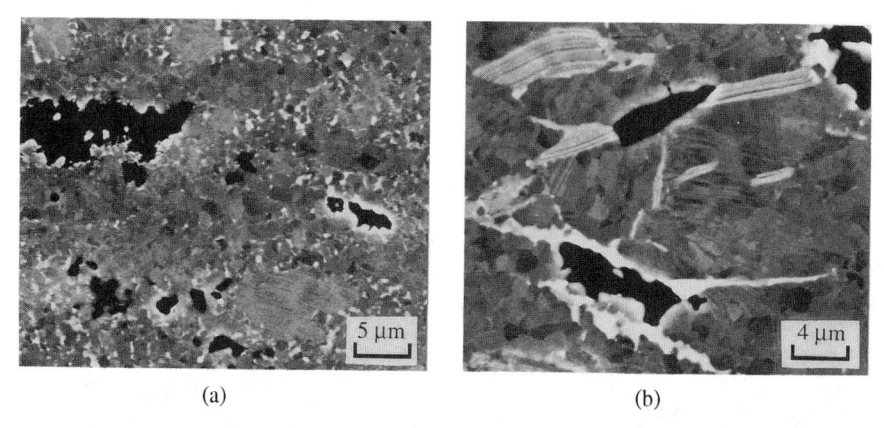

* = 12.7 mm gage length   ** = 6.35 mm gage length   # = α_2 + γ colony   ## = away from denuded zone

Fig. 3. Backscattered EI micrographs taken near the failure site of (a) AR or (b) HT specimens tested at 900°C. Tensile direction in these and following micrographs is along the horizontal.

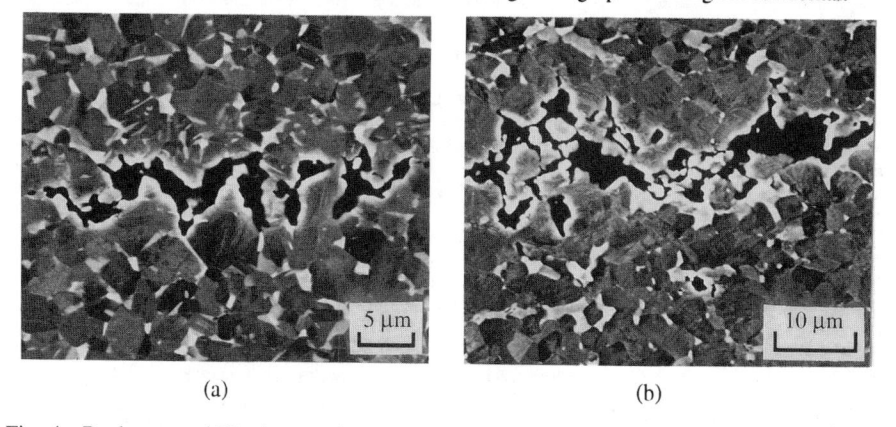

Fig. 4. Backscattered EI micrographs taken near the failure site of (a) AR or (b) HT specimens tested at 1000°C.

imaged white under the backscattered EI mode in the SEM and thus are believed to be α_2 which has formed at a temperature lower than the α_2 formed during rolling. The cavities were associated with both the lamellar colonies and the regions of recrystallized grains. Similarly, the HT specimens tested at 900°C contained an extensive recrystallized structure of γ grains and numerous cavities; the cavities in the HT material were associated predominantly with the lamellar colonies and γ/β interfaces (Fig. 3b). Unlike the AR material, however, the HT samples did not contain fine recrystallized α_2 grains.

After testing at 1000°C, the microstructures near the failure sites of both the AR and HT specimens exhibited a nearly completely recrystallized γ grain structure, some finer α_2 grains, and numerous voids (Fig. 4). In both cases, the voids were located along γ/γ and γ/α_2 interfaces. Unlike the 900°C behavior, voids initiated by the microfracture of a given grain were not found in HT samples tested at 1000°C. The microstructures observed in samples tested at 1100°C were similar to those tested at 1000°C with the exception that the amount of α_2 phase was greater at the higher temperature.

The nature of the microstructure and cavitation developed during testing at 1200°C was noticeably different from that observed at lower temperatures. Specifically, centerline porosity was observed near the failure site of the AR specimens tested at 1200°C (Fig. 5a). An examination of the structure along the centerline revealed a zone of larger γ grains (10 μm in diameter) which was denuded of the grain-growth inhibiting α_2 grains (Fig. 5b). Away from the centerline, the microstructure became more homogeneous, consisting of γ grains ($\approx$ 7 μm) and α_2 grains (7 to 10 μm). An examination of the AR sheet before tension testing and AR specimens pulled to failure at lower temperatures did not reveal any evidence of large scale microstructural inhomogeniety or porosity. Hence, the development of such porosity is most likely a result of the state of stress in the diffuse neck which may affect the local coarsening of the α_2 phase, the local γ grain growth rate, and hence the local cavitation behavior. However, a quantitative description of such phenomena cannot be given at this time. A similar centerline denuded zone with coarse γ grains was developed in the HT specimen tested at 1200°C (Fig. 5c, d).

Cavitation Kinetics

The measured cavity volume fractions were fit to the analytical expression proposed by Lian and Suery [3] :

$$C_V = C_{Vo} \exp(\eta \varepsilon_L)$$

C_V denotes the cavity volume fraction at an axial true strain (ε_L), C_{Vo} is the initial cavity volume fraction at $\varepsilon_L = 0$, and η is the cavity growth rate constant. Such a formulation assumes no nucleation of new cavities once deformation commences and no cavity coalescence. The research of Lian and Suery and Nicolaou, et al. [4] showed that the value of the strain rate sensitivity exponent (m) has a major influence on the strain to failure when η is small ($\eta \approx 1$ or less); in these cases, flow localization, not cavitation/fracture events, control the failure process. By contrast, when η is greater than approximately 3 the strain to failure is almost independent of m for $m \gtrsim 0.25$. Thus, a high m value is a necessary but not sufficient condition for large elongations to failure.

In the present work, because there probably was cavity nucleation during deformation as well as cavity coalescence, only an apparent cavity growth rate (η_{APP}) can be fitted to the measurements. Thus these values cannot be compared directly with those generated in Reference 3; however, qualitative statements about the trends found in this research can be made. The value of C_{Vo} was taken to be 10^{-4}. The values of η_{APP} for the AR and HT tension tests are summarized in Table II. For each test temperature, the value of η_{APP} for the AR material was less than that of the HT material. This trend is as was expected in view of the influence of grain size on flow stress and cavity initiation. Cavity initiation and growth would be expected to be slower for the fine-grained AR material because of the greater ability to relieve stress concentrations at grain triple points by diffusional processes for this material compared to the coarser-grained HT material. Secondly, the generally higher flow stresses of the HT material (see discussion below) would

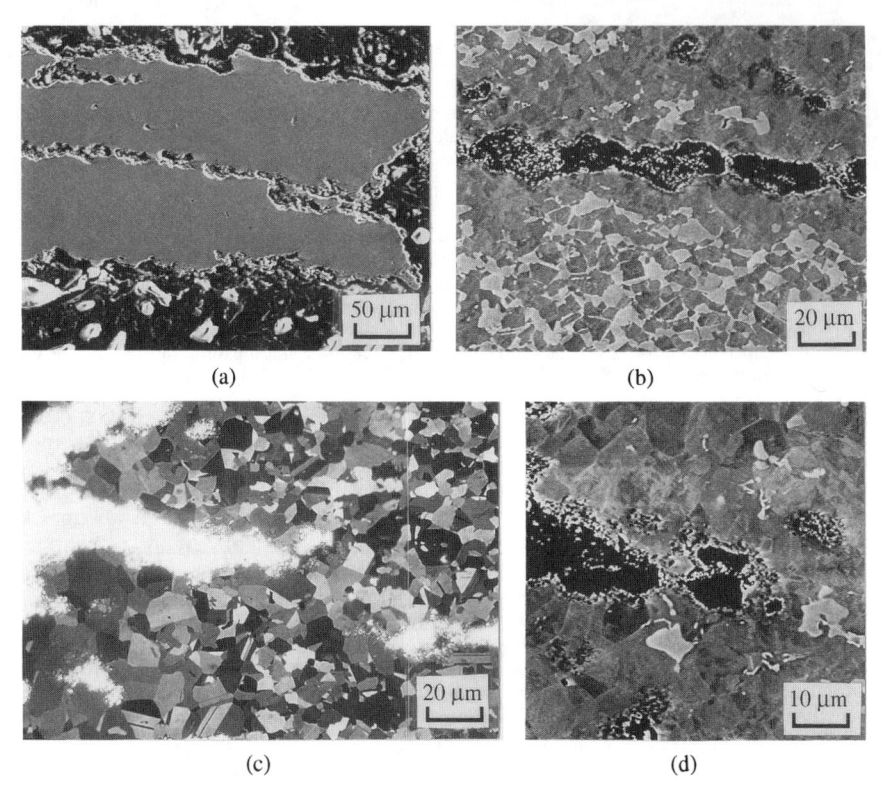

Fig. 5. Microstructures developed near the failure site of specimens tested at 1200°C. (a) AR Secondary EI (b) AR Backscattered (c) HT Polarized Light (d) HT Backscattered EI.

Table II. Apparent Cavity Growth Rates (η_{APP}) as a Function of Temperature

Temperature (°C)	As-Rolled (AR)	Heat Treated (HT)
900	3.1	8.0
1000	2.2	3.4
1100	2.3	2.9
1200	1.8	1.9

lead to higher local hydrostatic tensile stresses and hence higher cavity growth rates. Nevertheless, as the temperature was increased, the lower flow stresses and increased diffusion kinetics led to lower values of η_{APP} for both materials. In fact, the measured value of η_{APP} dropped to 1.8 to 1.9 for both the AR and HT materials, respectively. Thus, it can be concluded that as the temperature increased, flow localization began to affect the failure mode.

Flow Phenomenology
Mechanical property data from the uniaxial tension tests are summarized in Table III. Based on the peak stress and stress at 100 pct. elongation, the rate of decrease of engineering stress versus engineering strain is seen to be greater at low temperature for both material conditions. This more rapid drop at lower temperatures may have several sources such as the onset of localized necking within the diffuse neck and/or flow softening due to the spherodization of the remnant lamellar colonies (in the case of the AR material) or dynamic recrystallization of the γ phase.

Table III. Mechanical Property Data

Temperature (°C)	Condition	Peak Engineering Stress (MPa)	Engineering Stress at 100 pct. Extension (MPa)	Total Elongation (pct.)
900*	AR	357	-	72
1000*	AR	127	31.3	219
1100*	AR	51.2	22.1	350
1200**	AR	21.5	15.3	446
900*	HT	309	-	51
1000*	HT	169	-	104
1100*	HT	82.6	25.9	176
1200**	HT	25.0	18.6	244

* = 12.7 mm gage length sample   ** = 6.35 mm gage length sample

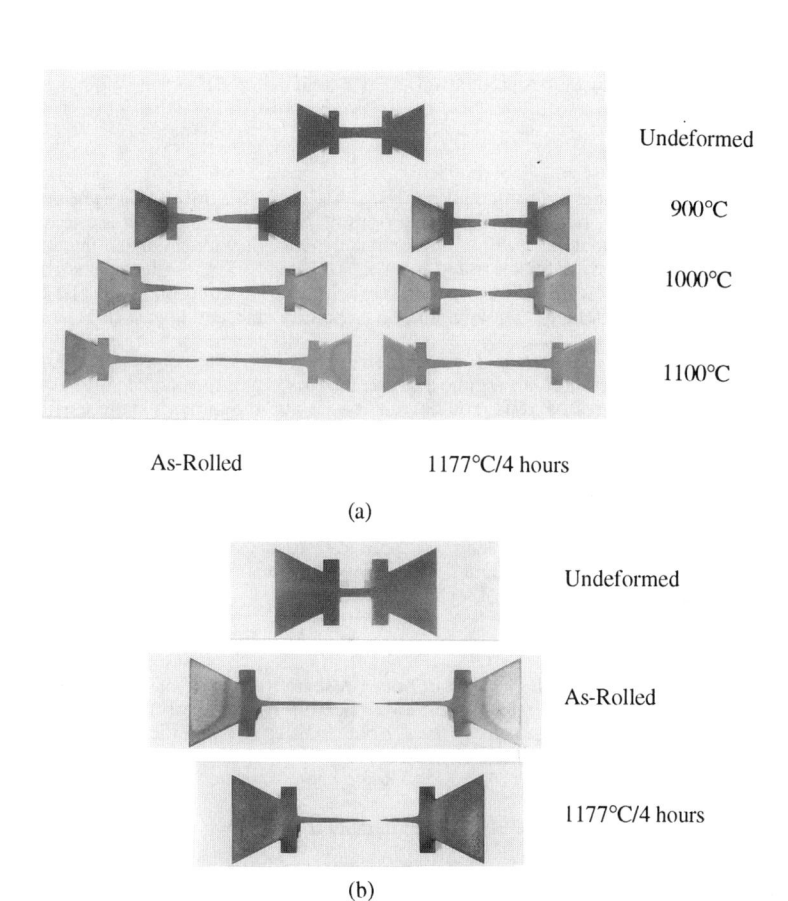

Fig. 6. Macrographs of failed test specimens with an initial gage length of (a) 12.7 mm gage or (b) 6.35 mm gage. In (b), the specimens were tested at 1200°C.

However, a rigorous explanation of the engineering stress-strain curves would require detailed numerical simulations such as those conducted by Nicolaou, et al. With regard to the magnitude of the peak stresses, the values for the HT material were higher than those for the AR except at 900°C; this is as expected in view of the coarser grain size of the HT material and deformation which has a large component of grain-boundary sliding. On the other hand, the higher peak stress of the AR material at 900°C may derive from deformation which has a large component of plastic flow dominated by slip.

Profiles of tested specimens showed various degrees of elongation and diffuse and localized necking (Fig. 6). Using a traditional definition of superplasticity as tensile elongation greater than 300 pct., the results in Table III reveal that AR material was superplastic at ≈ 1050°C and higher. While the HT material is not superplastic, it did exhibit extended ductility. A detailed interpretation of the failure mode (necking versus fracture controlled) must await measurements of the m values of the two materials. Comparison of m vs. elongation data for non-cavitating titanium and zirconium alloys from Lee and Backofen [5] with the current results allows an estimation of minimum m values for the program alloy at 1200°C; m_{AR} is at least 0.43 while m_{HT} is at least 0.31. Based on measured values of η_{APP} and assuming m is indeed of the order of 0.25 to 0.50, the results of Nicolaou, et al. suggest that failure of both materials was via a fracture-controlled process at 900°C to 1100°C and via a flow-localization process at 1200°C. Failure observations (Fig. 6) do indeed suggest fracture prior to localized necking at 900°C and localized necking almost to a point at 1200°C for both materials, which is in agreement with this broad conclusion.

Conclusions

1. The near-γ titanium aluminide alloy Ti-45.5Al-2Cr-2Nb exhibits superplasticity in the fine-grain as-rolled state, but not in the coarser-grain heat treated state. In the temperature range examined, as-rolled material exhibits approximately twice the elongations of heat treated material. The superplastic regime for as-rolled material is 1050-1200°C with the possibility of higher temperatures. For both materials, optimum sheet forming conditions are at 1200°C. At this temperature, maximum elongation, minimum cavitation rate, and lowest flow stresses are found.

2. Apparent cavity growth rates for the heat treated condition are consistently higher than those of the as-rolled material in the regime examined. This is as expected because as-rolled material has 1) lower flow stresses, thus reducing cavity growth due to hydrostatic tensile stresses, and 2) finer grain size and hence greater ease of accommodating stress concentrations/void initiation due to grain sliding. For both materials, the cavity growth rate decreases with temperature and approaches a single value at 1200°C indicating that cavitation plays a less important role while localized thinning becomes more important in causing final failure at higher temperatures.

3. At 1200°C, failure in both materials is seen to be associated with large γ grain regions which are devoid of α_2. The development of the denuded regions may be due to enhanced diffusion in local regions with higher levels of tensile stresses.

Acknowledgments

This work was done as part of the in-house research program of the Metals and Ceramics Division, Materials Directorate, Wright Laboratory, Wright-Patterson AFB, OH. The in-house program is supported in large part by the Air Force Office of Scientific Research (AFOSR).

References

1. C. M. Lombard, et al., Gamma Titanium Aluminides, eds. Y.-W. Kim, et al.; TMS; Warrendale, PA, pp. 579-586.
2. Seetharaman, V., et al., High Temperature Ordered Intermetallic Alloys IV, eds. L. A. Johnson et al.; MRS; Pittsburgh, PA, pp. 889-894.
3. J. Lian and M. Suery, Mat. Sci. Tech., Vol. 2 (1986), pp. 1093-8.
4. P. D. Nicolaou, et al., submitted to Met. Trans. A.
5. D. Lee and W. Backofen, Trans. Met. Soc. AIME, Vol. 239 (July, 1967), pp. 1034-1040.

LAMELLAR TO EQUIAXED GRAIN STRUCTURE BY TORSIONAL DEFORMATION OF Ti-6Al-2Sn-4Zr-2Mo ALLOY

G. Welsch, Case Western Reserve University, Cleveland, Ohio 44106
I. Weiss, Wright State University, Dayton, Ohio 45435,
D. Eylon, University of Dayton, Dayton, Ohio, 45469
F.H. Froes, University of Idaho, Moscow, ID 83843

Abstract

The effectiveness of torsional deformation at 925°C in transforming an alpha/beta lamella microstructure into an equiaxed grain structure was investigated by optical and electron microscopy on the Ti-6Al-2Sn-4Zr-2Mo alloy. The degree of the breakup of the lamellae into shorter grains, measured by the change of grains' length/width aspect ratio (AR), depends on the shear strain and recrystallization. In uniformly deformed regions. The grain aspect ratio decreased from an initial average AR of 12 to an AR of 3, as the shear strain was varied from zero to 1.5. The equiaxed grain structure was obtained by recrystallization of the deformed material. In shear bands, in which concentrated shear strain occurred, dynamic recrystallization led to equiaxed grains during the deformation processing. In moderately deformed regions of the microstructure, dynamic recrystallization results only in the formation of recrystallized grains (nuclei) at certain locations.

1. Introduction

The Ti-6Al-2Sn-4Zr-2Mo alloy is an alpha + beta alloy that exhibits the characteristic lamellar grain structure of beta-transformed titanium alloys after beta heat treatment. During cooling through the transus temperature the allotropic transformation proceeds from an all beta grain structure to the lamellar colony alpha+beta structure, eq.(1).

$$\beta super\text{-}transus \rightarrow f_\alpha \bullet \alpha + (1-f_\alpha) \bullet \beta sub\text{-}transus \tag{1}$$

where f_α is the volume fraction being transformed into the hexagonal alpha phase. The beta phase as it exists above the transus is considered chemically homogeneous and has the alloy's overall composition. The sub-transus beta phase, on the other hand, is enriched with the beta-stabilizing alloy elements and depleted of the alpha-stabilizing elements. The reverse is true for the alpha phase, thus balancing out the overall composition. The colonies grow in as many as twelve orientation variants from the initial high-temperature beta phase, Figure 1. They obey the Burgers orientation relationship, with the dense packed $\{011\}_\beta$ planes of the BCC beta phase forming the basis for dense packed $(0001)_\alpha$ planes. One of two $<111>\beta$ directions in each $(011)_\beta$ plane serves as the coincidence direction for a dense-packed $<11\bar{2}0>_\alpha$ direction. The relative orientations of alpha and beta lamellae and of their respective crystallographic slip planes, along which a colony can undergo shearing by dislocation glide, are illustrated in Figure 2.

The microstructural stability of the transformed lamella structure at sub-transus heat treatment temperatures is due to low alpha/beta interface energy. It is not possible to change this microstructure to an equiaxed one by heat treatment alone. An exception to this may be the so-called "Thermohydrogen Processing"[3]. The equiaxed grain structure is preferred for many applications in which cold formability, high fracture toughness and improved resistance against fatigue crack initiation are needed[4]. The only practical way to change the beta-transformed lamella structure into an equiaxed one is by forceful distortion of the orientation relationship. This is achieved by deformation processing, usually in the 800 to 1000°C temperature range. Forging, rolling, extrusion, swaging and drawing exert various degrees of homogeneous and inhomogeneous (redundant) deformation. On the mesoscale the material flow involves shear deformation along the principal shear stress planes. At the microscopic scale a variety of deformation mechanisms can occur. They include dislocation glide on those crystallographic slip planes that have orientations near a principal shear plane of the continuum. Consider the microstructure in Figure 2. Assume, the $(\bar{1}100)_\alpha$ and $(2\bar{1}1)_\beta$ are the primary slip planes. Shearing of the lamella packet can be effected by movement of $1/3\ [11\bar{2}0]_\alpha$ and $½\ [\bar{1}\bar{1}1]_\beta$ dislocations. Note, however, that the orientations of the respective slip planes and slip directions and the lengths of the Burgers vectors are somewhat different in the neighboring α- and β-phase lamellae. To avert a misfit in contiguity during deformation, additional mechanisms occur on the microscopic scale. They include dislocation glide on secondary and higher order slip systems, twinning, and diffusion-assisted creep. In a strain-hardening material the shear is relatively uniformly distributed. When strain-softening occurs, the deformation tends to be concentrated in shear bands.

Because torsion produces pure shear it is well suited to evaluate the effect of shear strain on the transformation from lamella to equiaxed grain structure. In previous work results were presented of crystallographic shear of α - lamellae, of shear bands through the lamellar aggregate and of the change of grain (length/width) aspect ratio with shear strain [11-19].

Objectives of present work
Microstructural detail of the transformation of the lamella to the equiaxed grain structure will be presented. Because individual lamellae or grains have width and length dimensions in the range of 1 to 50 μm, it was desirable to view relatively large areas, e.g. areas of tens of micrometers wide, and to have the ability of focusing on small regions, e.g. 1 μm wide, and to image microstructural details with good resolution. High voltage transmission electron microscopy on thin foils from deformation-processed alloy is a suitable method for this task. Argonne National Lab's high voltage electron microscopy facility [20] was made available for this purpose. All of the TEM micrographs shown in this paper were obtained at the Argonne HVEM facility at 1 MeV electron energy.

2. Experimental

Alloy Composition
The actual composition is listed in Table 1. The beta transus is approximately 993°C ± 5°C[22]. At the deformation temperature of 925°C the phases were in the majority alpha phase and the balance was beta phase. An estimate of the phase volume fraction from the quenched microstructure (e.g. see Figures 6 and 7) indicates the ratio of α/β was approximately (2/3) : (1/3).

Table 1. Chemical Composition in Weight Percent

Ti	Al	Sn	Zr	Mo	Si	Interstitial Elements
balance	5.58	1.91	3.0	2.37	0.09	Not determined

Microstructure
The specimen material was heat-treated for two hours at 1025°C for homogenization in the β-phase field, then air-cooled through the transus to generate a fine-lamellar α+β microstructure, as shown in figure 1. This microstructure represented the starting condition, labeled (A) in Figure 3, for subsequent deformation processing. The post-deformation heat treatment for one hour at 925°C in Figure 3, leading to condition (C) was for the purpose of completely recrystallizing the deformed materials. The present study focusses on the microstructural investigation of condition (B) only.

Deformation
Figure 3 shows a schematic of the procedure. A cylindrical sample having a gage length of 25.4 mm and a diameter of 7.9 mm was heated to 925°C, held for 12 minutes to ensure temperature uniformity, and then hot-torsion-tested at this temperature [21] at a shear strain rate of 1.73×10^{-1} s^{-1} in the near-surface layer. The duration of the test was approximately 10 seconds. The experimentally recorded torque versus twist curve is shown in Figure 4. Effective strain and effective strain rate are related to shear strain and shear strain rate by

$$\bar{\varepsilon} = \gamma/\sqrt{3} \quad (2a)$$

and

$$\dot{\bar{\varepsilon}} = \dot{\gamma}/\sqrt{3} \quad (2b)$$

Thus, the effective strain rate in the near-surface layer was 1×10^{-1} s^{-1}, and the total shear or effective strains were 1.73 or 1.0, respectively, in the near-surface layer. The strain in the torsion sample varies linearly with radius, r, from zero in the center of the sample to the maximum value at the outer surface, (Fig. 3),

$$\gamma = r \cdot \phi \quad (3)$$

where ϕ is the angle of twist (in radius) per unit length of the torqued cylinder.
By using thin foil samples taken from various distances, r, from the neutral axis of the

torsion specimen it was possible to investigate material that had experienced different total shear strains. For TEM investigation a radial thin foil section (different illustration in Fig. 4) was chosen. It had been deformed toa shear strain of 1.5 (effective strain = 0.87).

Electron Microscopy

Thin foil specimens were prepared following conventional electrolytic etch-polishing procedure [23]. High voltage transmission electron microscopy, that enabled the imaging of relatively wide (up to 100 μm) material sections, was carried out with a 1 MeV Kratos instrument at the HVEM facility at Argonne National Laboratory.

3. Results on Deformed Microstructure

Metallographic results of the torsionally deformed microstructure have been previously published [19], showing how the lamella aspect ratio varies with the deformation. After deformation to a shear strain of 1.7 the aspect ratio decreased from an initial average value of 12 to an average of 3. Additional heat treatment (1 hour at 925°C) resulted in complete recrystallization and an equiaxed grain structure.

The results in this paper show the deformed microstructure as it was obtained after water-quenching the material immediately following hot torsion. Figure 5 is an optical micrograph of the deformed microstructure in a near-surface region of the torsion sample. The viewing plane is that of a radial section as illustrated in the right-hand portion of Figure 4. Overall, the shear strain in this material was 1.5. Locally, however, there exist great variations in the magnitude of the shear strain. Within shear bands (evidenced by complete breakup of the lamellae) the strain was much higher than 1.5. In other regions, in which bent and otherwise distorted lamellae are still discernable, the strain was less than 1.5.

Figure 6 is a TEM micrograph showing the deformed and partially recrystallized grain structure in a shear band (top portion of Figure) and adjacent to a shear band (bottom portion of Figure). Dynamic recrystallization has resulted in a completely equiaxed grain structure in the shear band but only in occasional new grains in the lesser deformed bottom portion.

The transformation from the lamella structure to the equiaxed grain structure is shown at higher magnification in Figure 7. Figure 7(a) shows the initial stage of lamella breakup. Localized shearing has breached the continuity of the relatively thin β-lamellae and has enabled the formation of junctions between α-phase of neighbor lamellae. There are also nuclei of new recrystallized grains. The recrystallized α-grains have diameters (1.86 μm on average) that are considerably larger than the widths of the original α-lamellae (0.78 μm on average), see Figure 7(b). Likewise, the equiaxed β-grains have diameters (avg. dia. = 1.55 μm) that are considerably larger than the original β-lamella widths (0.39 μm average). The β-grains are easily identified in the micrograph by their feathery fine structure. It is due to orthorhombic martensite that formed in the β-grains during quenching.

4. Discussion

The experimental observations indicate several mechanisms in the dynamic transformation from the lamella α/β to the equiaxed $\alpha + \beta$ grain structure during hot torsion. They include the following:

Strain
Strain-softening caused the magnitude of localized shear strain to vary widely at the 100 μm scale between heavily strained shear bands and mildly deformed regions between shear bands. Only partial microstructural transformation occurred in regions that had less than 1.5 shear strain, and complete transformation to the equiaxed grain structure occurred in shear bands where the shear strain was greater than 1.5.

Strain plays several roles:

(1) High temporal defect (dislocation) concentrations in each the α-phase and the β-phase provide the driving force for nucleation and initial growth of recrystallized α-grains and β-grains. High dislocation concentration also facilitates diffusion that enables the redistribution of the alloy elements in the α and β-phases during recrystallization.

(2) Disruption of the continuity of β-phase ribbons that previously separated neighbor α-lamellae is the result of shear across lamella packets. Here, junctions form between the α-phase of neighbor lamellae. Upon recrystallization the β-phase may end up as islands within or between α-grains.

(3) Distortion of the crystallographic orientation relation of the as-grown α/β lamella structure. This has the effect of increasing the α/β interface energy. The partial reduction of this energy and the need of satisfying force balance at grain- and phase boundary triple junctions is a driving force behind the microstructure changes from lamellar to equiaxed.

In summary, strain provides the driving force for the microstructure transformation. The driving force is stored temporarily in newly generated lattice defects and in increased α/β interface energy caused by the destruction of the Burgers' orientation relationship between the phases.

B) Temperature and Time

Temperature and time are controlling parameters for the kinetics of the transformation from lamellar to equiaxed microstructure. Recrystallization of the deformed microstructure requires diffusion for the removal of lattice defects, for boundary migration and for redistribution of the alloy elements toward the new α- and β-grains, respectively. Diffusion of titanium and of atoms of the substitutional alloy elements proceeds by vacancy transport in the lattices of the α- and β-phases, along dislocation lines (pipe diffusion) and across and along phase and grain boundaries. Because of short-circuit diffusion along dislocations and boundaries it is difficult to estimate the effective diffusivities and diffusion distances. For the recrystallized microstructure in Figure 7(b) the diffusion distances must have been of the order of 1 μm. Diffusivities, D, of the substitutional elements at 925°C are listed in Table 2. The diffusion time, t, from the moment of deformation to the quenching event was of the order of 10 seconds. Approximate diffusions distances, $X = \sqrt{Dt}$, are also listed in Table 2 for assumed diffusion times of 10 seconds (quenched after torsion) and 3600 seconds (additional 1 hour anneal).

Table 2. Diffusivities of alloy elements in β- and α-titanium and approximate diffusion distances at the 925°C deformation temperature. Diffusion coefficients were excerpted from references [23, 24].

Diffusivity at 925°C [m²/s]	Diffusion distances ($X = \sqrt{Dt}$) in 10 sec	in 3600 sec
D_{Ti} in β-Ti ≈ 6 x 10^{-14}	0.8μm	14.7μm
D_V in β-Ti ≈ 2 x 10^{-14}	0.4μm	8.5μm
D_{Al} in β-Ti ≈ 2 x 10^{-13}	1.4μm	26.8μm
D_{Ti} in α-Ti ≈ 3 x 10^{-17}	0.02μm	0.3μm
D_v in α-Ti, no data	--	--
D_{Al} in α-Ti ≈ 10^{-15}	0.1μm	1.9μm

In the torqued and quenched material the diffusion distances, calculated for 10 seconds of diffusion time, are of the order of 1μm in the β-phase but much less in the α-phase. During and immediately following the deformation a temporal high dislocation density must have existed. Its presence would have caused the effective diffusivities to be much larger[26] than those listed in Table 2, and consequently the effective diffusion distances (for 10 seconds) would have been larger. For the recrystallized α-grains in Figure 7(b) the effective diffusion distance for Ti-diffusion in α-titanium must have been of the order of half the α-grain diameters, i.e., close to 1μm. The observations that only partial microstructural transformation occurred in the mildly deformed regions (bottom of Fig. 6 and Fig. 7(a)) and that complete transformation occurred after an additional 1 hour anneal[19] are consistent with the diffusion distances in Table 2. The fully dynamically recrystallized grain structure in a shear band, Figure 7(b), is consistent with increased effective diffusion by pipe diffusion along dislocations that existed temporarily in a high concentration in the shear bands.

In summary, the kinetics of microstructural transformation is controlled by the nucleation and growth of new α-grains and new β-grains[27]. As suggested by the results in Figure

7(a) and (b) the nucleation rate is dependent on strain, particularly on local strain. Nuclei of new grains appear in Figure 7(a) at certain locations where localized shearing has disrupted the continuity of α/β lamellae. New grains have nucleated and grown throughout the regions of former shear bands, Figure 7(b). This is similar to the observation of nucleation of recrystallized grains in Ti-Mo alloy[28]. The number density of disruptions of the continuity of α/β lamellae controls the number density of recrystallization nuclei. The diffusion distance for the growth of these nuclei to a fully recrystallized microstructure is approximately one-half the distance between the nuclei. The initial growth of recrystallized grains is driven by the annihilation of lattice defects. The growth kinetics is controlled by diffusion in the deformed α- and β- lattices. Short-circuit diffusion is believed to play a significant role. Further grain growth, including the formation of equiaxed α- and β-grains is driven by minimization of grain boundary and phase boundary energy. As the Burgers' coherency orientation relationship has been destroyed, and because it cannot be regained during sub-transus annealing, the path toward energy reduction is by reducing the α/β interface and the grain boundary areas in the α- and β-phases. This leads to the equiaxed grain structure. In the majority it consists of equiaxed α-grains and in the minority it consists of equiaxed β-grains. With diffusion in the α-phase being slower than in the β-phase it is likely that the redistribution of the alloy elements is controlled by diffusion in the α-phase. The complete recrystallization in Figure 7(b) that took place in some ten seconds was most likely through short circuit diffusion. A longer annealing time, e.g. 1 hour, would be necessary to enable complete recrystallization by lattice diffusion.

5. Conclusions

The microstructural transformation of lamellar α/β alloy (Ti-6Al-2Sn-4Zr-2Mo) is enabled through mechanical deformation (shear deformation in torsion in the present case) and through recrystallization. The deformation provides the driving force for the transformation. It is in the form of energy of mechanically generated defects and in the form of increased interface energy between α- and β-phases from distortion of the Burgers orientation relation. Schematically, an energy balance can be written as:

Applied Strain Energy → Heat dissipation
→ Stored energy in lattice defects, $\Delta G_{lattice}$
→ Stored energy in increased α/β interface energy, $\Delta G_{\alpha/\beta interface}$

A schematic free energy diagram of the mechanisms of microstructural transformation is given in Figure 8. The ascending branch of the free energy curve is made up of the stored energies from deformation. They provide the driving force for the microstructural changes

$$\text{Deformation} \rightarrow \Delta G_{\text{lattice defects in }\alpha} + \Delta G_{\text{lattice defects in }\beta} + \Delta G_{\alpha/\beta \text{ interface}} \qquad (4)$$

The descending branch of the free energy curve indicates the processes of recovery and recrystallization. The intermediate energy hills in the descending branch indicate reaction barriers, e.g. of diffusion, that are overcome by thermal activation. As the Burgers relation between α and β lamellae has been destroyed the next lowest recrystallized free energy state is that of an equiaxed α+β phase mixture.

Acknowledgements

The electron microscopy investigation was made possible through a grant for use of the HVEM-Tandem Facility at Argonne National Laboratory.

References

1. W.G. Burgers, Physica, vol. 1 (1934) p. 561.

2. T. Furuhara, H.J. Lee, E.S.K. Menon and H.I. Aaronson, Metall Trans, vol. 21A (1990) 1627-1643.

3. O.N. Senkov, J.J. Jonas and F.H. Froes, J. Of Metals, vol. 48, No. 7 (1996) 42-47.

4. K.J. Grundhoff and H. Schurmann, Mat.-Wiss. u. Werkstoff tech., vol. 20 (1989) 284-289.

5. G.R. Yoder, L.A. Cooley, and T.W. Crooker: 23rd Sructures, Structural Dynamics and Materials conference, New Orleans, LA, May 1982, pp. 132-36.

6. D. Eylon, M.E. Rosenblum, and S. Fujishiro: Titanium '80, 4th International Conference on Titanium, Kyoto, Japan, 1980, pp. 1845-54.

7. D. Eylon and C.M. Pierce: Metall. Trans. A, 1976, vol. 7A, pp. 111-21.

8. G. Welsch, W. Bunk and H. Kellerer, Z. Werkstofftechnik, vol. 8, pp.141-48.

9. Y.T. Lee, M. Peters, and G. Welsch, Metall. Trans. A, vol. 22A, (1991) 709-14.

10. I. Weiss, F.H. Froes, D. Eylon and G. Welsch, Metall. Trans. A, vol. 17A, (1986) pp. 1935-47.

11. I. Weiss, G. Welsch, F.H. Froes and D. Eylon, in Titanium Science and Technology Proceedings of 5th International Conference on Titanium, Deutsche Ges. f. Metallkunde, (1985) pp. 1503-10.

12. P. Dadras and J.F. Thomas, Jr., Metall. Trans. A, (1981) vol. 12A, pp. 1867-76.

13. S.L. Semiatin and G.D. Lahoti, Metall. Trans. A, (1981) vol. 12A, pp. 1705-17.

14. S.L. Semiatin and G.D. Lahoti, Metall. Trans. A, (1981) vol. 12A, pp. 1719-28.

15. S.L. Semiatin, J.F. Thomas and P. Dadras, Metall. Trans. A, vol. 14A (1983) 2363-74.

16. E.F. Rauch, G.R. Canova, J.J. Jonas and J.P. Michel, in 'Plastic Instability', Proceedings of Int'l. Symp. On Plastic Instability, Ecole Nat'l. Des Ponts Chaussées, Paris, (1985) 147-158.

17. T. Sheppard and J. Norley, Mat. Sci. & Technology, vol. 4 (1988) 903-908.

18. M. Peters, Y-T. Lee, K.-J. Grundhoff, H. Schurmann and G. Welsch, in Proc. on Microstructure/Property Relationships in Ti-Aluminides and Alloys, Y.W. Kim and R. Boyer (editors), The Minerals, Metals and Materials Society (1991), pp. 533-48.

19. G. Welsch, I. Weiss, D. Eylon and F.H. Froes, Sixth World Conf. On Titanium, P. Lacombe, R. Tricot and G. Beranger (editors), Soc. Française de Métallurgie, Paris (1988) pp. 1289-93.

20. C.W. Allen, E.A. Ryan and S.T. Ockers, Proc. of 46th Annual El. Micr. Soc. of America, G.W. Baily (editor), San Francisco Press (1988) 806-7.

21. E.F. Rauch, Ph.D. Thesis, McGill University, Montréal, Canada, (1984).

22. S. Nadiv, H.L. Gegel and J. Morgan, "Research to Develop Process Modeling for Producing a Dual Property Titanium Alloy Compressor Disk," AFW-TR-79-4156, AFWAL Materials Laboratory WP-AFB, OH 45433, (1979).

23. M.J. Blackburn and J.C. Williams, Trans TMS-AIME, vol. 239 (1967) 287-88.

24. U. Zwicker, Titan u. Titanlegierungen, Springer Verlag (1974) 102-113.

25. Z. Liu and G. Welsch, Metall. Trans. A, vol.19A (1988) 1121-25.

26. P.G. Shewmon, Diffusion in Solids, McGraw-Hill (1963) Ch.6.

27. R.W. Cahn, in Physical Metallurgy, R.W. Cahn, P. Haasen, editors, North Holland Physics Publ., Chapter 25 (1983) 1595-1671.

28. T.A. Schofield and A.E. Bacon, Acta Metall. vol.9 (1961) 653.

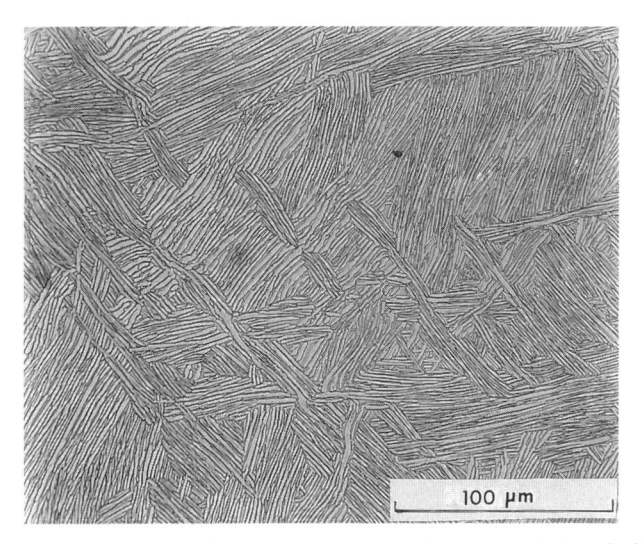

Figure 1 -   Alpha/beta lamella microstructure in beta-annealed and allotropically transformed Ti-6Al-2Sn-4Zr-2Mo alloy

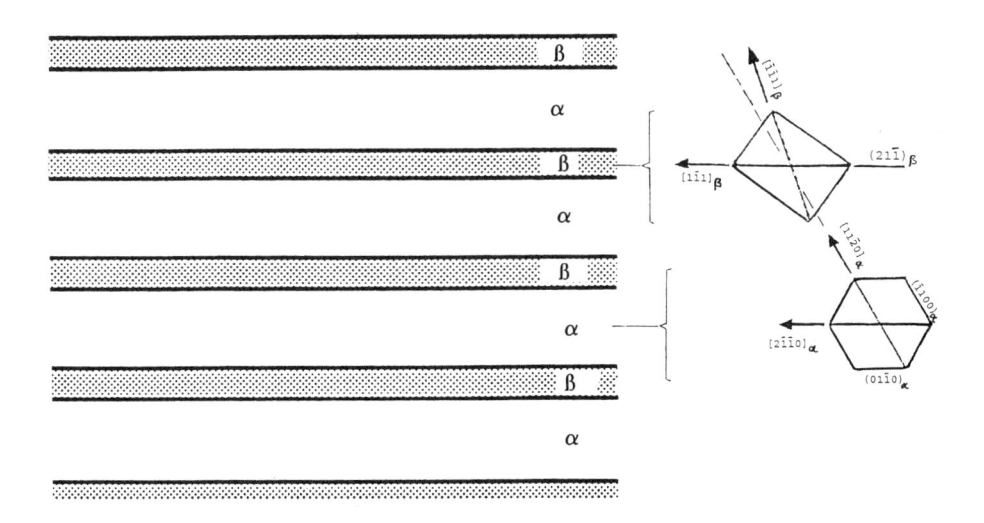

Figure 2 -

Illustration of the Burgers orientation relation between alpha and beta phase in a lamella colony. For the $[011]_\beta$ and $[0001]_\alpha$ zone axes shown, one of the close-packed $<11\bar{2}0>_\alpha$ directions will coincide with either the $[1\bar{1}1]_\beta$ or the $[\bar{1}11]_\beta$ directions. These are also the directions of the Burgers vectors of glide dislocations in the alpha and beta phases.

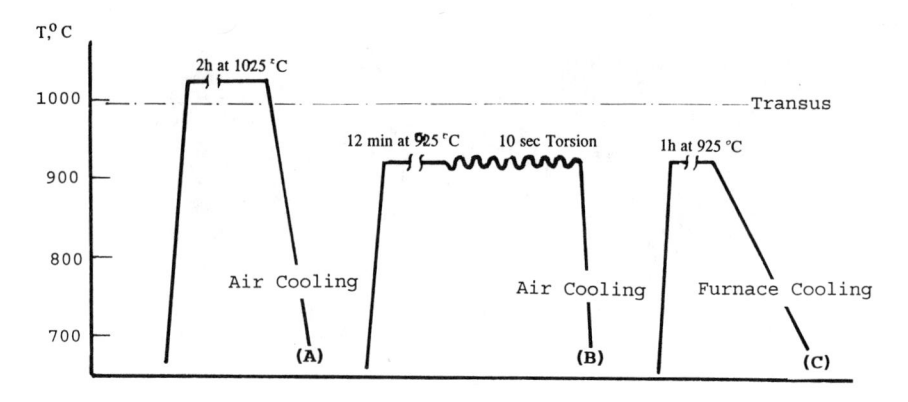

Figure 3 -   Schematic of the thermal-mechanical processing.

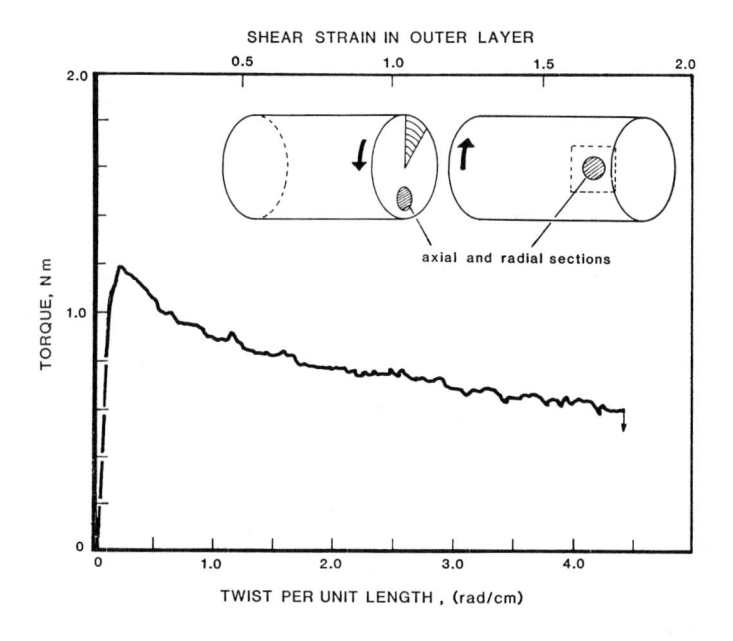

Figure 4 -

Torque versus twist test result from Rauch [16] of hot (925°C) torsion of a 7.9 mm diameter specimen at a shear strain rate of $\dot{\gamma} = 0.173$ s^{-1} in the near-surface layer. The general trend of strain softening is superimposed by small torque increments from localized, time-limited strain hardening. The starting and end points of the torsion-twist curve correspond to conditions (A) and (B) in Figure 3.

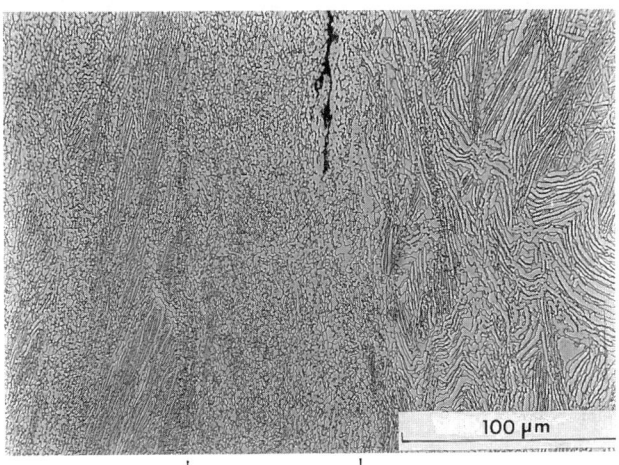

Figure 5 -

Optical micrograph of radial cross section of a torsionally deformed specimen, c.f. Fig. 4. The overall shear strain is approximately 1.5. Equiaxed grains have formed in a shear band. A surface crack (top) has partially penetrated into the same shear band. In regions of relatively uniform deformation the original lamella structure has been distorted but has not fully recrystallized, see details in Figures 6 and 7.

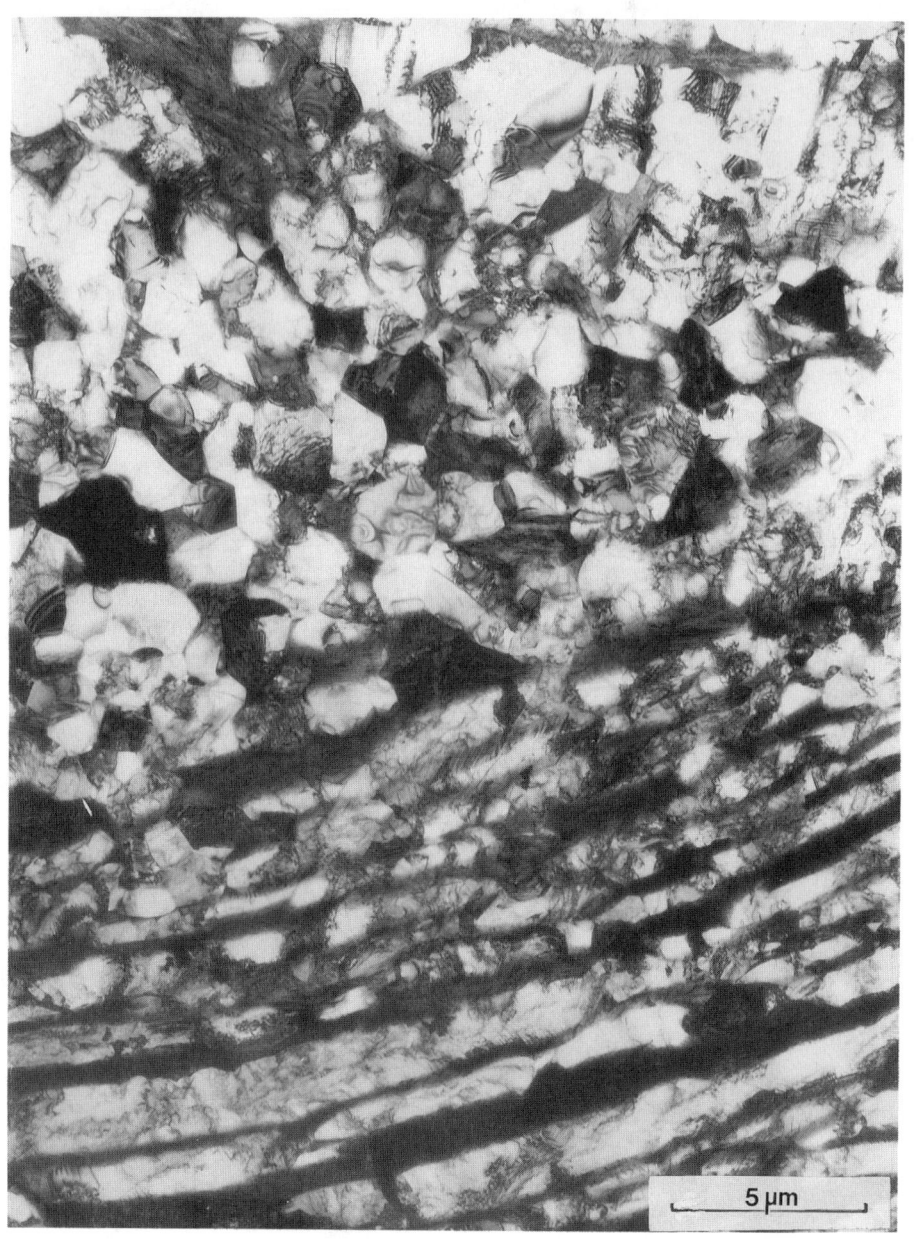

Figure 6 - TEM micrograph of deformed alloy (overall shear strain $\gamma \approx 1.5$. The upper portion of the micrograph shows an area within a shear band. Here, dynamic recrystallization has led to a fully equiaxed microstructure. In the bottom portion of the micrograph the original lamellae are still recognizable, although they have been distorted and are partially recrystallized into equiaxed grains. The TEM was carried out at 1 MeV, at the Argonne Nat'l. Lab. HVEM facility[20].

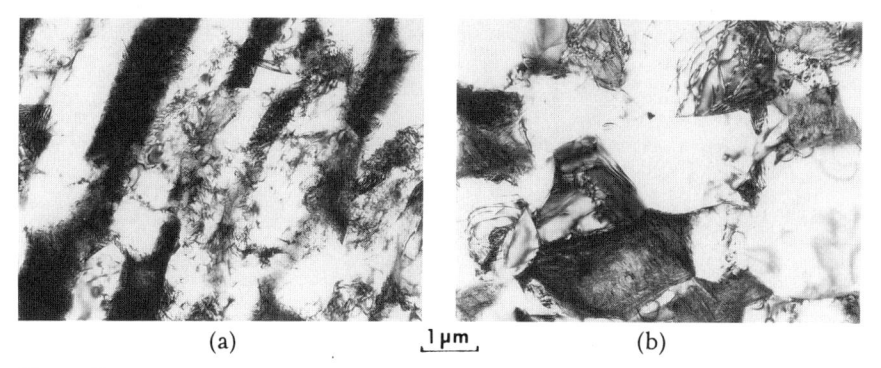

Figure 7 -
TEM micrographs of deformed alloy as in Fig. 6, showing progressive stages of the transformation from the lamellar to the equiaxed grain-structure by dynamic recrystallization.
a) Localized breakup of deformed lamellae which have been partially sheared as well as twisted and bent.
b) Fully transformed microstructure within a shear band.

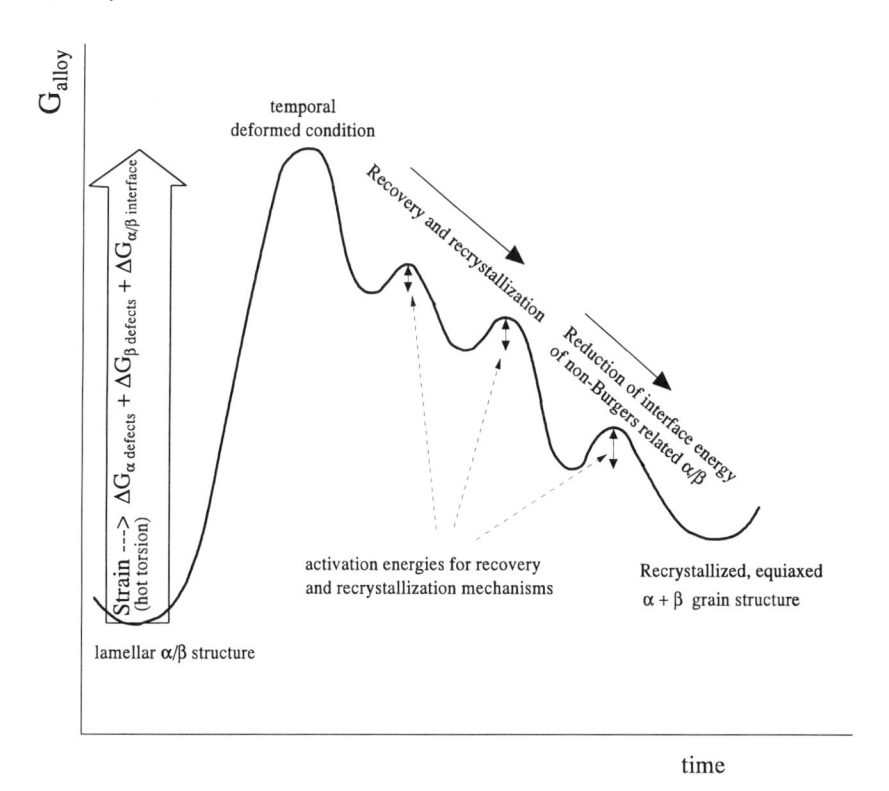

Figure 8 - Illustration of the mechanisms during hot torsion that lead to the transformation of the lamellar α/β to the equiaxed $\alpha+\beta$ grain structure.

INFLUENCE OF THERMOMECHANICAL PROCESSING ON

THE MATERIAL PROPERTIES OF TI-6%AL-4%V/TIC COMPOSITES

P. Wanjara, R.A.L. Drew and S. Yue

McGill University, 3450 University Street
Department of Mining and Metallurgical Engineering
Montreal, Quebec
Canada, H3A 2A7

Abstract

The influence of thermomechanical processing parameters on the flow stress behaviour, consolidation and microstructure of Ti-6%Al-4%V/TiC particulate reinforced metal matrix composites is presented. High temperature deformation processing was performed by compression testing at three constant strain rates in the range of 5×10^{-4} to 5×10^{-3} s^{-1} at various temperatures above and below the beta transus (850°C to 1000°C). Through systematic variations in the deformation processing parameters, the flow stress behaviour of the composites was modelled as a function of the deformation temperature and strain rate. Density measurements and microstructural analysis indicated that thermomechanical processing can reduce the consolidation temperature and time, while simultaneously improving the properties of the composite through elimination of the residual porosity. Furthermore, deformation also leads to modification of the matrix microstructure, offering the potential to control the mechanical properties of the composite.

Introduction

Interest in particle reinforced titanium composites has increased in recent years because these have the advantages of isotropic properties, lower processing costs, and the possibility of microstructural control by means of subsequent forming processes.[1] However, information on the high temperature deformation processing of these materials is still lacking. In the present study, the deformation response of titanium matrix composites reinforced with up to 20 vol.% TiC particles is considered.

Experimental Methods

Materials

Metal matrix composites, containing up to 20 volume percent of TiC particulate (TiC_p) reinforcement, were fabricated using the powder metallurgy technique of powder blending and cold pressing. The Ti-6%Al-4%V powder, used for the matrix, had angular morphology of mean particle size of 60 µm, and the TiC powder was both angular and facetted in appearance, with a mean particle size of 15 µm.[2] Pre-blended compositions of TiC and prealloyed Ti-6%Al-4%V powders were mixed in isopropanol using an attrition mill. After drying, the Ti-6%Al-4%V/TiC_p mixtures were uniaxially pressed at 170 MPa to form compression specimens. To remove the density gradients from axial pressurization, these were subsequently cold isostatically pressed at 210 MPa. These green compacts (height-to-diameter ratio of 0.8) of various Ti-6%Al-4%V/TiC_p mixtures, having an initial density of approximately 65%, were used for the present deformation investigation.

Deformation Technique

All high temperature deformation experiments were performed on a computerized Materials Testing System (MTS 810), adapted for hot compression tests.[3] The thermal cycle and deformation schedule of a single hit deformation test is shown in Figure 1. The compression specimen was placed between the upper and lower compression anvils. Boron nitride and thin sheets of mica were used to reduce the friction between the smooth faces of the specimen and compression anvils. To minimize oxidation, the specimen and anvils were enclosed by a quartz tube sealed with O-rings in which a high purity argon atmosphere was maintained. Each specimen was then heated, at a constant rate of 1.11°C·s⁻¹, to the deformation temperature, and held for 5 minutes to homogenize the temperature in the specimen. The specimen was deformed isothermally at a constant strain rate in the range of 5×10^{-4} to 5×10^{-3} s⁻¹ up to a total strain of 80%. Deformation temperatures were selected above and below the β transus (around 988°C for Ti-6%Al-4%V), ranging from 850°C to 1000°C. The specimen was then cooled to 600°C at a constant rate of 0.5°C·s⁻¹ to ensure uniform transformation conditions for all test schedules.

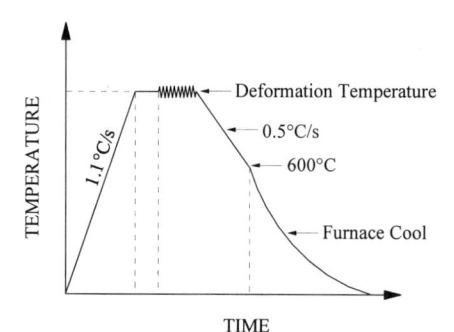

Figure 1. The heating and deformation schedule.

Results and Discussion

Deformation Behaviour

The compressive stress/strain behaviour of the Ti-6%Al-4%V alloy and Ti-6%Al-4%V/TiC$_p$ composites is shown in Figure 2. It should be noted that constant volume deformation (i.e. $v = 0.5$) was assumed even though the porous compacts densify with deformation. The flow stresses of the composites and the matrix alloy both decrease with increasing deformation temperature. This is primarily due to the role of thermally activated dislocation movement as well as increased recovery with increasing temperature. In addition, there is an effect of a phase transformation occurring in Ti-6%Al-4%V at a temperature below 988°C, since the transformation from α to β occurs exponentially with temperature in the following manner:[4]

$$V_{f_\alpha} = 0.925\left\{1 - e^{-0.0085(1261-T)}\right\} \qquad (1)$$

where V_{f_α} is the volume fraction of α at a certain temperature, T [K] and, for a given temperature, β, being bcc, has a lower flow stress than α (hcp).

Hence, plastic deformation below the β transus becomes increasingly difficult, contributing to the much higher flow stress at lower deformation temperatures (i.e. below 988°C). There is also a progressive rise in the flow stress with increasing TiC$_p$ volume fraction. The reasons for the observed difference in flow stress may be explained in two ways: (1) after densification, due to the differential thermal expansion coefficient between the reinforcing particles and the matrix (roughly 2×10^{-6} °C^{-1}), a high dislocation density is generated around the TiC particles which increases the flow stress during subsequent working, (2) the rise in flow stress must be related also to the constraint to plastic deformation caused by the presence of a large volume fraction of rigid TiC particles. Finally there is a continual increase in flow stress even at very high strains, due partly to the fact that the porous compact densifies with further plastic flow. This geometric hardening due to densification is difficult to separate from matrix hardening due to microstructural changes.

Analysis of the effect of strain rate indicates that the flow stress of the matrix alloy and composites increases when the strain rate is increased for all temperatures tested. Figure 3 shows this strain rate effect at 850°C.

Using the information from the stress-strain curves, the temperature and strain rate dependence of the flow stress at a given strain can be modelled using the following creep power law form of the constitutive equation:[5]

$$\dot{\varepsilon} = A\sigma^n e^{-Q/RT} \qquad (2)$$

where A is an empirical constant, Q is the activation energy and R is the universal gas constant. The stress exponent n is the slope of constant temperature points on a plot of log $\dot{\varepsilon}$ versus log σ. If the above expression is followed, the activation energy can be obtained from a plot of log σ versus $1/T$.[6]

On the plots of log σ versus log $\dot{\varepsilon}$ (Fig. 4), the slope, n, was calculated by a least squares regression of each isothermal line. The average values of n for the matrix alloy and composite materials are summarized in Table I. Previous work[6] on these materials indicates that log σ versus log $\dot{\varepsilon}$ plot is not linear over a wider range of strain rates and a modified form of the above creep power law (Eq. 2) can be used to obtain the stress exponent.[5] However, over the range tested the above relationship appears to hold.

Figure 5 shows the relationship between log σ versus 1/T. As this Arrhenius plot is linear, a single activation energy for each material (Table I) was obtained from the average slope, S_{av}, of the constant ε lines using the following relationship:

$$Q = 2.3 \, n \, R \, S_{av} \quad (3)$$

The increasing n value with increasing TiC_p volume fraction suggests that the composite materials are less sensitive to decreasing temperature and increasing strain rate. This agrees with the work of Johnson et al. in their findings of a lower strain rate sensitivity for Ti-6%Al-4%V-TiC_p composites as compared to the unreinforced matrix.[6] The apparent activation energy of flow is slightly higher than that for the self-diffusion of titanium in the β phase (250 kJ·mol[-1]). For this alloy, Briottet et al. reported an activation energy for deformation in the β phase to be 180 kJ·mol[-1] and calculated a value of 220 kJ·mol[-1] for the α phase.[7]

Previous work on this alloy[7] has shown that the activation energy for deformation in the α+β phase region can be substantially higher than that of the individual phases due to the influence of the transformation of α with temperature, and supports the values obtained in this work. In addition, there was a slight increase in the activation energy of deformation with increasing TiC volume fraction. Clearly, the examination of the dependence of Q on the temperature and phases present requires further investigation, over a wider range of strain rates and temperatures.

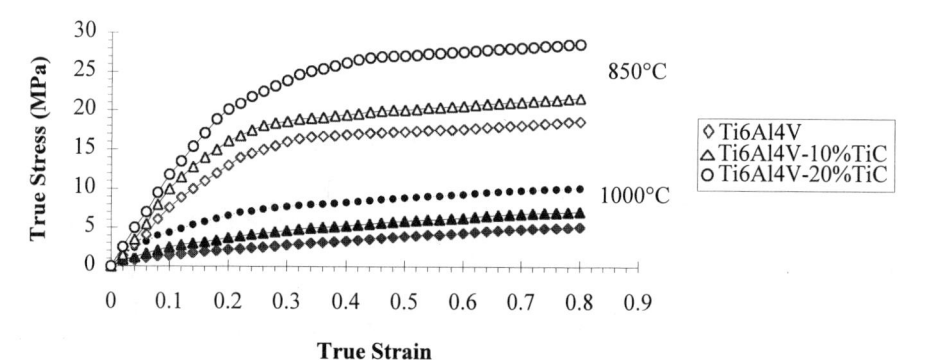

Figure 2. Stress-strain curves for the Ti-6%Al-4%V/TiC_p composites and Ti-6%Al-4%V alloy deformed at a strain rate of 0.0005 s[-1] above and below the β transus temperature.

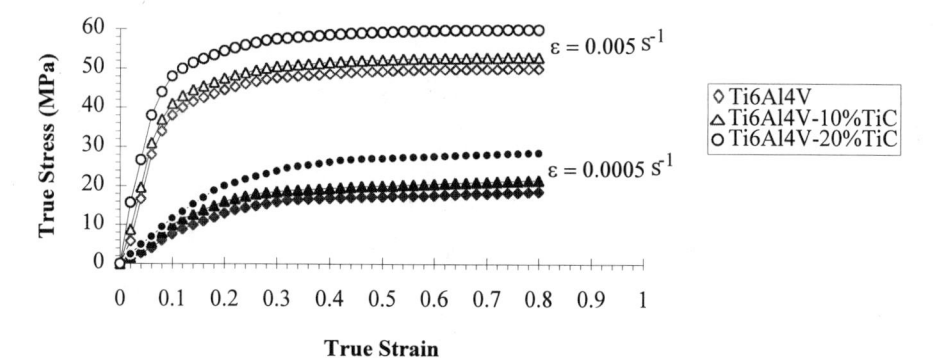

Figure 3. Stress-strain curves at 850°C for the Ti-6%Al-4%V/TiC_p composites and Ti-6%Al-4%V alloy at a strain rate of 0.0005 and 0.005 s[-1].

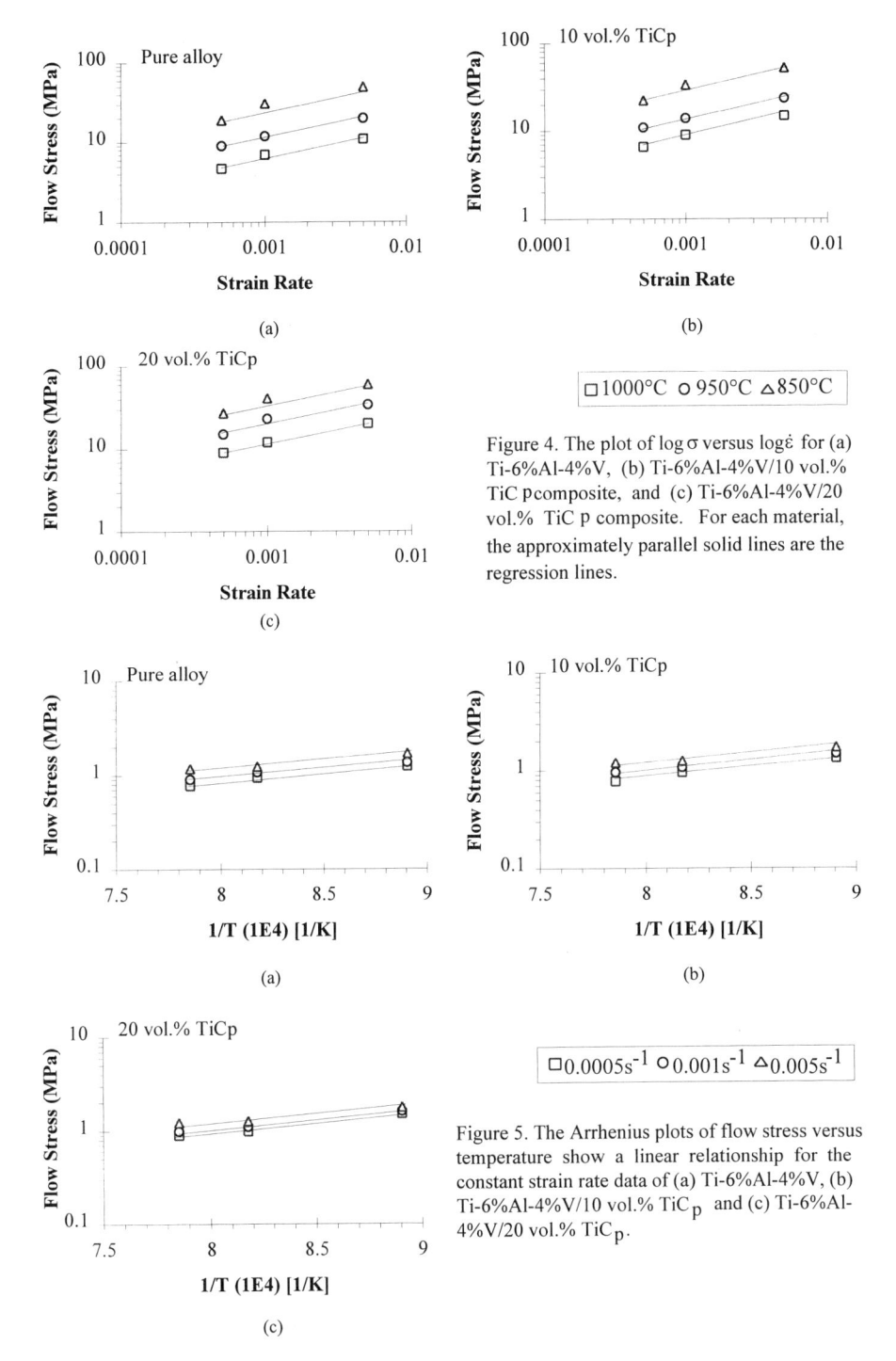

Figure 4. The plot of log σ versus log ἐ for (a) Ti-6%Al-4%V, (b) Ti-6%Al-4%V/10 vol.% TiC p composite, and (c) Ti-6%Al-4%V/20 vol.% TiC p composite. For each material, the approximately parallel solid lines are the regression lines.

□ 1000°C ○ 950°C △ 850°C

□ 0.0005s^{-1} ○ 0.001s^{-1} △ 0.005s^{-1}

Figure 5. The Arrhenius plots of flow stress versus temperature show a linear relationship for the constant strain rate data of (a) Ti-6%Al-4%V, (b) Ti-6%Al-4%V/10 vol.% TiC$_p$ and (c) Ti-6%Al-4%V/20 vol.% TiC$_p$.

Table I. Summary of the stress exponent (n) and activation energy (Q)

Material	n	Q (kJ·mol^{-1})
Ti-6%Al-4%V	2.74	260
Ti-6%Al-4%V/10 vol.% TiC$_p$	2.88	274
Ti-6%Al-4%V/20 vol.% TiC$_p$	2.96	282

Densification Behaviour

Previous work on the fabrication of Ti-6%Al-4%V/TiC$_p$ composites by sintering, alone, indicated that a high sintering temperature (1500°C) and long hold time (4 hours) were necessary to achieve near complete densification.[2] However, at temperatures above 1300°C there is extensive carbon loss from the TiC$_p$ into the matrix. Hence, lower temperature (< 1100°C) deformation assisted-sintering was pursued.

The densities of the deformed materials were determined by the Archimedes method and image analysis. Figure 6 illustrates the effect of temperature and total strain on densification of the composites containing up to 20 vol.% TiC. The curves indicate that the temperatures for near complete densification (> 95%) are substantially lower than for sintering alone (i.e. 1000°C versus 1500°C).[2] Hence the previously encountered problem of carbon loss from the reinforcement is minimized while simultaneously achieving conditions of near complete densification.

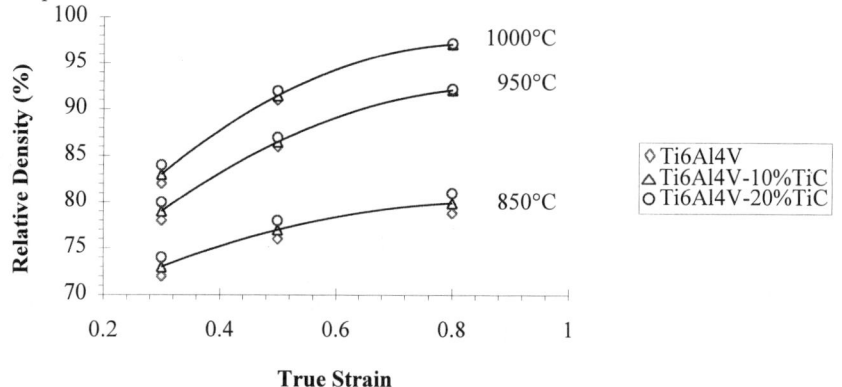

Figure 6. Increase in density as a function of the deformation temperature and total strain.

Microstructural Evolution

Besides enhanced densification, the importance of deformation processing in metallurgy is attributed to the ability to control the microstructure and properties by means of the deformation temperature, strain and strain rate. Preliminary investigation of the influence of the deformation temperature on the matrix alloy microstructure is shown in Figure 7. Deformation at 1000°C occurs in the single phase β region. Transformation of the deformed β grains upon cooling (<988°C) should produce α having an acicular or lamellar morphology, as observed in Figure 7a. The alternative microstructure for this alloy is the equiaxed structure which can only be formed by work hardening the lamellar structure (plastically deforming below the β transus) followed by recrystallization of the microstructure. The wavy nature of some of the lamellar α platelets (theoretically 64 vol.% of them, according to Eq. 1) in Figure 7b is probably indicative of work hardening, and heat treatment of this microstructure could lead to recrystallization.

The presence of the TiC$_p$, in the case of composite materials, leads to inhomogeneities in the straining of the matrix. During working, the dislocations generated at the interface between the non-deformable particles and the deforming matrix give rise to the formation of zones of relatively high dislocation density. The size and shape of the deformation zones are related to the size and shape of reinforcement particles, with angular morphologies (such as the TiC particles) leading to enhanced dislocation densities in the region closest to their stress-concentrating corners.[8] Also, since the matrix strain around the rigid particles is accommodated by diffusive plastic relaxation, the size of the deformation zone will decrease with increasing deformation temperature and decreasing strain rate. In other words, regions in the proximity of the particles lead to relatively high dislocation densities, which may recrystallize at the deformation temperature.

Observation of the effect of deformation on the microstructure of the Ti-6%Al-4%V/TiC$_p$ composite is shown in Figure 8 for temperatures above and below the beta transus. At 1000°C (Fig. 8a), the matrix microstructure appears to be similar to that of the unreinforced matrix alloy (i.e. lamellar morphology of α platelets with interplate β). However, the microstructure of the composite deformed at 850°C (Fig. 8b) differs significantly from the unreinforced matrix. The wavy lamellar α observed for the unreinforced matrix does not appear in the correspondingly processed composite. Instead, the α grains are randomly oriented and have a smaller aspect ratio. It is suggested that at 850°C, the presence of the angular TiC particles, through the formation of deformation zones of high dislocation density, may cause particle induced recrystallization, leading to a more equiaxed structure (Fig. 8b).

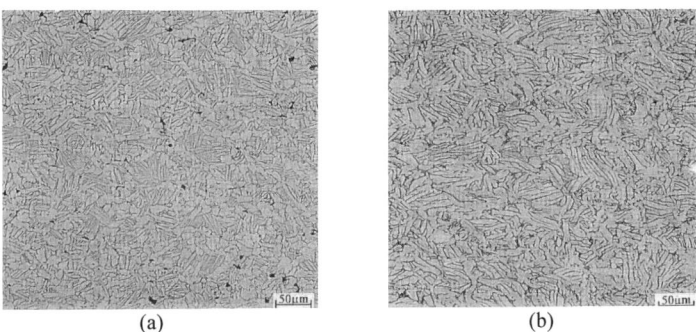

Figure 7. Optical micrographs of etched Ti-6%Al-4%V deformed at a strain rate of 0.0005 s^{-1} to a strain of 0.8 at (a) 1000°C and (b) 850°C.

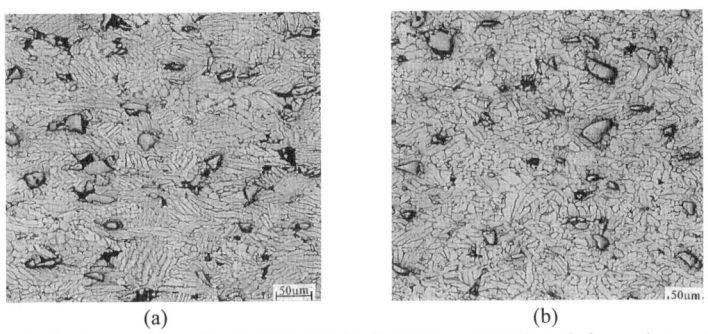

Figure 8. Optical micrographs of etched Ti-6%Al-4%V/10 vol.% TiC$_p$ deformed at a strain rate of 0.0005 s^{-1} to a strain of 0.8 at (a) 1000°C and (b) 850°C.

In general, the microstructures of the matrix and composite indicate that mechanical working has almost fully consolidated the porous compact, with the relative density being above 95% at a deformation temperature of 1000°C. For the composites, it has also improved the homogeneity of the material by breaking up clusters of reinforcement,[2] although these remain aligned in the working direction.

Conclusions

Investigation of the stress-strain behaviour of the matrix and composite materials indicated that the flow stress increased with increasing reinforcement volume fraction, strain and strain rate and decreasing temperature. Moreover, the deformation of the composite materials was found to be less sensitive to strain rate and temperature than the unreinforced matrix.

Analysis of the densification behaviour indicated that deformation processing can achieve near complete densification at temperatures at which sintering would not be effective. Hot deformation also modifies the matrix microstructure, especially in the presence of the TiC reinforcement particles.

Acknowledgements

This work was supported by a scholarship and research grants from the National Science and Engineering Research Council of Canada and the McGill Metals Processing Centre.

References

1. T.W. Clyne and H.M. Flower, "Titanium Based Composites," Titanium '92 Science and Technology, vol. 3, ed. F.H. Froes and I.L. Caplan (Warrendale, PA: TMS, 1993), 2467-2478.

2. P. Wanjara, R.A.L. Drew and S. Yue, "Characterization of Ti-6%Al-4%V/TiC MMCs Consolidated by Powder Metallurgy Processing," Presented at the Eighth World Conference on Titanium, Birmingham, U.K., October 22-26, 1995, in press.

3. P. Wanjara, R.A.L. Drew and S. Yue, "Influence of Thermomechanical Processing on As-Consolidated Properties of Ti-6%Al-4%V/TiC MMCs," Presented at the Eighth World Conference on Titanium, Birmingham, U.K., October 22-26, 1995, in press.

4. R. Castro and L. Seraphin, "Contribution à l'étude métallographique et structurale de l'alliage de titane TA6V," Mémoires et Etudes Scientifiques de la Revue de Métallurgie, 63 (1966), 1025-1058.

5. G.E. Dieter, Mechanical Metallurgy, Third Edition, (New York, NY: McGraw Hill Book Company, 1986), 306-307.

6. T.P. Johnson, M.H. Loretto, M.J. Walker and M.W. Kearns, "The Forging Characteristics of Two Ti-6Al-4V-Based Particulate Composites," Titanium '92 Science and Technology, vol. 3, ed. F.H. Froes and I.L. Caplan (Warrendale, PA: TMS, 1993), 2585-2592.

7. L. Briottet, J.J. Jonas and F. Montheillet, "A Mechanical Interpretation of the Activation Energy of High Temperature Deformation in Two Phase Materials," Acta Metall. Mater., in press.

8. F.J. Humpreys, "The Deformation Structure and the Recrystallization Behaviour of Two-Phase Alloys," Metallurgical Forum , 12 (1978), 123-135.

PRINCIPLES OF TITANIUM ALLOYS' STRUCTURE CONTROL WITH THE PURPOSE OF INCREASING THEIR MECHANICAL PROPERTIES

M. Brun, G. Shachanova

All - Russia Institute of Light Alloys
2, Gorbunov St. 121596, Moscow, Russia

Abstract

This paper discusses the general rules of $\alpha + \beta$ - Titanium alloys' structure formation during hot processing. They are presented in the form of diagrams describing the dependence of various structure parameters on deformation and heat treatment regimes. Such diagrams are used in industry for the development of semiproducts with regulated structure manufacture technology.

Introduction

For titanium alloys typical is a wide diversity of structures which is caused by the presence of several phases, differences in their morphology, a wide range of changes in volume fractions and sizes of various structural constituents. The diversity of these alloys' structure (especially of $\alpha + \beta$ - alloys) determines a significant change in their mechanical properties. Another important aspect of the " structure - properties " problem for titanium alloys is not the single influence of structure's type and parameters on different mechanical properties. Both factors require a preferable structure selection taking into account the specific component's service conditions.

This paper is devoted to the consideration of the general rules of titanium alloys structure formation under deformation and heat treatment, as a scientific basis for the designing of the optimum semiproducts structure. The same problem has been studied in many other researches[1-5], etc.. However, in contrast to the most of them in this research are established the quantitative dependencies of structure parameters on practically all processing regimes. Namely this is the most systematic way to understand and control structure change.

The data presented have been obtain on laboratory samples of the typical $\alpha + \beta$ - BT 8 alloy (Ti - 6.4 Al - 3.3 Mo - 0.27 Si) which were upset (in isothermal conditions) and annealed at various regimes. Similar investigations were carried out on the most industrial α and $\alpha + \beta$ - alloys. Besides, the generalisation thus obtained were tested in real processes of semiproducts manufacture.

As a rule these processes present a multy - stage deformation and a final heat treatment. But at the end they all come to formation of lamellar or globular structures.

1. General rules of lamellar structure formation

The main parameters of a lamellar structure determining its diversity are the value of initial and recrystallized β - grains (D_i, D_r), the degree of their inequiaxiality and recrystallization (μ, λ),the size of α - colonies (d), the thickness of α - lamellae (primary and secondary) and β - interlayers (b_1, b_2, b), the volume fraction of various α - and β - phase modifications (γ)[6]. The totality of the quantitative values of these parameters gives a complete conception about the concrete structure.

Titanium alloys have a lamellar structure in a cast state and after deformation or heat treatment in β - field. Deformation in β - field is used to refine coarse - grained cast stucture and heat treatment to obtain specific β - grains size. It is well known that for the refining of coarse - grained structure are necessary those conditions of deformation or / and annealing which favour the development of recrystallization and retard the growth of recrystallized grains.

At the usual deformation speeds of industrial processes recrystallization in titanium alloys develops during cooling after deformation (at the temperature range T_d - T_β) or during the following β - heating of the deformed material and depends on the deformation temperature (T_d), degree (ε) and speed ($\dot{\varepsilon}$), on the cooling speed (V_c), annealing temperature and time (T_a, τ), initial structure (D_β). Experimental results for the recrystallization kinetic are summarised in the form of recrystallization diagrams constructed in coordinates of the various treatment regimes and D_i values. Examples of such diagrams are shown in Fig. 1. Each diagram contains 2 systems of isolines: λ - constant and D_r - constant. Accordingly each point of the diagrams fully describes a grain structure of alloy at regimes given by point coordinates. For practical use the best are the diagrams in coordinates T_d - ε with isolines of $\tau = \tau^*$(Fig. 1c) at what the most fine - grained fully recrystallized structure ($\lambda = 1$) can be ensured (Fig. 2).

At the time τ^* size of β - grains is determined by relationship $D_r = D_i (1 - \varepsilon)$, that depends only on the deformation degree and initial structure. At the time $\tau < \tau^*$ structure is only partly recrystallized and accordingly is not fine enough. At $\tau > \tau^*$ structure is fully recrystallized but β - grains size is larger than D^* and can be coarser even than D_i.

The analysis of diagrams and other data shows that in terms of qualitative aspect the influence of deformation and annealing parameters on titanium alloys recrystallization is the same as in other metals: recrystallization degree is the greater the higher is T_d, ε and $\dot{\varepsilon}$, the lower is V_c and the finer is D_i. At the

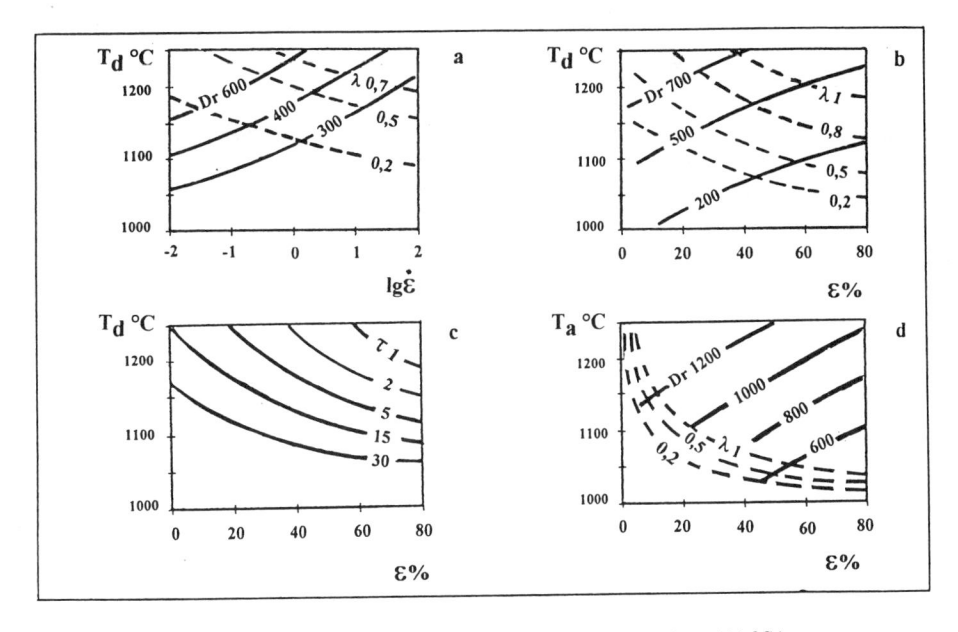

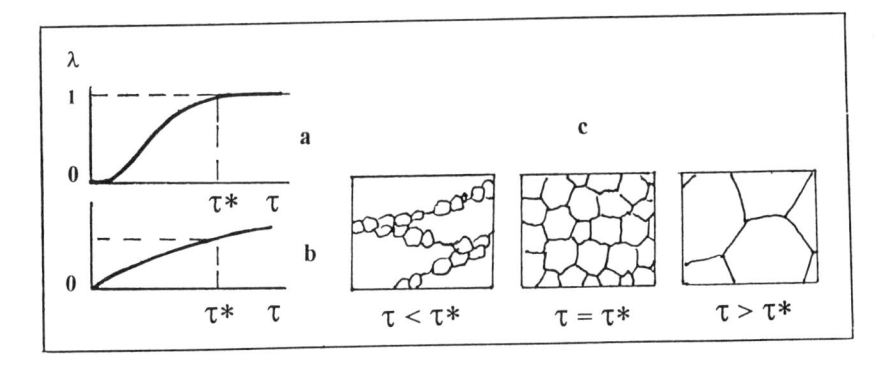

Fig. 1 Recrystallization diagrams of BT8 alloy (D_i = 1000 μm, $T_β$ = 980 °C).
a - T_d - lg $\dot{ε}$ at $ε$ = 40 %, $τ_d$ = 40 sec; b - T_d - $ε$ at $\dot{ε}$ = 10^2 sec^{-1}, $τ_d$ = 40 sec;
c - T_d - $ε$ at $\dot{ε}$ = 10^2 sec^{-1}, $τ_d$ = $τ^*$; d - T_a - $ε$ at $\dot{ε}$ = 10^2 sec^{-1}, T_d = 1100 °C,
$τ$ = 1 hour.

Fig. 2 Kinetic of $λ$ (a) and D_r (b) change and scheme of recrystallization development at $β$-treatment (c).

same time the recrystallization development in these alloys has some important features: 1 - At moderate $ε$ ($≤$ 40 - 50 %) and relatively low T_d ($≤ T_β$ + 100°C) a recrystallization develops inertly especially at low $\dot{ε}$ (at hydraulic presses). This is explained by the fact that only the initial $β$ - grains boundaries are the places of the recrystallized grains' nucleation (Fig. 2c). For a full recrystallization these new grains have to grow into their volume i. e. to overcome the distance of the deformed grains width; 2 - At such deformation conditions recrystallization is intensified significantly by preliminary $\alpha + \beta$ - deformation ($T_d \leq T_\beta$ - 20 - 30 °C, $\varepsilon \geq 5$ - 10 %), which promotes to nucleation of new grains in intrargranular volumes of the initial β - grains; 3 - At high ε and $\dot{\varepsilon}$ of one - orientated deformation the texture retarding of recrystallization is possible (in the case of sharp texture formation with low disorientation between neighbour initial β - grains impeding the formation of new grains); 4 - For the growth of recrystallized grains high speeds and nonuniformity are typical. In order to overcome these unfavourable features of titanium alloys' recrystallization in industrial β - processing mentioned below technological methods are used.

Unlike the parameters of grain structure, the parameters of intragranular structure are mainly regulated during the final heat teatment. To control them special diagrams are used (Fig. 3). Such diagrams describe the quantitative changes of the stated above stucture parameters in a wide range of the heat treatment regimes changes.

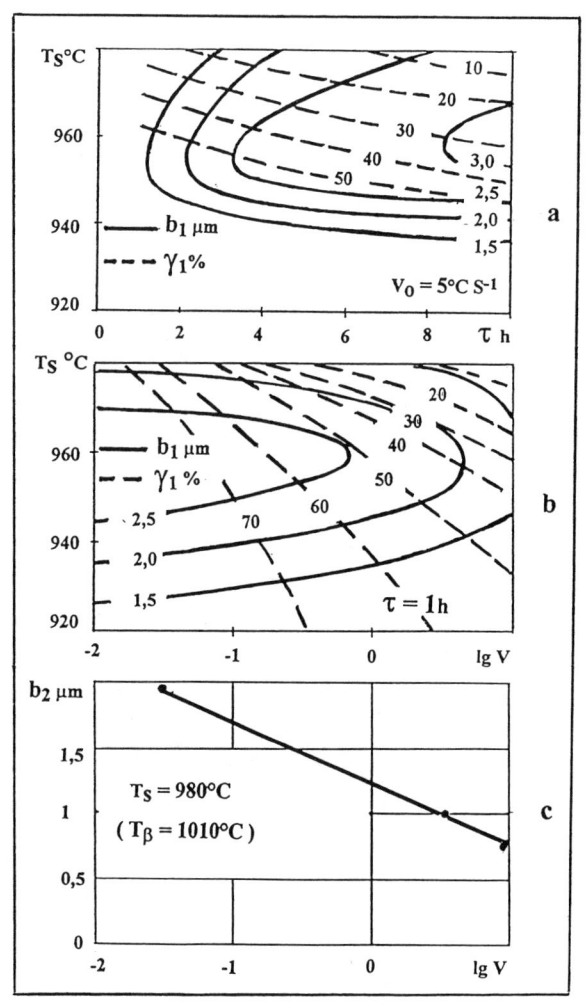

Fig. 3 Diagrams of BT8 alloy structure's parameters change at heat treatment

2. General rules of globular structure formation

The main parameters of this structure are the degree of globularization and anizotropy of α - lamellae (C, K_α), the value of β - grains (D_β), the size of globular α - particles (b_1), the thickness of secondary α - lamellae (b_2) and β - interlayers (b), the volume fraction of various α and β - phases modifications $(\gamma)^6$. The structure of globular type is formed during deformation and annealing at $\alpha + \beta$ - field temperatures. In this case both phases (α and β) participate in the deformation, and in both phases processes of recrystallization and poligonization can be developed. The initial structure of α and β - phases before $\alpha + \beta$.- deformation is always a lamellar one. In a result of deformation, poligonization, recrystallization and initiating by them spheroidization a lamellar structure changes into a globular structure (Fig. 4). The value C depends on all parameters of deformation and heat treatment regimes. The examples of such dependences are presented in Fig. 5.

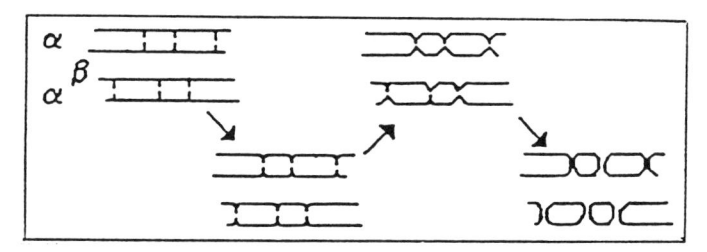

Fig. 4 The scheme of lamellar structure transformation to globular structure

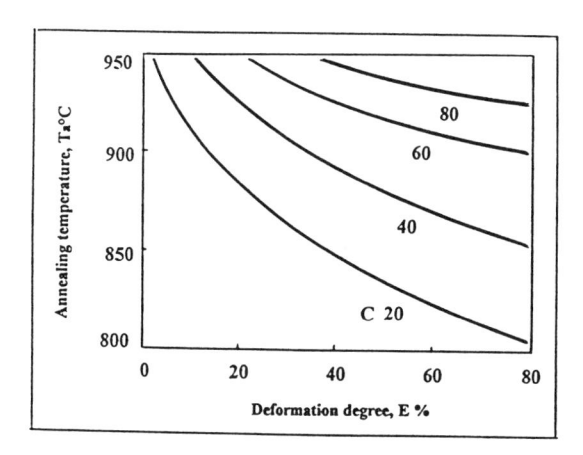

Fig. 5 Globularization diagram of BT8 alloy
($D_i = 2000$ μm, $b_1 = 3$ μm, $\dot{\varepsilon} = 10^0$ sec^{-1}, $\tau = 1$ h)

Their analysis together with the data of TEM reseaches have allowed to determine that for the active globularization development it is necessary to create the following conditions: 1 - The rather high ε (≥ 40 - 50 %) - for the most strongly deformed α - lamellae, which are parallel to the impressed stresses, and ≥ 70 -

80 % - for the least deformed perpendiculary oriented α - lamellae), how much possible low T_d and high T_α, that favours the development of polygonization, recrystallization and spheroidization, 2 - The thickness of of α - lamellae and β - interlayers in initial structure is not less than the equilibrium (at given regimes of deformation and heat annealing) sizes of subgrains or recrystallized grains (C_α, C_β), i.e. $K_\alpha = b_1 / C_\alpha \geq 1$, $K_\beta = b / C_\beta \geq 1$ (condition of recrystallization); 3 - Monolayer disposition of recrystallized grains or poligonized subgrains in lamellae and interlayers ($K_\alpha = 1$, $K_\beta = 1$) (condition of spheroidization). In thick lamellaes with multilayer disposition of grains (subgrains) the spheroidization does not develops.

The most perfect globular structure forms if the above conditions are realized similtaneously for α and β - phases. Otherwise transition structures with a different degree of lamellae' and interlayers' globularization form. This influence of the initial structure at the development of recrystallization, polygonization and spheroidization processes is shown by the scheme in Fig. 6. The states **A**, **B** and **C** differ for the thickness of the lamellae (interlayers) : $b_c < b_\alpha < b_\beta$. The sizes of C_α and C_β do not depend on the structure. They are determined only by the alloy's composition and by deformation and annealing regimes. In this case the character of K_α and K_β change depending on the annealing temperature corresponds to what shown in the scheme. On the basis of the above stated it is clear that at the same temperature in alloy with states **A**, **B** and **C** the various processes will be realized. The most perfect globular structure will form only in alloy with state **A** at a temperature T_2 in which $K_\alpha = 1$ and $K_\beta = 1$. For this reason the success of the globular structure formation depends not only on processing regimes but also on the preparing of the initial lamellae structure.

The final formation of globular structure parameters occurs during heat treatment and depends on the heating temperature and time, and also on the cooling speed. The regulation of these parameters is realized on the basis of diagrams which are similar to the ones shown in Fig. 3.

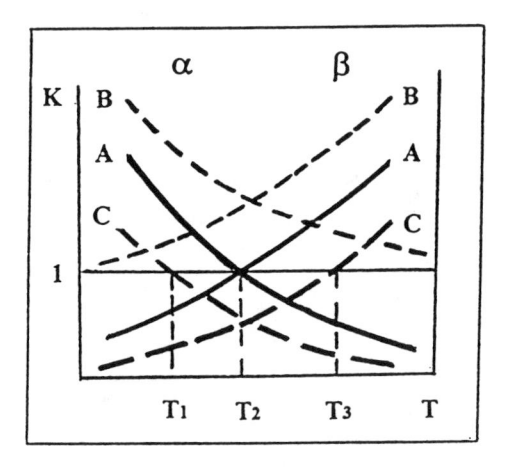

Fig. 6 The influence of initial structure at the processes of polygonization, recrystallization and spheroidization at annealing.

3. Technological methods of lamellar and globular structure regulation

Development of these methods is based on the use of the mentioned above quantitative dependences of structure's parameters on the deformation and heat treatment regimes. One of the main difficulty of optimum regimes realization at producing of real semiproducts (especially forgings) is connected with a zone nonuniformity of $\varepsilon, \varepsilon, T_d$ and V_c virtual values in their volume. It means that it is necessary to use such regimes which can garantee an uniform regulated structure in nonuniform processing's conditions.

In accordance with experimental data the influence of last three parameters' differences on β - recrystallization and $\alpha + \beta$ - globularization processes is decreased with increase of deformation degree. Therefore, the most important demand at β and $\alpha + \beta$ - deformation is ensuring of required ε value in all semiproduct volume (taking into account all other parameters).

At β - deformation it is achieved by use of the following methods: 1 - Combination of various deformation modes or regimes allowing to deform separately surface and deep zones of billet; 2 - Use of the special intermediate dies promoting to more efficient deformation of billet surface zones; 3 - Application for a billet of the so - called all - sided forging process including some operations of upsetting and drawing; 4 - Carrying out before final β - processing of $\alpha + \beta$ - deformation which even at very low degree ($\geq 5 - 10 \%$) promotes effectively to development of recrystallization.

At $\alpha + \beta$ - deformation ensuring of necessary summary reduction is achieved by the following methods: 1 - Carrying out of such deformation at the stage of an initial billet manufacture with creation of conditions for deformation of the whole billet volume; 2 - Combination of various deformation schemes at producing of billets and final semiproducts (for example, usial and radial - shearing rolling of bars or rolling of rings with the following die - forging of disks , etc) what due to change of deformation direction ensures the more favourable conditions for deformation and globularization of differently oriented α - lamellae; 3 - Using of multy - passage die - forging with application of the preliminary dies designing by method of computer simulation and creating conditions of the more uniform deformation in various zones of forgings. At the same time all other methods promoting to development of globularization are used. Among them there are preparing of initial grain and intragrain structure, choice of favourable deformation speed (for example β - forging on a hammer and $\alpha + \beta$ - forging on a press), deformation and annealing temperature, cooling speed (depending on alloys composition), etc.

All mentioned above measures allow to control by the processes of billets' and final semiproducts' structure formation and to ensure the necessary structure.

Conclusions

Structure is one of the major elements of $\alpha + \beta$ - alloys semiproducts' quality. The formation of the required structure and selection of the most efficient technological regimes of deformation and heat treatment can be based only on the processing - structure quantitative relationships.

References

1 - R. I. Jaffee," Overview on Titanium Development and Application ", Proceedings of the 4 th International Conference on Titanium, Kyoto, (1980),53 - 74.

2 - C. C. Chen, " Processing - Structure - Property Relationships in Commercial Titanium Alloys ", ",Proceedings of the 5 th International Conference on Titanium, Munich, (1984), 461 - 468.

3 - R. Tricot, " Termomechanical Processing of Titanium Alloys ", Proceedings of the 6 th International Conference on Titanium, Cannes, (1992), 23 - 26.

4 - M. Brun, G. Shakhanova, O. Ukolova, " The Relationship Inherent in the Formation of a Granular Structure in Titanium Alloys Subjected to Working and Heat treatment in β - field ", Titanium, 1995, 5, 37 - 42.

5 - M. Brun, G. Shakhanova, " Demands to Bimodal Structure with Optimal Combinations of Mechanical Properties and Regimes of Its Development ", Proceedings of the 8 th International Conference on Titanium Alloys, Birmingham, (1995), (in print 9).

6 - M. Brun, G. Shakhanova, " Titanium Alloys Structure and Parameters Defining Its Diversity ", Titanium, Moscow, (1993), 1, 24 - 29.

HOT WORKING OF CP TITANIUM

K. Ray, W.J. Poole, A. Mitchell and E.B. Hawbolt

Department of Metals and Materials Engineering
University of British Columbia
Vancouver, B.C., Canada V6T 1Z4

Abstract

A major source of concern in the breakdown of as-cast titanium ingots is the occurrence of surface cracks. Traditionally, the breakdown of as-cast ingots has been done by open-die forging, but recently there has been an interest in direct hot-rolling of the ingots. Constitutive equations for the hot working of commercially pure (C.P.) titanium in the alpha and the beta phase (700-950°C) have been developed by conducting axisymmetric compression tests on cylindrical specimens using a Gleeble 1500®* for strain rates appropriate for direct hot-rolling. A model of the breakdown rolling using DEFORM®**, a finite element software package, has been studied to identify the critical processing parameters.

---

* This is a registered product name for a thermomechanical simulator marketed by Dynamic Systems Inc., Troy, New York.
** This is a registered product name for a software designed by SFTC Inc., Columbus, Ohio.

Introduction

The breakdown of as-cast vacuum-arc remelted (VAR) and electron beam melted (EBM) titanium ingots into slabs and billets is traditionally done by open-die forging at relatively low strain-rates. The titanium industry has recently become interested in breakdown rolling instead of forging, since it increases the rate of production and promises better microstructural refinement of the as-cast structure. Titanium producers, lacking in-house facilities for breakdown rolling, typically roll ingots in mills normally used for steel. Before rolling, these ingots are soaked in soaking pits normally used for steels at temperatures between 1000-1100°C for 6-18 hours. It is desired that the ingots be initially rolled at a temperature in the beta phase, (i.e., above 882°C) and cool to the alpha phase during the course of rolling, to obtain the best combination of strength, microstructural properties, and processing ease. The soaking temperatures and heat-treating regimes vary from one company to another, and in most cases the procedures are proprietary. Problems can arise because cracks appear on the surface of the ingots during the initial rolling passes.

One of the objectives of this work was to develop a constitutive equation for the hot working of commercially pure titanium for processing conditions similar to those experienced during direct rolling. Another objective was to determine the effect of air-annealing, similar to that encountered at the surface during direct rolling, on the hot rolling properties of the CP titanium.

Experimental

Axisymmetric compression tests were conducted on cylindrical samples 10 mm in diameter and 15 mm in height from an extruded Grade II CP titanium rod having a composition (wt.%) of 0.11% O, 0.06% Fe, 0.01% H, 0.01% C, and the balance Ti. The effect of annealing in air was examined by heat treating the samples in air for 24 hours at 1100°C followed by air-cooling and compression testing. The aim was to develop an oxygen-rich solid solution layer on the specimen, and to test its susceptibility to cracking during deformation. After annealing, the oxide layer was removed, a platinum-platinum/10% rhodium thermocouple was welded to the mid-central surface of the sample for temperature measurement and control during deformation.

The samples were then deformed in axisymmetric compression to a strain of 1 at strain rates of 1 and $10s^{-1}$ at temperatures between 750°C and 950°C using the Gleeble 1500®. Tantalum foils were used at both ends of the sample, to prevent welding to the deformation anvils and to minimize friction during deformation.

Samples were also deformed in the as-received condition. These samples had an additional thermocouple at the mid-centre of the specimen inserted through a drilled hole. These samples served as control specimens to determine the presence of any temperature gradients, and to provide additional hot-working data for developing the constitutive response of CP Ti.

The finite-element software package DEFORM® was used to simulate the rolling process using material data available in the DEFORM Material Data-base and the constitutive behaviour of the titanium. The effect of the processing parameters, strain, strain rate and temperature has been examined.

Results

1) Constitutive Equations - The Alpha Regime

For the processing regime between 750°C and 875°C and strain rates between 1 and 10s^{-1}, the flow stress curves are shown in Figure 1. The flow stress achieves a steady state value at strains of approximately 0.1 when the strain rate is 1s^{-1}. For samples tested at strain rates of 10 s^{-1}, a steady state is reached for strains greater than 0.25. Note that the temperatures recorded in Figure 1 represent surface temperatures. The inner thermocouple recorded a higher temperature, the difference varying between 20-40°C due primarily to the thermal gradient. The measured stresses pertained to the power-law breakdown regime, as shown on the deformation-mechanism map for titanium [1]. In this regime, the steady-state stress values can be represented by a strain independent exponential law [2] of the form,

$$\dot{\varepsilon} = A\exp(n'\sigma)\exp\left(\frac{-Q}{RT}\right) \tag{1}$$

For the measured steady-state flow stresses of alpha C.P. titanium in our tests, the equation has the form,

$$\dot{\varepsilon} = 4.2 \times 10^6 \exp(0.1\sigma)\exp\left(\frac{-216500}{RT}\right) \tag{2}$$

The activation energy, Q, resulting from the analysis was similar to that reported for lattice controlled diffusion, 242 kJmol^{-1} [1].

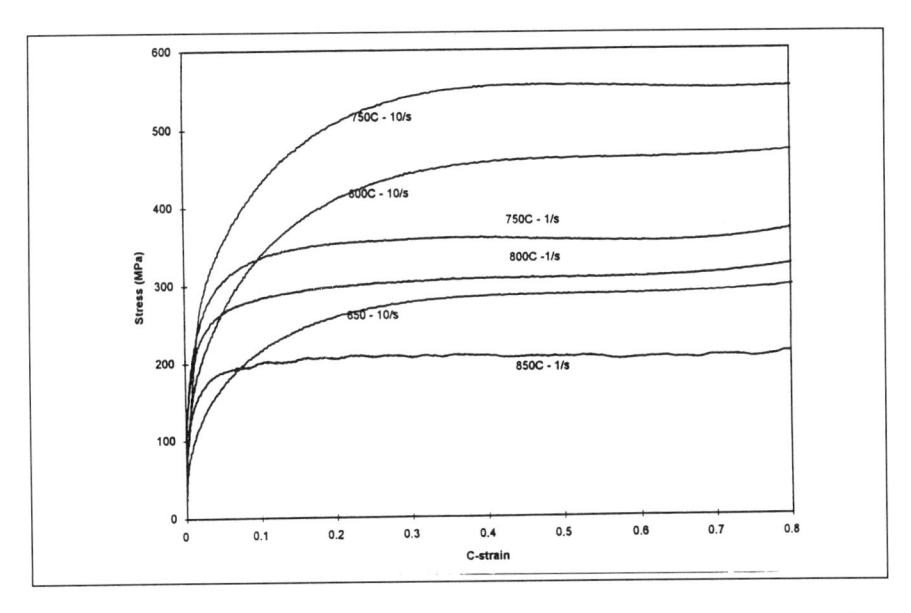

Figure 1: Flow Stress Curves for Alpha Titanium obtained by Axisymmetric Compression Testing.

The Beta Regime

The flow stress curves obtained for samples tested in the beta phase are shown in Figure 2. In this case, the overall flow stress levels were considerably lower than those obtained for the alpha phase and the material also showed an increased rate of strain-hardening, as compared to curves obtained in the alpha regime. Hence, for the beta phase, the flow stress values were calculated at different strains, using equation (1) in a similar manner to that reported by Rao & Hawbolt [2]. The Q, n and A values in Table I were calculated independently and at each fixed strain. The Q values in this case do not have physical significance, but also describe the form of the stress-strain data.

Table I

Strain	Q (kJmol^{-1})	n	A
0.1	1188	0.48	7.52x10^{41}
0.2	613	0.20	7.42x10^{21}
0.3	434	0.14	1.96x10^{15}
0.4	408	0.12	2.2x10^{14}
0.5	305	0.11	1.83x10^{10}
0.6	217	0.10	4.22x10^{6}

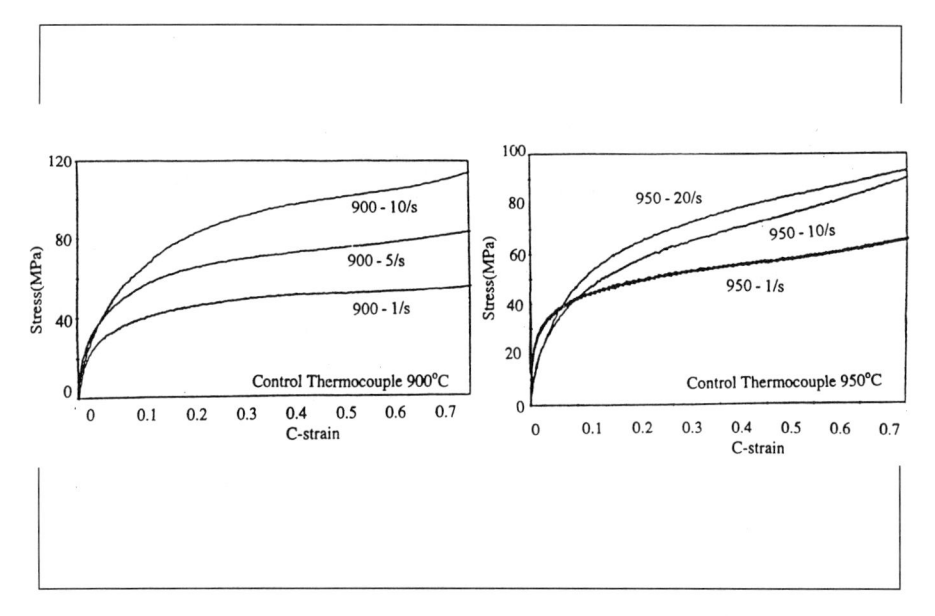

Figure 2: Flow Stress Curves for Beta Titanium obtained by Axisymmetric Compression Testing.

2) Modelling Results

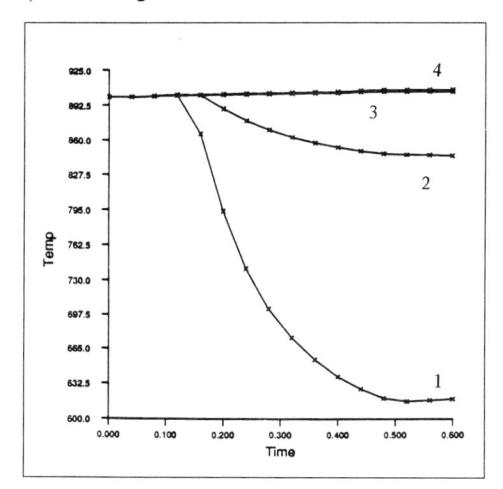

Figure 3: Temperature vs. Time in the Roll Gap for Positions 1,2,3 and 4, Showing the Chilling Effect.

Using the finite element software DEFORM®, the thermal and deformation conditions in the slab were examined. Figure 3 shows the model results for the temperature as a function of time for four position in the slab, i.e. (1) was at the surface of the slab; (2) was 10 mm below the surface; (3) was midway to the centre of the slab (3) and (4) was at the centre of the slab. The results indicate that a roll-chilled layer extends to about 10 mm below the surface. Figure 4 and 5 also illustrate that the strains and strain rates experienced at the surface are larger than those at the centre.

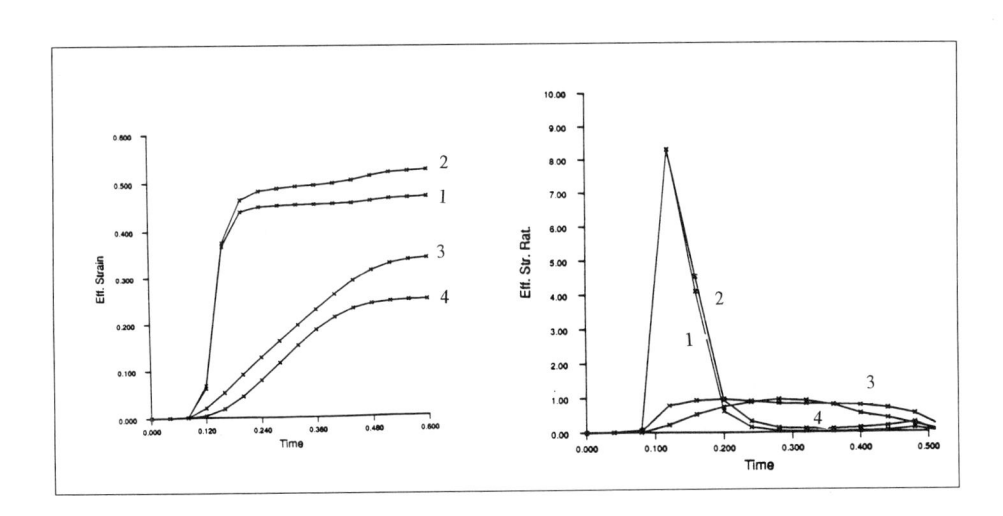

Figure 4: Effective Strain vs. Time in the Roll Gap at Positions 1, 2, 3 and 4.

Figure 5: Effective Strain Rate vs. Time in the Roll-Gap at Positions 1,2, 3 and 4.

3) Deformation of the Air-Annealed Samples

When samples which had been previously air-annealed (i.e. 24 hours at 1100 °C) were compressed at strain rates of 1-10 s^{-1}, evidence of surface cracks was readily observed. In contrast, samples which were annealed in vacuum or samples taken from the as-received extruded rod showed no evidence of surface cracking. Examination of the microstructure for the air-annealed samples (see Figure 6) shows that a stabilized alpha layer was formed during air annealing. After deformation, cracks were observed in this surface layer as shown in Figure 7.

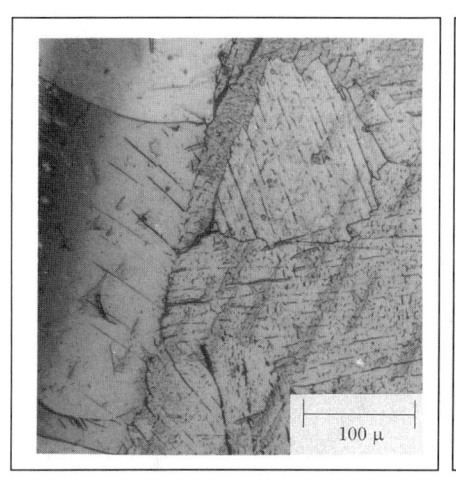

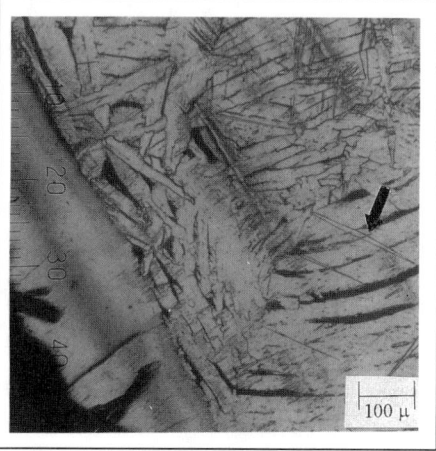

Figure 6: Micrograph of the sample before deformation, after annealing at 1100°C for 24 hours.

Figure 7: Micrograph of sample deformed at 850°C at 10s-1 to a strain of 1. Cracks are indicated by the arrows. indicated by the arrow. (100X)

A more detailed examination of the cracks indicates that cracks are also observed between the Widmanstatten platelets under the oxygen stabilized alpha layer. Deformation twins in the Widmanstatten platelets are also visible, consistent with previous reports [3]. However, it is important to note that cracking of the Widmanstatten structure was noticed only in the oxygen-contaminated material with large grain-sizes. The role of oxygen content on the formation of Widmanstatten structures is complicated but it is known to promote the lamellarity and definition of the alpha platelets [4]. This may have significant effects on the deformation behaviour of these structures. Clearly, there is complex interaction between oxygen content, microstructure and deformation behaviour (both plastic deformation and fracture). A complete understanding of these interactions requires a more detailed examination.

Finally, differences in the stress strain behaviour of air annealed and vacuum annealed samples show important differences. The overall flow stress levels are much higher for oxygen contaminated samples suggesting the importance of dissolved oxygen on the flow stress behaviour (possibly as a solid solution hardener). The stress-strain curve of the oxygen contaminated specimen showed a steadily rising stress up to a peak point, after which the stress decreased with increasing strain, as seen in Figure 8. The peak stress was associated with attainment of the strain to failure ($\varepsilon_f \sim 0.2$), for these test conditions, resulting in a loss of load bearing capacity of the material.

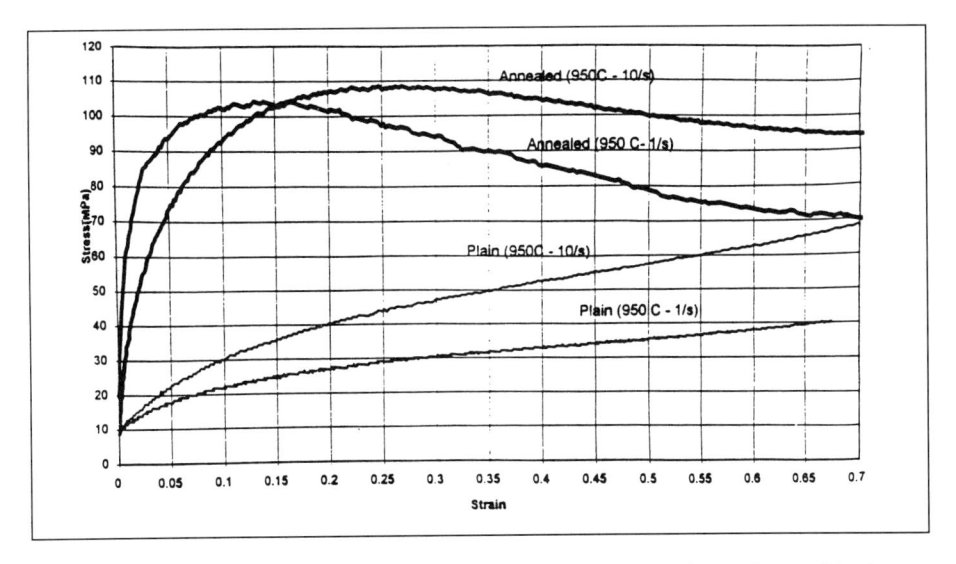

Figure 8: Flow curve of air-annealed sample compared against control sample tested in the beta region. Similar curves were obtained in the alpha region.

Summary

During hot-working of CP Ti at strain rates comparable to those employed in industrial rolling practice, the material exhibited substantial strain hardening at temperatures in the beta phase and the attainment of a steady state flow stress at temperatures in the alpha phase. The activation energy, Q, calculated from the constitutive response in the alpha regime is similar to the value for lattice controlled diffusion.

The modelling of the roll pass using the DEFORM software indicated that the surface layer of the rolled slab experienced the most severe strain, strain-rate and chilling conditions, which tended to enhance surface cracking. Air-annealed samples tested in similar conditions of strain, strain rate and temperature, showed evidence of surface cracking. The oxygen-stabilized alpha layer at the surface of these specimens cracked when subjected to strains of approximately 0.2. It was also observed that cracks developed in the Widmanstatten pattern. When the samples with pronounced Widmanstatten structures were subjected to high strain rate deformation, the plates separate from each other.

In conclusion, it is believed that the air annealing of C.P. titanium produces an oxygen stabilized alpha layer at the surface and a Widmanstatten structure near the surface. Both of these structures show a decreased level of ductility leading to an increased propensity for cracking during hot rolling.

Acknowledgments

We would like to acknowledge the funding we received from the Natural Sciences and Engineering Research Council of Canada for this work. We would also like to thank Mr. Binh Chau and Dr. Xiande Chen, who were responsible for performing the compression tests on the Gleeble®.

References

1. H.J. Frost and M.F. Ashby; Deformation-Mechanism Maps: The Plasticity and Creep of Meals and Ceramics, (Oxfordshire, New York; Pergamon Press, 1982), 52.
2. K.P. Rao and E.B. Hawbolt; Development of Constitutive Relationships using Compression Testing of a Medium Carbon Steel, Journal of Mater. and Tech., vol. 114, January 1992, 116-123.
3. M. Hayashi, H. Yoshimura, M. Ishii, and H. Harada; Recrystallisation Behaviour of Commercially Pure Titanium during Hot Rolling, Nippon Steel Technical Report No. 62, July 1994, 64-68.
4. M.K. McQuillan; Phase Transformation in Titanium, Metallurgical Reviews (8), 1963, 41-43.

COLD WORKING

COLD FORMING OF TITANIUM ROUNDS AND FLATS

by CHARLES PEPKA
RENTON COIL SPRING CO.

THE MANUFACTURE OF SPRINGS AND SHEET METAL COMPONENTS IS CHANGING EACH YEAR. EXOTIC MATERIALS OF THE 60'S AND 70,S ARE NOW COMMON, AND NEW MATERIALS ARE COMING TO THE MARKET YEARLY. FORMABILITY OF MATERIALS HAS BEEN A MAJOR CONCERN, AND COLD FORMABILITY AT RENTON COIL SPRING IS ONE OF THE MOST IMPORTANT FACTORS IN PRODUCTION. COLD FORMING OF TITANIUM ROUNDS AND FLATS WILL COVER, MATERIALS, TYPES OF SPRINGS, FORMING ,MODULUS, MECHANICAL PROPERTIES AND CHANGES DURING THE FORMING OPERATION

ROUND WIRE SPRINGS

THE MATERIAL SIZE RANGE FOR COLD FORMING "BETA C" WIRE (AMS 4957) IS FROM .009 IN. TO .5 IN. DIAMETER . THIS IS LIMITED BY THE DRAW SIZE OF STARTING STOCK. THE TITANIUM BAR (AMS 4958) RANGE FROM .375 IN. TO 2.5 IN. DIA. THIS IS THE STA CONDITION OF "BETA C". SHEET MATERIALS IN 15-3-3-3 (AMS 4914)SIZE RANGE FROM .002 IN. TO .75 IN. 15-3-3-3 NORMALLY IS USED IN SIZES UP TO .125 IN.

BEFORE THE PRODUCTION PHASE BEGINS, RENTON COILS SPRING'S ENGINEERING STAFF HAS CONSULTED ON THE DESIGN, COMPLETED A COMPUTER MODEL OF THE SPRING, AND CHECKED FINISH REQUIREMENTS . COMPUTER MODELING IS USED THOUGH OUT THE PRODUCTION PROCESS. DESIGN PROBLEMS ARE FIXED BEFORE ACTUAL PRODUCTION BEGINS.

SPRINGS FROM LARGE DIAMETER BAR STOCK

TYPES OF SPRINGS COLD FORMED FROM ROUND WIRE OR BAR.
COMPRESSION
EXTENSION
TORSION
TORSION BARS
WIRE FORMS.

COMPRESSION SPRINGS ARE THE MOST COMMON COLD WORKED TITANIUM SPRING. EXTENSION AND TORSION SPRINGS ARE COILED AND LEGS OR HOOKS ARE FORMED. TORSION BARS AND WIRE FORMS ARE FORMED PER PRINT REQUIREMENTS.

THE MODULUS OF EACH TITANIUM ALLOY IS SLIGHTLY DIFFERENT. IT IS NORMAL FOR TITANIUM SPRINGS TO HAVE ABOUT 1/2 THE COILS OF STEEL SPRINGS

THE MODULUS OF TITANIUM		
	E	G $X=10^6$
BETA C TI 3-8-6-4-	14.5	5.6
LCB TI 6.8-4.5-1.5	16.4	6.3
TI 15-3-3-3	15.5	

ROUND WIRE SPRINGS USE THE SPRING INDEX AS A FORMING LIMIT . THE MINIMUM DIAMETER LIMIT COMES FROM THE MANDREL SIZE BEING SMALLER THAN THE WIRE SIZE . THE OUTER LIMIT COMES FROM PRACTICAL LIMITS OF COILERS AND STABILITY OF SPRINGS. THE EFFECT OF THE SPRING INDEX ON FORMING IS THAT THE SMALLER THE SPRING INDEX THE GREATER THE SECTION CHANGE DURING COILING. THE POWER REQUIREMENTS IN COILING ARE MUCH HIGHER FOR SMALL INDEX SPRINGS .

SPRING INDEX

<u>MEAN SPRING DIAMETER</u>
WIRE DIAMETER

MINIM INDEX >= 3
MAXIMUM <=22

SPRING-BACK DURING COILING OF COLD WOUND SPRINGS.

SPRING-BACK AFTER FORMING IS CAUSED BY TITANIUM'S LOW MODULUS AND HIGH TENSILE STRENGTH. WHEN FORMING, THE FINAL DIAMETER OF A SPRING IS ALWAYS LARGER THAN THE MANDREL USED FOR COILING. IN RENTON COIL SPRING'S MANUFACTURING PROCESS THE USE OF COMPUTER MODELING CAN PREDICT THE STARTING MANDREL SIZE . THIS PREDICTION IS IMPORTANT BECAUSE VARIATIONS LARGER THAN 3% CAUSE US TO LOOK AT THE WIRE FOR PROBLEMS.
DURING THE MANUFACTURE OF SPRINGS IT IS IMPORTANT TO GET THE EARLIEST TEST OF UNIFORM PROPERTIES OF THE MATERIAL. DURING THE COILING PROCESS, A LARGE VARIATION IN DIAMETER WOULD INDICATE SIGNS OF TENSILE CHANGES IN THE MATERIAL.

SPRING-BACK DURING COILING
OF COLD WOUND SPRINGS

Da = ARBOR DIAMETER
q = UTS. OF WIRE
E = MODULUS OF ELASTICITY
d = WIRE DIAMETER
OD = OUTSIDE DIAMETER

$$DA = \frac{1.02d}{d/(OD - d) + 1.85 q/E}$$

THE SPRING RATE CAN BE DETERMINED USING THE FOLLOWING FORMULA. SPRING RATE CAN ALSO BE MODELED IN THE COMPUTER AND COMPARED WITH ACTUAL RESULTS FOR AN ADDITIONAL OVER CHECK ON MATERIAL PROPERTIES. ONCE A TEST IS MADE SPRING RATE IS CORRECTED BY CHANGING THE ACTIVE NUMBER OF COILS.

SPRING RATE

d = WIRE SIZE
D = MEAN DIAMETER
G = MODULUS TORSION
N = ACTIVE COILS

$$\text{RATE} = \frac{Gd^4}{8D^3 n}$$

DESIGN CONSIDERATIONS FOR SHAPED WIRE COIL SPRINGS

FOR SOME APPLICATIONS THERE IS A NEED FOR NON-ROUND SECTIONS. THE PRODUCTION OF SHAPED WIRE PRESENTS SOME UNIQUE PROBLEMS. THE QUALITY OF STARTING STOCK HAS TO BE CLOSELY CONTROLLED, AND THERE IS ONLY LIMITED SCALE REMOVAL. CROSS SECTIONS OF THE MATERIAL HAVE TO BE FACTORED INTO THE DESIGN, BECAUSE OF KEYSTONING DURING THE COILING PROCESS. RECTANGULAR MATERIAL CAN BE COILED ON EDGE (HARD WAY) OR ON FLAT (EASY WAY). ENVELOPE REDUCTION IS THE MAIN REASON TO GO TO NON-ROUNDS. AS AN EXAMPLE TORSION SPRINGS CAN BE MANUFACTURED WITH A SHORTER OVERALL LENGTH.

THIS IS AN EXAMPLE OF A KEYSTONE SECTION IN A RECTANGULAR WIRE, FROM A SPRING COILED ON EDGE (HARD WAY). THE MOUNT ON THE TOP SIDE IS THE STARTING SECTION. THE SECTION ON THE BOTTOM IS AFTER COILING. NOTE THE CHANGES, WITH THE WIDEST AREA BEING THE INSIDE DIAMETER OF THE SPRING. THIS VIEW ALSO SHOWS CROWNING IN THE INSIDE AND OUTSIDE DIAMETERS. THIS IS CONSISTENT WITH ALL BENDING IN RECTANGULAR AS WELL AS FLAT MATERIALS. THIS KEYSTONING CHANGES THE SHAPE AWAY FORM THE SIMPLE BEAM STRESS FORMULA.

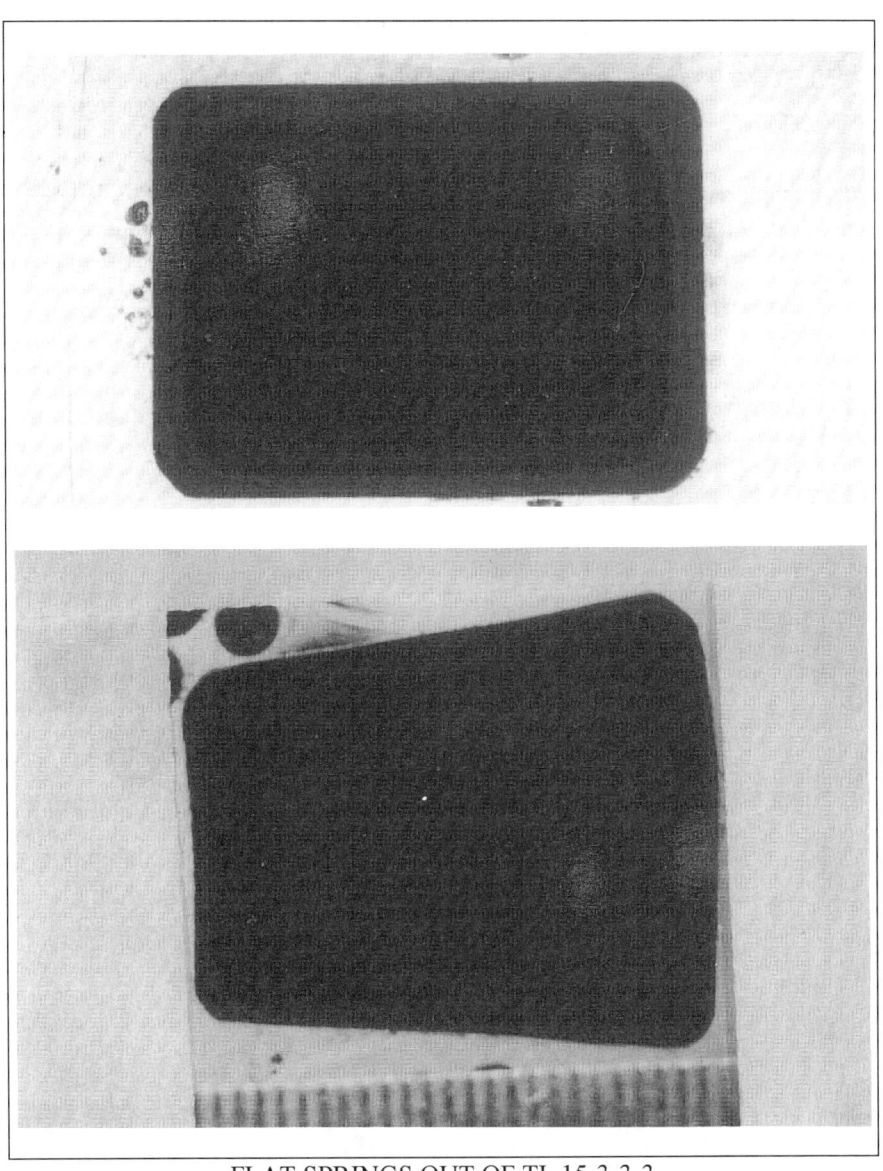

FLAT SPRINGS OUT OF TI. 15-3-3-3
FLAT SPRINGS MANUFACTURED OUT OF 15-3-3-3 SHEET OR STRIP
MATERIALS ARE USED IN MANY COMMERCIAL AIRCRAFT . THE USE OF
15-3-3-3 HAS ALLOWED RENTON TO COLD FORM SOME VERY COMPLICATED
SHAPES. THERE IS FORMABILITY DOWN TO A 2T RADIUS. FLAT PARTS ARE
ALSO BEING FORMED ON AUTOMATIC 4 SLIDE MACHINES. STAMP AND FORM

PARTS ARE REPLACING MANY STEEL AND STAINLESS STEEL PARTS WITH IMPROVED PERFORMANCE AND LESS WEIGHT. IN MANY DESIGNS TITANIUM CAN RUN WITH NO SECONDARY FINISH, ONLY STAMP AND FORM. THIS METHOD HAS MANY ADVANTAGES WHEN COMPARED TO THE COST OF FINISHING STEEL PARTS. THERE IS GOOD COMPATIBILITY OF TITANIUM WITH GRAPHITE EPOXY COMPOSITES .THE EXPECTED WEIGHT SAVING IN SMALL FLAT PARTS WOULD BE 33.% LIGHTER THAN STEEL .

```
            15-3-3-3 MATERIAL PROPERTIES

   ST  :  UTS 102-130 KSI
          YS   100-121 KSI
          %E   12 MIN

   STA:  UTS  150 KSI MIN
          YS   140 KSI MIN
          % E   7 MIN
```

SPRINGS MANUFACTURED OUT OF TITANIUM 15-3-3-3

FLAT SPRING TYPES
SPIRAL TORSION , POWER (CLOCK TYPE), BEAM, CANTILEVER CONSTANT FORCE, SPRING WASHERS.

THIS IS THE ENTIRE RANGE OF FLAT SPRING PRODUCTS.
THE SPIRAL TORSION AND POWER SPRINGS ARE USUALLY FORMED FROM
STRIP MATERIAL AND THE REST ARE STAMPED AND FORMED.

POWER (CLOCK TYPE SPRINGS) OUT OF 15-3-3-3

THIS IS A TOOLING PROGRESSION FROM FLAT PATTERN TO FINISH PART

KEY ISSUES

AS YOU CAN SEE COLD FORMING FOR COIL SPRINGS IS A MATURE TECHNOLOGY. COMPUTERS ASSIST IN THE DESIGN OF SPRINGS AND MODELING THIS ELIMINATES WASTED TIME AND MATERIAL .
TITANIUM FLAT SPRINGS IS A NEW MARKET WITH A LOT OF POTENTIAL. RENTON COIL SPRING CO IS CONTINUING ITS TITANIUM RESEARCH AND DEVELOPMENT. WE ENJOY WORKING ON NEW IDEAS .

THANKS TO ALAN CADDEY, JIM MILLER, SPRING MANUFACTURES INSTITUTE AND THE DYNAMET, TIMET, AND RMI COMPANIES, FOR THEIR HELP IN THE PRODUCTION OF THIS PAPER.

Ti-22Al-23Nb ALLOY FOILS COLD ROLLED FROM STRIP

CAST BY THE PLASMA MELT OVERFLOW PROCESS

T.A. Gaspar, and I.M. Sukonnik*

Ribbon Technology Corporation, P.O. Box 30758, Columbus, Ohio 43230;
*Texas Instruments, 34 Forest Street, MS 10-28, Attleboro, MA 02703

Abstract

Direct cast strips of a Ti-22Al-23Nb alloy were cold rolled to 0.125-mm-thick foils with and without intermediate annealing. Cold rolling refined the relatively coarse cast microstructure to a fine three phase microstructure consisting of equiaxed primary α-2 grains, orthorhombic platelets with the B2 phase along the grain boundaries. The cast and cold rolled foils were compared to ingot metallurgy foils.

Introduction

The production of titanium alloy foils by ingot metallurgy techniques involves casting ingots, forging slabs, hot rolling, thermal treatments, pickling, grinding and trimming to produce strip for cold rolling to foil gauge. Production of wrought strip is complicated and expensive because of the operations involved and process losses. Cold rolling of wrought titanium aluminides to foil gauge using the isobaric rolling process has been previously described (1).

One alternative to wrought titanium strip on a laboratory scale is to cast titanium strip to near-net-shape thereby eliminating all of the operations described above. The plasma melt overflow process is a single-chill roll technique for direct casting thin strips of titanium alloys. Titanium alloys were melted in a water-cooled-copper crucible with a plasma arc torch. After a melt pool was established, the crucible was rotated about the same axis of rotation as the chill roll, to overflow liquid metal onto the chill roll circumference. Strips nominally 0.5-mm-thick, 100-mm-wide and 3-m-long to 4-m-long were cast in each batch.

Isobaric rolling of direct cast strips to foil gauge has been reported for 98% pure titanium (2), and Ti-6Al-2Mo-4Zr-2Nb alloy (3). This investigation focuses on isobaric cold rolling of a Ti-22Al-23Nb (at %) orthorhombic Ti-aluminide cast strip.

Experimental Procedures

The Ti-22Al-23Nb alloy master alloy was produced by induction skull melting and cast into 150-mm-diameter cylindrical molds. The alloy was re-melted for 13 minutes 20 seconds in the plasma melt overflow furnace using a helium plasma with an average power of 75 kW. The strip was cast by rotating the water-cooled-copper hearth about the same axis of rotation as the chill roll to overflow the liquid alloy onto a molybdenum chill roll with a surface speed of 1 m/s and a tilting time of 3.3 seconds.

The precursor strip materials used in the cold rolling investigation were a direct cast strip of Ti-22Al-23Nb intermetallic alloy and an ingot metallurgy Ti-22Al-23Nb 0.5-mm-thick wrought strip. Primary breakdown for both cast and wrought strips consisted of 50 percent cold work and intermediate annealing at 996°C for one hour under vacuum followed by cold rolling to the gauge of approximately 0.125mm [4]. The foils were vacuum heat treated at 996°C for one hour after cold rolling.

In addition some of the cast strips were ground from both sides from an initial thickness between 0.7-mm-thick and 0.9-mm -thick to thickness between 0.6-mm-thick and 0.8-mm-thick to remove surface features. The direct cast and ground Ti-22Al-23Nb alloy strip was successfully cold rolled from approximately 0.7-mm-thick to a final gauge of 0.125 mm without intermediate annealing with a total cold reduction of approximately 82 percent.

Results

The as-cast strips and foils were evaluated using optical microscopy, microhardness measurements, scanning electron microscopy (SEM), and transmission electron microscopy (TEM).

SEM micrographs of the as-cast surfaces are shown in Figure 1. Both the chill cast surface and free cast surface exhibited equiaxed grains. The diameter of the grains was approximately 50 microns. The free surface exhibited a dendritic structure while the chill cast surface exhibited fully transformed grains. Intergranular porosity was evident on the free-cast surfaces as a result of solidification. Some of the cast samples were ground before cold rolling to remove surface irregularities.

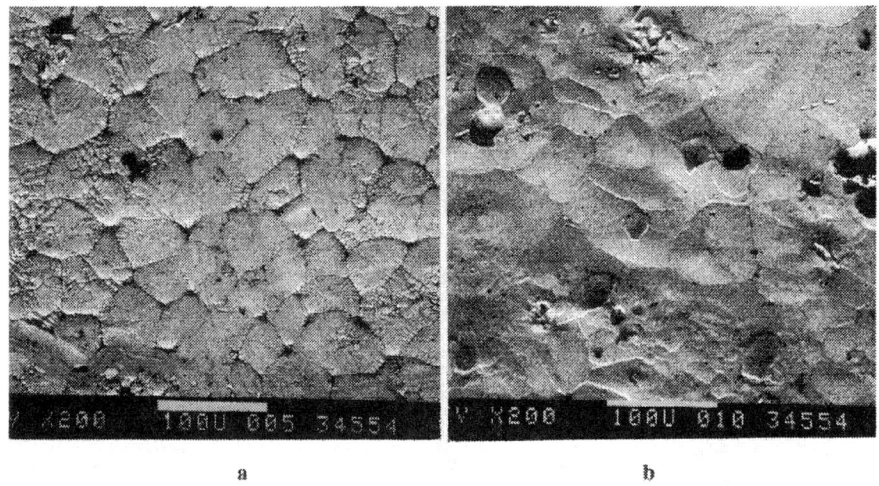

Figure 1: SEM micrographs of as-cast Ti-22Al-23Nb strip material: (a) free surface; (b) roll side surface.

Examination of the cross-sections of the direct cast precursor strip revealed that 25 to 40 percent of the thickness of the strip had a dendritic structure, as shown in the box micrograph, Figure 2a.

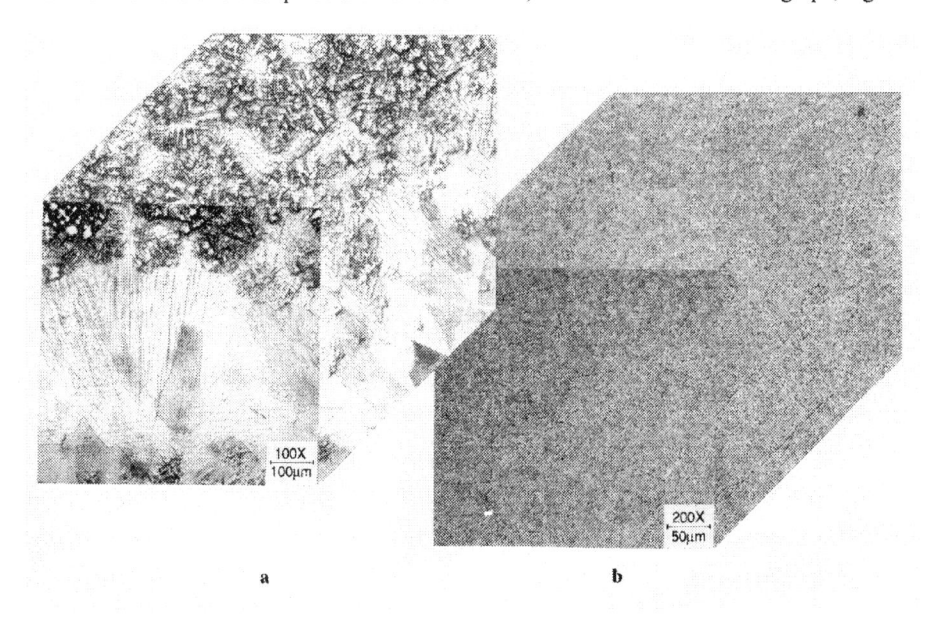

Figure 2: Optical block micrographs of precursor strip: (a) Cast; (b) Ingot Metallurgy. Rolling direction is horizontal.

The strip solidified from the chill-roll surface to the free cast surface which resulted in the microstructure being aligned across the thickness of the strip. Along the chill cast surface, a transformation to a coarser grain structure appeared to have taken place. By contrast, the wrought precursor had a finer, more equiaxed microstructure as shown in Figure 2b.

The relatively coarse microstructure of the cast precursor strip was transformed after cold reduction of approximately 82% followed by heat treatment (Figure 3a). Cold rolling and annealing refined the relatively coarse cast microstructure to a very fine, transformed, three-phase microstructure with some banding parallel to the rolling direction. By contrast, the microstructure of ingot metallurgy foil, shown in Figure 3b, resembled the microstructure of the wrought strip precursor, shown in Figure 2b, but had coarsened after cold rolling and heat treatment.

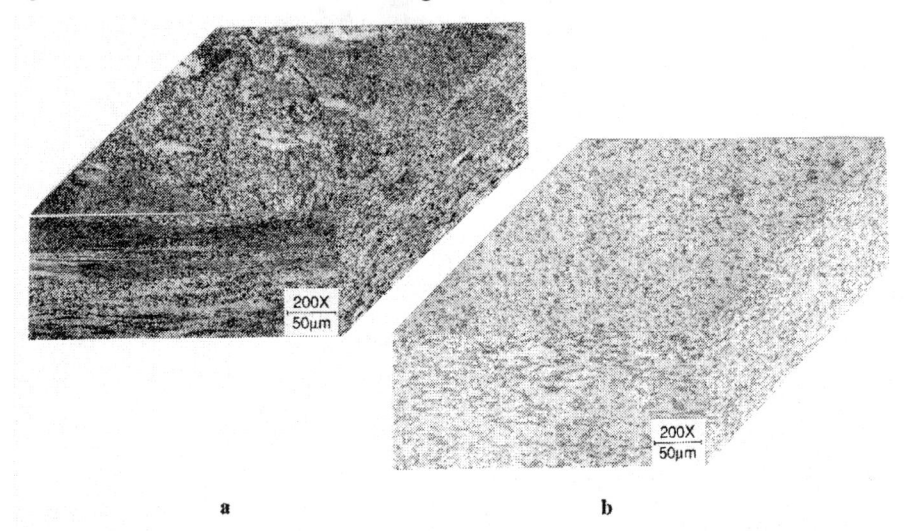

Figure 3: Optical micrographs illustrating microstructures developed in cold rolled/annealed Ti-22Al-23Nb material: (a) Cast; (b) Ingot metallurgy. Rolling direction is horizontal.

Results of microhardness tests along the primary rolling direction (L) and transverse direction (T) of both the wrought and cast precursor strips and cold rolled foils are summarized in Table 1. There was not much variation in microhardness between the longitudinal and transverse test directions of each sample. There was a large variation in microhardness between the wrought and direct cast strips but after cold rolling, the cast and wrought foils exhibited little variation in microhardness. Similar behavior has been observed in other materials which have been heavily cold rolled.

Table 1. Microhardness of Ti-22Al-23Nb Alloy Strips and Foils (VHN, 100g load)

Condition	Longitudinal	Transverse
Wrought Strip	336 - 376	n/a
Direct Cast Strip	391 - 413	387 - 407
Wrought Foil	333 - 356	334 - 352
Direct Cast Foil	337 - 343	322 - 327

TEM analysis of thin foil specimens of both the ingot metallurgy foils (Figure 4a) and the cast and cold rolled foils (Figure 4b) revealed a three phase microstructure. The phases were identified by energy dispersive spectroscopy (EDS) as equiaxed primary α2 grains; orthorhombic platelets; with the B2 phase along the grain boundaries.

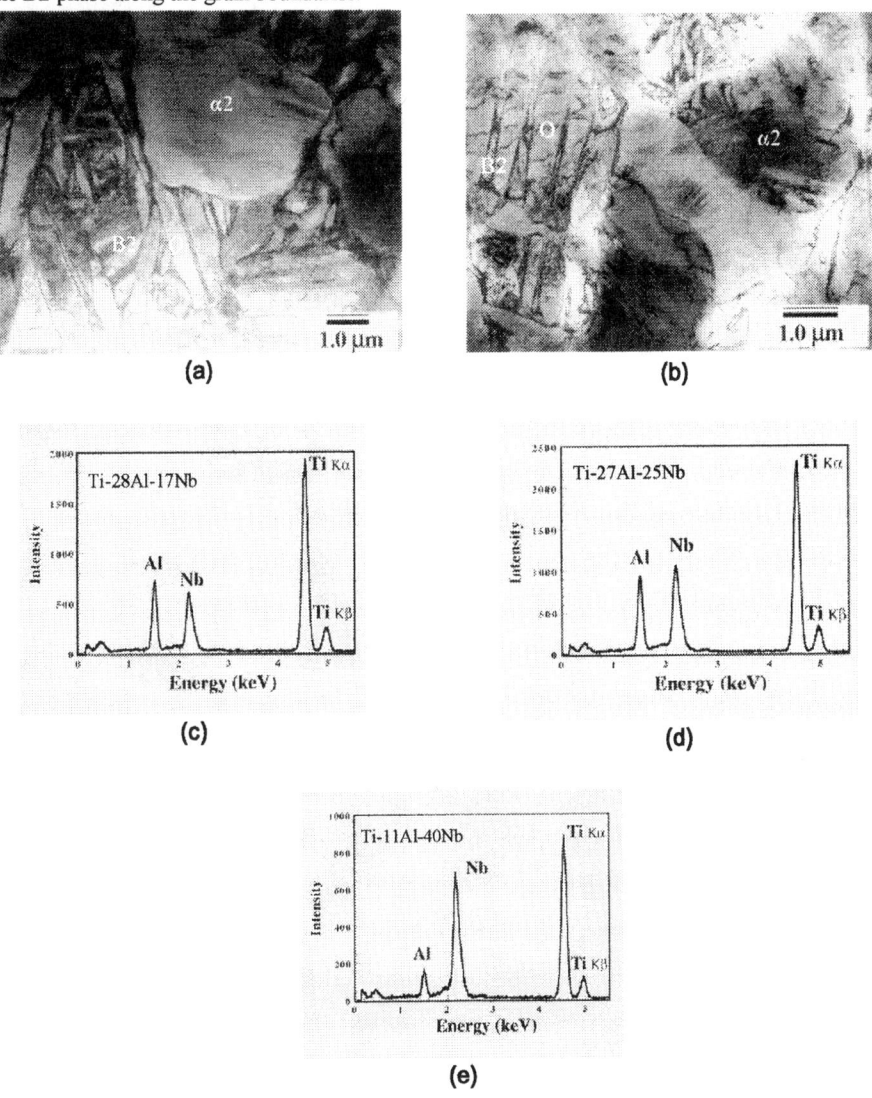

Figure 4. Transmission Electron Micrographs of (a) ingot metallurgy foil and (b) cast and cold rolled foils revealed a three-phase microstructure. EDS analysis identified the phases as (c) equiaxed primary alpha-2; (d) orthorhombic platelets and (e) B2 phase along the grain boundary.

Conclusions

The feasibility of direct casting strip of Ti-22Al-23Nb Ti-aluminide to nominally 0.7-mm-thick strip was demonstrated. Both direct cast and ingot metallurgy strips were cold rolled to foil gauge of 0.125-mm-thick undergoing the same thermal and mechanical treatments, then compared. The microstructure of the as-cast strip was much coarser than the hot worked ingot metallurgy strip. After rolling, the direct cast foils exhibited a finer microstructure than the wrought foils. The microstructure of cast foil was elongated into a banded structure along the rolling direction. Both cast and ingot metallurgy foils exhibited a three-phase microstructure consisting of equiaxed primary $\alpha 2$ grains; orthorhombic platelets; with the B2 phase along the grain boundaries.

It was possible to cold roll the direct cast strip to foil gauge without intermediate annealing if the irregularities in the cast surfaces were removed by grinding, with a total cold deformation of approximately 82 percent. By contrast, the wrought material required intermediate annealing to withstand this much reduction. This was attributed to greater ductility in the direct cast strip.

References

1. I.M. Sukonnik, S.L. Semiatin, and M. Haynes, " Effect of Texture on the Cold Rolling Behavior of an Alpha-Two Titanium Aluminide", *Scripta Metallurgica et Materialia*, 26, (1992) 993-998.

2. T.A. Gaspar, I.M. Sukonnik, R.K. Bird and W.D. Brewer, "Feasibility of Cold Rolling Titanium Strip Cast by the Plasma Melt Overflow Process", from *Synthesis/Processsing of Lightweight Metallic Materials*, ed. by F.H. Froes, et. al. (TMS, Warrendale, PA, 1995), 119-128.

3. T.A. Gaspar, I.M. Sukonnik, and A.J. Kummnick, "Titanium Foils Cold Rolled from Direct Cast Strip", (paper presented at the 8th World Conference on Titanium, 25 October 1995) in print.

4. I.M. Sukonnik, et. al., "Advances in Thermo-Mechanical Processing of Ti-Aluminide Orthorhombic Foils", Proceedings from Titanium Matrix Composites Workshop, Orlando, FL, Nov. 1991, Report WR-TR-92-4035, Wright Laboratory, Wright-Patterson AFB, OH , 98-115.

ADIABATIC SHEAR BANDS FORMED BY PUNCHING IN Ti-6Al-4V

Yoji Kosaka and John E. Kosin

Oregon Metallurgical Corporation
530 34th Ave. S.W. , Albany OR 97321

Abstract

The formation of adiabatic shear bands along fracture surface was investigated in punching tests with a moderate strain rate in Ti-6Al-4V bar and plate. Microstructure and strength of specimens were changed by heat treatment. Cross-sectional observation of fractured surface revealed adiabatic shear bands formed along fracture surface. This is due to a rapid release of strain energy at the final stage of punching, even though the initial strain rate is not extremely fast. The width of shear bands was measured as 10 to 30 μm. Both of fine dimples and stretched dimples were observed by SEM, that indicates the occurrence of fracture at elevated temperatures and the involvement of localized shear deformation. Shear strength of bar samples punched toward rolling direction appeared to be higher than that of through thickness shear stress of plates having equivalent hardness.

Introduction

It is generally recognized that adiabatic shear bands (ASB's) are products created by high speed shear deformation such as ballistic impact and drop hammer forging. In this circumstance shear deformation concentrates in narrow bands, which derives a local temperature rise in bands. Eventually, unstable fracture takes place along ASB's due to softening of these bands. According to TEM observation of ASB in Ti-6Al-4V [1], ASB's consist of martensitic transformation products with finer grains due to rapid heating and cooling cycles.

It was reported that ASB's were able to be introduced by laboratory scale tests such as dynamic punch [2], dynamic torsion [3] and dynamic compression [4,5] tests. These testing methods will offer high strain rate deformation that is realized to be necessary to introduce ASB's. However special equipments are required to generate high speed deformation. Recent studies by D. Makel [6,7] demonstrated that localized melting surface was observed at shear lips of tensile specimen pulled with quasi-static strain rate in Ti-6Al-4V. The result suggests dynamic tests may not be an absolute requirement for the occurrence of ASB's. In this study, a slow speed punch test was selected to examine if the test could generate ASB's and provide any useful information on actual dynamic failures such as ballistic fracture. Microstructural effects on shear strength obtained by the punch test were also studied.

Advances in the Science and Technology of
Titanium Alloy Processing
Edited by I. Weiss, R. Srinivasan,
P.J. Bania, D. Eylon, and S.L. Semiatin
The Minerals, Metals & Materials Society, 1997

Experimental Procedures

Test Method

Baldwin Testing Machine with a capacity of 120,000lbs was used for the punch tests. Photo.1 (a) shows a close view of a bottom die holder, an insert die and a pin for the test. The die holder was machined from AISI4140, insert die with the diameter of 0.72" was machined from A2 Steel then heat treated to get enough hardness. Pins were machined from Dowell Pin. Photo.1 (b) shows an example of punched sample. A size of a typical disc specimen used for the punch test is 2.0"dia. x 0.2"thick. A typical cross-head speed of the punch test was estimated as 56mm/in. which is converted to strain rate as 1.6×10^{-1}/sec.

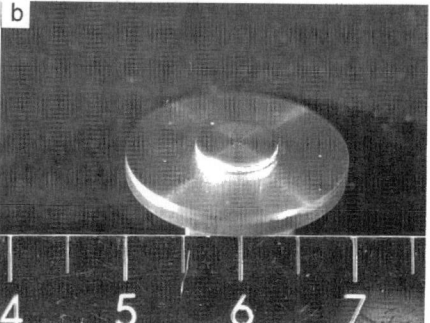

Photo.1 - Appearance of equipment for punch test and sample. (a) Die Holder, Die and Pin (b) Sample after punching.

Materials

A preliminary test was carried out on Ti-6Al-4V 2"diameter bar of which chemistry is given as A in Table 1. Disc specimens were cut perpendicular to the rolling direction that gives shear deformation toward longitudinal direction. After completing a preliminary test of bar samples, disc specimens were cut from Ti-6Al-4V plate samples parallel to the rolling surface in order to evaluate through thickness property by the punch test. Again microstructure and strength were changed by various heat treatments.

Table 1. Chemistry of Ti-6Al-4V plate used for punch test. (wt%)

ID	Prod.	Size	Al	V	Fe	O	C	N
A	Bar	2"Dia.	6.30	3.88	0.159	0.18	0.02	0.01
B	Plate	3"Thick.	6.36	4.00	0.177	0.19	0.01	0.01

Heat Treatment

Heat treatment conditions used for Ti-6Al-4V Alloy in this study are summarized as follows,

RA : 1775 F x 1 hr. --- Air Cool
STA : 1775 F x 1 hr. --- Water Quench, 900 F x 4 hrs. --- Air Cool
BFC : 1900 F x 1 hr. --- Furnace Cool
BAC : 1900 F x 1 hr. --- Air Cool
BQA : 1900 F x 1 hr. --- Water Quench, 900 F x 4 hrs. --- Air Cool

Results

Microstructure

Photo.2 shows microstructure of Ti-6Al-4V plates heat treated differently. Among three α/β micrographs, MA gives highest volume of primary alpha, while STA gives lowest volume of primary α which is estimated as approximated 30%. On the other hand, three kinds of β heat treatment created different morphology of transformed β products as shown in Photo.2(d) thru 2(f). BQA exhibits decomposed α' martensite in coarse β grains, although TEM work was not carried out in this work. BFC and BAC show similar microstructure except that BFC consists of thick α/β laths and grain boundary α. Similar microstructure was observed in bar samples if heat-treated by the same condition as plate samples.

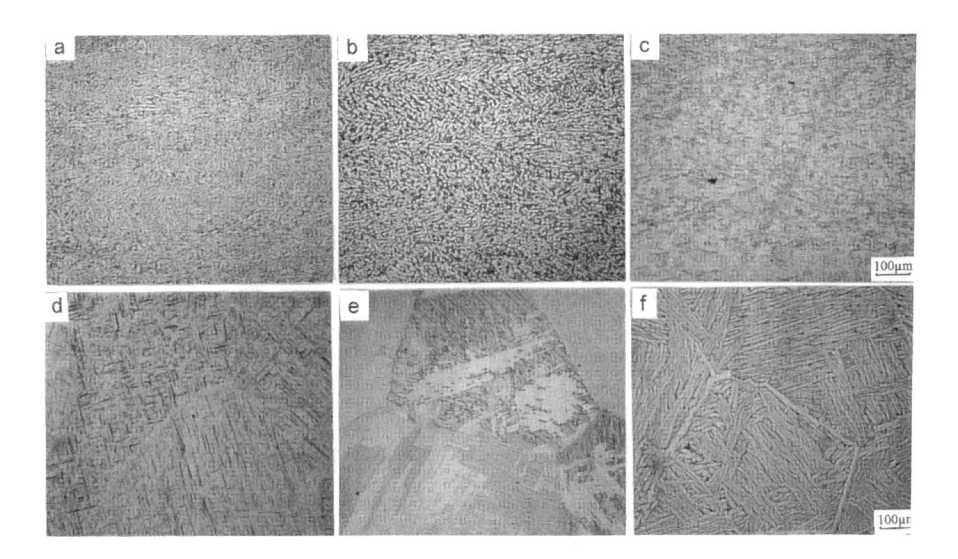

Photo. 2 - Micrographs of Ti-6Al-4V plate samples heat-treated by (a) Mill Anneal, (b) Recrystallize Anneal, (c) Solution Treatment and Age, (d) Beta Quench and Age, (e) Beta Anneal/Air Cool , and (f) Beta Anneal/Furnace Cool.

Shear Strength

Table 2. shows the results of maximum shear strength of mill annealed Ti-6Al-4V bar obtained by punching with two different strain rates. There was no significant change in shear strength in a range of strain rate varied in this study.

Table 2. Effect of cross head speed on shear strength measured by punch test.

Cross Head Speed mm/min.	Strain Rate 1/sec.	Maximum Shear Strength ksi
0.3	0.00083	88.9
56	0.16	88.1, 90.3

Table 3. shows the maximum shear strength for Ti-6Al-4V bar heat-treated variously. Hardness value and estimated tensile strength from the hardness are listed in the table as well. STA or BQA gives higher shear strength than annealed specimens. It should be noted that BFC gives higher strength than RA although their hardness values are equivalent.

Table 3. Maximum shear strength for heat treated Ti-6-4 bar.

Heat Treatment	Hardness (HRC)	Estimated Tensile Strength (ksi)	Maximum Shear Strength (ksi)
Recrystallize Anneal (RA)	31.3 31.0	143 142	87.7 87.5
Solution Treatment And Age (STA)	37.3 38.2	168 172	96.0 97.0
Beta Quench And Age (BQA)	38.3 37.7	173 170	98.9 99.1
Beta Anneal (BFC)	31.4 32.9	143 149	93.1 92.3

Maximum shear strength and hardness values for Ti-6Al-4V plate samples are listed in Table 4 with estimated tensile strength by hardness. Among α/β heat-treated samples, STA gives highest shear strength. Similarly, BQA shows highest shear strength in β heat-treated samples. Again, it should be noted that BFC shows higher shear strength than α/β samples having equivalent hardness.

Table 4. Maximum shear strength of through thickness direction in Ti-6Al-4V heat-treated variously.

Heat Treatment	Hardness (HRC)	Estimated Tensile Strength (ksi)	Maximum Shear Strength (ksi)
Beta Anneal Furnace Cool(BFC)	32.3 32.7	147 148	85.9 88.6
Beta Anneal Fan Cool (BAC)	35.3 34.7	158 156	81.7 81.8
Beta Quench (BQA)	38.0 37.0	171 166	91.1 94.3
Mill Anneal (MA)	34.0 34.7	153 156	82.0 84.3
Recrystallize Anneal (RA)	32.3 32.3	147 147	79.9 82.1
Solution Treatment and Age (STA)	36.3 36.3	163 163	85.6 82.4

Formation of Adiabatic Shear Bands

After the punch tests, the cross-section of fractured surfaces was observed by an optical microscope. A portion of fracture surface had been often damaged by rubbing between inner and outer pieces, however, ASB's were observed in all of the specimens.
Photo.3 shows an example of ASB observed in mill-annealed bar sample. Thickness of the ASB

was measured as approximately 30 μm. A void is observed in the ASB of Photo.3, although it is not very common in ASB's created by the punch test. This observation suggests that thermal softening and possibly local melting took place in the ASB [8].

All of specimens exhibited ASB as a part of fracture surface regardless their heat treatment conditions. Photo.4 shows an example of ASB observed in three plate specimens heat-treated differently. The width of the ASB's was measured as 10 to 20 μm. A distorted microstructure adjacent to fracture surface indicates the occurrence of localized shear deformation prior to the formation of ASB. Fig.1 shows a hardness distribution across ASB in the STA sample. The temperature of the ASB is considered to have exceeded beta transus, but no hardness change was detected.

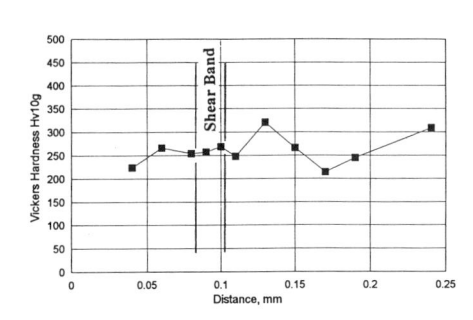

Fig. 1 - Micro-hardness distribution across adiabatic shear band formed by punch test.

Photo. 3 - Adiabatic shear band observed in punched sample of mill annealed bar.

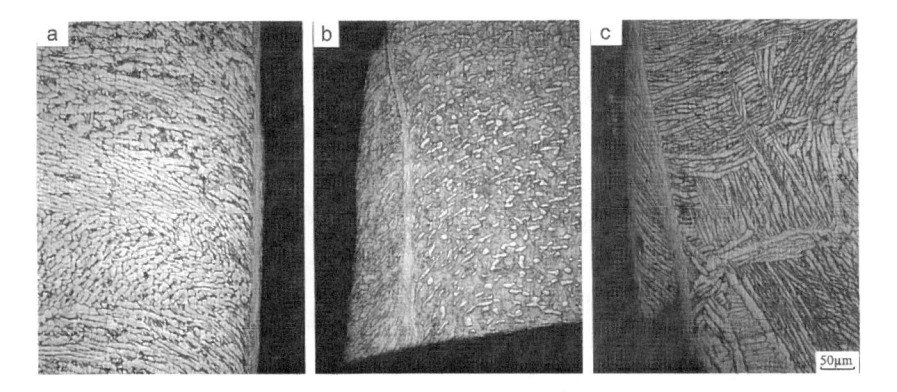

Photo. 4 - Adiabatic shear bands observed in (a) Mill Anneal (b) Solution Treatment and Age and (c) Beta Anneal/Furnace Cool samples.

Fracture Surface

Photo.5 shows fractographs of sample BAC. In lower magnification, Photo.5(a), a combination of three fracture types are observed, i.e. fine ductile dimples Photo.5(b), elongated dimples Photo.5(c), and smooth surface. Elongated dimples suggests the occurrence of fracture by shear mode. The smooth surface is thought to be resulted from extremely large shear deformation or partially melt surface. Photo.6 shows fractographs of sample MA. These two fractographs are predominantly seen on the fractured surface. Photo. 6(a) shows ductile dimples, while Photo.6(b) shows more elongated dimples. Shapes of dimples of the fracture surfaces appear to be more irregular than those in Photo.5.

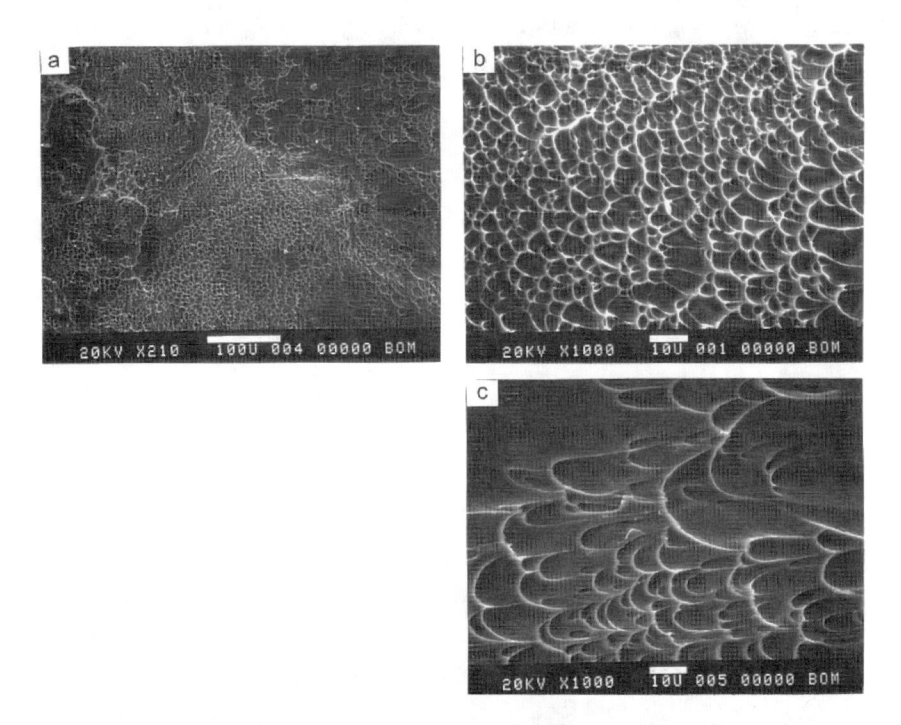

Photo. 5 - SEM fractographs of punched surface of Beta Anneal / Air Cool sample. (a) Combination of fine dimples, elongated dimples and flat surface. (b) Fine dimples with higher magnification. (c) Elongated dimples with higher magnification.

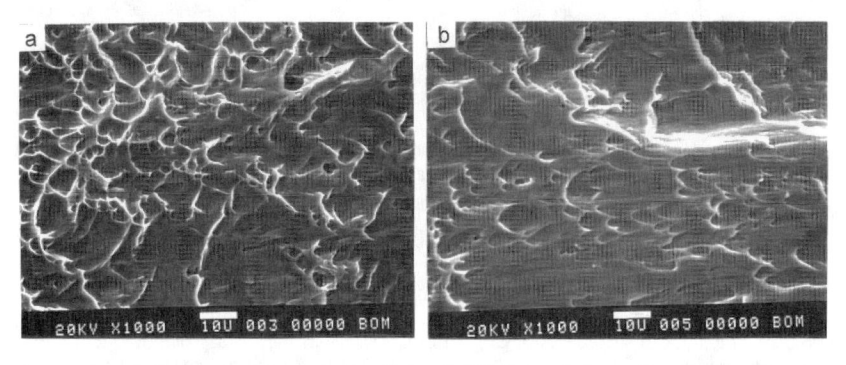

Photo. 6 - SEM fractographs of punched surface of Mill Annealed sample. (a) Fine dimples. (b) Elongated dimples.

Discussion

Effect of Strain Rate on the Formation of ASB

It is generally understood that ASB's are formed under the deformation with extremely higher strain rate. The strain rate used in the present punch test was 0.16/sec. which is considered to be slow strain rate compared with a dynamic condition such as drop hammer forge, machining, and ballistic impact. Although the strain rate of the present test is not extremely fast, an elastic energy stored in a punch specimen is rapidly released at the final stage of the fracture. This results in a fracture with extremely high strain rate regardless of the speed of the initial strain rate. This phenomena was also observed by D.Makel et.al. [6,7] in tensile test, and by S.V.Kailas et.al. [4] in compression test.

Besides strain rate effect, the other important factor in punching is that deformation is very limited to a narrow volume between a die and a pin. In addition, the deformation is limited to a shear mode due to a geometric restriction. The concentration of shear strain at limited area induces significant amount of slips within a short period, that results in a temperature rise at deformed bands. The temperature rise causes softening of a shear band and eventually triggers the formation of ASB's and fracture.

Shear Strength Measured by Punch Test

Fig.2 shows the correlation between shear strength and hardness, HRC for Ti-6Al-4V alloy heat-treated variously. The figure includes both bar and plate samples heat-treated at both α/β and β regions. Bar samples show higher shear strength than plate samples at the same hardness. Also, β microstructure gives higher shear strength than α/β microstructure at the same hardness except for BAC. This result indicates that the longitudinal shear deformation of bars is harder than through thickness shear deformation of plates. The difference of shear strength between bar and plate is considered to be primarily due to the difference of crystallographic orientation, i.e. texture. In round bar, basal axis tends to align parallel to rolling direction. While transverse texture is the most common in Ti-6Al-4V plates, although texture measurement was not carried out in this study. Higher shear strength of beta microstructure is resulted from the higher resistance of crack propagation due to coarse microstructure. Cracks are forced to direct cylindrical side surface by geometric restriction, although β microstructure has a preferential crack path such as β grain boundaries or colony boundaries. A disagreement of these two directions causes crack branching or irregular surface as often seen in fracture toughness specimens having coarse beta microstructure. Photo.7 is a micrograph showing secondary crack and irregular surface in β heat-treated sample.

Fig. 2 - Correlation between hardness, HRC, and maximum shear strength in Ti-6Al-4V bar and plate samples heat-treated variously.

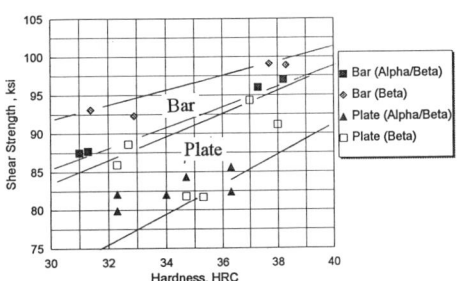

Photo. 7 - Cross-sectional view of fracture surface in beta anneal, BFC, sample. (Note rough surface and secondary crack indicated by arrow.)

Conclusions

1. Adiabatic shear bands were formed along fracture surfaces by slow punch test. This is due to the occurrence of fracture with extremely high strain rate in a localized band resulted from a rapid release of strain energy at the final stage of fracture.

2. The width of adiabatic shear bands is in a range of 10 to 30 μm. There was no hardness change between matrix and shear bands.

3. Higher hardness plate gives higher shear strength in α/β heat-treated plates.

4. Bar sample shows higher shear strength than plate samples having equivalent hardness. This is considered to be due to the difference of texture in two types of products.

5. Furnace-cooled sample from the β region exhibited higher shear strength than α/β heat-treated samples with an equivalent hardness. This is due to the occurrence of crack branching and the formation of irregular surface.

Acknowledgements

The authors would like to acknowledge the assistance of Michael B.Daggett who aided mechanical tests and metallographic works.

References

1. H.A.Grebe, H-R. Pak, and M.A.Meyers, "Adiabatic Shear Localization in Titanium and Ti-6 Pct Al-4 Pct Alloy," Met. Trans., 16A(1985), 761-775

2. A.K.Zurek, "The Study of Adiabatic Shear Band Instability in a Pearlitic 4340 Steel Using a Dynamic Punch Test," Met. Trans., 25A(1994), 2483-2489

3. K.Cho, Y.C.Chi, and J.Duffy, "Microscopic Observations of Adiabatic Shear Bands in Three Different Steels," Met. Trans., 21A(1990), 1161-1175

4. S.V.Kailas, Y.V.R.K.Prasad, and S.K.Biswas, "Flow Instabilities and Fracture in Ti-6Al-4V Deformed in Compression at 298 K to 673 K," Met. Trans., 25A(1994), 2173-2179

5. L.W.Meyer and E.Staskewitsch, "Adiabatic Shear Failure of the Titanium Alloy Ti6Al4V under Biaxial Dynamic Compression/Shear Loading," in "Titanium '92 science and Technology", 1939-1946, edited by F.H.Froes et.al. in 1993

6. D.D.Makel and H.G.F.Wilsdorf, "An Investigation of Unusual Surface Features Caused by Adiabatic Shear During Tensile Separation," Scripta Met., 21(1987), 1229-1234

7. D.D.Makel and D.Eylon, "The Effect of Microstructure on Localized Melting at Separation in Ti-6Al-4V Tensile Samples," Met.Trans., 21A(1990), 3127-3136

8. S.P.Timothy and I.M.Hutchings, "Initiation and Growth of Microfractures along Adiabatic Shear Bands in Ti-4Al-4V," Mater. Sci. and Tech. , 1(1985), 526-530

The Effect of Residual Work on the Aging Response of Beta Alloys

P. G. Allen and P. J. Bania

TIMET, Henderson Technical Laboratory
P O Box 2128, Nevada, USA

Abstract

When metastable beta alloys undergo deformation processing, particularly at elevated temperatures, the stored residual work is usually utilized to drive subsequent recrystallization. However, due to other considerations such as grain size, the recrystallization process may not always be driven to near completion. Occasionally an intermediate condition of sub-grain formation or polygonization is developed. Whether it is simple incomplete recrystallization or sub-grain formation/polygonization, the lack of complete recrystallization leaves added driving force that can significantly influence the precipitation process during subsequent ageing. This paper will review some experiences with metastable beta alloys relating to the effects of incomplete recrystallization on aging kinetics and mechanical properties.

Introduction

Metastable beta alloys are differentiated from their near-alpha or alpha-beta counterparts by their deep hardening potential, heat treatability and, in the solution treated condition, their cold workability. The aging reaction in metastable beta alloys simplistically consists of hardening via the partial decomposition of the metastable β matrix to a finely dispersed alpha precipitate. In reality the ageing reactions are complex and may involve intermediate phase formation such as β' or ω. The kinetics and mode of this decomposition are heavily dictated by the chemical composition of the alloy. Chemistry is only one factor however and for a given chemistry many other factors can influence both the kinetics and morphology of the α precipitation. These factors include, but are not confined to, :-

- Cooling rate from solution treatment.
- Ageing temperature
- Heating rate to the ageing temperature.
- The presence of precipitates, inclusions or other preferred alpha nucleation sites, such as dislocation sub-structure contained in unrecrystallized grains.

The focus of this paper will be the latter factor. Specifically the effects of strain (retained deformation) from prior thermo-mechanical processing in the unrecrystallized grains of wholly or partially recrystallized structures on the ageing behavior of various metastable beta alloys. Information is drawn from previously unpublished data on TIMET produced strip alloys such as Ti-15V-3Al-3Cr-3Sn and Ti 13V-11Cr-3Al and also published data on alloys such as Beta C™ and Beta III.

Polygonization in Ti-15V-3Al-3Cr-3Sn

The example of polygonization or sub-grain formation in the above alloy is drawn from a TIMET internal study[1] aimed at defining optimum processing parameters for cold rolled strip. The input stock for the experiment was 10 mm thick, hot rolled plate. The processing used to manufacture the plate consisted of cross-rolling 75 mm plate from 954°C on a cascading temperature basis to the 10 mm thickness. The input stock was then Vacuum Creep Flattened (VCF'd) for 8 hours at 760°C, slow cooled and subsequently solution treated for 1 hour at 788°C and air cooled.

Examination of the plate in this condition, to characterize the input stock for the experiment, revealed an equiaxed grain size of 117μ (ASTM 3). No sub-structure was evident in this metallurgical condition. To assess the true degree of recrystallization, a decoration age of 30 minutes at 482°C was applied. The principle of decoration aging is that the retained work in an unrecrystallized grain will accelerate the precipitation kinetics of the alpha phase, while the truly recrystallized grains exhibit no optically distinguishable precipitation. This precipitation results in darkening of the unrecrystallized grains upon subsequent etching, readily distinguishing them from their recrystallized counterparts. Examination of the plate in this condition revealed an extensive network of sub-grain formation as shown in Figure 1. The sub-grain boundaries being preferred sites for alpha phase nucleation.

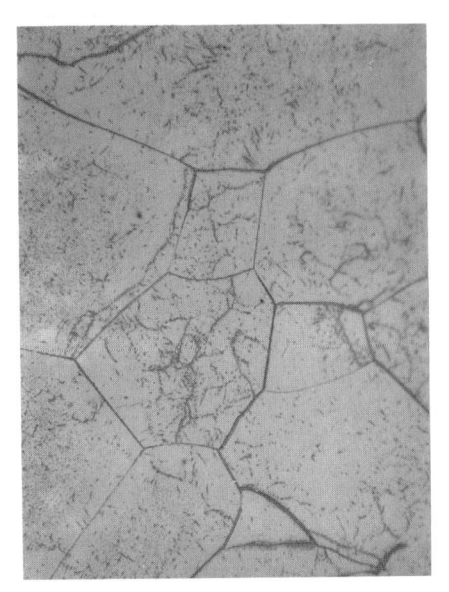

Figure 1: Sub-structure in hot rolled and VCF'd Ti-15-3-3-3 plate after decoration ageing. (x500)
50µ ├─────────┤

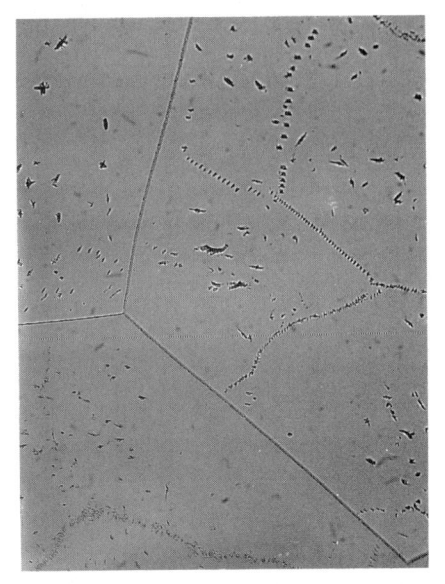

Figure 2: Discontinuous particles along prior sub-boundaries after 1 hour at 1093°C and decoration ageing. (x500)
50µ ├─────────┤

A program of annealing trials encompassing a matrix of times and temperatures between 788°C and 954°C, for up to 24 hours, was undertaken to try and eliminate this polygonized network. In all cases the sub-structure was still evident after the decoration age. Finally, using an anneal cycle of 1 hour at 1093°C and air cooling, the sub-grains were no longer evident upon decoration ageing. However, a distribution of discontinuous inclusion like particles still persisted along the prior sub-boundaries (refer Figure 2). These particles were shown by EDS analysis to be rich in silicon, phosphorus and sulfur. Precipitation of these particles could only be suppressed by water quenching from above 1038°C.

The above experiment demonstrates the role that sub-boundaries play in influencing alpha phase precipitation upon ageing. It also illustrates how thermally stable such polygonized grains can become once formed. Despite the apparent thermal stability of these polygonized grains, subsequent processing involving relatively low cold roll reductions (10%) and 788°C, 30 minute anneal cycles produced true recrystallization eradicating the sub-grain network.

Recrystallization and Property Relationships in Ti-13V-11Cr-3Al

This example of the influence of degree of recrystallization is drawn from a TIMET internal study[2] of the influence of several process variables on the mechanical properties of aged Ti-13V-11Cr-3Al. The study encompassed variables such as varying reductions in the hot rolled starting stock, final cold roll reduction, final solution treatment temperature and final ageing cycle. The processing matrix is illustrated schematically in Figure 3. The final aged products were characterized in terms of microstructure (degree of recrystallization and grain size) and L & T mechanical properties. Results were then analyzed statistically to determine the ranking importance of these process variables on aged properties.

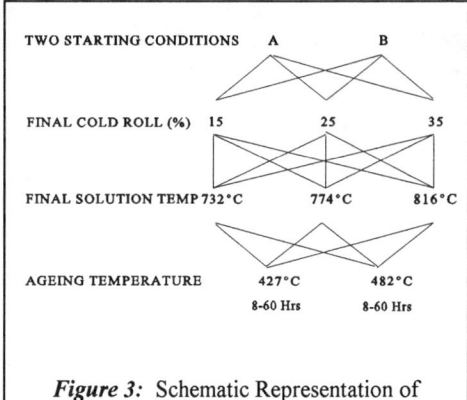

Figure 3: Schematic Representation of Experiment Processing Martix.

In terms of microstructural characterization, 3 distinct structural categorizations influencing the mode of ageing were noted. These can be summarized as follows:-

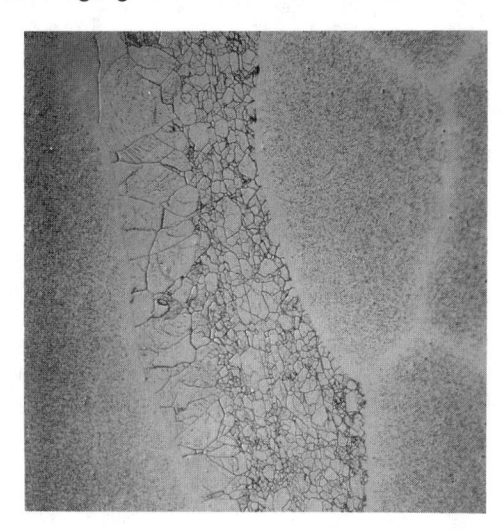

Figure 4: Co-existence of Type 1 and Type III Microstructural Categories. (x500) 50μ

1) *Type I* - the development of a fine sub-structure within the unrecrystallized grains. As might be expected from the previous discussion on Ti-15-3, these structures produced the fastest ageing response.

2) *Type II* - Partially recrystallized structures where the alpha precipitation in the recrystallized grains exhibited slower alpha precipitation kinetics (than Type 1) and where alpha precipitation was non-uniform within the recrystallized grains.

3) *Type III* - Partially recrystallized structures where the alpha precipitation within the recrystallized grains was uniform but exhibited a "lesser aged" band along the boundaries. These grains exhibited the slowest mode of alpha precipitation.

An example of the co-existence of Type I and III structures is depicted in Figure 4.

For a given ageing cycle, analysis of the mechanical property data concluded that the processing variables that *decreased* the aged strength were:-
- Increasing solution treatment temperature
- Greater reductions in the prior hot-roll processing sequence
- Greater reductions in the final cold roll sequence

All the above factors have the tendency to *increase* the degree of recrystallization in the final product prior to ageing.

To test the concept that, for a specific ageing cycle, overall degree of recrystallization was the overriding influence on product strength, average yield strength (L & T) was plotted against degree of recrystallization for all processing variables. An example of one such plot for an age cycle of 24 hours at 427°C is depicted in Figure 5. This plot illustrates that, despite many process variations, there is an excellent correlation between the resultant degree of recrystallization and strength.

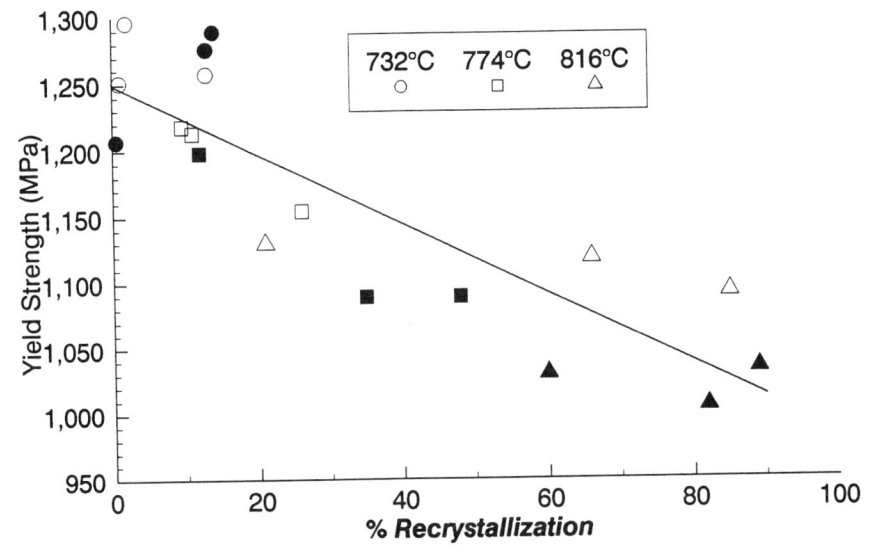

OPEN SYMBOLS - STARTING CONDITON A
SOLID SYMBOLS - STARTING CONDITION B

Figure 5: Yield Strength vs. Recrystallization for all Process Conditions Aged for 24 hours @ 427°C.

Effect of Solution Treatment Temperature on the TTT Curve for BETA III

This example of the effect of residual work on subsequent ageing kinetics is drawn from previously published work on the alloy BETA III[3], Ti-11.5Mo-6Zr-4.5Sn. This work studied the correlation between microstructure and age hardening response.

Two ingots of similar alloy content, but differing oxygen contents (0.17% and 0.28%), were hot rolled, annealed at 649°C and lightly cold rolled prior to solution treatment. The resultant product was then solution treated at temperatures above and below transus, recognizing the influence of the varying oxygen contents. TTT curves for the precipitation of ω and α phases during subsequent ageing were developed by a combination of hardness, optical microscopy and X-ray studies.

The TTT diagrams obtained for the low and high oxygen variants are reproduced in Figure 6. In both cases, the displacement to shorter times of the nose of the alpha precipitation curve for the lower solution treatment temperature is testament to the accelerating effect of retained work in unrecrystallized grains. There is little time displacement on the nose of the ω curve, temperature displacements being attributed to phase composition effects of the super versus sub transus solution treatments.

The referenced paper includes an excellent TEM replica portraying the variation in the coarseness of alpha precipitation in adjacent recrystallized and unrecrystallized grains. The substantially finer alpha precipitation in the unrecrystallized grain illustrates the increased nuclei provided by the retained deformation.

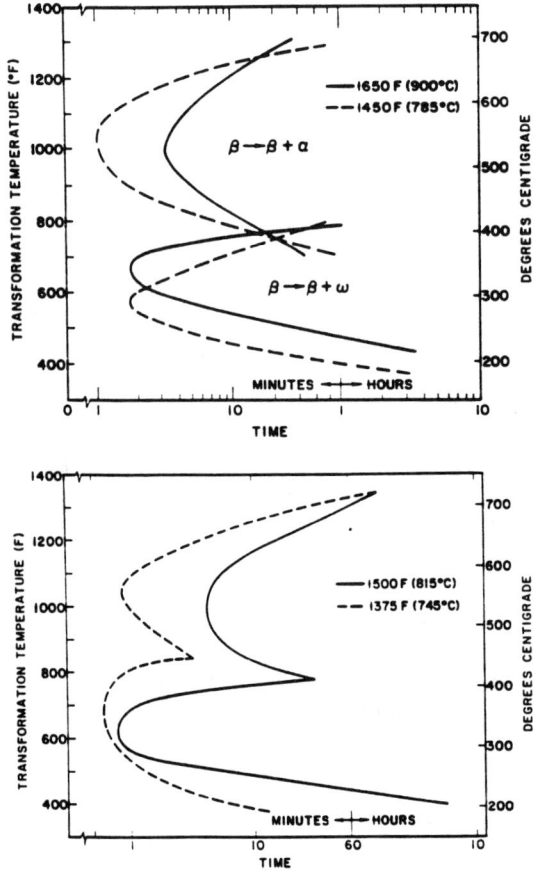

Figure 6: Effect of Solution Treatment Temperature on the TTT curves of Beta III with Oxygen contents of 0.28% (top) and 0.17% (bottom).

Effect of Degree of Recrystallization on the Properties of Beta C™.

This example is drawn from previously published work that defines the effects of differing solution treatment temperatures on the degree of recrystallization and subsequent aged properties of 75 mm diameter Beta C™ bar[4]. Various solution treatment temperatures were evaluated for degree of recrystallization. Two were down selected; one at 788°C, 1 hr, AC which produced a predominantly unrecrystallized structure, the other at 843°C, 1 hour, AC which was predominantly recrystallized. Tensile properties for these two solution treatment conditions were studied after various age cycles. For the purposes of comparison here, the tensile properties after ageing are compared for only two conditions, 24 hours @ 482°C, AC and 24 hours @ 552°C, AC.

Figure 7 presents ultimate and yield tensile data for both L & T tests for the two solution-treatment/age-cycle combinations. The data clearly demonstrates the strengthening effect derived from the finer alpha precipitation resultant from the stored energy present in the unrecrystallized grains that predominate at the lower solution treatment temperatures. Ductilities are not presented here, since without consideration of the differing strength levels, direct comparisons of ductility have little meaning. The reader is referred to the original paper where such strength/ductility relationships are presented.

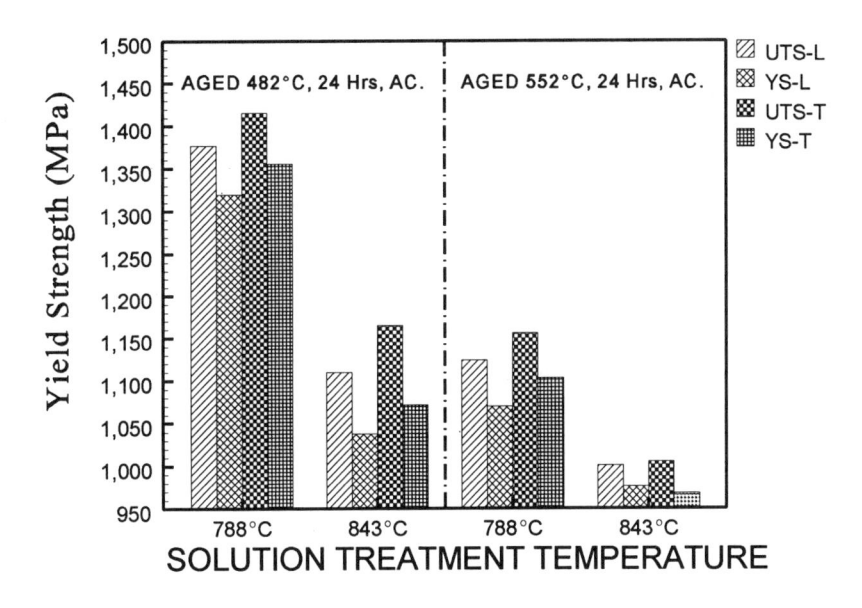

Figure 7: Yield Strength of Beta C™ bar for two solution treatment and age combinations

Summary

Examples of the effects of residual work in the unrecrystallized grains of partially recrystallized structures on subsequent alpha phase nucleation and growth have been reviewed for various metastable beta alloys. The common aspects of all of these examples can be summarized as follows:-

a) Residual work in unrecrystallized grains can significantly effect both the kinetics and morphology of alpha precipitation by virtue of providing more driving force and nucleation sites. The mechanical properties of metastable beta alloys resulting from a given age cycle are thus heavily influenced by the proportions of recrystallized and unrecrystallized grains.

b) Recovery and polygonization in the unrecrystallized grains of hot worked alloys can also influence the ageing kinetics. The homogeneity and accelerating effects of alpha precipitation diminish as the polygonized grains grow. Such polygonized structures, once formed, can become very thermally stable, and are difficult to eradicate without further deformation processing.

c) Accelerating effects of retained work can be used to advantage in metastable beta alloys, particularly those with sluggish ageing response. Residual work will promote fine alpha nucleation, and avoid grain boundary alpha films or inhomogeneous nucleation that can produce poor strength/ductility relationships in these alloys. Disadvantages are the potential development of strong textural effects and anisotropic properties; wide property scatter due to difficulty of reproduceability of partially recrystallized structures; and adverse effects on cold formability.

References

1. S. W. Sorchik and P. J. Bania, "Ti-15-3 Recrystallization Studies", Technical Report No. 3, March 1984, _Unpublished_

2. A. J. Hatch, "Effect of Prior Processing History on 800°F and 900°F Aging Characteristics of Ti-13V-11Cr-3Al", Progress Report No. 3, June 6 1963, _Unpublished_

3. F.H. Froes et al., "The Relationship Between Microstructure and Age Hardening Response in the Metastable Beta Titanium Alloy Ti-11.5Mo-6Zr-4.5Sn (Beta III)", Metallurgical Transactions (11A) (1980), 21-31.

4. Bella et al., "Effects of Processing on Microstructure and Properties of Ti-3Al-8V-6Cr-4Mo-4Zr (Beta C™)", Microstructure/Property Relationships in Titanium Aluminides and Alloys, ed. Y-W. Kim and R. R. Boyer, TMS, 1991, 493-510

COLD AND WARM WORKING OF LCB TITANIUM ALLOY

I. Weiss*, R. Srinivasan*, M. Saqib*, N. Stefansson*
A. Jackson**, and S.R. LeClair**

*Mechanical and Materials Engineering Dept.
Wright State University, Dayton, Ohio 45435

**Manufacturing Technology Branch,
Wright Laboratories/Materials Directorate,
Wright Patterson AFB, Ohio 45433,

Abstract

A study was undertaken to evaluate the cold and warm forming characteristics of the low cost beta titanium alloy Ti-6.8Mo-4.5Fe-1.5Al wt% (Timetal® LCB). Compression tests were conducted at room temperature and at warm working temperatures in the range of 150 to 300°C at strain rates in the range of $0.01s^{-1}$ to $5s^{-1}$. During room temperature deformation, the material work hardens at slow strain rates, while deformation heating and localization causes the material to flow soften at fast strain rates. The temperature rise has been estimated to be as high as 500°C. Increasing the initial deformation temperature decreases the tendency for flow softening. At temperatures above 150°C the materials work-hardens during deformation. Above starting temperature of about 300°C, alpha phase precipitation is observed in the deformed specimens. This paper discusses the cold and warm deformation of LCB titanium alloy in terms of flow curves, optical microscopy and transmission electron microscopy (TEM) analysis.

Introduction

There has been increased interest in titanium alloys for automotive applications in recent years for the purpose of increasing the fuel economy of vehicles. Titanium alloys have high specific strength, specific stiffness, and excellent corrosion resistance, making them potentially useful to the automotive industry. A new low cost beta titanium alloy, Timetal® LCB (Ti-6.8Mo-4.5Fe-1.5Al wt%), which utilizes the beta stabilizing elements molybdenum and iron, added in the form of a ferro-moly compound, has been recently introduced into the market. This method of adding alloying elements has greatly reduced cost of making the material, in comparison to conventional titanium alloys [1]. The LCB alloy, in the solution-heat-treated and quenched condition, has a bcc beta phase structure at room temperature. The alloy, therefore, possesses excellent workability compared to other titanium alloys that contain the hcp alpha phase. Aging the LCB alloy at a temperature 155° to 185°C below the transus temperature (T_β = 804°C, 1480°F) causes the alpha phase to precipitate. In this condition, the alloy has a yield strength greater than 1100 MPa (160 ksi) and a tensile elongation of about 10% [2,3], which compares very favorably with high strength steels.

In this paper we discuss the results of a study on the cold and warm deformation of Timetal® LCB titanium alloy. Compression tests were carried out at different strain rates over a temperature range from room temperature 25°C (77°F) to 370°C (700°F). Optical microscopy and TEM analysis were performed to characterize the microstructural changes due to the deformation.

Materials and Experimental Procedures

The as-received LCB titanium alloy had been hot rolled to a 51 mm (2 in.) diameter bar, and then aged at 788°C (1450°F) for 30 minutes, followed by air cooling. The as-received material was solution heat treated above T_β at 816°C (1500°F) for 30 minutes and air cooled at 5°C/s (9°F/s). The microstructure of the as-received material, shown in Figure 1a, consists of fine alpha phase particles in a beta phase matrix. The microstructure after heat treatment (Fig. 1b) consisting of beta grains in the 50 to 200 µm size range. Specimens with a length to diameter (aspect) ratio of 1.125, with a 10.2 mm diameter and 11.5 mm height (0.4 in. by 0.45 in.) were machined by EDM, surface ground, and then heat treated. As shown in an earlier study, this material is prone to shear band formation, and reducing the aspect ratio decreases the tendency for flow localization [4].

Compression tests were carried out on an MTS hydraulic test frame, controlled by the MTS TestStar system. Constant strain rate compression tests were carried out at strain rates of 0.01, 0.1, 1, and 5s^{-1}, at 25°, 150°, 205°, 290°, and 370°C (77°, 300°, 400°, 550°, and 700°F, respectively). A graphite base lubricant, DeltaForge 31 (Acheson Chemical Co.), was used between the specimen and die surfaces. Data was recorded in the form of applied load and ram displacement as a function of time. This data was then converted to nominal true stress - true strain flow curves.

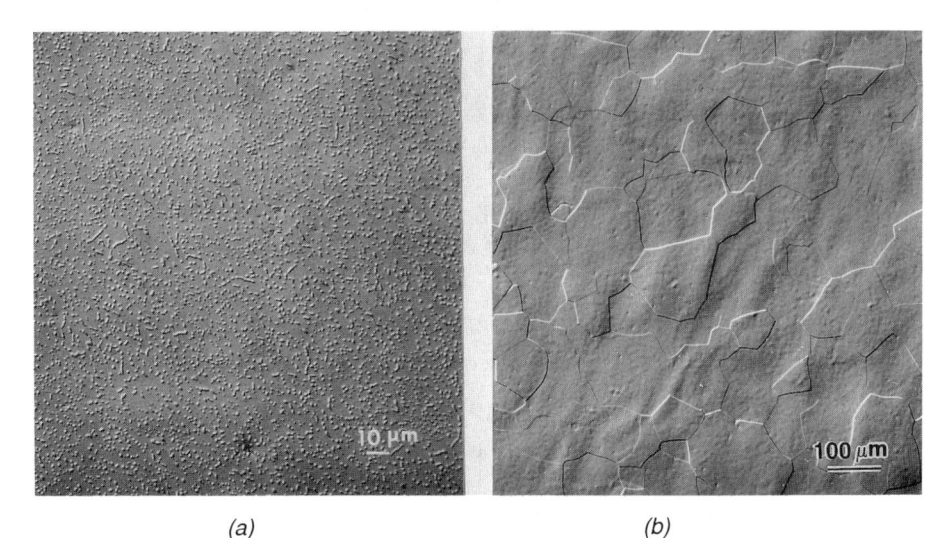

(a) *(b)*
Fig. 1: Microstructure of LCB titanium alloy (a) As-received, (b) After heat treatment at 816°C (1500°F) for 30 min. and cool at 5°C/s (9°F/s).

Results and Discussion
Macroscopic Deformation

Figure 2 shows macrographs of specimens deformed to a reduction in height of approximately 75% at a strain rate of 1s^{-1} and at different temperatures. At room temperature, distinct shear bands form early in the deformation, and subsequently spread into the bulk of the specimen [4]. The tendency for flow localization decreases as temperature increases. However, due to friction, at all temperatures, the deformation is non-uniform. This results in the bulging of specimens and the formation of dead-metal zones near the die contact surfaces. Though the deformation was non-uniform, no macroscopic failure of the specimens was observed. This type of non-uniform deformation and shear band formation has been previously reported in other titanium alloys [5,6].

Flow Curves at Room Temperature

Flow curves for the deformation of LCB titanium alloy, deformed to approximately 75% reduction in height, at 25°C (77°F), are presented in Figure 3a. These curves show the variation of the average true stress (σ = P/A) as a function of nominal true strain (ε = ln[h$_0$/h]), are presented in Figure 3a. Though the yield stress of the alloy increases with increasing strain rate, deformation at high strain rates of 1 and 5s^{-1} results in flow softening after yielding. At slower strain rates of 0.1 and 0.01s^{-1}, the flow curves show work hardening. The temperature rise during a compression test was estimated assuming 95% of the work of deformation was converted to heat, with a fraction of the heat being retained to uniformly heat up the specimen [7]. As seen in Figure 3b, the average temperature of the specimen rises quite rapidly at strain rates $\geq$ 1s^{-1}, reaching approximately 150°C by a strain of 0.2 and almost 500°C, by the end of deformation. As seen in Figure 4a, since the flow stress decreases considerably with increasing

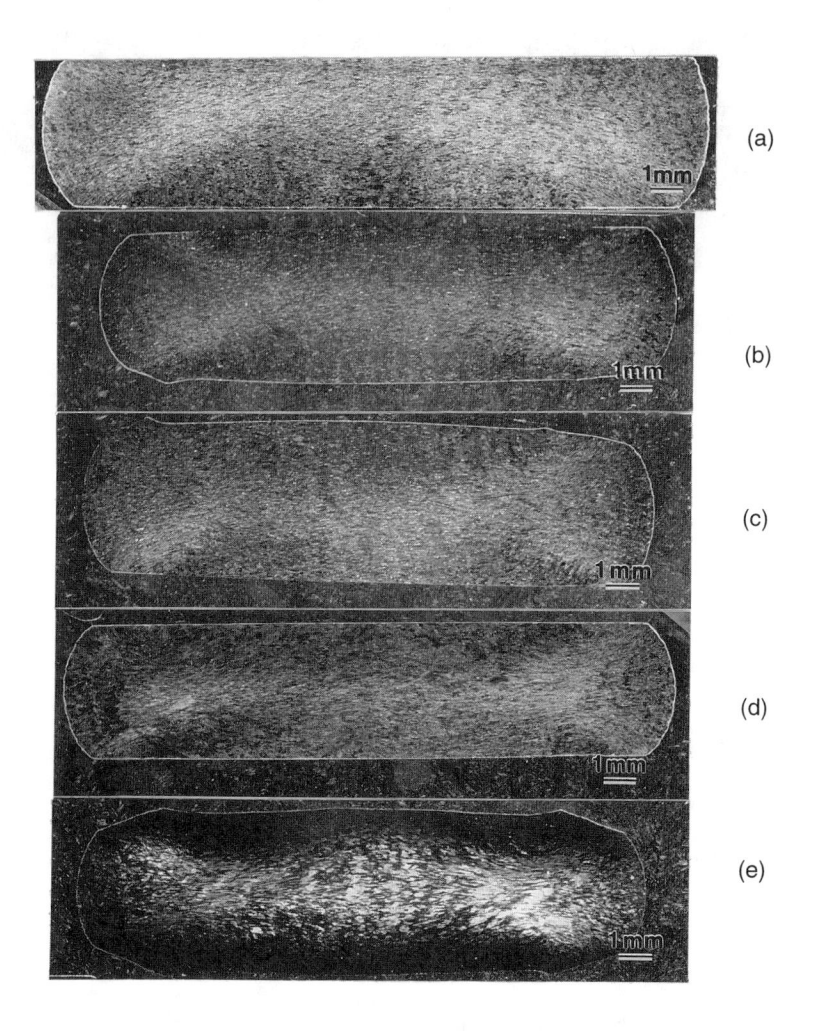

Fig. 2: Macrophotographs of specimens deformed at 1s^{-1} at different temperatures to a true strain of approximately 1.3. (a) 25°C (room temperature), (b) 150°C, (c) 205°C, (d) 290°C, and (e) 370°C.

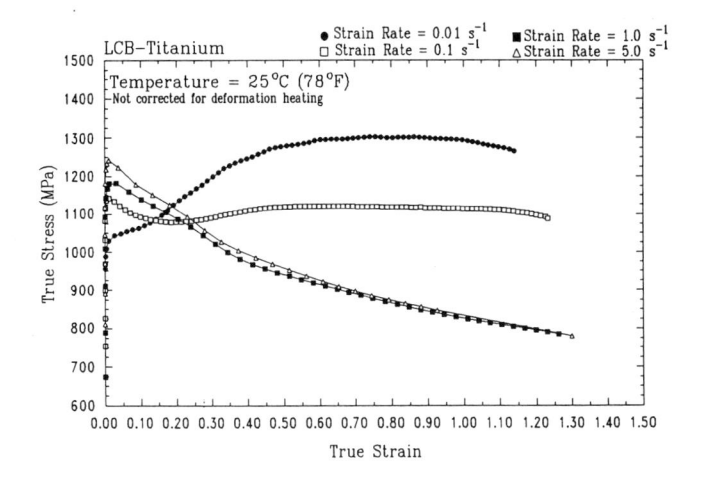

Fig. 3(a): Flow curves from compression tests at 25°C (77°F).

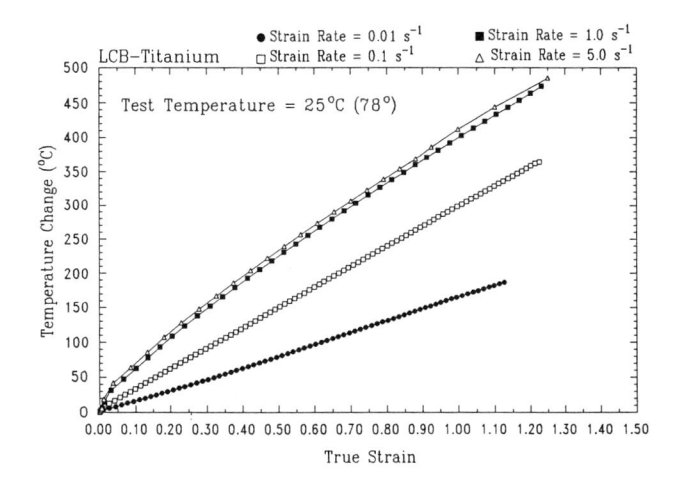

Fig. 3(b): Estimated temperature rise for tests conducted at 25°C (77°F).

temperature, flow localization is further enhanced at high deformation rates in this material.

Flow Curves for Warm Deformation

Figure 4a shows flow curves for compression tests carried out at 150°, 205°, 290°, and 370°C (300°, 400°, 550° and 700°F) at a strain rate of $1s^{-1}$, in comparison to room temperature deformation. There is continuous flow softening at room temperature. As the initial deformation temperature is raised, there is a decrease in the flow stress level, as well as a change in the work hardening behavior. The flow curve at 150°C shows a distinct yield point followed by almost constant flow stress, whereas at 205°C there is a small amount of work hardening followed by deformation with little change in flow stress. At 290°C, however, there is continuous work hardening. At a starting temperature of 370°C, the yield stress of the material is higher than that observed at 205° and, the material work hardens to a higher stress level. As seen in Figure 4b, even when the initial test temperature is increased, the estimated temperature rise is in the range of 350° to 500°C for deformation to a strain of 1.3. The yield stress at a strain rate of $1s^{-1}$ decreases as temperature increases up to a temperature of 290°C, and then increases at 370°C, as shown in Figure 5.

Work Hardening Behavior

Figure 6 shows the normalized work hardening parameter $\gamma = (1/\sigma)(d\sigma/d\varepsilon)$, for the LCB alloy as a function of plastic strain at a strain rate of $1s^{-1}$ for different deformation temperatures. At room temperature, γ is negative beyond a strain of about 0.02. The normalized work hardening parameter is as low as -0.75 at strain of 0.28. The rate of softening decreases (γ becomes less negative) with increasing strain. Had the work hardening parameter become positive, stable deformation of the material would have occurred. Instead, the work hardening parameter becomes more negative beyond a strain of about 0.15. When the deformation temperature is raised to 150°C, there is initial work softening. Steady state behavior ($\gamma = 0$) is observed beyond a strain of 0.15. At 205°C, deformation at a strain rate of $1s^{-1}$ shows a positive work hardening rate which decreases with increasing strain. The stress reaches a steady state beyond a strain of about 0.5, which results in a zero work hardening rate. At 290°C, the work hardening rate is higher than at 205°C, and flow stress value exceeds that for a specimen initially at 205°C at strains greater than about 0.6 (Fig. 4a). The highest initial work-hardening rate is observed for a starting temperature of 370°C. Subsequently, the rate decreases, reaching steady state deformation at strains greater than about 0.5. This phenomenon may be related to microstructural instability of the material above about 250°C which results in the precipitation of the alpha phase [8].

Microstructural Changes During Cold and Warm Deformation

The Starting Microstructure

The as-received material was annealed above the beta transus temperature at 816°C for 30 minutes and cooled at a rate of 5°C/s to room temperature. The microstructure at the resolution level of optical microscope consists of equiaxed beta grains of 50 to 200μm size, as shown in Figure 1b. TEM examination of the as-starting material reveals the presence of the omega phase in the metastable beta phase matrix. Figure 7a shows the

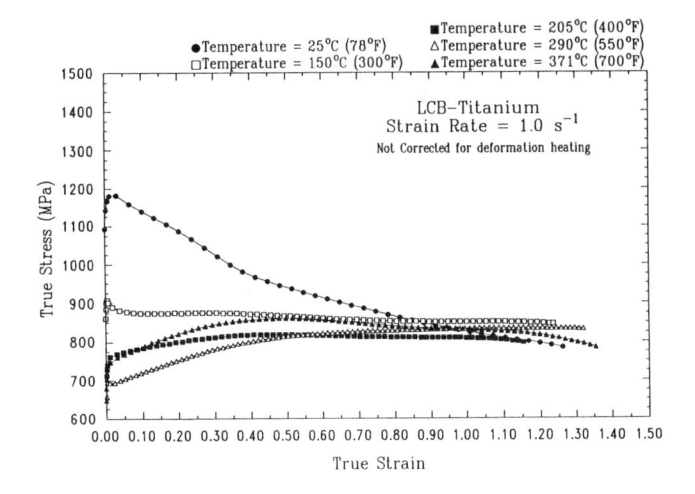

Fig. 4(a): Flow curves from compression tests at a strain rate of $1s^{-1}$.

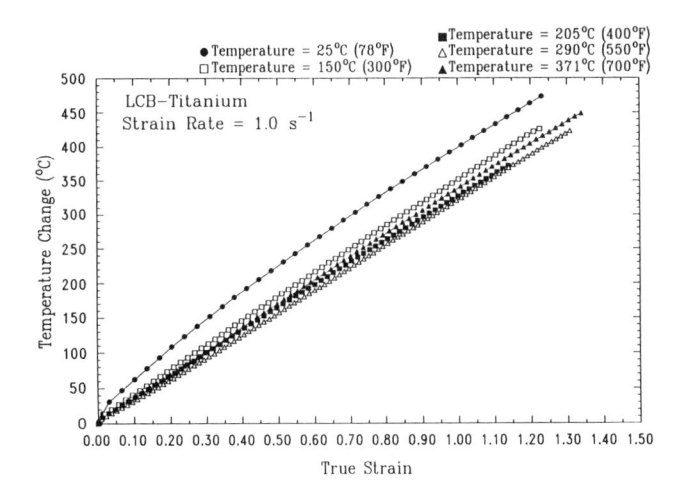

Fig. 4(b): Estimated temperature rise for tests conducted at a strain rate of $1s^{-1}$.

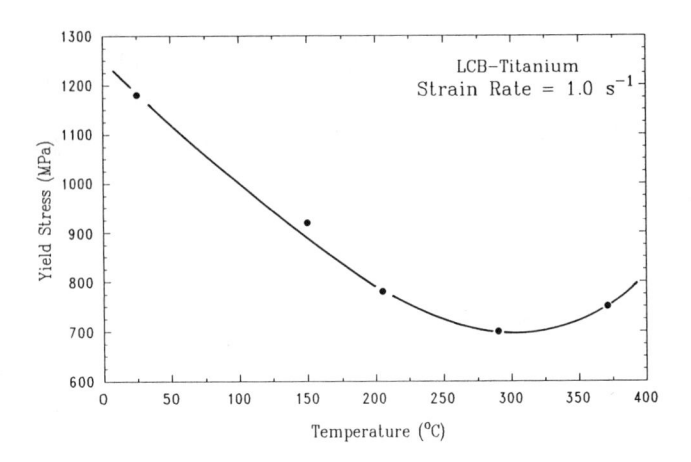

Fig. 5: Variation of yield stress with temperature at a strain rate of 1s^{-1}.

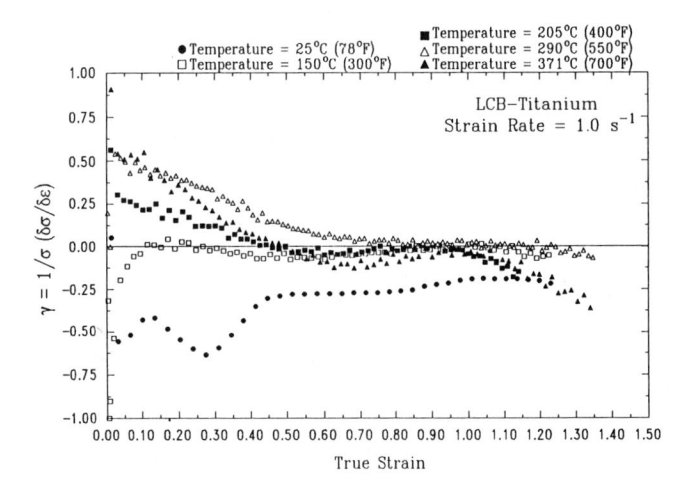

Fig. 6: Variation of work hardening parameter γ with strain at different temperatures and at a strain rate of 1s^{-1}.

bright field image from a beta grain and the corresponding diffraction pattern obtained after tilting the specimen to [110]β zone axis (Fig. 7b). The presence of the extra spots at 2/3(111)β positions indicate the presence of the hexagonal omega phase. The presence of the omega phase in the LCB alloy and many other metastable beta titanium alloys is well established [4,9,10]. The omega phase typically forms athermally during rapid cooling from above the beta-transus temperature, and has been shown to have little effect on the mechanical properties of beta titanium alloys in the as-quenched condition [11]. However, the presence of the athermal omega phase may affect the microstructure, the kinetics and mode of alpha phase precipitation, and consequently the mechanical properties of the alloy [12].

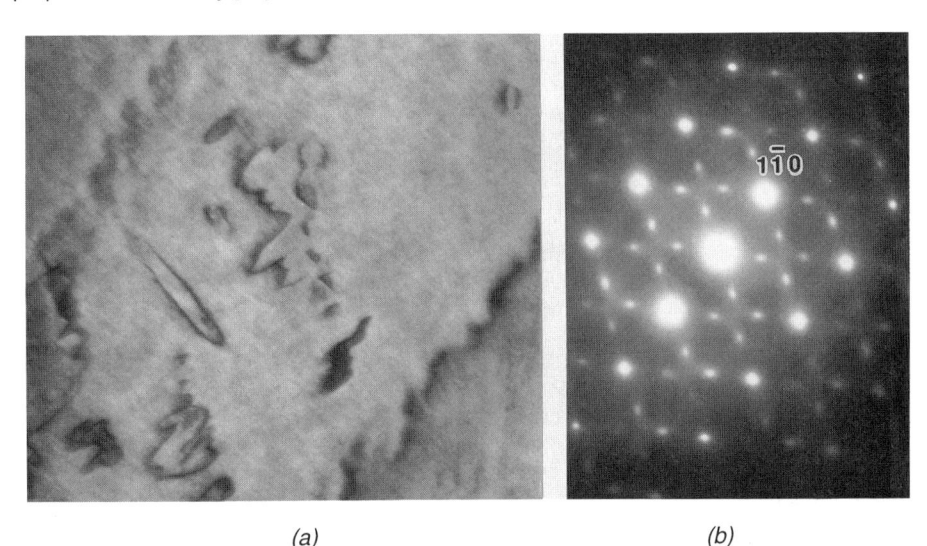

(a) *(b)*
Fig. 7: (a)Bright field image from a beta grain, and (b) the corresponding diffraction pattern.

Microstructural Changes during Cold Deformation

The microstructural changes during room temperature deformation are discussed in detail in another paper [4]. The deformation at room temperature occurred predominantly by slip. The slip character changed progressively from single slip at small strain levels (<0.06) to multiple slip and wavy glide at higher strain levels. A progressively increasing level of macrostructural inhomogeneity and formation of the shear bands in the specimens were observed with increasing strain rate and strain. Some features of the microstructural changes during deformation at a strain rate of $1s^{-1}$ are presented here for comparison with the elevated temperature deformation under similar conditions.

Room temperature deformation at a rate of $1s^{-1}$ was inhomogeneous. Optical microscopic examination at low magnification revealed the presence of shear bands at a strain of 0.26. The grains within the shear band were elongated, whereas the grains outside the band appeared to be equiaxed. TEM examination revealed the occurrence of predominantly single slip in regions outside the shear band. Whereas in areas within the shear band, multiple slip was the dominant mode of deformation. At large strains (~ 1.2), the shear bands spread through the entire specimen, as seen in the macrograph of Figure 2a, and the micrograph of Figure 8a.

Microstructure During Warm Deformation

The specimens deformed at 150°, 205°, 290° and 370°C at a rate of $1s^{-1}$ to a strain of about 1.2, were examined using optical microscopy. The deformation at elevated temperatures was more uniform than deformation at room temperature. Optical micrographs from the longitudinal sections of the specimens deformed to a total strain of about 1.2 at 150°, 205°, 290° and 370°C are presented in Figures 8b, c, d, and e, respectively. The microstructural features seen in all of these micrographs, except Figure 8e, are quite similar. The grains are elongated in the plane perpendicular to the compression direction. The micrographs show that, like at room temperature, deformation at these temperatures also occurred predominantly by multiple slip, with a small fraction of the grains deforming by single slip. The microstructure of the specimens deformed at 370°C is distinctly different from the specimens deformed at lower temperatures. The micrograph in Figure 8e shows straight and wavy lines (slip lines) with dark contrast. These features were present in about half of the grains. While the remaining grains do not show any contrast caused by the slip. It is worth noting that the yield strength decreases up to a temperature of 290°C, but then increases from 290°C to 370°C. This change in yield behavior and the microstructure are probably related to the presence of alpha precipitates in the alloy prior to the start of deformation. Previous work on the LCB indicates that alpha phase precipitation in the metastable beta phase begins during isothermal aging at temperatures of about 300°C [4]. In the present study, the specimens were soaked at 370°C for about 15 minutes before compression. Therefore, it is likely that precipitation of alpha phase started in the specimens before the deformation.

Transmission electron microscopy was carried out on specimens deformed at 205°, 290° and 370°C. The microstructures of specimens deformed at 205° and 290°C were similar

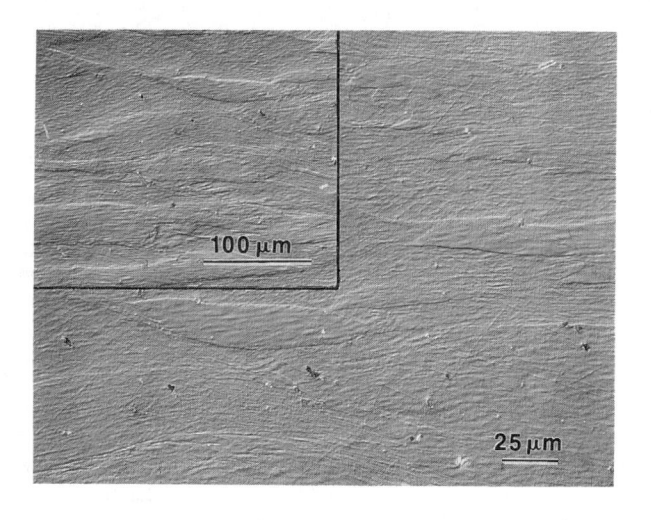

(a) 25°C
Fig. 8: Micrographs of specimens deformed at $1s^{-1}$.

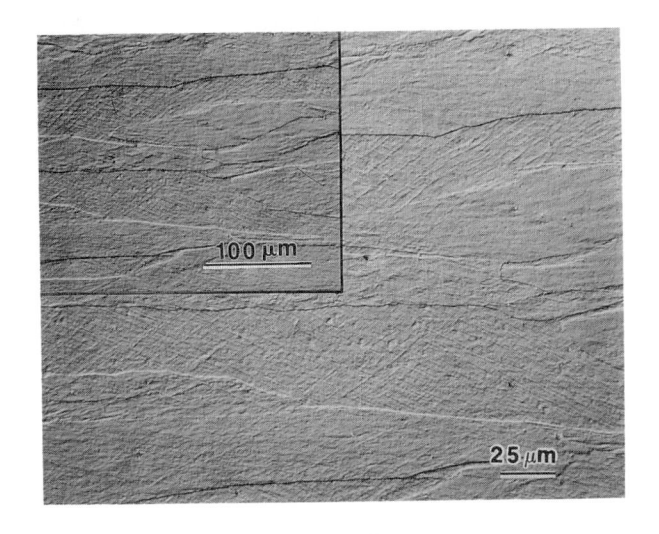

(b) 150°C

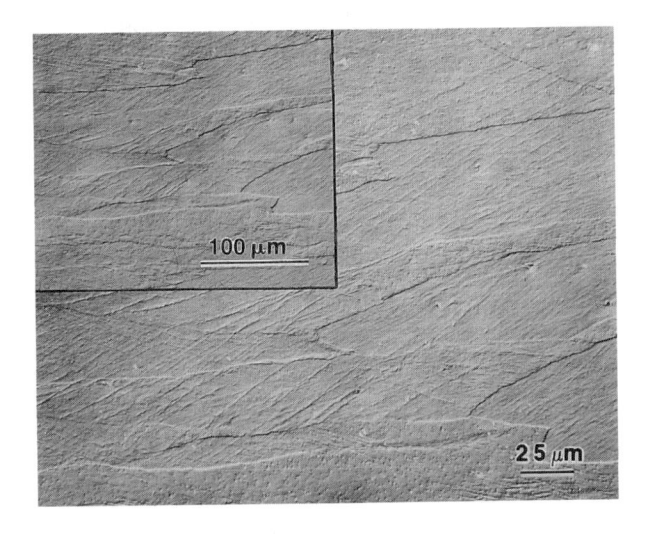

(c) 205°C
Fig. 8:(cont.) Micrographs of specimens deformed at $1s^{-1}$.

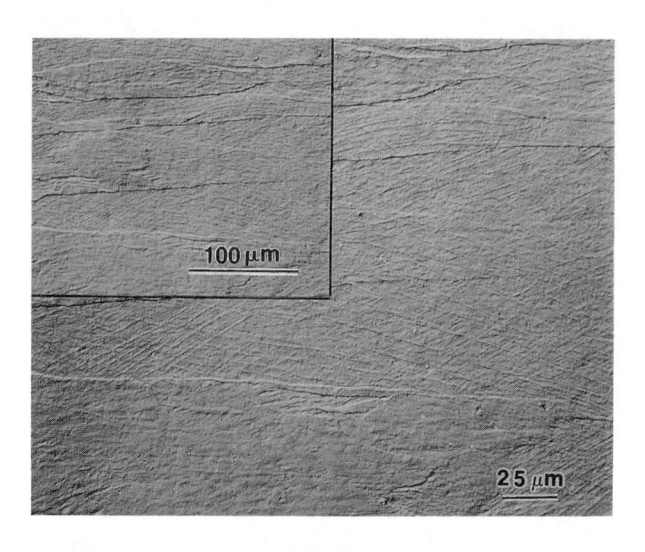

(d) 290°C

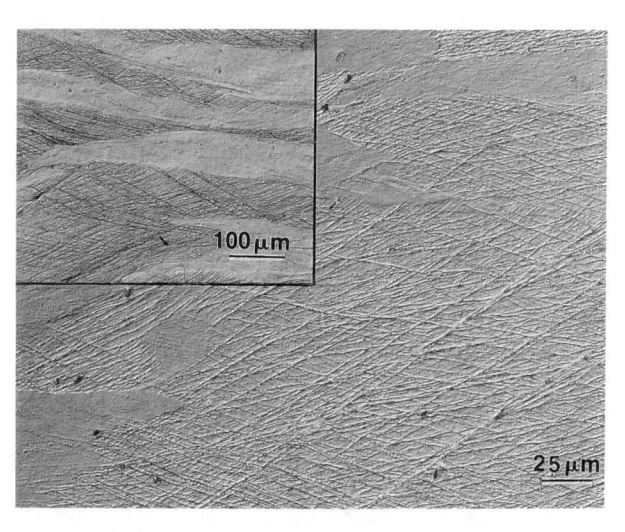

(e) 370°C

Fig. 8: (cont.) Micrographs of specimens deformed at $1s^{-1}$.

(a) Multiple slip region

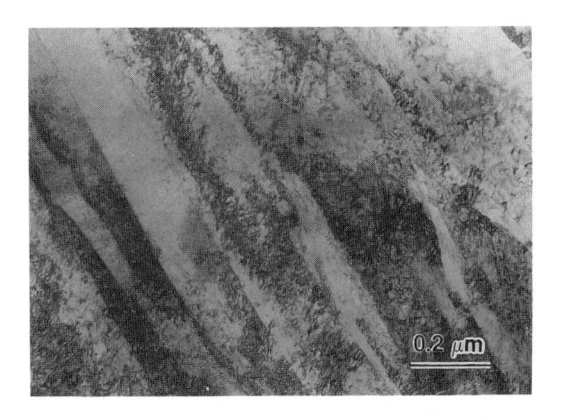

(b) Single slip region

Fig. 9: TEM image from a specimen deformed at $1s^{-1}$ at 205°C.

at the TEM level also. Figure 9 shows microstructures observed in the areas of multiple slip (Fig. 9a) and single slip (Fig. 9b), after deformation at 205°C, as seen in the optical micrographs (Fig. 8c). The diffraction patterns obtained from these specimens (not shown here) indicate the presence of finely dispersed precipitates of omega phase in the metastable beta phase matrix. Some of the regions which had deformed by multiple slip, also show the presence of alpha precipitates. Figure 10 a, b and c show the bright field image, dark field image using a diffraction spot (marked a) from the alpha phase, and corresponding diffraction pattern, respectively. The streaking in the diffraction pattern and extra spot due to the presence of omega phase can also be seen in Figure 10c. Furthermore, it seems that the precipitates of alpha phase nucleate preferentially at the slip lines, where the dislocation density is high. The volume fraction of the regions containing alpha precipitates was very small in the specimens deformed at 205° and 290°C. It is worth noting that the estimated temperature of the specimens increases above 630°C due to deformation heating by the end of the compression test. This temperature is well above the reported stability range of omega phase in other beta titanium alloys [12]. Apparently, the time for which a specimen is at a temperature of 630°C is too short to dissolve all the omega precipitates, or the omega precipitates seen after deformation were formed during cooling after deformation.

Figure 11 shows TEM information obtained from a specimen deformed at 370°C at a rate of $1s^{-1}$ to a strain of 1.3. The images shown in Figure 11 represent the areas with multiple slip, as seen in the optical micrograph (Fig. 8e). The criss-cross morphology seen in the bright field image in Figure 11a is similar to multiple slip. Figure 11b is a higher magnification image from one of the slip bands, marked X in Figure 11a. Instead of the presence of a high density of dislocations, these areas show high density of precipitates. Figure 11c and d show the high magnification dark field image and corresponding diffraction pattern from the area marked Y in Figure 11b. The diffraction

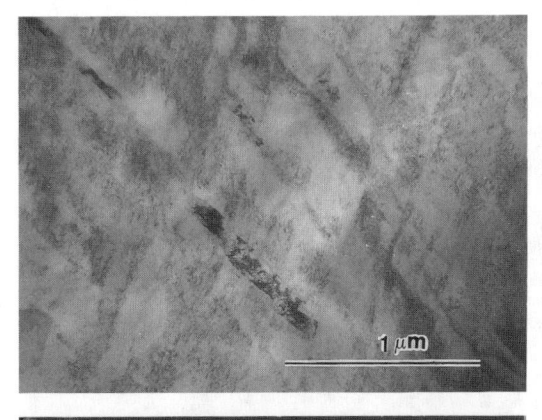

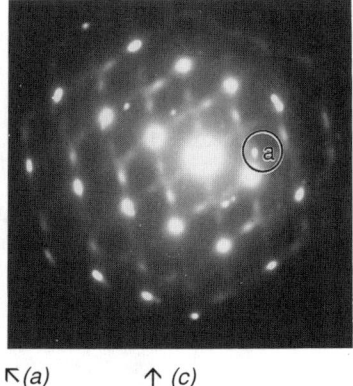

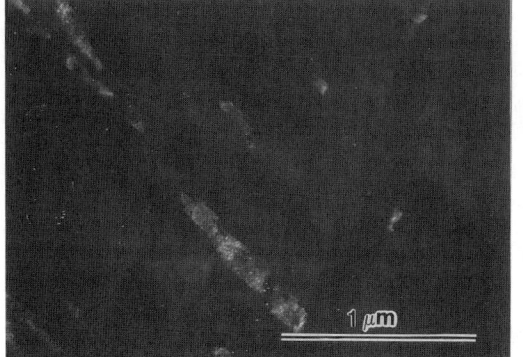

↖(a) ↑(c)
←(b)

Fig. 10: TEM images and diffraction patterns showing the presence of the alpha phase in regions of multiple slip. (a) Bright field (b) Dark field, and (c) Diffraction pattern.

rings in Figure 11d are consistent with a polycrystalline alpha + beta structure. The dark field image obtained using one of the rings from the alpha phase lightens up the α precipitates (Fig. 11c). Figure 12a is a higher magnification bright field image from a region marked Z in Figure 11a. The diffraction patterns from the beta phase region marked X in Figure 12a, after tilting the specimens to [110]β (Fig. 12b) and [111]β (Fig. 12c) zone axes, show spots in addition to those expected from the beta phase. The mottled contrast in the bright field image (Fig. 12a) is similar to the omega precipitates, but the diffraction patterns do not match with the alpha phase, or the hcp omega phase seen in the undeformed material and specimens deformed at lower temperatures (Fig. 7b and 9c). Further analysis to characterize this phase is in progress.

In addition to the grains with dark contrast features, about one half of the grains in the specimen deformed at 370°C were featureless (Fig. 8e). A representative TEM image and diffraction pattern from such areas are shown in Figure 13. The image shows that the dislocations are arranged in cellular type morphology, typical of recovery following

(a) Bright field image *(b) B.F. Image from region X in Fig 11a*

(c) Dark field image *(d) Diffraction pattern*

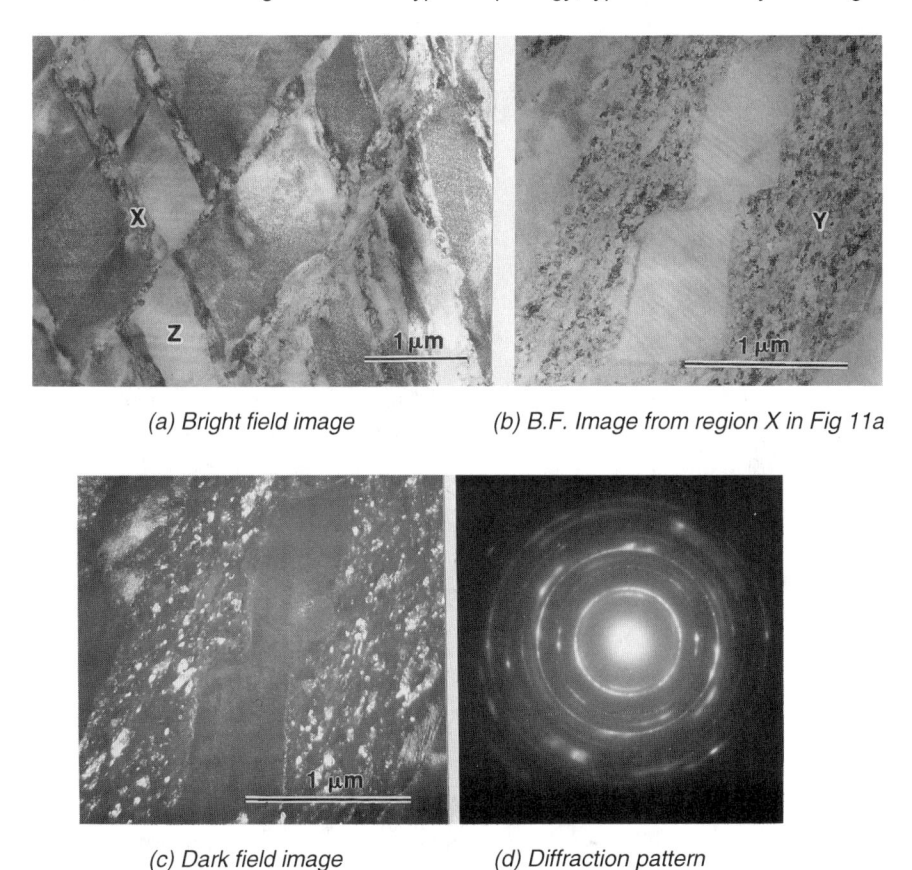

Fig. 11: TEM images and diffraction pattern from a specimen deformed at 1s^{-1} at 370°C.

deformation (Fig. 13a). Furthermore, the alpha precipitates are not present in these areas. The diffraction pattern shows that the metastable beta phase contains the typical omega phase seen in the as starting material, and not the unidentified phase seen in the regions of multiple slip (Fig. 12).

A plausible explanation of microstructure produced by the deformation of LCB alloy at 370°C at 1s^{-1} is as follows. Alpha precipitates form in the material during soaking at

Fig. 12: TEM images From region marked Z in Fig. 11a. (a) Bright field image, (b) Diffraction pattern after tilting to [110]β zone axis (c) Diffraction pattern after tiling to [111]β zone axis.

←(a)
↙(b) ↓(c)

(a) Bright field image *(b) Diffraction pattern*

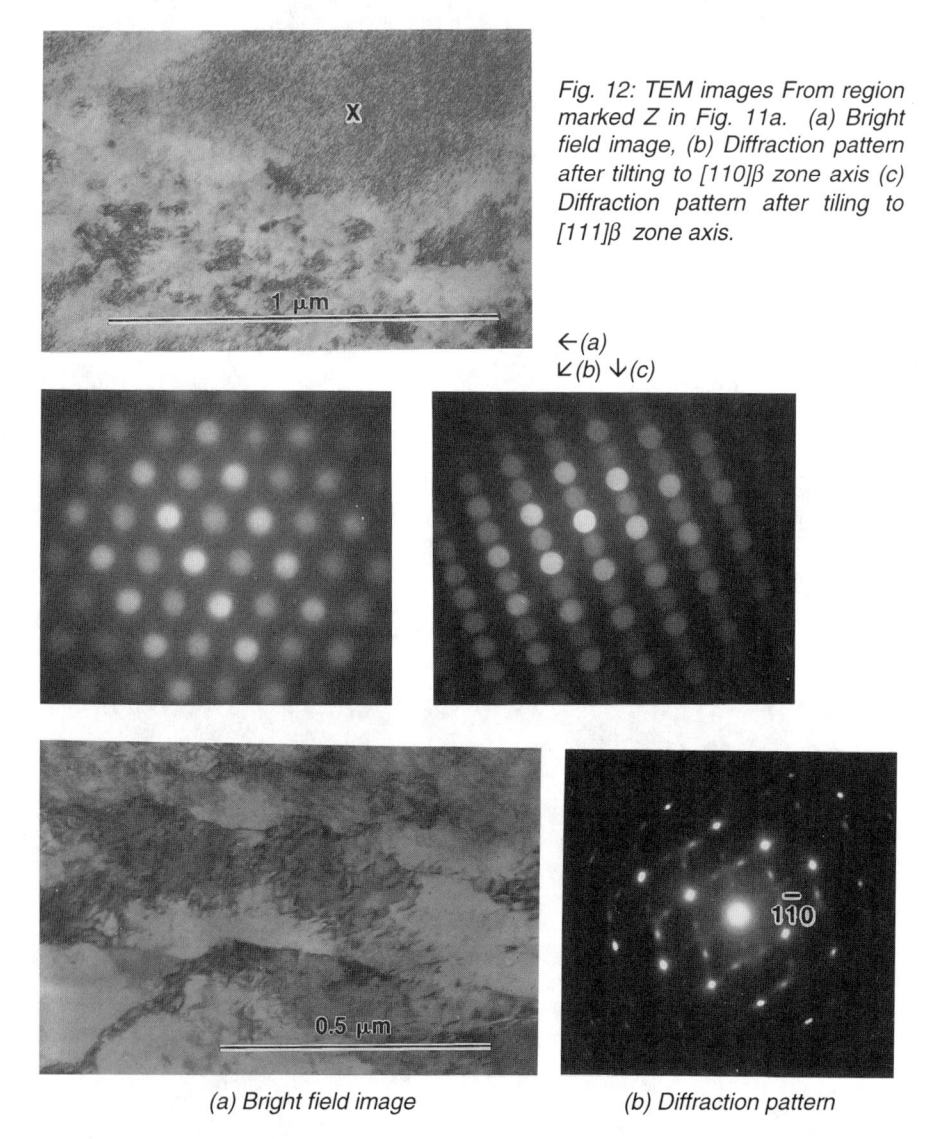

Fig. 13: TEM image from a region without alpha precipitates showing the omega phase and the unidentified phase.

370°C prior to deformation. The uniformly distributed alpha precipitates are expected to offer resistance during deformation and may be responsible for the increased yield strength of the material. As the deformation continues, additional slip lines are produced and more alpha precipitates form preferentially on the slip lines. The rate of nucleation is expected to be higher where larger number of nucleation sites are available i.e. the regions with multiple slip. The rate of growth of the alpha precipitates should also be higher in the regions with multiple slip as compared to the areas with single slip because the temperature rise due to deformation heating is expected to be higher in regions of high deformation. The continued deformation provides the opportunity of growth for the existing alpha precipitates. The growth of the alpha precipitates will also cause the annhilation of the dislocations present in the slip lines. Towards the end deformation when the strain is about 1.3, the deformation induced heating raises the temperature of the specimen to above 825°C, which is higher than the beta transus temperature of this alloy. Above the beta transus, alpha precipitates are not stable. The precipitates may dissolve completely in the areas where the opportunity of nucleation and growth of the precipitate was smaller, i.e. the areas with lesser slip activity. Whereas, in the areas of multiple slip, the precipitates might dissolve only partially. During cooling, following the deformation, the grains with metastable beta phase with no alpha precipitates show the recovery and arrangement of dislocations in a cellular type morphology. In these regions, the omega phase forms during cooling. Whereas, in the grains with large number of alpha precipitates, an unidentified phase forms in the metastable beta matrix. The inhomogeneity of the precipitation, the dissolution of alpha phase, the formation of hcp omega, and the unidentified phase during cooling are reflected in the microstructure observed after deformation.

Summary and Conclusions

The deformation of the low cost beta titanium alloy Ti-6.8Mo-4.5Fe-1.5Al wt% (Timetal® LCB) at room temperature and at warm working temperatures (150 to 370°C) was studied.

Deformation at room temperature at a strain rate of $0.01 s^{-1}$ is characterized by work hardening, whereas, flow softening occurs at a fast strain rate of $1.0 s^{-1}$. At an intermediate strain rate of $0.1 s^{-1}$, following an initial drop in stress upon yielding, little work hardening is observed, and plastic deformation occurs at almost constant flow stress. The yield stress of the material at room temperature is high (1000 to 1250 MPa) over the strain rate range of 0.01 to $1 s^{-1}$. After deformation to a strain between 1.15 and 1.30, the estimated temperature rise, due to deformation heating, is in the range of 200 to 500°C. Despite this large rise in temperature, no alpha phase precipitation was observed. The slip characteristics during deformation change from single slip to multiple slip and then to wavy glide with increasing strain.

Upon increasing the initial specimen temperature between 150 and 370°C (300 to 700°F), stable deformation associated with work hardening is observed at a strain rate of $1.0 s^{-1}$. Below 290°C the slip characteristics are similar to those observed at room temperature. But, above this temperature alpha phase precipitates were observed in the deformed specimens. The precipitates formed predominantly on slip lines. The distribution of alpha precipitates was not uniform. More alpha phase precipitates were observed in grains with multiple slip, while some grains showed no precipitation at all. In addition, precipitates of an unidentified phase were observed in samples deformed at 370°C.

Acknowledgment

Support of this work by the U.S. Air Force Wright Laboratory-Materials Directorate is acknowledged. The authors would also like to thank Drs. Paul Bania and Paul Allen of Timet Corp. for stimulating discussions during the course of this work.

References

1. P.J. Bania, Beta Titanium Alloys in the 1990's, edited by D. Eylon, R.R. Boyer and D.A. Koss, The Minerals, Metals & Materials Society, Warrendale, PA, p 3, 1993.
2. P. G. Allen and A. J. Hutt, Proceedings of the 1994 International Titanium Conference, Titanium Development Association, Boulder, CO, p 397, 1994.
3. "Low-Cost Beta Titanium," Journal Announcement, Mechanical Engineering, p 64, July 1993.
4. I. Weiss, et al., J. Materials Engineering and Performance, Vol. 5(3), p335, 1996.
5. S.L. Semiatin and G. Lahoti, Metall. Trans. A, Vol 12A, p. 1705, 1981.
6. S.L. Semiatin and G. Lahoti, Metall. Trans. A, Vol 13A, p. 275, 1982.
7. P. Dadras and J.F. Thomas, Jr., Metall. Trans. A, Vol 12A, p. 1867, 1981
8. P.A. Blankinsop, Titanium Science and Technology, edited by G. Luetjering, U. Zwicker and W. Bunk, DGM Publishers, Oberursel, FRG, p 2323, 1985.
9. K. Chaudhuri and J. H. Perepezko, Metall. Trans. A, Vol. 25A, p 1109, 1994.
10. T. W. Duering, G.T. Terlinde, and J.C. Williams, Beta Titanium Alloys in the 1980's, edited by R.R. Boyer and H.W. Rosenberg, TMS, Warrendale, PA, p 1219, 1984.
11. T.W. Duering and J.C. Williams, Beta Titanium Alloys in the 1980's, edited by R.R. Boyer and H.W. Rosenberg, TMS, Warrendale, PA, p 24, 1984.
12. A.W. Bowen, Beta Titanium Alloys in the 1980's, edited by R.R. Boyer and H.W. Rosenberg, TMS, Warrendale, PA, p 85, 1984.

COLD FORMABILITY OF TIMETAL® 21S SHEET MATERIAL

J. Reshad,* I. Weiss,* R. Srinivasan,* T. F. Broderick,‡ and S. L. Semiatin‡

* Mechanical and Materials Engineering Department, Wright State University, Dayton, OH 45435

‡ Materials Directorate, Wright Laboratory, WL/MLLN, Wright-Patterson Air Force Base, OH 45433-7817

Abstract

The cold formability of the titanium sheet alloy Timetal® 21S, in the solution-treated and solution-treated-and-aged conditions, was determined through uniaxial tension testing. In the solution-treated (ST) condition, the deformation was essentially all quasistable with negligible uniform elongation. This behavior was explained on the basis of a finite, positive strain rate sensitivity of the flow stress and negligible strain-hardening rate. Failure occurred by ductile fracture within a diffuse neck. Total elongation varied from approximately 20 percent at low strains to 10 percent at higher rates approaching those of conventional sheet forming practice, a trend attributable to the effect of deformation heating on flow localization kinetics. In contrast to the ST behavior, the deformation response of solution-treated-and-aged (STA) material was characterized by measurable uniform elongation, little to no post-uniform quasistable flow, and a brittle, "woody"-type fracture. The total elongation's of the STA material were markedly lower than those of the ST material and showed less of a dependence on deformation rate. The implications of the laboratory formability results on commercial forming practice were examined.

Introduction

The development of advanced aircraft such as the High-Speed Civil Transport (HSCT) requires the introduction of skin materials which can withstand temperatures approaching or exceeding 500°C. Conventional aluminum-base skin alloys are unsuitable for such service conditions; therefore, attention is now focusing on various lightweight, titanium-base alloys. Due to cost considerations, most of the interest in titanium is centered on beta alloys which can be manufactured into coil product using relatively inexpensive continuous cold rolling operations. A second major cost consideration relates to the ability to use conventional sheet forming practices, such as those commonly used by the aerospace, automotive, and appliance industries to stamp aluminum and steel sheet metals. Initial prototype production experience on beta titanium alloys[1] suggests that the cold formability of these materials is limited, and careful selection of sheet forming processes and/or processing conditions are required.

The objective of the present work was to develop a fundamental understanding of the effect of microstructure, for the beta titanium alloy Timetal® 21S, on cold formability. For this purpose, uniaxial, tension tests were conducted. The failure phenomenology and mechanisms were determined through analysis of the engineering stress-strain behavior and macro- and micro-fractography.

Materials and Procedures

Materials

The material used in this investigation was 1.52 mm-thick cold rolled sheet of Timetal® 21S. The chemical composition (in weight percent) of this alloy consists of 13 Mo, 3 Al, 3 Nb, .2 Si, and the balance titanium. The beta transus temperature of the alloy is 807°C. The mechanical properties of the alloy were measured in solution treated (ST) and solution treated and aged (STA) conditions. The solution treatment is comprised holding at 845°C for 30 minutes followed by air cooling. The ST microstructure consisted primarily of 30 μm equiaxed beta grains (Figure 1a) with a small volume fraction of unrecrystallized, elongated grains located at the midplane of the rolled sheet (Figure 1b). In addition, air cooling following the solution treatment gave rise to a very fine second-phase precipitate, presumably silicides or athermal omega, that was not resolvable by optical microscopy.

The STA condition was produced by aging of the ST material at 540°C for 4 hours followed by air cooling. The resulting beta grain structure was identical to that in the ST condition, with a relatively homogenous precipitation of acicular alpha-phase particles (Figure 1c).

Tensile Test Procedures

Material flow behavior was determined by tension tests on ASTM standard E-8 specimens which had been electric-discharge machined from sheets in the two heat treatment conditions with their tension axes along either the rolling or the transverse

(a)

(b)

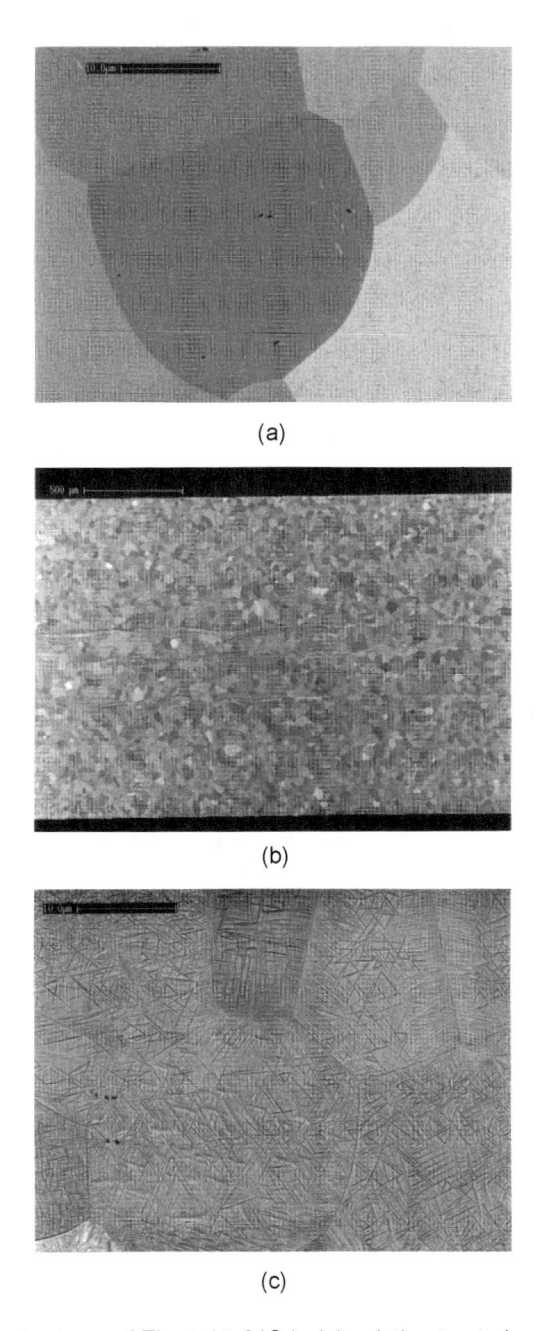

(c)

Figure 1. Microstructures of Timetal® 21S in (a) solution-treated condition (b) solution-treated condition at low magnification and (c) solution-treated-and-aged condition.

direction. Each sample had a reduced section 25 mm long. Tension tests were conducted to failure using an Instron 4500 screw driven machine operated at constant crosshead speeds of .508, 5.08, and 50.8 mm/min., corresponding to initial strain rates ($\dot{\varepsilon}$) of 3.4 x 10^{-4}, 3.4 x 10^{-3}, 3.4 x 10^{-2} s^{-1}, respectively. A 25 mm extensometer was attached to each sample during testing to allow the measurement of the stress-strain behavior, leading to determination of the strain hardening and strain-rate sensitivity exponents. After tension testing, the fracture surfaces of the broken tension specimens were examined using a secondary electron imaging technique in a Leica Stereoscan 360FE scanning electron microscope.

Results and Discussion

The results of this work are summarized in terms of flow behavior and failure observations.

Stress-Strain Behaviors

Typical engineering stress-strain curves for the ST and STA conditions are shown in Figure 2 and 3; other mechanical properties are summarized in Table I. The engineering stress-strain curves exhibited different behaviors for the two heat treatment conditions.

The ST material flow curves (Figure 2) were characterized by a peak load at the onset of plastic flow (with no region of uniform elongation), a period of quasistable flow in which the load dropped slowly with increasing extension, and lastly a region of rapid load drop leading to sample failure. Such observations suggest that the strain hardening exponent $n (= (\partial \log \bar{\sigma} / \partial \log \bar{\varepsilon})|_{\dot{\varepsilon},T})$ is essentially zero, but the strain rate sensitivity exponent $m (= (\partial \log \bar{\sigma} / \partial \log \dot{\bar{\varepsilon}})|_{\bar{\varepsilon},T})$ at the yield stress is finite and positive.[2,3,4] Using the peak load data corresponding to the onset of plastic flow, or data which would not need correction for any deformation heating effects, the strain rate sensitivity exponent was calculated to be approximately 0.025.

The magnitude of quasistable flow observed for the ST condition can be interpreted in terms of previous experimental and theoretical work on the effect of strain rate sensitivity on sheet tension behavior. The magnitude of the quasistable flow for a sheet tension sample with m = 0.01 or 0.03 tested under isothermal conditions is predicted to be approximately 10 or 25 percent respectively[3]. The validity of such predictions was confirmed by comparison to isothermal tension test data for low carbon steel which exhibits m values of this magnitude. For the Timetal® 21S material, testing at 3.4 x 10^{-4} s^{-1} would correspond to isothermal conditions. Hence the observed level of quasistable flow at this strain rate (20 percent) is certainly very reasonable in view of the m value 0.025 and the previous theoretical calculations.

The observed decrease in the amount of quasistable flow with increasing strain rate for the ST material (Figure 2, Table I) can also be interpreted with reference to previous research on the effect of deformation heating on flow localization kinetics. For example, Ayres and Semiatin[5,6] have shown that

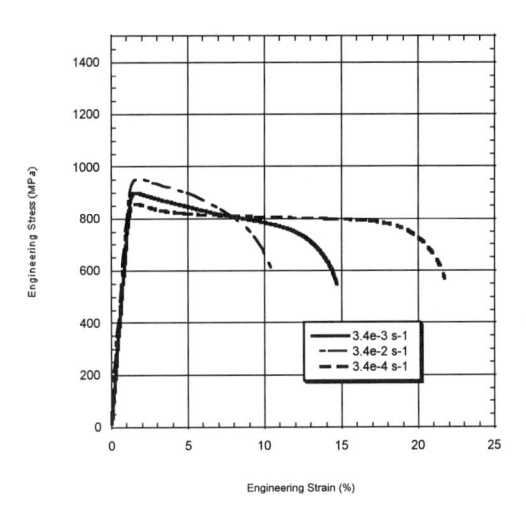

Figure 2. Engineering stress-strain curves for transverse-direction Timetal® 21S specimens tested in the solution-treated condition.

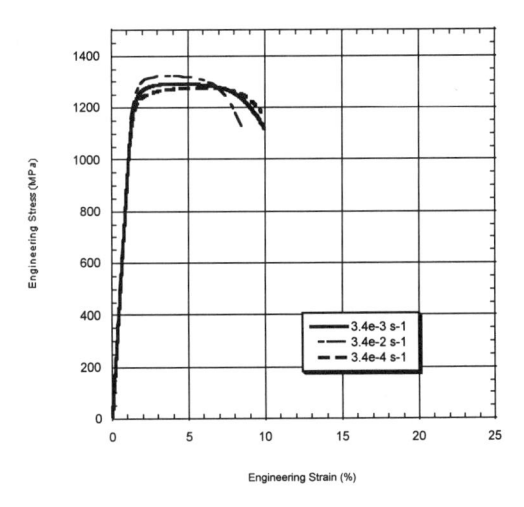

Figure 3. Engineering stress-strain curves for transverse-direction Timetal® 21S specimens tested in the solution-treated-and-aged condition.

deformation heating effects can reduce the amount of quasistable flow by a nominal strain of 10 percent. These former results were for low carbon steel sheet specimens tested at 5.8×10^{-2} s^{-1} in air compared to samples tested at the same strain rate in water (i.e.,under isothermal conditions). This behavior was explained on the basis of the decrease in flow stress with increasing temperature. Strain localization leads to higher temperatures at the center of a diffuse neck, thus resulting in further localization. For Timetal® 21S, a similar qualitative dependence of quasistable flow on strain rate can be expected. However, a quantitative explanation would require detailed calculations (such as those in reference 6 for low carbon steel) to determine the influence of the flow stress levels, thermal properties, etc. that pertain specifically to the titanium alloy.

In contrast to the results for the ST material, the engineering stress-strain curves for solution-treated-and-aged Timetal® 21S showed evidence of nominally uniform flow (increasing true stress-strain curves / broad load maxima) but relatively little post-load-maximum quasistable flow (Figure 3). Thus, it can be concluded that the STA material condition is characterized by a finite strain-hardening exponent ('n') and a rather small strain-rate sensitivity exponent ('m'). In fact, a log-log plot of peak stress-strain rate data in this case yields an m value of only 0.008, thus explaining the limited quasistable flow. As with the ST material, the tensile ductility decreases with increasing strain rate, most likely because of the effect of deformation heating on strain localization kinetics as well. Moreover, the tensile ductilities in the STA condition are only about one-half those of the ST.

Failure observations

The ST and STA samples also exhibited noticeably different failure modes. As would be predicted by the form of their stress-strain curves, examination of the broken ST samples revealed that diffuse necking occurred prior to the fracture event (Figure 4a). The ductile failure mode was accomplished by a microscopically ductile fracture mode as well (Figure 5). Fracture was seen to be controlled by the formation, growth, and coalescence of voids; the fracture surfaces lying at 45° to the tension axis suggest some form of void sheet mechanism. The void-initiating particles could not be identified, but are presumably the second-phase particle (silicides or omega-phase) that formed on air cooling from the solution treatment. It is well documented that the omega phase is known to promote localized slip through particle shearing. This deformation characteristic would decrease the opportunity to promote work hardening in the material yielding low n values. Therefore, the ST flow behavior with approximately zero strain hardening could have been caused by the presence of the omega phase. Identification of the void-initiating particles is in-progress and will be included in a later publication.

The broken STA specimens revealed essentially no diffuse necking, an observation in line with the low m value and near-absence of quasistable flow (Figure 4b). The fracture mode for STA samples was also quite different from that of the ST samples. The STA specimens failed in a brittle manner (Figure 6), and this type of fracture is often referred to as having a "woody" appearance.[7] It may be hypothesized that the woody fracture appearance is likely related to the acicular alpha phase. In addition, the macroscopic fracture at approximately 45° to the

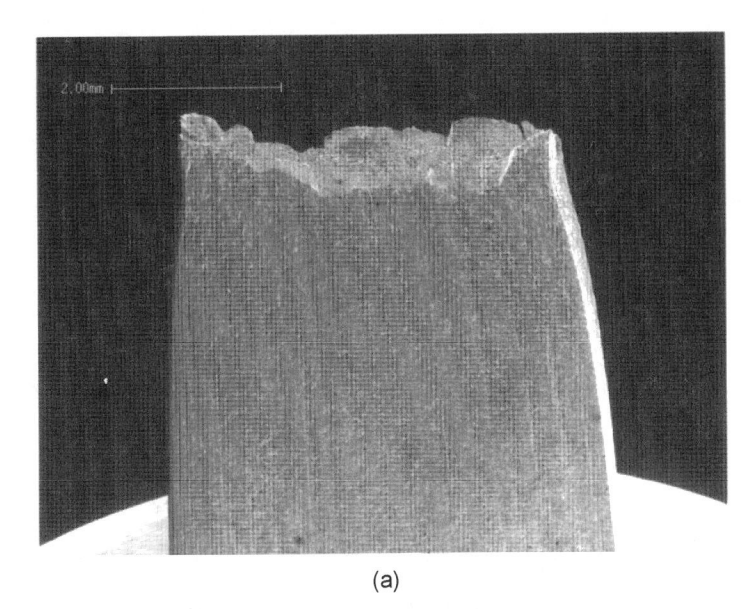

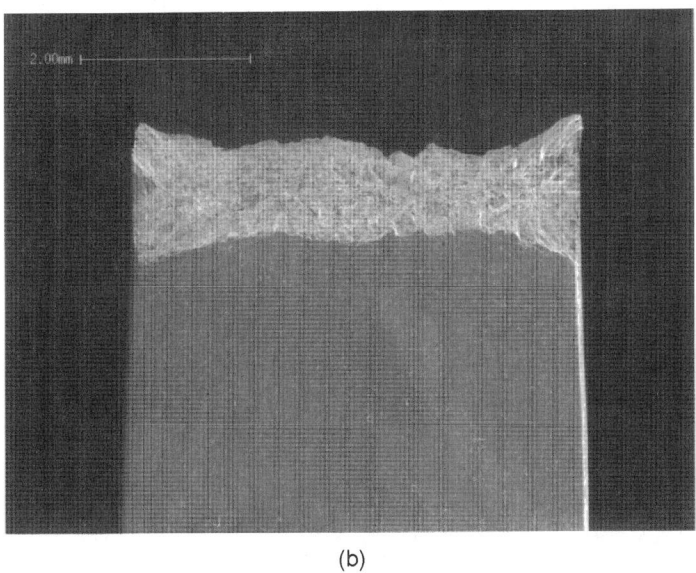

Figure 4. Macrographs of failed transverse-direction Timetal® 21S specimens in (a) solution-treated and (b) solution-treated-and-aged conditions, tested at $\dot{e} = 3.4 \times 10^{-4}$ s^{-1}.

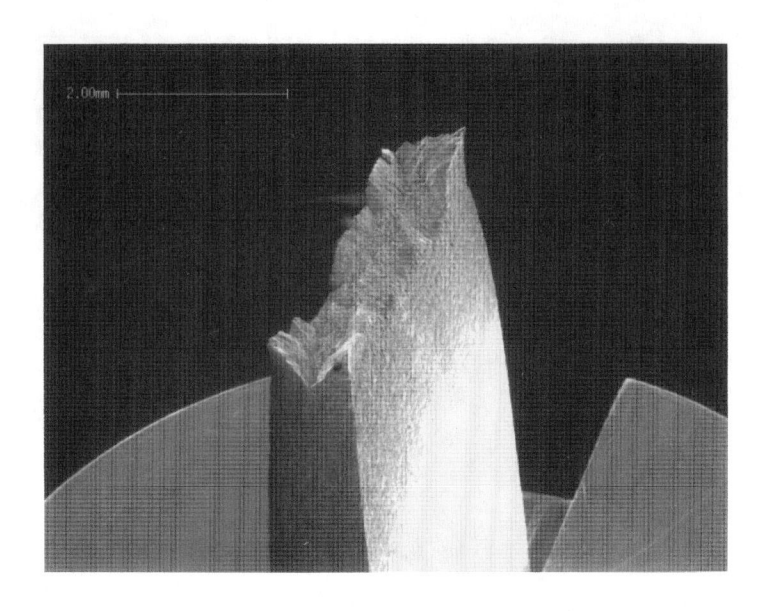

Figure 5. SEM fractographs of a failed transverse-direction ST Timetal® 21S specimen tension tested at $\dot{e} = 3.4 \times 10^{-4}$ s^{-1}.

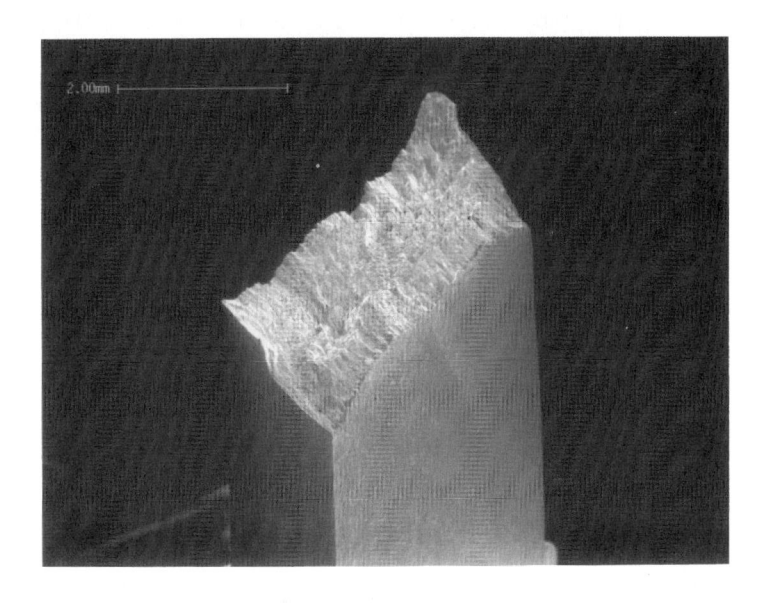

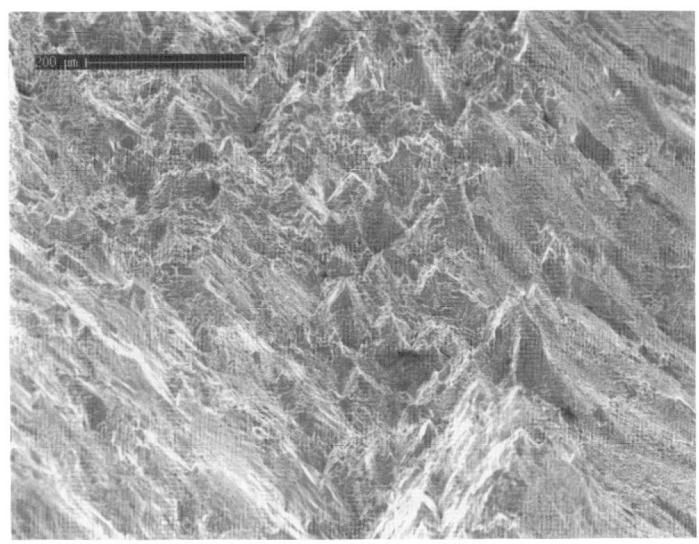

Figure 6. SEM fractographs of a failed transverse-direction STA Timetal® 21S specimen tension tested at $\dot{e} = 3.4 \times 10^{-4}$ s^{-1}.

tension axis for the STA material suggests the importance of stress state as well as microstructure in controlling fracture. In this regard, because of the multiple orientations in which the alpha plates may precipitate within a given beta grain, the macroscopic "ledge" formed during fracture (shown in Figure 6) suggests that crack propagation occurs along those directions most favorably oriented relative to a specific alpha plate orientation within each beta grain.

Production Implications

The present formability results demonstrate that Timetal® 21S has limited cold formability. At strain rates typical of production stamping operations (~0.1 to ~1.0 s^{-1}), the solution-treated condition has only approximately 10 percent elongation, and the solution-treated-and-aged condition exhibits only about 6 percent elongation. These ductilities are considerably less than those of conventional aluminum and steel sheet metals (i.e., ~25 and ~40 percent elongation, respectively) which are stamped. Hence, the Timetal 21S material may be suitable only for the most mild of stretching-type operations or relatively large bends (> 10 x thickness). Moreover, additional testing (e.g., forming limit diagram determination, deep drawing evaluations) for Timetal® 21S should be conducted to define the formability of this alloy in more detail.

The limited ductility of Timetal 21S in the ST and STA conditions suggests that other processing treatments should be applied in order to improve formability. One possible approach would involve solution treatment, cold rolling to a 4 to 8 percent reduction, and partial aging or overaging treatments. By this means, a low density homogeneous distribution of fine (or coarse) spheroidal alpha-phase precipitates may be developed. Such microstructures may enhance a more uniform elongation which is lost in the ST condition due to the presence of athermal omega phase. The spheriodal alpha-phase precipitate morphology may also enhance post-uniform, quasistable flow relative to that obtained in the conventional STA material condition with an acicular alpha phase microstructure.

Summary

The formability of Timetal® 21S sheet was determined by uniaxial tension tests on material in the solution-treated (ST) or solution-treated-and-aged (STA) conditions. The ST material had a tensile ductility between 10 and 20 percent, the higher values obtained at low strain rates. At all strain rates, quasistable, post-load maximum flow accounted for the majority of the elongation. The magnitude of the quasistable flow was explained on the basis of the strain-rate sensitivity of the flow stress and deformation heating effects. In contrast, the STA ductilities were controlled primarily by the magnitude of the uniform elongation. The fracture mechanisms (ductile for the ST condition and brittle for the STA conditions) mirrored the tensile ductility observations.

Acknowledgments

This work was part of a joint research project between NASA Langley, the Wright Laboratory Materials Directorate, and Wright State University. The authors thank D. Dikus of NASA Langley and H. Lipsitt of Wright State University for helpful discussions and D. Hoffman of Boeing Aircraft Corporation for the supply of test materials. One of the authors (JR) was supported by the Southern Ohio Council for Higher Education.

Table I. Mechanical Properties of Timetal® 21S Sheet Material

Material Condition*	Nominal Strain Rate (s^{-1})	Test Direction	Yield Strength (MPa)	UTS (MPa)	Uniform Elongation (pct.)	Total Elongation (pct.)
ST	3.4×10^{-4}	RD	874.8	882	~0	15.1
ST	3.4×10^{-4}	TD	849.9	870.9	~0	21.5
ST	3.4×10^{-3}	RD	909.6	912.9	~0	10
ST	3.4×10^{-3}	TD	897.4	902	~0	14.7
ST	3.4×10^{-2}	RD	894.2	998.6	~0	10.7
ST	3.4×10^{-2}	TD	954.3	955.3	~0	9.8
STA	3.4×10^{-4}	RD	1226	1327	5.2	10.5
STA	3.4×10^{-4}	TD	1190	1279	5.7	9.8
STA	3.4×10^{-3}	RD	1289	1362	3.2	8.4
STA	3.4×10^{-3}	TD	1219	1295	3.9	9.7
STA	3.4×10^{-2}	RD	1344	1393	3.1	6.7
STA	3.4×10^{-2}	TD	1281	1324	3.7	8.1

* ST: solution treated 845°C/30 minutes + air cooled
STA: ST + aged 540°C/4 hours + air cooled

References

1. D. Dikus, NASA Langley Research Center, Hampton VA, unpublished research, 1994.

2. A.K. Ghosh, *Metall. Trans. A*, 1977, vol. 8A, p.1221.

3. S.L. Semiatin, A.K. Ghosh, and J.J. Jonas, *Metall. Trans. A*, 1985, vol. 16A, p. 2291.

4. S.L. Semiatin and J.J. Jonas, *Formability and Workability of Metals: Plastic Instability and Flow Localization*, ASM, Metals Park, OH, 1984.

5. R.A. Ayres, *Metall. Trans. A*, 1985, vol. 16A, p. 37.

6. S.L. Semiatin, R.A. Ayres, and J.J. Jonas, *Metall. Trans. A*, 1985, vol. 16A, p. 2299.

7. G. F. Voort, Metals Handbook® Ninth Edition, 1987, vol. 12, p. 104.

COLD EXTRUSION OF TITANIUM ALLOYS AND TITANIUM - MMCs

Hans Wilfried Wagener and Joachim Wolf

Metal Forming Laboratory, Department of Mechanical Engineering
University of Kassel, 34109 Kassel, Germany

Abstract

For the optimum use of titanium alloys and titanium metal matrix composites (MMCs) as structural materials, the disadvantage of high producer prices has to be overcome by engaging a manufacturing process, i.e. cold extrusion, which saves high price material (net shape forming) and improves the properties of the components produced: These are high accuracy, excellent surface quality, improved hardness by strain hardening, and good fatigue properties.
The following metals were tested by the basic cold extrusion processes:
TiAl6V4 (grade 5), TiV13Cr11Al3 (UCA), TiAl3V8Cr6Mo4Zr4 (ß-C) and the two MMCs Ti 99.9 +10% TiC and TiAl6V4+10% TiC.
For comparison purposes, extrusion experiments were carried out with Ti 99.9 (grade 1) and the austenitic stainless steel X2CrNiMo1812 (AISI, TP, 316 L).

Introduction

The application of titanium, titanium alloys, and titanium MMCs as structural materials is very promising due to their excellent mechanical (strength/density ratio) and corrosion properties. Because of the high producer price for these materials it is necessary to employ processing methods for the production of structural components without any waste of material. Compared with manufacturing these titanium based components by machining, 20% to 40% of the material can be saved by applying cold or hot forging processes, i.e. near net shape or net shape forging. Depending on the final configuration and on the batch sizes of the component to be produced, net shape forging can be achieved by cold extrusion if the production sequence, the lubricational system used, and the tool design are optimum. Cold extrusion of these materials includes the additional advantages of high accuracy production combined with very good surface quality of the components produced and of improved fatigue peoperties and increased strength by strain hardening. The successful application of cold extrusion of titanium alloys and titanium matrix MMCs gives a new approach and a new dimension in designing components for highly loaded but light-weight machine elements. The problems concerning cold extrusion of unalloyed commercial pure titanium Ti 99.9 of galling, i.e. to stick and fret and also cold weld with the high strength steel tools used in cold extrusion, were solved by introducing special lubricational systems [1].

The greater technological and economic advantages of cold extrusion of titanium alloys, because of the larger tonnage used for industrial purposes, but also the advantages of titanium matrix MMCs, were not used so far for mass production and industrial application. The greater values of the flow stress and the reduced cold workability (brittle behaviour) of some alloys and of the MMCs reduce the chances of the production of crack-free components by cold extrusion.

The high flow stress of the alloys and MMCs requires tool materials and tool design, e.g. extrusion punch and extrusion die of outstanding hardness and resistance against wear. It requires high pressure lubricational systems and a complicated tool configuration to improve the workability of the material by a highly compressive state of stress in the forming zone [2, 3, and 4].

Tested Metals

The main purpose of the research is the determination of the forming parameters and characteristics of cold extrusion of the titanium alloys TiAl6V4 (grade 5), of the older ß-alloy TiV13Cr11Al3 (UCA), of the ß-alloy TiAl3V8Cr6Mo4Zr4 (ß-C), and of two types of titanium matrix MMCs Ti 99.9+10% TiC and TiAl6V4+10% TiC. For comparison purposes, the unalloyed Ti 99.9 (grade 1) and the austenitic stainless steel X2CrNiMo1812 (AISI, TP, 316 L) were also tested.

The chemical composition is listed in Table I and the mechanical properties in Table II.

Table I: Chemical composition of the tested metals.

Metal weight [%]	Fe	C	O	N	H	Al	V	Ni	Cr	Zr	Mo	Ti
Ti 99.9	0.04	0.01	0.04	0.01	0.002							bal.
TiAl6V4	0.13	0.02	0.18	0.01	0.005	6.44	4.0					bal.
TiAl3V8Cr6Mo4Zr4	0.08	0.02	0.08	0.01	0.01	3.30	7.8		5.8	3.7	3.7	bal.
TiV13Cr11Al3	0.16	0.02	0.16	0.02	0.015	3.35	14		11			bal.
X2CrNiMo1812	bal.	0.01				0.02		13.6	17.4		2.63	

The main forming characteristics of a metal are determined by the flow curve (true stress vs. true strain diagram). The flow curves of the tested materials are shown in Figure 1. They are experimentally obtained by the uni-axial compression test.

These diagrams demonstrate that the two titanium MMCs (curves 3 and 7) have a very small ductility, i.e. the maximum strain is only ϵ_h = 0.2 and 0.27 respectively.

The MMCs could be classified as brittle metals because the fibrous and particle strengthening act as

Table II: Mechanical properties of the tested metals.

Mech. properties Tested Metal	$R_{P0.2}$ [MPa]	R_m [MPa]	A_s [%]	HV10 [-]	n [-]	Z [%]
Ti 99.9	156	370	42	137	0.33	72
Ti 99.9 + 10 Vol.% TiC	645	700	2.7	331	0.011	4
TiAl6V4	989	1060	16	297	0.09	40
TiAl6V4 + 10 Vol.% TiC	910	924	1.5	407	0.05	< 1
TiAl3V8Cr6Mo4Zr4	813	815	15	274	0.11	46
TiV13Cr11Al3	931	933	32	317	0.09	51
X2CrNiMo1812	244	591	55	168	0.34	48

material separations in the matrix. The alloy TiAl6V4 is of medium ductility and the two ß-alloys, unalloyed Ti 99.9, and the stainless steel have a limit strain which is greater than $\epsilon_h = 1.0$.
The strengths of TiAl6V4 + 10% TiC (curve 3) and of the alloy TiV13Cr11Al3 (curve 1) are nearly of the same value. The metals start yielding at a flow stress $\bar{\sigma} = 1,200$ MPa. The flow stress of Ti 99.9 + 10% TiC (curve 7) corresponds to the value for the stainless steel and is approx. 45 % greater than the flow stress of the unalloyed titanium Ti 99.9 (curve 6).
These are important data for designing components of these metals for industrial commercial application.

The work hardening exponents (n = $\lg\Delta\bar{\sigma}/\Delta\lg\epsilon_h$) of the metals are listed in Table II. The n-values also give certain information about the ductility of a metal.

Cold Extrusion Processes and Test Set-up

Experiments were carried out utilizing the following standard extrusion processes:
1. Forward bar extrusion (FBE)
2. Forward cup extrusion (Hooker; FTE)
3. Backward can extrusion (Cupping, BCE)

The extrusion tests were performed on a 1,000 kN hydraulic press or on an eccentric press of a nominal force of 2,500 kN.

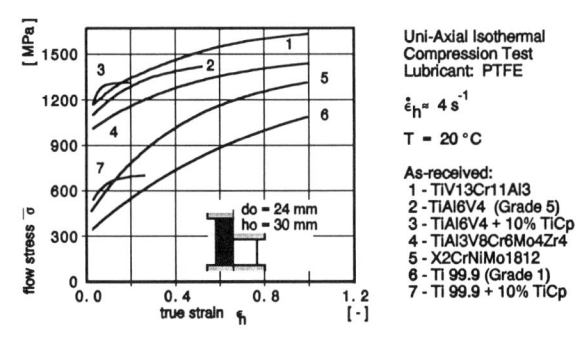

Figure 1. Flow curves of the tested metals.

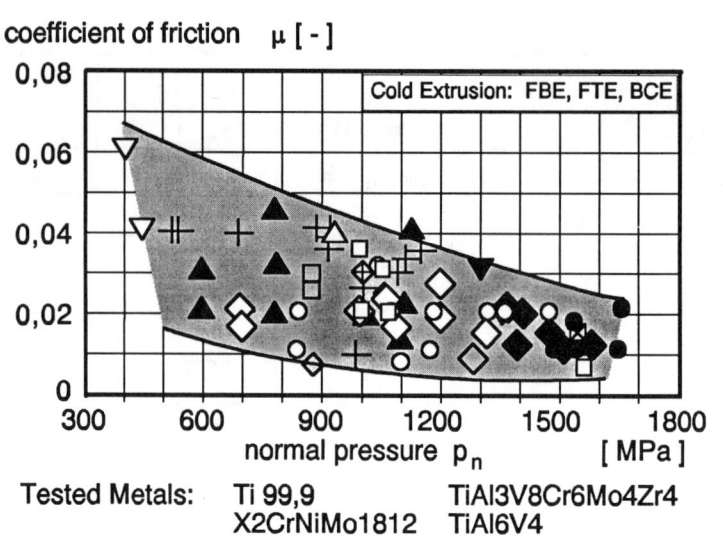

Figure 2. Influence of normal pressure on the coefficients of friction

The surface of the cylindrical specimens (slugs) is coated by means of specific lubricational systems (Table III) used for the different types of materials under test to ensure the avoidance of galling and to obtain minimum frictional forces during the cold extrusion process.

The punches and the dies are made of high-speed steel of HRC 63. The dies are reinforced by double shrink rings. Figure 3 demonstrates the principel of an experimental tool set-up for FBE.

One objective of the extrusion experiments is to find the extrusion parameters where the load on the tools is minimum to guarantee maximum tool life. For this reason, in Table III the values of the coefficient of friction between the coated slugs and the tool surface are also mentioned. They were determined experimentally by the usual extrusion tests. It was found that the μ-values for the extrusion of different metals are a function of the normal pressure acting on the tool surface. The values are quite close and decrease with increasing normal pressure p_n, see Figure 2. The lubrication system Galvanic-Cu and PVD-Cu combined with MoS_2 have the smallest coefficients of friction $\mu = 0.015$ to 0.02. The μ-values for the system PVD-Fe with zinc phosphate and zink soap (BONDER) are around 0.045. Since the chemical equipment for the surface treatment of the slugs with zinc phosphate is available in all cold forging shops, the final solution to the problem of applying cold extrusion to all metals and alloys is to produce a thin PVD layer of iron (Fe) on the surface of the slugs.

Extrusion Force vs. Punch Travel Diagrams

The main characteristics and forming data of the cold extrusion processes can be studied by means of the extrusion force vs. punch travel diagram. This type of diagram deliveres the required data and information to the metal forming expert to select the optimum press, concerning nominal force and work capacity, and to design the tool strength for the processing of these metals by cold forging.

Forward Bar Extrusion (FBE)

In Figures 4 to 7 diagrams for FBE of TiAl6V4, of TiV13Cr11Al3, and of the two titanium-MMCs are plotted. Figure 4 shows the force vs. travel diagram and, for comparison with the other metals, the required extrusion pressure of FBE of TiAl6V4. At the beginning of the extrusion operation, there is a steep increase of the extrusion force, then the period of steady-state flow can be observed. For

Figure 3. Tool set-up for FBE.

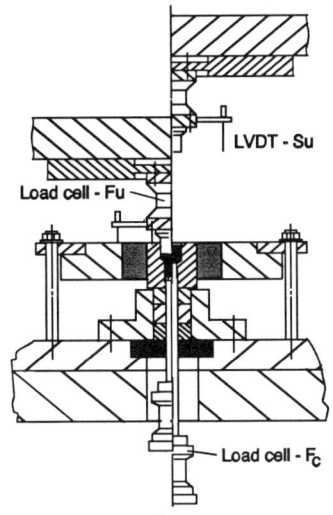

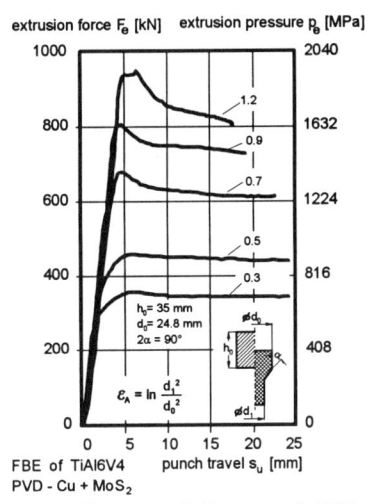

Figure 4. Force-travel diagrams of FBE of TiAl6V4

greater values of extrusion strain ϵ_A, a reduction of the force takes place after the force maximum due to decreasing frictional forces and increasing temperature of the material in the forming zone.
For these experiments a maximum extrusion pressure $p_e = 1,940$ MPa is required.
For the ß-alloy TiV13Cr11Al3 the corresponding diagrams are shown in Figure 5a. Additionally to the normal force-travel curves three dotted curves are plotted. These are the results of extrusion experiments making use of the method of "thick-film lubrication", i.e. the slug is covered on its whole surface with a 0.5 mm thick coating of MoS$_2$ paste. This method simulates "hydrostatic extrusion" to

Table III: Lubricational systems and coefficients of friction.

Lubricational systems	Metal	Coefficient of Friction
Oxide (300 °C, 4 h) + MoS$_2$	Ti 99.9 TiAl3V8Cr6Mo4Zr4	0.11
PVD-Copper+MoS$_2$ approx. 6 μm thickness of Cu layer	Ti 99.9 TiAl6V4 TiAl3V8Cr6Mo4Zr4 X2CrNiMo1812	0.025 0.02 0.02 0.02
PVD-Iron+zinc phosphate+zinc soap approx. 24 μm thickness of Fe layer	Ti 99.9 TiAl6V4 TiAl3V8Cr6Mo4Zr4 X2CrNiMo1812	0.04 0.025 0.02 0.03
Galvanic-Copper+MoS$_2$ approx. 15 μm thickness of Cu layer	Ti 99.9+10% TiC X2CrNiMo1812	0.015
Zinc phosphate+zinc soap (Bonder)	TiAl6V4+10% TiC Mild steels	0.045

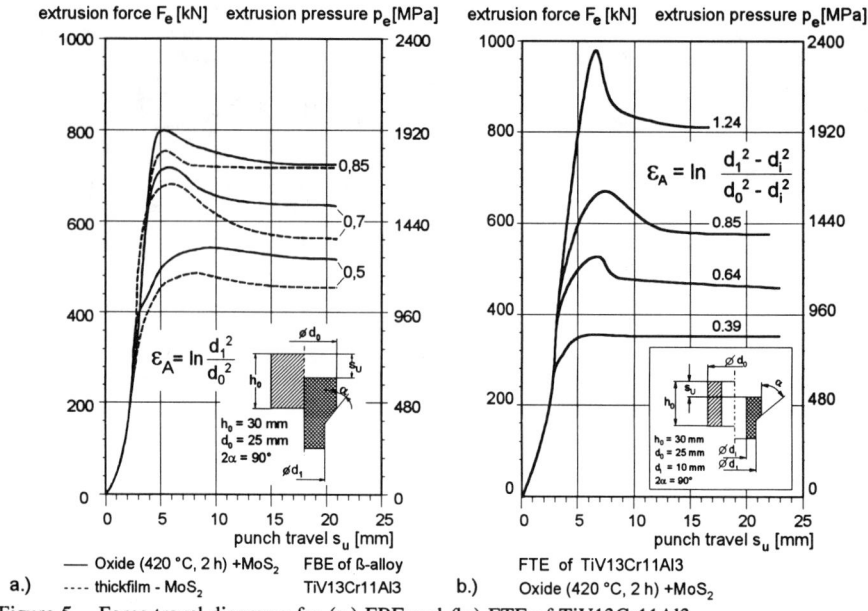

Figure 5. Force-travel diagrams for (a.) FBE and (b.) FTE of TiV13Cr11Al3.

a certain extent. By the reduction of friction between the acting surfaces of the extrusion die and the component under extrusion, the required values of extrusion force and extrusion pressure are decreased by approx. 10%. Because of the high load put on the extrusion punches used for cold extrusion of titanium alloys, punch live will be positively influenced by a reduction of 10% of the usual load.

It is generally understood that the titanium matrix MMCs could not be cold extruded due to their small

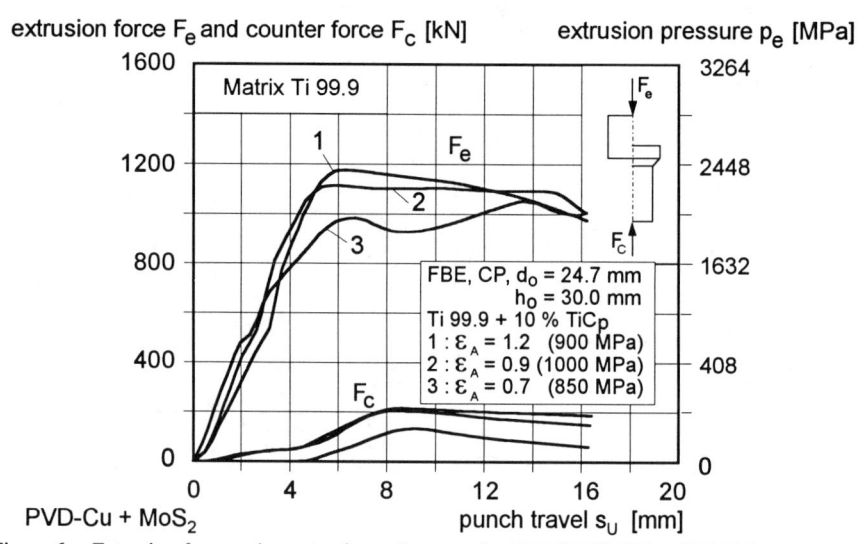

Figure 6. Extrusion force and counter force diagrams for FBE,CP Ti 99.9+10% TiC.

value of maximum strain, i.e. its brittle behaviour. In the case of cold extrusion of brittle materials, the most promising measure is the employment of a counter force by means of a counter punch during the extrusion operation, because the workability of the metal is improved if the state of stress in the forming zone is more to the compressive side, i.e. the hydrostatic pressure in the forming zone should be maximum [4].

Due to this fact, the employment of a hydrostatic counter pressure for crack prevention seems most suitable for the extrusion of titanium MMCs as well. In Figure 6, the diagrams for FBE of Ti 99.9+10% TiC and in Figure 7 of TiAl6V4 +10% TiC are plotted. Compared with the force-travel diagrams for FBE of unalloyed Ti 99.9 [1], the maximum value of extrusion pressure (p_e = 2.440 MPa) of the MMC Ti 99.9 +10% TiC is approximately 50 % greater, if the influence of the counter pressure is omitted, see curves 6 and 7 in Figure 1. This extrusion operation was carried out with an extrusion strain of ϵ_A = 1.2 and a counter pressure of p_c = 900 MPa (Figure 6).

In the case of the MMC TiAl6V4+10% TiC, cold forged with an extrusion strain ϵ_A = 0.9, an extrusion pressure of p_e = 2,320 MPa is required to produce crack-free components if additionally a counter pressure of p_c = 800 MPa is applied, see Figure 7. The extrusion pressure for the ordinary alloy TiAl6V4 is p_e = 1,430 MPa, only if a counter pressure of p_c = 200 MPa is introduced into the forming zone.

Forward Tube Extrusion (FTE)

The extrusion force vs. punch travel diagrams of FTE are quite similar to those of FBE, see Figure 5, but the maximum of the extrusion force is more clearly defined with increasing values of extrusion strain. This phenomenon is due to the fact that in connection with incre-asing extrusion strain more extrusion work is done, thereby establishing increasing temperature in the forming zone due to the low heat conductivity of the titanium materials.

For this reason, because of the conversion of the deformation work into heat, the flow stress of the material being a function of temperature is reduced as a function of strain. After the maximum, the extrusion force remains nearly constant (Fig. 5b.). This can be explained by the fact that in FTE, besides the frictional forces acting on the inner die surface, additional frictional forces have to be overcome which act between material and mandrel in the forming zone and remain constant during the whole forming operation.

Backward Cup Extrusion (BCE)

Figure 8 represents the force travel diagrams of BCE of the two ß-type alloys. The character of the curves of BCE, which is a non-steady forming process, is different compared with the diagrams for FBE and FTE, which are quasi-steady extrusion processes. At the beginning of the backward extrusion process, when the punch enters the slug, the extrusion force increases to a maximum value and decreases to the end of the deformation process.

The negative slope of the force travel curves after the maximum increases with increasing extrusion strain. This is also due to the increasing forming temperature in the forming zone.

Determination of Extrusion Force by Computer Simulation

The extrusion force vs. punch travel diagrams, which must be available for the selection of the forming press and for tool design, can be determined in advance in the phase of tool design by computer simulation. The mathematical background for the main extrusion processes FBE, FTE, and BCE is well known [1].

In the case of unalloyed titanium, of titanium alloys, but also of stainless steels, a special approach must be taken. The temperature of these metals in the forming zone increases quite rapidly due to the small values of heat conductivity. Hence, the flow stress of the metal is speedily reduced, because of the temperature increase.

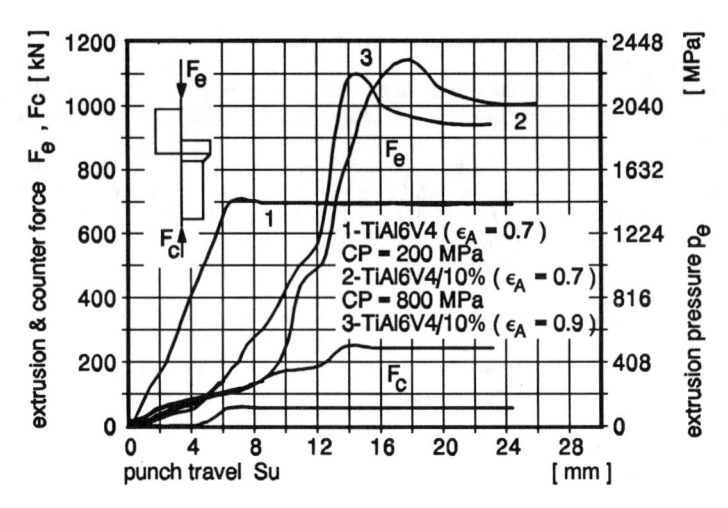

Figure 7. Extrusion-force and counter force diagrams for FBE,CP of TiAl6V4 & TiAl6V4+10%TiC.

Correct diagrams will be obtained only, if adiabatic conditions are assumed, i.e. in the equations, the adiabatic change of forming temperature and flow stress are considered, and the correct values of the coefficient of friction are used. Figure 9 demonstrates the accuracy of computer simulation of FBE of the ß-alloy TiAl3V8Cr6Mo4Zr4. Between the experimentally obtained diagram and the result of computer simulation, there is only a difference of 4 %.

a.) TiAl3V8Cr6Mo4Zr4
PVD - Cu + MoS$_2$

b.) TiV13Cr11Al3
PVD - Cu + MoS$_2$

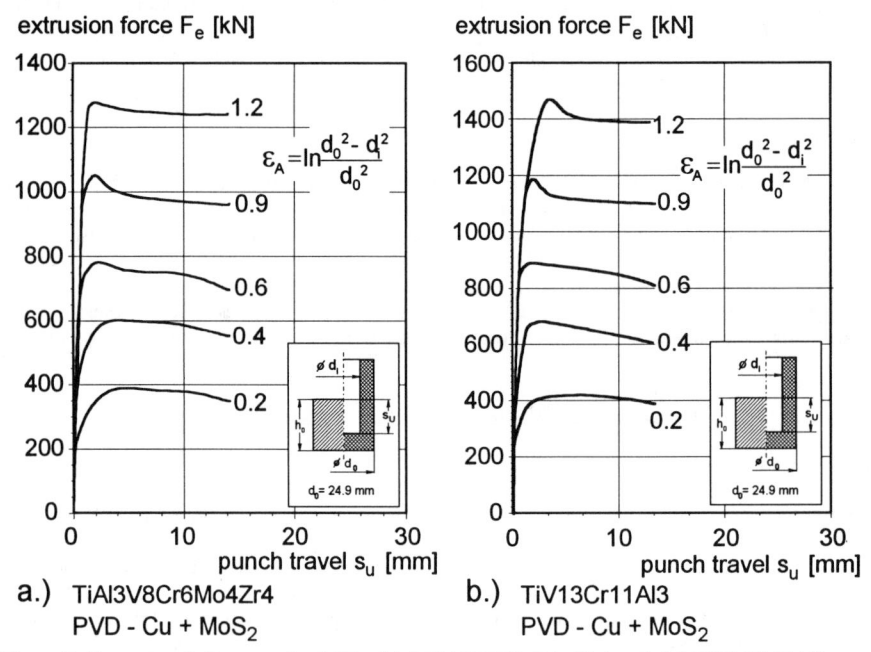

Figure 8. Force-travel diagrams for BCE of (a.) TiAl3V8Cr6Mo4Zr4 and (b.) TiV13Cr11Al3.

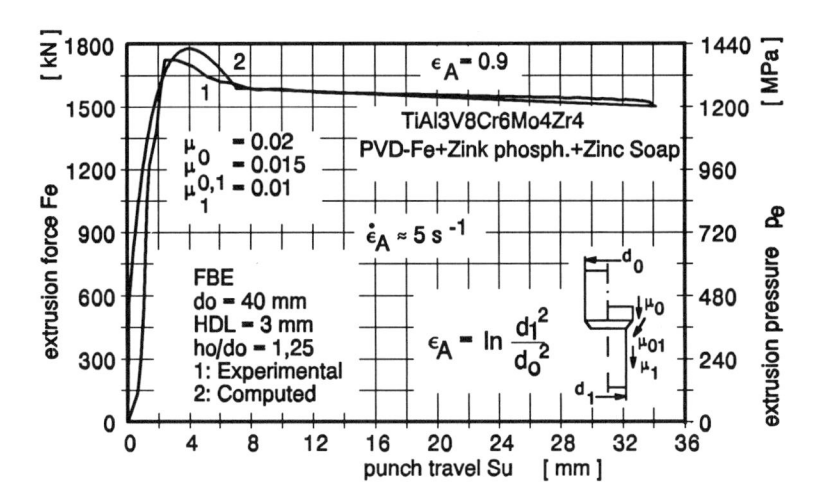

Figure 9. Comparison of experimental and computed force travel diagrams for FBE of TiAl3V8Cr6Mo4Zr4.

Comparison of Extrusion Pressures

For the design of extrusion tools, but mainly for the selection of the tool material and its working hardness, it is necessary to know the magnitude of the extrusion pressure for the different extrusion processes. In Figure 10, the maximum values of extrusion pressure of the three different extrusion processes are plotted as a function of the extrusion strain. The curves for the quasi-steady processes FBE and FTE are quite similar. The increase of the p_e-values is almost linear with increasing strain. For FBE, the maximum value of extrusion pressure is 2,500 MPa.

The values for the non-steady BCE are approximately two times greater than the values for FBE and FTE. There is a minimum at an extrusion strain of $\epsilon_A = 0.6$. In the case of BCE of the ß-alloy TiV13Cr11Al3, the maximum value of extrusion pressure measured for this series of experiments is

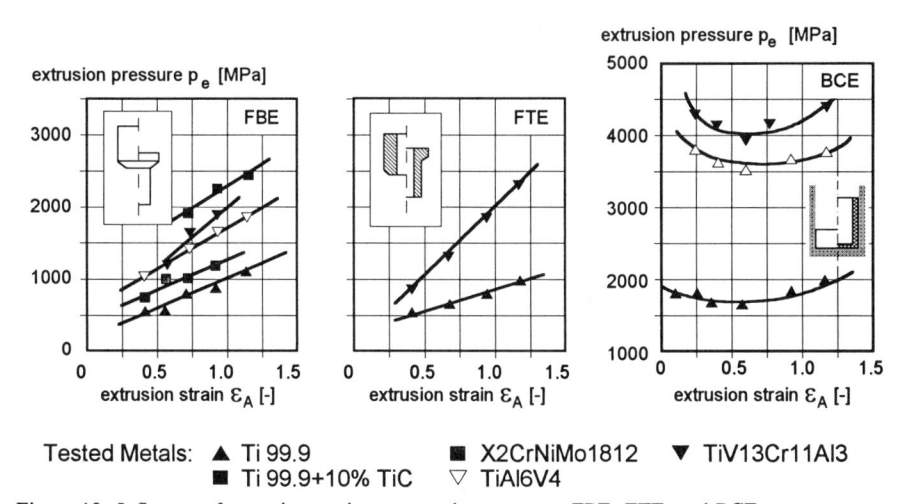

Figure 10. Influence of extrusion strain on extrusion pressure, FBE, FTE, and BCE.

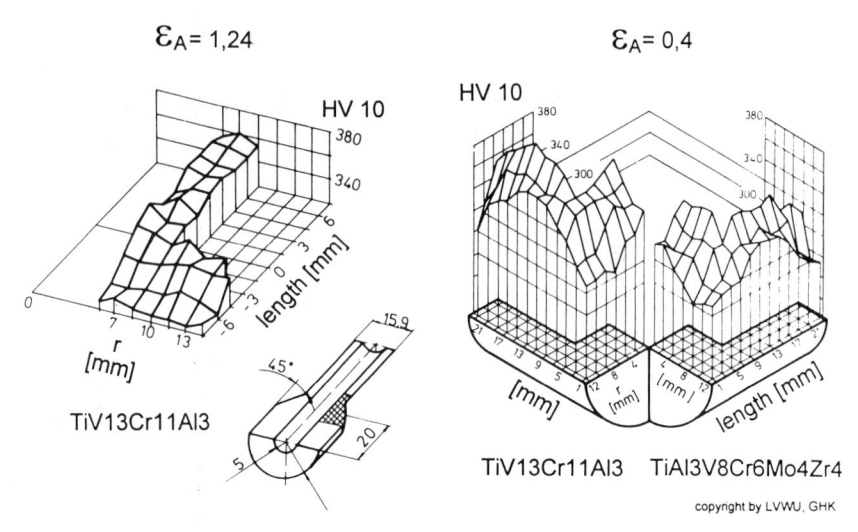

Figure 11. Hardness distribution of cold extruded components.

p_e = 4.400 MPa. This great value of extrusion pressure explains why cupping is the most difficult extrusion process. The punch is maximum loaded and the lubricational system, too.

Mechanical Properties of Extruded Components

To prove that components made of titanium alloys or of titanium MMCs produced by cold extrusion can be applied as structural components, several aspects were under study.
Additionally to the prerequisite that the workpieces be under intensive inspection with regard to surface quality and, more important, to cracks, hardness tests and tensile tests were carried out to determine to what extent strain hardening took place during extrusion. The phenomenon of strain hardening is one of the main advantages of cold extrusion of metals, because the strength of the workpiece can be inreased by this method partially or as a whole.

Hardness Tests

In Figure 11a., the increase of hardness of the part of the material which is in the conical forming zone of FBE of the ß-alloy TiV13Cr11Al3 is shown. This three-dimensional diagram indicates that the hardness increases by FBE due to the extrusion strain ϵ_A = 0.85 from 300 HV10 to a value of 405 HV10. A similar behaviour is demonstrated in the diagrams for BCE of the two ß-alloys (Figure 11b.). For an extrusion strain ϵ_A = 0.4, the hardness of the strain-hardened TiV13Cr11Al3 increases up to 360 HV10 and the hardness of the alloy TiAl3V8Cr6-Mo4Zr4 to 350 HV10. The inhomogenous character of the hardness plane indicates that in the case of non-steady BCE the strain distribution is not uniform.

Table IV: Values of hardness HV0.2 & tensile strength R_m of MMC of TiAl6V4+10% TiC.

Extrusion strain ϵ_A	0	0.5	0.7	0.9
Vickers hardness HV0.2 [-]	407	460	476	514
Tensile strength R_m [MPa]	924	/	1031	1264

In Table IV, the hardness HV0.2 before and after extrusion of the TiAl6V4+10% TiC specimens is listed. The hardness increases from 407 to 514 (26 %) due to the extrusion strain $\epsilon_A = 0.9$.

Tensile Tests

Tensile tests were carried out with the TiAl6V4+10% TiC specimens (Table IV). The strength of the alloy TiAl6V4 (Table II) is slightly greater than the value of the MMC in the as-received condition. But due to cold working, the MMC is strain-hardened up to R_m = 1264 MPa (37 %). Table IV proves, in accordance with the results in Figure 11, that the phenomenon of strain hardening also plays an important role in the process of industrial application of titanium MMCs as structural metals.

Micrographs

Figure 12 demonstrates that in TiAl6V4+10% TiC the ß-matrix and the TiC particles are orientated by extrusion. The general character of the materials structure is more uniform after extrusion than before. These micrographs also prove that "brittle" MMCs can be deformed in such a way that crack-free components can be produced if a highly compressive state of stress is dominant.

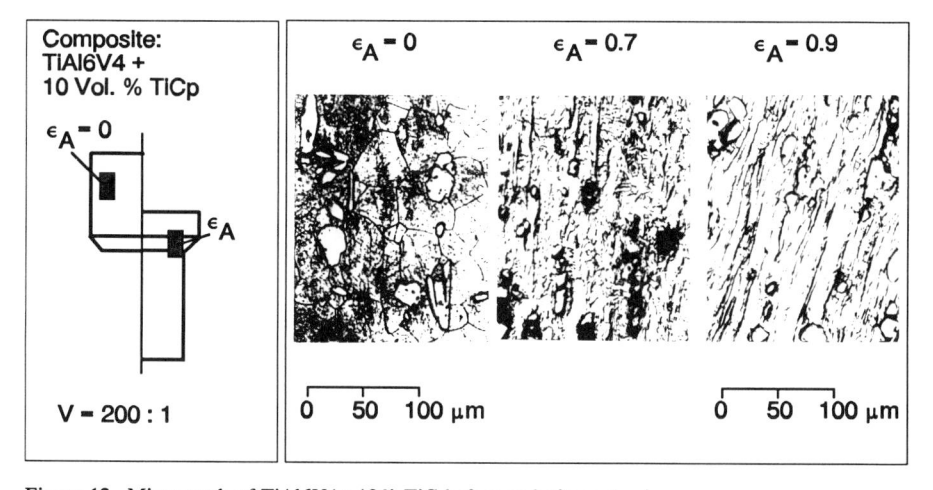

Figure 12. Micrograph of TiAl6V4+10% TiC before and after extrusion.

Conclusions

For the application as structural materials, titanium alloys and titanium MMCs can be plastically deformed at room temperature by the standard cold extrusion processes thereby producing crack-free components. This includes the advantages of near net shape or net shape manufacturing. The brittle materials, the most frequently applied alloy TiAl4V6 and the titanium MMCs, can be successfully subjected to cold extrusion by making use of a counter pressure, by which a high hydrostatic pressure is introduced into the forming zone. The main mechanical properties and the strength of the cold extruded components are increased due to strain hardening.

References

1. H. W. Wagener and K. H. Tampe: Cold Extrusion of Titanium (in German) (Düsseldorf, VDI-Verlag, 1985) 162.

2. H. W. Wagener, J. Haats, and J. Wolf: "Increase of Workability of Brittle Materials by Cold Extrusion". Journal of Materials Processing Technolgy, 32 (1992) 451-460.

3. H. W. Wagener and J. Wolf: "Cold Forging of MMCs of Aluminium Alloy Matrix". Journal of Materials Processing Technology, 37 (1993) 253-265.

4. H. Stenger: Influence of the State of Stress on the Ductility of Homogenous Metals (in German). Dr.-Ing. Thesis, Technol. University of Aachen 1965.

5. H. W. Wagener, J. Wolf, and K. Thoma: Coefficient of Friction in Cold Forging (in German), (Oberursel VDM-Informationsgesellschaft 1996) 162.

Development of Processing Methods for Ti-13Nb-13Zr

E. W. Robare [*], C. M. Bugle [*], J. A. Davidson [+], and K. P. Daigle [+]

[*] Dynamet Incorporated
Research and Development
195 Museum Road, Washington, PA 15302

[+] Smith & Nephew Orthopaedics
1450 Brooks Road, Memphis, Tennessee 38116

Abstract

Ti-13Nb-13Zr (Ti-13-13™) is a new titanium alloy which has a unique combination of properties that may make it an attractive choice for a variety of applications. Originally developed for use in biomedical implants, this alloy combines a low elastic modulus, high strength, excellent hot and cold workability, superior corrosion resistance, and the ability to be surface hardened to improve wear properties.

Research on this age-hardenable alloy has shown that the mechanical properties can be controlled over a significant range through hot working, heat treatment and cold working. The elastic modulus of Ti-13-13 can be varied between approximately 41 and 83 GPa, and strengths as high as 1330 MPa have been achieved. Some typical combinations of properties are presented.

Further, the material does not seem to exhibit some of the typical processing difficulties encountered in some age-hardenable beta Ti alloys. For example, in wire drawing trials, Ti-13-13 yielded exceptionally good surface finish in the "as-drawn" condition with over 85% cold work. Some examples of processing behavior are discussed.

Introduction

Ti-13Nb-13Zr (Ti-13-13™) is a new titanium alloy which was originally developed for medical implant applications. It has shown promise with respect to various applications for biomedical devices. Its properties make it an attractive option for applications such as total hip replacements, orthopaedic fixation devices, dental components, and devices that come into contact with blood such as heart valve housings and stents.

It can be said that an "ideal" material for implantable applications should possess the following properties: excellent biocompatibility, low modulus, and good corrosion and wear properties. Strength is important also, with different strength levels required for different applications. Ti-13-13 was developed with the goal of optimizing performance in these areas for biomedical applications.

Some typical materials currently used for such applications include Ti-6Al-4V, Co-Cr-Mo alloys, commercially pure Ti, and carbon coated graphite. Examples of how Ti-13-13 may provide a potential improvement can be seen by examining some of these areas.

Biocompatibility

Biocompatibility is a major consideration in the selection of an implantable alloy. Applications in which the component must operate in direct contact with the bloodstream are especially demanding. The chemical composition of Ti-13-13 is shown in Table I. The main components of the alloy are Ti, Nb, and Zr, which are three of only four metals (along with Ta) that have been shown to meet all the criteria for optimum biocompatibility (1). Ti-13-13 has been shown to exhibit excellent biocompatibility (2,3), and has performed well in blood-compatibility testing including protein, platelet, and bacterial attachment tests (4). As a metal alloy, it exhibits much higher strength than the graphite material typically used for heart valves.

Table I: Chemical Composition of Ti 13Nb-13Zr

Element	Wt %
Nb	12.00-14.00
Zr	12.00-14.00
C	0.08 max
O	0.15 max.
N	.0.05 max.
Fe	0.25 max.
H	.012/.015

Elastic Modulus

An important concern in total hip replacments, in particular press-fit designs, is the pain and eventual potential for loosening of the implant that has been associated with stress shielding due to modulus mismatch between the implant and surrounding bone (3,5). Ti-13-13 in the ST or STA condition has a significantly lower modulus than either Ti-6Al-4V or Co-Cr-Mo. The Nb in the alloy contributes to this reduction in the elastic modulus (5).

A somewhat unusual behavior which has been observed in Ti-13-13 is that the modulus can be varied by heat treatment and cold working. This phenomenon is believed to be due to a stress-induced phase transformation and has interesting potential ramifications.

Corrosion/Wear

Corrosion and wear properties are important for both orthopaedic applications as well as other uses. The alloy must be resistant to corrosive attack in the presence of body fluids, and it must be able to resist abrasion and wear resulting from relative motion between the implant and bone or articulation against another surface. Although the biocompatibility of Ti alloys is well documented, they are also known for their generally poor wear properties.

Due mainly to the Nb and Zr content of Ti-13-13, it is possible to produce a hard, continuous surface oxide layer by heat treating the alloy in a controlled atmosphere (6). The oxide layer that forms is smooth, adherent and is black in color. This "diffusion hardening" treatment has been shown to result in dramatic improvements in the wear properties of Ti-13-13 (2).

Process Development

Process development was done to develop methods for production of bar, rod and wire from Ti-13-13. Typical product forms and conditions that are of interest are large diameter bar (25 mm and over) and rod and coil (6-12 mm diameter) in the solution treated or STA condition for applications such as heart valve housings, orthopaedic implants, fixation rods and other devices. Coil and fine wire (0.5-2.5 mm diameter) in the cold worked condition are of interest for both orthopaedic and dental applications, and very fine wire (under 0.25 mm) for applications where wire is woven into mesh or cloth, for example, in stents.

Processing-property relationships were developed for solution treated and cold worked conditions. Wire finishing techniques for producing a clean, smooth surface were also developed.

Microstructure

The micrographs in Figure 1 show typical examples of the structures representing various heat treatment conditions. Figure 1a shows a solution treated and water quenched structure, which has been shown to consist of α' hcp martensite (5). Figure 1b shows this structure after cold drawing plus annealing, which results in a finer dispersion of alpha plus beta. With multiple cycles of cold work and annealing, the structure can be refined to an extremely fine alpha-beta dispersion, as shown in Figure 1c.

Mechanical Properties and Cold Work

Some typical combinations of mechanical properties are shown in Table II. Particularly interesting to note are the range of strengths obtainable and the amount of cold work that is possible. This alloy was able to be cold drawn to a final reduction of over 90%, much more than is possible with traditional Ti alloys such as Ti 6Al-4V.

Table II: Typical Mechanical Properties of Ti-13Nb-13Zr

Condition	UTS, MPa	0.2%YS, MPa	RA (%)	El (%)	Young's Modulus, GPa
ST+WQ Bar	725	475	70	27	65
STA Bar	1005	875	40	13	81
ST+WQ Coil	670	450	68	30	51
STA Coil	860	760	60	15	76
ST+Cold Worked Coil (62%)	930	800	53	13	41
ST + Cold Worked + Aged Coil	950	820	68	19	83
Annealed + Cold Worked Wire (87%)	1330	980	*	3.0 (in 10")	53

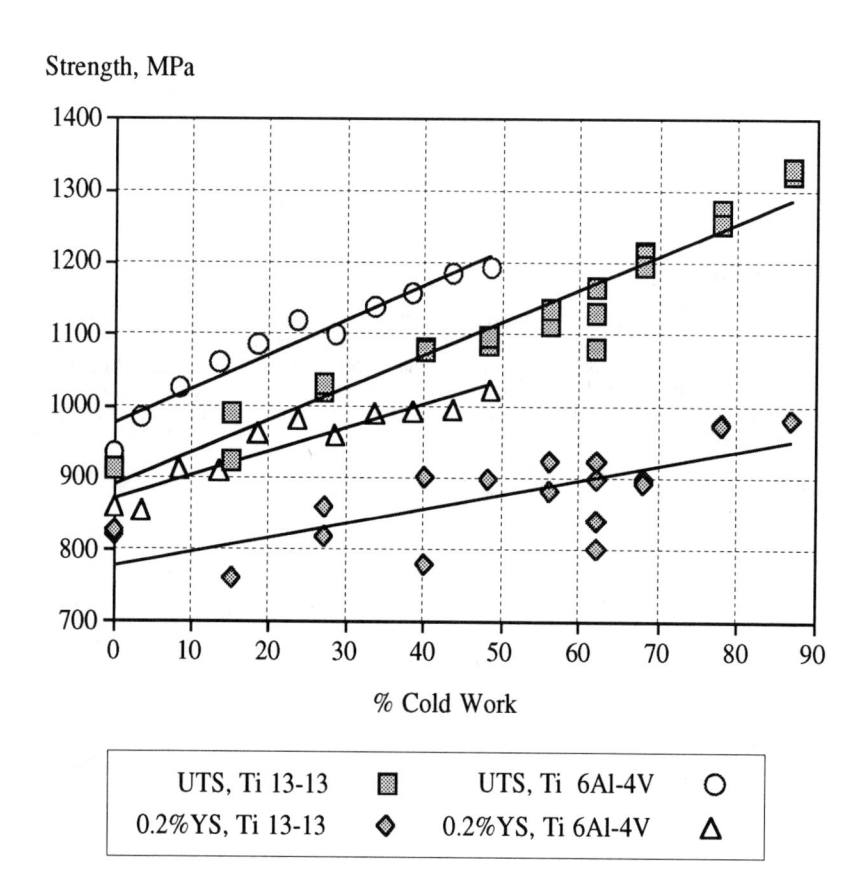

Figure 2: Cold Work vs. Strength for Ti-13-13 and Ti-6Al-4V

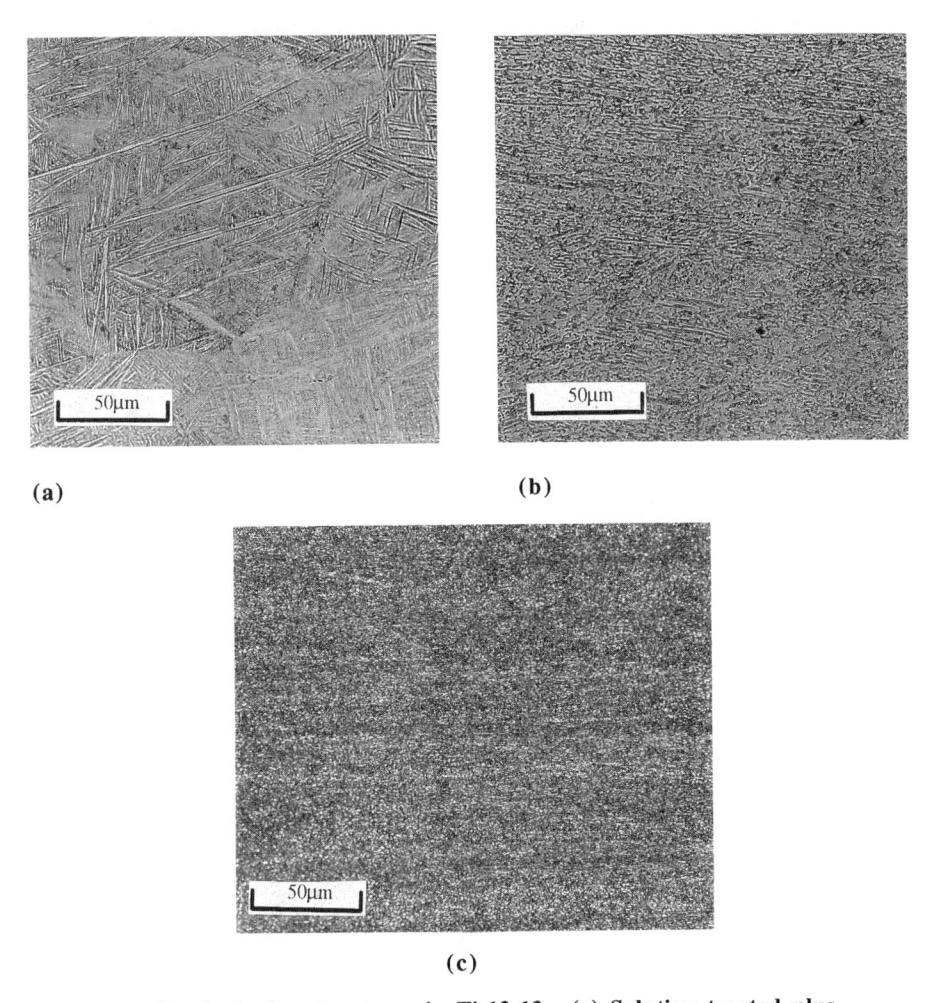

Figure 1: Typical microstructures in Ti-13-13. (a) Solution treated plus water quenched; (b) Cold worked and annealed; (c) Cold worked and annealed - multiple cycles.

The cold work vs. strength relationship is illustrated in Figure 2. This chart compares the cold working behavior of annealed Ti 13-13 to that of annealed Ti 6Al-4V. The chart shows that, while the Ti-13-13 starts out at a lower strength than Ti 6Al-4V, because of its excellent cold workability it can be worked to higher strengths than are possible for Ti 6Al-4V. By tailoring the anneal temperatures and amount of cold work in this way, a wide range of properties can be achieved in Ti-13-13.

Another unique feature of Ti-13-13, apparent in Table 2, is the fact that the elastic modulus also varies with material condition and cold work. This behavior is not typical of other Ti alloys and is believed to be due to a stress-induced phase transformation. The variation in the modulus with heat treatment and cold work is illustrated in Figure 3. In cold worked fine wire, moduli as low as 35 Gpa were obtained in some cases. In the solution treated condition, the modulus of 50-60 Gpa is less than half that of Ti 6Al-4V (110 GPa) and 75% lower than Co-Cr-Mo (227 Gpa). This low modulus can be very advantageous for applications such as orthopedic implants, where a lower modulus material may help to lessen the problems associated with modulus mismatch between implant and bone.

For some applications, such as orthodontic wire in particular, the combination of high strength and low modulus would be desirable. It has not been shown, however, that the low elastic (Young's) modulus observed in cold worked wire corresponds to low bending and/or bulk moduli, which can be more important parameters for orthodontic wire applications.

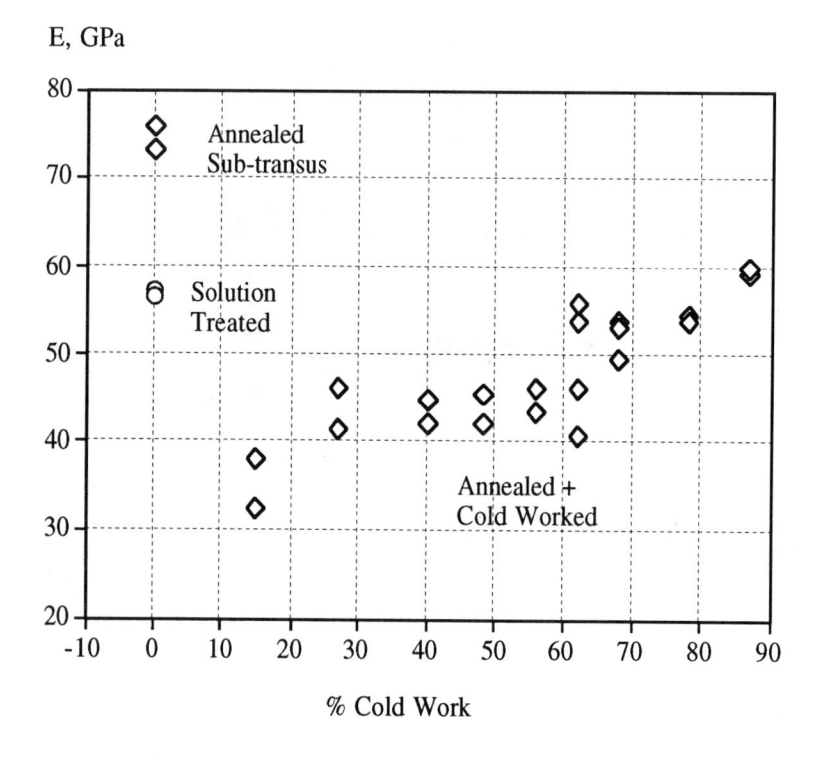

Figure 3: Cold Work and Material Condition Condition For Ti-13Nb-13Zr

Wire finishing - Surface Effects

Strengthening by cold working is most appropriate for small diameter wire. Since the type of products made from fine wire are usually used in contact with soft tissues in the body, a smooth surface is preferred to minimize tissue irritation. It was found that an exceptionally smooth surface finish could be obtained on Ti-13-13. Figure 4 shows scanning electron micrographs of a typical surface on CP Ti wire, and on Ti-13-13 wire that was processed using techniques that were developed in this study. The Ti-13-13 was able to be prepared so that it had virtually no irregularities on the surface.

Another feature of the Ti-13-13 alloy is that due to the presence of Nb and Zr, a tightly adherent, hard oxide layer can be formed to improve the wear resistance of surfaces. This oxide is black in color and contains TiO_2, ZrO_2, and Nb_2O_5. The oxide layer is approximately 0.8 microns thick, and underneath, a hardened metal layer of about 2-3 microns is formed (6). This diffusion hardening process is of great interest for applications requiring excellent wear resistance, such as orthopedic implants.

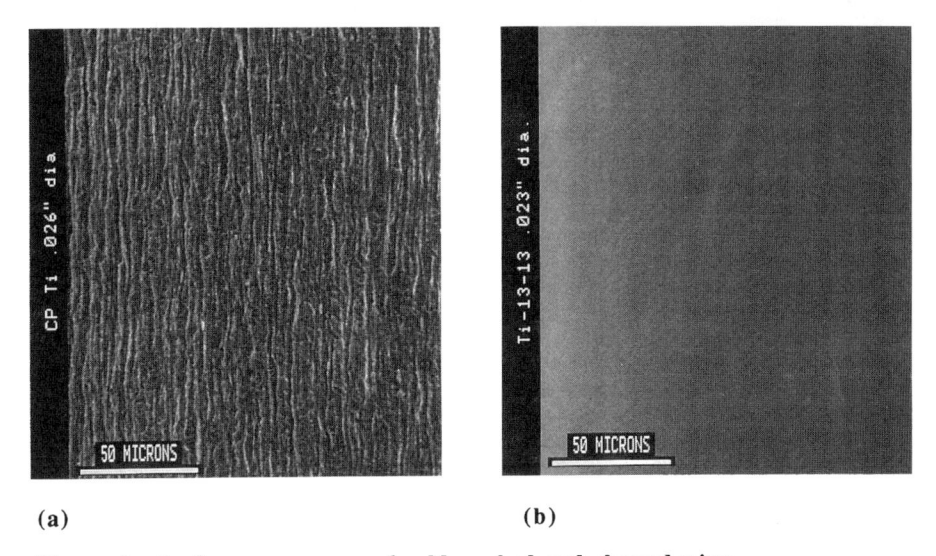

(a)   (b)

Figure 4: Surface appearance of cold worked and cleaned wire; (a) a typical CP Ti surface finish, and (b) surface finish obtained on Ti-13-13.

The material's affinity for oxygen, however, can result in special requirements for processing wire. Annealing, solution treating, or even aging in air can make subsequent cold reduction difficult. The hardened layer which forms is brittle, and cracks when cold worked, leading to a poor surface finish on the wire. Figure 5a shows the surface of a wire that was air annealed and cold drawn, resulting in the formation of numerous cracks on the surface. A cross-section of the same wire is shown in Figure 5b. The cracks propagate to a depth of approximately .001" and, interestingly, they are blunt at the tip and did not result in failure of the material during drawing.

Because of the way the material behaves in this condition, it would be possible to air anneal and cold draw without a wire failure, unknowingly creating the type of surface shown in Figure 5a. For this reason, it is recommended that all heat treating processes be performed in vacuum or inert atmosphere.

(a)   (b)

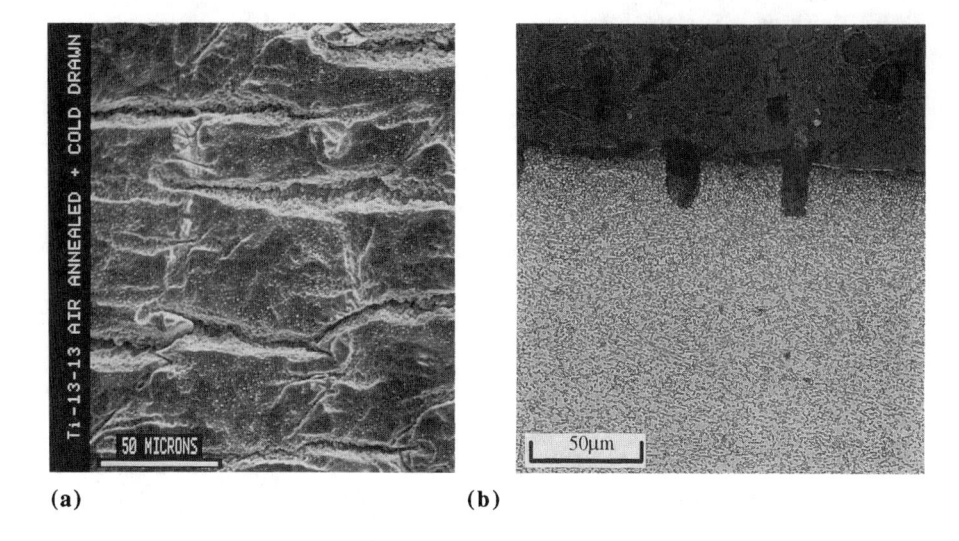

Figure 5: Air annealed plus cold drawn Ti-13-13; (a) surface appearance (SEM) and (b) transverse cross-section through the cracked surface.

Conclusions

Processing methods for Ti-13Nb-13Zr bar and wire products were developed. It was found that the properties of the alloy could be varied significantly by adjusting the cold work and anneal cycles. The properties were quite predictable, which would allow for "tailoring" of individual products to meet specific requirements.

In addition, it was observed that the elastic modulus was not only lower than that of Ti 6Al-4V and Co-Cr-Mo, it also varied with material condition and could be changed by cold working. This feature may be able to be exploited for orthopaedic implants.

The wire was able to be processed so as to produce an exceptionally smooth surface finish. This is an attractive feature for dental and wire weaving applications.

Some potential processing pitfalls were also identified. The most significant of these is that the alloys's high affinity for oxygen makes it important to utilize a protective atmosphere during all heat treatment operations.

References

1. P. Kovacs and J. A. Davidson, "Chemical and Electrochemical Aspects of the Biocompatibility of Titanium and its Alloys," Medical Applications of Titanium and its Alloys: The Material and Biological Issues, ASTM STP 1272, ed. S. A. Brown and J. E. Lemmons (Philadelphia, PA: American Society for Testing and Materials, 1995).

2. A. K. Mishra, J. A. Davidson, R. A. Poggie, P. Kovacs, and T. J. Fitzgerald, "Mechanical and Tribological Properties and Biocompatibility of Diffusion Hardened Ti-13Nb-13Zr," Medical Application of Titanium and its Alloys: The Material and Biological Issues, ASTM STP 1272, ed. S. A. Brown and J. E. Lemmons (Philadelphia, PA: American Society for Testing and Materials, 1995).

3. S. B. Goodman, J. A. Davidson, V. L. Fornasier, and A. K. Mishra, "Histological Response to Cylinders of a Low Modulus Titanium Alloy (Ti-13Nb-13-Zr) and a Wear Resistant Zirconium Alloy (Zr-2.5Nb) Implanted in the Rabbit Tibia," Journal of Applied Biomaterials, 4 (1993), 331-339.

4. K. P. Daigle, P. Kovacs, A. Mishra, and J. A. Davidson, "Development of a New Medical Implant Alloy," Proceedings of the 1994 International Conference on Titanium Products and Applications (Boulder, CO: Titanium Development Association, 1994).

5. J. A. Davidson, A. K. Mishra, P. Kovacs and R. A. Poggie, "New Surface-Hardened, Low-Modulus, Corrosion-Resistant Ti-13 Nb-13 Zr Alloy for Total Hip Arthroplasty," Biomedical Materials and Engineering, 4 (3) (1994), 231-243.

6. R. A. Poggie, P. Kovacs, and J. A. Davidson, "Oxygen Diffusion Hardening of Ti-Nb-Zr Alloys," Materials and Manufacturing Processes, II (2) (1996).

TITANIUM ALUMINIDE FOIL PROCESSING

C.C.Wojcik, Teledyne Wah Chang, P. O. Box 460, Albany, Or 97321
R.Roessler and R.Zordan, Allison Engine Co., P. O. Box 420, Indianapolis, In 46206-0420.

Abstract

Orthorhombic titanium aluminides based on the Ti_2AlNb composition are of interest for high temperature composite applications because of their excellent strength and compatibility with SiC fibers. The performance improvements attainable with this new type of titanium alloy will only be realized if orthorhombic alloy foil and SiC fibers can be produced economically. Fabrication of these alloys into thin (e.g., 0.10 mm thick) foil presents some difficult problems. We will report on our experience fabricating Ti-22Al-26Nb into foil. The effects of hot working and annealing temperature have profound influences on microstructure and properties. Additionally, this alloy is sensitive to cooling rate which may be used to quench in metastable structures that are more workable than ordered orthorhombic or ordered alpha phases. The effects of heat treatment on microstructure, hardness and cold workability will be discussed.

Introduction

Recently various researchers[1-3] have reported on the properties of several orthorhombic alloys for potential use as matrix materials in foil/SiC fiber composites. Their results have shown that these alloys offer major performance improvements over existing commercial titanium alloys. In consideration of these reported studies, which were performed mainly with small laboratory size quantities and by means that were not economical from a production perspective, our objective has been to develop a low cost manufacturing process for Ti-22Al-26Nb (atomic percent) alloy in foil form for high volume use.

It is important to realize that most of the common high strength titanium alloys in use today are not available in thin foil form due to limited cold workability. In fact, the most common titanium alloy (Ti-6Al-4V) was modified many years ago to a less alloyed version Ti-3Al-2.5V (weight %) specifically so that foil and seamless tubing could be economically manufactured by cold working. Because ordered alloys can not simply be "leaned down" without causing major shifts in phase boundaries it is imperative that thermomechanical behavior and microstructural features be well understood to optimize processing capability as well as the materials final properties. It is also important to recognize that in foil/fiber composites usually the composite consolidation step involves hot pressing which may be the most controlling factor in determining the final microstructure of the metal foil. Our work therefore was mainly focused on developing a practical manufacturing process for foil rather than optimizing the final mechanical properties.

Experimental

The initial ingot used for this work was vacuum arc cast in three consecutive melts into a final ingot measuring 380 mm diameter X 1.3 meters long. The ingot was machined on all surfaces, and forged into a rectangle measuring ~100 mm thick X 400 mm wide at 1150°C, which was well into the beta phase field. After forging, small trial samples of material were hot rolled a total of 3:1 area reduction at nominal temperatures of 1045°C and 1015°C. Reheating was performed after every pass to ensure that the material passed into the two phase (α_2+ B2) phase field between each rolling pass. Reductions were 15% per pass. After examining the microstructures produced by the rolling trials, the remaining forging was hot rolled 3:1 using a furnace temperature of 1020°C, reheating after every pass, and 15% reductions. A small fraction of the plates (~ 60kg) was subsequently hot rolled again at a later date to 5.5 mm thick X 450-550 mm wide using the same rolling reductions and temperatures. After appropriate mechanical and chemical surface cleaning these pieces were cold rolled 40-50% on a four high rolling mill having work rolls measuring ~250 mm in diameter. A series of annealing experiments was then carried out to determine intermediate heat treatments between successive stages of cold rolling. We continued a sequence of cold rolling and annealing to achieve the desired thickness. Rolling under tension was started at a thickness of 1.5 mm, on a 300 mm wide Sendzimer mill.

Results and Discussion

Chemical analysis of the cast ingot is shown in Table I. In this alloy, aluminum content is the most difficult element to accurately control, due to the high vapor pressure of aluminum relative to the melting temperature of the alloy. This alloy melting temperature is also elevated by the high niobium content. Aluminum levels were lowest at the top of the ingot. Oxygen content was kept as low as possible using commonly available raw materials. This decision was made as a result of ongoing work with two small heats that were cast into rectangles by cold wall induction melting. One of these heats was found to have poor fabricability at room temperature due to high levels of oxygen, which were measured at 1100-1500 ppm by weight. The other heat that had an oxygen content of 880 ppm was fabricable.

Table I Chemical Analysis of
Ti-22Al-26Nb (Ti-43.9Nb-10.8Al Wt %) Ingot Ht.#878879

Element	Ingot Position		
Wt %	Top	Middle	Bottom
Al	10.2	10.4	10.3
Nb	44.5	44.3	44.2
O	820 ppm	840	950
N	34 ppm	37	39
C	80 ppm	249	97
Fe	282 ppm	740	340
Ta	570 ppm	590	560
Ti	bal.	bal.	bal.

After successfully forging the ingot in air at 1150°C three small sections were rolled at 1045°C, 1015°C, and 1015°C with an intentional delay to cool the surface 60-70°C before the piece entered the rolling mill. The cooling delay was intended to simulate the heat loss anticipated while transferring material from a roller hearth furnace to the rolling mill. Reheating was performed after every pass. Reductions per pass were 15% and total area reduction was 3:1. After the last pass the material was annealed for 10 minutes at the rolling temperature. The resulting microstructures are shown in Figure 1. Rolling at the higher temperature, 1045°C, produced a lower volume fraction and less uniform distribution of the second phase α_2 particles as compared to hot rolling at 1015°C. It appeared that the cooling delay prior to entering the rolling mill, had little effect on the microstructure. At both rolling temperatures the prior beta grains are elongated in the rolling direction. Based on these observations and our prior rolling experience, with a small induction cast slab we continued all hot rolling using a preheat temperature of 1020°C. Final thickness of the hot rolled plates after removing the surface oxide was ~5 mm.

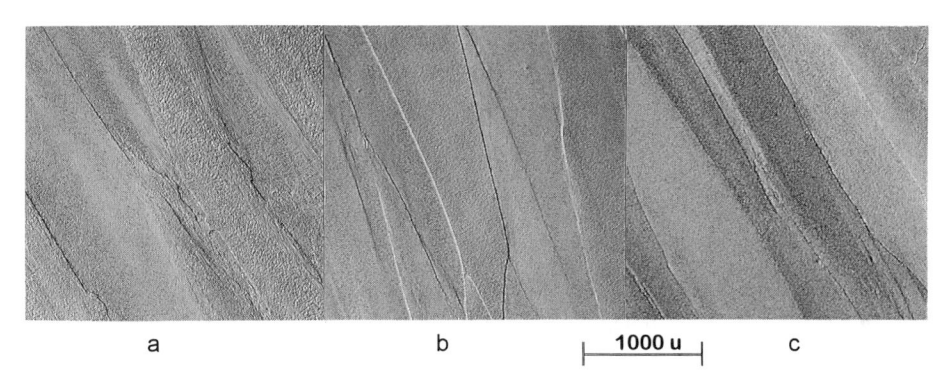

Figure 1 Resulting microstructures after hot rolling 3:1 at (a) 1045°C, (b) 1015°C, and (c) 1015°C with a cooling delay. 50X

Figure 2 Hot rolled plate after cold rolling 50% to 2.5 mm thick. 200X

Cold rolling was somewhat complicated by the fact the plates had more crown (i.e. thicker in the middle of the sheet width) than other materials commonly rolled on our hot rolling mill. To address this problem, crowned work rolls were used in addition to a hydraulic roll bender on a four high cold rolling mill. Cold rolling becomes increasingly difficult as the material work hardens to ~380 DPH. Microstructure of the sheet after rolling 50% is shown in Figure 2. Interstitial content of the sheets at this stage in the processing is summarized in Table II.

Table II
Intersticial Content of 2.5 mm Thick Sheet

Oxygen	870 ppm	880
Nitrogen	26 ppm	34
Hydrogen	26 ppm	34

Annealing the 2.5 mm thick sheet proved to be the most challenging step in the thermomechanical processing. It was not considered practical to anneal this thickness of material in air at 1020°C because of surface oxidation. Additionally we could not perform annealing in a continuous protective atmosphere furnace because of temperature limitations. An alternate method envisioned was vacuum annealing, which is commonly used for other refractory metals that are rolled into foil. It was however known from previous work, that vacuum annealing the Ti-22Al-26Nb alloy at 1020°C would actually increase the hardness, due to precipitation of the orthorhombic phase during cooling under a vacuum.

The temperature at which the orthorhombic phase precipitates during cooling was determined by running samples in a differential scanning calorimeter. Samples which had previously been cold rolled and then annealed in air at 1020°C at a thickness of 2.3 mm were heated and cooled at a rate of 5°C per minute. The results of these scans are shown in Figure 3. It appears that the orthorhombic phase change occurs during cooling at ~ 840°C. During heating a smaller peak is shown at ~ 930°C. We believe this peak corresponds to dissolution of the orthorhombic phase into the B2 matrix phase, however more microstructural analysis is necessary. Samples heat treated at 1000°C and air cooled do not have any orthorhombic phase visible by light microscopy.

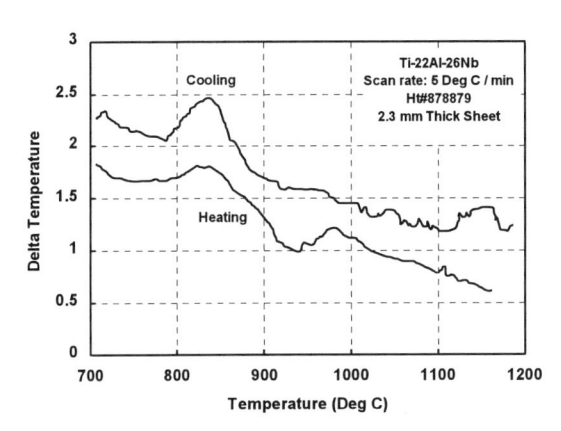

Figure 3 Differential calorimeter scan for heating and cooling

To further understand the heat treatment response of this alloy a series of annealing and aging treatments were performed. Using a section of the material that was cold rolled 50%, one third was air annealed 15 minutes at 1015°C, another third was air annealed in the beta phase for 15 minutes at 1080°C. The remaining third was left in the cold rolled condition. Small sections from each of the three pieces were then encapsulated in quartz capsules, evacuated, and heat treated for one hour at 600°C -1000°C. The microstructure and microhardness of each piece was evaluated.

As shown in Figure 4, the microhardness reaches a maximum at ~ 700°C. This hardening peak is due to precipitation of the orthorhombic phase within the B2 phase. At higher temperatures, the hardness decreases as the orthorhombic phase becomes coarser. The coarsest orthorhombic phase was observed in the material which was previously given a beta heat treatment and contained no visible α_2 phase prior to aging. The finest orthorhombic structure was observed in the samples that were heat treated in the cold worked condition. No orthorhombic phase was observed in any of the samples that were heat treated at 1000°C, Figure 5. These samples contained the B2 phase and the α_2 phase. Several more samples of the cold worked material were heat treated at 1010°C, 1030°C and 1050°C for 15 minutes. As shown in Figure 6 the volume fraction of the α_2 phase rapidly decreases as the temperature increases.

Figure 4
**Hardness Response to Heat Treatment
Ti-44Nb-10.8Al Sheet**

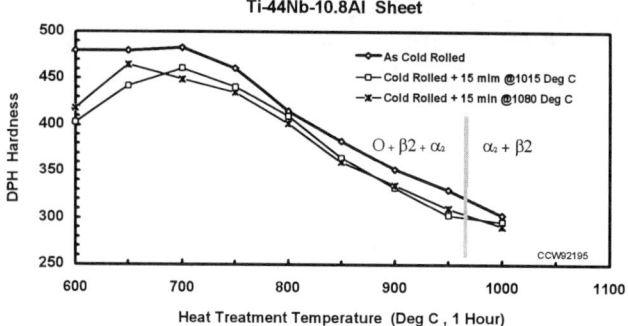

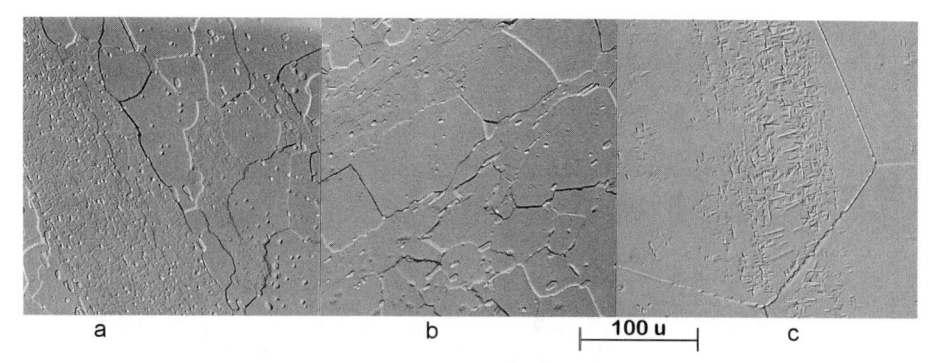

$O + \beta 2 + \alpha_2$ $\alpha_2 + \beta 2$

Figure 5 Microstructure of samples heat treated for one hour at 1000°C after (a) cold rolling, (b) cold rolling and air annealing 15 min. @ 1015°C, (c) cold rolling and air annealing 15 min. @ 1080°C. 200X

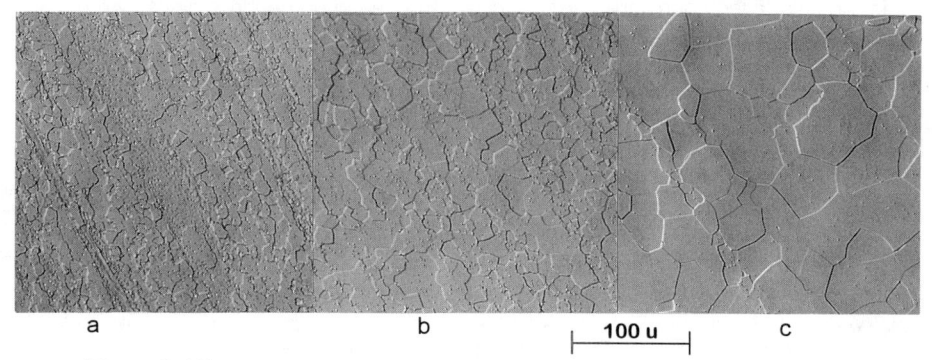

Figure 6 Microstructure of cold rolled samples after air annealing for 15 minutes at (a) 1010°C, (b) 1030°C, and (c) 1050°C. 200 X

Since it was known that vacuum annealing from 1020°C actually hardened the material due to precipitation during slow cooling, a third set of samples were heat treated in vacuum below the B2 to orthorhombic phase transformation temperature. Pieces of the cold worked material were heat treated for 90 minutes at 910°C, 930°C and 950°C. The resulting microstructures are shown in Figure 7. The cooling rate observed in vacuum was ~ one order of magnitude slower than the cooling rate in ambient air. Microhardness of these samples was 323, 318, and 330 DPH respectively. Small pieces (~ 100 mm X 100 mm X 2.5 mm thick) given these heat treatments were cold rolled 30-40% before edge cracking was observed. When a 90 minute, 930°C heat treatment was performed on the full size pieces it was found that these pieces did not roll well. In most cases severe cracking occurred after only several rolling passes. It was also observed that the large pieces cooled ~ half as fast as the small trial pieces that were vacuum annealed in a much smaller laboratory furnace.

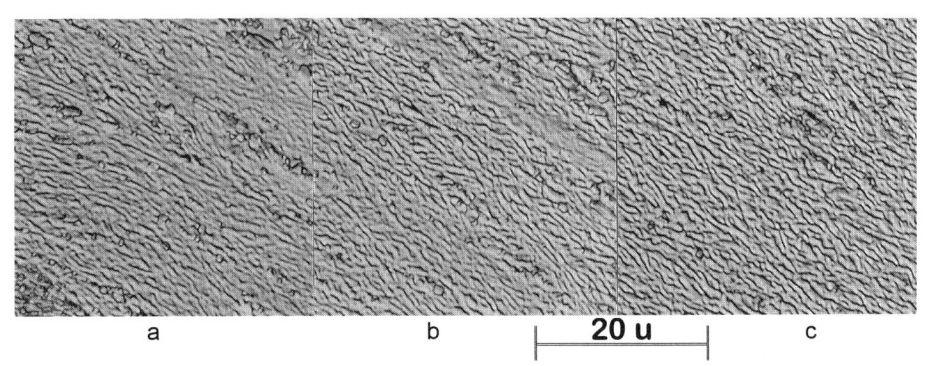

Figure 7   Microstructure of cold rolled material after vacuum annealing for 90 minutes at (a) 910°C, (b) 930°C, (c) 950°C. 1500X

The material was successfully "repaired" by annealing at 1020°C in air and appropriately cleaned to remove all surface oxide. Cold rolling was again performed on a four high mill to a thickness of 1.3 mm. After annealing in air again, and rolling under tension on a sendzimer mill, the material measured 0.75 mm thick. Subsequent annealing was performed under vacuum with rapid cooling by argon gas convection. The material was then cold rolled on a sendzimer mill 50% to a thickness measuring 0.37 mm.

Conclusions

In summary, we have demonstrated that the Ti-22Al-26Nb alloy should be fabricable into foil on a production scale if certain processing requirements are met. Specifically, it should be recognized that this alloy is very sensitive to cooling rate. At high cooling rates, from the B2 or the B2+α_2 phase field, the orthorhombic phase transformation can be avoided. This produces a metallurgical structure that is easily cold rolled. At slower cooling rates, such as experienced during vacuum annealing, the orthorhombic phase precipitates, and a less ductile structure results. The temperature range for annealing this alloy in the B2+α_2 phase field appears to be quite small ~ 1000-1050°C. Above 1050°C the α_2 phase is barely present. This leads to rapid beta grain growth, which would not be acceptable for thin gage foil.

This work represents the first attempt to fabricate production size quantities of the Ti-22Al-26Nb alloy into both hot and cold rolled sheet products. Considering our degree of success, it is encouraging to think that this new class of titanium aluminides may be the first type of aluminides that are readily fabricable into thin sheet at room temperature.

Acknowledgments

The authors wish to thank Dr. Paul Smith and program monitor Mr. Jim Terry for helpful discussion and encouragement throughout the course of this project. This work is sponsored by WL/POTC Wright Labs Aero Propulsion and Power Directorate under contract F33615-94-C-2410. We also thank Paul Danielson for his metallographic work which was crucial toward understanding this alloy system.

References

1. R. G. Rowe et al., "Tensile and Creep behavior of Ordered Orthorhombic Ti2AlNb-Based Alloys," MRS Proceedings, Vol. 213 (1991), 703-708.

2. P. R. Smith, J. A. Graves, and C. G. Rhodes, "Comparison of Orthorhombic and Alpha-Two Titanium Aluminides as Matrices for Continuous SiC-Reinforced Composites," Metall. Trans. A, 1994, vol. 25A, 1267-1283.

3. P .R. Smith, and J. A. Graves, "Monotonic Behavior of Neat High Temperature Alloys for Use as Composite Matrices," Scripta Metall., 1995, vol 32, 695-700.

ABRASIVE WATERJET CUTTING OF TITANIUM VENT SCREENS FOR THE F-22, NEXT-GENERATION AIR SUPERIORITY FIGHTER

Henry R. Phelps
Lockheed Martin Aeronautical Systems
86 South Cobb Dr
Marietta, Georgia 30063

Abstract

The F-22, Next-Generation Air Superiority Fighter, has 22 titanium screens per aircraft for ventilation, exhaust, and drainage. These screens contain between 100 and 27,000 densely packed, diamond shaped holes. The hole configurations precluded the use of conventional machining practices. This paper compares the potential methods considered for producing these screens, and explains why precision, abrasive waterjet (AWJ) was selected as the preferred method. Issues associated with the use of the AWJ technology are discussed.

Introduction

The F-22, Next-Generation Air Superiority Fighter, has 22 titanium screens per aircraft, for ventilation, exhaust and drainage. Titanium was selected due to its excellent fatigue resistance. The screens are between 2.54 and 6.35 mm (0.100 and 0.250 In.) thick and contain between 100 to 27,000 diamond shaped holes. Hole openings are typically 2.29 mm (0.090 In.) with nominal web thicknesses of 0.64 mm (0.025 In.). The hole configuration was selected to allow an increased level of porosity over that obtainable with round holes. To improve flow, the holes in some screens are set at an angle to the panel surface. The Air Cooled, Fuel Cooler (ACFC) screen, located in the mid-fuselage, is the largest and most complex of the vent screens on the F-22 (Figure 1). It contains three different sizes of vent holes, cut at a 45 degree angle to the screen surface. Locations of the titanium screens on the aircraft are shown in Figure 2.

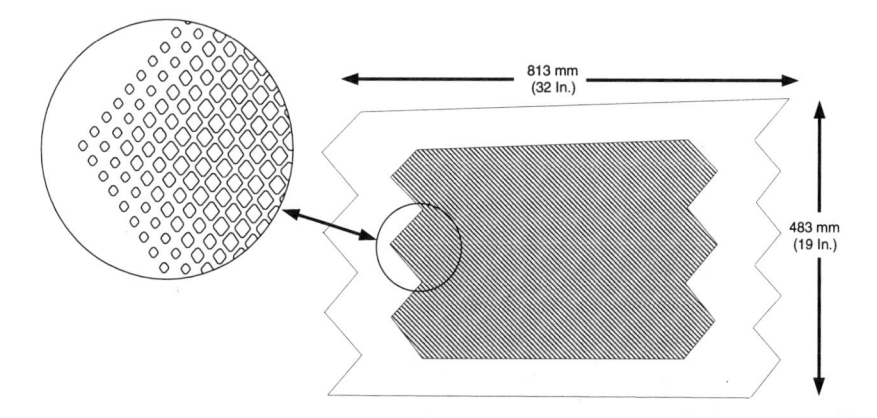

Figure 1 - The largest and most complex of the titanium vent screens on the F-22 is the Air Cooled, Fuel Cooler (ACFC) screen located in the mid-fuselage. This screen has 27,000 diamond shaped holes cut at 45 degrees to the surface.

Potential Fabrication Methods

There are three potential methods for producing the diamond hole patterns in the screens. Electrical Discharge Machining (EDM), Laser Cutting, and precision, Abrasive Waterjet (AWJ) Cutting. Plunge EDM uses a graphite electrode to penetrate the metal and produce a hole with the shape of the electrode. This process produces a small, uniform heat affected zone (HAZ) approximately .051 mm (0.002 In.) thick (Ref 1). The arithmetical average, surface roughness produced with this method is typically 5.7 µm (225 µ In.) (Ref 1). Electrodes can be gang mounted to produce several, highly accurate, holes simultaneously. Plunge EDM is an extremely slow and costly process.

Laser cutting uses either a carbon dioxide (CO_2) or neodymium:yttrium aluminum garnet (Nd:YAG) laser to vaporize the metal and form the hole. A high pressure, inert gas flowing through the nozzle acts to eject the vaporized metal from the hole as well as to protect the system optics. In addition, the gas provides some shielding of the front surface during cutting. Air mixes with the inert gas causing an exothermic reaction with the heated titanium, which increases the cutting action. A nonuniform, HAZ develops which is largest at the bottom of the cut. Chemical etching is required to remove the re-solidified material (called recast layer) as well as the underlying HAZ from the cut surface. The recast layer and HAZ are typically 0.152

mm (0.006 In.) thick when cutting titanium less than 5.1 mm (0.200 In.) thick. If the HAZ is not removed, the fatigue resistance of the material can be reduced by a factor of 3 (Ref 2). Laser cutting is significantly faster than EDM. Thermal build up during cutting, however, can cause part distortion and affect hole quality. Hole surface quality can be variable, but is typically better than 9.14 µm (360 µ In.) (Ref 1).

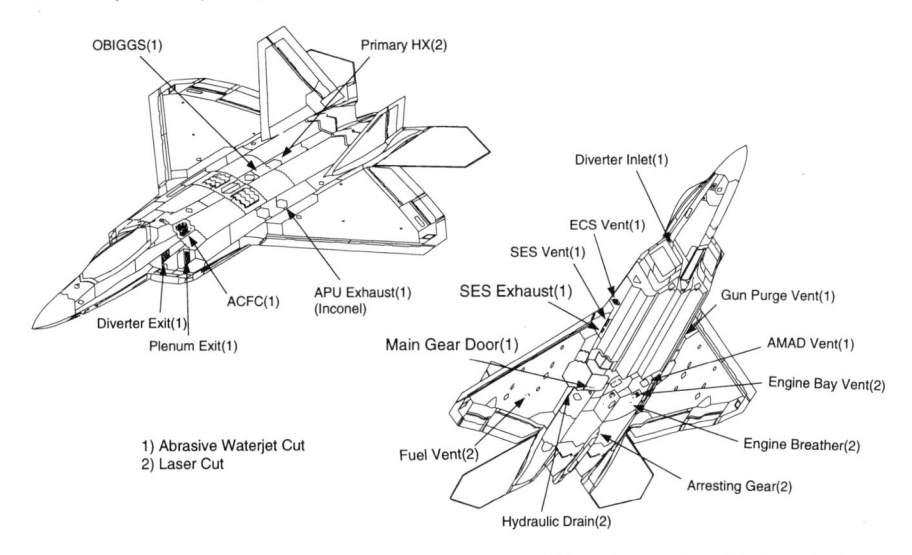

Figure 2 - Locations of Titanium Screens on the F-22 Next-Generation, Air Superiority Fighter.

Abrasive Waterjet Process

The basic waterjet process has existed for more than 25 years. High pressure water, at 340 MPa (50 Ksi), is forced through a 0.254 mm (0.010 In.), sapphire or diamond orifice and channeled down a tungsten carbide focusing tube to form a uniform jet, typically 0.86 mm (0.034 In.) in diameter. In order to cut metals, a garnet abrasive is added to the jet. The venturi effect of the high pressure stream passing through the mixing chamber forms a vacuum which pulls the abrasive into the nozzle. Some systems augment the grit feed with either positive or vacuum pressure to improve uniformity of the abrasive delivery. Figure 3 shows a sketch of the elements of a waterjet nozzle assembly (Ref 4).

The cut rates with AWJ are slower than those achievable with laser cutting. The quality of an AWJ cut edge, however, is significantly better than a laser cut edge. Surface roughness of the cut is typically 3.81 µm (150 µ In.) or better.

Representations of hole cross sections produced with each of the candidate cutting processes are shown in Figure 4. Abrasive waterjet cutting offered several key advantages. Edge quality is better than that produced with a laser or plunge EDM. No HAZ is developed by the AWJ process. In addition, thermal distortion of the work piece during cutting is not a problem with this method. When post processing costs are considered, the AWJ process is cost competitive with laser cutting and significantly less costly than EDM. Precision AWJ cutting was subsequently selected as the preferred cutting method for the F-22 screens.

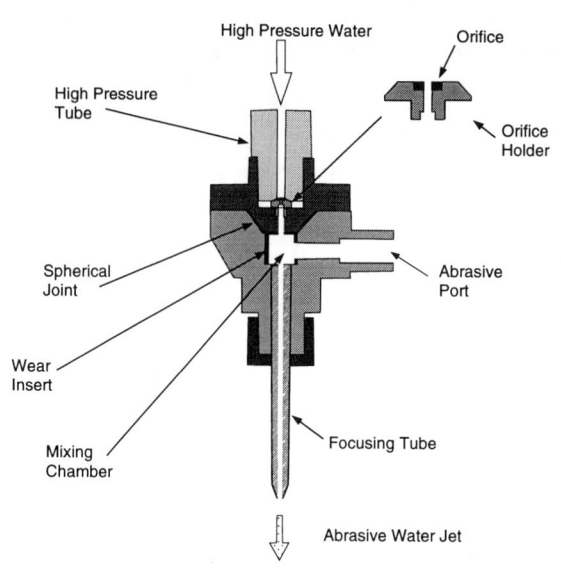

Figure 3 - Basic Elements of Abrasive Waterjet Nozzle Assembly

Critical AWJ Process Parameters

The F-22 Team and selected vendors are working to optimize the AWJ process for production of the vent screens. Table I shows the process parameters and equipment capabilities that need to be considered when optimizing the manufacturing process. The cut characteristics affected by these parameters are briefly discussed below. Detailed discussions are contained in references 4 and 5.

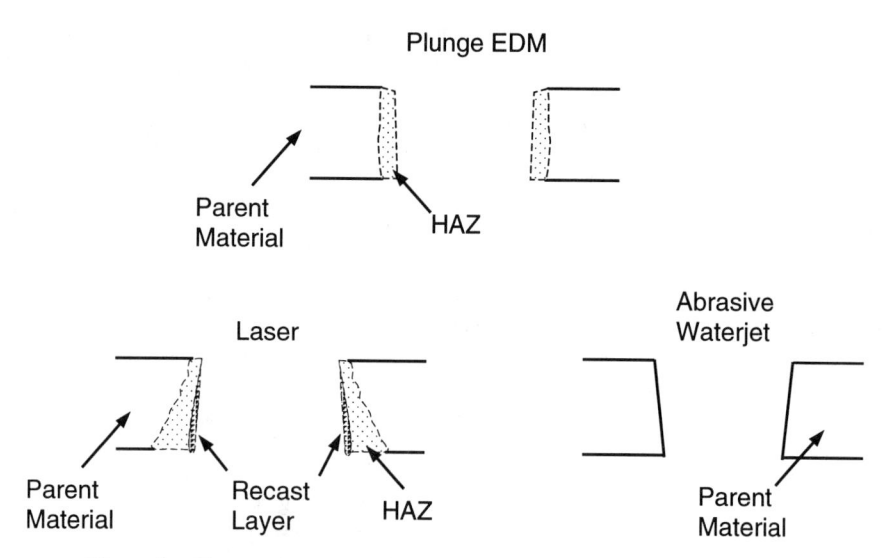

Figure 4 - Characteristics of Holes Produced with EDM, Laser, and AWJ Cutting Methods

304

	These Attributes Can Be Optimized for Specific Application			These Attributes Established by System Capabilities	
Parameter	Cut Depth	Edge Taper	Cut Edge Roughness	Positioning Accuracy	Repeatability
Abrasive					
Type & Grit Size	X		X		
Feed Rate	X	X	X		
Cut Rate	X	X	X		
Nozzle					
Orifice Dia	X		X		
Focusing Tube Dia	X	X	X		
Nozzle-to-Work Dist	X				
Inprocess Replacement					X
Water Pressure	X	X	X		
Robot					
Gantry or Manipulator				X	X

Table 1 - Critical Parameters for Precision, Abrasive Waterjet Cutting

Depth of Cut

The depth of material that can be cut is a function of the power that is delivered by the AWJ. This is established by the water pressure, orifice and focusing tube diameters and grit size of the abrasive. High cut rates can limit the cutting depth. The effective depth of angle cut holes needs to be considered when optimizing process parameters.

Edge Taper

Figure 5 show representations of the possible edge conditions produced by the AWJ. The cut edge normally exhibit some degree of negative kerf, i.e.. less material is removed from the exit side than the entry side. The taper can be minimized by slowing the cut rate. This also improves the surface roughness of the cut edge, but increases part cycle time and therefore end item cost. At extremely slow cut rates, it is possible to form a positive kerf as the spreading jet erodes the bottom layers of the material.

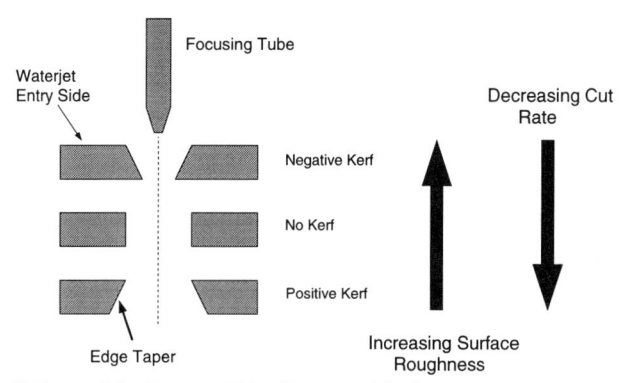

Figure 5 - Effects of Cut Rate on Edge Taper and Surface Roughness

Surface Roughness

Figure 6 shows the surface conditions produced for 3 different AWJ cut rates. The slowest rate produced the best edge quality, 1.0 µm (38 µ In.), while a rate approximately 10x faster produced an edge quality of 3.8 µm (150 µ In.). Surface roughness varies from entry to exit side and is highest near the exit side of the cut. The morphology of the waterjet cut surface includes striations, and randomly distributed fine pits. Edge striations run perpendicular to the cut edge. This differs from a conventionally machined edge. A tail forms near the exit side because the jet lags slightly behind the traversing head . The effect is more pronounced at high cut rates and can significantly affect cut- corner geometry. The pits are more prevalent near the exit side of the cut and may be due to momentary imbedding of particles in the wall.

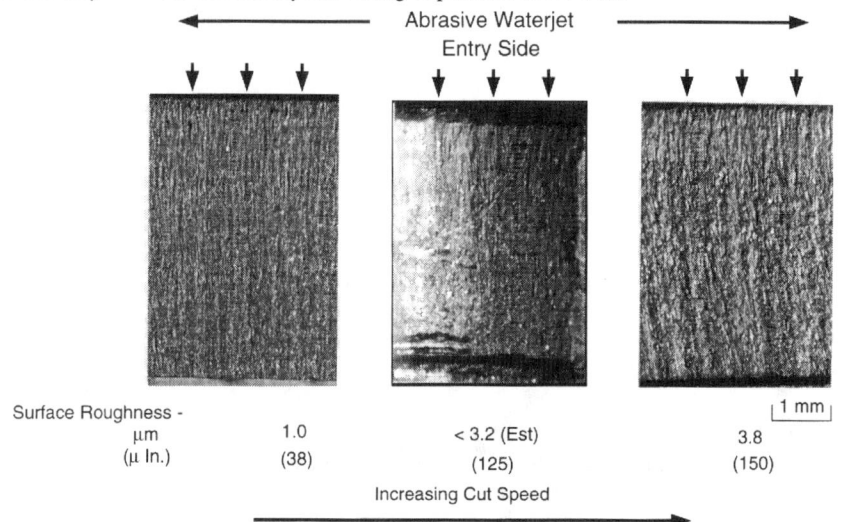

Figure 6 - Comparison of AWJ cut edges produced with different cut rates.

Fatigue Characteristics - Figure 7 compares the effect of edge, surface roughness on fatigue resistance. Two different edge qualities, 1.0 and 3.2 µm (38 and 125 µ In.), were evaluated. The results are compared against data from reference 3 for conventionally prepared specimens. The results indicate the AWJ produced edge degrades fatigue resistance. With the possible exception of high cycle performance, a better edge quality did not appear to translate into an improvement in fatigue resistance of AWJ cut specimens.

The reduced fatigue resistance may be attributable to two factors. First, the AWJ forms striations that run perpendicular to the cut edge. Conventional machining produces cutter marks that run parallel to the cut edge. The perpendicular marks increase the effective stress concentration factor (Kt) of the cut edge. Second, conventional machining tends to cold work the cut edge. The cold work leaves a residual compressive stress in the edge material, that improves tension-tension fatigue resistance.

Positional Accuracy and Repeatability

Though there are numerous AWJ systems currently in use, most are not capable of producing the F-22 screens. Precision, AWJ systems, with positional accuracy of 0.076 mm (0.003 In.) and repeatability of 0.038 mm (0.0015 In.) or better, are being used to insure all design requirements are met.

Equipment Issues

Screens fabrication will put unusual demands on the waterjet equipment. The large screens will require several days of continuous waterjet cutting. Power failures during screen cutting could lead to the scrapping of a high value part. Sufficient system positioning repeatability is required to allow restarting the cutting operation, without shifting the pattern, following such a process interruption.

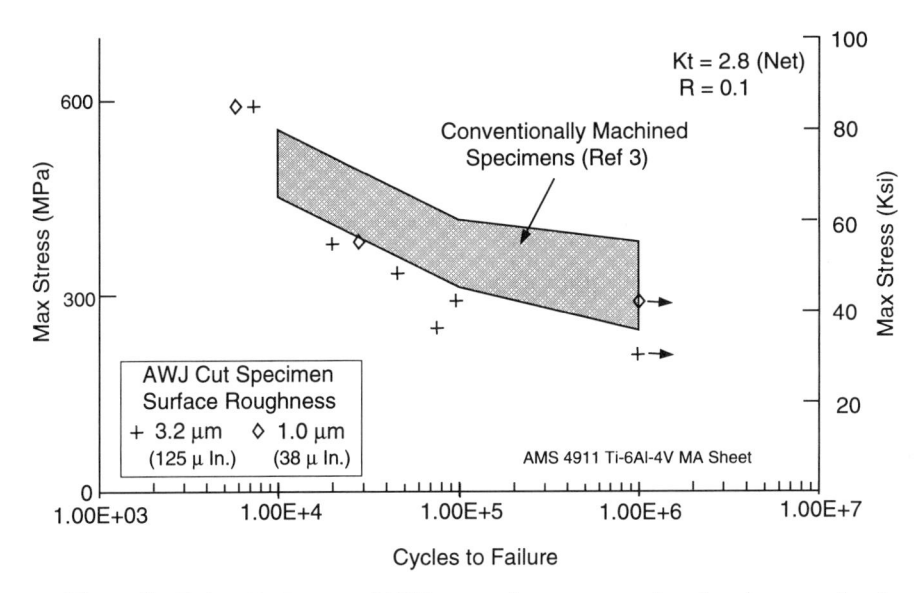

Figure 7 - Fatigue Resistance of AWJ cut specimens compared against the scatter band of data for conventionally machined specimens.

The long run times and high cycle rates (multiple cycles per minute) will put unusual demands on the high pressure pumps and valves. These items are high maintenance items under less demanding usage. Though a pump malfunction will shut the system down until repaired, water leaking through a high pressure valve can have more severe consequences by damaging the screen surface as the nozzle traverses from one hole to the next.

Another concern is water from the jet migrating up the abrasive feed lines when the high pressure valves are cycled on and off. This can wet the grit and clog the feed line, leading to irregular or uncut features. Some systems have sensors that detect when clogging has occurred and allow it to be cleared before starting the next cut.

Over time, the abrasive flow will cause the focusing tube to wear and the jet diameter to increase. This decreases cutting efficiency and increases the cut hole size. Slowing the cut rates can make up for reduced efficiency and offsets in the cut path can correct some of the hole features. The hole corner radii, however, are set by the jet diameter. Once the wear becomes excessive or uneven, the focusing tube must be replaced. The frequency of replacement must be increased if tight radius control is required and can impact the end item cost.

The equipment issues discussed above can be addressed through rigid adherence to periodic maintenance schedules and by having an inventory of critical spares. In addition, the precision AWJ systems used for the F-22 vent screens are highly automated. Extensive use of sensors and in-process inspection are used to detect and make corrections for anomalous situations before they cause a part to be scraped.

Summary

The F-22 requirements for vent screens required the extensive use of titanium and precluded conventional manufacturing processes.

Precision, abrasive waterjet cutting was selected as the preferred method for the F-22 screens because it offered the best combination of cost and part quality.

The characteristics of a AWJ produced edge are different than those produced by conventional machining. These differences need to be understood at the design phase to insure adequate fatigue resistance.

Acknowledgment

The Author wishes to thank the F-22 System Program Office for allowing the presentation and publication of this material.

References

1) Gary Giessmann, Meeting with author, Hi-Tek Manufacturing, Inc. Mason, Ohio, 10 August 1994

2) J. Huber and W. Marx "Production Laser Cutting", (Proceedings of ASM Applications of Lasers in Materials Processing Conference, Washington DC, 18 April, 1979, Pages 273-290)

3) Military Handbook 5G, Metallic Materials and Elements for Aerospace Vehicle Structures. 1 November 1994, Vol 2, Page 5-74.

4) M. Hashish, "Optimization Factors in Abrasive-Waterjet Machining," Journal of Engineering for Industry, 113 (1991) 29-37

5) M. Ramulu and M. Hashish, "Machining Characteristics of Advanced Materials", (Paper presented at the Winter Annual Meeting of the American Society of Mechanical Engineers, San Francisco, California, December 10 - 15, 1989

TEMPERATURE DISTRIBUTION IN CRYOGENIC MACHINING OF TITANIUM BY FINITE ELEMENT ANALYSIS

Shane Y. Hong
Yucheng Ding
Department of Mechanical & Materials Engineering
Wright State University
Dayton, Ohio

Abstract

Titanium and its alloys are characteristic of a poor tool life in conventional machining due to an intense temperature rise incurred at the tool-chip interface. The use of liquid nitrogen (LN2) as a replacement for common cutting fluids has been explored in industry and universities worldwide to improve the machinability for this type of difficult-to-machine materials. A general investigation into the cutting temperature distribution is required to determine a cost-effective LN2 supply for this cooling technology. This paper presents a finite element simulation of the cutting temperatures for the LN2 cooled machining processes. The effect of the LN2 coolant jets on the cutting processes is evaluated by using an experimentally estimated convective heat transfer, which is assigned to boundaries in the cutting zone based on the cutting strategy used. The results from the finite element analyses are found to be in a reasonable agreement with experimental measurements for the temperature picked up at the tool insert rake by an imbedded thermocouple. The discussion will be focused on Ti6-4.

Introduction

Titanium (or its alloys, e.g. Ti6-4) has long been classified as a difficult-to-machine materials due to an intense temperature rise generated at the tool-chip interface. The high temperature can promote various tool wear mechanisms, leading to poor tool life. This material is also notorious for a strong chemical reactivity with almost all tool materials available, especially at cutting temperatures higher than 500°C[1]. This promotes its alloying tendency with the tool materials, often resulting in galling, welding and smearing, along with premature tool failure. Therefore, current machining practices in industry are forced to limit the cutting speed to lower than 1m/s (or 200ft/min) in order to maintain a reasonable tool life.

Liquid nitrogen (LN2) has been investigated as an alternative to conventional cutting fluids (usually various oils or chemicals) for metal cutting processes. It has been reported[2-9] that this cryogenic coolant can increase tool life, improve chip breaking and refine workpiece surfaces. Another important fact that favors the use of the LN2 coolant is that this cryogen is not hazardous to the environment. It evaporates into the atmosphere and entirely eliminates the necessity of often expensive disposal. On the other hand, the application of the LN2 coolant can be a considerable expense. This includes the cost for acquisition, transportation, storage, and on-the-spot delivery of LN2. Therefore, it is a challenge to minimize LN2 consumption while maintaining an adequate improvement to the machinability of the cryogenic cutting processes. Since the cutting temperature plays a critical role in determining the machinability for titanium machining processes, an insight into the temperature distribution in the cutting zone, as affected by LN2 coolant, is required to design a cost-effective LN2 supply and cooling strategy.

Numerous attempts [10-17] have been made by other researchers to analytically predict the cutting temperatures by finite element methods (F.E.M.) or other analytical approaches, but almost exclusively for dry cutting conditions. The data accumulated thus far does not apply to the cryogenic machining processes, where LN2 is to be used as a coolant and sprayed toward a zone localized near the major tool cutting edge. A research by Childs *et.al.*[18] was devoted to the influence of waterjet coolant on the cutting temperatures and gave an estimated value for the convective heat transfer by the coolant.

This paper presents a finite element (F.E.) simulation of the temperature distributions for LN2 cooled cutting processes. The effect of LN2 jets on the cutting processes is taken into account by introducing a convective heat transfer coefficient, estimated experimentally. Assignment of the boundaries of the heat convection depends on the LN2 jet configurations. The F.E. analysis results are compared to the temperatures measured experimentally at specific location of the tool insert by an imbedded thermocouple.

Finite Element Simulation

If the cutting process is an orthogonal one and the chip width is much larger than the chip thickness, the problem is considered as a two-dimensional steady heat transfer in a domain as shown in Figure 1. Obviously the 2D problem domain shown should be taken perpendicularly to the major tool cutting edge. In order to determine the model geometries, the shear angle and deformed chip thickness can be measured under a metallographic microscope while the tool-chip contact length can be obtained from visual inspection of the tool wear marks under a tool microscope scaled in fillar unit.

The boundary can be categorized into three portions, that is, S_t, where the temperature is known, S_o, which can be assumed to be thermally insulated and S_h, which is exposed to heat transfer by the LN2 jet. The classification of boundaries depends on the configuration of the coolant jets. For a dry cutting process (therefore without any coolant), the boundaries are classified as dealt with in many other studies[10, 12, 18]

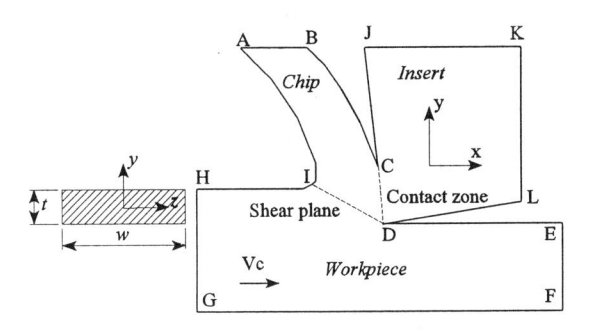

Figure 1: finite element model

$$\begin{aligned}(a) \ & S_t \ni HG, GF \\ (b) \ & S_o \ni HI, IA, AB, BC, CJ, JK, KL, LD, DE, EF \end{aligned} \quad (1)$$

For cryogenic machining, two cooling strategies are discussed here: one LN2 jet applied to the tool rake or two LN2 jets applied to the tool rake and flank simultaneously. If the LN2 coolant is delivered only to the tool rake face in jet, the following relationships apply:

$$\begin{aligned}(a) \ & S_t \ni HG, GF \\ (b) \ & S_o \ni HI, IA, AB, JK, KL, LD, DE, EF \\ (c) \ & S_h \ni BC, CJ \end{aligned} \quad (2)$$

and with a second LN2 jet sprayed toward the tool flank additionally, S_t, S_o and S_h can expressed in the following form:

$$\begin{aligned}(a) \ & S_t \ni HG, GF \\ (b) \ & S_o \ni HI, IA, AB, JK, KL, EF \\ (c) \ & S_h \ni BC, CJ, LD, DE \end{aligned} \quad (3)$$

Mathematically, the following boundary conditions are imposed on each boundary class:

$$\begin{aligned}(a) \ & T = T_s \quad \text{on boundary} \ S_t \\ (b) \ & -k\frac{\partial T}{\partial n} = 0 \quad \text{on boundary} \ S_o \\ (c) \ & -k\frac{\partial T}{\partial n} = h(T - T_\infty) \quad \text{on boundary} \ S_h \end{aligned} \quad (4)$$

where T_s is the temperature specified along the boundary S_t. T_s can be equated to room temperature (25°C) if S_t is located far enough from the tool-chip interface and the shear plane. T_∞ is the ambient temperature and is taken as the LN2 temperature for LN2 cooled cuttings. n stands for the outward normal to all boundaries. Finally, h represent the convective heat transfer coefficient on those LN2 cooled boundaries.

By merging the tool and chip boundaries at the tool-chip interface, it has been implicitly assumed that the tool and chip share the same temperature at this contact area. This is justified by the fact that the high normal contact stress produces only a negligible thermal resistance in the tool-chip interface. Two heat sources are considered: the deformation heat at the primary shear zone ID and the friction heat at the tool-chip interface CD. For simplicity, in practical finite element computations these two heats are considered to be volumetric and lumped uniformly along ID and CD, respectively. The heat generation rates for these two heat sources can be obtained from the cutting force components on-line measured by a 3D dynamometer.

The detailed F.E. formulation for the heat transfer can be found in various publications[10, 18]. It should be indicated here that the information on the temperature gradient in the tool is required to estimate the heat convection, which is crucial to the F.E. modelling developed above. It is technically difficult, however, to measure the tool temperature distribution directly during cutting processes due to the limited size and the high temperature gradients near the tool-chip interface. Therefore, a setup, as sketched in Figure 2, was designed to estimate the heat convection by the LN2 coolant sprayed toward a flat face. The heat convection obtained was then used as an approximation of what was actually produced in the cutting zones.

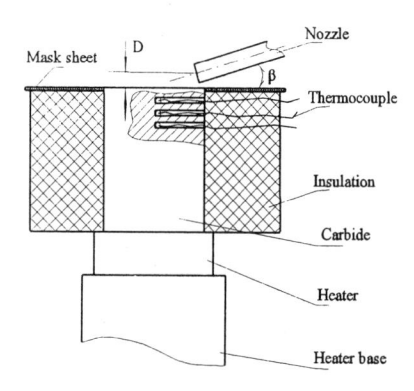

Figure 2: experimental set-up for heat transfer estimation

As shown in Figure 2, a carbide (K-68) block was mounted to the heater through a thermally conductive cement and insulated thermally from the ambient air, with only its top face exposed to the LN2 coolant spray. The LN2 was applied at an inclining angle, β, and a spraying distance, Δ, to the top face of the carbide block. These two variables were equated with those determined from the nozzle subassembly of the LN2 delivery system. Three miniature thermocouples (each with a junction diameter of 0.003") were embedded beneath the top face of the carbide block. The temperatures picked up from the thermocouples can be used to compute the gradient and extrapolate the surface temperature of the carbide block, both of which are needed to derive the convective heat transfer by the LN2 jet.

Experiment & Temperature Measurements

Cutting tests were performed for turning operations of Ti6-4 with different configurations of the LN2 jet. The tool insert CNMA432-K68 (from Kennametal Inc.), with a negative rake angle (*i.e.* -5°), was used for both materials. A volumetric flowrate of 0.20GPM was used for each activated jet. This corresponded to an estimated heat transfer ranging from 1.013×10^4 to

1.206×10^4 Btu/hr-ft^2-°F, depending on the temperature difference between the tool faces and LN2. The cutting speed ranged from 200 to 500ft/min (or from 1.02 to 2.54m/s). A fixed feed rate (0.01in./rot) and cutting depth (0.05in.) were applied throughout the current research.

The temperature measurements were performed to test the reliability of the F.E. simulation of the cryogenic cutting processes. The measurement pick-up spot was located on the rake face, close to the major cutting edge. A thermocouple (with a junction diameter of 0.003 in.) was planted into an EDMed hole through the tool insert. The hole was filled with a fine silicon cement of high thermal conductivity so that a good heat transfer could be produced between the tool materials and the thermocouple junction. The thermocouple junction was flushed even with the tool rake face. The location of the hole (0.015in. in diameter) was selected so that the thermocouple junction fell entirely into the tool-chip contact zone. The exact location of the thermocouple with respect to the major cutting edge was measured under a ruler microscope and associated with a corresponding F.E. node. All these cares were taken to insure that the detected temperature was in close approximation to what occurred at a certain point on the tool rake. The sensed temperature was typical for the purpose of a comparison with the F.E. prediction, although it may not be the maximum tool-chip interface temperature (which should be of the interest for our discussion).

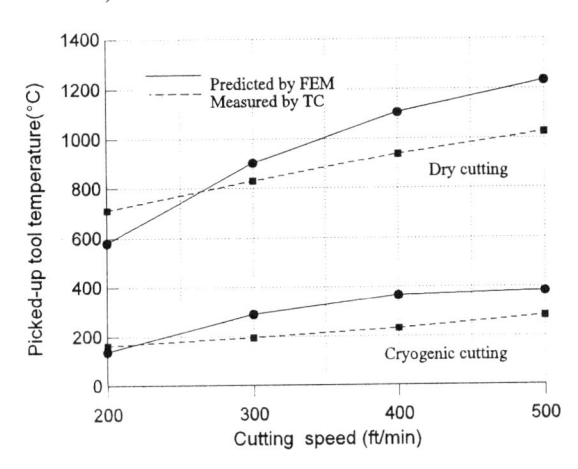

Figure 3: experimentally measured and F.E predicted tool temperature

Figure 3 shows the experimental and analytical results for the temperatures at the mentioned pick-up point. For cryogenic machining, the temperature was measured for the cutting processes where only one LN2 jet was applied to the tool rake (one nozzle configuration). The F.E. predicted temperature deviated from the measured temperature by less than 18% for dry cuttings and less than 30% for the cryogenic cuttings over the cutting speeds used. This is quite acceptable in view of the intricacies involved in the heat transfer of the cryogenic cutting processes. The deviations can be attributed to the errors in the thermocouple temperature measurements caused by the large temperature gradient and the inaccuracy in associating the location of the thermocouple junction with a finite element node. The comparatively large deviation for cryogenic machining can also be due to inaccuracy in the estimated heat transfer coefficient and the assumption of a unique heat transfer on all the boundaries affected by the LN2 jets. Fortunately, the F.E.M. tends to provide a conservative estimate of the tool-chip interface temperature, compared to the experimental measurement.

Temperature Distribution & Cooling Conditions

The F.E. analyses were performed for the turning operations of Ti6-4, with different jet configurations. Figure 4 shows the isothermal contours for dry cutting of Ti6-4 with a surface cutting speed of 500ft/min (depth of cut of 0.05in. and feedrate of 0.01in.). The maximum temperature induced occurred close to the cutting edge on the tool-chip interface and reached a level of 1200°C at this cutting speed. Such a high temperature was attributable to the small tool-chip contact length, the high cutting force and the poor thermal conductivity of the titanium alloy. It should be indicated that most of the tool wear mechanisms such as abrasion, adhesion and diffusion (or fusion) are highly dependent on the cutting temperature. For instance, Ti6-4 begins to exhibit a strong chemical reactivity to most tool materials (including the carbide K68 used here) even at a temperature of 500°C [1], impairing the wear resistance of the tool faces. Therefore, the dry cutting of Ti6-4 is not advisable because of the unacceptable tool life, especially at high cutting speeds.

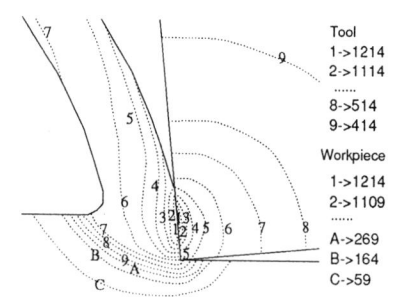

Figure 4: isothermal contours for dry cutting (500ft/min)

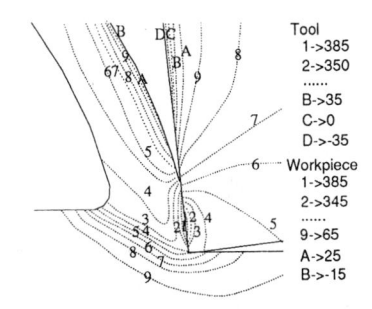

Figure 5: isothermal contours for one LN2 jet applied to rake only (500ft/min)

Figure 5 shows that by applying only one LN2 jet to the tool rake the maximum tool-chip interface temperature was reduced to 385°C. At this temperature the adhesion and diffusion of Ti6-4 to the tool materials generally become negligible [1]. A further decrease in tool temperature can reduce abrasion by maintaining the tool hardness and strength. Fig.6 indicates that the application of a second LN2 jet to the tool rake causes an additional reduction of more than 28% in the maximum tool-chip interface temperature. The temperature in the tool flank is also reduced significantly by the direct application of LN2 to this region, which is important for reducing the wear rate at the tool flank.

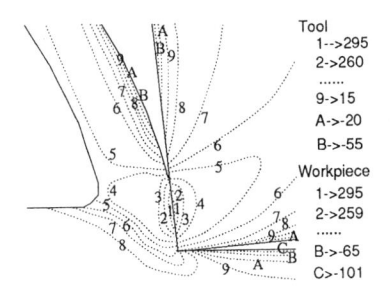

Figure 6: isothermal contours for two LN2 jets applied to tool rake & flank(500ft/min)

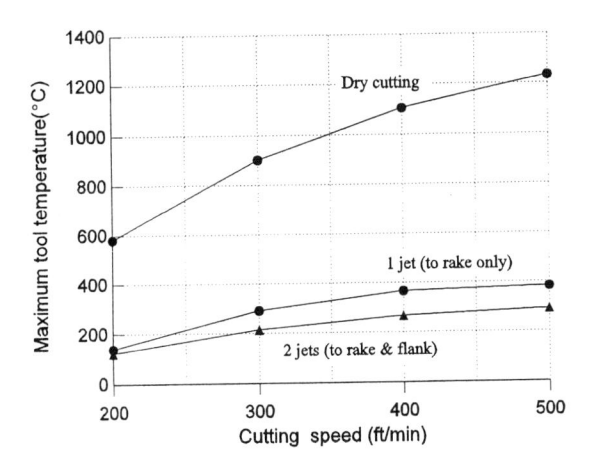

Figure 7: maximum tool temperature vs. cutting speed

Figure 7 shows the maximum tool-chip interface temperature predicted by F.E.M. versus cutting speed for different cooling conditions. At the cutting speed 200ft/min, simultaneous application of the second LN2 jet to the tool flank does not seem to make a distinguishable difference to the maximum tool-chip interface temperature, compared to application of one LN2 jet to the tool rake alone. The second LN2 jet to the tool flank is more effective in cooling the tool, in terms of the percentage reduction in the tool-chip interface temperature, for high cutting speed. Since use of the second LN2 jet increases the total LN2 consumption, it is only preferable for higher cutting speeds.

Conclusion

The use of LN2 coolant can generally decrease the temperature in the cutting zone significantly. For Ti6-4 cuttings, compared to dry cutting, the application of one LN2 jet at a volumetric flowrate of 0.20GPM only to the tool rake results in a reduction in the tool-chip interface temperature from 1200°C to 385°C. The application of a second LN2 jet to the tool flank can produce a further temperature reduction in the cutting zone, but depending on the cutting speed. The second LN2 jet applied to the tool flank tends to be more effective at a high cutting speed. Therefore, application of LN2 jets to the tool rake and flank simultaneously is especially advisable for high cutting speeds.

Acknowledgements

The authors wish to express their gratitude for the support of the cryogenic machining project by The Edison Materials Technology Center, for the joint partnership from Cincinnati-Milacron, General Motors-Delco-Chasis, Timken Company, Kennametal Inc., GE Aircraft Engines, BOC Group, Vortec Co., A.F. Leis Co., Enginetics Co., Abrasive-Form, Inc., and Gem City Engineering. Thanks are also due to Mr. Atul Saksena, Kevin Ryan and Mamaud Hanafi for their valuable involvement in the cutting experiments for the current project.

References

[1] Machado, A.R., Wallbank, J., "Machining of Titanium and Its Alloys—A Review", Proc. Instn. Mech. Engrs., Vol.204, No.B1, 1990.

[2] Uehara, K., Kumagai, S., "Chip Formation, Surface Roughness and Cutting Force in Cryogenic Machining", Annals of CIRP, Vol.17(1), 1968.

[3] Uehara, K., Kumagai, S., "Characteristics of Tool Wear in Cryogenic Machining", J. Japan. Sco. Precis. Engrs., Vol.35, No.9, 1969.

[4] Jainbajranglal, J.R., Chattopadhyay, A.B., "Role of Cryogenics in Metal Cutting Industry", Indian J. of Cryogenics, Vol.9, No.1, 1984.

[5] Fillipi, A.D., Ippolite R., "Facing milling at -180°C", Annals of CIRP, Vol.19(2), 1971.

[6] Dillon, O.W., De Angels, R.J., Lu, W.Y., Gunasekera, J.S. and Deno, J.A., "Some Results from Cryogenic Cutting Processes", J. Mater. Shap. Technol., Vol.8, No.1, 1990.

[7] Zhao, Z., Hong, S.Y., "Cooling Strategies for Cryogenic Machining from Materials Viewpoint", J. of Mater. Eng. & Performance, Vol. 1, No.5, 1992.

[8] Hong, S.Y., "Economical Cryogenic Machining for High Speed Cutting of Difficult-to-Machine Materials", Proc. of 1st Int. Conf. on Manufacturing Technology, Hong Kong, Dec. 27-29, 1991.

[9] Zhao, Z., Hong, S.Y., "Cryogenic Properties of Some Cutting Tool Materals", J. of Mater. Eng. & Performance, Vol.1, No.5, 1992.

[10] Tay, A.O., Stevenson, M.G. and G.de Vahl Davis, "Using Finite Element Method to Determine Temperature Distributions in Orthogonal Machining", Proc. Inst. Mech. Engrs., Vol.188, No.55, 1974.

[11] Usui, E., Shirakashi, T. and Kitagawa, T., "Analytical Prediction of Three Dimensional

Cutting Processes", Trans. of ASME, J. of Eng. for Ind., Vol.100, pp236-243, 1978.

[12] Murarka, P.D., Barrow, G. and Hinduja, S., "Influence of the Process Variables on the Temperature Distribution in Orthogonal Machining Using Finite Element Method", Int. J. of Mech. Science, Vol.21, pp445-456, 1979.

[13] Stevenson, M.G., Wright, P.C. Chow, J.G., "Further Development in Applying the Finite Element Method to Calculation of Temperature Distribution in Machining and Comparisons with Experiment", Trans. of ASME, J. of Eng. for Ind., Vol.105, pp.149-154, 1983.

[14] Childs, T.H.C., Maekawa, K., "A Computer Simulation Approach towards the Determination of Optimum Cutting Conditions", Proc. ASM Int. Conf. on Strategies for Automation of Machining, April 15-17, 1987, Orlando, Florida.

[15] Lin, J., Lee, S.L., Weng, C.I., "Estimation of Cutting Temperatures in High Speed Cutting", J. of Eng. Materials and Technology., Vol.144, pp.289-296, 1992.

[16] Tay, A.O., Stevenson, M.G., G.de Vahl Davis and Oxley, P.L.B., "A Numerical Method for Calculating Temperature Distributions in Machining, from Force and Shear Angle Measurements", Int. J. Mach. Tool Des. Res., Vol.16, pp.335-349, 1976.

[17] Strenkowski, J.S., Moon, K.J., "Finite Element Prediction of Chip Geometry and Tool/Workpiece Temperature Distributions in Orthogonal Metal Cutting", Trans of ASME, J. of Eng. for Ind., Vol.112, pp.313-318, 1990.

[18] Childs, T.H.C., Maekawa, K., Maulik, P., "Effects of Coolant on Temperature Distribution in Metal Machining", Material Science & Technology, Vol.4, pp1006-1019, 1988.

[19] B. J. Moniz, "Metallurgy", American Technical Publishers, Inc., 1992.

Manufacture of Fasteners and Other Items in Titanium Alloys

Viacheslav A. Volodin, Igor A. Vorobiov

NORMAL CORPORATION

74, Litvinova St., Nizhny Novgorod, 603600, RUSSIA

ABSTRACT

The manuscript proposes for consideration some data relating physical and mechanical properties of titanium alloys which are used for fastening parts production by cold and hot plastic deformation, as well as relating strength features of screws at different temperature operating conditions

Titanium and its alloys is a very attractive material for commercial applications. It is widely used in the aerospace industry to manufacture fuselages, engines and fastening parts. It is also used in chemical, nuclear, nautical and medical industries where a higher anti-corrosion features are required.

At present, titanium has found its application in sporting equipment and bijouterie and it has a huge potential in the automotive engineering.

Both Russian and foreign experts state that titanium market will keep on growing and that there will be a constant demand for titanium alloy products especially in the engineering, oil and gas industries. The reason to this is that titanium and its alloys offer a higher strength, corrosion resistance, reduced weight of a structure in comparison with a similar one in steel; in certain cases, titanium alloys offer a lower cost, higher reliability and service life of structures operating in a high corrosion environment.

At present, Russian industry produce various fastening systems using a wide range of Titanium brands and alloys like BT1-00, BT1-0, OT4-1, BT14, BT15, BT16, BT22, BT3-1 and some other. Table 1 shows mechanical properties of rods made in some of the above mentioned types.

Figure 1 displays a number of fasteners in titanium alloys made by cold heading. These parts have been successfully utilized for several various applications.

The parts have been produced to various standards including DIN and ISO standards. The fasteners can also be manufactured according to specially developped drawings.

An undeniable leader among other alloys turns out to be the high strength alloy BT16 strengthened by hardening and ageing up to $\sigma t \leq 1470$ MPa. This alloy contains a small amount of Al, which allows to raise the strength and heat-resistance and retain a higher plasticity in the annealled and hardened conditions.

Due to a relatively high content of molybdenum and vanadium, BT16 can effectively be strengthened during hardening and ageing processes. The temperature of $\alpha + \beta \rightleftarrows \beta$ transition flactuates within the range of 820-870 °C.

BT16 is a two-phase alloy (structure $\alpha + \beta$).

Parts made in BT16 have a higher reliability and service life which make them advantageously distinguished from other titanium alloys. Parts in BT16 having thread up to M12 are manufactured by cold heading process, whereas the parts over M12 are made by hot heading.

Basing on this point, it could be very curious to compare the stress-strain properties of BT16 with a well known Western titanium alloy Ti 6 Al 4 V used for a similar application.

Table 1. Mechanical Properties of Rods in Titanium Alloys

Alloy	Rod dia. mm	Thermal treatment	σ_B, MPa	Elongation, δ_5, %	Contraction, ψ, %	Impact Strength KCU J/cm^2	Shear Resistance, τ_{cp}, MPa
1	2	3	4	5	6	7	8
BT1-00	25	annealing	295 - 440	25	55	120	250
BT1-0	25	annealing	390 - 540	20	50	100	300
OT4-1	10-12	annealing	580 - 655	20	60	70	430
BT6p	5 - 10	annealing	981	17	57	40	519
BT6p	10	hardening + ageing	1155	10	25	15,7	646
BT16	5 - 10	annealing	830 - 950	22	65	160	617
BT16	10	hardening + ageing	1080 - 1470	16	60	80	706
BT3-1	5 - 10	isothermal annealing	1050	14	40	45	720
BT22	25	isothermal annealing	1250	13	40	35	700
BT30	2mm sheet	annealing	981 - 1079	14 - 17	45 - 60	-	-

321

Table 2. Comparative physical and mechanical properties of Titanium Alloys Ti 6 Al 4V and Ti 5Mo 5V 2,5Al (BT16)

Designation	Ti 6Al 4 V	Ti 5Mo 5V 2,5Al (BT16)
1	2	3
1. **Chemical Composition** in % (of max.value, if the range not specified)	0,8C; 0,25Fe; 0,05N$_2$; 0,02O$_2$; 0,125H$_2$; (5,5-7,0) Al; (3,5-4,5)V	0,10C; 0,25Fe; 0,15 Si 0,30Zr; 0,15O$_2$; 0,05N$_2$; 0,015H$_2$; (1,8-3,8)Al; (1,0-5,0)V; (4,5-5,5)Mo
2. **Mechanical Properties**		
Temporal Resistance, MPa	1078 / 931	1080 - 1470 / 880
Yield Strength, MPa	1029 / 857	1050 - 1225 / 770
Elongation, %	10/10	16/22
Contraction, %	25	60/65
Creepage	$\sigma_{0,2/100}$=539 at 350°C	$\sigma_{0,2/100}$=392 at 350°C

Table 2 continued

1	2	3
Charpy V-notch Impact strength at 20°C (ft/lb)	10550 (1, 57 kgf·m/cm^2)	$\frac{53760}{107500}\left(\frac{8}{16}\text{ kgf·m/cm}^2\right)$
K_{1c}, MPa·mm$^{1/2}$	1160-1420 / 770	1890/1260
Shear Resistance, MPa	646/519	706/617

3. **Physical Properties**

Conversion Temperature, (F +25)	1796 (980°C)	1544 -1616 (840-880°C)
Density, kg/m^3	4350	4680
Melting Point, approx. (F)	3200 (1530-1630)°C	3100 (1704)°C

Table 2 continued

1	2	3
Specific Electric Resistance at 20°C, microohm·cm	166	111
Tensile Modulus, MPa	112700	110340 / 108570
Annealing Temperature, (F) - Dead Annealing	1380-1470 (750-800)°C 45 minutes, air cooling	1440±50 (780±10)°C 1 hour, with furnace cooling
- for Stress Relaxation	1110-1200 (600-650)°C 60 minutes, air cooling	1020-1200 (550-650)°C 1 hour, with furnace cooling
Rolled Stock	Bar, wire, forging	Bar, wire, forging, sheet
Typical Application	Fasteners: screws, bolts, nuts, studs	Fasteners: screws, bolts, nuts, studs
Industrial Specifications	AMS 4928 F	TU1-92-3-74 (dia. 4,5-10)mm TU1-9-623-87 (dia.2,5-8,5 mm)

NOTE: hardened in Numerator
annealed in Denominator

Comarative data of temperature impact on mechanical properties of Ti-6 Al 4V and BT16 in as-received condition and after thermal treatment are shown on Figures 2 and 3. Throughout the hole range of temperature researched, BT16 alloy has a higher strength and plasticity features which provide a higher operating life of any product design made in this alloy. Such advantage is especially vital when the parts operate for bending and tension cyclic loads.

Various fastening parts of complex shape are usually produced by cold plastic deformation with the use of a special processing technology in grinded or machined bars in BT1-00, BT1-0, BT16, BT30 as-received or after vacuum annealing condition which both get a special technological coating.

Cold forging process can be done with multi-positon cold headers to take advantage of work harderning effect which allows to obtain a product of a required strength avoiding hardening thermal treatment.

Fastening parts in BT6p, BT3-1, BT9, BT8M are produced by hot forging with a special equipment. Parts (like sheet anchor nuts) in BT30 alloy, which is supplied as sheet material, are manufactured by deep extrusion stamping.

Tables 3,4 show typical calculated breaking loads at normal temperature for fastening parts and bolted joints in titanium alloys.

Table 3

Calculated Breaking Loads to Rupture Screws and Bolts in BT16 after strain hardening.

Temperature, °C	MR4	MR5	MR6	MR8	MR10
	Calculated breaking loads for rupture, H (kgf)				
20	8500 (865)	13600 (1390)	19400 (1980)	35000 (3570)	55300 (5640)
130	7200 (730)	11600 (1180)	16600 (1690)	29800 (3040)	47800 (4870)
160	6860 (700)	11070 (1130)	15880 (1620)	28910 (2950)	46060 (4700)
200	6800 (690)	10800 (1100)	15500 (1580)	28000 (2860)	44100 (4500)

Table 4 shows Ultimate Tensile Stress of Bolts in various alloys at ambient temperature when tested for static tension with a skew positioned washer and a straight positioned washer. Alloys BT16 and BT6p (pure) undergone hardening and ageing, alloys BT3-1 and BT8M were isothermally annealled, alloy BT9 was twice annealled. Although the incoming bars in heat-resistant alloys possess a significantlly higher strength in comparison with bars in BT16 (Table 6), bolts made in these alloys lose to a great extent such advantage if compared with strain hardened bolts in BT16.

When bolts are tested for tensile strength with a skew positioned washer, they get a drop in strength which is averagely equal to 34, 35, 22, 31, 41% for BT16, BT6p, BT3-1, BT9 and BT8M alloys. The sensitivity to the skewness of bolts in BT16, BT6p, BT9 is approximately identical to each other to make about one third of a load. A different case with bolts in BT3-1, sensitivity of which is considerably lower if compared with the other abovementioned series of alloys. As to BT8M , the sensitivity of this alloy is slightly greater.

Table 4

Breaking Load of Bolts in Titanium alloys under Static Tests* at 20°C

Bolt Type	Breaking Load (kN) for Alloys as					
	BT16	BT6p	BT3-1	BT9	BT8M	BT30
M5	13,90 ----- 9,17	13,80 ----- 9,30	14,00 ----- 12,12	13,80 ----- 9,48	14,30 ----- 8,40	--
M6	19,80 ----- 13,26	19,30 ----- 14,52	21,00 ----- 15,13	21,80 ----- 13,24	20,20 ----- 12,20	25,0 ---- 19,75
M8	35,70 ----- 22,70	34,20 ----- 17,65	33,50 ----- 24,95	32,70 ----- 24,72	38,30 ----- 22,30	47,5 ---- 26,0

Note: * test results without obliquity (skewness) in Numerator;
8 degrees angle obliquity (skewness) in Dominator

Breaking Load of Bolts in BT6p, BT3-1, BT9 and BT8M tested for Shear is slightly greater than the one for bolts in BT16 (Table 5). Here we may underline that Bolts M5 for the majority of alloys appear to have a stronger shear resistance in comparison with Bolts M8.

Table 5

Shear Resistance of Bolts at ambient Temperature

Bolt Type		Shear Resistance (MPa) for Alloys as				
	BT16	BT6p	BT3-1	BT9	BT8	BT30
M5	650-680	750	740	719	690	--
M6	650-680	665	700	640	700	713
M8	650-680	668	720	672	634	736

Although, mechanical properties of bolts made in heat-resistant alloys are higher at ambient temperature than those of bolts in BT16 (Table 4), the difference in properties is not that great to doubt the application expediency of the latter to produce fastening parts. In fact, BT16 has a significantly higher processing capabilities and it is more cost-efficient.

Table 6

Strength features of M8 Bolts in Titanium Alloys at higher temperatures

Test Type	BT6p		BT3-1		BT9		BT8M	
	\multicolumn{8}{c}{Temperature, °C}							
	350	400	400	450	400	450	400	450
Static Tension without Skewness, CBL,kN	26,2	25,9	32,5	31,0	29,8	29,0	--	29,8
8 Degree Skewness Static Tension, CBL,kN	18,2	18,0	32,3	--	--	28,0	--	28,7
Shear Test, SBL,kN	18,5	16,9	30,2	--	--	27,8	--	28,1
LCF, L=0,6CBL; L=0,1CBL, number of cycles	11050- -30024	4070- -5280	7700- -11315	--	--	3500- -5000	--	6420- -4810

CBL- Calculated Breaking Load
SBL- Calculated Breaking Load for Shear
LCF- Low-Cycle Fatigue
L- Load

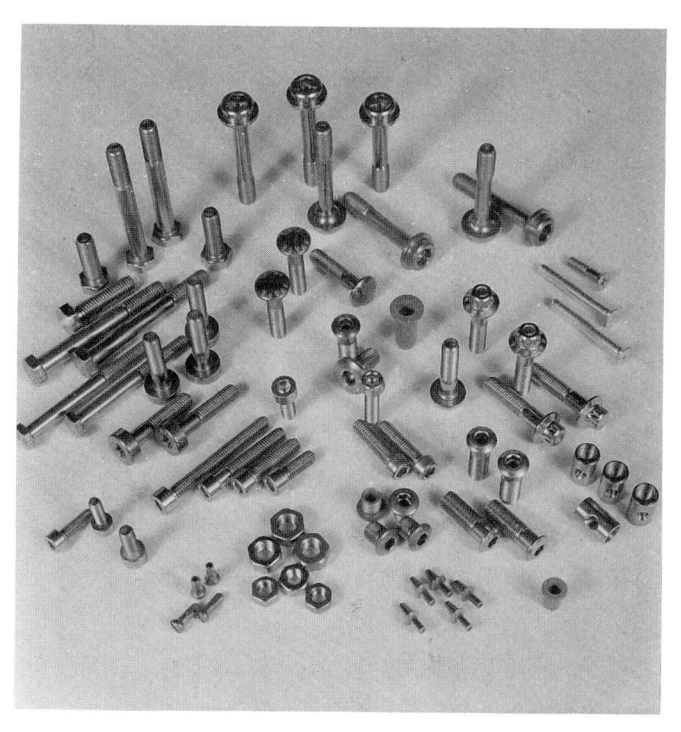

Figure 1. Fastening Parts in Titanium Alloys

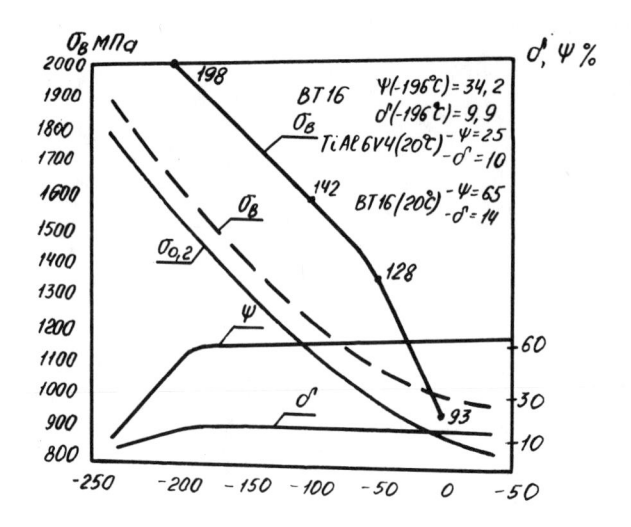

Figure 2. Temperature features of TiAl6V4 and BT 16 alloys

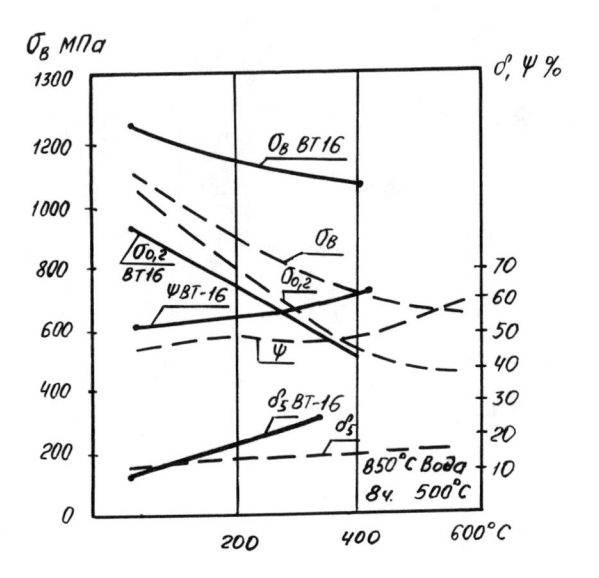

Figure 3. Temperature impact on TiAl6V4 and BT 16 alloys' mechanical properties

Hydrogen Technology as New Perspective Type of Titanium Alloy Processing

Boris A.Kolachev, Alexander A.Ilyin, Vladimir K.Nosov

Moscow State University of Aviation Technology named after K.Tsiolkovsky
Metals Science and Heat Treatment Technology Department
Petrovka Str. 27, K-31, Moscow, Russia, 103767

Abstract. Hydrogen alloying of titanium alloys gives several effects which may be used for the purpose of improvement of technological properties during production. These effects created the basis of titanium alloys hydrogen technology including: a) thermohydrogen treatment; b) hydrogen plastification; c) compacthydrogen processing; d) mechinohydrogen (mechanical treatment using hydrogen as temporary alloying element) treatment. The paper deals with description of technological possibilities which hydrogen technology may secure.

Introduction

In the case of investigations of metal-hydrogen problem, it was found that hydrogen not only causes hydrogen embrittlement but also gives several effects which may be used for the purpose of improvement of processing technology of metals, including titanium alloys. Hydrogen caused stress flow lowering, increasing of ultimate deformation limit before the first crack occurrence, structure and phase transformations, enhanced adhesion phenomena, hydrogen influence on properties being important in technological aspects may be attributed to these effects [1-8]. Above mentioned effects created the basis of titanium alloys hydrogen technology [9] which suggests hydrogenation of metal within certain concentrations, technological operations to use favourable hydrogen effects and vacuum annealing to reduce hydrogen content in a metal to safe concentrations which is not followed by hydrogen embrittlement of components and structures during operation.

Hydrogen technology includes the following technological processing [9]: a) thermohydrogen treatment, b) hydrogen plastification; c) compacthydrogen operations; d) machinohydrogen treatment. These technologies may be used as independent operations of titanium alloy processing or as one of a component of united technology which begins with hydrogenation of an ingot and ends in vacuum annealing of a finished product. Practical use of hydrogen technology requires design of specialised metal hydrogenation equipment only, other operations are carried out using standard equipment. Inasmuch as hydrogenation of a metal is an additional and not safe operation, hydrogen technology is the most reasonable in that cases when standard technological processes are low efficient or impossible at all.

Hydrogen technology is based on reversible hydrogen alloying of titanium alloys [10]. Hydrogen is the only alloying element providing efficient reversible alloying. In view of unusually high diffusion mobility, hydrogen may be easily introduced into the metal as well as removed from it after technological carrying out of all procedures at conditions where hydrogen causes the most favourable influence on properties determined by technological properties of alloys. Technological perspectives of different operations of hydrogen technology are described below.

Thermohydrogen treatment

Thermohydrogen treatment is based on the possibility of control of phase transformations and structure forming as a result of the significant hydrogen influence on amount, composition and properties of α- and β-phase.

The structure control of titanium alloys by means of thermohydrogen treatment is based on the following effects [5,7,11]:

1. Hydrogen is enough powerful β-stabiliser and causes significant $\alpha+\beta/\beta$ transition temperature decrease, increasing β-phase amount in alloys at annealing, and solution treatment temperatures.

2. Hydrogen increases heat treatment hardening during ageing in quenched $\alpha+\beta$-alloys; hydrogen alloying of pseudo- α-alloys converses them into a thermally hardened class of $\alpha+\beta$-alloys.

3. Enhancing of β-phase stability, hydrogen decreases the critical cooling rate that leads to harden penetration increase.

4. Hydrogen results not only in the β-phase amount increase but also in alloying elements redistribution between α- and β- phases in some alloys. For example hydrogen introduction in VT23 alloy (Ti-4.5Al-4.5V-2Mo-1Cr-0.6Fe) causes decreasing of all alloying element concentrations in β-phase and increasing of β-phase quantity at the same time.

5. Hydrogen has the appreciable influence on the lattice periods, phase volume fracture and volume effect at $\alpha \Leftrightarrow \beta$ transformation. In non-hydrogenated $\alpha+\beta$ alloys the atomic volume of β-phase is less that of α-phase one. Hydrogen increases β-phase volume fracture and has a little effect on α-phase one because of its better solubility in β-phase. As result of the introduction of a definite hydrogen concentration volume fracture of α- and β-phases become equal and volume effect of $\alpha \Leftrightarrow \beta$ transformation turns into zero. It secures the nucleation of α-phase in a form close to spherical one, the formation of globular structure and the improvement of plastic and fatigue properties of alloys with rough lamella initial structure.

6. Hydride precipitation becomes possible at high enough hydrogen concentrations. Titanium hydride has much higher volume fracture than α- and β- phases with the result that during thermocyclic treatment, including direct and reverse hydride transformation, the phase hardening occurs enough for α- and β-phase recrystallisation and essential refinement of not only intergrain microstructure but also macrograin too.

The most numbers of published papers [12-15] was devoted to refinement of intergrain microstructure by means of hydrogen doping of titanium alloys and following vacuum annealing. This technology was called "hydrovac" [12]. From mentioned above data it follows that "hydrovac", as the means of microstructure refinement, is one of directions of more general technology - themohydrogen treatment.

It will be noted that phase and structural transformations described above occur during realisation of any mentioned processing of hydrogen technology. More over the microstructures ensuring the optimum conditions of technological processing realisation may be formed by the suitable choice of thermohydrogen treatment. In this sense thermohydrogen treatment is the base of overall hydrogen technologies [12].

Hydrogen plastification

Hot deformation of titanium alloys in conditions of hydrogen plastification is based on hydrogen induced lowering of metal flow stresses and increasing of ultimate upsetting degree until the first crack occurrence [1-4]. This effect was found at first by Zwiker and Schleicher in 1956 year [1] but was small studied for a long time. Investigations carried out in Moscow aviation technological institute named after K.Tsiolkovsky in the early 70-s showed that hydrogen plastification (as this phenomena was called by us) may be used effectively for improving hot deformation technology of titanium alloys especially hard-to-deform ones.

Hydrogen plastification takes place due to the following causes [1-4, 6, 9]: a) β-stabilising action of hydrogen leading to formation of the structure with high enough amount of technological β-phase at lower temperatures as compared to non-hydrogenated metal; b) facilitated realisation of dynamic equilibrium (softening-strengthening) during hot deformation; c) β-phase leaning with β-alloying elements; d) microstructure refinement as a result of thermohydrogen treatment effect; e) enhanced dislocation generation and its increased mobility (this effect was experimentally proved at temperatures close to room one).

In paper [16] we drew conclusion that hydrogen plastification is an unusual picture of usual superplasticity. This unusual effect features its occurrence under lower temperatures and more strain rates as compared with superplasticity. Foreign investigators [17-19] found dramatic improvement in superplastic forming of titanium alloys by hydrogen additions, but did not put this effect in touch with hydrogen plastification. At present we believe than hydrogen plastification is more general phenomena that hydrogen-induced superplasticity. Hydrogen does not change the superplasticity mechanism but displaces its development to lower temperatures. Hydrogen-induced superplastic forming corresponds to deformation conditions of hydrogen plastification being the most pronounced.

The conditions of the most pronounced development of hydrogen plastification during tensile tests with initial strain rate $\dot{\varepsilon}=1.4*10^{-3}$ s^{-1} and its characteristics are presented in Table I.

Figure 1 illustrates in detail hydrogen influence on mechanical properties of VT6 (Ti-6Al-4V) alloy. At the temperatures close to 800°C yield strength of Ti-6Al-4V alloy with 0.3%wt. H is about two times lower than at initial hydrogen content. Elongation of Ti-6Al-4V alloy at the same temperature 800°C is equal to 50% at 0.005wt.% and 105% at 0.3wt.% H.

It is the most reasonably to carry out hot deformation of titanium alloys with using hydrogen plastification in isothermal conditions. At present isothermal forming of titanium alloys is conducted at the temperatures 925-952 °C requiring using tools from superalloys of GS-6 type difficult machined by cutting. Hydrogen plastification allows us to carry out isothermal forming at the temperatures at which stamps from comparatively easy machined superalloys of nimonic type have enough endurance leading to facilitating of stamp tool machining. At the same time tool metal consumption reduces in some times; endurance of stamp tool rises, energy expense decreases.

Hydrogen alloying allows us to reduce in about two times number of passes during rolling of sheets from elevated-temperature VT18U alloy and an experimental high-temperature alloy based on titanium aluminide Ti$_3$Al. Hydrogenated sheets of these alloys were cold rolled from the thickness 2 mm to 1 mm.

Table I The conditions of hydrogen plastification development of titanium alloys and its characteristics (initial strain rate $\dot{\varepsilon}=1.4*10^{-3}$ s^{-1}) [4]

Alloy	Conditions		Characteristics	
	wt.% H	t, °C	UTS,MPa*	El,%*
VT1-0 (99.5Ti)	0.3	700	69 / 22	92 / 98
OT4 (Ti-4Al-1.5Mn)	0.3	800	(165) / (56)	51 / 34
VT18(Ti-6.5Al-2.5Sn-4Zr-1Nb-0.7Mo-0.15Si)	0.3	800	(140) / (35)	-
VT20(Ti-6.5Al-1Mo-1V-2Zr)	0.25	800	100 / 50	-
VT6(Ti-6Al-4V)	0.3	900	(94) / (45)	50 / 100
VT3-1(Ti-6Al-2.5Mo-1.5Cr-1.5Fe-0.3Si)	0.3	850	(110) / (47)	68 / 112
VT22(Ti-5Al-5Mo-5V-1Cr-1Fe)	0.25	800	86 / 58	100 / 122
VT14(Ti-5Al-3Mo-1V)	0.15	750	47 / 34	406 / 464
ST4	0.3	900	(272) / (116)	51 / 100

Footnotes : * in numerator - non-hydrogenated samples; in denominator - hydrogenated ones; in brackets - yield strength

After hot deformation with using of hydrogen plastification, hydrogen should be removed from metal to avoid hydrogen embrittlement development of details and structures during operation. Properties of semiproducts may be changed in a proper direction by means of thermohydrogen treatment. As a result mechanical properties of titanium alloys after deformation with using of hydrogen plastification and vacuum annealing are not worse that of ones being produced by standard processing.

Hydrogen alloying of annealed α+β alloys with a great amount of β-phase and quenched pseudo-β-alloys results in technological plasticity increase at room temperatures. This effect occurs due to hydrogen induced depressing of ω-phase formation in subcritical composition alloys during quenching [6] and β- phase amount increase in β- rich α+β- titanium alloys with accompanying leaning of β-phase with β-stabilisers [6, 7].

The effect described above were used for improving of the technology of fasteners production from VT16 (Ti-2.5-4.5) [6]. At present cold heading of titanium VT16 alloy bolts up to 9-12 mm is mastered in industry scale whereas 16 mm diameter bolts and those with larger diameter are manufactured by hot heading at 800-850 °C. Reversible hydrogen alloying of VT16 (Ti-2.5Al-4.5V-5Mo) alloy allows to manufacture bolts up to 16 mm in diameter by means of cold heading resulting in labour productivity increase in 10-12 times during heating stage.

The effectiveness of new processing was shown not only at cold heading but also at cold forming. Cylindrical self-safety nuts were manufactured from VT30 (beta-III) alloy by this processing. At present these nuts from titanium alloys can not be made; instead of titanium alloys alloyed steels are used for these purposes.

Compacthydrogen processes

Compacthydrogen processes include diffusion bonding and compacting of hydrogenated powders and granules with subsequent their conversion in semiproducts according to current technology [9]. Adhesion effects and deformation in contact adjacent surfaces play a significant role in diffusion bond formation. As a result of hydrogen induced enhanced adhesion and hydrogen plastification it is possible to obtain qualitative diffusion bonds of hydrogenated specimens under much lower temperatures (100-150ºC lower) as compared with standard technology. Current diffusion bonding temperatures to obtain hydrogenated blanks joints allow to reduce specific upsetting forces by 50-70% and decrease process duration.

High temperatures of diffusion bonding create essential difficulties in tooling manufacture. For diffusion welding it is necessary to provide hard-to-machine GS6 superalloy tooling. Lower temperatures allow to use tooling made of nimonic type alloys as specific forces during welding of hydrogenated metal are low.

Reversible hydrogen alloying of titanium alloys favours substantial decrease of compacting temperature of titanium alloy powder and granules components without mechanical properties loss [9].

Machinohydrogen Treatment

Machinohydrogen treatment intends to improve titanium alloys machining by cutting [9, 20]. Hydrogen alloying of titanium and its alloys results in cutting area temperatures decrease and reduction of cutting forces; chip becomes fragile at that. These effects lead to 2-10 fold life increase depending on alloy grade. Favourable hydrogen influence upon machining of titanium alloys by cutting occurs in the definite range of hydrogen concentrations in the metal being machined (Figure 2) and enhances with lowering of hydrogenation temperatures and decreasing of cutting rates.

The improvement of machining of titanium alloys by reversible hydrogen alloying is due to refinement of metal structure, decrease at metal flow stresses and toughness, increase of thermal conductivity of titanium and change in character of the tool wear [24]. During cutting of titanium alloys with the initial hydrogen content (0.003-0.005%wt.) cutting wear occurs both along the front face and along the clearance face. During machining of hydrogenated metal tool clearance face wear prevails.

Hydrogen technology of pressing production from titanium chips without its remelting

Favourable in technological aspects hydrogen alloying effects of titanium alloys were used in design of hydrogen technology of pressing production from titanium chips without its remelting [21].

This technology includes cold compacting of chips, heating of chip brick combined with hydrogenation, hot compacting with subsequent hot isothermal extrusion and thermohydrogen treatment with finishing vacuum annealing. The proposed hydrogen technology of pressing production from titanium chips without its remelting may be considered to be the further development of diffusion bonding at which large number of surfaces are diffusionly bonded not two. At the hydrogen action consists in following: a) adhesion enhances; b) metal flow stresses reduce due to hydrogen plastification; c) oxidation of chip decreases because of protective action of

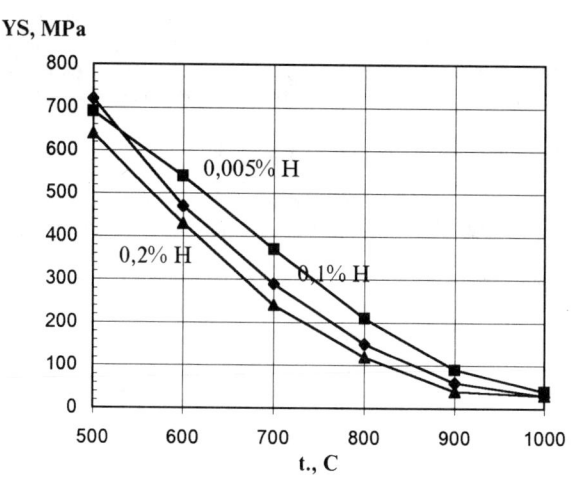

Figure 1 - Temperature effect on tensile yield strength of VT6 (Ti-6Al-4V) alloy with different hydrogen content; values on curves are wt.%H.

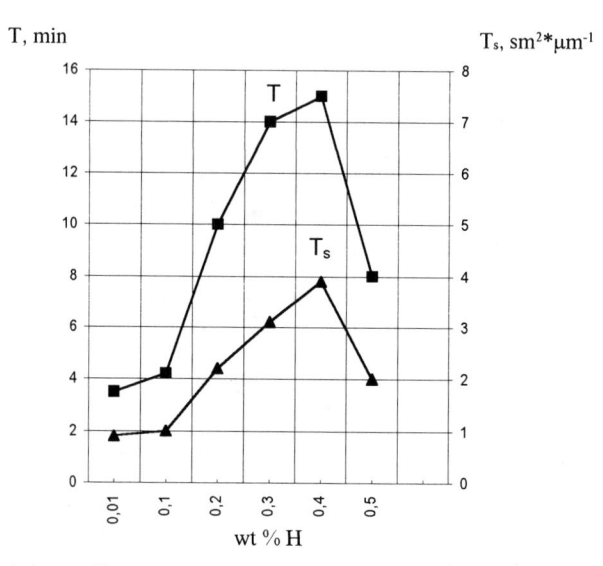

Figure 2 - Hydrogen effect on VK8 hard alloy specific size endurance Ts and tool life T, evaluated by clearance wear of 0.3 mm, during VT6 (Ti-6Al-4V) turning at the rate of about 60 m*min^{-1}; feed of 0.2 mm*rev.$^{-1}$; cutting depth of 1 mm.

hydrogen within chip brick; a formation of chemical contact between chip elements during annealing becomes more easily because of thermohydrogen effects.

It will be noted that interchip boundaries in structure of pressing remain after hot extrusion of chip brick. Recrystallisation occurs during thermohydogen treatment which leads to disappearance of interchip boundaries [25, 26].

Short time mechanical properties of rods produced from chips by hydrogen technology meet to technical specifications of standard production [21] (Table II).

Not only round rods but also corners, tubes, strips were produced by hydrogen technology from chips without its remelting. These semiproducts are cheaper than standard ones became of low cost chips and excluding of such expensive operation as vacuum arc melting.

Table II Mechanical properties of rods produced by hydrogen technology from chips of titanium and its alloys without remelting [21].

Alloy	Mechanical properties of rods from chips				Technical specifications			
	UTS, MPa	El, %	RA %	KCU* J/cm^2	UTS, MPa	El, %	RA, %	KCU* J/cm^2
						not less		
Ti (99.5%)	490	27	51	120	345-540	≥ 15	≥ 40	≥ 70
Ti-6Al-2.5Mo-1.5Cr-0.5Fe-0.3Si (VT3-1)	1180	10	29	29	980-1230	≥ 8	≥ 25	≥ 25
Ti-6Al-4V (VT6)	1100	10	25	40	885-1050	≥ 8	≥ 20	≥ 25

Footnotes: Technical specification for rods of usual grade,
RA- Retardation area ,* - impact strength.

Conclusion

In conclusion it will be noted that hydrogen technology allows to increase a part-to-scrap ratio, decrease in working hours of processing and reduce of production costs of difficulty workable titanium alloys, improve operation properties. The proposed developments may be used in many fields-major titanium alloys users, and in civil industries semiproducts manufacture.

References

1. U.Zwicker, H.Schleicher. Titanium Alloys Deformability Improvement Technique during Hot Pressure Shaping. (USA patent N2892742, grade 148-11,5; 1959).
2. B.A.Kolachev, V.A.Livanov, V.K.Nosov. "Hydrogen Influence on Deformability of Titanium Alloys of Different Phase Compositions". (Titanium. Physical Metallurgy and Technology. Proc.3rd Intern. Conf. Moscow. VILS, v.3, 1978) 61-68.
3. B.A.Kolachev, V.K.Nosov. "Hydrogen Plastification in Hot Deforming of Titanium Alloys. Titanium Science and Technology". (Proc.5th Intern.Conf. Munich. v.1, 1984 - 1985) , 625-632.

4. Vladimir K.Nosov, Boris A.Kolachev, Hydrogen Plastification in Hot Deforming of Titanium Alloys (Moscow, Metallurgia, 1986), 118.
5. A.A.Ilyin. Mechanism and Kinetics of Phase and Structure Transformations in Titanium Alloys (Moscow, Nauka, 1994), 304.
6. B.A.Kolachev, et al. "The effect of Hydrogen Alloying on workability of Titanium Alloys". (Titanium' 92. Science and Technology Proc. of 7th World Conf. on Titanium. San Diego, California, USA, 1992), 861-870.
7. A.A.Ilyin, B.A.Kolachev, A.M.Mamonov. "The Phase and Structure Transformations in Titanium Alloys during Thermohydrogen Treatment". (Titanium' 92. Science and Technology Proc. of 7th World Conf. on Titanium. San Diego, California, USA, 1992), 941-948.
8. B.A.Kolachev et al. "The Investigations of Thermophysical Properties of Ti-H-alloys". Metals RAN, 5 (1994), 85-91.
9. B.A.Kolachev, V.D.Talalaev, "Hydrogen Technology of Titanium Alloys", Titanium, 1 (1993), 43-46.
10. B.A.Kolachev,"Reversible Hydrogen Alloying of Titanium Alloys". Metallovedenie i termicheskaya obrabotka (Russia), 10 (1993), 28-32.
11. A.A.Ilyin, A.M.Mamonov, V.K.Nosov. "Specific Principles, Technology and Perspectives of Applications of Thermohydrogen Treatment of Titanium Alloys", (Science. Production and Applications of Titanium in Conversion Conditions. Proc. 1st Int.Conf. on Titanium SNG-countries. Moscow, VILS, 1994), 500-527.
12. W.R.Kerr e.a. "Hydrogen as an Alloying Element in Titanium (Hydrovac)". (Titanium'80. Science and Technology. Proc. 4th Intern. Conf. on Titanium. Kyoto, 1980), 2477-2486.
13.W.R.Kerr, "The Effect of Hydrogen as a temporary Alloying Element on the Microstructure and Tensile Properties of Ti-6Al-4V", Met.Trans. 16A, (1-6), (1985), 1077-1087.
14. D.Eylon, F.H.Froes, W.J.Barice. "Effects of Treatments on Mechanical Properties of Titanium Alloy Castings", (2nd Intern. SAMPE Metals and Metals Process Conf. Dayton) v.2 (1988), 28-36.
15. Zhang Shaoqing, Pan Feng, "Hydrogen Treatment of Cast Ti-6Al-4V Alloy". Chin.J.Met.Sci.Technol., v.6 (1990), 187-192.
16. B.A.Kolachev, V.K.Nosov, "Hydrogen Plastification and Superplasticity of Titanium Alloys". Physica metallov i Metallovedenie (Russia). v. 57, 2(1984), 288-297.
17. R.J.Lederich e.a. "Advanced Processing Methods for Titanium", TMS-AIME, (Warrendale ,1982), 115.
18. R.J. Lederich, S.M.L.Sastry, J.E.O'Neal. "Microstructural Refinements for Superplastic Forming Optimisation in Titanium Alloys". (Titanium Science and Technology. Proc. 5th Intern. Conf. on Titanium. Munich) v.2 (1984), 695-702.
19. L.R.Zhao, S.Q.Zhang, M.G.Yan. "Improvement in the Superplasticity of Ti-6Al-4V Alloy by Hydrogenation". (Superplasticity and Superplastic Forming. Proc. Intern. Conf. TMS. Blain, Washington. 1988) 459-464.
20. B.A.Kolachev e.a. "About Favourable Influence of Hydrogen on Machinability of titanium Alloys". (Science, Production and Applications of Titanium in Conversion Conditions. Proc. 1st Intern. Conf. on Titanium SNG-countries. Moscow. VILS, 1994) 873-882.
21. B.A.Kolachev e.a. "Hydrogen Technology of Production of Pressing from Titanium Chip". (Science, Production and Applications of Titanium in Conversion Conditions. Proc. 1st Intern. Conf. on Titanium SNG-countries. Moscow. VILS, 1994) 283-291.

HYDROGEN INFLUENCE ON MACHINING OF TITANIUM ALLOYS

B.A. Kolachev, Y.B. Egorova
Moscow State University of Aviation Technology after K. Tsiolkovsky
Petrovka 27, 103767 Moscow K-31, Russia
V.D. Talalaev
Scientific-Production Association "AVITOM"
Ulanskyi pereulok 16, 101000 Moscow, Russia

Abstract

The paper deals with the influence of hydrogen in a metal being machined on cutting forces, cutting area temperature and tool life during turning of titanium alloys of various classes; VT1-0, VT5-1 (α - alloys); VT20, VT25 (pseudo α - alloys); VT3-1, VT6, VT8 ($\alpha+\beta$ - alloys). It was shown that a certain hydrogen concentration interval exists for each alloy within which hydrogen increases tool endurance essentially. Within this hydrogen content interval, tool endurance rises 2 - 10 times depending on the alloy type. The improvement of machinability of titanium and its alloys by hydrogen alloying may be due to microstructure refinement, hydride precipitations, decrease of metal flow stresses and impact ductility, increase of thermal conductivity, chips being more fragile, and changes in character of tool wear.

INTRODUCTION

The favourable influence of hydrogen on machining of titanium and its alloys was discovered in investigations described in papers [1,2], which was the foundation of one of the directions of hydrogen technology i.e. machinohydrogen treatment. Mechanohydrogen treatment includes: a) metal hydrogenation; b) machining of hydrogenated metal; c) vacuum annealing. This treatment may be used both as an independent technological process and in combination with other directions of hydrogen technology (hydrogen plasticization, termohydrogen treatment, hydrogen procedures of powder and granular metallurgy).

The object of the present work is a more detailed investigation of hydrogen influence on machinability of titanium alloys of different classes and closer definition of mechanisms of the favourable hydrogen machining effect.

MATERIALS AND EXPERIMENTAL PROCEDURE

Investigations were carried out on forged rods of 40 - 50 mm diameter. The chemical compositions of alloys investigated are given in table I.

Table I. The chemical compositions of alloys investigated.

Alloys	Chemical compositions, wt. %							
	Al	Sn	Zr	Mo	V	Cr	Fe	Si
VT1-0 (99.5)	-	-	-	-	-	-	-	-
VT5-1	5.2	2.4	-	-	-	-	-	-
VT20	6.1	-	1.9	0.9	1.6	-	-	-
VT25	6.8	1.3	1.7	2.0	1.1	-	-	0.18
VT3-1	6.3	-	-	2.6	-	1.9	0.6	0.24
VT6	6.3	-	-	-	4.3	-	0.07	0.01
VT8	6.7	0.8	1.1	3.4	-	-	-	0.22

Rod stocks were hydrogenated in molecular hydrogen at temperature 800°C during 10 hours. Turning experiments were conducted on machine 16K20 with using cutters with the mechanical support of VK8 and VK60M hard alloy plates. Dynamometer UDM600 was used to measure cutting forces. The cutting area temperature was determined by the natural thermocouple method.

According to the experimental results tool life T and specific size endurance T_s were determined: $T_s = 1 \cdot S / h$, where h is clearance face wear of cutter, l is the length of cutting path, S is feed. Specific size endurance characterizes the surface area which the cutter is able to machine per unit of wear. The accelerated method of endurance tests [3] was used to estimate the optimum conditions of the favourable hydrogen influence on machinability of titanium as full endurance tests require great amount of metal conversion into chips and large consumption of expensive hard alloy tool.

Metallographic analysis was carried out on the microscope "Metaval", X-radiography investigations were conducted on "DRON-2" apparatus, using cobalt radiation, at angle interval $\theta = 19 - 23$ and $31 - 35°$ where the X-ray diffractions from α, β - phases and hydrides are observed.

THE EXPERIMENTAL RESULTS

The favourable hydrogen influence on machinability was observed both for titanium and all alloys investigated. Hydrogen leads to reduction of cutting area temperature during turning and milling of VT1-0 titanium at that the extent of hydrogen influence on this temperature decreasing with the increase of cutting rate. The introduction of 0.3 - 0.6 wt.% H results in the reduction of cutting area temperature θ by 100 - 150°C at cutting rates 40 -50 m·min⁻¹. Along with that hydrogen causes the decrease of cutting forces P_x, P_y and P_z. At rates 40 - 60 m·min⁻¹ force P_z during turning VT1-0 titanium decreases from 380 - 330 N to about 230 - 200 N with the increase of hydrogen content in metal from 0.003 to 0.1 - 3 wt. %. Specific size endurance T_s at hydrogen alloying of metal being machined increases about three times. The greatest tool life and specific size endurance are observed at hydrogen contents in metal equal to 0.1 - 0.3 wt. %.

The hydrogen influence on cutting forces and cutting area temperature θ during turning of VT5-1 alloy is shown in figure 1.

Cutting area temperature change small with hydrogen content increase to 0.5 wt. %. Cutting forces rise in some extent up to hydrogen content equal to 0.2 wt. %, then they reduce and become lower as compared with the initial billet at concentration 0.5 wt. % H. Further increase of hydrogen content results in great rise of cutting forces and cutting area temperature, which causes corresponding effect in VK60M tool wear during turning of VT5-1 alloy billet with different hydrogen content. The greatest tool life and the least tool wear are observed during turning of billet with hydrogen concentration equal to 0.5 wt. % at that tool life T increases about 10 times (figure 1). A great amount of titanium hydrides was revealed in the structure of this sample. Results shown in figure 1 were obtained at cutting rate equal to 100 m·min^{-1} which is about 3 - 5 times higher than ones used in industrial conditions (20 - 30 m·min^{-1}). The favourable hydrogen effect on machiability of VT5-1 alloy is enhances with cutting rate decreasing (figure 2).

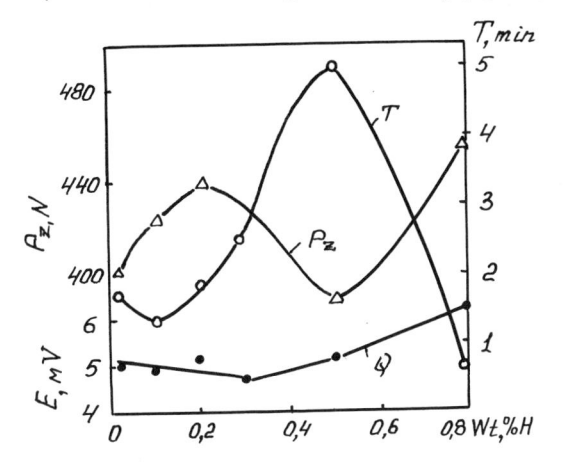

Figure 1 - Hydrogen effect on cutting area temperature (thermo E), cutting force P$_z$ and tool life T at h = 0.3 mm during turning of VT5-1 alloy by VK60M tool at cutting rate 100 m·min^{-1}, feed S = 0.2 mm/rev and cutting depth 1 mm.

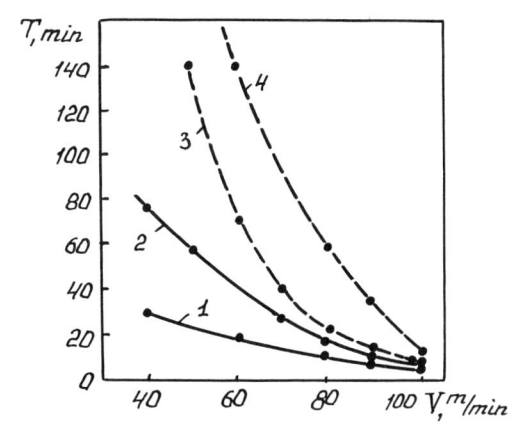

Figure 2 - The dependence of VK60M cutter life T evaluated from clearance face wear h = 0.3 mm on cutting rate during turning of VT5-1 alloy with hydrogen content 0.004 (1,2) and 0.5 wt. % (3,4) at cutting conditions: 2,4 - S = 0.2 mm per rev, t = 1 mm; 1,3 - S = 0.2 mm per rev, t = 2 mm.

Chips become more fragile during turning of hydrogenated metal. However chip crushing was not observed during turning of hydrogenated samples by fourthfaceted plates from VK8 alloy, but crushing of hydrogenated chips occured during turning by threefaceted plates with chip-crushing ditches, extent of crushing being more pronounced with hydrogen content increase.

The maximum tool endurance during turning of pseudo α-alloy VT20 was observed at hydrogen content equal to 0.1 - 0.35 wt. %. The favourable hydrogen influence on machinability of this alloy enchanced with reduction of cutting rate and depth. Microstructure of VT20 alloy with initial hydrogen content was presented by transformed prior beta grains with lamellar intergrain alpha morphologies and alpha phase segregations along prior beta boundaries. The introduction of 0.3 wt. % H into metal leads to transformation of lamellar structure into globular one and the amount of beta phase increase. If the amount of beta phase in the initial alloy was equal to 5%, then at 0.5 wt. % it was 48%.

The minimum cutting forces during turning of VT25 alloy are observed for billets with hydrogen content 0.6 - 0,7 wt. %. Thermo E is little changed with hydrogen content increase. The maximum tool endurance during turning of VT25 alloy billets is observed in two concentration intervals: 0.15 - 0.2 and 0.7 - 0.75 % H. The extent of hydrogen-induced cutting forces reduction and tool endurance increase becomes less pronounced with rise of hydrogenation temperature from 750 to 850°C.

Microstructure of VT25 alloy billet being investigated has basket morphology without distinct alpha colonies of similar oriented alpha platelets. Surfaces of alpha platelets become toothed with hydrogen contents increase and then platelets crush although do not transform into globules. The great segregation of alpha phase along prior beta grain boundaries was found in structure of billet with 0.3 wt, % H, which corresponds to the least tool endurance.

Alloy of α + β - type VT6 (Ti-6Al-4V) exhibits the best machinability by turning at hydrogen concentrations 0.3 - 0.4 wt. % at which cutting area temperature is 50 -70°C lower and force P_z is about 1, 4 times less than during turning of metal with initial hydrogen content. The maximum tool life and specific size endurance of tool are found at the same hydrogen concentrations. Endurance of cutters increases with cutting rate reduction, the extent of favourable hydrogen influence on machinability being maintained (figure 3).

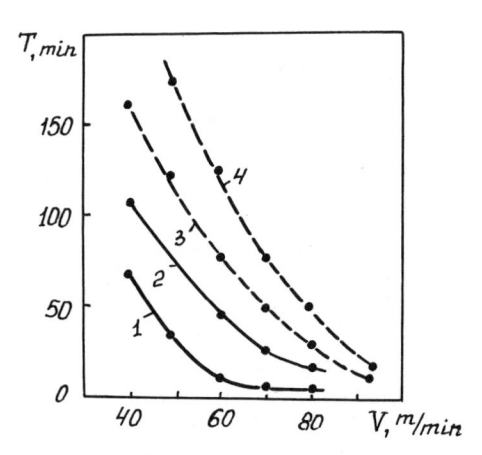

Figure 3. - The dependence of VK8 cutter life T estimated by clearance face wear h = 0.3 mm on cutting rate during turning of Ti-6Al-4V alloy with hydrogen content 0.003 (1,2) and 0.3 wt. % (3,4): 1,3 - S = 0.2 mm per rev, t = 1.0 mm; 2,4 - S = 0.2 mm per rev, t = 0.5 mm.

Hydrogen-induced improvement of machinability was found for α + β - alloys VT3-1 and BT8 too. This effect is the most pronounced at hydrogen contents 0.3 - 0.5 wt. %

for VT3-1 alloy and 0.7 - 0.8 wt. % for VT8 alloy. The extent of favourable hydrogen influence on machinability of VT3-1 and VT8 alloys by turning increases with cutting rate decrease and hydrogenation temperature lowering (figure 4).

Microstructure of initial billets of $\alpha + \beta$ alloys was lamellar with clearly formed prior beta grain boundaries, alpha colonies and alpha platelets. Crushing of alpha platelets and their transformation into globules occur with hydrogen content increase along with that amount of beta increase. While the amounts of beta phase are 12, 23 and 12 % in initial billets of VT6, VT3-1 and VT8 alloys they are 56.75 and 41 % after introduction of 0.75 - 0.8 wt. % H accordingly. Microstructure of VT3-1 alloy with 0.8 wt. % H corresponds to structure of annealed pseudo β - alloys and was presented by β - retained phase with alpha phase inclusions of globular form. Structure becomes rougher and beta phase amount increases slightly with hydrogenation temperature raise. The significant amount of hydrides was found in the structure of alloys with hydrogen contents 0.45 - 0.5 wt. % and more.

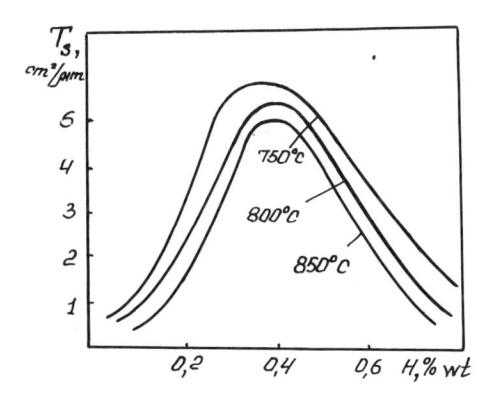

Figure 4. - Hydrogen influence on specific size endurance of VK8 cutter during turning of VT3-1 alloy hydrogenated at different temperatures.

DISCUSSION

The following regularities are characteristic for results discribed above:
1) The reduction of cutting forces and the increase of wear resistance occur in the definite range of hydrogen concentrations in the material being machined; lowering of cutting area temperature is observed in some cases;
2) The greatest effect of favourable hydrogen influence corresponds to certain hydrogen concentrations depending upon alloy composition;
3) The increase of hydrogenation temperature from 750 to 850°C results in the deterioration of machinability of titanium alloys by cutting;
4) With the decrease of cutting rate, feed and cutting depth the effect of favourable hydrogen influence on machinability is enhanced;
5) Chips become more brittle with hydrogen content increase; hydrogenation of α - alloys leads to chip crushing.

That the optimum hydrogen contents corresponding to the more pronounced effect of machinohydrogen treatment are different for various alloys (figure 5). However obvious dependence of effectiveness of mechanohydrogen treatment on titanium alloys classes is not observed.

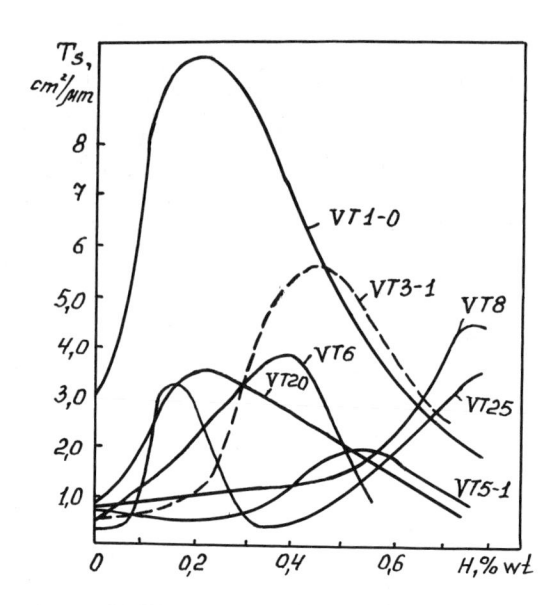

Figure 5 - The dependence of VK8 cutter specific size endurance on hydrogen content during turning of titanium alloys (accelerated tests).

The optimum hydrogen concentrations at which effects of machinohydrogen treatment being the most pronounced are given in table 2.

Table 2. Hydrogen Influence on Machinability Characteristics of Titanium Alloys.

Alloy	Optimum Hydrogen Concentrations, % (wt.)	The Change of Machinability Characteristics of hydrogenated Metal as Compared to Initial one		
		Decrease of Cutting Area Tempetature C*	Reduction of P_z, number of times	Increase of Specific Size Resistance T_s, number of times
VT1-0 (Ti - 99.5)	0.25	100 - 150	1.6 - 1.7	3.5
VT5-1(Ti-5Al-2.5Sn)	0.45 - 0.6	S	1.1	2 - 2.5
VT20	0.2 - 0.3	-	-	3.5
VT3-1	0.3 - 0.5	100 - 120	1.4	8 - 12
VT6 (Ti-6Al-4V)	0.3 - 0.4	50 - 70	1.4	7 - 8
VT8	0.7 - 0.8	S	1.1	4 - 5
VT25	0.15 - 0.2	S	1.3	8 - 10
VT25	0.7 - 0.75	S	1.3	8 - 10

Remark: * S - temperature changes are small.

Machinohydrogen treatment of titanium alloys with the optimum hydrogen contents allows to increase cutting rate 1.5 - 2 times while retaining other cutting parameters or to use higher feeds and cutting depths without changing rates.

The favourable hydrogen influence on machinability of titanium alloys may be due to following:
1) Structural and phase transformations caused by hydrogen alloying of titanium alloys;

2) Hydrogen alloying caused changes of mechanical properties determining machinability;
3) Hydrogen induced changes of thermophysical properties;
4) Changes in extent of titanium adhesion on tool as a result of hydrogen alloying of metal.

Machinability of α and pseudo α - alloys by turning improves with hydride precipitations which leads to metal being brittle and fragile, which facilitates chip removal. General features of pseudo α - and $\alpha + \beta$ - alloys are beta phase amount rise with hydrogen content increase, structure refinement in a definite hydrogen concentration interval, hydride precipitations at hydrogen contents above solibility limit and formation of the structure characteristic to pseudo β - alloys at high enough hydrogen concentrations. From the comparison of structure parameters of pseudo α - and $\alpha + \beta$ - titanium alloys alloyed by different hydrogen contents with the dependences of cutting characteristics on hydrogen content it follows that alloys have the best machinability at finegrained structure. Hydrogen content increase in $\alpha + \beta$ - alloys above the optimum concentration leads to beta phase amount raise up to quantities characteristic to pseudo β - alloys, which are machined poorly because of high toughness of material and great adhesion of material being machined on tool. Diminishing of machinability with hydrogenation temperature raise from 750 to 850°C is probobly due to the formation of more coarse-grained structure and beta phase amount increase.

Hydrogen-induced phase and structural changes are accompanied by changes of mechanical properties determining machinability: impact ductility reduction, metal flow stresses decrease at cutting area temperatures, increase of susceptibility to brittle fracture at temperatures close to room one [3, 4].

The increase of thermal- and temperature-conductivity at temperatures characteristic to cutting area was found for titanium [6]. This effect must accelerate heat conduction from cutting area and as a consequence to cutting area temperature lowering and tool endurance improvement. However this effect cannot be dominant. The favourable hydrogen influence on machinability was observed for alloys too essential cutting area temperature drop not being observed. It will be noted that hydrogen effect on thermophysical properties was studied for titanium only; but not for alloys.

The experimental data devoted to hydrogen influence on titanium adhesion upon tool was not published. It will be noted however that the character of tool wear changes because of hydrogen alloying of titanium alloys. During cutting of titanium alloys with the initial hydrogen contents (0.003-0.005%) tool wear occurs both along the front face due to friction of tool against chip and along the clearance face due to friction of tool against the surface being machined. During machining of hydrogenated billets tool clearance face wear prevails. Such change of wear character caused by hydrogen alloying of titanium alloys doubtless makes a certain contribution into the improvement of their machinability by cutting.

CONCLUSION.

1. Hydrogen alloying of titanium and its alloys results in the increase of tool life 2-10 times depending upon alloy composition.
2. Favourable hydrogen influence on machinability of titanium alloys by cutting occurs at the highest extent at certain hydrigen concentrations depending upon alloy composition and increases with the decrease of hydrogenation temperature and the decrease of cutting rate.
3. The improvement of machinability of titanium and its alloys by hydrogen alloying may be due to microstructure refinement, hydride precipitations, decrease of metal

flow stresses and impact ductility, increase of thermal conductivity, brittleness of chips and changes in tool wear characteristics.

REFERENCES.

1. B.A. Kolachev, V.D. Talalaev, "Hydrogen Technology of Titanium Alloys", *Titanium*. №1 (1993), 27 - 30.
2. B.A. Kolachev et. al., "About Nature of the Favourable Hydrogen Influence on Machinability of Titanium Alloys by Turning," Science, Production and Applications of Titanium in Conversion Conditions. Proc. 1st Intern. Conf. on Titanium in SNG-Countries, vol. 2 , Moscow, VILS, (1994), 873 - 882.
3. A.D. Makarov: Optimization of Machining (Moscow, Mashinostroenie, 1976) 278.
4. B.A. Kolachev et. al. "Hydrogen Influence on Structure and Mechanical Properties of Titanium Alloy VT3-1," Metallovedenie i termicheskaja obrabotka metallov, Moscow, № 1 (1992) 32 -33.
5. Vladimir K. Nosov and Boris A. Kolachev . Hydrogen Plasticization at Hot Deformation of Titanium Alloys. (Moscow, Metallurgia, 1986), 118.
6. B.A. Kolachev et al. "Investigation of Thermophysical Properties of Ti-H Alloys', Metalli, (RAN), № 5 (1994), 85 - 91.

HEAT TREATMENT

Heat Treatment of Titanium Alloys: Overview

R.R. Boyer[1] and G. Lütjering[2]

1 - Boeing Commercial Airplane Group
P.O. Box 3707, MS 6H-CJ
Seattle, WA 98124 USA
2 - Technical University Hamburg-Harburg
21071 Hamburg, Germany

Abstract

Heat treatment serves many purposes, e.g., stress relief, a means of strengthening, or a means of optimizing certain properties such as creep or fracture toughness. The values of some properties, such as crack growth- or creep-resistance, can be changed by an order of magnitude, and more, through heat treatment variation. Thermomechanical processing and heat treatment are the two key processing variables which control the microstructure and thus the properties combinations obtained from titanium alloys. Heat treatment alone can not guarantee that the proper microstructure will be achieved. If the prior working history has not been done properly, heat treatment will not correct the situation; it is a key factor in achieving the proper microstructure to control the properties of interest.

The different types of heat treatments used for titanium alloys, the rationale for their use, and their effect on properties will be discussed. The effects of heat treat parameters such as heating rate, cooling rate, temperature selection, etc. will also be presented

Introduction

Heat treatment is a powerful tool used to obtain the desired set of properties for a specific alloy for a specific application. In some cases the heat treatment requirements are not very restrictive, for instance, for a stress relief or anneal, the temperature and time limits may be quite broad. In other cases, where tight microstructure controls are required to optimize a given property, the heat treat requirements may be quite restrictive. For example, to obtain the maximum creep resistance in IMI 834 the solution treatment temperature must be controlled to within ±5°C.[1] As the properties requirements become more demanding, the heat treatment controls become tighter.

Ti-6Al-4V provides an excellent example to demonstrate the impact heat treatment variations can have on properties. In going from a mill anneal to a β anneal, the fracture toughness can be changed by a factor of almost 2, fatigue strength by a factor of ~3, and the change in fatigue crack growth resistance can be greater than a factor of 10 as illustrated in Figure 1.[2-5] The anneal times in Figure 1 were 2 hours for the sub-transus anneals and 20 minutes for those annealed above the beta transus (β_t), each followed by an air cool. For material annealed above 870°C, they were followed by a 730°C/2 hour mill anneal, again followed by an air cool. The fatigue lives are log average lives of notched fatigue specimens, K_t = 2.53, tested at a maximum stress of 345 MPa at R = 0.1. The crack growth rate data are measurements at a ΔK of 16.5 MPa√m (R = 0.05).[2]

The heat treatments appropriate for the various alloy types and how they affect the properties will be discussed in the following sections. In addition, the influence of variations in heat treatment parameters, such as heating rate, cooling rate, temperature and heat treat sequence on the final properties are presented. Specific times and temperatures used for most of the conventional heat treatments for the more common alloys can be found in references 1,6,7.

Conventional Heat Treatments

Conventional types of heat treatments include:
   • Stress Relief (SR) - used to remove residual stresses present from prior fabrication steps, or even prior heat treatment operations. This could be an intermediate step to relieve residual stresses from previous operations, such as welding, or a final heat treat operation prior to being put into service. Removal of these stresses can be very important if high residual stresses are present during machining of the parts. The machining operation could create unbalanced residual stresses, which could result in part distortion. The removal of the residual stresses is also critical for alloys susceptible to H_2 embrittlement, particularly if they operate at elevated temperatures where diffusion is enhanced. Commercially pure (CP) Ti ducts which operated at temperatures in the range of 120-175°C and had residual stresses on the order of 480 MPa from welding, caused H_2 embrittlement with bulk hydrogen contents on the order of only 30-50 ppm.[8]

For the α and α/β alloys the SR temperature will be in the range of 480-815°C. The more creep resistant alloys/heat treat conditions will require higher temperatures. Times will be a strong function of temperature. Boeing experience indicates that for CP Ti, for instance, a full stress relief is achieved after 6 hours at 525°C, while only 30 min. is required at 650°C. Mill Annealed (MA) Ti-6Al-4V is stress relieved in 30 minutes at 675-790°C; the temperature must be increased by ~55°C for β-annealed material.

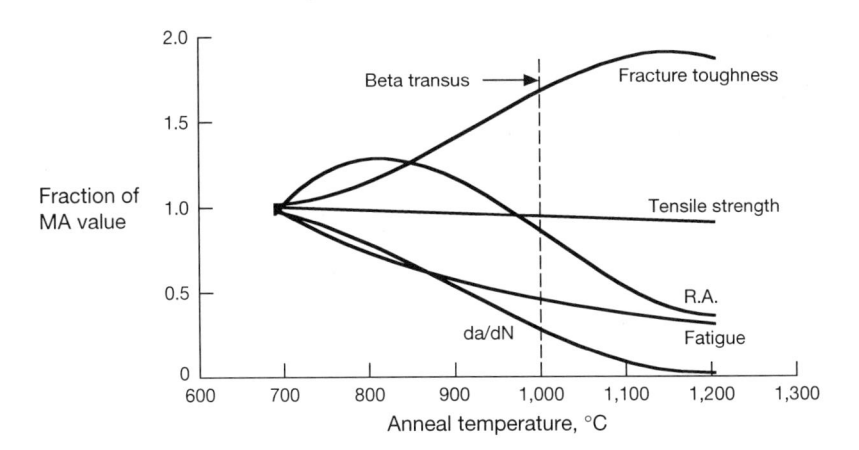

Figure 1. Properties of Ti-6Al-4V as a function of annealing temperature. The properties are ratioed to the properties of a standard mill anneal (730°C/2 hr). See text (Introduction) for further details.

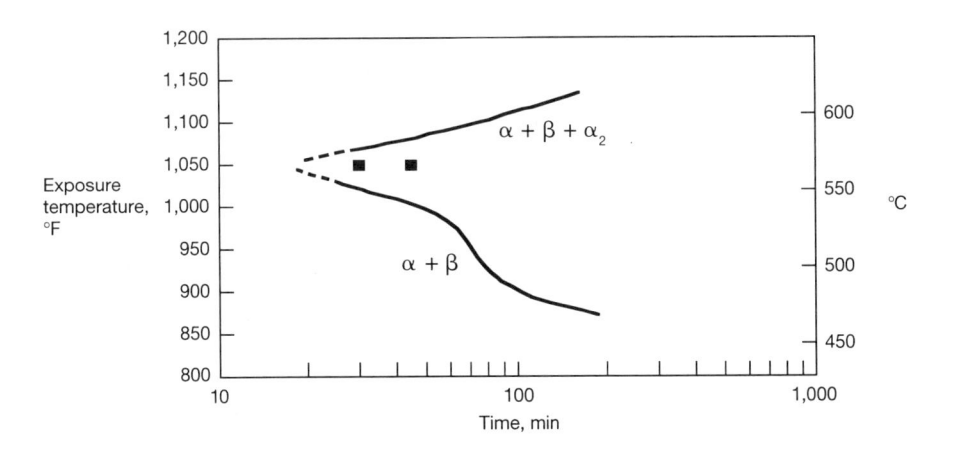

Figure 2. T-T-T diagram indicating the time-temperature-transformation parameters for the onset of ordering within the α-phase within 2 hr as detected by a reduction of the stress-corrosion cracking threshold (from Ref. 11).

The cooling rate from the SR can be important for these types of alloys if the alloy has an Al equivalent greater than ~6.5. The Al equivalent, as defined by Rosenberg (9) is:

Al Equivalent = %Al + 1/3(%Sn) + 1/6(%Zr) + 10(O_2)

Slow cooling, at rates of ~150°C/hour and slower [10] can result in a marked degradation of the stress-corrosion threshold due to the formation of fine α_2 precipitates which can occur due to the increased residence time in the $\alpha+\alpha_2$ phase-field during cooling. (At the time of the cited study it was felt that the use of the ELI grade took care of the problem. However, recently many more complex machined components have been designed where a slow cool from the anneal temperature was requested to minimize distortion. It was felt that the stress-corrosion issue should be re-assessed prior to proceeding.) This later study indicated the critical temperature regime for the ordering to occur in Ti-6Al-4V is in the range of 480-595°C, considering exposure times up to 2 hours, as illustrated in Figure 2.[11] The boundaries of Figure 2 were determined by a reduction in the stress-corrosion threshold due to the ordering reaction. The nose of the curve must be projected as the data points shown at 565°C, (30 and 60 minutes) exhibited ordering so the nose of the curve has not yet been defined. This temperature regime will vary somewhat from alloy to alloy. Evans and Broderick, for instance, have observed ordering in Ti-6Al-2Sn-2Zr-2Mo-2Cr at an anneal temperature of 650°C. [12] On the other hand Lütjering and Jaffee [13] utilized the formation of fine α_2 precipitates to provide additional strength, improving the fatigue performance of Ti-6Al-4V.

The SR treatment could also result in aging of all alloy types. As described above, the α could be aged by the precipitation of α_2. If there is metastable β present, α precipitation would occur, providing strengthening.

Care must be exercised for SR of β-alloys to assure that it does not interfere with the final age-hardening treatment. It would normally be done sub-transus for material not in the heat treated condition, far enough below the β_t to prevent extensive grain boundary α formation. If the material is in the final aged condition (for solution treated and aged material), the SR temperature must be maintained at or below the age temperature - above the age temperature would reduce the strength. Aging time-temperature-strength characteristics must be determined to assure that a SR after heat treatment will not degrade the strength requirements.

Anneal - The anneal is similar to a SR except it is usually accomplished at higher temperatures. In addition to removing residual stresses, an anneal is generally intended to remove all vestiges of prior work, i.e., obtain minimum hardness. This is often the final heat treat condition used for α- and α/β-alloys as a good properties balance is achieved using a wide processing window - the material is *forgiving*. This condition is commonly referred to as a mill anneal (MA), and the requirements can be quite broad. Mil-H-81200 [6], a titanium heat treat specification, permits MA temperatures from 705-870°C for times ranging from 15 to 60 min. Care must be taken in using a β-alloy in the annealed condition. If it is used at elevated temperatures, it may not be thermally stable, particularly if there is any remnant deformation structure. Service temperatures up to ~400°C could result in ω or a very fine α formation within the metastable β-phase which could result in embrittlement. Service temperatures in the range of ~400-600°C could age harden the material, resulting in a strengthening and loss of ductility. Annealed β-alloys used at higher service temperatures would have very low tensile and creep strengths at temperature.

Duplex Anneal (DA) - This heat treatment, used for α- and α/β-alloys, involves an initial anneal, high in the α/β phase-field, followed by a mill anneal, to provide thermal stability - the cooling rate following each of these steps is most often an air cool. This type of heat treatment is used for Ti-6Al-2Sn-4Zr-2Mo to improve the creep resistance. The higher the initial annealing temperature, the more transformed β (lamellar α) is present in the final microstructure, and the higher the creep resistance (and fracture toughness). The effect of annealing temperature on an array of properties of Ti-6Al-4V is illustrated in Figure 1. The cooling rate from the initial anneal temperature (air cooling was used in Figure 1) can also be critical. Faster cooling rates will result in a finer transformed structure, which affects essentially all material properties. Figure 3 illustrates the influence anneal or solution treat temperature and cooling rate has on microstructure, while Figure 4 shows the resultant tensile and fatigue strengths and creep resistance. The differences in microstructure between the air cool and water quench are relatively small in Figure 3, as it was constructed using sheet material, and air cooled sheet has a fairly high cooling rate. Larger cross-sections would show greater differences between the air cooled and water quenched microstructures. The largest microstructural changes occur at the higher temperatures. No discernible microstructural change is observed for the lowest temperature as the temperature is not high enough to revert some of the α-phase to β, and there was not enough work in the material to cause recrystallization. At the higher temperatures the faster cooling rates result in a finer transformed structure and/or more retained β, depending on the alloy, anneal temperature, and cooling rate. The finer transformed structure provides material with an inherently higher strength, while higher strengths can be achieved in the latter via α-precipitation during aging of retained β. The faster the cooling rate, the higher the volume fraction of the retained β-phase, and the greater the strengthening potential. Figure 4a illustrates the effect of anneal or solution treat temperature on strength in Ti-6Al-2Sn-2Zr-2Mo-2Cr [14], and Figure 4b the effect of cooling rate on tensile strength of the same alloy.[15] The improvement in fatigue strength with the faster cooling rates (Figure 4c) in Ti-6Al-2Sn-4Zr-2Mo [16] is attributed to the higher yield strength due to the finer transformed structure, and hence enhanced resistance to crack nucleation. The creep strength is also strongly affected by the fineness of the transformed structure as can be seen in Figure 4d.[17] At cooling rates in the range of ~ 100-300°C/min., increasing the cooling rate (finer transformed structure) decreases the slip distance, increasing the creep resistance. However, further microstructural refinement by cooling at a faster rate, in the range of 300-3000°C/min., resulted in decreased creep resistance. It is speculated that this could be due to dislocation annihilation at lamellae boundaries controlling the creep rate; the finer structure associated with the faster cooling rates will have a higher density of boundaries

Recrystallize Anneal (RA) - this is a specialized DA used for Ti-6Al-4V which is done at a temperature high enough in the α/β phase field to ensure recrystallization, and then cooled slowly to result in a high volume fraction of equiaxed α with islands of retained β at the triple points and some interfacial β at α/α boundaries. This heat treatment, used for Ti-6Al-4V fracture critical applications on the B-1 and B-2 bombers, provides high damage tolerance properties (fracture toughness, crack growth resistance, stress-corrosion threshold) with a modest fatigue debit in comparison to MA.

Beta Anneal (BA) - Annealed at a temperature above the β_t with a subsequent MA. This heat treatment, used for fracture critical components on the F-22 fighter, maximizes the damage tolerance properties of α and α/β alloys. However, as previously mentioned, this improved damage tolerance is associated with a very substantial debit in fatigue performance, as illustrated in Figure 1; notched fatigue strength is on the order of 30% of the MA condition.

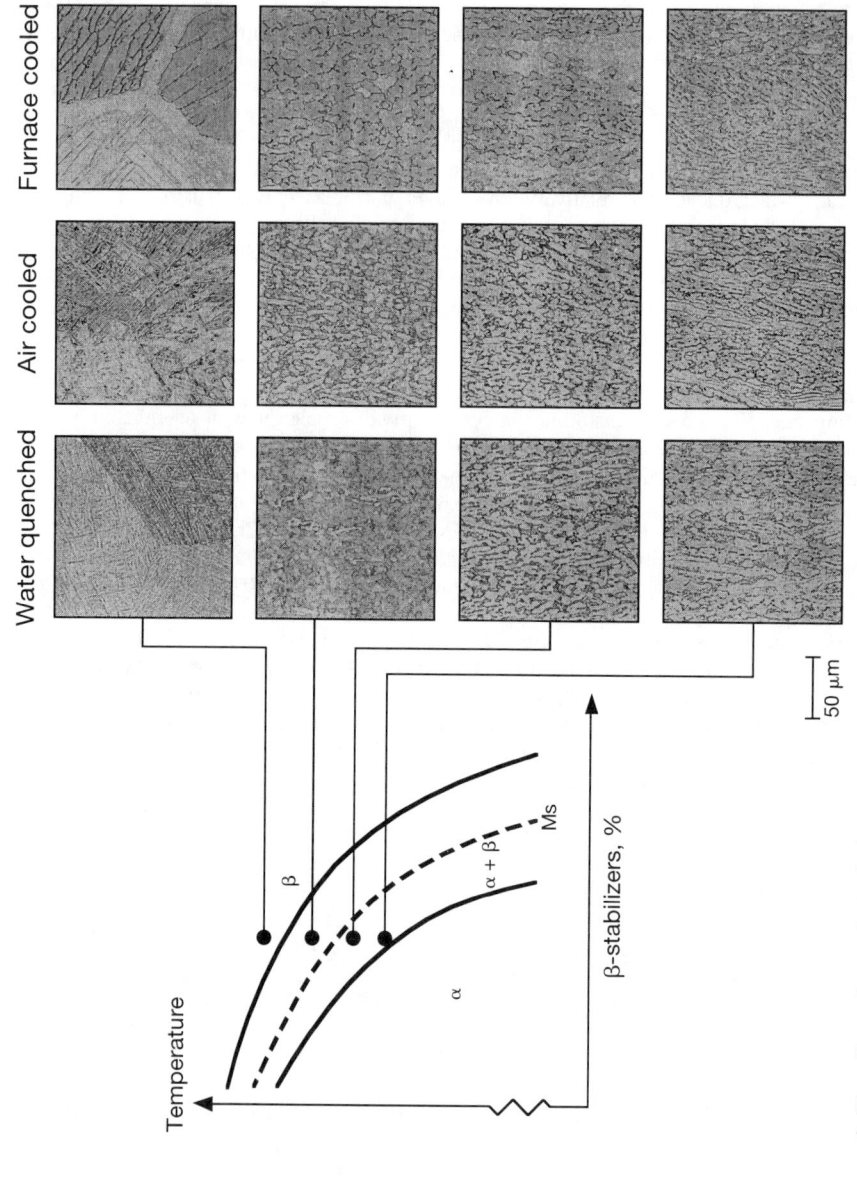

Figure 3. The effect of anneal or solution treat temperature and cooling rate from that temperature on the microstructure of Ti-6Al-4V.

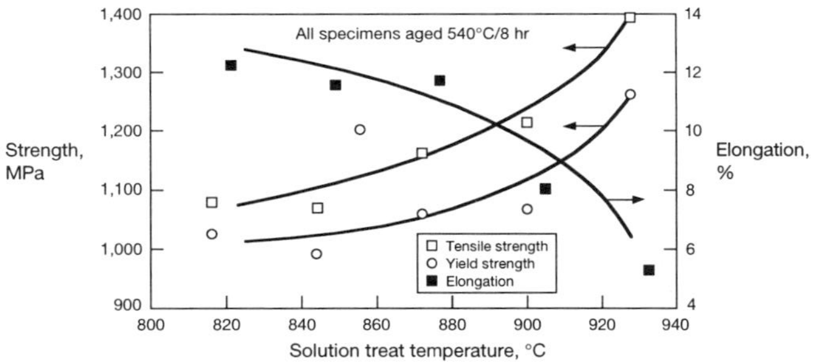

a.) Effect of solution treat temperature on the strength of Ti-6Al-2Sn-2Zr-2Mo-2Cr (from Ref. 14).

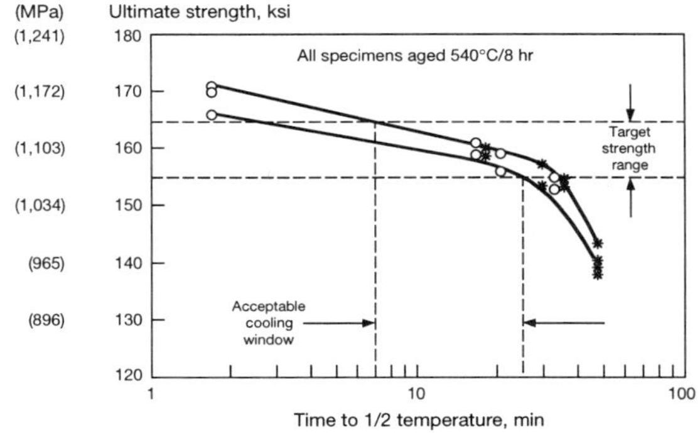

b.) Effect of cooling rate on tensile properties of Ti-6Al-2Sn-2Zr-2Mo-2Cr (from Ref. 15).

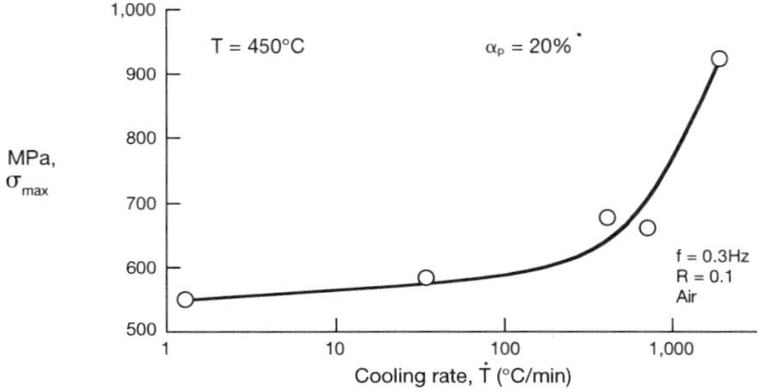

c.) Effect of cooling rate on the LCF strength (10^4 cycles) of Ti-6Al-2Sn-4Zr-2Mo at 450°C (R=0.1)(from Ref. 16).

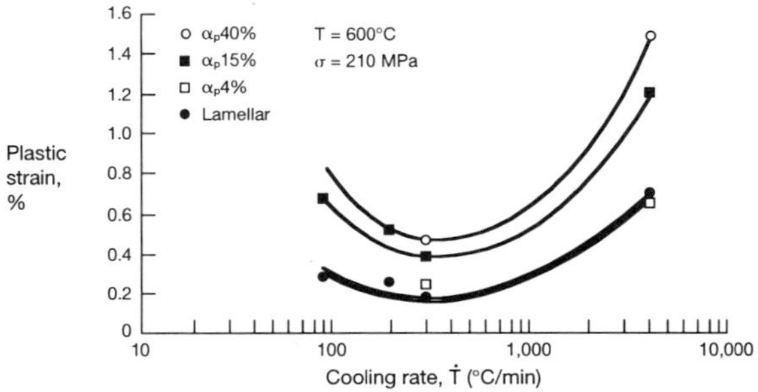

d.) Effect of cooling rate on the creep strain of IMI 834 after 100 hr of exposure for both lamellar and bi-modal microstructures. α_p: Volume fraction of primary α-phase (from Ref. 17).

Figure 4. Effects of solution treat temperature and cooling rate on mechanical properties.

The time and temperature above the transus should be minimized to prevent excessive grain growth. Again, cooling rate is very important. Higher cooling rates, with the finer transformed structure, will have higher tensile and fatigue strength than material cooled at a slower rate, but all of the damage tolerance properties will be reduced.

- Solution Treat (ST) - this is a heat treatment used to revert a desired amount of α to β–phase followed by cooling at a rate sufficient to retain the β-phase for subsequent aging by fine α precipitates or to form martensite, depending on the alloy type. The cooling could be an air cool for metastable β-alloy sheet - such as Ti-15V-3Cr-3Al-3Sn - or a water quench for thicker sections or less stable alloys. While β–alloys are normally solution treated above the β_t (except for Ti-10V-2Fe-3Al) the ST of α and α/β alloys is normally done below the β_t. For most β-alloys the object of the solution treatment is to recrystallize the structure and retain 100% β to attain maximum hardenability. Ti-10V-2Fe-3Al is sub-transus solution treated to control the morphology of the primary α-phase to optimize the strength-ductility-toughness combination which can be achieved.[18,19] Beta alloy sheet is normally supplied in the ST condition, which will be minimum strength and 100% β, to provide maximum cold formability. This may also be referred to as the solution annealed condition, when an air cool is employed.

Solution Treat and Age (STA) - used to provide increased tensile, and, sometimes, increased fatigue strength. The factors controlling the strength which can be achieved by this heat treatment can be very complex, especially for β-alloys - ST temperature and cooling rate, aging temperature and time, heating rate, and aging sequence must all be taken into account.

The first consideration must be the alloy and alloy type. The α-type alloys are generally considered non-heat treatable from the standpoint of increasing the strength, i.e., not age-hardenable. The α/β-alloys are more capable of being strengthened by aging, with the strengthening capability increasing as the β-stabilizer content increases. The strength of an α/β–alloy such as Ti-6Al-4V, for example, can be increased by ~ 200 MPa in sections up to about 25 mm thick. The strength of a β-alloy such as Ti-3Al-8V-6Cr-4Mo-4Zr can be increased by >400 MPa through heat treatment in comparable section sizes.

The next consideration must be the ST temperature; the higher the temperature, up to the β_t, the greater the strengthening capability. The rationale for this is straightforward; as the ST temperature increases the amount of β at temperature progressively increases, increasing the volume fraction of material which can be subsequently aged (precipitation hardened via α precipitation), increasing the achievable strength. For those alloys which transform martensitically on cooling the volume fraction of martensite will increase with increasing ST temperature, hence increasing the achievable strength. Based on this premise, it is obvious that the cooling rate from the ST temperature is also an important variable, as has already been discussed. These points are demonstrated in Figure 4. In addition, the heating rate to the solution treatment temperature can be a consideration as will be discussed later under specialized heat treatments

Quench delay, the time between when the part is extracted from the furnace and is immersed into the quenchant, is a further consideration. Longer quench delays will result in reduced strength. Ishikawa et al [20] demonstrated the effect of quench delay on the tensile properties of Ti-6Al-4V and SP-700. The reduced strength with the longer quench delays (Figure 5) is ascribed to the formation of coarse acicular α prior to immersion in the quench medium.

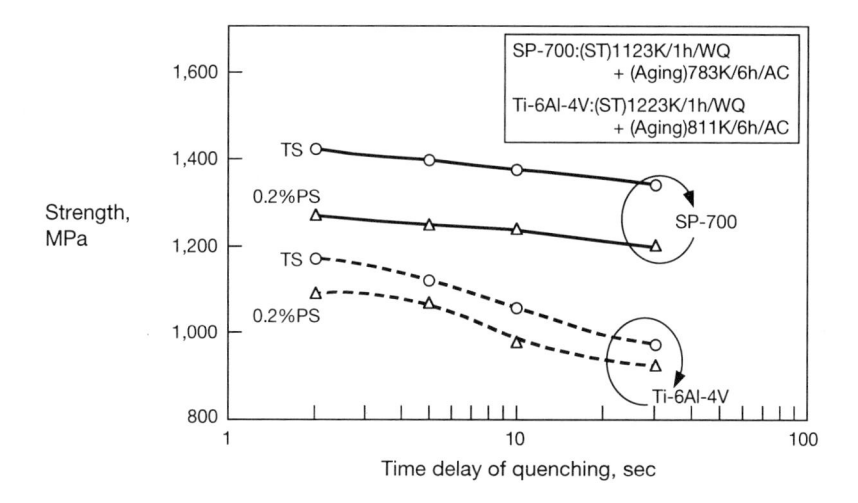

Figure 5. Effect of quench delay during solution treatment on the strength of Ti-6Al-4V and SP-700 (from Ref. 20).

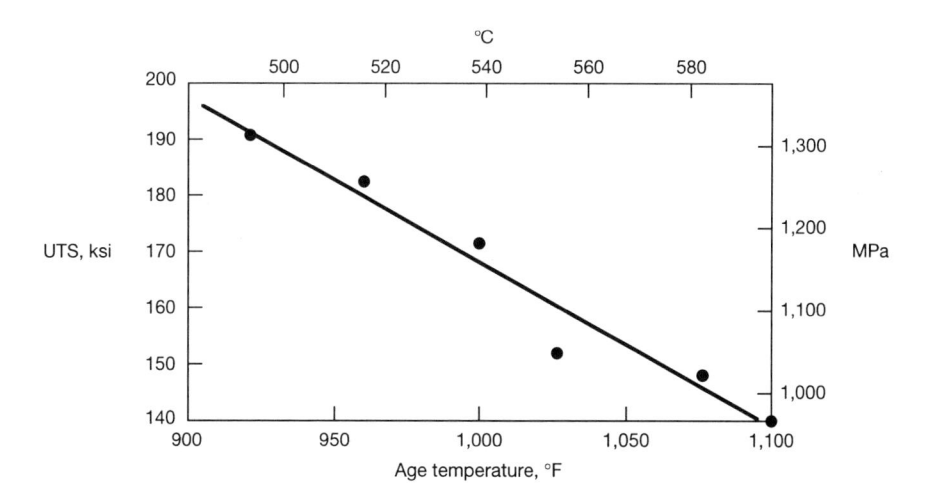

Figure 6. Effect of aging temperature on the tensile strength of Ti-10V-2Fe-3Al. All specimens were solution treated at 750°C/2 hr/WQ followed by an 8-hr age at the indicated temperature (from Ref. 18).

The aging temperature controls the size and volume fraction of the aged α-precipitates and thus the tensile strength. (Other precipitates can be formed, depending on the alloy and aging times and temperatures involved, such as ω, $α_2$, or $β_1+β_2$, but these are not generally used in production practice as they are impractical in terms of the aging times involved and/or often result in embrittlement.[21]) Generally, the lower the aging temperature the higher the strength. This of course assumes that the aging temperature is high enough to provide adequate diffusion to achieve an acceptable aging response in reasonable time periods. This is illustrated in Figures 6 and 7. The shapes of the curves in Figure 7 for aging temperatures from 480 to 595°C are consistent in that the higher the aging temperature, the coarser the α precipitates and the lower the strength. The 425°C aging kinetics are slower in that precipitation of the α is now diffusion rate controlled. One would expect, however, that it would ultimately have a higher strength than the higher temperature ages, given sufficient aging time

The heating rate to the aging temperature is another factor in metastable β-alloys. Slower heat-up rates can result in the formation of ω-phase. This ω-phase, a transition precipitate, normally forms as a very fine precipitate, with a very high density of precipitates for production alloys. This ω-phase serves as a pre-cursor for subsequent α-precipitation. The α precipitates near the ω, resulting in a higher density of precipitates than would be observed in the absence of the ω-phase.[23] As a specific example, discrepancies have been identified in the tensile strengths achieved in Timetal®21S sheet when comparing material heat treated in a vacuum furnace (slow heat-up rate), vs. that heat treated in an air furnace, with the material being inserted into the furnace heated to the aging temperature.[24] For a given aging temperature, the material aged in the vacuum furnace exhibited a higher strength. To achieve similar strength levels, material heat treated in the vacuum furnace had to be aged at a temperature 55°C higher than the material aged in the air furnace. The higher density of α-precipitation resulting from the slow heating rate (due to the ω precursor) caused the higher strength; aging at a higher temperature was thus required to coarsen these precipitates to make the strength commensurate with that from the material aged in the air furnace, which had a lower density of α precipitates.

The ω precursor can also control the morphology of the α.[23] If the ω to α transformation occurs at a temperature close to the ω-solvus, where it will co-exist with the α for a significant time period the presence of the residual ω will constrict the growth of the α needles, resulting in a stubby morphology. If there is no ω precursor, or if the aging temperature is sufficiently above the ω-solvus such that the ω quickly reverts, the α forms with the expected acicular morphology (Figure 8).

Azimzadeh [23] also demonstrated that the path taken to the aging temperature can affect the morphology and distribution of the α-phase. He studied Timetal®LCB quenched (1) from the ST temperature into the aging bath (direct aging), and (2) into room temperature water and then re-heated to the aging temperature (indirect aging), the more common commercial practice. The path taken to the aging temperature and the actual aging temperature influence the kinetics of α-precipitation and athermal ω ($ω_{ath}$) formation, as indicated in Figure 9. The $ω_{ath}$ will form if it is quenched to a temperature below the ω start temperature. If $ω_{ath}$ is present at the onset of aging, whether the material is direct or indirect aged, the aging response is rapid as the $ω_{ath}$ is rapidly converted to isothermal ω ($ω_{iso}$), resulting in immediate hardening with no incubation period (345°C aging curve in Figure 10a.). If, however, the material is direct aged to a temperature above the ω-start temperature so no $ω_{ath}$ is present (or

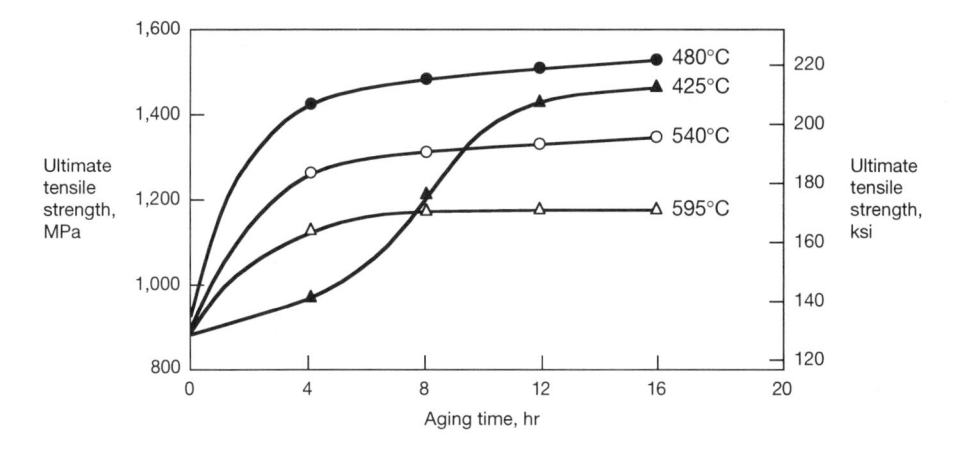

Figure 7. Effect of aging temperature and time on the tensile strength of Beta-C. All specimens were solution treated at 815°C (from Ref. 22).

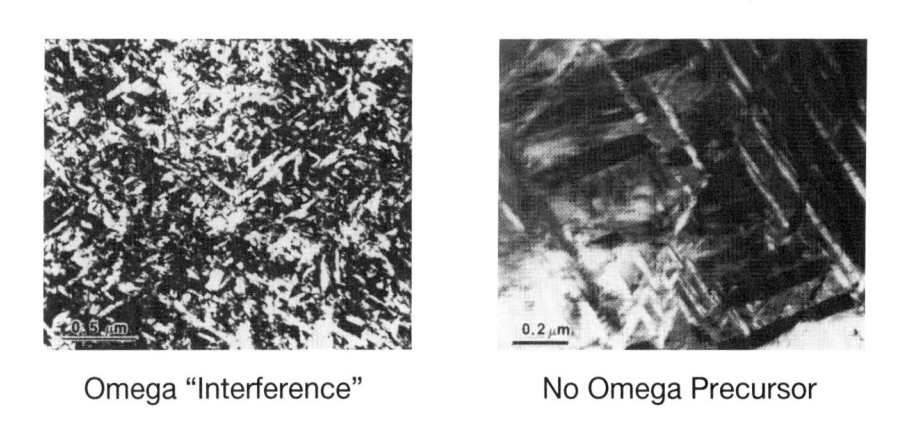

Omega "Interference"　　　　No Omega Precursor

Figure 8. TEM photomicrographs illustrating the effect of the presence of ω-phase on α-morphology. If a high volume fraction of ω is present at time of α-precipitation, the constraint of the ω-phase results in a "stubby" α (a.). If no ω is present or the temperature is high enough that it reverts quickly, the α-phase is more acicular (b.) (from Ref. 23).

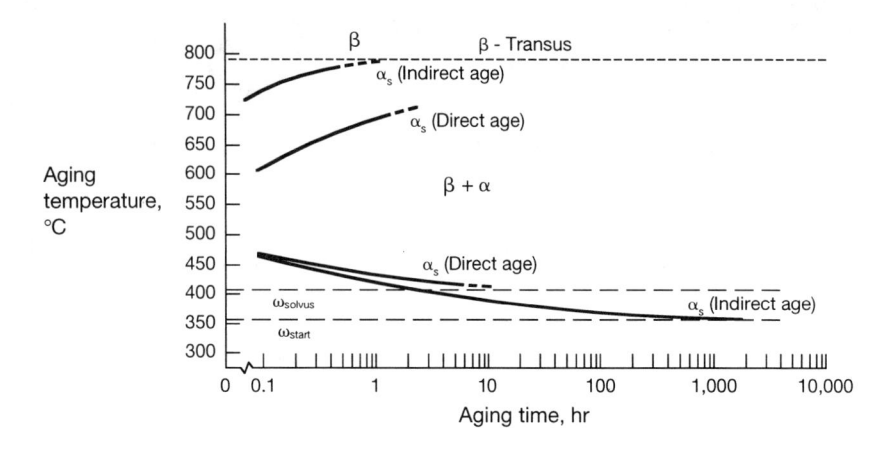

Figure 9. Comparison of indirect and direct aging α-precipitation domains and the ω solvus and start temperatures for TimetalwLCB (from Ref. 23)

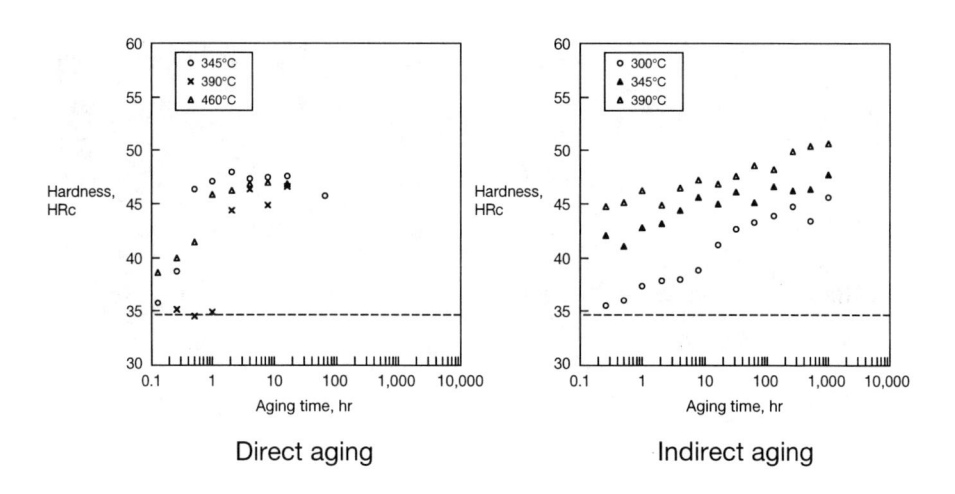

Figure 10. Aging response curves for Timetal® LCB illustrating the effects of heat treatment route (direct versus indirect age) and age temperature. The dotted line is the hardness of as-quenched material (from Ref. 23).

if it is indirect aged at a temperature significantly above the ω-solvus so the ω reverts to β prior to α-precipitation) the kinetics are much slower, and an incubation period, related to the nucleation and initial growth of the α-phase, may be observed (390°C aging curve in Figure 10a.). The 460°C age in Figure 10a. again has a rapid aging response as growth of the α-phase is rapid at this temperature.

Similar data for indirect aged material is illustrated in Figure 10b.). None of these indicate an incubation period as ω_{ath} is present in all cases prior to aging. The response rate of the low temperature age is much slower due to reduced diffusion rates at the lower temperature. Aging at higher temperatures than those shown in Figure 10 resulted generally in lower hardness and overaging within an hour.

The solution treatment of α/β alloys may be conducted above the β_t in some cases. This is sometimes used for Ti-6Al-4V castings, for instance, which already have a fully transformed β structure. A beta ST is sometimes used to refine the transformed structure obtained from the casting and HIP processes to improve the high-cycle fatigue performance. This would be referred to as a beta solution treat and age (BSTA) or beta solution treat and overage, depending on the age (or anneal) temperature used.

Triplex Heat Treatments - These could be triplex-anneals or double-solution treat and age treatments depending on the intention of the heat treatment. A triplex anneal has been used for Ti-6Al-2Sn-4Zr-2Mo to provide slight strength and creep resistance improvements. This heat treatment consists of 900°C/2.5 hr + 785°C/.25 hr + 595°C/2 hr.[25]

A double-solution treat and age was developed to improve the damage tolerance properties of Ti-6Al-2Sn-2Zr-2Mo-2Cr forgings. Initially it was felt that the best fabrication method for these parts, for the Boeing/Lockheed F-22, was to β-forge and α/β-solution treat and age. Chakrabarti et al [26] felt that more uniform properties would be obtained throughout the forging using α/β forging and β-heat treating. An extensive TMP study indicated that α/β forging followed by a β-solution treat, an α/β-solution treat, and an age provided the best properties combination.

Double solution treatments below the β_t could be utilized to optimize damage tolerance properties. The first solution treatment, high in the α/β phase-field, would be followed by a slow cool to develop a coarse lamellar α-phase for improved toughness and crack growth resistance. The second solution treatment, 30 to 50°C below the first one, is followed be a more rapid cooling rate to establish the hardenability for the subsequent aging cycle. Boyer et al [27] demonstrated the advantage of the double solution treatment for the SP 700 alloy.

Specialized Heat Treatments

Wagner and Gregory [28] found that a duplex aging treatment could provide improved tensile and fatigue strengths in Beta-C. Single step aging resulted in non-uniform α nucleation, with β regions not containing α precipitation. Slip was concentrated in these softer regions providing sites for early fatigue crack initiation. Duplex aging treatment involves an initial age at a lower temperature where α–nucleation kinetics are favorable to form a high density of α-precipitates, followed by a higher temperature age to coarsen the precipitates to achieve the desired strength. This resulted in a more uniform precipitation, improved fatigue strengths at comparable strength levels, and higher strengths for duplex aging as compared to single step aging using the same final age temperature. They also reported that the duplex age improved both the strength and the toughness at a given strength level for the Ti-3Al-8V-6Cr-4Mo-4Zr alloy.[29] Boyer et al [30] observed similar behavior regarding the fatigue properties of Ti-15V-3Cr-3Al-3Sn castings; limited data using an extended aging time to improve the homogeneity of α precipitation improved the fatigue performance.

In contrast to this Niwa et al [31] used a high-low aging temperature sequence to improve the strength ductility relationship of heavily cold-worked Ti-15V-3Cr-3Al-3Sn. The initial high temperature age of heavily cold worked material effected a partial recovery; the resultant dislocation network provided a higher density of nucleation sites for α precipitation with subsequent low temperature aging.

Yet another approach was used by Ouchi et al [32] to obtain very high strength and ductility in Ti-15V-3Cr-3Al-3Sn, schematically illustrated in Figure 11. The thickness reduction in the first cold rolling step (CR(1), see the figure) must be sufficient to provide a very fine recrystallized β grain structure following the first solution treatment, ST(1). This fine grain size enhances the uniformity of the dislocation distribution during the 2nd cold rolling step - CR(2)- which is limited to 5-20% reduction. The strength is also strongly related to the rate of heating to the solution treatment temperatures (Figure 12). The slower heating rates would result in larger amounts of recovery (or recrystallization depending on the circumstances) which would then diminish the nucleation sites, thus reducing the density of α precipitates and the strength. This also explains limiting the amount of reduction during the CR(2) step. Greater reductions would result in recrystallization and the loss of the nucleation sites. The ductility achieved at these strength levels, on the order of 10% elong. at a tensile strength of over 1860 MPa is ascribed to the very fine β grain size. Grain boundary α, which will always be present to some degree, has a negative effect on ductility. However, the effect of the grain boundary α is minimized for very small β grain sizes.

Makino et al [33] studied the microstructures of cold-worked and aged Ti-15V-3Cr-3Al-3Sn and determined that more α variants were present in material which was rolled, recovered and aged than in the one which was rolled and direct aged. This could imply that better and more uniform properties would be observed in the former case with the additional variants.

Gray et al reported a unique thermomechanical treatment to achieve a good combination of creep/damage tolerance properties with high fatigue crack initiation resistance.[34] The lamellar microstructure is desirable for maximum creep resistance; but it would have inherently poor fatigue crack initiation resistance. They shot peened the surface and then annealed the material; the cold work on the surface from the shot peening was sufficient to drive recrystallization, resulting in a fine-grained equiaxed microstructure on the surface (Figure 13). This then provides an essentially *ideal* microstructure where both creep resistance (or fracture toughness) and fatigue crack initiation resistance are both optimized. The fine grained microstructure on the surface to provide maximum fatigue crack nucleation resistance, and the lamellar microstructure in the interior with the maximum crack growth resistance and fracture toughness. This provided about a 50 MPa improvement in the fatigue strength of Ti-6Al-2Sn-4Zr-2Mo at 550°C as shown in Figure 14. Similar results have been reported on Beta-C cold worked in the solution treated condition in conjunction with a specially developed aging treatment as seen in Figure 15.[35] They have reported using both shot peening and surface rolling to impart the cold work. The fatigue improvement in the β-alloy was ascribable to preferential aging at the cold-worked surface, and, depending on the aging temperature, it may be beneficial or detrimental relative to the as-rolled material. Heat treatment SSA 2 (SSA = selective surface aging) provided an improvement, while SSA 1 had

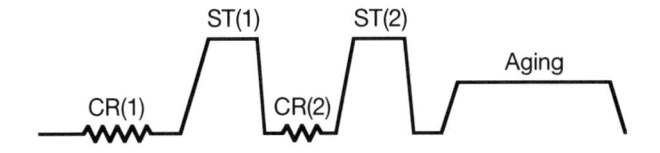

Figure 11. Schematic of TMP route used by Ouchi et al. [32] to obtain very high strength and ductility in Ti-15V-3Cr-3Al-3Sn.

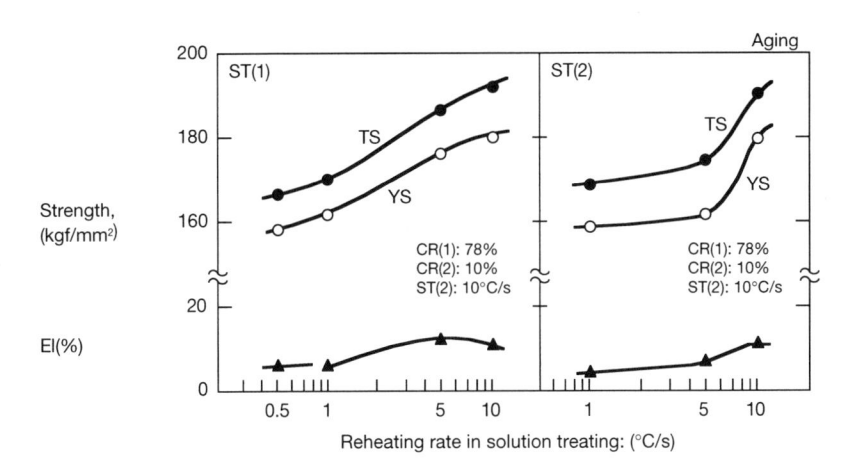

Figure 12. The influence of heating rate to the ST(1) and ST(2) temperatures on the aged strength (from Ref. 32).

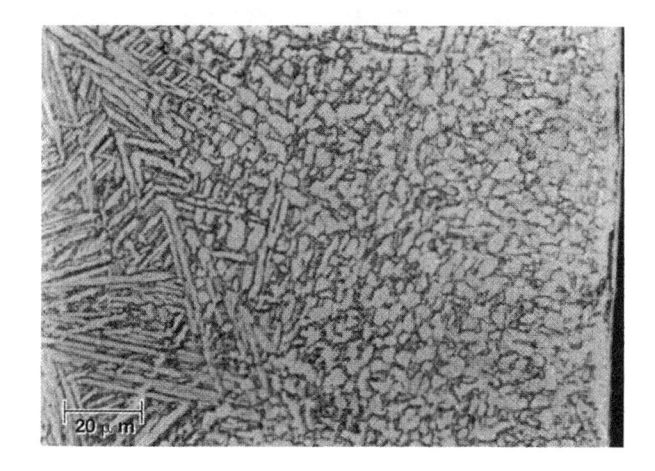

Figure 13. Ti-6242 with a lamellar core and equiaxed surface microstructure.

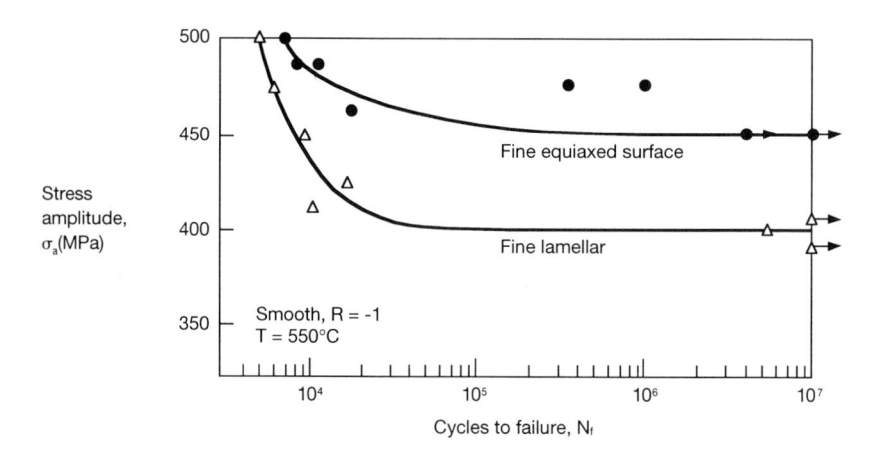

Figure 14. S-N curves for Ti-6242 at 550°C with a fine lamellar microstructure with and without a thermomechanical surface treatment (from Ref. 34).

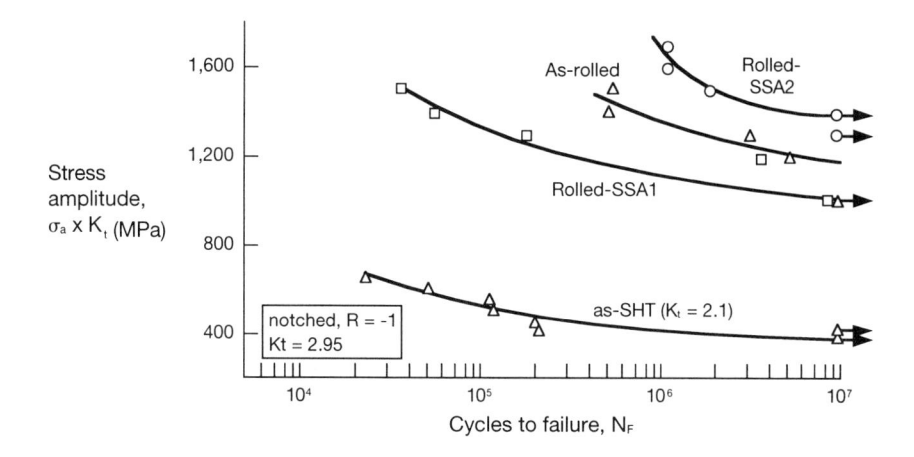

Figure 15. S-N curves for Ti-3Al-8V-6Cr-4Mo-4Zr in the as-SHT and after rolling both with and without subsequent aging treatments (from Ref. 35).

a fatigue debit. They speculated that SSA 1 was at a high enough temperature to relieve the residual compressive stresses, which mitigated any gain provided by the harder surface, resulting in a fatigue debit. It could simply be related to differences in hardness (strength) at the surface because of the different aging times and/or temperatures.

Summary

The heat treatment of titanium alloys can provide a very broad range of properties combinations. As the properties requirements become more critical, the heat treatment parameters, and prior processing history, must be more tightly controlled. All aspects of the heat treatments (which can be very complex) used must be understood to have a viable, robust, repeatable process. Consideration must be given not only to the times and temperatures used, but to the prior working history, heating rates, cooling rates, processing sequences, etc.

Heat treatments can be utilized to improve or maximize a given set of properties, but this almost always means a compromise in another property, e.g., fatigue crack nucleation resistance is sacrificed to maximize creep resistance. Specialized heat treatments are available which can mitigate these compromises in certain instances.

Acknowledgments

R. Boyer would like to thank L.M. Micona, G.E. Trepus and L.M. Gammon for their contribution in the construction of Figure 3. He would also like to thank Prof. H.J. Rack from Clemson University for extensive discussions on details of their work.

References

1. Materials Properties Handbook: Titanium Alloys, (Materials Park, OH: ASMI, 1994), 439-444
2. R.R. Boyer, R. Bajoraitis and W.F. Spurr, "The Effect of Thermal Processing Variations on the Properties on Ti-6Al-4V", Microstructure, Fracture Toughness and Crack Growth Rate in Ti Alloys, (Warrendale, PA: TMS, 1987), 149-1703
3. I.W. Hall and C. Hammond, "Fracture Toughness, Strength and Microstructure in $\alpha+\beta$ Titanium Alloys," Titanium and Titanium Alloys, (New York, N.Y., Plenum Press, 1982), 601-613
4. U. Zwicker, M. Blass and U. Hofmann, "Effects of TiAl6V4 Metallurgical Structures on Fatigue Properties," ibid., 2027-2045
5. G. Lütjering and A. Gysler, "Critical Review: Fatigue," Titanium Science and Technology, (Oberursel, Germany: DGM, 1985), 2065-2083
6. Mil-H-81200, Military Specification, Heat Treatment of Titanium and Titanium Alloys
7. Specification AMS 2801, Heat Treatment of Titanium Alloy Parts
8. E.R. Barta, R.R. Boyer and G.H. Narayanan, "Delayed Hydrogen Embrittlement of C.P. Titanium," Proceedings of the International Symposium for Testing and Failure Analysis, (Materials Park, OH, ASMI, 1988), 387-396
9. H.W. Rosenberg, "Titanium Alloying in Theory and Practice," The Science, Technology and Application of Titanium, (London, U.K., Pergamon Press, 1970), 851-859
10. R.R. Boyer and W.F. Spurr, "Effect of Composition, Microstructure, and Texture on Stress-Corrosion Cracking in Ti-6Al-4V Sheet," Met Trans A, 9A (10) (1978), 1443-1448
11. R.D. Briggs, "Effects of Cooling Rate from Mill Annealing Temperature on Stress Corrosion Cracking Threshold of Titanium 6Al-4V ELI Beta Annealed," Proceedings of this conference
12. D.J. Evans and T.F. Broderick, "The Role of Intermetallic Precipitates in Forged T--62222S," (Paper Presented at the 8th World Conference on Titanium, Birmingham, England, Oct. 22-26, 1995)
13. R.I. Jaffee, L. Wagner and G. Lütjering, The Effect of Cooling Rate From the Solution Anneal on the Structure and Properties of Ti-6Al-4V Alloy," Sixth World Conference on Titanium, (Cedex, France: Les Éditions de Physique, 1989), 1501-1506
14. R.R. Boyer and A.E. Caddey, "The Properties of Ti-6Al-2Sn-2Zr-2Mo-2Cr Sheet," Titanium '92 Science and Technology, (Warrendale, PA: TMS, 1993), 1647-1652
15. H.R. Phelps and J.R. Wood, "Correlation of Mechanical Properties and Microstructures of Ti-6Al-2Sn-2Zr-2Mo-2Cr-0.25S Titanium Alloy," Titanium '92 Science and Technology, (Warrendale, PA: TMS, 1993), 193-199
16. S. Saal, L. Wagner, G. Lütjering, "Effect of Cooling Rate on Creep and Low Cycle Fatigue Resistance in Ti-6242," Z. Metallkde, 90, (1990), 535-540
17. C. Andres, A. Gysler and G. Lütjering, "Correlation Between Microstructure and Creep Behavior of the High-Temperature Ti-Alloy IMI 834," Titanium '92 Science and Technology, (Warrendale, PA: TMS, 1993), 311-318
18. R.R. Boyer and G.W. Kuhlman, "Processing Properties Relationships of Ti-10V-2Fe-3Al," Met. Trans. A, 18A (12) (1987), 2095-2103

19. G. Terlinde, H-J. Rathjen and K-H. Schwalbe, "Microstructure and Fracture Toughness of Aged β-Titanium Alloy Ti-10V-2Fe-3Al," Met. Trans A, 19A (4) (1988), 1037
20. M. Ishikawa, O. Kuboyama, M. Niikura and C. Ouchi, "Microstructure and Mechanical Properties Relationship of β-rich α–β Titanium Alloy; SP-700," Titanium '92 Science and Technology, (Warrendale, PA: TMS, 1993), 141-148
21. T.W. Duerig and J.C. Williams, "Overview: Microstructure and Properties of Beta Titanium Alloys," Beta Titanium Alloys in the 1980's, (Warrendale, PA: TMS, 1983), 19-67
22. Materials Properties Handbook: Titanium Alloys, (Materials Park, OH: ASMI, 1994), 826
23. S. Azimzadeh, "Phase Transformations in Metastable β Titanium Alloy Timet LCB (Ti-6.8Mo-4.5Fe-1.5Al)," (M.S. thesis, Clemson University, 1995)
24. R.R. Boyer, Unpublished Boeing Research
25. Materials Properties Handbook: Titanium Alloys, (Materials Park, OH: ASMI, 1994), 374
26. A.K. Chakrabarti, R. Pishko, V.M. Sample and G.W. Kuhlman, "TMP Conditions - Microstructure - Mechanical Properties Relationships in Ti-6-22-22S Alloy," Titanium '92 Science and Technology, (Warrendale, PA: TMS, 1993), 209-216
27. R.R. Boyer, K. Minikawa and A. Ogawa, "Strength/Toughness Properties of SP-700," (Paper Presented at the 8th World Conference on Titanium, Birmingham, England, Oct. 22-26, 1995)
28. L. Wagner and J.K. Gregory, "Improvement of Mechanical Behavior in Ti-3Al-8V-6Cr-4Mo-4Zr by Duplex Aging," Beta Titanium Alloys in the 1990's, (Warrendale, PA: TMS, 1993), 199-209
29. R.R. Boyer and J.A. Hall, "Microstructure-Property Relationships in Titanium Alloys (Critical Review)", Titanium '92 Science and Technology, (Warrendale, PA: TMS, 1993), 77-88
30. R.R. Boyer, D.A. Wheeler and R.G. Vogt, "Effects of Weld Repair on the Fatigue Performance of Ti-15-3 Castings," (Paper Presented at the 8th World Conference on Titanium, Birmingham, England, Oct. 22-26, 1995)
31. N. Niwa, A Arai, H. Takatori and K. Ito, "Mechanical Properties of Cold-Worked and High-Low Temperature Duplex Aged Ti-15V-3Cr-3Al-3Sn Alloy," ISIJ Intl., 31 (1991), 856-862
32. C. Ouchi, H. Suenaga and Y. Kohsaka, "Strengthening Mechanisms of Ultra-High Strength Achieved by New Processing in Ti-15%V-3%Cr-3%Al-3%Sn Alloy," Sixth World Conference on Titanium, (Cedex, France: Les Éditions de Physique, 1989), 819-824
33. T. Makino, R. Chikaizumi, T. Nagaoka, T. Furuhara and T. Maki, "Microstructure Development in a Thermomechanically Processed Ti-15V-3Cr-3Al-3Sn Alloy," (Paper Presented at the SAMPE International Symposium on Metallurgy and Technology of Titanium Alloys, Tokyo, Japan, Sept. 25-28, 1995)
34. H. Gray, L. Wagner and G. Lütjering, "Effect of Modified Surface Layer Microstructures Through Shot Peening and Subsequent Heat Treatment on the Elevated Temperature Fatigue Behavior of Ti Alloys," Shot Peening, (Oberursel, Germany: DGM, 1987), 467-475
35. L. Wagner and J.K. Gregory, "Improve the Fatigue Life of Titanium Alloys, Part II," AM&P, July, 1994, 50HH-50JJ

SELECTION OF HEAT TREATMENT OPTIMUM TECHNOLOGIES

FOR INTRICATELY SHAPED TITANIUM ALLOY ARTICLES

Dr Alexey N. Lozhko, Dr Grigory Z. Malkin

State Metallurgical Academy of the Ukraine
Gagarin Ave 4, Dnepropetrovsk, Ukraine

Abstract

Cooling of intricately shaped Ti-alloy blanks during the thermal treatment has been described. The determined optimum cooling schedules provide for the required microstructure of the articles due to controlled and variable cooling of the sections of the articles featuring different mass. A search of the specific technology of cooling (heating) of an intricately shaped article has been carried out on a computer model which visualized the distribution of temperatures, distribution of heating (cooling) rates, thermo-sresses, gas saturation layer and scale layer. An original mathematical apparatus based on conformal transformations has been used for the simulation of the non-stationary temperature field.

Introduction

The purpose of heat treatment is improvement of a metal microstructure. It requires to fulfill a determined temporary - temperature mode. Usually these modes are well known from laboratory experiments. They are received in result of the microstructural analysis of small metal pieces. Data of such experiments easily to use for development of heat treatment technology for products of the regular form (parallelepiped, cylinder, sphere). However, there are following problems for products of a irregular form:

-it is impossible to ensure identical rate of cooling in all product's volume because its parts has various heat build-up,

-at intensive cooling, it is necessary to take into account thermo-stress, and if their level is high, residual stress too (this phenomenon is present for any forms, but the products with the regular form resist temperature deformation better),

-on a irregular surface the control of oxidation and gas saturation processes is dificult (it is connected to various cooling rate of various parts of the irregular form product).

Main Factors

Thus the problems of uniform heat treatment of irregular form products is problems of complexity of the summation of many known factors and search their optimal relation. Problem can be resolved by development of adequate mathematical model of heat treatment, which takes into account the product form non-regularity. Object of simulation are:

-a temperature field of product under arbitrary distributed conditions of cooling on surface,
-rate-temperature field of product, for control of a formed microstructure,
-a field of temperature stresses in view of elastic and plastic deformation possibility,
-distribution of residual stresses in a product after its heat treatment,
-distribution of oxydated and gas saturated layers depth on a surface of a product after its heat treatment.

Mathematical Statment

The mathematical statement of a task is simulation of a non-stationary fields of temperatures and stresses. This process is completely described systems of differential equations

$$\begin{cases} \dfrac{\partial t}{\partial \tau} c(t)\rho(t) = \lambda(t)\cdot\left(\dfrac{\partial^2 t}{\partial r^2}+\dfrac{1}{r}\dfrac{\partial t}{\partial r}+\dfrac{\partial^2 t}{\partial z^2}\right), \\ \dfrac{\partial^2 S}{\partial r^2}+\dfrac{1}{r}\dfrac{\partial S}{\partial r}+\dfrac{\partial^2 S}{\partial z^2} = -\dfrac{E(\sigma,t)\alpha}{1-\upsilon(\sigma)}\left(\dfrac{\partial^2 t}{\partial r^2}+\dfrac{1}{r}\dfrac{\partial t}{\partial r}+\dfrac{\partial^2 t}{\partial z^2}\right), \\ \dfrac{S}{\partial r\cdot\partial z} = -\left(\dfrac{\partial^2 \sigma_{rz}}{\partial r^2}+\dfrac{1}{r}\dfrac{\partial \sigma_{rz}}{\partial r}+\dfrac{\partial^2 \sigma_{rz}}{\partial z^2}\right), \\ S = \sigma_r + \sigma_z, \end{cases} \quad (1)$$

under boundary conditions:

$$\lambda(t)\dfrac{\partial t}{\partial n} = \alpha(r,z,t)\cdot\left[t_\alpha(n) - t_s(n)\right], \quad (2)$$

$$\dfrac{\partial S}{\partial k} = -\left(\dfrac{\partial \sigma_{rz}}{\partial n} + P_n\right), \quad (3)$$

r, z - cylindrical coordinates,
τ - time,
t = t(r,z,t) - temperature in a point r,z at the moment of a time t,
$\sigma_r = \sigma_r$ (r,z) - stresses in a point r,z along a axis r,
$\sigma_z = \sigma_z$ (r,z) - stresses in a point r,z along a axis z,
$\sigma_{rz} = \sigma_{rz}$(r,z) - shear stresses in a point r,z,
$\lambda(t)$, $c(t)$, $\rho(t)$ - thermal characteristics of a metal (heat conduction coefficient, heat capasity coefficient, strength of a metal),

$E(\sigma,t)$, $\upsilon(\sigma)$, α_T - mechanical characteristics of a metal (integrated coefficient of deformation, integrated coefficient of shift, coefficient of thermal expansion),
n, k - direction, which is normal to surface, direction, which is parallel to surface,
P_n - external force.

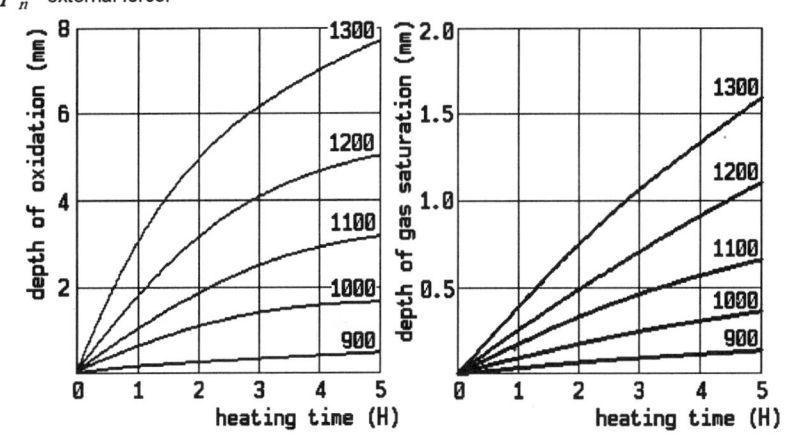

Figure 1. - The dependens of oxydation and gas saturation processes on a temperature state

Distributions of cooling rates, residual stresses, gas saturation and oxidation are received by the decision of the given equations. This decision was compared with experimental datas about dependence of these parameters on temperature in a surface and time of heating. Large statistical information about these processes was examined. The result of this exploration is many empirical dependences, which are various for different alloys, and conditions of heat treatment. For example, on a fig.1 dependenses of oxydation and gas saturation depthes for one of Ti- alloys are shown. These dependenses are formalized by algebraic expression such as

$$L_{oxd}^{aT+b} = K_o \exp\left(-\frac{\beta}{t_s}\right)\tau, \quad L_{gst}^{cT+b} = N_o \exp\left(-\frac{\gamma}{t_s}\right)\tau, \qquad (4)$$

L_{oxd} - depth of oxidated layer,

L_{gst} - depth of gas saturated layer,

t_s - temperature of a sample surface,

t - time of thermal action,

(K_O, a, b, β) and (N_O, c, d, γ) - coefficiets, which are defined by the analysis of experimental data.

Solving

Joint solving of equations under specific boundary conditions is a dificult-solving problem. The problem was decided by use of a original mathematical method, based on orthogonal conformal transformation. A ccording to these method, the computer builds a new system of coordinates for each geometrical form. In a new system of coordinates the search of the decision is considerably simplified. It occurs because the system of simulative equations has no derivative on mixed directions. From the mathematical theory it is known, that such derivative are the main reason of instability of a decidion search procedure. Derivative on a mixed direction occur in the decision if the task is decided with the help of a design net, which is constructed in the geometrical area with a irregular border. However, if it is require to fulfill a condition of orthogonality (to save orthogonality of design net lines), the system of equations will be so simple to decide, as for rectangular area. But the construction of a coordinate system, which has such property, is a very important question.

Solving Transformation

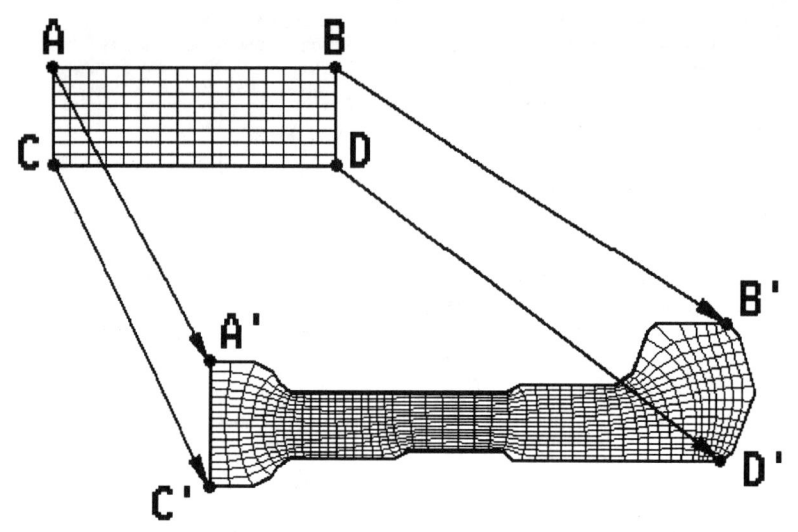

Figure 2. - The example of curvelinear orthogonal coordinates system construction for a formed disk

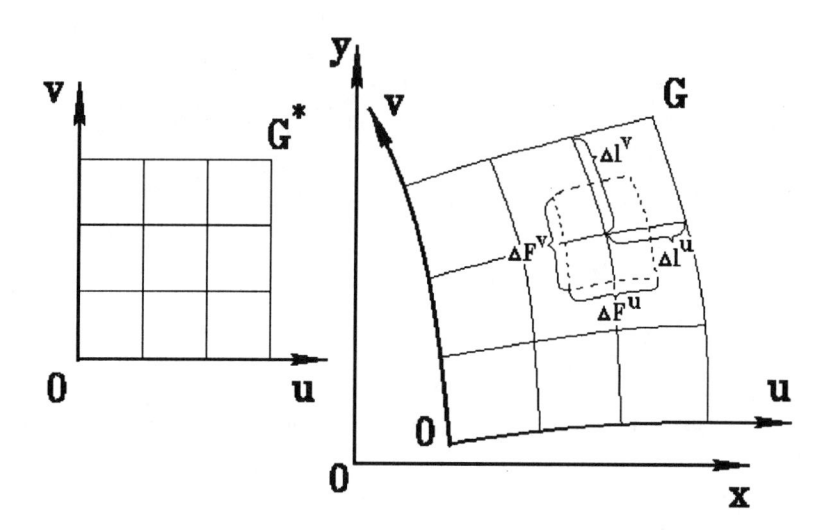

Figure 3. - Transformation for curvelinear orthogonal coordinates system

There is a example of rectangular design area transformation to real formed blank in fig.2. It is visible, that all lines, forming design net of formed blank, are crossed with one another orthogonally. There is a geometrical interpretation of the decision in fig.3. There is the rectangular area G in a plane w, and in plane Z there is single-connected area G^*. It is require to find such complex function,

$$Z = f(\omega) = X(u,v) + iY(u,v), \qquad (5)$$

which can to transform the area G to the area G^* and to save rectangularity between small sections (condition of orthogonality):

$$\Delta Z_u = [Z(u+\Delta u, v), Z(u, v)],$$
$$\Delta Z_v = [Z(u, v), Z(u+\Delta u, v)].$$

If thus a rectangular net is consrtucted in area G it will be reflected on curvelinear orthogonal net of the area. And condition of orthogonality has a kind:

$$\frac{\partial Z}{\partial u} = -\frac{\partial Z}{\partial v}.$$

Arrangement of curvelinear orthogonal net points strongly depends on the geometrical features of the area G^*. The design practice has shown, that it is necessary to have parameter, which drive on density of net points distribution. It let to maked net more dense there, where more exact account (for example at a surface) is necessary. Here why the final equation inclucds distributed parameter $K_{uv}(x, y)$. Thus the a local coordinate system construction is the decision of a equation

$$\frac{1}{K_{uv}} \frac{\partial}{\partial u} \left(\frac{K_{uv}}{\frac{\partial Z}{\partial u}} \right) = K_{uv} \frac{\partial}{\partial v} \left(\frac{1}{\frac{\partial Z}{\partial v}} \right). \qquad (6)$$

Software

Described approach gave opportunity to creat the series of software, which optimize a cooling (heating) technology of titanium products in view of all factors, accompanying to this process. The purpose of development there are utmost automation of selection. It is useful to technologists which are not professional mathematicians or programmers.
Analysis of technology usually has three stages
   -input of a type, the geometrical form and material of a product,
   -choice of heat exchange conditions for various zones of surfaces of a product and change of these conditions at time,
   -the control of temperatures, rates of cooling (heating), thermo and residual stresses, depthes of oxidated and gas saturated layer.

Examples

The search of cooling tecnology of a titanium valve of locking crane, which works under large pressure in a gressive environment with abrasive inclusions was examined as example.

According to technology "A" the locking valve is cooled on the air. Rate of cooling on a lateral surface insufficient for formation of a qualitative microstructure. It does not provide resistance of a valve to effect of abrasive inclusions on a stem. Valve, received by such technology is quickly destroyed. According to the technology "B" the same valve, after heating up to temperature $950\,^\circ C$, colled by a water, that provides required high rate of cooling. Mechanical properties of a metal on surfaces, which come into contact with abrasive environment is satisfactory. However thus arise high thermal stresses. The valve can be destroyed or to lose the initial form under effect of residual stresses. According to technology "C" the valve was exposed to selective cooling of the most bulky parts. Thus the cooling rate on a surface of a stem is sufficient for formation of qualitative microstructure, but the thermal and the residual stresses do not exceed a critical level.

The figure 4 shows temperature fields of valve after 20 minutes of cooling on technologies "A", "B" and "C". Their comparison gives representation about rate of cooling in different parts of a product. There are appropriative thermal stress fields in figure 5, and field of residual stresses after total cooling in figure 6. Figure 7 illustrates distributions of oxidation and gas saturation. They concern to this valve, but are received at analysis of technology of its heating for heat treatment.

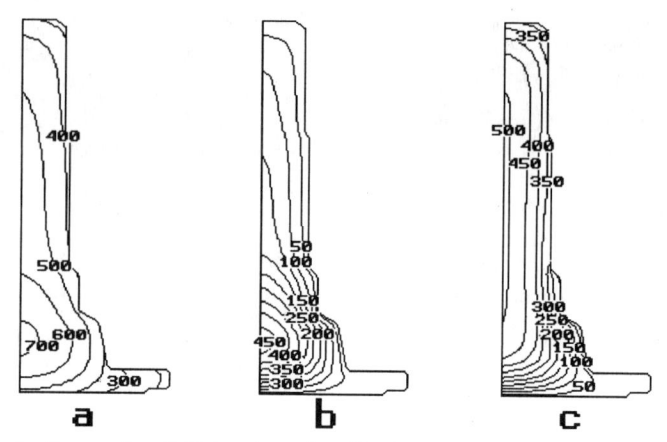

Figure 4. - Temperature fields in valve after 20 minutes of cooling at various heat treatment technologies

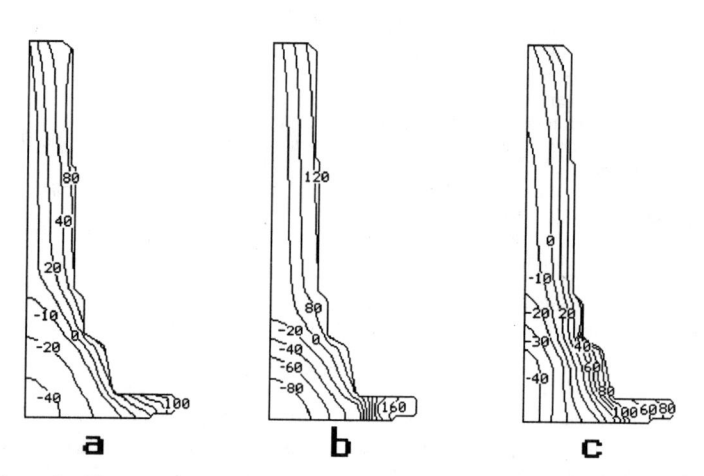

Figure 5. - Thermo-stresse fields in valve after 20 minutes of cooling at various heat treatment technologies

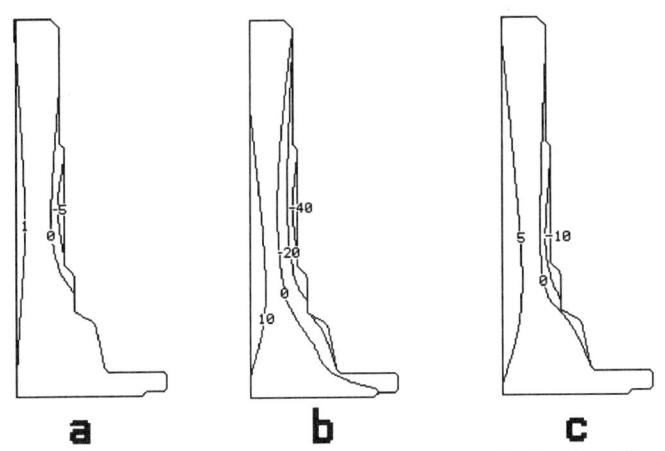

Figure 6. - Distribution of residual stresses in a valve after heat treatment by various technologies

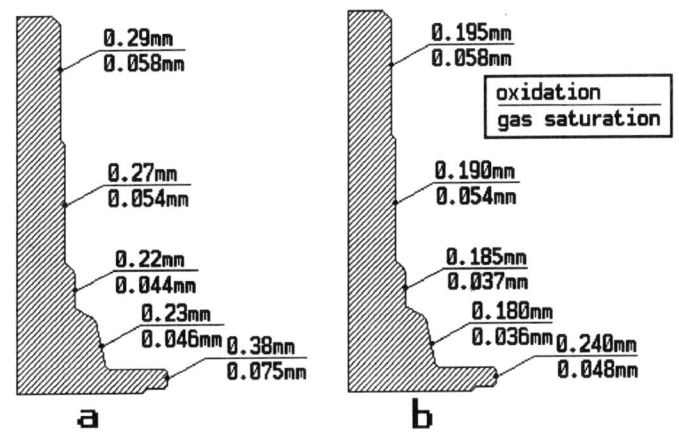

Figure 7. - The depth distribution of oxydation and gas saturation layerst at valve heating for heat treatment

Adequacy

The adequacy of results strongly depends on completeness of thermal and mechanical characteristics of a material. If these variables sizes are given completely, the results difference between docision and experiment is near 15%.

Optimization

However the visualization of heat treatment process is not always causes to fast choice of technology. The requirements, which touches upon cooling rate and mycrostructure uniformity in all product, frequently run counter one another.

At those case the technology should be optimum as ti is possible. For choise technology in similar cases a method of optimization, which takes into accoµnt all parameters of heat treatment in a complex, is used. For solving of this task concept "quality of heat treatment" is entered. It is functional, the minimum of which corresponds to optimum process.

Optimization Example

For example, the optimum technology of cooling, which provides:
 -a minimum deviation from given cooling rate in all volume a product,
 -non-exiding of determined temperature difference between a most cold and hottest point of a product,
 -absence of residual stresses.
Was choised with the help of such criterion:

$$c = \int_0^{\tau_k} \left\{ b_1 \int_\Omega \left(f^* - \frac{\partial t}{\partial \tau} \right)^2 d\Omega + b_2 \int_\Omega \left(\Delta t^* - t_{max} + t_{min} \right)^2 d\Omega + b_3 \int_\Omega \left(\sigma_E - \sigma \right)^2 d\Omega \right\} d\tau, \qquad (7)$$

Ω - considered geometrical area,

τ_k - time of cooling,

$t = t(x,y)$ and $\sigma = \sigma(x,y)$ - temperature and stresses in area Ω,

t_{max}, t_{min} - maximum and minimum temperature of a body at the moment of a time τ,

σ_E - limit of product material elasticity,

f^* - optimum rate of cooling,

Δt^* - allowable difference of temperatures,

b_1, b_2, b_3 - coefficientes, which account subjectiv importance of each factors.

Figure 8 shows the dependence of criterion C to speed of a air flow, which was created by a fan, for product cooling. The minimum of curve is seen well.

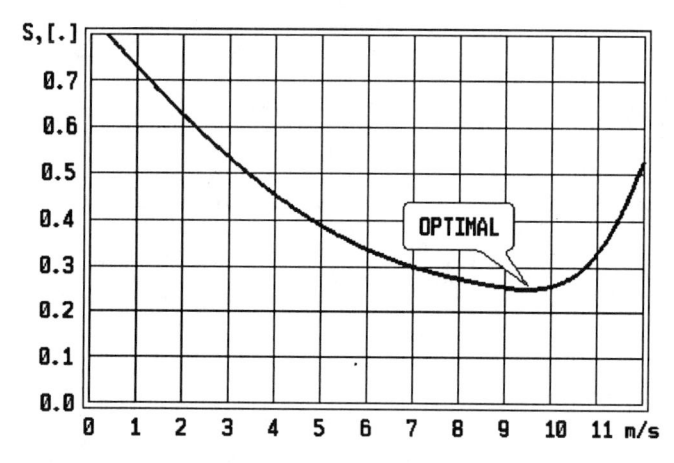

Figure 8. - Dependence of criterion C from air speed flow

Development of method

The developed method of tecnology selection permits to decide some non-standard tasks too.
figure 9 represents the   residual radial stresses in a disk. These stresses are created so, that at rotation of a disk a total level of stresses, connected with centrifugal forces, is reduced.

figure 10 shows distribution of the axial stresses in a valve, the stem of which should be in agressive abrasive environment. The reason of valve stem fragility in such conditions are microcrack on its surface. Such microcracks become the centers of destructions. If recisdual compression stresses are created on the stem surface, microcracks don`t became deeper. The valve stem surface acquires a resitance to chemical and mechanical destrutures. Resistance of such valve is considerable higher.

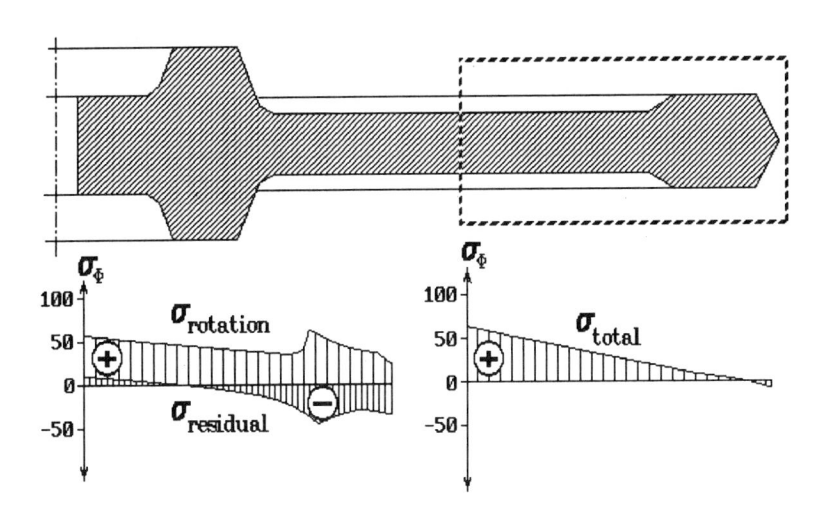

Figure 9. - Distribution of residual stresses, which reduces mechanical load at fast rotation of a disk

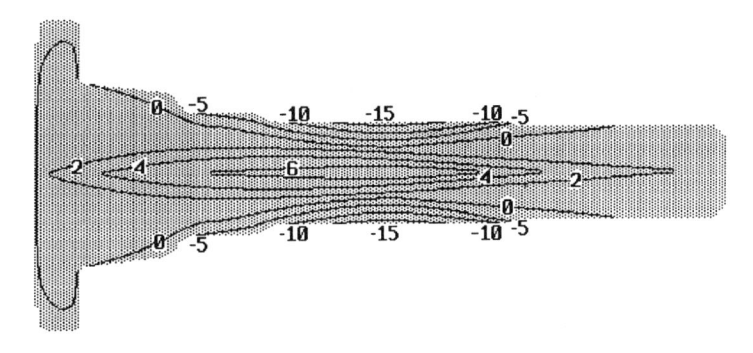

Figure 10. - Distribution of residual stresses in a valve, which increase it wear- resistanse

PROCESSING, MICROSTRUCTURE, AND PROPERTIES OF ß-CEZ

J.O. Peters[*], G. Lütjering[*], M. Koren[**], H. Puschnik[**], and R.R. Boyer[***]

[*]Technical University Hamburg-Harburg, Germany
[**]Böhler Schmiedetechnik GmbH, Kapfenberg, Austria
[***]Boeing Materials Technology, Seattle, USA

Abstract

Two distinctly different microstructures were established in ß-CEZ material by different processing routes, a bi-modal microstructure by conventional α+ß processing and a necklace microstructure by through-transus forging. Comparing the mechanical properties of the two microstructures at the same yield stress level of 1200 MPa, the bi-modal microstructure exhibited better HCF and LCF properties whereas the necklace microstructure showed a higher fracture toughness value. With respect to fatigue crack propagation, small surface cracks (microcracks) propagated slower in the bi-modal microstructure whereas for the large cracks (macrocracks) the propagation rates were similar in both microstructures. The results on crack propagation can be explained by taking into account the differences in the crack front profiles of the two microstructures.

Introduction

The ß-CEZ alloy, recently developed by CEZUS, France, shows potential for application at high strength levels [1-3]. Especially the so-called necklace type of microstructure exhibits a very good combination of strength and fracture toughness [3-6]. This necklace microstructure can be produced by through-transus deformation resulting in round α particles at ß grain boundaries instead of a continuous α-layer [5-8].

In a recent investigation [9], the high fracture toughness values of the necklace microstructure reported in the literature were only obtained in the longitudinal direction of forgings, whereas especially in the short transverse direction much lower fracture toughness values were obtained. It was therefore suggested that the necklace type of microstructure might be useful for application in thin plates or sheets for which the properties in the short transverse direction are not important. So-called bi-modal microstructures, consisting of equiaxed primary α particles in a fine grained ß matrix, exhibited also fairly low but isotropic fracture toughness values. The advantage of the bi-modal microstructure is that it can be produced fairly easily by conventional α+ß processing [9].

The main purpose of the present work was to evaluate in addition to the tensile and fracture toughness properties the fatigue behavior (S-N curves, da/dN-ΔK curves for small and large cracks) for the two microstructures (bi-modal and necklace). Some fatigue properties are reported in the literature for the necklace type of microstructures [10,11] but none for the bi-modal microstructure.

Experimental Procedure

The ß-CEZ material produced by CEZUS, France, had the composition (wt.-%) Ti-5.0Al-1.9Sn-4.5Zr-3.9Mo-2.2Cr-1.1Fe-0.1 O and a ß-transus of about 900°C. The material was delivered by CEZUS in an α+ß forged condition as round bar with a diameter of 245 mm. The microstructure of this round bar in the as-received condition contained a homogeneous distribution of equiaxed primary α particles (Figure 1). Two different types of thermomechanical treatments were used to generate the two microstructures of interest (bi-modal and necklace).

For the bi-modal microstructure blocks of the as-received material with a thickness of about 30 mm (40x40x30 mm) were rolled in the α+ß region at 850°C with a deformation degree of $\varphi = 1.3$ followed by air-cooling. For the necklace microstructure a 45 mm thick slab was produced by forging. The 245 mm round starting bar was first α+ß forged at 850°C to a diameter of 115 mm using an open die hammer. This 115 mm round bar was annealed in the ß region at 920°C and then forged through the ß-transus by flattening into a slab of 45 mm thickness on an open die hammer. The deformation degree φ was 0.9 and the finishing temperature was about 850°C. After forging the slab was air-cooled. After processing, specimens blanks were heat treated in the α+ß phase field. For the bi-modal microstructure the blanks were first heated to 870°C for 1 hour followed by cooling at a rate of 1.7 K s^{-1} (100°C/min). The rest of the heat treatment was identical for both microstructures: Annealing at 820°C for 1 hour followed by cooling at a rate of 1.7 K s^{-1} (100°C/min) and aging at 580°C for 8 hours followed by air-cooling.

The microstructure was investigated by light microscopy (LM) and by transmission electron microscopy (TEM). Tensile tests were performed using an initial strain rate of 8×10^{-4} s^{-1}. Fracture toughness tests according to ASTM E-399 were performed on 3-point bend specimens (thickness 8 mm). S-N curves were measured using electrolytically polished hourglass shaped rotating beam specimens at a frequency of about 50 Hz. The propagation rates of small surface cracks (microcracks) at a stress amplitude of 750 MPa were measured using LM and interrupting the fatigue test for each measurement.

The propagation rates of large cracks (macrocracks) were measured on CT-specimens (thickness 8 mm) using a stress ratio of $R = 0.1$ and a frequency of 30 Hz. The fracture surfaces were investigated by scanning electron microscopy (SEM) and the crack front profiles by sections perpendicular to the crack propagation direction using LM. For the mechanical tests the stress direction was parallel to the rolling direction or to the major forging direction (L-direction). The crack propagation direction for the macrocracks (3-point bend specimens and CT-specimens) was the LT-diretion.

Results and Discussion

1. Microstructure

The bi-modal microstructure is shown in Figure 2. As compared to the as-received starting structure (Figure 1) the volume fraction of α_p is reduced by the annealing treatment of 1h at 870°C after the rolling process. This annealing treatment served also as a recrystallization treatment for the ß phase. The resulting average size of the equiaxed ß grains which is determined by the distance between α_p was about 40 µm. During the subsequent second annealing treatment at 820°C for 1 hour coarse α plates precipitated within the ß grains (Figure 2). At some ß grain boundaries continuous α layers developed.

For the so-called necklace type of microstructure the relatively large equiaxed ß grains (size about 450 µm) became pan-cake shaped during the through-transus deformation process. The through-transus deformation parameters were controlled in such a way, that α precipitated at the ß grain boundaries during the deformation process decorating the pan-cake shaped grain boundaries by round α particles (necklace structure), but that the deformation time was short enough to avoid the precipitation of round α particles within the ß grains [9]. This necklace type of microstructure can best be pictured after the deformation process (Figure 3). Using a deformation degree of $\varphi = 0.9$ the dimensions of the pan-cake ß grains were about 1000 x 650 x 160 µm. The forged plate used in this investigation for the mechanical tests exhibited small recrystallized ß grains at some of the elongated ß grain boundaries (Figure 4). The coarse α plates formed during the annealing treatment of 1 hour at 820°C reached in some cases a length of 100 µm or more for this necklace type of microstructure (Figure 4).

The fine α platelets formed during the aging treatment of 8 hours at 580°C were similar for both microstructures (Figure 5) because of the identical final annealing treatment at 820°C and the identical cooling rate of 1.7 K s^{-1}.

2. Tensile Properties

The results of the tensile tests are shown in Table 1. It can be seen that the yield stress $\sigma_{0.2}$ was about the same for the bi-modal and the necklace microstructure because the yield stress is determined mainly by the fine α platelets from the aging treatment. On the other hand, the tensile ductility was much higher for the bi-modal microstructure (RA = 34 %) as compared to the necklace microstructure. The reason for the superior tensile ductility is the small equiaxed ß grain size of about 40 µm for the bimodal microstructure. The lower ductilitiy (16 %) of the necklace microstructure is a result of the partial fracture along ß grain boundaries observed for this microstructure.

3. Fracture Toughness

The fracture toughness values measured for the two microstructures are also listed in Table 1. It can be seen that the bi-modal microstructure exhibited a fracture toughness of 37 MPa m^{½}, compared to 68 MPa m^{½} for the necklace microstructure.

Table 1: Tensile and Fracture Toughness Properties

Structure	$\sigma_{0.2}$ MPa	UTS MPa	σ_F MPa	El. %	RA %	K_{Ic} MPa m^{½}
Bi-Modal	1200	1275	1660	13	34	37
Necklace	1190	1275	1480	10	16	68

Taking the tensile ductility as the primary parameter determining fracture toughness the necklace microstructure should show lower K_{Ic} values than the bi-modal structure. It is obvious from Table 1 that besides tensile ductility a second parameter has to be effective explaining the high fracture toughness of the necklace microstructure. Figure 6 shows the fracture surfaces at the transition from pre-cracking to unstable fracture for the bi-modal structure and the necklace structure and Figure 7 the corresponding crack front profiles in the region of unstable fracture. It can be seen that the bi-modal microstructure exhibited a relatively smooth fracture surface (Figure 6a) and that the roughness of the crack front profile (Figure 7a) was limited by the small ß grain size to a maximum value of about 40 µm. On the other hand, the fracture surface of the necklace structure (Figure 6b) was very serrated and the roughness of the crack front profile (Figure 7b) was very pronounced. Apparently, the crack tried to follow the ß grain boundaries which, due to the pancake shaped grains, were aligned predominantly parallel to the stress axis and therefore very unfavorably for crack propagation inducing steps in the crack front profile as large as 300 µm. This serrated crack front profile hindering crack extension can therefore explain the high fracture toughness value of 68 MPa m^{½} (Table 1) for the necklace microstructure.

4. S-N Curves

The S-N curves from rotating beam tests are shown in Figure 8. It can be seen that the 10^7 cycles HCF strength of the bi-modal structure was about 625 MPa, about 50 MPa higher than the value for the necklace structure tested in L-direction (575 MPa) although the yield stress values were about the same (Table 1). This higher resistance against crack nucleation can be explained by the smaller maximum slip length of the bi-modal microstructure due to the smaller ß grain size (40 µm) as compared to the larger dimensions of the pancake grain structure of the necklace microstructure even in the thickness direction (about 160 µm). Examples for crack nucleation sites are shown in Figure 9. It can be seen that the crack in the bi-modal structure was transcrystalline and was limited in size by the ß grain size (Figure 9a) whereas the crack in the necklace structure developed along a fairly coarse α plate with a length of about 80 µm. The difference in S-N curves between the two microstructures was maintained up to high stress amplitudes in the LCF regime in which in addition to the resistance against crack nucleation the resistance against crack propagation becomes important.

5. Microcrack Propagation

The propagation rates of the small surface cracks shown in Figure 9 were measured for both microstructures at a stress amplitude of 750 MPa. The results are shown in Figure 10 as da/dN-ΔK plots. In addition to the ΔK axis, the surface crack length 2c is shown in Figure 10. It can be seen that the microcracks are propagating much faster in the necklace microstructure as compared to the bi-modal microstructure. This can be explained by the much smaller ß grain size of the bi-modal microstructure, which is also the reason for the higher ductility as compared to the necklace microstructure. The higher number of grain boundaries serving as strong obstacles for the advancing crack is retarding fatigue crack propagation of microcracks for the bi-modal structure.

6. Macrocrack Propagation

The propagation behavior of large cracks (macrocracks) was measured on CT-specimens with a specimen width (crack front length) of 8 mm. The resulting da/dN-ΔK curves are also shown in Figure 10 for the two different microstructures. It can be seen that the macrocracks propagated at a much slower rate as compared to the microcracks and that there was no measurable difference in propagation rates of the macrocracks for the two microstructures. That means that increasing the crack front length from the microcrack level to more than a few millimeters (8 mm for the CT-specimens) had a stronger retardation effect on fatigue crack propagation for the necklace microstructure as compared to the bi-modal structure.

The fracture surfaces of the macrocracks in the threshold region are shown in Figure 11 for the two microstructures and the corresponding crack front profiles in Figure 12. It can be seen that even in this region where the plastic zone size ahead of the crack tip is relatively small the necklace microstructure exhibited a rougher crack front geometry than the bi-modal microstructure due to the coarser α plate structure. The da/dN-ΔK curves for the macrocracks of the two microstructures deviated from each other in the fast crack propagation regime in agreement with the different fracture toughness values observed for the two microstructures (Table 1). It should be pointed out that in this region of the da/dN-ΔK curves the roughness of the fracture surfaces and the crack front profiles developed the pronounced differences already discussed for the fracture toughness tests (Figures 6 and 7).

Since the fatigue crack propagation curves of the macrocracks were measured at the low stress ratio of R = 0.1, crack closure may occur. The crack closure levels were determined using back-face strain measurements and an evaluation method described elsewhere [12]. A relatively small crack closure effect of about 11 % was measured for both microstructures in the threshold region, thereby changing the R value from 0.1 to an effective R value of 0.21 and reducing the ΔK values by a factor of 1.1 to the corresponding ΔK_{eff} values.

Conclusions

Comparing for the ß-CEZ alloy the fatigue and fracture properties of the two quite different microstructures (bi-modal and necklace) at the same yield stress level of about 1200 MPa the following conclusions can be drawn:

- The resistance against fatigue crack nucleation (HCF strength) was higher for the bi-modal microstructure due to the much smaller ß grain size.
- For the same reason (smaller ß grain size), the resistance against fatigue crack propagation of small surface cracks (microcracks) was higher for the bi-modal microstructure leading in combination with the higher resistance against crack nucleation also to a higher LCF strength as compared to the necklace microstructure.
- The resistance against fatigue crack propagation of macrocracks, measured on CT-specimens with a thickness of 8 mm, was about the same for both microstructures, that means that with increasing crack front length the necklace microstructure exhibited a stronger retardation effect due to the development of a rougher crack front profile even for a relatively small plastic zone size ahead of the crack tip.
- The difference in crack front profile became very pronounced for a large plastic zone size (fracture toughness tests) resulting in a much higher K_{Ic} value for the necklace microstructure (68 MPa m$^{1/2}$) as compared to the bi-modal microstructure (37 MPa m$^{1/2}$).

References

1. B. Champin, B. Prandi, P.-E. Mosser, and Y. Honorat, Proc. 1991 AAAF Meeting, (1991), 55-76.
2. Y. Combres and B. Champin, Beta Titanium Alloys in the 1990's, TMS, (1993), 27-38.
3. Y. Combres and B. Champin, Beta Titanium Alloys in the 1990's, TMS, (1993), 477-484.
4. B. Prandi, J.-F. Wadier, F. Schwartz, P.E. Mosser, and A. Vassel, Titanium 1990, Products and Applications, TDA, (1990), 150-159.
5. D. Grandmagne, Y. Combres, and D. Eylon, Beta Titanium Alloys in the 1990's, TMS, (1993), 227-236.
6. Y. Combres and B. Champin, Advanced Materials, SAMPE, (1993), 1744-1749.
7. B. Champin and B. Prandi, European Patent Office, EP 0 514 293 Al, (1992).
8. Y. Combres and B. Champin, Beta Titanium Alloys, Editions de la Revue de Métallurgie, (1994), 319-325.
9. J.O. Peters, G. Lütjering, M. Koren, H. Puschnik, and R.R. Boyer, Symposium-Metallurgy and Technology of Ti-Alloys, 4th Japan Intern. SAMPE Conf., Tokyo (1995), Materials Science and Eng. A, Elsevier (1996).
10. H.G. Suzuki, W.J. Porter, and D. Eylon, Metallurgy and Technology of Practical Titanium Alloys, TMS, (1994), 119-126.
11. E. Awadé, J. Mendez, and J.M. Rongvaux, Beta Titanium Alloys, Editions de la Revue de Métallurgie, (1994), 143-150.
12. H. Hargarter, G. Lütjering, J. Becker, and G. Fischer, Aluminum Alloys, ICAA4, Vol.II, Georgia Inst. of Techn., Atlanta, (1994), 420-427.

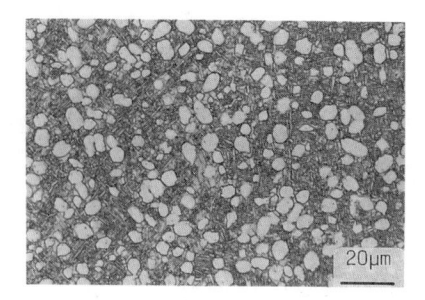

Figure 1: As-Received Structure (LM)

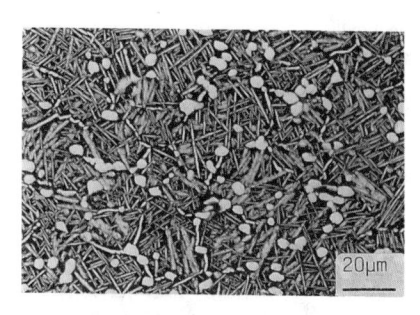

Figure 2: Bi-Modal Structure (LM)

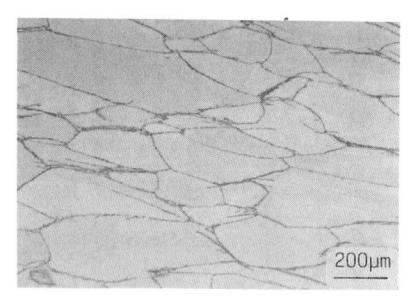

Figure 3: As-Forged Necklace Structure (LM)

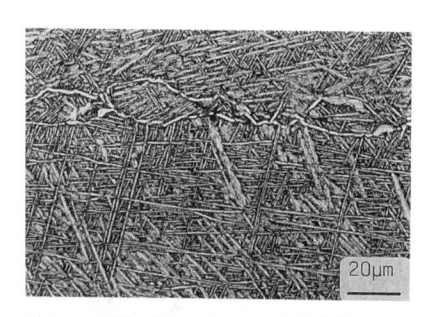

Figure 4: Necklace Structure (LM)

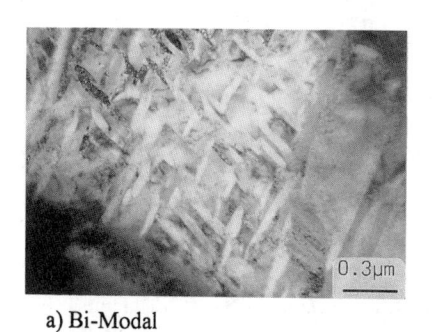

a) Bi-Modal 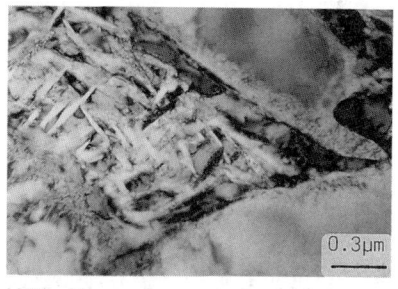 b) Necklace
Figure 5: Microstructures (TEM)

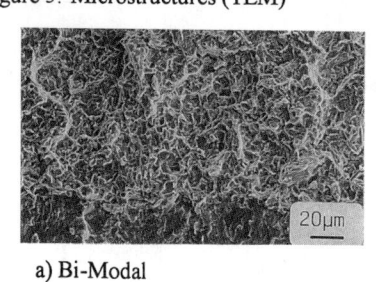

a) Bi-Modal 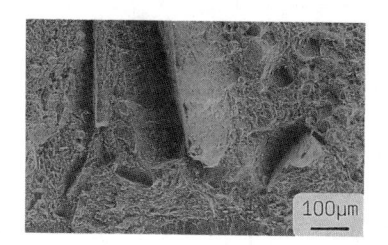 b) Necklace
Figure 6: Fracture Surfaces (SEM), K_{Ic} Specimens

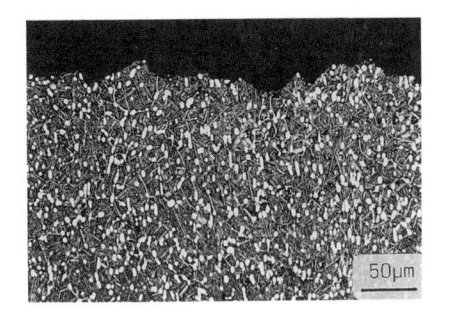

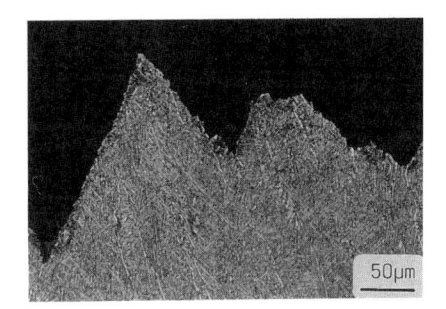

a) Bi-Modal

b) Necklace

Figure 7: Crack Front Profiles (LM), K_{Ic} Specimens

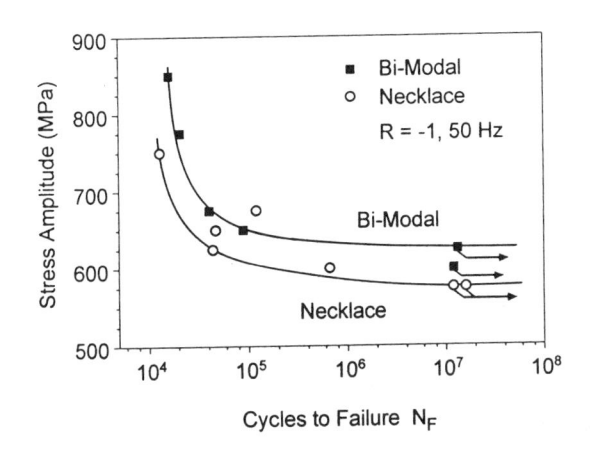

Figure 8: S-N Curves

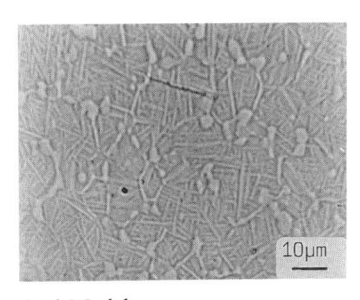

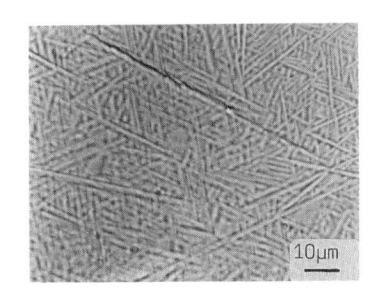

a) Bi-Modal

b) Necklace

Figure 9: Fatigue Crack Nucleation (LM)

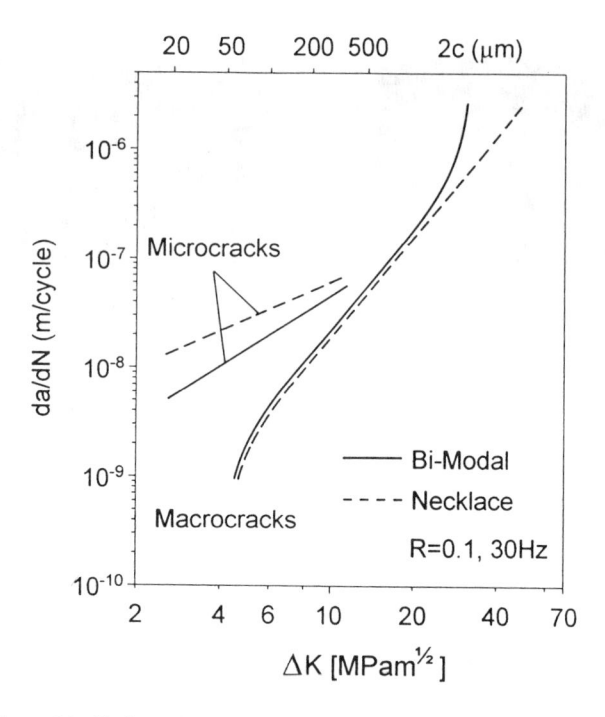

Figure 10: Fatigue Crack Propagation of Micro- and Macrocracks

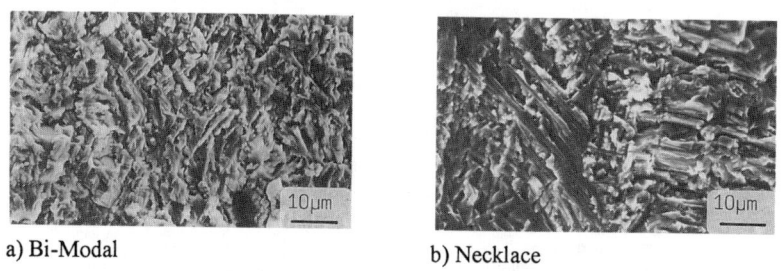

a) Bi-Modal b) Necklace
Figure 11: Fracture Surfaces (SEM), da/dN-ΔK Specimens

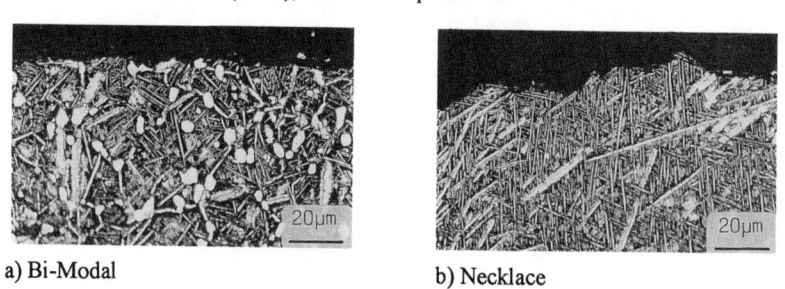

a) Bi-Modal b) Necklace
Figure 12: Crack Front Profiles (LM), da/dN-ΔK Specimens

PROCESSING OF NANOSTRUCTURED GAMMA TiAl BY MECHANICAL ALLOYING AND HOT ISOSTATIC PRESSING

F.H. Froes[1], C. Suryanarayana[1], N. Srisukhumbowornchai[1], X. Chen[1],
D.K. Mukhopadhyay[1], M.L. Öveçoğlu[2], K. Brand[3], and J. Hebeisen[4]
[1]Institute for Materials and Advanced Processes (IMAP)
University of Idaho, Moscow, ID 83844-3026
[2]Istanbul Technical University, Faculty of Chemistry & Metallurgy
Maslak 80626, Istanbul, TURKEY
[3]Institute für Werkstoffwissenschaft Technische Universität
Dresden, 01062 Dresden, GERMANY
[4]IMT, Inc, 155 River Street, Andover, MA 01810-5-85

Abstract

This paper discusses the processing of nanostructured gamma TiAl by mechanical alloying (MA) and hot isostatic pressing (HIP'ing). The extremely small size (≤100 nm) of nanograined materials makes them inherently unstable and susceptible to grain growth on elevated temperature exposure. The present paper will report on a study of the compaction of amorphous gamma (TiAl) titanium aluminides Ti-55Al and Ti-47.5Al-3Cr (at.%) produced as powder by mechanical alloying (MA). Compaction was achieved by HIP'ing at temperatures ranging from 725°C to 975°C and pressures from 105 to 310 MPa, to full density. A fine grain size was present in the HIP'd material, for example in the Ti-47.5Al-3Cr alloy the gamma phase has a grain size of about 25-50nm after HIP'ing at 725°C and 100-200nm at 975°C. The subsequent grain growth behavior was studied and will also be discussed.

Introduction

In recent years it has been clearly demonstrated that powder metallurgy (P/M) approaches can lead to substantial departures from equilibrium allowing novel combinations of constitutional and microstructural effects to be achieved (1,2). This in turn can lead to significant enhancements in mechanical and physical behavior compared to material produced by the more conventional ingot metallurgy (I/M) technique. Amongst the P/M processes allowing this departure from equilibrium are synthesis methods such as rapid solidification, mechanical alloying, and physical vapor deposition. All of these approaches have the additional advantage that they are scaleable processes rather than being laboratory curiosities yielding only gram quantities of material.

However, inherent to the P/M approach is the necessity to compact the powder produced into a solid article allowing use in device or structural components. To accomplish this consolidation step a thermal excursion is required, with a loss in the novel constitutional and microstructural effects being directly related to the extent of the thermal exposure. Thus constitutional effects such as novel phases (microcrystalline or amorphous) and extension of solubility limits can be lost during an elevated temperature process. Equally, microstructural refinement will be removed as features such as second phase particles and grains grow with increasing temperature and time. In general, as the time-temperature exposure increases, the desirable features built-in to the material by the "far from equilibrium" are removed (Fig. 1) (3); recognizing that if these features are modified in a controlled fashion they can still give novel characteristics and enhanced behavior in the compacted article.

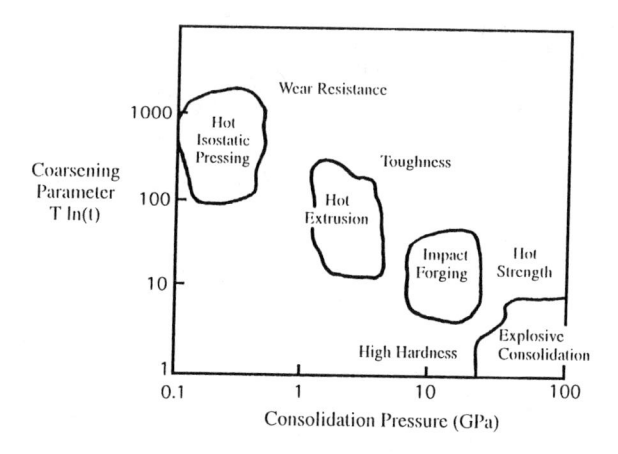

Fig. 1. Effect of time-temperature exposure on non-equilibrium material.

Nanostructured Materials

Nanostructured materials are materials with at least one dimension in the nanometer range (1nm = 10^{-9}m), generally $\leq$ 100 nm. Because of their novel combinations of mechanical, physical, and magnetic properties they have received considerable attention in the past few years (2,4,5). The nanostructures can be one-dimensional (layered), two-dimensional (fibrous) or three-dimensional (crystallites) (5). However, the vast majority of work to date has been on the third type which is generally produced using a P/M approach; and this paper will only consider this type of nanostructured material.

A schematic representation of a nanostructured crystallite material is shown in Fig. 2 (4). The large fraction of atoms located in the grain boundary regions results in novel processing possibilities and enhanced combinations of mechanical, physical and magnetic behavior compared to material with a more conventional grain size (> 1µm). The extremely fine grain size of nanostructured material offers an intriguing avenue for exploration for ductility enhancement in the normally brittle intermetallics, recognizing that creep performance is likely to be degraded, unless creative grain boundary engineering can be employed.

The goal of the present work was to define whether HIP'ing allowed retention of the nanometer-sized grains (6,7), and to evaluate subsequent grain growth behavior.

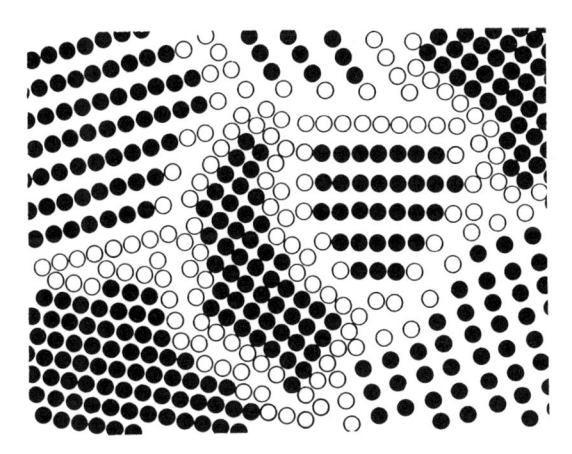

Fig. 2. Schematic representation of a nanostructured material. The black circles represent atoms in normal lattice positions within the grains; the white circles indicate atoms that are "relaxed" in grain boundary regions (courtesy H. Gleiter).

Hot Isostatic Pressing of Nanograined Gamma TiAl

Preliminary work designed to retain nanograins in MA'd Ti-24Al-11Nb (at.%) after consolidation was conducted using dynamic explosive techniques (8). The compacts produced exhibited less than 100% density, with the presence of intra- and interparticle porosity. When very high consolidation pressures were used, by increasing the impact velocity using a larger amount of explosive charge, evidence of melt pores was detected. A detailed evaluation of the microstructure revealed that this technique retained small nanostructured grains about 15nm in diameter. Thus while the goal of retaining the nanograins was achieved, less than a fully dense compact was obtained.

In the present work, mechanical alloying (Fig. 3), which results in a grain size decrease (Fig. 4), was used to produce powder with a grain size in the nanometer range.

Although Fig. 1 predicts that the time-temperature exposure characteristic of the HIP process is considerably more extensive than that for dynamic techniques, it was decided to investigate the potential of retaining nanograins by this compaction method, while achieving full density. Work on gamma TiAl powder with a conventional grain size indicated that a

HIP temperature in excess of 1100°C was required to achieve full density (9). In the present work two gamma TiAl compositions were investigated - Ti-55Al(at.%) and Ti-47.5Al-3Cr(at.%)

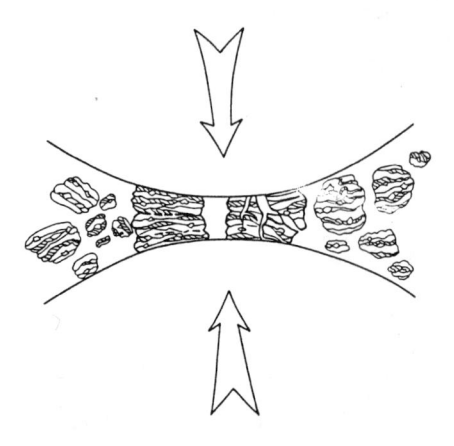

Fig. 3. Schematic of mechanical alloying process.

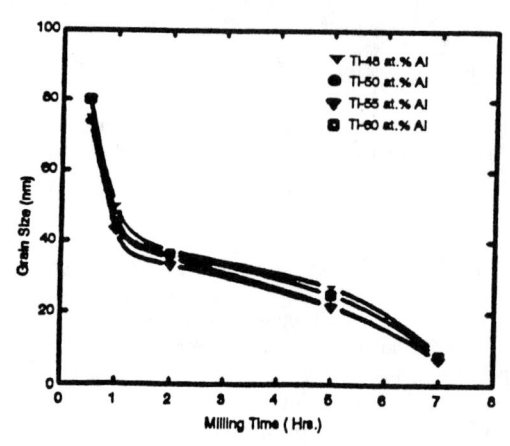

Fig. 4. Variation of grain size as a function of milling time in mechanically alloyed Ti-Al (gamma) alloys.

Ti-55Al(at%) Alloy. Ti-55 at%Al powder was MA'd in a Szegvari 01-HD attritor for 115h under an argon atmosphere cover using a ball-to-powder ratio of 20:1 and 1wt% stearic acid as a PCA. After MA the powder was in an amorphous state as shown by the broad peak in the X-ray diffraction pattern, Fig. 5. This material was compacted using the HIP parameters shown in Table I.

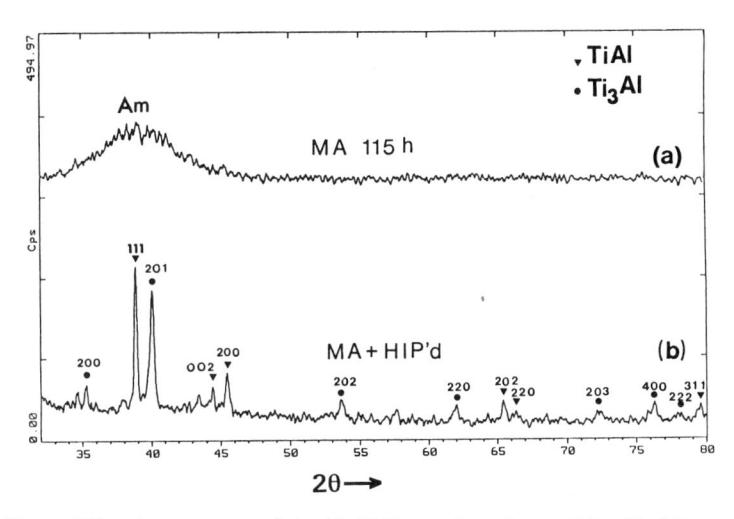

Fig. 5. X-ray diffraction patterns of the Ti-55Al powder mixture (a) milled in an attritor to the amorphous state, and (b) after HIP'ing at 975°C/207MPa/2h showing that the amorphous phase has crystallized to a mixture of γ-TiAl and α_2-Ti$_3$Al phases.

The compacts were not removed from the can and thin slices were cut from the consolidated specimens using a low speed diamond saw. A diamond drill was used to cut 3mm disks which were then dimpled, ion-milled, and thin foils prepared by electropolishing in an electrolyte consisting of 5% H_2SO_4 + 95% methanol cooled to -25°C and at 175mA and 25V. Transmission electron microscopy was conducted using either a Philips EM420T operating at 120kV or a JEOL 2010 operating at 200kV. Grain sizes were determined using a linear intercept method (Table I).

Table I. HIP Consolidated Mechanically Alloyed Ti-55at.%Al Powder

Consolidation Method	Consolidation Parameters (Temp(°C)/Pressure(MPa)/Time(h))	Grain Size, by TEM (nm)
As-Milled	---	Amorphous
HIP	800/207/2	8-90
HIP	900/207/2	30-140
HIP	975/207/2	85 TiAl 220 Ti$_3$Al

Fig. 6 is a high resolution electron micrograph of a Ti-55Al sample HIP'd at 900°C, with one of the grains in good contrast and the lattice planes imaged. The boundary between two grains (marked by arrows) is sharp and devoid of any second-phase particles or contamination.

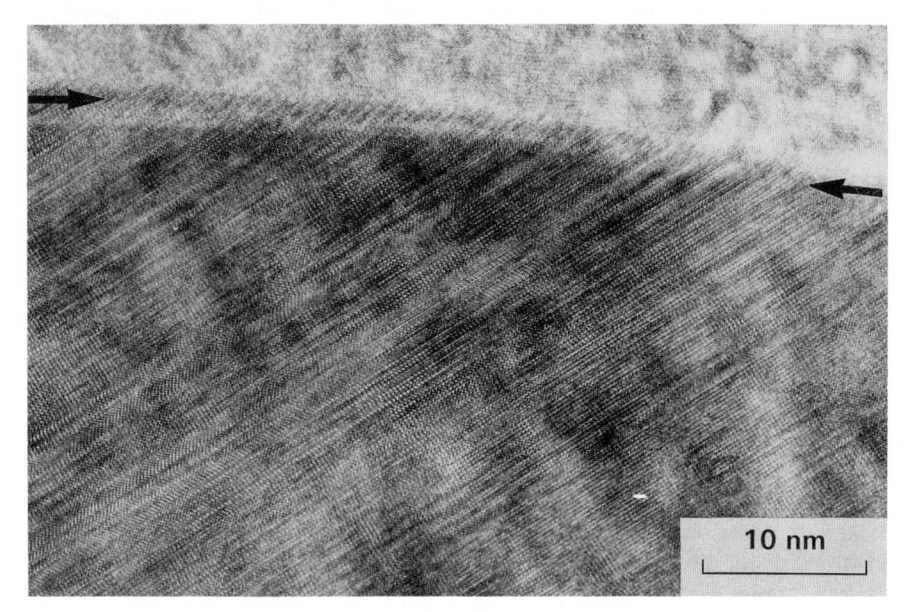

Fig. 6. High resolution electron micrograph of MA'd Ti-55Al after HIP'ing at 900°C/207MPa/2h.

Ti-47.5Al-3Cr(at.%) Alloy. The as-received gas-atomized powder was milled in a SPEX 8000 mill under an argon atmosphere using a 10:1 ball-to-powder weight ratio and 1 wt.% stearic acid as a PCA. Loading and unloading of the powder was conducted in an argon-filled glove box. Milling was continued for 15 hours without opening the vial. At this point loose powder was then discarded, leaving the powder which was clad to the sides and bottom of the vial and the balls in place. This approach was taken to minimize the contamination of the powder by Fe and Cr from the vial and the balls. Subsequent X-ray diffraction patterns of the MA'd powder indicated that it was in an amorphous state, similar to that shown in Fig. 5.

In preparation for HIP consolidation the MA'd powder was vibration packed into commercially pure CP titanium HIP canisters. Filled canisters were packed with a porous plug of CP titanium and leak checked using Helium mass spectrometry. After an elevated temperature out-gas at 300°C, the tubes were hermetically sealed by hot forge crimping.

Four of the sample canisters were HIP'd in a 20cm diameter HIP unit under conditions of 105MPa/4h at 975°C and 1100°C, and five additional canisters were HIP'd in a 41cm diameter pressure unit under conditions of 310 MPa/4h at 725°C, 850°C and 975°C giving a total of 9 conditions (Table II).

Samples suitable for examination in the TEM were prepared in a similar fashion to that described for the Ti-55Al(at.%) material.

Table II. Hot Isostatic Pressing Conditions for Ti-47.5Al-3Ct(at.%)*

Powder Condition	725/310	850/310	975/310	975/105	1100/105
GA[1]	--	✓	✓	✓	✓
MA[2]	✓	✓	✓	✓	✓

*°C/MPa
[1] Gas Atomized
[2] Mechanically Alloyed

Fig. 7 shows the grain size of the MA'd Ti-47.5Al-3Cr alloy after HIP'ing at various temperatures, which have been determined to date; although there are also areas in which larger grains are present probably a result of incomplete MA'ing. An estimate of the grain sizes from a number of electron micrographs shows that the grain sizes of MA'd material vary from as small as 25nm (725°C HIP) up to 200 nm (975°C HIP), Table III.

As expected, the grain size of gas-atomized material is much larger than that of MA'd material HIP'd under the same conditions. There is a strong dependence of the grain size of the MA'd material on the HIP temperature, but the HIP temperature did not have much effect on the grain size of the gas-atomized powder. Thus this work has clearly demonstrated that nanometer sized grains can be retained in gamma compositions after HIP'ing.

In order to determine the density of the HIP'd compacts, polished sections were examined in the optical and scanning electron microscopes. Minimal porosity was observed in the MA'd material, even with a 725°C HIP. In contrast, the GA'd material HIP'd at 850°C exhibited about 3-5% porosity (7).

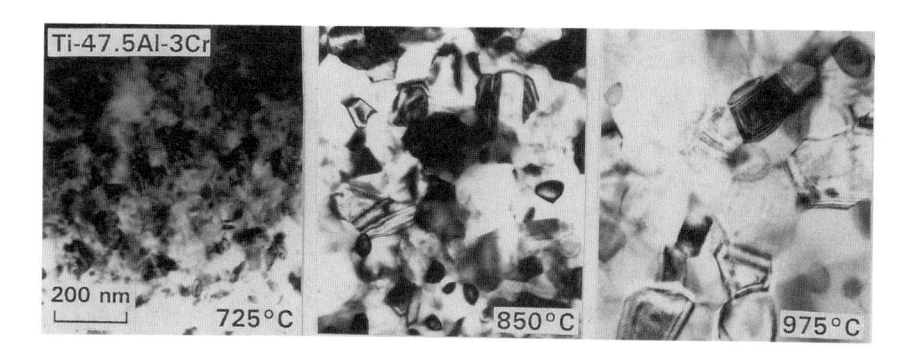

Fig. 7. Grain size of MA'd Ti-47.5Al-3Cr after HIP'ing at various temperatures.

Grain Growth Behavior

Ti-47.5Al-3Cr (at.%) samples HIP'd at 975°C/105MPa/4 hours were heat treated at 975°C to study grain growth behavior. The grain sizes were measured using the transmission electron microscopy, with thin foils prepared after heat treatment.

Table IV and Fig. 8 show that the grain size is about 100 nm in the as-HIP'd condition, and on annealing for 100 and 200 hours, the average grain sizes are 313 and 412 nm, respectively. Studies are in progress to evaluate the activation energy for grain growth in this material.

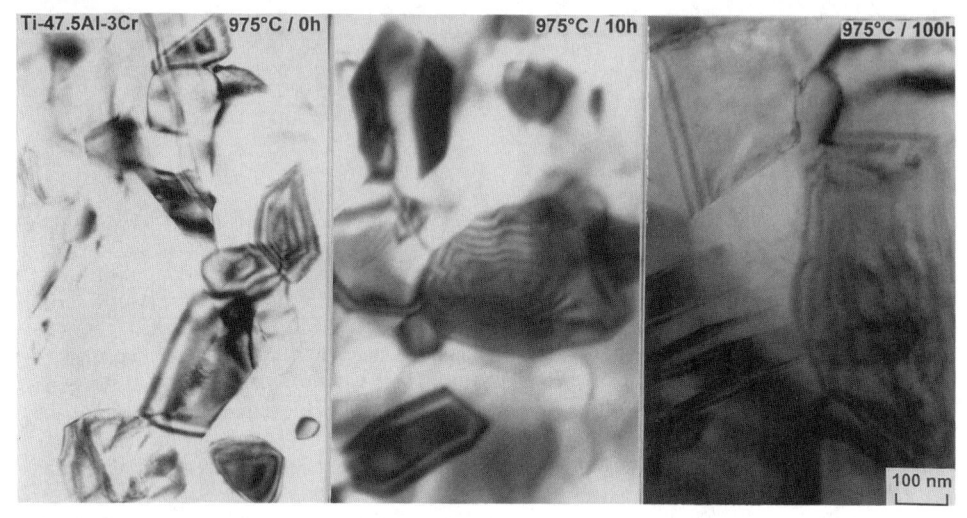

Fig. 8. Grain growth behavior of MA'd Ti-47.5Al-3Cr annealed at 975°C.

Table III. Grain size of MA'd Ti-47.5Al-3Cr after HIP'ing at various temperatures

Condition	Gas-Atomized	SPEX-milled
HIP'd 725°C/310 MPa/4h	---	25-50 nm
HIP'd 850°C/310 MPa/4h	1-2 μm	60-100 nm
HIP'd 975°C/105 MPa/4h	---	100-200 nm
HIP'd 975°C/310 MPa/4h	1-2 μm	---

Table IV. Grain sizes of MA'd Ti-47.5Al-3Cr (at.%) heat treated at 975°C

Time (h)	Grain Size (nm)
0*	100-200
100	313±32
200	412±77

* as - HIP'd at 975°C (Table III).

Conclusions

A study was conducted on the effect of HIP'ing temperature on the grain size of gamma TiAl alloys. It was shown that full density could be achieved at temperatures as low as

725°C at a pressure of 310 MPa while retaining grains in the nanometer range (≤ 100 nm) at temperatures up to 850°C. Thus the HIP temperature can be reduced by as much as 400°C by starting with amorphous MA'd material rather than conventional gas atomized product. The work demonstrates the applicability of the HIP compaction technique to consolidation of nanometered materials. Preliminary grain growth studies at 975°C are also reported.

Acknowledgements

The authors would like to acknowledge useful discussions with Dr. Sarah Dillich and Dr. Garrett Hyde. They would also like to thank D. Zick and P. Tylus for help in the HIP experiments. In addition, the assistance of Mrs. Susan Goetz and Miss Jonica Johnson in typing and formatting the manuscript is greatly appreciated.

References

1. F.H. Froes, C. Suryanarayana, K.C. Russell and C.M. Ward-Close, "Far from Equilibrium Processing of Light Metals," in Novel Techniques in Synthesis and Processing of Advanced Materials, eds. J. Singh and S.M. Copley, TMS, Warrendale, PA (1995), 1-21.

2. F.H. Froes and P.D. Desai, "Recent Developments in Powder Metallurgy Processing Techniques", MIAC/CINDAS, University of Purdue, West Lafayette, IN, Oct. 1994.

3. C.M. Adam, and R.E. Lewis, "High Performance Aluminum Alloys", in Rapidly Solidified Crystalline Alloys, eds. S.K. Das et al., TMS, Warrendale, PA (1985), 157-183.

4. H. Gleiter, "Nanocrystalline Materials", Progress in Materials Science, 33 (1989), 223-315.

5. C. Suryanarayana, "Nanocrystalline Materials", Internat. Mater. Rev., 40 (1995), 41-64.

6. J.P. Hebeisen, Tylus, D. Zick, D.K. Mukhopadhyay, K. Brand, C. Suryanarayana and F.H. Froes, "Hot Isostatic Pressing of Nanometer Sized γ-TiAl Powders," MPIF, ADDA Conference, ed. F.H. Froes, 1995, MPIF, Princeton, NJ, to be published.

7. F.H. Froes, C. Suryanarayana, D.K. Mukhopadhyay, G. Korth and J. Hebeisen, "Compaction of Nanograined Gamma Titanium Aluminide," (1995 Int. Conf. and Exhibition on PM and Particulate Materials, MPIF, Seattle, WA, 1995), Proceedings to be Published.

8. G.E. Korth, "Dynamic Consolidation of Mechanically Alloyed Nanocrystalline Al-Fe and Ti-Al-Nb Alloys", in Advanced Synthesis of Engineered Materials, eds. J. Moore, F. H. Froes and E. Lavernia, ASM INT., Materials Park, OH (1993), 81-86.

9. J.C. Beddoes, W. Wallace and M.C. deMalherbe, "Densification of γ-TiAl Powder by Hot Isostatic Pressing", Int. J. PM 28 (1992), 313-325.

Effects of Heat Treatment on Matrix Microstructure, Interfacial Reactions and Fatigue Crack Growth Resistance of Titanium Metal Matrix Composites

S.V.Sweby & P.Bowen.

School of Metallurgy and Materials / IRC in Materials for high performance applications, The University of Birmingham, Edgbaston, B15 2TT. UK.

Abstract

The fatigue crack growth resistance of two titanium matrix alloys, IMI318 (Ti-6Al-4V (wt.%)). and IMI834 (Ti-5.8Al-4Sn-3.5Zr-0.7Nb-0.5Mo-0.3Si (wt.%)) reinforced with 35% volume fraction of silicon carbide monofilament 'Sigma' fibres has been assessed at room temperature. Two fibre coating systems have been employed. These consist of, a 1µm inner carbon coating surrounded by a 1µm outer boron rich TiB_2 coating (SM1240) and a 4.5µm carbon coating (SM1140+). Work presented here involves the effects of heat treatment on the matrix microstructure, fatigue crack growth resistance and interfacial reactions of IMI318 matrix composites. Also a comparison between the fatigue crack growth resistance of both IMI318 and IMI834 matrix composites is made in the as received condition. A fully transformed beta microstructure may confer improved fatigue crack growth resistance provided that excessive fibre matrix interactions do not occur.

Introduction

Continuous fibre reinforced metal matrix composites (MMCs) have received much attention due to their excellent specific strength and stiffness for potential use in gas turbine aero-engines. Degradation of composite performance has been observed after exposure at, and above, consolidation temperatures (≈900°C) of the titanium MMCs. Improvements in the fatigue crack growth performance of monolithic titanium alloys can be achieved by beta or near beta heat treatments. However, the temperatures required for these heat treatments are above the processing temperatures for titanium MMCs. IMI318 is the standard commercial alloy used by DRA Sigma for production of Titanium MMCs as it is readily processed into large thin foils. Other alloys most notably metastable beta alloys have been investigated[1]. Near alpha alloys such as IMI834 are more difficult to process into thin foils, but it is envisaged this alloy may improve the performance of the composites at temperatures above 300°C.

Experimental

Materials

Two titanium matrix alloys were chosen for this study, IMI318 a standard alpha beta alloy and IMI834, a near alpha alloy normally used at elevated temperatures. The matrix alloys were reinforced with 35% by volume of 100μm diameter silicon carbide 'Sigma' fibres. Two different fibre coating (SM1240 and SM1140+) were evaluated in different 8 ply composite panels. The SM1240 fibre has an 1μm inner carbon coating and an outer 1μm boron rich titanium diboride coating, while the SM1140+ has a single 4.5μm carbon coating. The composites were processed using the foil-fibre-foil technique[2].

Heat Treatment and Materials Characterisation

The principal objectives of the heat treatment studies were to determine the beta transus approach curve for the IMI318 composites, and to assess fibre matrix reactions. All heat treatments were completed in a standard muffle type furnace, and a heating rate of 10°C per minute was used up to the pre determined hold temperature. The specimens (dimensions 75x4.5x1.5mm^3) were held at the pre determined hold temperature for ten minutes and then air cooled. To avoid oxidation during heat treatment the specimens were wrapped in tantalum foil and encapsulated in an evacuated silica tube, and back filled with argon prior to heat treatment. After heat treatment specimens were either sectioned for metallographic analysis or retained for fatigue crack growth tests. Temperature induced changes in phase volume fraction were assessed optically and scanning electron microscopy with micrographs subsequently being analysed using a Leica Quantimet image analyser. Fracture surfaces were studied with the use of an Hitachi S4000 field emission gun scanning electron microscope. Metallographic studies of the extent of fibre matrix interactions and elemental analysis was completed on a JEOL 840A fitted with both energy and wavelength dispersive analysis facilities.

Fatigue Crack Growth Studies

Fatigue crack growth tests were carried out on the IMI318 and IMI834 matrix composites in the as processed conditions. In addition for the IMI318 matrix composites fatigue crack growth tests were performed following heat treatment at 950, 1030 and 1070°C. All tests used specimens of dimensions 75x4.5x1.5mm^3, under three point bending with a half span of 30 mm. All testing was completed in air with an R ratio of 0.1 using an Instron servo-hydraulic test machine with a 10 kN load cell operating at a frequency of 5 Hz, and using a constant load range. Both IMI318 and IMI834 matrix composites were tested at ambient temperature. Notches were cut using a 0.15 mm thick diamond blade to produce an initial unbridged notch, a_0, to width, W, ratio of $a_0/W \approx 0.25$. Crack length was monitored continuously during testing using the direct current potential drop technique with crack growth rates being calculated using an incremental five point secant method. The methodology used in defining crack arrest conditions has been given in detail elsewhere[3]. For the purpose of this work crack arrest is assumed to have occurred when the crack growth rate in the mode I direction drops below 1×10^{-8} mm/cycle.

Results and Discussion.

Microstructural Characterisation

The microstructure of the IMI318 matrix composites in the as processed condition consists of large primary alpha grains with a discontinuous network of retained beta phase around the primary alpha phase boundaries. The heat treatment response (of the matrix) of the composite was similar in nature to monolithic IMI318 in terms of the phases present namely primary alpha and transformed beta products. The effects of microstructural manipulation of the IMI318 matrix composites through thermal exposure has been documented elsewhere[4]. In summary the heat treatment response of the IMI318 matrix material is modified by the presence of the silicon carbide fibre reinforcement through diffusion of the alpha stabilising elements from within the fibre coating layers during thermal exposure. The beta transus curve for the composite (matrix) is shifted to higher temperatures compared to conventionally processed IMI318, and the transus temperature for the composite (matrix) is predicted to be 1060°C +/- 5°C. After thermal exposure at a temperature of 1070°C for ten minutes the IMI318 matrix consists of a relatively coarse grained, fully transformed beta microstructure. At lower temperatures of thermal exposure (900-1000°C) the alpha phase volume fraction was significantly greater in the carbon coated fibre composite, SM1140+. This is consistent with the presence of the outer boron rich coating in the SM1240 acting as a diffusion barrier to the diffusion of carbon, hence the build up of carbon at these lower levels of thermal exposure would be expected to be significantly less. This is consistent with WDS measurements of carbon profiles across the fibre matrix interface and is also supported by the presence of blocky titanium carbides in regions of the matrix remote from the fibres (figure 1)

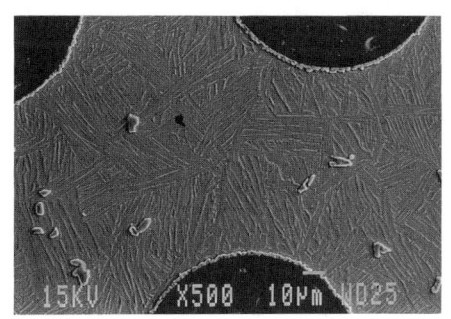

Figure 1, Scanning electron micrograph showing titanium carbides remote from the fibres in SM1140+ fibre reinforced IMI318 after heat treatment at 1070°C

Fibre-Matrix Reactions (IMI318)

Typical scanning electron micrographs illustrating the difference in morphology of the reaction products formed at the interface between the fibre and matrix in the SM1240 and SM1140+ are reproduced in figure 2

In both cases distinct reaction layers were observed, and as the exposure temperature was raised the extent of the reaction zone increased significantly. In all conditions examined, including the as processed condition, interactions between the matrix and the SM1140+ fibre were characterised by a continuous blocky titanium rich carbide layer which extended around the carbon coating layer and into the metal matrix. Kinetic data relating to the effects of temperature on the depletion of the carbon layer is reproduced in figure 3 for both SM1140+ and SM1240 fibres.

Each data point represents approximately thirty data points randomly selected from fibres across the full thickness of the specimen. As indicated a progressive reduction in the carbon layer thickness was observed for both fibre types with increasing temperature. A progressive increase in the reaction layer was also observed as the temperature of thermal exposure was increased. The average thickness of the SM1140+ reaction layer increased from 0.82 μm after exposure at a temperature of 950°C up to approximately 2.90 μm at 1100°C. This corresponded with average reductions in the carbon layer thickness of 0.15 and 0.57 μm respectively. Quantitative data relating to the reduction in carbon layer thickness for the carbon and titanium boride coated SM1240 fibre reinforced IMI318 are presented in figure 3.

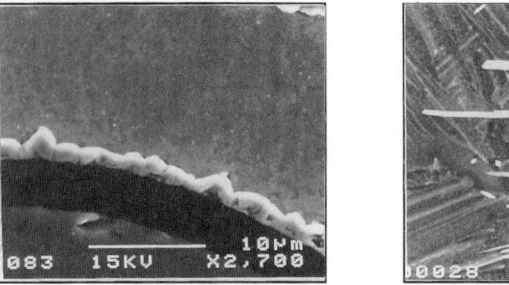

Figure 2(a)   **Figure 2(b)**
Figure 2, SEM micrographs showing the interfacial reaction layers after heat treatment at 1100°C in (a) SM1140+ fibre reinforced IMI318, and (b) SM1240 fibre reinforced IMI318.

A number of detailed[5,6,7] studies have been conducted relating to the reaction layers formed on the SM1240 type of fibre during thermal exposure and a good data base has developed regarding both the kinetics of the fibre / matrix reaction[6] and the morphology[7] of the reaction products. The present results suggest the development of a duplex reaction product, (figure 2); consisting of an inner titanium diboride layer surrounding the carbon layer and extending into the matrix from the diboride layer a network of interconnecting needles. To characterise these reaction products, thermodynamic considerations[8] together with EDS and WDS analysis suggest that these layers consist of a mixture of TiB_2 and TiB titanium borides.

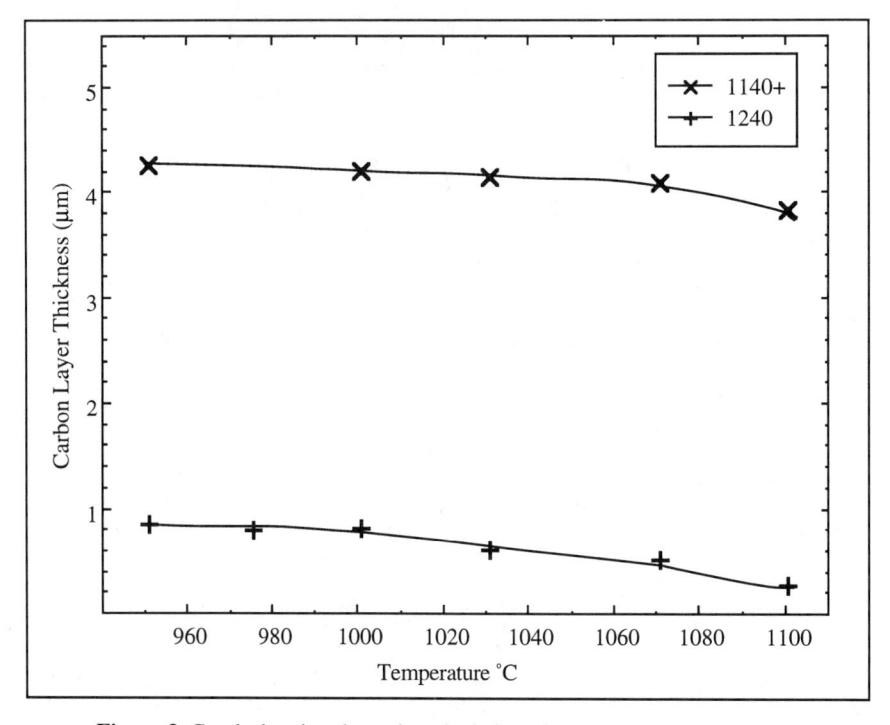

Figure 3, Graph showing the carbon depletion after exposure at temperature

As indicated in (figure 2), the reaction zone was observed to develop primarily by the extension of the outer layer of monatomic TiB needles into the metal matrix. Penetration into the metal matrix increased from approximately 2.25μm after exposure at a temperature of 950°C to in excess of 20μm at a temperature of 1100°C. In contrast the inner TiB_2 layer exhibited very little growth, and reached an average thickness of approximately 1.70μm after exposure at 1100°C.

Fatigue Crack Growth (IMI318)
As Received and After Heat Treatment

Fatigue crack growth studies were conducted on both as received and heat treated IMI318 reinforced with SM1240 and SM1140+ fibres. Typical fatigue crack growth curves, da/dN Vs crack length curves indicating the effect of initial ΔK are reproduced in figure 4. In general at lower levels of initial applied stress intensity factor range result in a decrease in crack growth rate, da/dN, with increasing crack length despite the increase in nominal level of ΔK. This fatigue crack growth response is characteristic of continuous fibre reinforced MMCs and is a consequence of fibre bridging in the wake of the fatigue crack. This will result in a lower driving force (ΔK) for crack growth in the matrix. If load bearing fibres do not fail when the

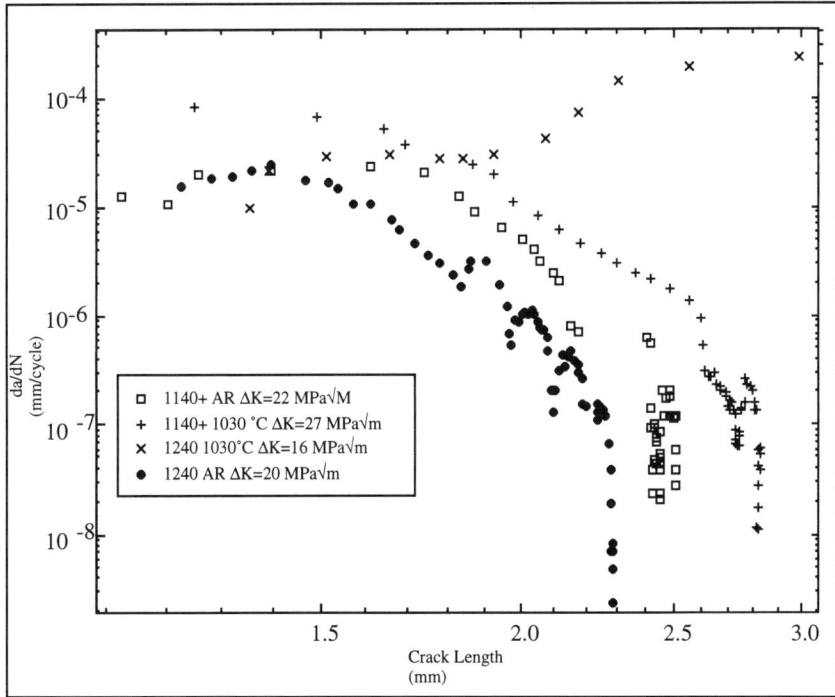

Figure 4, Typical fatigue crack growth resistance curves for continuous fibre reinforced metal matrix composites.

crack grows around them, the fibres will effectively bridge the crack. As the driving force for crack extension is increased larger crack tip opening displacements will result in an increased probability of fibre failure Figure 5 indicates, based on the limiting values of initial ΔK applied values, that for the SM1140+ system a progressive increase in fatigue crack growth resistance is obtained from the as received condition to that obtained after heat treatment at temperatures of 1030 and 1070°C, with crack arrest now being recorded at initial ΔK levels up to 31 MPa√m for the latter heat treatment. In sharp contrast such elevated heat treatment temperatures degrade the fatigue crack growth resistance of the SM1240 system significantly. Now catastrophic failure has been observed at ΔK values of 14 MPa√m in less than 25000 cycles. The initial stress range, Δσ, has been included to figure 5 to give some indication of the possible stress range limits in testpieces containing notches(Δσ values have not included the stress concentration factor due to the notch and are calculated from the net section area at the position of maximum bending moment). They also highlight the effects of heat treatment in reducing or promoting such applied stress range limits to give crack arrest In this series of tests the fatigue crack growth resistance of the SM1140+ system appears to be superior to the SM1240 system in all conditions . The acute differences in the crack growth response of the SM1140+ and SM1240 material following heat treatment at and above temperatures of 1030°C appear to be best interpreted in terms of both the interfacial reaction layer between the fibre and the matrix (figure 2), the extent of matrix modification and earlier fibre failure. A characteristic of the SM1240 material observed was the development of a duplex reaction layer consisting of an inner TiB_2 layer surrounded by a network of interlocking TiB needles extending into the matrix. These needles appear to initiate interfacial cracks, which have been shown to result in matrix crack growth away from the fibre and prior to the arrival of the main fatigue crack growing from the unbridged notch[4]. The main fatigue crack that initiated and grew from the notch is characterised by striated growth. Some further work is in progress to quantify the reasons for the poor fatigue crack growth resistance of the SM1240 system after heat treatment at temperatures of 1030°C and above. Possible factors include: (a) a reduction of fibre strength, (b) an increase in interfacial strength and (c) matrix crack growth from the reaction zone reducing the local matrix load bearing area. Evidence suggests that the improvements in crack growth behaviour observed in the SM1140+ system may be attributed to temperature

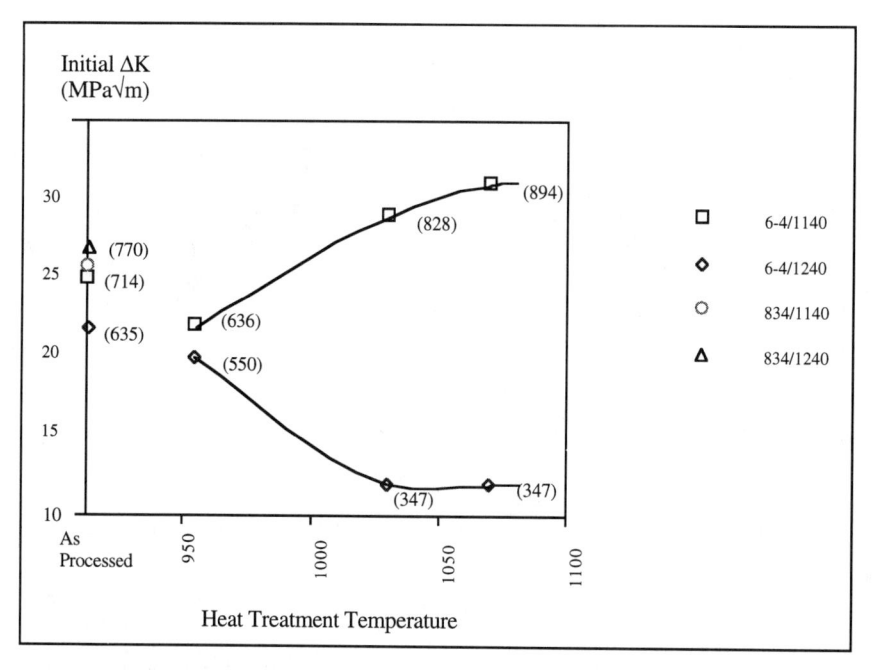

Figure 5, Measurement of crack arrest limits for fatigue crack growth in terms of initial ΔK for IMI318 and IMI834 reinforced with SM1240 and SM1140+ 'Sigma' fibres. Figures in parentheses indicate the initial stress range, Δσ, applied in MPa.

induced microstructural changes. Heat treatments resulting in reductions in the volume fraction of primary alpha and coarsening of the transformed beta would be expected to have a major influence on the crack growth characteristics of the metal matrix. In the present instance heat treatment at a temperature of 1030 and 1070°C were found to promote crack branching and secondary cracking within the metal matrix and at the fibre matrix interface.

Fatigue Crack Growth(IMI318 and IMI834)
As Received Condition.

Fatigue crack growth tests were conducted on the IMI318 and IMI834 matrix composites and were carried out as function of fibre type and initial ΔK. Based on the limiting values of initial ΔK applied, figure 5 shows the transition, in terms of initial ΔK applied, for crack arrest / catastrophic failure for the IMI318 and IMI834 matrix composites reinforced with the SM1140+ and SM1240 coated fibres. The results indicate that there is a significant difference not only in the crack arrest limit values of ΔK applied(or initial stress range) to the different alloys but also in the response to the different fibre types within these alloys. The SM1140+ fibre reinforced IMI318 and IMI834 have very similar crack arrest limits and also fractography revealed a very similar fracture mode consisting of, long 'delamination' lengths. The SM1240 reinforced IMI318 and IMI834, however, show a significant difference in the crack arrest limit values of initial ΔK. These observations are still in the early stages of being analysed, They can be tentatively explained by the reduction in reactivity of the IMI834 to the fibre coating elements of carbon and more importantly boron compared to IMI318. A great deal more work is required to completely understand the differences in the fatigue crack growth crack arrest limits of the IMI318 and IMI834 reinforced composites especially when reinforced with SM1240 fibres.

CONCLUSIONS

Microstructural development of IMI318 reinforced with SM1140+ and SM1240 continuous 'Sigma' fibres are identical in nature to monolithic IMI318 in terms of the phases present, namely primary alpha and transformed beta products. However, quantitative studies have revealed large shifts in the beta transus approach curve for the composite materials with the beta transus temperature increased to 1060°C +/- 5°C.

The morphology of the fibre matrix reaction products after processing and heat treatment, for the IMI318 matrix composites, are dependent critically on the fibre coating elements within each system. The SM1140+ reinforced material was characterised by blocky titanium carbides extending around the carbon coating layers at the interfaces with the matrix. In contrast the SM1240 reinforced material was characterised by the development of a duplex reaction layer, consisting of a compact titanium diboride layer surrounded by a network of interlocking titanium boride needles which extend into the matrix.

SM1140+ reinforced IMI318 composites exhibited superior fatigue crack growth resistance to the SM1240 reinforced IMI318 composites. As the heat treatment temperature was raised to 1030°C the fatigue crack growth performance of the SM1240 reinforced material deteriorated and that of the SM1140+ reinforced material was enhanced. The deterioration in the fatigue crack growth performance of the SM1240 reinforced material after being heat treated at or above a temperature of 1030 °C could be caused by several factors; a reduction in fibre strength, an increase in interfacial strength and matrix crack growth from the reaction zone, and/or a combination of these factors. The fatigue crack growth resistance of the SM1140+ reinforced material when heat treated at or above a temperature of 1030°C appears to be enhanced by the promotion of crack tip deflection and bifurcation within the matrix. In the as received condition the fatigue crack growth resistance of the IMI834 and IMI318 composites reinforced with SM1140+ fibres is very similar. However, there is a marked improvement in the fatigue crack growth resistance of the SM1240 reinforced IMI834 composites when compared with all other composites in the as processed condition.

ACKNOWLEDGEMENTS

The authors would like to thank the Department of Trade and Industry for their support, and the support of Dr M. Winstone, the programme manager, Structural Materials Centre, Defence Research Agency. The work was carried out predominately within the I.R.C. at The University of Birmingham, England, and which is supported by E.S.P.R.C. British Crown Copyright 1996/DERA.

REFERENCES

1) K.M.Fox, PhD Thesis, The University of Birmingham, 1995

2) J.G.Robertson, AGARD Report 796, 'Characterisation of Fibre Reinforced Titanium Matrix Composites', 7, (1993)

3) S.V.Sweby, MPhil Thesis, The University of Birmingham, 1995.

4) S.V.Sweby, A.Dowson & P.Bowen, ICCM10, Metal Matrix Composites, Vol. 2, pp 513-521, 1995

5) D.B.Gundel & J. E. Wawner, Scr. Met., Vol. 23, (1991), pp437-441

6) C.M.Warwick & R.A.Shatwell, B.P.Composites, Report nos, 137,999 & 138, 535.

7) F.Brisset and A.Vassel, "Compatibility of SiC and Al_2O_3 Continuos fibre with Ti Aluminides matrices', 1st Int. Symp. on Structural Intermetallics, Severn Springs, TMS 1993, pp 739-747.

8) J.J.Valencia, Mater. Sci & Eng., Vol. A144, 1991, pp25-36.

FLOW BEHAVIOR OF A MECHANICALLY ALLOYED AND HIPPED NANOCRYSTALLINE γ-TiAl

R. S. Mishra, A. K. Mukherjee
Department of Chemical Engineering and Materials Science
University of California, Davis, CA 95616
and
D. K. Mukhopadhyay, C. Suryanarayana and F. H. Froes
Institute for Materials and Advanced Processes
University of Idaho, Moscow, ID 83844-3026

Abstract

Nanocrystalline γ-TiAl alloy (25-50 nm) was obtained by hot isostatic pressing of mechanically alloyed powders at 998 K and 310 MPa. The flow behavior of the nanocrystalline γ-TiAl was evaluated in compression at 1023-1223 K and in the strain rate range of 10^{-5}-10^{-3} s^{-1}. The stress exponent was ~5-6 at all the test temperatures used in this study. Based on the superplasticity in micron size grained γ-TiAl alloys, a stress exponent of 2 in this temeprature and strain raterange is expected. In spite of the high stress exponent, the nanocrystalline γ-TiAl showed no edge cracking when deformed to a true compressive strain of 1.0 at 998 K and a strain rate of 1×10^{-3} s^{-1}. The flow stresses for nanocrystalline γ-TiAl are significantly lower than a coarse grained γ-TiAl (~10 μm). The possible mechanisms for a grain size dependent flow behavior coupled with high stress exponent will be discussed.

Introduction

Nanocrystalline materials have attracted considerable interest because of the exceptional properties (1-4). Most of these properties relate to the microstructural scaling of the applicable phenomenon. Nanocrystalline microstructure provides opportunity to examine the applicability of microcrystalline mechanisms at a much finer scale. The possibility of low temperature superplasticity in nanocrystalline materials has been dicussed (5-6). This is particulary attractive for high temeprature materials for which the superplastic temperatures are usually high. Gamma titanium intermetallic alloys are potentially attractive for elevated temperature applications (7-9). A number of studies have shown that superplasticity is possible in γ-TiAl alloys (10-20). The superplastic temperature are usually quite high. Figure 1 shows the variation of superplastic temperature with grain size for a number of TiAl alloys. It is important to establish whether the scaling law extends to nanocrystalline γ-TiAl regime or the flow behavior changes. Recently, nanocrystalline γ-TiAl alloys have been synthesized by hot isostatic pressing (HIP) of mechanically alloyed (MA) Ti-47.5 Al-3 Cr (at.%) powders (21-22). The purpose of this study was to evaluate the possibility of observing low temperature superplasticity in this nanocrystalline alloy. By determining the stress exponent for flow, it should be possible to comment on the micromechanism of deformation in a nanocrystalline intermetallic alloy.

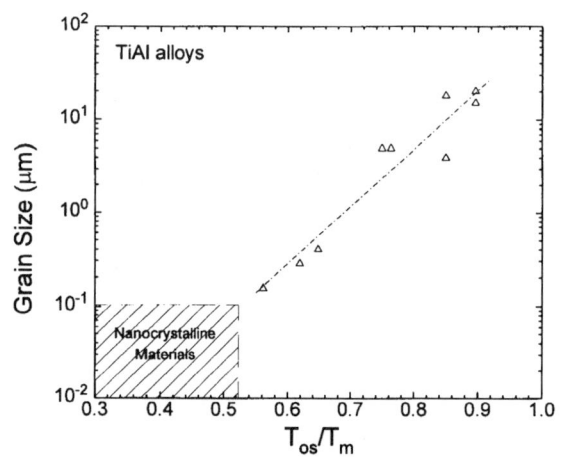

Figure 1. The variation of optimum homologous superplastic temperature with grain size in a number of γ-TiAl alloy (10-20).

Experimental

Two TiAl alloys have been used in this study: (a) a coventionally processed duplex TiAl alloy with a grain size of ~10 μm and (b) a mechanically alloyed & HIP'ed Ti-47.5-3Cr alloy. The processing and microstructural conditions of the wrought TiAl alloy are reported by Schwenker and Kim (23). The as-received gas atomized powder was mechanically milled in a SPEX 8000 mill under a nominal argon atmosphere using a 10:1 ball to powder weight ratio and 1 wt.% stearic acid as a process control agent. Loading and unloading of the powder was conducted in an argon-filled glove box. Milling was continued for 15 h without opening the vial. X-ray diffraction (XRD) patterns of the milled powder indicated that the powder was in an amorphous state (22). The MA powder was vibration packed into commercially pure CP titanium HIP canisters. After outgassing at 573 K, the tubes were HIP'd at 998 K and 310 MPa for 4 h. Compression samples of approximately 3.6 mm x 3.3 mm x 2.5 mm size were cut by a low speed diamond saw. The compression testing was carried at a constant strain rate using a MTS closed loop servo hydraulic test machine. Both single strain rate and strain rate jump tests were used to determine the stress-strain rate behavior. The specimens were coated with BN spray to minimize the friction between SiC compression rods and specimen. Transmission electron microscopy (TEM) and X-ray line broadening (XRLB) techniques were used to evaluate the grain size before and after testing. The density of as compacted specimen was determined by Archimedes' method using a Mettler H51AR balance with resolution of 0.01 mg.

Results and Discussion

Figure 2 shows a bright field TEM micrograph of the MA'd Ti-47.5 Al-3 Cr alloy in the as HIP'd condition. An estimate of the grain size was made from a number of bright and dark field micrographs. The average grain size from TEM micrographs is in the range of 25-50 nm. From this grain size estimate it can be noted that the as consolidated microstructure is indeed nanocrystalline. The XRD pattern of as HIP'd alloy is shown in Figure 3. The presence of a small volume fraction of Ti_3Al phase can be noted. Also, a number of small unidentified peaks can be seen. These peaks are not consistent with the standard peaks of β-phase or Ti_2AlCr phase. The density of the as consolidated specimen was 3.92 g/cc. The as consolidated specimen was polished and examined using optical and scanning electron microscopy for residual porosity. No porosity was detected in the as consolidated specimen. The as consolidated specimen appears to be fully dense, although a direct comparison with the theoretical density is not possible for this alloy because of the unknown volume fraction of second phase. It can be noted that the theoretical density of stoichiometric TiAl is 3.91 g/cc (24).

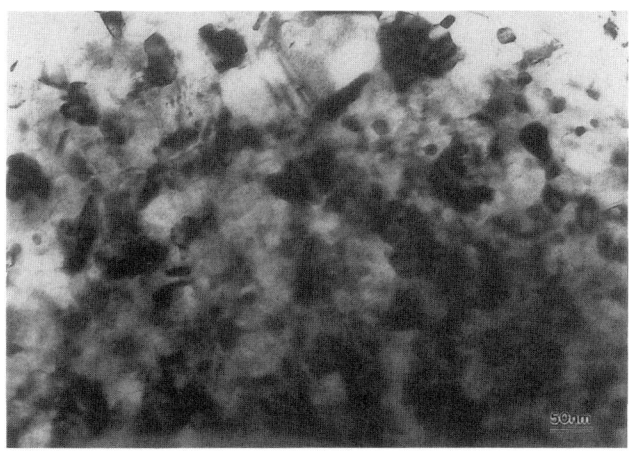

Figure 2. A bright field micrograph of mechanically alloyed and HIP'ed Ti-47.5Al-3Cr alloy.

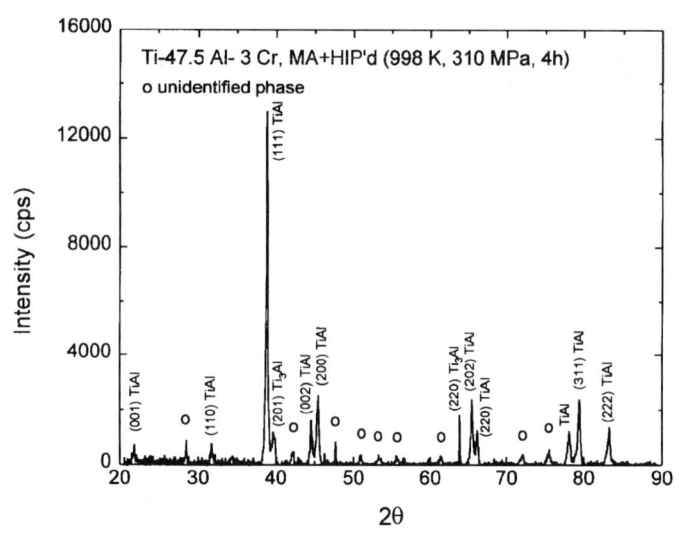

Figure 3. X-ray diffraction of the mechanically alloyed and HIP'ed Ti-47.5Al-3Cr alloy.

The stress-strain behavior at 1×10^{-4} s^{-1} and 1023 K is shown in Figure 4 for both grain sizes. The flow stress for nanocrystalline alloy is stable and after a true compressive strain of ~0.9, no edge cracking was observed. The flow stress for microcrystalline alloy is significantly higher. The stress-strain rate behavior is shown in Figure 5. The stress exponent is ~6 in the strain rate range of 1×10^{-5}-1×10^{-3} s^{-1}. It is important to note that the stress exponent is not 2. It appears that the nanocrystalline microstructure does not result in a shifting of the grain boundary sliding mechanism to lower temperatures in the investigated strain rate range. Recently, Mishra et al. (19) have noted that the superplastic data on various γ-TiAl alloys with grain sizes ranging from 0.4 μm to 20 μm can be represented by a generalized equation

$$\dot{\varepsilon} = 2 \times 10^{12} \left(\frac{D}{d^2}\right) \left(\frac{\sigma}{E}\right)^2 \quad (1)$$

where $\dot{\varepsilon}$ is the strain rate, D is the diffusivity, d is the grain size, σ is the applied stress and E is the Young's modulus. It is interesting to compare the superplastic deformation results on γ-TiAl alloys with different grain sizes. The results of Mishra et al. (19) on three γ-TiAl alloys with ~15-20 μm grain size showed optimum superplasticity at 1553 K and in the strain rate range of 3×10^{-5}-5×10^{-4} s^{-1}. On the other hand the results of Imayev et al. (13) showed a reduction in the superplastic temperatures (1073-1173 K) for fine grained γ-TiAl alloy with 0.4 μm grain size, while the strain rate range (10^{-4}-10^{-3} s^{-1}) was quite similar to other γ-TiAl alloys with coarser grain sizes. Bohn et al. (25) observed n=2 regime in γ-TiAl alloy with grain size of 150-250 nm in compression at 973-1073 K. Based on these results (10-20) it was expected that grain boundary sliding should be a dominant mechanism at the present test conditions.

The variation of flow stress with test temperature in nanocrystalline γ-TiAl alloy is shown in Figure 6. Also, the expected flow stress using equation (1) for 35 nm grain size is plotted. The present flow stresses are higher than the expected flow stresses for grain boundary sliding

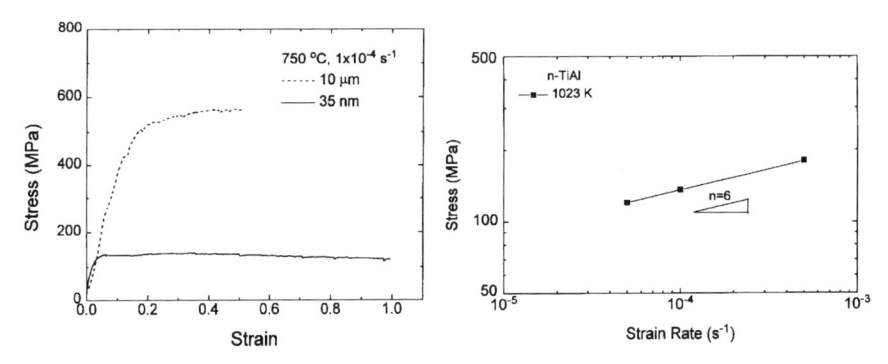

Figure 4. The flow behavior of a microcrystalline and a nanocrystalline γ-TiAl alloys.

Figure 5. The variation of flow stress with strain rate at 1023 K.

mechanism. This suggests two possibilities, (a) the grain boundary sliding mechanism in nanocrystalline γ-TiAl alloy is different from the micron- and submicron-grained γ-TiAl alloys and/or (b) mechanically alloyed γ-TiAl has second phase particles that alter the flow behavior. Further experiments are required to find the optimum strain rate for grain boundary sliding mechanism at 1023-1223 K. It is possible that the optimum superplastic strain rates have shifted to higher values, similar to the observation of high strain rate superplasticity in fine grained aluminum alloys (26).

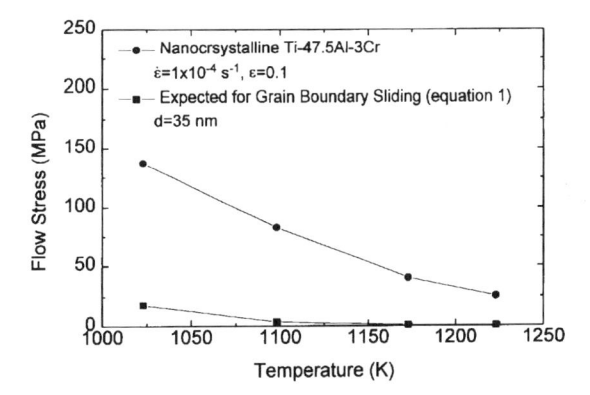

Figure 6. The variation of flow stress with temeprature for the nanocrystalline Ti-47.5Al-3Cr alloy. The expected flow stress for grain boundary sliding mechanism is also plotted using equation (1).

A stress exponent of ~6 means that climb controlled dislocation mechanism is operative. This is interesting because the change of Hall-Petch slope in the nanocrystalline grain size range is attributed to the change in micromechanism of dislocation generation and movement (27,28), although the exact mechanism is not clear. Further experiments, particularly TEM of the deformed specimens is required to understand the micromechanism of dislocation climb controlled process in nanocrystalline TiAl.

Conclusions

1. The stress exponent for deformation of a nanocrystalline Ti-47.5 Al-3 Cr alloy is ~6 in the strain rate range of 10^{-5}-10^{-3} s^{-1} and 1023 K.
2. The flow stresses for mechanically alloyed γ-TiAl are higher than that expected for grain boundary sliding mechanism.

Acknowledgment

This work was partly (at UCD) supported by National Science Foundation through a grant NSF-DMR-93-00217. The authors are grateful to Dr. Y-W. Kim for the micrograined γ-TiAl alloy.

References

1. H. Gleiter, Prog. Mater. Sci., 33, 223 (1990).
2. H. Gleiter, Nanocrystalline Materials, 1, 1 (1992).
3. C. C. Koch, Nanostructural Materials, 2, 109 (1993).
4. J. Karch, R. Birringer and H. Gleiter, Nature 330, 556 (1987).
5. R. W. Siegel, Superplasticity of Metals, Ceramics and Intermetallics, edited by M. J. Mayo, M. Kobayashi and J. Wadsworth, p. 184, MRS, vol. 196, (1991).
6. C. Altstetter, Mechanical Properties and Deformation Behavior of Materials Having Ultra-Fine Microstructures, edited by M. Nastasi, p. 381, Kluwer Academic Publishers (1993).
7. Y-W. Kim, JOM, 41, 24 (July 1989).
8. Y-W. Kim and D. M. Dimiduk, JOM, 43, 40 (August 1991).
9. Y-W. Kim, Acta Metall. Mater., 30, 1121 (1992).
10. N. Masahashi, Y. Mizuhara, M. Matsuo, T. Hanamura, M. Kimura and K. Hashimoto, ISIJ International, 31, 728 (1991).
11. R. M. Imayev, O. K. Kaibyshev and G. A. Salishchev, Acta Metall. Mater., 40, 581 (1992).
12. S. C. Cheng, J. Wolfenstine and O. D. Sherby, Metall. Trans. A, 23A, 1509 (1992).
13. R. M. Imayev, V. M. Imayev and G. A. Salishchev, J. Mater. Sci., 27, 4465 (1992).
14. H. S. Yang, W. B. Lee and A. K. Mukherjee, Structural Intermetallics, edited by R. Darolia, J. J. Lewandowski, C. T. Liu, P. L. Martin, D. B. Miracle and M. V. Nathal, p. 69, The Minerals, Metals and Materials Society, Warrendale, PA, (1993).
15. H. S. Yang, W. B. Lee and A. K. Mukherjee, First International Conference on Processing Materials for Properties, edited by H. Henein and T. Oki, p. 233, The Minerals, Metals and Materials Society, Warrendale, PA, (1993).
16. W. B. Lee, H. S. Yang, Y-W. Kim and A. K. Mukherjee, Scripta Metall. Mater., 29, 1403 (1993).
17. M. Nobuki, D. Vanderschueren and M. Nakamura, Acta Metall. Mater., 42, 2623 (1994).
18. W. B. Lee, H. S. Yang, and A. K. Mukherjee, Mater. Sci. Eng. A, A192/193, 733 (1995).
19. R. S. Mishra, W. B. Lee, A. K. Mukherjee and Y-W. Kim, *International Symposia on Gamma Titanium Aluminides*, edited by Y-W. Kim, R. Wagner and M. Yamaguchi, TMS, in press (1995).

20. C. M. Lombard, A. K. Ghosh and S. L. Semiatin, *International Symposia on Gamma Titanium Aluminides*, edited by Y-W. Kim, R. Wagner and M. Yamaguchi, TMS, in press (1995).
21. J. Hebeisen, P. Tylus, D. Zick, D. K. Mukhopadhyay, K. Brand, C. Suryanarayana and F. H. Froes, ADDA, edited by F. H. Froes, MPIF, Princeton, NJ, in press (1995).
22. F. H. Froes, C. Suryanarayana, D. K. Mukhopadhyay, K. Brand, G. Korth, D. Zick, P. Tylus and J. Hebeisen, PM2TEC'95, edited by M. A. Phillips and J. Porter, MPIF, Princeton, NJ, in press (1995).
23. S. W. Schwenker and Y-W. Kim, *International Symposia on Gamma Titanium Aluminides*, edited by Y-W. Kim, R. Wagner and M. Yamaguchi, TMS, in press (1995).
24. N. S. Stoloff, High-Temperature Ordered Intermetallic Alloys, edited by C. C. Koch, C. T. Liu and N. S. Stoloff, MRS vol. 39, MRS, PA, 1985, p 3.
25. R. Bohn, M. Oehrig, Th. Pfullmann, F. Appel and R. Bormann, *Processing and Properties of Nanocrystalline Materials*, edited by C. Suryanarayana, F. H. Froes and J. Singh, TMS, in press (1995).
26. R. S. Mishra, T. R. Bieler and A. K. Mukherjee, Acta Metall. Mater., 43, 877 (1995).
27. J. R. Weertman, Mater. Sci. Eng., A166, 161 (1993).
28. R. W. Siegel and G. E. Fougere, NanoStructured Materials, 6, 205 (1995).

Effect of Cooling Rate from Mill Annealing Temperature on Stress Corrosion Threshold of Titanium 6Al-4V ELI Beta Annealed.

Robert D. Briggs
Boeing Materials Technology
Boeing Commercial Airplane Group
Seattle, Wa. USA

Slow cooling from the mill annealing temperature can result in a significant reduction in the stress corrosion threshold of beta annealed Ti 6Al-4V ELI. This report compares the stress corrosion threshold resulting from several continuous cooling rates from the mill annealing temperature. A Time-Temperature-Transformation curve based on the stress corrosion susceptibility of Ti 6Al-4V ELI beta annealed plate is also presented. The presence of Ti_3Al, an ordered phase which forms during slow cooling, was found to cause the loss in stress corrosion threshold.

INTRODUCTION

The stress corrosion threshold of beta annealed Ti 6Al-4V Extra Low Interstitial (ELI) was found to decrease 50% when slow cooled from the mill annealing temperature. The stress corrosion threshold for several different continuous cooling rates was determined. Iso-thermal exposure was used to establish a Time-Temperature-Transformation curve (TTT) based on the reduction in stress corrosion threshold.

PROCEDURE

A titanium 6Al-4V ELI extrusion was selected for initial testing. The chemical composition is shown in Table I. The material was extruded at a temperature above the beta transus (988C), air cooled, then mill annealed at 730C for 2.5 hours followed by air cooling. Seven specimen blanks were cut from the extrusion. Each blank was individually mill annealed at 730C for 1.5 hours in a vacuum furnace followed by controlled cooling from 730C to 480C at rates shown in Table II. Blanks were machined to the specimen configuration shown in Figure 1 using a T-L orientation.

An additional 17 stress corrosion specimens from 28 mm thick Ti 6Al-4V beta annealed ELI plate were fabricated in the T-L orientation. Composition of the plate is also shown in Table I. These specimens were exposed to various combinations of time and temperature to determine the effects of specific time-temperature exposures on the stress corrosion threshold, enabling construction of a TTT curve.

TABLE I CHEMICAL COMPOSITION

Material	V	Al	Fe	N	C	O	H
Extrusion	4.4	6.2	.12	.015	.032	.12	16
Plate	4.1	6.1	.23	.011	.012	.11	85

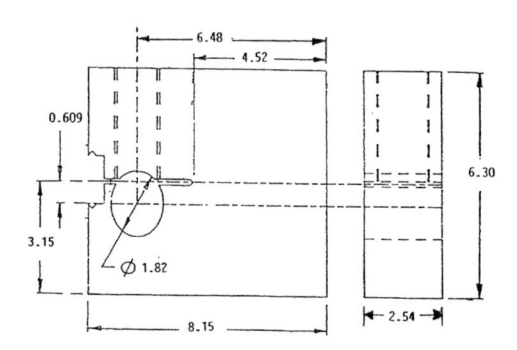

Figure 1. Specimen Configuration. Dimensions in cm.

Specimens were fatigue pre-cracked 2.5 mm then loaded to an initial stress intensity of 30 MPA√m, a clip type displacement gage was used to measure the crack opening displacement (COD). Loading was performed with salt water in the pre-crack. Specimens were then immersed in a 3.5 percent salt water solution. After 24 hours of immersion the specimens were examined for visible crack extension beyond the pre-crack. If no visible crack extension occurred the stress intensity was increased by 10 MPA√m and the specimens immersed for an additional 24 hours. This process was repeated until visible crack extension occurred, at which point the specimens were returned to the to the salt water and the crack allowed to propagate. After the crack had arrested, the specimens remained in the solution for a minimum of 50 hours to ensure the threshold was achieved. The specimens were then removed from the salt water, the COD was measured and the load required to produce that COD determined. The specimens were then fractured in liquid nitrogen to enhance crack length measurement. The K_{ISCC} threshold was calculated using equation (1).

$$K_{ISCC} = (P \cdot C) \div (B \cdot \sqrt{a}) \qquad (1)$$

Where "P" is the crack opening load, "B" is the specimen thickness, "a" is the final crack length and "C" is the function in equation (2).

$$C = 30.96 \cdot [a/W] - 195.8 \cdot [a/W]^2 + 730.6 \cdot [a/W]^3 \qquad (2)$$
$$- 1186.3 \cdot [a/W]^4 + 754.6 \cdot [a/W]^5$$

Where "a" is the final crack length and "W" is specimen width.

RESULTS

Stress corrosion thresholds for Ti 6Al-4V cooled from 730C to 480C at various rates are shown in Table II and plotted in Figure 2. As can be noted, the stress corrosion threshold is progressively reduced as the cooling rate is decreased below approximately 250C per hour.

TABLE II
STRESS CORROSION THRESHOLD VS COOLING RATE

Cooling Rate from 730C to 480C	Stress Corrosion Threshold MPA√m
25C per hour	43.7
33C per hour	44.7
55C per hour	54.3
125C per hour	63.7
250C per hour	91.2
500C per hour	90.0
6200C per hour	98.9

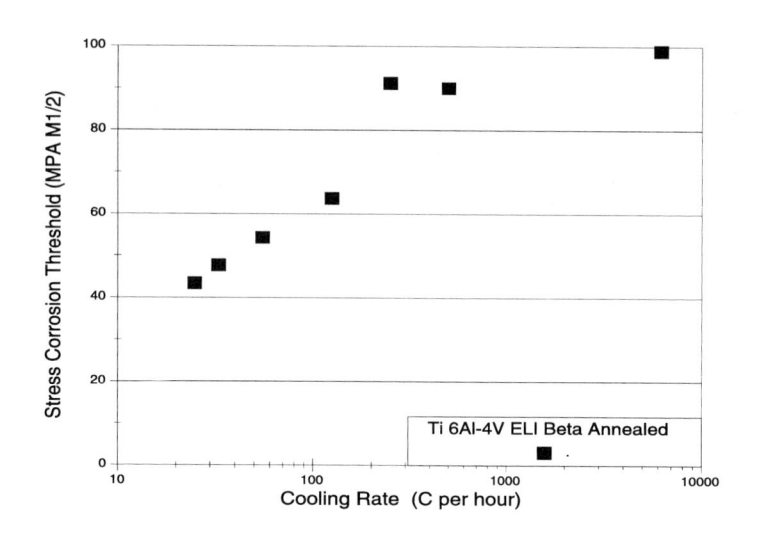

Figure 2. Stress Corrosion Threshold Versus Cooling Rate From Mill Annealing Temperature.

The stress corrosion thresholds measured after isothermal exposures at various temperatures for different length of time are presented in Table III. The data is used to produce a TTT curve by determining if the reduction in stress corrosion threshold relative to the as-received material exceeded the test tolerance of ~10 MPA$\sqrt{m}$. Those specimens which exhibited a stress corrosion threshold below 75 MPA $\sqrt{m}$ were determined to have a stress corrosion threshold reduction due to the elevated temperature exposure resulting in precipitation of Ti_3Al. Thus, this data could be used to define the TTT curve for the onset of ordering. The plot is shown in Figure 3.

TABLE III
STRESS CORROSION THRESHOLD VERSUS ISOTHERMAL TEMPERATURE

Temperature	Time	Stress Corrosion Threshold MPA$\sqrt{m}$
As-received	none	84.4
425C	4 hours	84.5
450C	3 hours	88.0
480C	1 hour	81.2
480C	1.5 hours	67.6

480C	2 hours	74.6
480C	3 hours	56.8
540C	1 hour	76.2
540C	1.5 hours	69.7
540C	2 hours	63.3
565C	30 minutes	63.6
565C	45 minutes	59.9
565C	1.5 hours	58.8
590C	1 hour	77.9
590C	1.5 hours	73.7
590C	2 hours	61.8
650C	2 hours	88.9

Figure 3 TTT Curve based on Stress Corrosion Threshold.

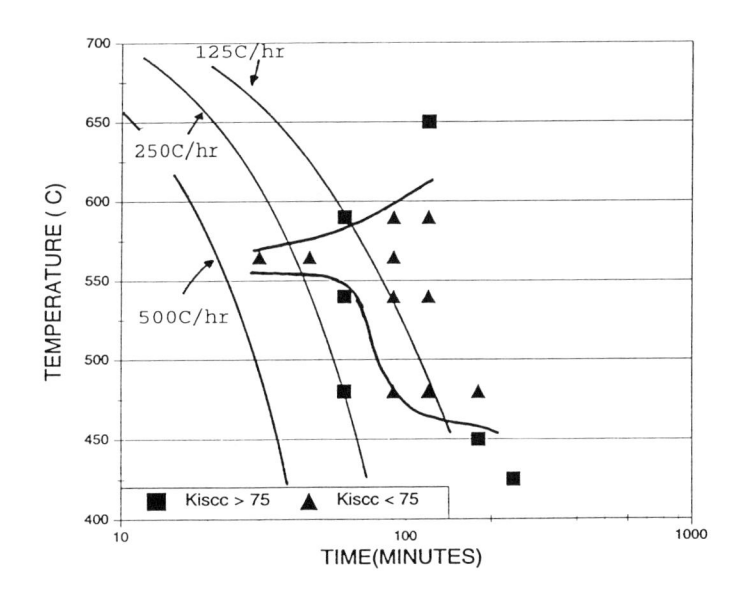

Thin foil transmission electron microscopy was used to follow the microstructural changes that resulted from cooling the specimens from mill anneal temperature at various rates. The electron diffraction pattern of the alpha phase from a specimen continuously cooled from 730C to 480C at 55C per hour is shown in Figure 4. Also shown in Figure 4 is a computer generated diffraction pattern for alpha titanium with and without Ti₃Al. The precipitation of Ti3Al was not detected in specimens cooled at a rate of 250C per hour or higher.

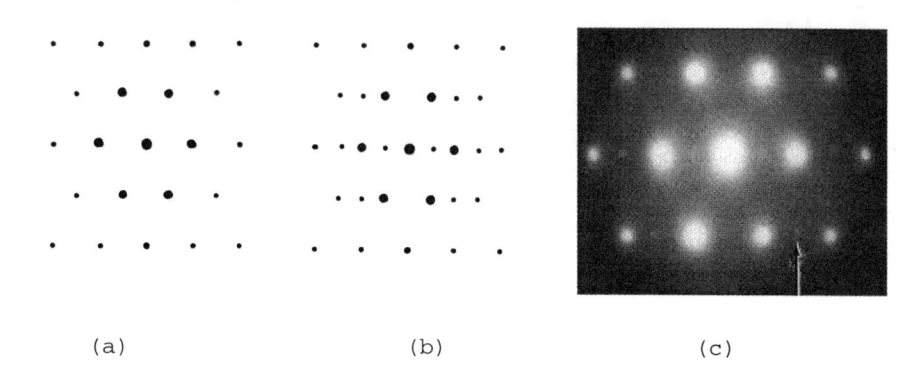

(a) (b) (c)

Figure 4. (a) Computer Generated Diffraction Pattern of Alpha Ti, (b) Alpha Ti plus Ti₃Al, (c) Diffraction pattern of specimen cooled at 55C per hour. Zone Axis [1̄2̄13].

DISCUSSION

Titanium-Aluminum alloys can form Ti₃Al as shown by a titanium aluminum phase diagram. The phase diagram indicates that a minimum of 6 percent aluminum is required to form Ti₃Al at 500C (1). Increasing the aluminum content increases the amount of Ti₃Al which can form. The aluminum content of the alpha phase of beta annealed Ti 6Al-4V was measured above the 6 percent bulk composition due to partitioning. Partitioning occurs when crystallographically oriented alpha phase precipitates from the beta creating the lamellar alpha/beta microstructure. As the alpha forms it rejects beta stabilizers. The beta phase that remains is higher in the beta stabilizer-vanadium and the alpha phase which forms is higher in the alpha stabilizer-aluminum. Using the thin foil specimens, the aluminum content of the alpha phase was measured to be 7 to 8 percent and the vanadium content to be 2 to 3 percent. The higher aluminum content in the alpha phase increases the propensity of Ti₃Al formation. Vanadium content increased in the beta phase to 20 to 25 percent while the aluminum dropped to 1 to 2 percent.

The presence of Ti₃Al depends not only on chemistry but also on temperature i.e. must be within the alpha plus Ti₃Al phase field. At the mill annealing temperature of 730C, and an alpha phase aluminum content of 7 percent, Ti₃Al would not expected to be present as the phase diagram indicates a condition above the alpha plus Ti₃Al phase field. Cooling below the mill annealing temperature is required for the formation of Ti₃Al.

Since the phase diagram represents an equilibrium condition, cooling sufficiently fast may not allow time to form Ti_3Al. Cooling at 55C per hour provided sufficient time to precipitate Ti_3Al as evidenced by the diffraction pattern in Figure 4.

The kinetics of precipitation of Ti_3Al is a function of the driving force due to supersaturation and diffusion. For a given composition, the higher the temperature within the Ti plus Ti_3Al phase field, the lower the degree of supersaturation, resulting in a lower driving force for precipitate nucleation. At lower temperatures, on the other hand, the kinetics of precipitate growth, which are controlled by diffusion are retarded due to slower diffusion, thus requiring longer time to induce the same degree of precipitation (2). The net result is that the maximum rate of precipitation occurs at some intermediate temperature, which gives rise to the well known C-curve.

A correlation of the microstructural changes observed by TEM with the stress corrosion threshold results suggest that the reduction in the measured stress corrosion thresholds of Ti 6Al-4V is associated with precipitation of Ti_3Al. The increased presence of Ti_3Al results in a progressive degradation in stress corrosion threshold.

Ti_3Al is an ordered phase, where the lower free energy state is created by the titanium and aluminum atoms occupying particular positions in the hexagonal close packed unit cell. During deformation the interaction of dislocations with these ordered Ti_3Al precipitates results in intense planar slip. The shearing of the ordered precipitates by dislocations causes local disordering along the slip plane, which makes movement of subsequent dislocations on the same slip plane more favorable. Consequently, slip tends to be confined to a few well defined crystallographic planes and little or no cross slip occurs. Under conditions of planar slip, dislocation pile-ups will occur at grain boundaries and colony boundaries. Such dislocation pile-ups result in local stress concentration at the head of the pile-ups (3). Cleavage fracture occurs when the adjacent grain or colony is oriented such that the cleavage plane is subjected to a stress in excess of the threshold normal stress required to induce cleavage fracture. This threshold stress will be reduced in the presence a of 3.5 percent salt water solution. If the slip plane in the adjacent grain or colony is more favorably oriented to the stress a ductile fracture occurs (4,5).

The TTT curve shown in Figure 3 provides the time and temperature range where the influence of the ordering reaction on stress corrosion threshold is most significant. At temperatures below 480C, the formation of Ti_3Al is sufficiently slow that even though supersaturation is present, diffusion does not occur fast enough to cause the stress corrosion threshold reduction associated with ordering. The combination of driving forces due to supersaturation and diffusion appears to be maximum at 565C. Above 565C the driving force for precipitation decreases as the equilibrium amount of Ti_3Al

deceases. Above 590C, as the solvus of Ti$_3$Al is approached the driving force is sufficiently low, resulting in no loss of stress corrosion thresholds within the time frame of these tests. The stress corrosion results from the continuous cooling tests are consistent with the C-curve in Figure 3. As can be noted, at cooling rates of 250C per hour or slower, the cooling curves cut through the C-curve leading to precipitation of Ti$_3$Al.

CONCLUSIONS

1. Slow cooling from the mill annealing temperature through the alpha plus Ti$_3$Al phase field, reduces the stress corrosion threshold of Ti 6Al-4V ELI.

2. The decrease in stress corrosion threshold is related to the extent of formation of the ordered phase Ti$_3$Al, which promotes planar slip due to shearing of the precipitates, dislocation pile-ups and localized stress concentration which in turn increases the propensity for cleavage fracture of the alpha phase in the presence of a corrosive environment.

3. Within 90 minutes, material exposed in the critical temperature range 480C to 590 experienced a loss in stress corrosion threshold, with exposure at 565C the most severe.

REFERENCES

1. T. Massalski, *Binary Alloy Phase Diagrams*, (Metals Park OH, American Society of Metals 1986), 175.

2. M.J. Blackburn, "Ordering Transformation in Titanium-Aluminum Alloys Containing up to 25 at. pct Aluminum", *AIME*, 239 (1967), 1200-1208.

3. R.E. Curtis and W.F. Spurr, "Effect of Microstructure on the Fracture Properties of Titanium and Titanium Alloys in Air and Salt Solution", *Met. Trans.*, 61 (1968), 115-127.

4. R.E. Curtis, R.R. Boyer, J.C. Williams, "Relationship Between Composition, Microstructure and Stress Corrosion Cracking (in Salt Solution) in Titanium Alloys", *Met. Trans.* 62, (1969) 457-469.

5. R.R. Boyer and W.F. Spurr, " Effect of Composition, microstructure and Texture on Stress Corrosion Cracking of Ti 6Al-4V Sheet", *Met. Trans.*, 9A (1978) 1443-1448.

ACKNOWLEDGMENTS

The author wishes to express his gratitude to R.R. Boyer and Dr. G. H. Narayanan for their technical expertise, and to Dr. A.K. Eikum for many stimulating discussions.

HEAT TREATMENT OF TITANIUM ALLOYS

J. R. Wood and P. A. Russo

RMI Titanium Company
1000 Warren Avenue
Niles, Ohio 44446

Abstract

Titanium and titanium alloys are heat treated for a variety of purposes. Examples are: stress relieving to relieve residual stresses during fabrication; process annealing to produce an acceptable combination of ductility, machinability and dimensional stability; and solution heat treating and aging to optimize strength, ductility, fracture toughness, fatigue strength, creep resistance, etc. A review of these processes and a discussion on specialized heat treating processes are described in this paper with particular emphasis on alpha-beta titanium alloy Ti-6Al-2Sn-2Zr-2Mo-2Cr-Si (Ti-6-22-22) and metastable beta titanium alloy (Ti-3Al-8V-6Cr-4Mo-4Zr (Beta-C^{TM}).

Introduction

Titanium alloys are heat treated to provide specific microstructures and mechanical properties for a given application. Heat treatments may also be coupled with a specific thermomechanical processing sequence to further enhance or optimize microstructures and properties. The response of titanium alloys to heat treatment depends on the composition of the particular alloy and the effect of alloying elements on the alpha-beta phase balance. Some titanium alloys are classified as alpha-beta alloys and respond to various heat treating times and temperatures and cooling rates in a different manner than do the more heavily alloyed metastable beta titanium alloys. Not all heat treating cycles are applicable to all titanium alloys because of differences in composition and microstructure. Therefore, certain alloys may have unique heat treatments designed for specific property enhancement such as creep resistance, fatigue strength, fracture toughness, high strength, etc. A review of various heat treatments for alpha-beta and metastable beta alloys is presented herein with particular emphasis on alpha-beta alloy Ti-6Al-2Sn-2Zr-2Mo-2Cr-Si (Ti-6-22-22) and metastable beta alloy Ti-3Al-8V-6Cr-4Mo-4Zr (Beta-C).

General Heat Treatments

Heat treatments generally applicable to all titanium alloys are classified as follows:

- <u>Stress Relieving</u> - to relieve residual stresses present after heat treating, welding, forming, machining, etc.
- <u>Annealing</u> - to produce an acceptable microstructure and combination of properties to enhance fabricability, machinability, dimensional stability and service life.
- <u>Solution Treating and Aging</u> - to optimize strength, ductility, fracture toughness, fatigue strength, creep resistance, etc., for the intended application.

Heat Treatments for Alpha-Beta Titanium Alloys

Typical heat treatments for alpha-beta titanium alloys are as follows:

- Stress Relief Anneal (SR): 480/705°C (900/1300°F) for 30 min. to 4 hrs.; air cool or slower.
  <u>Microstructure</u>: Essentially no change in prior microstructure.
  <u>Property Enhancement</u>: Relieves residual stresses from heat treatment, welding, fabrication, etc.

- Mill Anneal or Full Anneal (MA): 650/815°C (1200/1500°F) for 30 min. to 4 hrs.; air cool or slower.
  <u>Microstructure</u>: Incompletely recrystallized (Rx) alpha (α) and beta (β) phases.
  <u>Property Enhancement</u>: Minimum strength and maximum ductility.

- Recrystallization Anneal (RA): Beta transus (β_T) minus 25/55°C (50/100°F) for 1 to 4 hrs.; fce. cool to 760°C (1400°F); air cool.
  <u>Microstructure</u>: Rx coarse alpha with small amounts of β phase.
  <u>Property Enhancement</u>: Fracture toughness (K_{Ic}) and fatigue crack growth resistance (da/dn) at moderate strength and ductility.

- Duplex Anneal (DA): β_T minus 25/55°C (50/100°F) for 1 hr.; air or fan cool; overage or stabilize at 600/760°C (1100/1400°F) for 2 to 4 hrs.; air cool.
  Microstructure: Equiaxed primary alpha plus platelet alpha in transformed beta matrix.
  Property Enhancement: Strength, creep resistance, K_{Ic}, fatigue.

- Beta Anneal (BA): β_T plus 15/30°C (25/75°F) for 30 min.; air cool; overage at 600/760°C (1100/1400°F) for 2 to 4 hrs.; air cool.
  Microstructure: Widmanstätten or basket weave alpha + beta colonies.
  Property Enhancement: Maximum K_{Ic} and da/dn with lower fatigue strength and ductility.

- Alpha-Beta Solution Treat and Age (STA): β_T minus 15/55°C (25/100°F) for 1 hr.; air, fan, or water quench; age at 480/650°C (900/1200°F) for 4 to 8 hrs.; air cool.
  Microstructure: Equiaxed primary alpha plus Widmanstätten alpha + beta plus tempered alpha prime (α').
  Property Enhancement: Highest strength with moderate ductility and lower K_{Ic} and da/dn. Good combination of properties such as fatigue and creep.

- Alpha-Beta Solution Treat and Overage (STOA): Similar to STA except for higher aging temp. of 650/760°C (1200/1400°F).
  Microstructure: Similar to STA.
  Property Enhancement: Intermediate strength between MA and STA with improved K_{Ic} and da/dn over STA.

- Beta Solution Treat and Age (β STA): β_T plus 15/30°C (25/75°F) for 30 min.; water quench or fan cool; age at 480/650°C (900/1200°F) for 4 to 8 hrs.; air cool.
  Microstructure: Fine Widmanstätten alpha + beta plus tempered α'.
  Property Enhancement: Good K_{Ic} and strength combination with lowest ductility.

- Beta Solution Treat and Overage (β STOA): Similar to β STA with higher aging temp. of 650/760°C (1200/1400°F).
  Microstructure: Similar to β STA.
  Property Enhancement: Highest K_{Ic} with lower strength and better ductility than β STA.

- Beta Plus Alpha-Beta Solution Treat and Age (TRIPLEX STA): β_T plus 15/30°C (25/75°F) for 30 min.; air or fan cool; β_T minus 25/55°C (50/100°F) for 1 hr.; air or fan cool; age at 480/590°C (900/1100°F) for 8 hrs.; air cool.
  Microstructure: Widmanstätten alpha-beta structure with coarse alpha platelets.
  Property Enhancement: Optimum combination of UTS, K_{Ic} and da/dn.

Commonly heat treated alpha-beta titanium alloys include the following:

- Ti-6Al-4V (Ti-6-4)
- Ti-6Al-2Sn-4Zr-2Mo (Ti-6242)
- Ti-6Al-2Sn-4Zr-6Mo (Ti-6246)
- Ti-6Al-2Sn-2Zr-2Mo-2Cr-Si (Ti-6-22-22)
- Ti-5Al-2Sn-2Zr-4Mo-4Cr (Ti-17)
- Ti-6Al-6V-2Sn (Ti-662)

The microstructural changes that can be brought about through thermal treatment of Ti-6-4 bar are shown in Figure 1 for four different temperatures and three different cooling rates from each temperature[1]. The tensile properties which accompany each of the 12 thermal treatments are shown in Table 1 before and after aging at 538°C (1000°F)[1]. The presence of nonequilibrium phases, such as alpha-prime or metastable beta, results in substantial increases in tensile and yield strength properties following aging. The tensile data show that 1) no response to aging occurs on furnace cooling from solution temperatures; 2) only a slight response occurs on air cooling; and 3) the greatest response is experienced with a water quench from the solution temperature. Good response to aging occurs on water quenching from the beta field; however, ductility values are quite low. The best combination of properties can be produced by solution treating at temperatures relatively high in the alpha-beta field.

Titanium alloy Ti-6-22-22 is used in heavy sections up to 102 cm (4 in.) in which a good combination of tensile strength and fracture toughness is desired for fracture critical applications. The effect of cooling rate from beta forging and alpha-beta solution heat treating has a pronounced effect on alpha platelet size and morphology and tensile properties as shown in Figure 2[2]. A comparison of beta forging plus alpha-beta STA; alpha-beta forging plus beta STA; and alpha-beta forging plus TRIPLEX STA is shown in Figure 3, in which the best combination of strength and fracture toughness properties is achieved with the TRIPLEX STA.

For sheet products of Ti-6-22-22, moderate strengths can be achieved through alpha-beta STA with highest strengths achieved at temperatures approaching the beta transus of 954°C (1750°F), as shown in Figure 4[3].

Heat Treatments for Metastable Beta Titanium Alloys

Typical heat treatments for metastable beta alloys are as follows:

- Solution Treating - Puts material in softened condition for optimum formability. Material may be subsequently aged to increase strength.

- Subtransus Solution Treating - Generally does not recrystallize material, which promotes uniform aging in solute-rich alloys at the potential adverse effect on transverse properties. Results in the presence of alpha in solute lean alloys (e.g., Ti-10-2-3).

- Supratransus Solution Treating - Recrystallizes or partially recrystallizes structure. May result in nonuniform aging in solute-rich alloys (e.g., Beta-C).

- Direct Aging - Performed directly from hot working. Promotes uniform aging with potential decease in ductility in transverse direction if material has substantial microstructural texture.

- Solution Treat/Single Age - Solution treat followed by aging at a temperature and time that achieves required mechanical properties (e.g., 480-650°C (900-1200°F) for 4 to 24 hours).

- Solution Treat/Duplex Age - Solution treat plus a low temperature age applied first to precipitate alpha or beta prime in the case of solute-rich alloys. This is followed by a higher temperature age which coarsens alpha and establishes final mechanical properties (e.g., 260-370°C (500-700°F) for 4 to 8 hours plus 480-650°C (900-1200°F) for 8 to 24 hours).

- **PASTA** - PreAge Solution Treat and Age for Beta-C. A preage is applied to cold worked material to precipitate alpha uniformly. A solution treatment is applied at temperatures below the silicide solvus which precipitates silicides at alpha sites prior to the dissolution of alpha. These silicides provide nucleation sites for the precipitation of alpha during the aging process.

Typical commercial heat treated metastable beta titanium alloys include the following:

Alloy	Relative Beta Stability Molybdenum Equivalency[1]
Ti-10V-2Fe-3Al (Ti-10-2-3)	9.5
Ti-15V-3Cr-3Sn-3Al (Ti-15-333)	11.9
Ti-3Al-8V-6Cr-4Mo-4Zr (Beta-C)	16.0
Ti-13V-11Cr-3Al (Ti-13-11-3)	23.3

1. Mo Eq. = 1.0 wt.% Mo + 0.67 wt.% V + 2.9 wt.% Fe + 1.6 wt.% Cr - 1.0 wt.% Al.

The relative beta stability of these alloys is assessed by using calculated molybdenum equivalency (Mo. Eq.).

A schematic pseudo-binary equilibrium diagram indicating the relative positions of several metastable beta titanium alloys based on calculated stability is shown in Figure 5[4]. Some of the phases present include the following:

- β phase - Body centered cubic.
- α phase - Hexagonal closed packed.
- ω phase - Forms in solute lean metastable beta alloys athermally and during low temperature aging.
- β' - Solute lean beta phase. Can form in solute-rich alloys during aging.

Typical temperature ranges for heat treatment of metastable beta alloys is shown in Figure 6[4].

The effect of solution annealing and aging of Beta-C bar at various temperatures is shown in Figure 7[5]. The effects of incomplete recrystallization is evident at 788°C (1450°F) in which residual warm work and substructure produced more extensive aging and highest strength and hardness, whereas 843°C (1550°F) produced a recrystallized structure with less uniform aging and lower strength.

An example of the PASTA heat treatment is shown in Figure 8 in which cold pilgered pipe of Beta-C is uniformly aged with an intermediate preage heat treatment at 621°C (1150°F) to take advantage of residual cold work and rapid aging response at 621°C (1150°F) prior to STA[6].

Effect of high solution heat treatments on strength-toughness trends is shown in Figure 9 for beta extruded and heat treated Beta-C[7]. The best combination of strength and fracture toughness was achieved with a relatively high temperature beta solution heat treatment at 927°C (1700°F).

Finally, the effect of duplex aging on Beta-C plate is shown in Figure 10[8]. Here it is shown that long-time aging at a low temperature followed by a shorter-time age at a higher temperature achieves a more uniform distribution of alpha and superior tensile and fatigue strengths over simplex aging at a single temperature.

Summary

Many titanium alloys are used in service in the mill annealed or stress relieved condition without the need for further heat treatment.

Some α-β titanium alloys are given various heat treatments to develop the desired microstructure for enhancement of certain key properties. Such heat treatments include RA, DA, BA, STA, STOA, β STA, β STOA, and TRIPLEX.

Metastable β titanium alloys can achieve the highest strengths through proper selection of TMP routes coupled with solution treating and/or age hardening treatments.

References

1. RMI Titanium Company Brochure, "Metallography," 1994, 22-23.

2. H. R. Phelps and J. R. Wood, "Correlation of Mechanical Properties and Microstructures of Ti-6Al-2Sn-2Zr-2Mo-2Cr-0.25S Titanium Alloy," Titanium '92 Science and Technology, vol. I, (Warrendale, PA: The Minerals, Metals & Materials Society, 1993), ed. F. H. Froes and I. L. Caplan, 193-199.

3. R. C. Bliss, "Evaluation of Ti-6Al-2Sn-2Zr-2Cr-2Mo-.23Si Sheet," ibid., 201-208.

4. S. Ankem and S. R. Seagle, "Heat Treatment of Metastable Beta Titanium Alloys," Beta Titanium Alloys in the 1980's, (Warrendale, PA: The Metallurgical Society of AIME, 1984), ed. R. R. Boyer, H. W. Rosenberg, 107-126.

5. J. G. Ferrero, J. R. Wood and P. A. Russo, "Microstructure/Mechanical Property Relationships in Bar Products of Beta-C (Ti-3Al-8V-6Cr-4Mo-4Zr)," Beta Titanium Alloys in the 1990's, (Warrendale, PA: The Minerals, Metals & Materials Society, 1993), ed. D. Eylon, R. R. Boyer and D. A. Koss, 211-226.

6. R. W. Schutz and S. R. Seagle, "Method for Improving Aging Response and Uniformity in Beta-Titanium Alloys," RMI Titanium Company, Niles, OH, U. S. Patent No. 5,201,967.

7. P. A. Russo et al., "High Strength Structural Applications for Large Diameter Forgings of Beta-C (Ti-3Al-8V-6Cr-4Mo-4Zr)," Beta Titanium Alloys in the 1990's, (Warrendale, PA: The Minerals, Metals & Materials Society, 1993), D. Eylon, R. R. Boyer, D. A. Koss, 361-374.

8. L. Wagner and J. K. Gregory, "Improvement of Mechanical Behavior in Ti-3Al-8V-6Cr-4Mo-4Zr by Duplex Aging," ibid., 199-209.

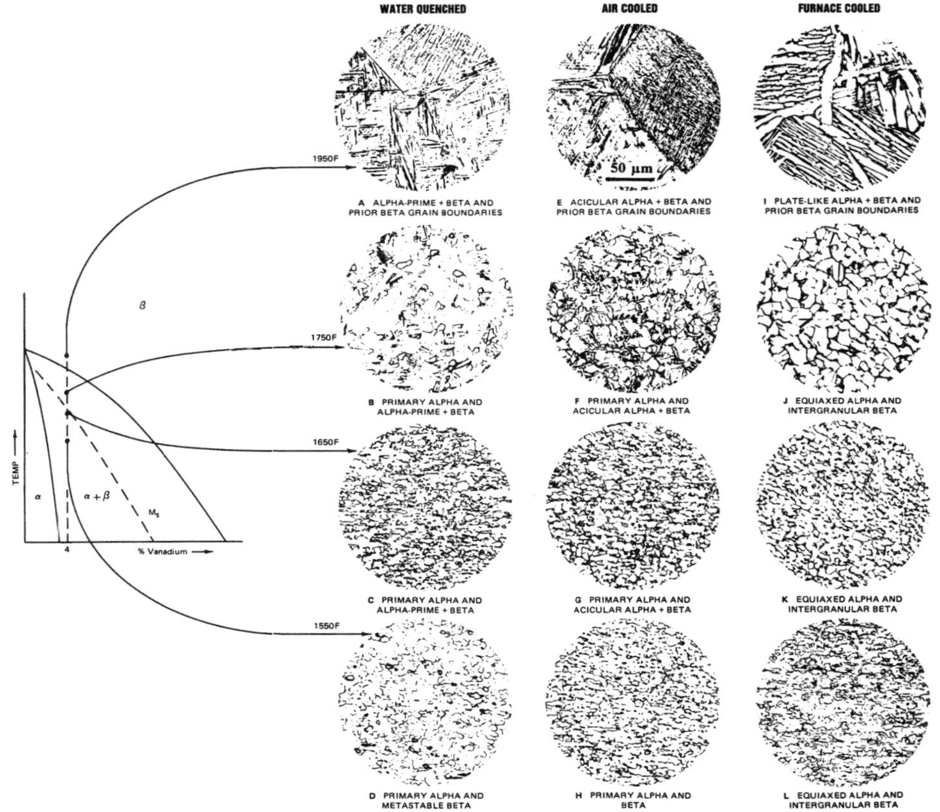

Figure 1 - Effect of thermal treatment on microstructure of Ti-6-4, 15.8 mm (.625 in.) bar.

Table I. Tensile Properties of Ti-6-4 Bar Shown in Figure 1

Micro	Treatment*	UTS KSI	.2% YS KSI	Elongation %	RA %
A	1950F/WQ After Aging	160.7 169.7	138.3 153.3	7.7 8.5	19.2 19.2
B	1750F/WQ After Aging	162.3 171.6	138.3 155.0	17.0 16.5	60.2 56.4
C	1650F/WQ After Aging	162.0 162.0	134.0 147.0	15.2 15.3	53.9 47.5
D	1550F/WQ After Aging	146.4 156.3	112.0 141.7	20.0 16.5	54.7 48.8
E	1950F/AC After Aging	153.7 153.7	137.0 136.3	7.0 9.8	10.3 16.0
F	1750F/AC After Aging	144.3 148.0	122.7 130.3	17.8 16.1	54.1 45.7
G	1650F/AC After Aging	145.3 149.3	126.0 136.0	17.5 17.3	54.7 50.2
H	1550F/AC After Aging	148.0 150.3	127.3 135.0	17.8 16.8	47.7 46.9
I	1950F/FC After Aging	151.0 146.6	136.0 136.0	10.5 9.5	15.6 15.4
J	1750F/FC After Aging	136.3 140.3	121.3 128.0	18.8 18.2	46.0 49.1
K	1650F/FC After Aging	139.6 139.6	124.0 127.0	16.5 16.8	43.3 48.3
L	1550F/FC After Aging	144.6 154.0	134.0 138.3	17.3 17.0	48.9 49.6

*Aging in all cases: 1000F/4 Hrs.; Air Cool

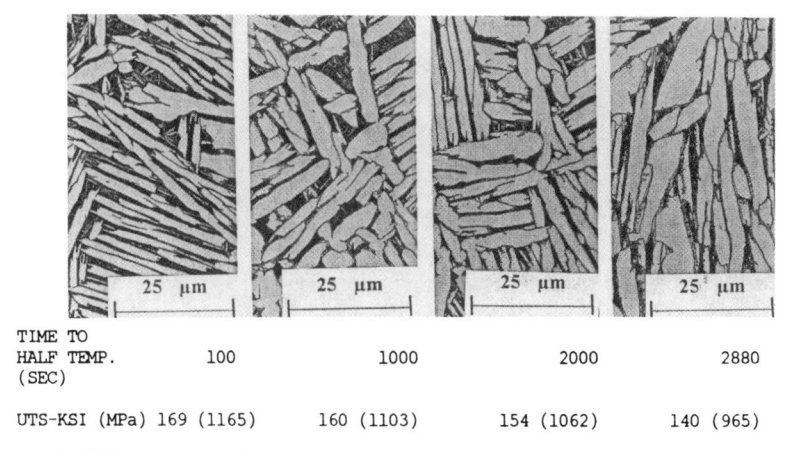

TIME TO HALF TEMP. (SEC)	100	1000	2000	2880
UTS-KSI (MPa)	169 (1165)	160 (1103)	154 (1062)	140 (965)

Figure 2 - Microstructure/mechanical property/cooling rate correlation for Ti-6-22-22 102 cm (4 in.) plate heat treated at 926°C (1700°F).

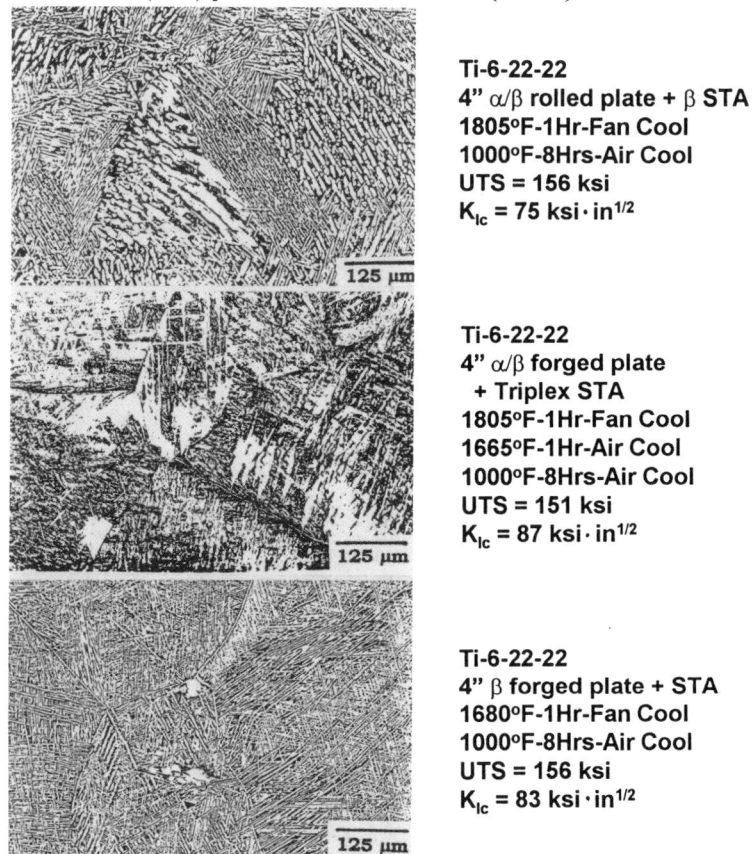

Ti-6-22-22
4" α/β rolled plate + β STA
1805°F-1Hr-Fan Cool
1000°F-8Hrs-Air Cool
UTS = 156 ksi
K_{Ic} = 75 ksi·in$^{1/2}$

Ti-6-22-22
4" α/β forged plate
+ Triplex STA
1805°F-1Hr-Fan Cool
1665°F-1Hr-Air Cool
1000°F-8Hrs-Air Cool
UTS = 151 ksi
K_{Ic} = 87 ksi·in$^{1/2}$

Ti-6-22-22
4" β forged plate + STA
1680°F-1Hr-Fan Cool
1000°F-8Hrs-Air Cool
UTS = 156 ksi
K_{Ic} = 83 ksi·in$^{1/2}$

Figure 3 - Ti-6-22-22 plate microstructure and mechanical properties in three heat treated conditions.

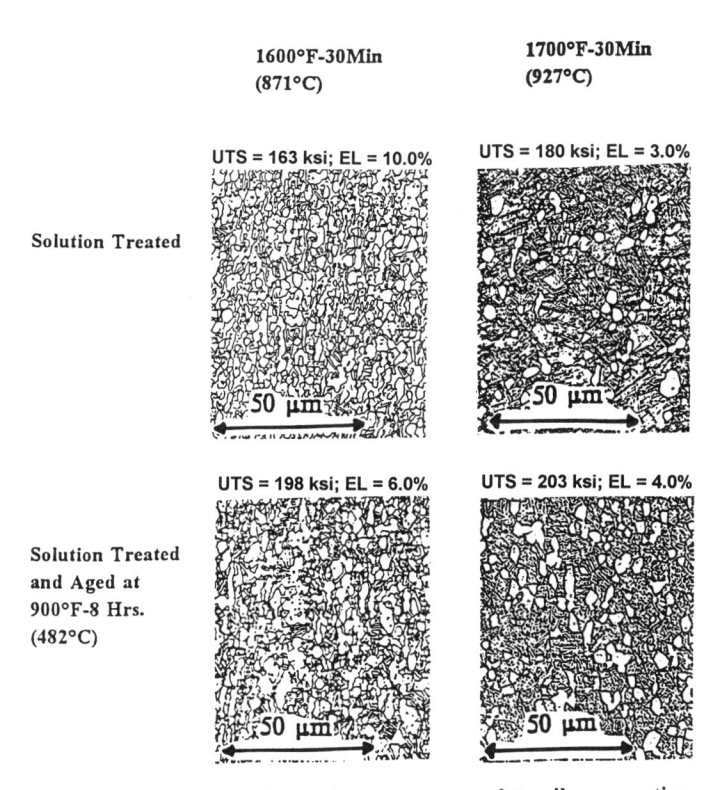

	1600°F-30Min (871°C)	1700°F-30Min (927°C)
Solution Treated	UTS = 163 ksi; EL = 10.0%	UTS = 180 ksi; EL = 3.0%
Solution Treated and Aged at 900°F-8 Hrs. (482°C)	UTS = 198 ksi; EL = 6.0%	UTS = 203 ksi; EL = 4.0%

Figure 4 - Ti-6-22-22 sheet microstructure and tensile properties in the ST and STA conditions.

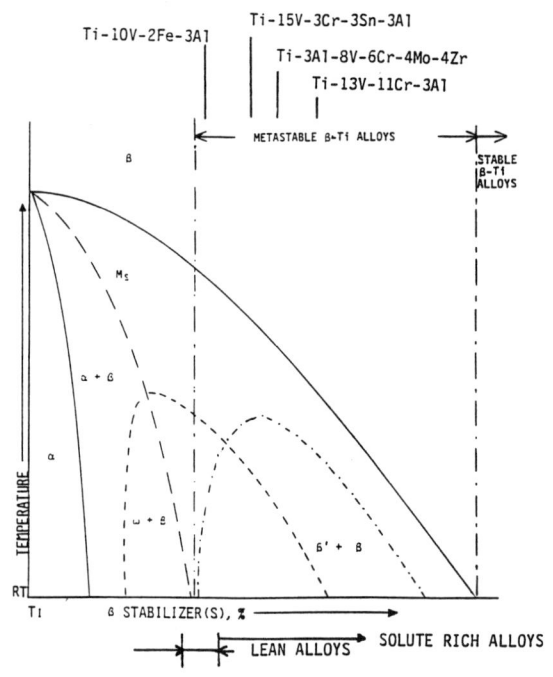

Figure 5 - A schematic pseudo-binary equilibrium diagram showing relative positions of common metastable beta titanium alloys.

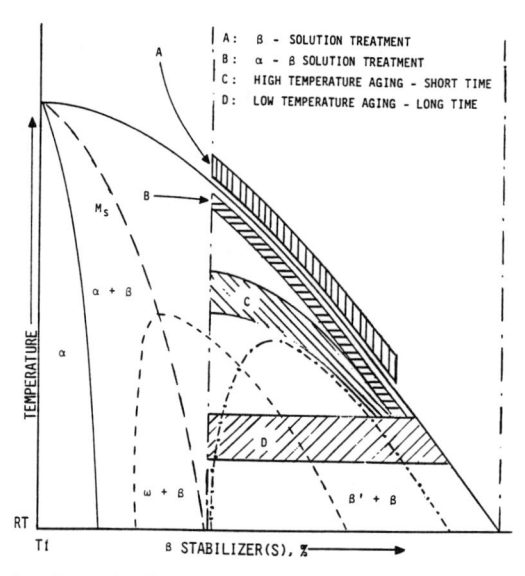

Figure 6 - A schematic diagram showing the ranges of solution and aging heat treatments for metastable beta titanium alloys.

1450°F-1Hr-AC + 900°F-24Hr-AC
192 ksi YS   8% El

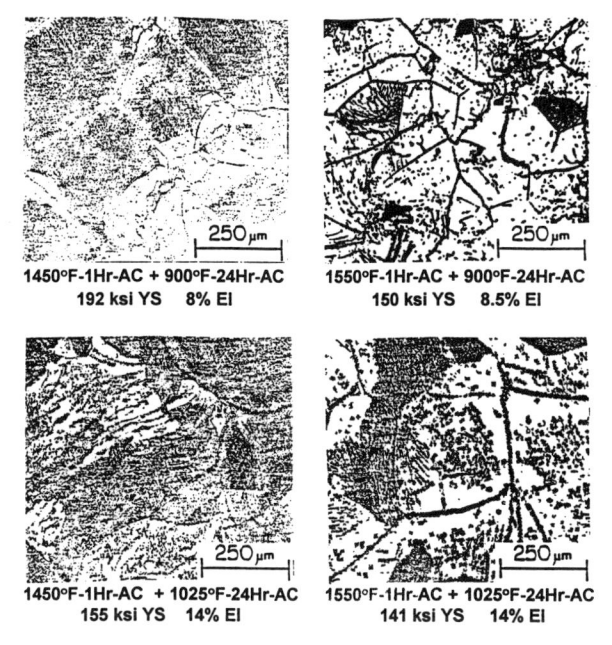

1550°F-1Hr-AC + 900°F-24Hr-AC
150 ksi YS   8.5% El

1450°F-1Hr-AC + 1025°F-24Hr-AC
155 ksi YS   14% El

1550°F-1Hr-AC + 1025°F-24Hr-AC
141 ksi YS   14% El

Figure 7 - Effect of solution annealing and aging temperatures on microstructure of Beta-C bar.

(STA)
HEAT TREATMENT:
1500° F-30 Min-AC +
1050° F-24 Hrs-AC
(100X)

137 ksi YS   19% El.

(PASTA)
HEAT TREATMENT:
1150°F-1Hr-AC +
1500°F-30 Min-AC +
1050°F-24 Hr-AC
(100X)

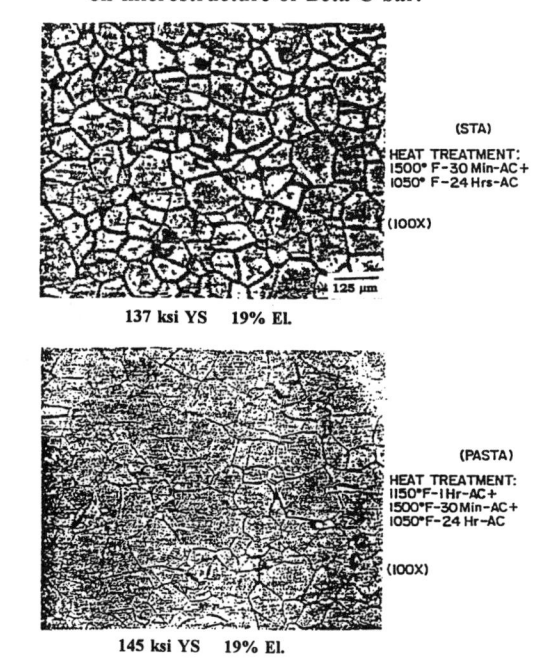

145 ksi YS   19% El.

Figure 8 - Beta-C cold pilgered pipe microstructures in STA and PASTA heat treated conditions.

431

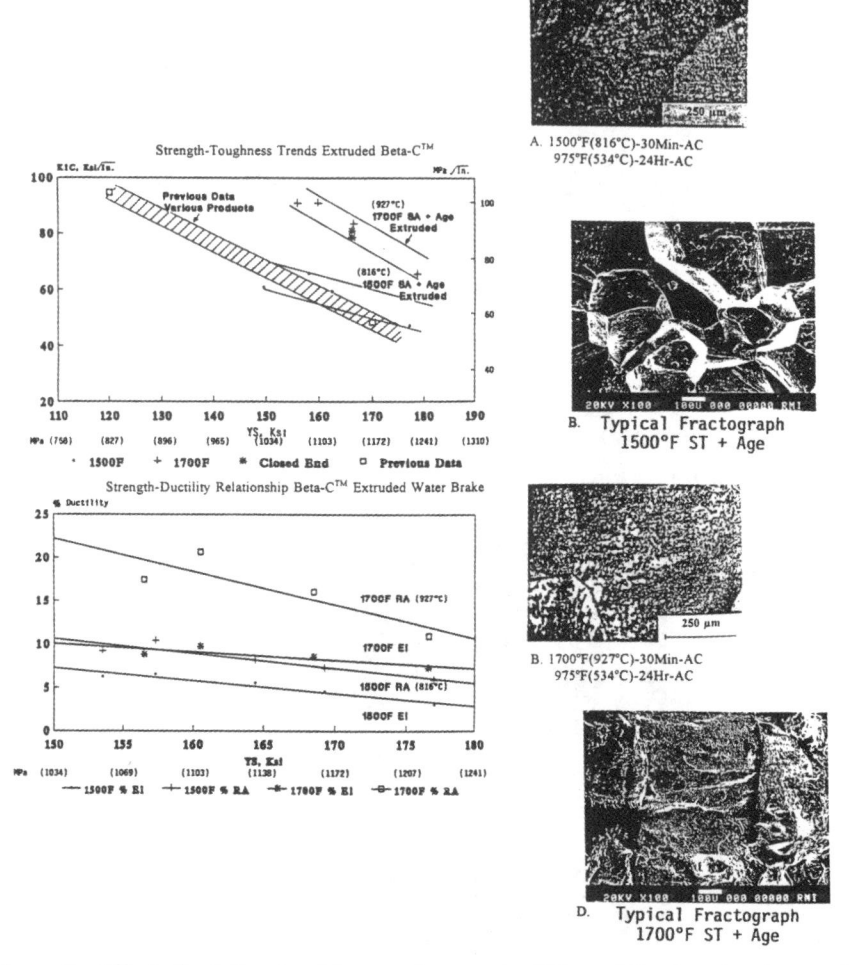

Figure 9 - Effect of solution treat temperature on ductility and fracture toughness of extruded Beta-C.

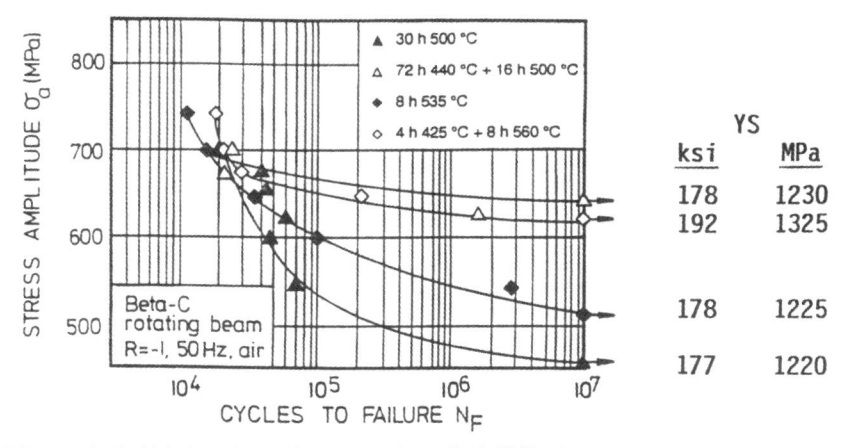

S-N curves for the high strength conditions, rotating beam, R=-1, 50 Hz, air

Fully aged condition after 30 h 500 °C

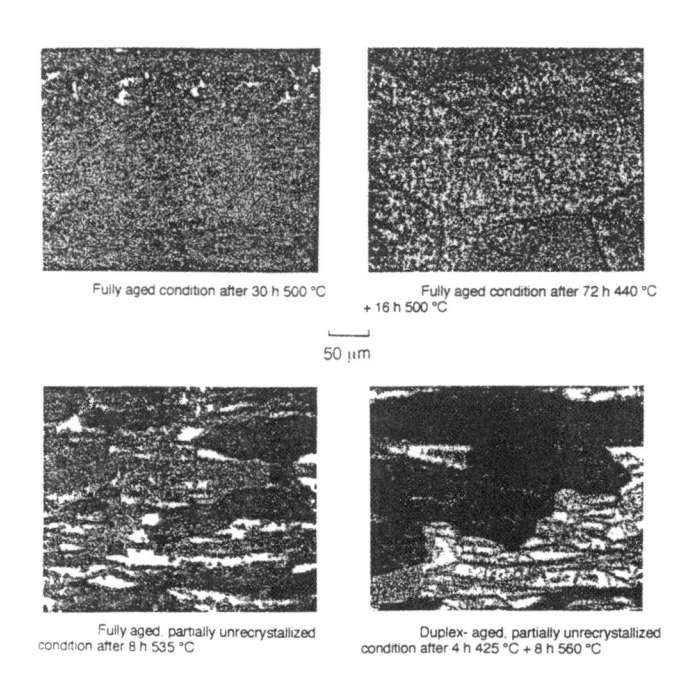

Fully aged condition after 72 h 440 °C + 16 h 500 °C

50 µm

Fully aged, partially unrecrystallized condition after 8 h 535 °C

Duplex- aged, partially unrecrystallized condition after 4 h 425 °C + 8 h 560 °C

Figure 10 - Effect of heat treatment and microstructure on fatigue properties of Beta-C plate.

Role of oxygen on Transformation Kinetics
in Timetal-21S Titanium Alloy

M. A. Imam and C. R. Feng
Materials Science and Technology Division
Naval Research Laboratory
Washington, DC 20375-5343

Abstract

The Timetal-21S, Ti-15Mo-3Al-2.8Nb-0.2Si, is one of the newest beta titanium alloy which has shown excellent resistance to corrosion and heat compared to other beta titanium alloys. Interstitial elements such as oxygen, nitrogen, carbon, and hydrogen affect the transformation characteristics of the Timetal-21S resulting in variations of the mechanical properties. Although the stable phase at room temperature is alpha, the metastable beta is retained even after slow cooling from above the beta transus temperature. The transformation of beta phase is sluggish and can be controlled to produce desired properties. The structure of the transformed products and their distribution depends on the aging conditions and interstitial contents. The present investigation deals with the role of oxygen on the phase transformation kinetics and establish a time-temperature and transformation curves of Timetal-21S. Oxygen contents in the alloy was systematically varied from 0.09 to 0.33 wt%.

Introduction

Titanium alloys, in general, offer low density, high specific room and elevated-temperature strength, and resistance to corrosion. Among titanium alloys, beta titanium alloys have received considerable attention in the aerospace industry because of their cold-formability and age-hardening characteristics. Timetal-21S (also called Beta 21S) is one of the newest beta titanium alloys. This alloy has shown excellent mechanical properties and resistance to corrosion [1-9] as compared to other beta titanium alloys. However, to exploit the mechanical behavior of this alloy, in general and for specific applications, it is crucial to understand the kinetics of phase transformations because they dictate the microstructure, and thus the properties. Phase transformation kinetics may be related to variables such as alloy chemistry, processing conditions, and thermomechanical treatments. For example, alloying additions can stabilize certain phases. Molybdenum stabilizes the beta phase. Alternately, oxygen is known to stabilize the alpha phase thereby raising the beta transus temperature and making transformation kinetics sluggish. The amount of oxygen will affect the transformation characteristics of this alloy, and therefore, will affect the microstructure and mechanical properties, too. Controlling the cooling rate from the beta-phase field can change the morphology, size, volume fraction and distribution of phases resulting in wide variations of mechanical properties.

The present investigation deals with the role of oxygen on the phase transformation kinetics and establish a time-temperature and transformation curves of Timetal-21S, Ti-15Mo-3Al-2.8Nb-0.2Si. Oxygen contents in the alloy was systematically varied from 0.09 wt% to 0.33 wt%.

Experimental Procedure

The materials for this study were obtained from two sources. It is noted that oxygen is an intentional addition in this alloy. First, materials with an oxygen content of 0.14 wt% were furnished by Texas Instruments, Boston, MA in the form of 0.5 mm (0.02 inch) thick rolled sheet. Second, materials with oxygen contents varying from 0.09 to 0.33 wt% were supplied by Timet Metal Corporation, Hendersen, NV in the form of 1.2 mm (0.05 inch) thick plate. The alloy chemistry in weight percent (the balance is Ti) is shown in Table 1 for each material provided. For the present work, study is concentrated on oxygen contents of 0.14 wt% and 0.23wt%.

TABLE I - Chemical Analyses of Timetal-21S

Identification	Mo	Nb	Al	Si	Fe	O	N
TI (Ave.)*	14.9	2.80	3.23	0.23	0.065	0.14	0.008
Timet V-7297	15.1	2.85	3.06	0.18	0.081	0.09	0.003
V-7504	14.5	2.68	3.05	0.23	0.083	0.13	0.010
V-7505	14.4	2.67	3.08	0.23	0.088	0.18	0.010
V-7506*	14.7	2.69	3.08	0.23	0.086	0.23	0.008
V-7518	15.2	2.80	3.14	0.20	0.054	0.33	0.004

* Predominant alloy compositions used in this study

Samples for evaluating the phase transformation kinetics were prepared as follows. Several pieces of material, approximately 2 cm x 2 cm, were encapsulated in Pyrex glass under vacuum to avoid oxidation during subsequent heat treatment operations. The encapsulated materials were heat treated at 950°C (above the beta transus temperature) for

10 minutes and then quenched in water. This was done to assure solutionization and homogenization of the alloying elements and to have a single phase as the starting microstructure for subsequent aging. Aging treatments were performed at temperatures ranging from 204°C to 760°C (400°F to 1400°F) at intervals ~111°C (200°F) apart for times ranging from 15 minutes to 100 hours.

Initially efforts were also made to use dilatometry to investigate the phase transformation behavior of the alloy but ended up with not much success. Details of dilatometry works are given in reference #10. Finaly the transformation characteristics of aged materials were studied by using light and transmission electron microscopy (TEM), and by hardness measurements. Standard metallographic techniques were used to prepare the samples for optical evaluation. The thin films for TEM were produced by electrolytic thinning using a solution composed of 12.5% methanol, 31% butanol and 6.5% of 70% perchloric acid at ~14 volts and temperatures between -40 to -50°C. The hardness measurements were performed using a Buehler Micromet II (Vickers) hardness tester with a 100 gm load. Tension tests were also conducted on the 1.2 mm thick plate materials with oxygen contents varying from 0.09 to 0.33 wt%. Tensile samples were fabricated from rectangular blanks 1.90 cm by 13.97 cm (0.75 in by 5.5 in) with a reduced gage section 0.635 cm (0.25 in) wide by 2.54 cm (1 in) long.

Results and Discussion

All materials for aging treatment were initially homogenized at 950°C for 10 minutes and water quenched. For the sake of clarity, discussion will be limited to only two oxygen contents, namely 0.14 wt% and 0.23 wt%, unless it is necessary to refer to other oxygen contents. The microstructure of the heat treated and quenched sample with an oxygen content of 0.14 wt% is shown in Figure 1. Figure 1a is an optical metallograph whereas 1b is a TEM micrograph. The optical micrograph shows polycrystals of single-phase beta, and the TEM micrograph shows dislocations with helical and loop configurations in a beta-phase matrix. These dislocations are characteristic of samples which have had excessive vacancies annihilated. Similar dislocation configurations were observed in other solution-treated titanium alloys [11]. Since the TEM foils were thinned by conventional electropolishing techniques, the planar features exhibited in the background of some figures were believed to form during the thinning process [12] and be a result of the relaxation of bulk constraints. This is a common observation of the specimens of metastable bcc titanium alloys. The planar features have the appearance of stacking faults, narrow twins, thin plates, or martensite. Similar features were noted upon observation of the specimens with varying oxygen concentrations in the quenched condition.

Timetal-21S may be classified as a lean beta titanium alloy since isothermal omega phase is formed during aging [1] (see Figures). The behavior of selected, aged samples with the two oxygen contents will be compared and contrasted. Micrographs of the specimens aged at 204°C (400°F) for five hours are shown in Figure 2. The micrograph of the specimen with 0.14 wt% oxygen content, Figure 2a shows dislocation loops and helical dislocations with planar features in the background of beta phase similar to those seen in the quenched specimens. No omega phase was observed in this specimen whereas the TEM micrograph of the specimen with 0.23 wt% oxygen content, Figure 2b, shows possible omega phase formation.

Micrographs of the specimens aged at 315°C (600°F) for 64 hours (0.14 wt% oxygen) and for 5 hours (0.23 wt% oxygen) are shown in Figure 3. The micrographs of the specimens,

Figures 3a and 3b, show omega phase in a beta phase matrix. Note that the specimen with 0.14 wt% oxygen content required a much longer aging time to produce approximately the same amount of omega phase as the specimen with 0.23 wt% oxygen.

Figure 4 presents micrographs of the specimens aged at 427°C (800°F) for 1 hour. Figure 4a is a TEM micrograph of the specimen with 0.14 wt% oxygen content showing a small percentage of omega phase whereas the TEM micrograph of the specimen with 0.23 wt% oxygen content, Figure 4b, shows only beta phase.

Micrographs of the specimens aged at 649°C (1200°F) for 0.5 hour are shown in Figure 5. Indications of small percentages of alpha phase in the grains are also evident from the TEM micrograph, Figure 5a. The TEM micrographs of the specimen with 0.23 wt% oxygen content, Figures 5, show the presence of large amounts of alpha phase.

Finally, micrographs of the specimens aged at 760°C (1400°F) for 5 hours are shown in Figure 6. Figure 6a shows a low volume fraction of alpha phase whereas Figure 6b, of the specimen with 0.23 wt% oxygen content, shows a large volume fraction of alpha phase. Also note that the relatively large density of alpha phase laths in Figure 6b are much coarser than those found in Figure 6a, indicating that the higher oxygen content promotes alpha formation.

During continuous heating or cooling in the dilatometer, there were no apparent length change patterns that could be associated with first order transformations. First order transformations can usually be detected and followed by the dilatometric technique because of the relatively large length changes that occur with these transformations. Second order phase changes are much more difficult to detect as they produce changes in the coefficient of expansion rather than in the length itself. However, at slower cooling rates there did appear to be some structure to the curves which might warrant further analysis. The experimental length change versus temperature data were fit with polynomial equations. The derivatives of such equations are in fact proportional to the coefficient of thermal expansion. The details of the temperature dependence of the length change derivatives are given in reference 10. These data suggest the possibility that two or perhaps three transformations may have initiated, each in a different temperature range, during cooling at the slower cooling rates (i.e. longer cooling times). Except for the case of ~700°C aging, the specimens which were quenched from 950°C to various temperatures ranging from 300°C to ~700°C, isothermally aged for 95 hours, and finally cooled to room temperature, did not show any conclusive results in the plots of length versus temperature. There is the indication of transformation at 701°C for aging times of 2 minutes and 4 hours. To investigate further, another set of specimens were given aging treatments at 701°C for 2 minutes. These samples were metallographically examined. The micrograph, shown in Figure 7, revealed the presence of alpha phase at the beta grain boundaries for the 2 minute aged specimens.

Hardness testing results showed a systematic change in hardness with increasing time at temperature as a result of microstructural modifications as shown in Figures 8 and 9 for specimens with oxygen contents of 0.14 wt% and 0.23 wt%, respectively. Aging at 204°C did not evolve microstructural changes, and as such, the hardness remained constant. The propensity for omega phase formation is very high in the material aged at 427°C, with long times resulting in large increases in hardness. When omega phase formation is limited, such as when aging at 315°C, the increase in hardness is smaller as compared to the 427°C aged sample. The sharp rise in the hardness vs. time plot and the subsequent slow rise is consistent with microstructural evolution of the material aged at 538°C. Aging at 538°C for short times (less than one hour) did not promote new phase formation except the possible precipitation of alpha phase at grain boundaries. Longer annealing times showed omega phase formation followed by alpha phase. Aging at 649°C and 760°C formed alpha phase in a beta matrix. Since the 760°C annealing temperature is close to the beta transus temperature, the extent and morphology of alpha phase formation is different compared to the aging temperature of 649°C. The alpha phase evolved in the 649°C aged sample was smaller in size and lesser in amount compared to the material aged at 760°C. The microstructural changes for the two aging temperatures and the corresponding hardness vs. time curves are consistent. A detailed analysis of the material microstructure and related properties is described elsewhere [13].

The tensile test results, for specimens in the "as-received" condition with oxygen contents varying from 0.09 to 0.33 wt%, are shown in Figure 10. As expected, the yield stress and tensile strength increase with increasing oxygen content whereas the percentage elongation remains constant within experimental error initially and then decreases. The elastic modulus values are shown in Figure 11. It is interesting to note that the modulus changes by approximately 30% with the increase in oxygen content from 0.09 to 0.33.

The microstructural modification as found by different aging conditions is evaluated for creation of time-temperature-transformation curves. The percentages of phases present at different aging conditions were estimated using micrographs obtained by optical and transmission electron microscopy. Based on the information of the quantitative analysis obtained by optical and transmission electron microscopy and dilatometric results, an attempt was made for the creation of time-temperature-transformation curves of Timetal-21S. Time-temperature-transformation curves are shown in Figures 12 and 13 for the two oxygen contents. It is noted that the plot is based on a limited number of aging conditions. The information of grain boundary alpha formation which comes out at an early stage of aging was obtained from samples used for dilatometric experiments.

Conclusions

1. The propensity for omega phase formation is very great at an aging temperature of 427°C for lower oxygen content alloys resulting in large increases in alloy hardness and decreases in ductility. Higher oxygen contents suppress the formation of omega phase at 427°C.
2. Alpha phase precipitates much earlier at the beta grain boundary during aging than inside the grains. Under the same aging conditions, increased oxygen contents promote alpha formation.
3. Results obtained from dilatometric technique are not easy to interpret (as compared to steel for instance) because of smaller length changes during phase transformations in Timetal-21S.
4. The alloy microstructure is stable below the aging temperature of 200°C.
5. Yield stress and tensile strength increase with increasing oxygen content, while percentage elongation decreases with increasing oxygen content.
6. The elastic modulus can increase by as much as ~30% with increase in oxygen content from 0.09 to 0.33 wt%.
7. Time-temperature-transformation (TTT) curves show a narrow zone of three phase field.

Acknowledgments

The authors gratefully acknowledge the assistance of Dr. R. A. Vandermeer and W. E. King (late) for dilatometric experiments and analysis, to C. L. Vold (late) for x-ray analysis, and to K. W. Robinson for microscopic sample preparation.

References

1. Duerig, T. W., Terlinde, G. T. and Williams, J. C., *Metall. Trans.*, **11A**, 1980, p. 1987.
2. *Aviation Week and Space Technology*, May 3, 1993, p. 36.
3. Advanced Aerospace Alloys, *Materials Engineering*, August 1991, p. 26.
4. Titanium Alloy for Composites, *Materials Engineering*, June 1991, p. 10.
5. *Aviation Week and Space Technology*, April 15, 1991, p. 69.
6. Grauman, J. S., "A New High Strength, Corrosion Resistant Titanium Alloy," TDA International Conference Presentation, 1990, Orlando, FL.
7. Wallace, T. A., Wiedemann, and Clark, R. K., "Oxidation Characteristics of Beta 21S in Air in the Temperature Range 600 to 800°C," Proc. of 7th World Conference on Titanium, 1992, p. 2117, San Diego, CA.
8. Bania, P. J. and Parris, W. M., "Beta 21S: A High Temperature Beta Titanium Alloy," TDA International presentation, 1990, Orlando, FL.
9. Mahoney, M. W., Martin, P. L. and Hardwick, D. A., "Microstructural Stability of Beta 21S," Proc. of 7th World Conference on Titanium, 1992, p. 161, San Diego, Ca.
10. Imam, M. A., Feng, C. R. and Everett, R. K., in Proc. of Titanium Metal Matrix Composite II (WL-TR-93-4105), LaJolla, CA, June 2-4, 1993.
11. Fujii, H. and Suzuki, G. G., *Metall. Trans. JIM*, **34**, 373, 1993.
12. Spurling, R. A., Rhodes, C. G. and Williams, J. C., *Metall. Trans.*, **5**, 2597, 1974.
13. Imam, M. A. and Feng, C. R., "Study of Transformation Kinetics in Timetal-21S Titanium Alloy," Proceedings of "Eight World Conference on Titanium", 22-26 October 1995, Birmingham, U. K., to be published.

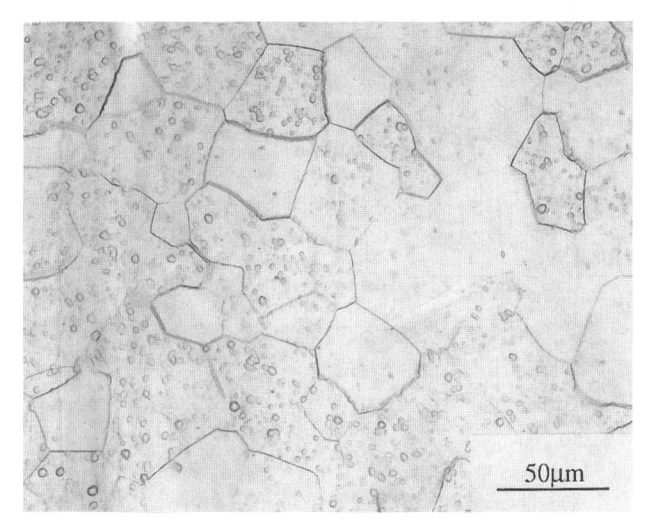

Figure 1(a) - Optical Micrograph of 0.14 wt% oxygen content material heat treated and quenched

Figure 1(b) - TEM Micrograph of 0.14 wt% oxygen content material heat treated and quenched

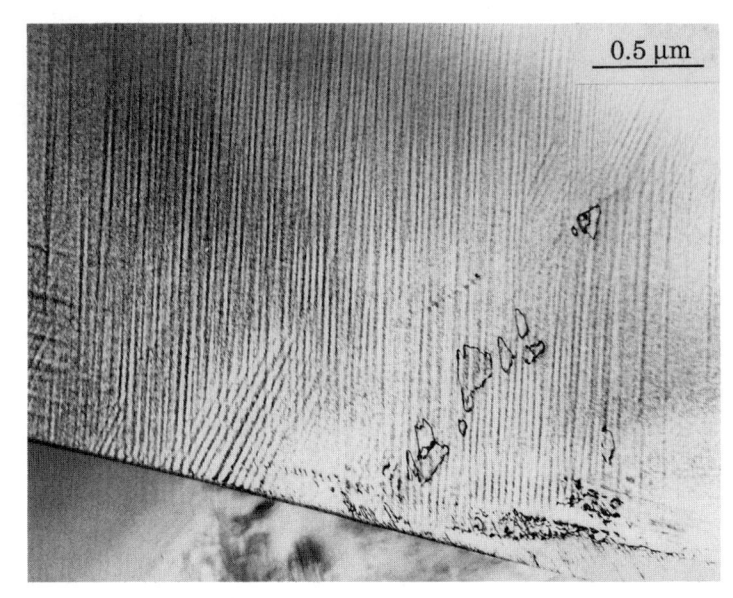

Figure 2a - TEM micrograph of 0.14 wt% oxygen content material aged at 204°C for 5 hours.

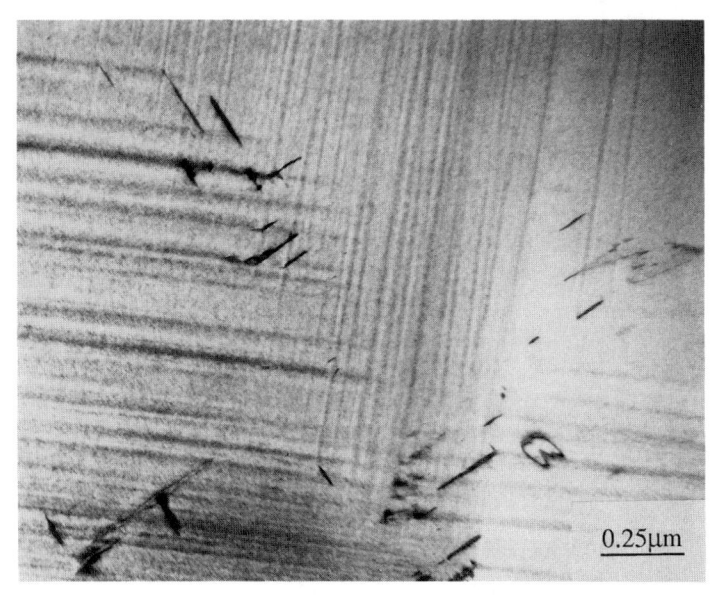

Figure 2b - TEM micrograph of 0.23 wt% oxygen content material aged at 204°C for 5 hours.

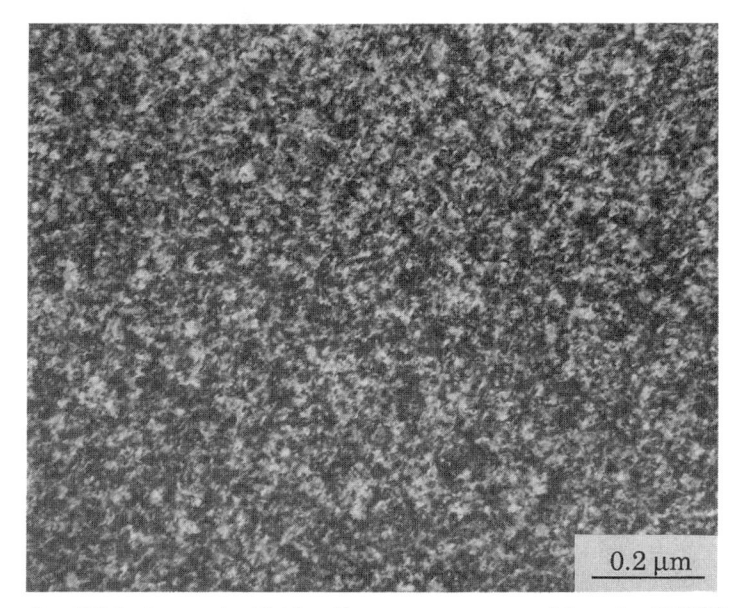

Figure 3a - TEM micrograph of 0.14 wt% oxygen content material aged at 315°C for 64 hours.

Figure 3b - TEM micrograph of 0.23 wt% oxygen content material aged at 315°C for 5 hours.

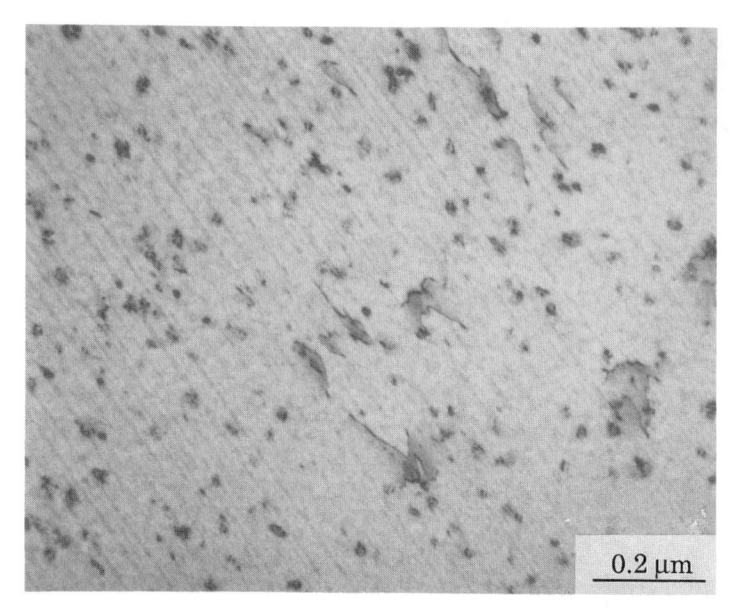

Figure 4a - TEM micrograph of 0.14 wt% oxygen content material aged at 427°C for 1 hour.

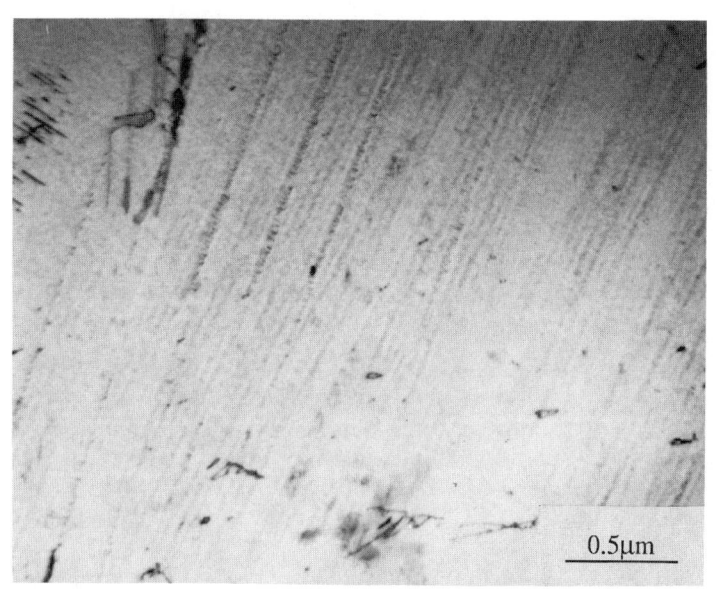

Figure 4b - TEM micrograph of 0.23 wt% oxygen content material aged at 427°C for 1 hour.

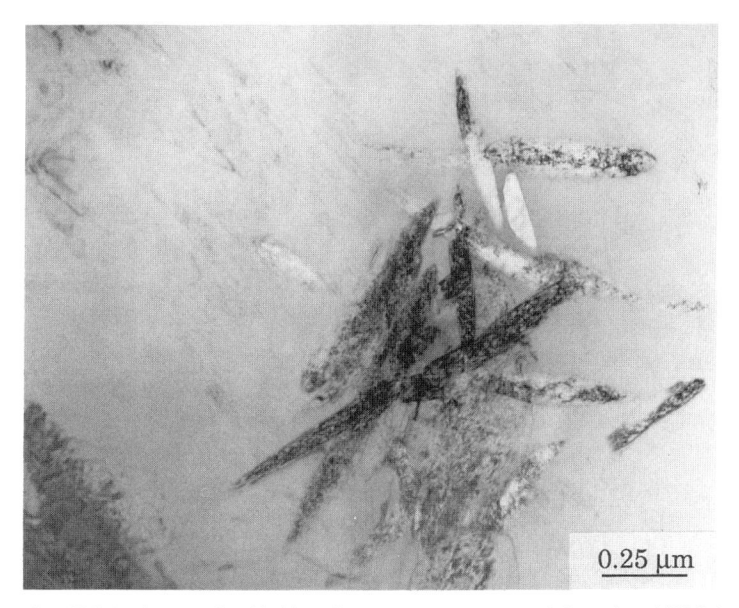

Figure 5a - TEM micrograph of 0.14 wt% oxygen content material aged at 649°C for 0.5 hour.

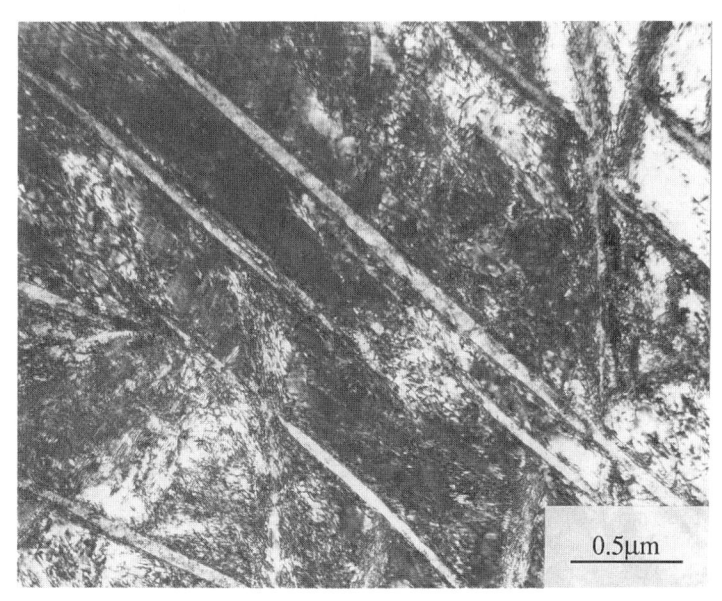

Figure 5b - TEM micrograph of 0.23 wt% oxygen content material aged at 649°C for 0.5 hour.

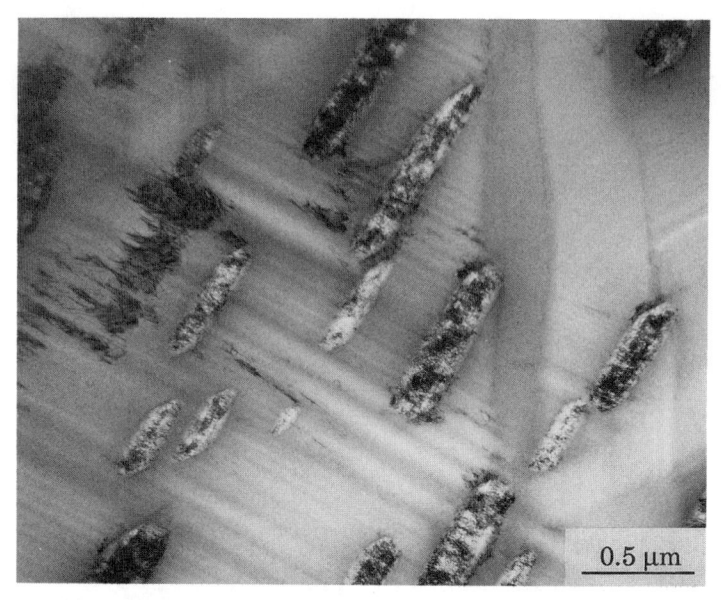

Figure 6a - TEM micrograph of 0.14 wt% oxygen content material aged at 760°C for 5 hours.

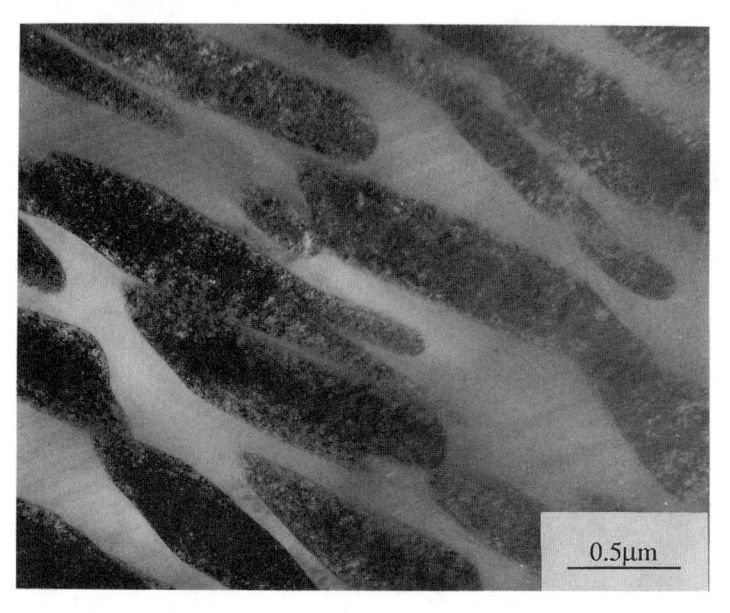

Figure 6b - TEM micrograph of 0.23 wt% oxygen content material aged at 760°C for 5 hours.

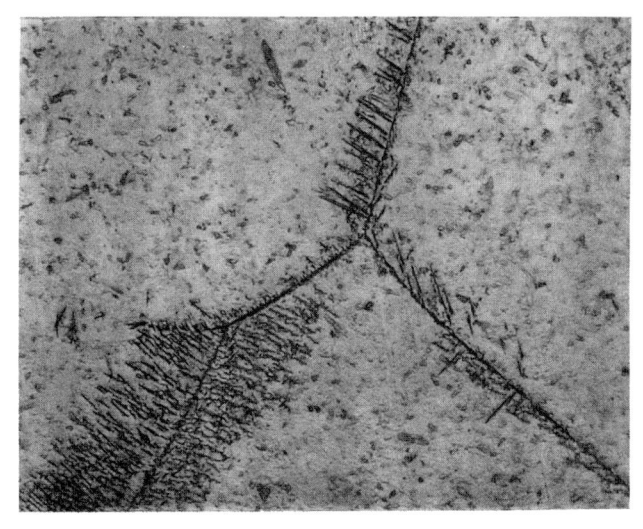

Figure 7 - Optical micrograph of 0.14 wt% oxygen content material after 950°C homogenization, water quench, and 701°C aging for 2 minutes showing some alpha phase at the grain boundaries (and also etch pits in the grains).

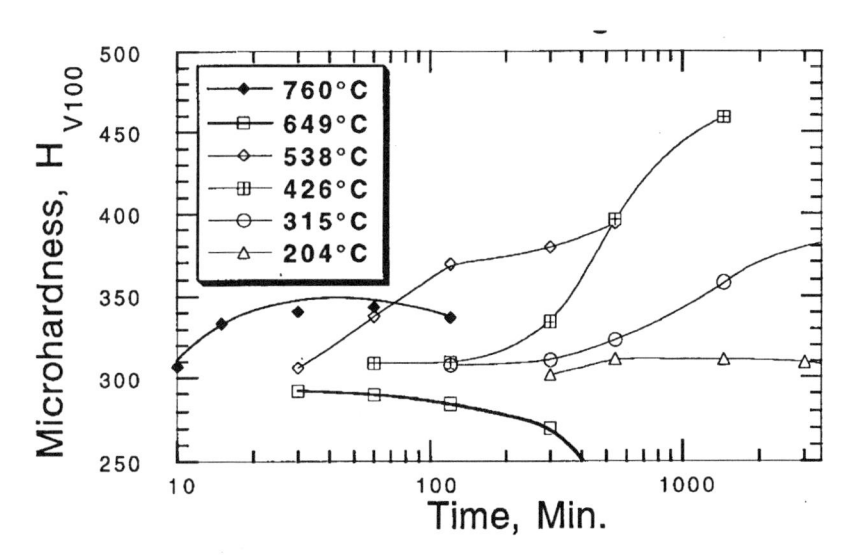

Figure 8 - Average hardness of Timetal 21S as a function of isothermal aging conditions for 0.14 wt% oxygen material.

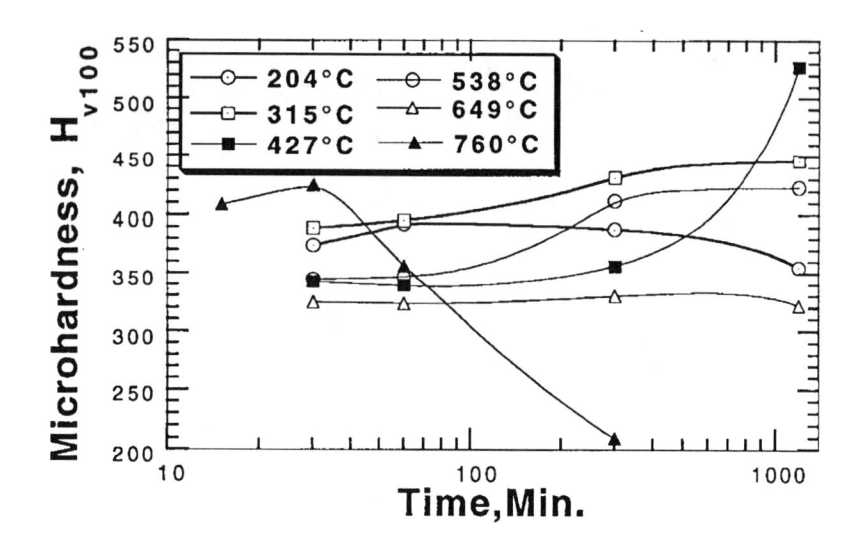

Figure 9 - Average hardness of Timetal 21S as a function of isothermal aging conditions for 0.23 wt% oxygen material.

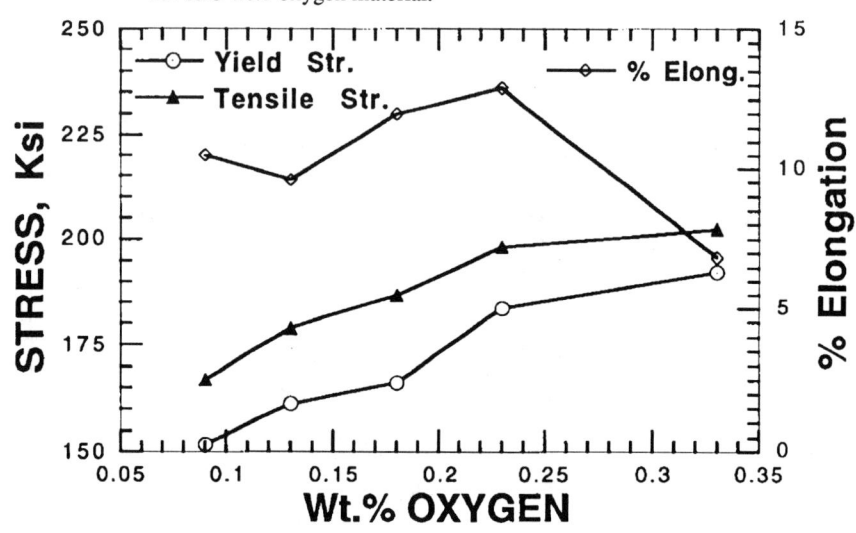

Figure 10 - Yield strength, tensile strength, and percentage elongation as a function of oxygen content.

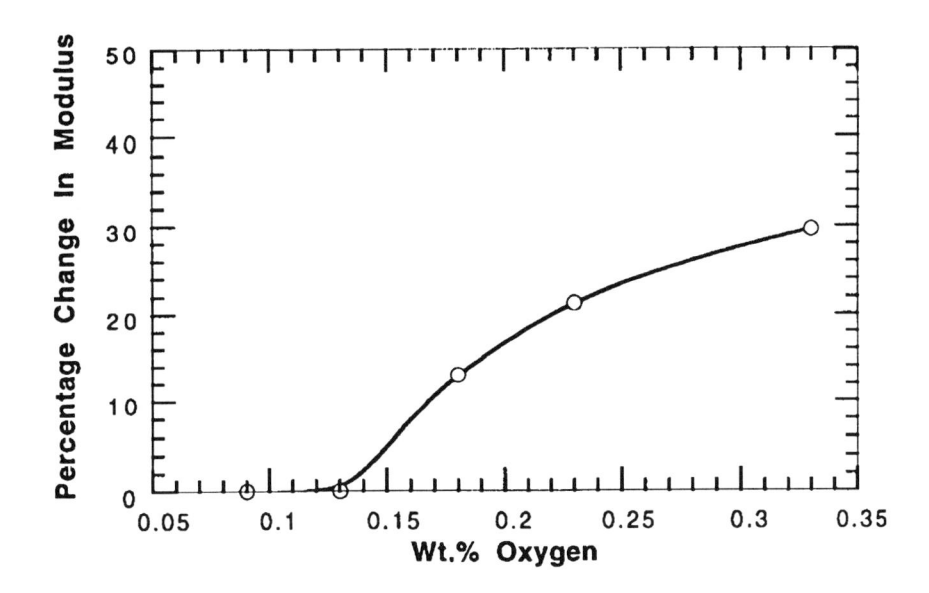

Figure 11 - Modulus change as a function of oxygen content.

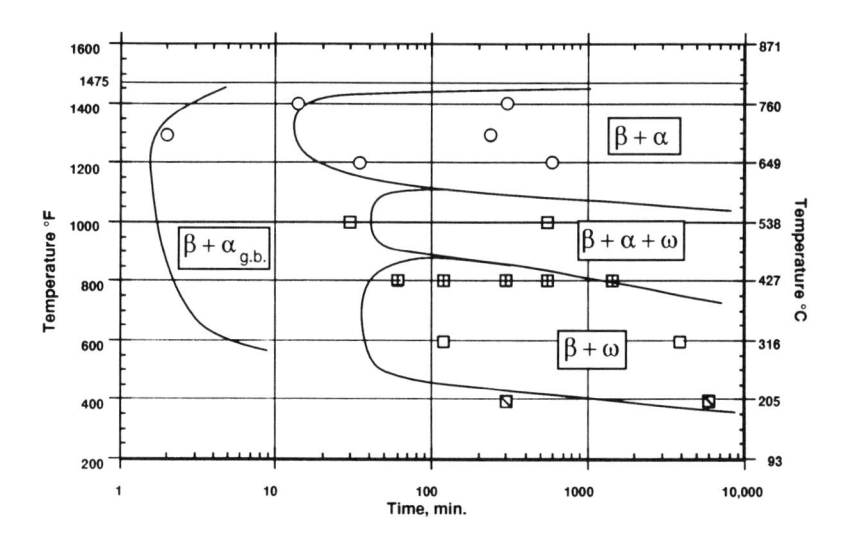

Figure 12 - Time-temperature-transformation curve of Timetal-21S for 0.14 wt% oxygen material.

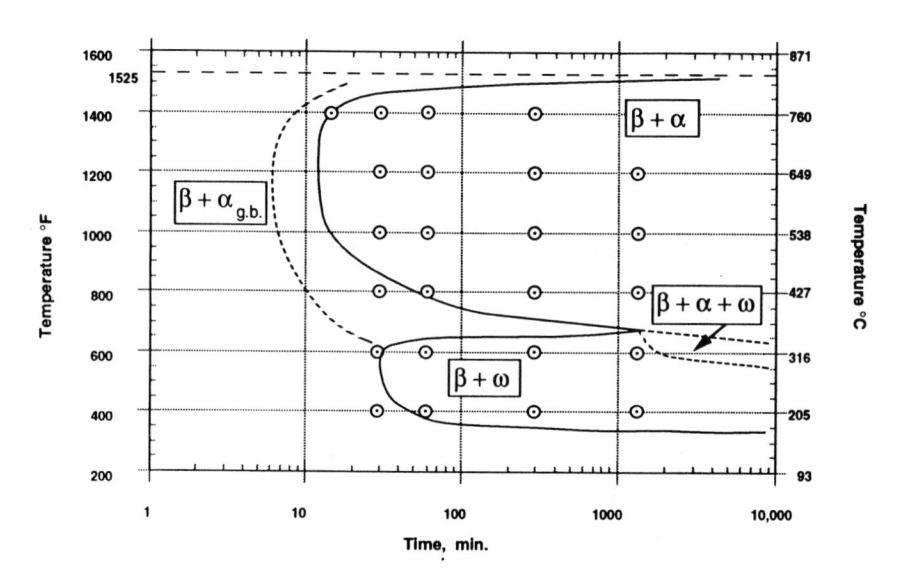

Figure 13 - Time-temperature-transformation curve of Timetal-21S for 0.23 wt% oxygen material.

PROCESSING - MICROSTRUCTURE - PROPERTY RELATIONSHIPS IN Ni MODIFIED CORONA 5

P.L. Martin and J.C. Fanning*

Rockwell Science Center
1049 Camino Dos Rios
Thousand Oaks, CA 91360

*Timet Henderson Technical Laboratory
P.O. Box 2128
Henderson, NV 89015

Abstract

Three alloys, with compositions based on Corona 5 (Ti-4.5Al-5Mo-1.5Cr), were defined in an attempt to modify the relative strength and volume fractions of the α and β phases. The Al and oxygen levels were altered to affect the α strength while Mo, Cr and Ni were added to increase the volume fraction of β. Ingots were double vacuum arc remelted (VAR), β forged and finally α/β forged to produce slab. Sections of the slab were cross-rolled at temperatures both above and below the beta transus. The microstructure at each stage of processing was documented using backscattered electron metallography on electropolished sections. Multi-step heat treatments, beginning both above and below the β transus, were used to modify the primary and secondary α morphology in an attempt to maximize the strength/toughness trade-off. The β processed material tended to have higher combinations of strength and toughness. The alloys were found to have superior strength and toughness compared to typical values for Ti-10-2-3 and the maximum data measured for Corona-5. Ambient temperature plane strain fracture toughness values as high as 84 MPa$\sqrt{m}$ were measured at a corresponding UTS of 1120 MPa. The maximum strength was 1340 MPa with a corresponding toughness of 58 MPa$\sqrt{m}$.

This work was sponsored by Boeing Commercial Aircraft Group.

Introduction

The impetus for this study was to improve upon the work that resulted in the formulation of the moderate strength - high fracture toughness α/β titanium alloy called Corona-5 [1-3]. In a series of programs the ingot melting scale-up, oxygen sensitivity, thermomechanical processing, cold rollability, heat treatment, SPF/DB, tensile strength, toughness, fatigue and crack growth properties of Corona-5 were thoroughly investigated. Ingots as large as 5,000 pounds were VAR melted prior to processing to mill product for closed die forgings and bar stock for sheet rolling. The alloy was found to be more forgiving than lower volume fraction α/β alloys to oxygen content (as high as 0.2 wt%) and thermomechanical processing conditions. Furthermore, Corona-5 could be produced to one microstructure for secondary processing, such as fine equiaxed grains for SPF, and then heat treated to the transformed β higher toughness microstructure as a final step. Very high combinations of tensile strength and fracture toughness were found following β working and subtransus annealing. The best properties were obtained in a microstructure that consisted of high aspect ratio "primary" α needles in a matrix of transformed β containing finer α precipitates.

However, the Corona-5 composition was never used as a base for subsequent alloy "optimization"; it was selected from a group of 16 alloys studied in the original program. There is every reason to suspect that improved properties are possible by changing the relative strengths and volume fractions of the α and β phases. Using SP700 and β-CEZ as examples, an increase in the Mo equivalent of Corona-5 to approximately 5.0 (excluding the effect of oxygen) seemed appropriate. To avoid decreasing the SPF properties, it seemed best to accomplish this through the use of the anomalous diffusing species Fe, Cr or Ni. Using this strategy, intermetallic compound formation (e.g. TiFe, $TiCr_2$ or Ti_2Ni) might be a problem unless there was a sufficient quantity of isomorphous β stabilizer (the refractory metals). Since Corona-5 was already rich in Mo, most of the additional β stabilizer could be of the eutectoid type in order to maximize the potential SPF benefit.

Preliminary studies of small induction skull melted ingots showed that alloys with Ni additions had a promising combination of strength and toughness [4]. This paper will describe the initial data on alloys based on the composition: Ti-4.5Al-5.5Mo-2Cr-1Ni.

Experimental Procedure

Double melted VAR ingots weighing 36 kg were produced at Timet. The ingots were side forged from 20 cm in diameter to 10x15 cm, beginning at a temperature of 1093°C (2000°F). The second forging, from 10 to 5 cm thick, was done in the α+β field beginning at 871°C (1600°F). Using pieces from the 5 cm thick forgings, metallographic examination was used to determine the actual β transus temperature through heat treatment at successively higher temperatures at 12°C increments. Two portions, each approximately a third of the 5 cm thick forging, were cross-rolled from 28°C above and below the β transus in order to evaluate the effect of working in the single versus the two phase field (hereafter referred to as either β rolled or α/β rolled). This operation resulted in plate slightly over 1.9 cm thick. All subsequent evaluations were conducted on this material.

Heat treatments were defined with reference to the early Corona-5 literature and, whenever possible, were conducted at constant values below T_β. All specimens for high temperature heat treatment were wrapped in Ta foil and encapsulated in quartz after partially backfilling with Ar. Air cooling was used for all treatments except those done at or above T_β where a fan air cool was used. Similar precautions were used in heat treating the mechanical property specimens. Electropolished surfaces (unetched) were examined in a CamScan SEM using backscattered electron images (BEI). With this technique, etching artifacts are avoided and magnifications higher than those possible with optical metallography are possible. All metallographic data presented will be from the longitudinal section of the rolled plate.

All tensile testing was conducted at a strain rate of 8×10^{-4} sec^{-1} in accordance with ASTM E-08. Fracture toughness testing used compact tension specimens and ASTM E-399 procedures. All fracture toughness values reported are valid K_{1c}.

Results and Discussion

The chemical compositions of the 3 alloys are listed in Table I. They closely approximate the aim compositions except for the slightly higher than desired oxygen in heat V-7819.

Table I. Compositions of 80 pound VAR ingots used (wt.%).

Heat #	Al	Mo	Cr	Ni	C	Fe	O	N	H
V-7817:									
Top	4.49	5.40	1.99	0.96	0.022	0.039	0.158	0.002	0.0037
Bottom	4.56	5.41	1.98	0.96	0.023	0.041	0.108	0.004	0.0054
V-7818:									
Top	4.99	5.41	2.00	0.99	0.012	0.038	0.117	0.004	0.0035
Bottom	5.17	5.62	2.04	0.96	0.018	0.038	0.112	0.007	0.0058
V-7819:									
Top	4.48	5.37	1.96	0.96	0.017	0.042	0.185	0.033	0.0077
Bottom	4.60	5.49	2.06	1.00	0.012	0.042	0.218	0.039	0.0086

Examination of these compositions shows that V-7818 and V-7819 have a higher content of α stabilizing elements relative to the base alloy V-7817; V-7818 has higher Al (at constant oxygen) and V-7819 has higher oxygen (at constant Al). This intentional variation will highlight the effects of strengthening the α phase relative to the β and the efficacy of accomplishing this with Al or oxygen.

Up-quenching experiments for the 3 alloys revealed that V-7817 and 7818 had similar beta transus temperatures; 910°C for both alloys. However, the high oxygen/nitrogen variant, V-7819, showed a transus of 950°C due to their potent effect on α stability. In addition, the α particles produced by aging just under T_β in the high oxygen alloy were non-uniformly distributed relative to the other compositions.

The as-rolled microstructures of the 3 alloys were similar. The plate rolled from above T_β had large grains, ≈200μm, that were somewhat elongated in the rolling direction. The air cool following the final rolling pass resulted in the precipitation of fine α needles, Figure 1(a). The unrecrystallized β grain boundaries were decorated with a layer of α less than 1μm thick. In contrast, the plate rolled from below T_β had a very fine grain size, ≈5μm, with elongated primary α, Figure 1(b).

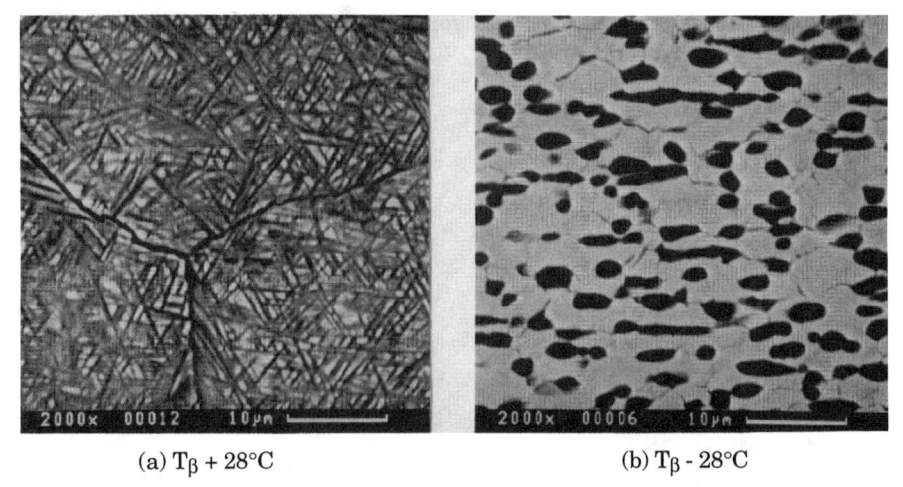

(a) $T_\beta + 28°C$ (b) $T_\beta - 28°C$

Figure 1. BEI metallography of V-7817 as-rolled from above and below T_β.

Microstructure Development

A variety of heat treatments were applied to the plate material. In all cases, their rationale was based on experience with Corona 5 [1-3,5]. The β rolled material was given a high temperature anneal for 2 hours at approximately 50, 100 and 150°C below T_β followed by an age at 350°C below T_β for 8 hours. The α/β rolled plate was solution treated for 15 minutes at $T_\beta+5°C$ followed by a fan air cool and then given the same 2-step subtransus annealing treatments. In all cases, the goal was to obtain a bimodal distribution of high aspect ratio α particles in a tough transformed β matrix. The large volume of microstructural data precludes its complete presentation in this paper, but the microstructures for all 3 alloys were similar. Typical examples will be shown for each combination of working operation and annealing treatment.

Figure 2 compares the microstructure of both β and α/β rolled V-7817 in the coarsest microstructure evaluated. In both cases, the volume fraction, size and aspect ratio of the primary α needles are virtually identical. The grain boundary α layers are similar, but there is some tendency for incomplete coverage of the boundaries in the β worked plate. The most noticeable difference is the shape of the primary β grain boundaries; they are serrated in the subtransus aged plate and extremely smooth in the α/β rolled and solution treated material. The α precipitates formed during the low temperature (final) age are too fine to be observed at this magnification.

Figure 3 makes a similar comparison of β and α/β rolled microstructure for V-7818 in the finer microstructure formed with the lower temperature anneal cycle. As seen before, the primary α morphology is identical; only the grain boundary shape is different as revealed by the thin α layer.

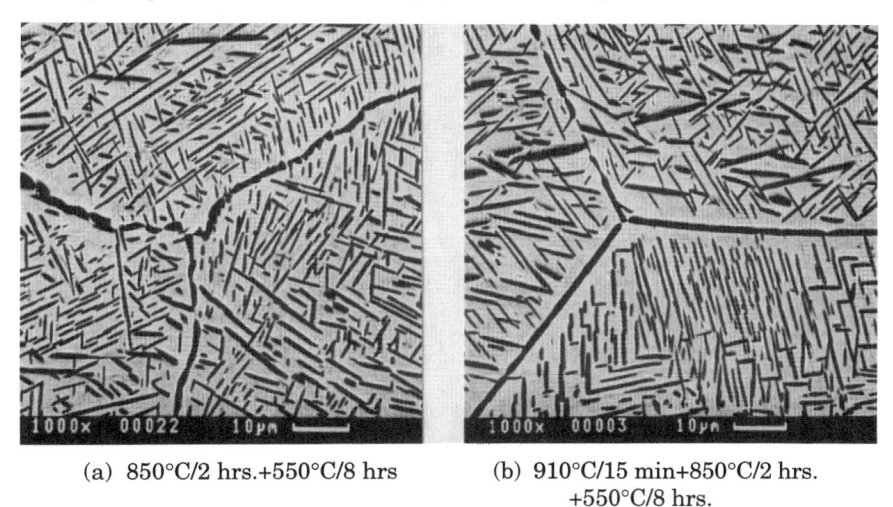

(a) 850°C/2 hrs.+550°C/8 hrs

(b) 910°C/15 min+850°C/2 hrs. +550°C/8 hrs.

Figure 2. BEI images of heat treated V-7817 plate; (a) β rolled, (b) α/β rolled.

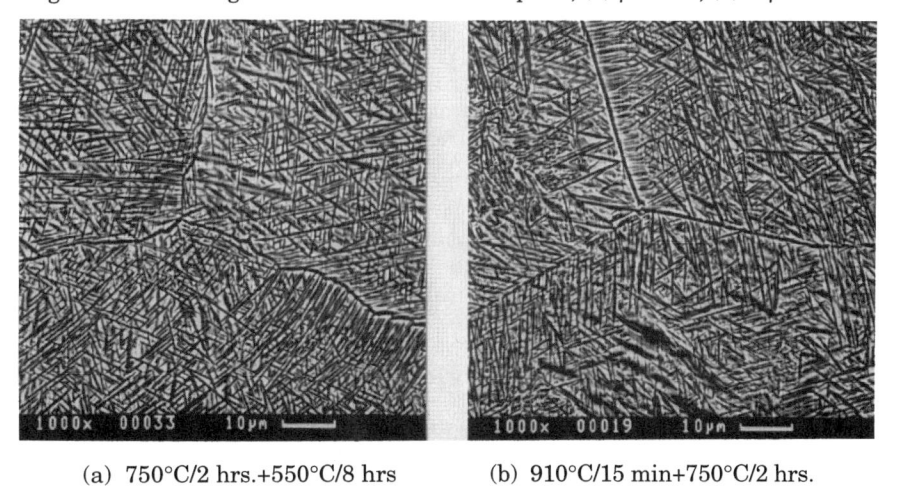

(a) 750°C/2 hrs.+550°C/8 hrs

(b) 910°C/15 min+750°C/2 hrs. +550°C/8 hrs.

Figure 3. BEI images of heat treated V-7818 plate; (a) β rolled, (b) α/β rolled.

The only exception to the comparability of the microstructures obtained with heat treatments defined with respect to the β transus occurred for the high oxygen alloy, V-7819. In the coarsest microstructure obtained with the high subtransus anneal, the high volume fraction of the secondary α was observable, Figure 4, as faint contrast in the matrix. Comparison to Figure 2 will highlight the effect of the higher oxygen on the secondary α. Morphologies of the grain boundaries and their ubiquitous coverage by α were the same as noted for the other alloys.

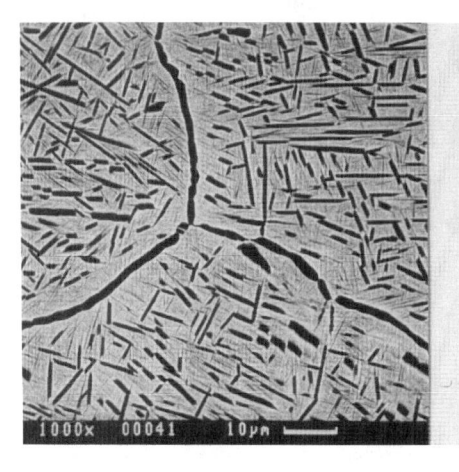

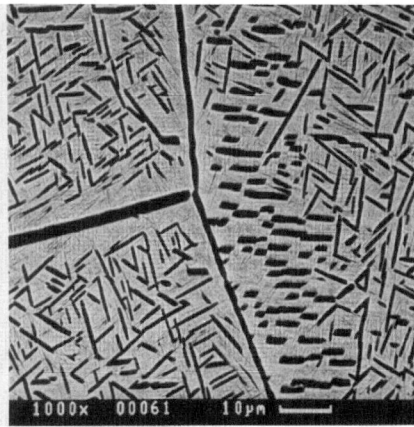

(a) 890°C/2 hrs.+590°C/8 hrs (b) 950°C/15 min+890°C/2 hrs. +590°C/8 hrs.

Figure 4. BEI images of heat treated V-7819 plate; (a) β rolled, (b) α/β rolled.

Mechanical Properties

Tensile and plane strain fracture toughness values for these alloys, processing routes and microstructures are listed in Table II. While not shown, the transverse tensile and TL fracture toughness properties were consistent with those tabulated and little anisotropy was observed.

Comparison of the effect of heat treatment within each alloy group indicates that higher temperature primary annealing treatments result in higher tensile strengths and reduced ductility. Since the final aging treatment was kept constant, the higher temperature anneal produced a coarser primary α needle morphology; compare Figure 2(a) to 3(a). This initial treatment also resulted in more solute rich β phase and therefore a higher volume fraction of fine α (unresolved in the SEM micrographs shown) that was precipitated in the final age. Generally, the fracture toughness varied in the opposite direction to the tensile strength; lower strength led to higher fracture toughness within a given alloy/process history. The only exception to this was for α/β rolled V-7919.

Comparison of the effect of hot rolling temperature on properties for each alloy also reveals some consistent trends. The ultimate tensile strength is not dramatically affected by the rolling temperature while the ductilities are slightly lower in the α/β rolled and solution treated condition. The most dramatic effect is noted in the fracture toughness. The β rolled plate consistently shows higher fracture toughness at nearly the same strength levels compared to the α/β rolled plate. The normally observed trade-off between tensile strength and ductility is illustrated in Figure 5 which plots UTS versus K_{1c} for this data set. Since the α volume fraction and aspect ratio, as observed at the SEM magnification level,

Table II. Tensile and fracture toughness data; average of 2 tests for tensile data and single specimens for toughness. Longitudinal orientation for the tensile specimens and LT crack propagation plane for the toughness tests.

Heat Treatment	0.2% YS (MPa)	UTS (MPa)	Elong. (%)	RA (%)	K_{1c} (MPa√m)
V-7817, β rolled:					
850/550	1135	1235	8	16	64.8
750/550	1063	1121	11	26	84.8
V-7817, α/β rolled:					
910/850/550	1166	1259	6	11	58.9
910/750/550	1083	1128	9	16	72.2
V-7818, β rolled:					
850/550	1190	1280	7	14	62.3
750/550	1107	1170	11	21	73.8
V-7818, α/β rolled:					
910/850/550	1239	1342	4	8	58.2
910/750/550	1114	1163	9	18	69.3
V-7819, β rolled:					
890/590	1225	1294	9	19	62.9
790/590	1170	1211	13	24	71.6
V-7819, α/β rolled:					
950/890/590	1180	1259	5	10	62.7
950/790/590	1121	1159	10	18	56.9

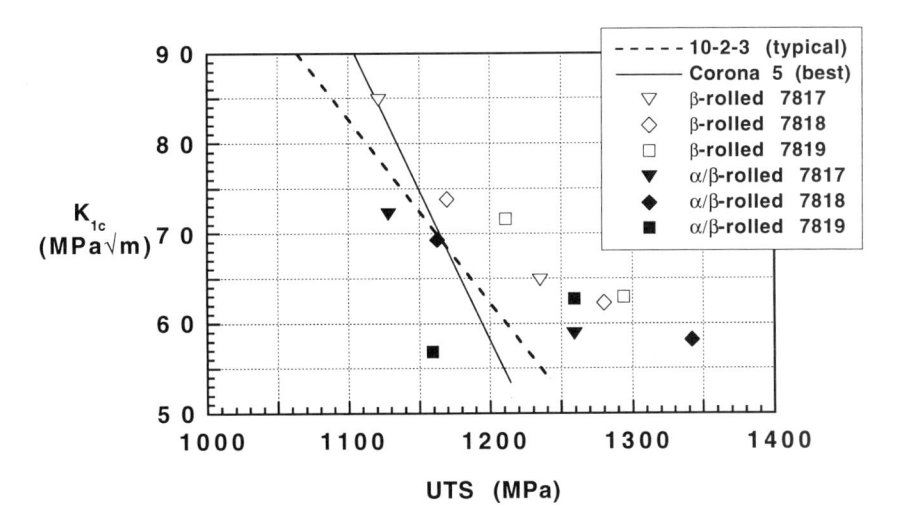

Figure 5. UTS versus K_{1c} plotted for the data in Table II.

are virtually identical for this processing and heat treatment schedule, the only observed difference is the grain boundary morphology. The serrated nature of the prior β boundaries is distinctly changed by the solution treatment; compare Figures 2, 3 and 4.

Comparison of alloy element effects shows that the base composition, V-7817, developed the highest fracture toughness values. V-7818 and V-7819 contained elements that would strengthen the α relative to the β phase. It is not surprising, therefore, that these alloys showed generally higher strengths.

It should be remembered that the heat treatment strategy used in this study was based on the extensive work on Corona 5. This limited series of heat treatments should not be construed as representing the optimized values for any of these compositions. Balancing the volume fraction, morphology, distribution and relative strengths of the α and β phases is difficult and warrants further study.

Conclusions

1. Promising strength/fracture toughness properties are possible in Ni modified Corona 5.

2. In all 3 alloys, β rolled plate exhibits better fracture toughness properties. The β grain boundary morphology is the most obvious difference in the microstructures evaluated.

3. Aluminum and oxygen can increase the strength of the base alloy, albeit with a decrease in the toughness.

References

1. R.G. Berryman, F.H. Froes, J.C. Chesnutt, C.G. Rhodes, J.C. Williams and R.F. Malone, "High Toughness Titanium Alloy Development", TFD-74-657 Naval Air Systems Command contract N00019-73-C-0335, July 1974.

2. R.G. Berryman, F.H. Froes, J.C. Chesnutt, C.G. Rhodes, J.C. Williams and R.F. Malone, "High Toughness Titanium Alloy Development", TFD-75-640 Naval Air Systems Command contract N00019-74-C-0273, July 1975.

3. R.G. Berryman, J.C. Chesnutt, F.H. Froes and C.G. Rhodes, "High Toughness Titanium Alloy Development", TFD-76-471 Naval Air Systems Command contract N00019-75-C-0208, June 1976.

4. P.L. Martin, "Toughness-Strength Trend in Alloys Derived from Ti-4.5Al-5Mo-1.5Cr", to be published in *Proceedings of the 8th World Conference on Titanium*, edited by M. Loretto et al., Institute of Materials, London, 1996.

5. G.R. Keller, J.C. Chesnutt, W.T. Highberger, C.G. Rhodes and F.H. Froes: "Relationship of Processing/Microstructure/Mechanical Properties for the Alpha-Beta Titanium Alloy Ti-4.5Al-5Mo-1.5Cr (Corona-5)," *Titanium '80 Science and Technology,*' H. Kimura and O. Izumi, eds., TMS, Warrendale, 1980, pp1209-1220.

A STUDY OF GRAIN BOUNDARY NUCLEATED α PRECIPITATES IN AN (α+ß) TITANIUM ALLOY

Kei Ameyama, Hiroshi Fujiwara, Hajime Kawakami and Allec Mitchell*

Dept. Mechanical Engineering, Ritsumeikan University,
Kusatsu, Shiga, Japan 525-77
*Dept. Metals & Materials Engineering, University of British Columbia,
Vancouver, B.C., Canada V6T 1Z4

Abstract

Morphology and crystallography of grain boundary α (HCP) precipitates nucleated at ß (BCC) matrix grain boundaries have been studied with TEM and SEM in a Ti-22V-4Al alloy. The α precipitates show a globular allotriomorph or an elongated film morphology. Their shape is an elongated lath with a growth direction lying close to the grain boundary plane. Although intragranular α precipitates are very close to a Burgers orientation relationship (OR), the grain boundary α precipitates show different crystallographical features. According to Kikuchi pattern analysis of α precipitates at β grain boundaries, they are able to be classified in three types; Type A: a near-Burgers OR to one β grain and no specific OR to another one. Type B: a near-Burgers OR to one β grain and a near-$\{1\bar{1}01\}\alpha$ // $\{110\}\beta$ OR to another one. Type C: a near-$\{1\bar{1}01\}\alpha$ / $\{110\}\beta$ OR to both β grains. Since a near-Burgers OR is a particular case of a near-$\{1\bar{1}01\}\alpha$ //$\{110\}$ β OR, the near-$\{1\bar{1}01\}\alpha$ / $\{110\}\beta$ OR is a common crystallographical feature in all types of the grain boundary α precipitates. This $\{1\bar{1}01\}\alpha$ plane is confirmed to be parallel to one of the twelve $\{110\}\beta$ planes in the adjacent β grains, which has the minimum angle with the grain boundary plane. Therefore, the variants of an α precipitate at a β grain boundary are selected from 24 (12 possible variants in each β grain) to 2 at the most, and it reduces to only one in case of good coherency to the opposite β grain.

Introduction

Grain boundaries play an important role in many diffusional transformations in alloys as nucleation sites of precipitation. The grain boundary precipitates have, in general, a unique orientation relationship (OR) with, at least, one of the adjacent matrix grains [1-11]. Furthermore, in many cases, restriction of variants of the grain boundary precipitate are observed. The variant selection is found to be due to effect of grain boundary plane orientation on the OR between the grain boundary precipitate and the matrix grains in Fe-C alloys and ($\alpha+\gamma$) two phase stainless steels [6-9]. Single variant precipitates grow to form film morphology by the coalescence of each other, while multiple variant precipitates form globular morphology by impingement.

In the case of the ($\alpha+\gamma$) two phase stainless steels [6, 8, 9], variants of the precipitates (γ) are restricted by a rule that the close packed {110} plane of matrix grain (α) which is the most close to parallel to the grain boundary plane is chosen for a Kurdjumov-Sachs OR between the precipitate and the matrix grain. Recently, Furuhara et al [11] indicated the dependence of morphology of grain boundary precipitates on the variant selection in an ($\alpha+\beta$) Ti alloy. They concluded that a critical nucleus with a particular variant which has an elongated shape along the grain boundary plane is preferred, since a large grain boundary area will be destroyed by such a nucleus formation. These results are consistent with the calculation done by Lee and Aaronson [12, 13] that a precipitate at the grain boundary should have its low energy interface inclined very close to the grain boundary to minimize the activation energy for critical nucleus formation.

In the present study, the crystallography and morphology of α (HCP) phase nucleated at β (BCC) grain boundaries in a near-β Ti-22V-4Al alloy is described. The variant selection of the grain boundary α precipitates is discussed from a standpoint of the orientation of grain boundary plane. Kikuchi pattern analysis for both HCP and BCC phases are employed to carry out the accurate orientation analysis.

Experimental Procedure

A near-β Ti-22V-4Al alloy containing V: 22.2, Al: 4.2, Fe: 0.14, O: 0.11, N: 0.008, C: 0.008 and H: 0.017 (mass%) was prepared by vacuum melting. The β transus and α precipitation nose temperatures of the alloy are approximately 1000 K and 873 K, respectively, so that the alloy was solution treated at 1573 K for 18 ks and aged at 923 K for 400 ks under He gas atmosphere followed by furnace cool to obtain ($\alpha+\beta$) microstructure. Prior to the solution treatment, the specimen was 90 % cold rolled to make the grain boundary plane normal to the specimen foil by recrystallization.

Specimens for TEM were prepared by mechanical polishing and ion milling technique and examined in a JEOL 2010 electron microscope at 200 kV. The relative orientation relationship of adjacent β grains and α precipitates was determined by Kikuchi pattern analysis. The error associated with this technique was assessed by analysis of annealing twins in FCC crystals. This gave an average error less than 0.10 degree. Grain boundary plane was determined by trace analysis.

Results and Discussion

Morphology of grain boundary α phase

Figure 1 shows (a) SEM back scattered image and (b) SEM grain boundary fractograph of the specimen aged at 923 K for 400 ks. The specimen shown in (b) was fractured approximately at 77 K. As seen in (a), α precipitation took place mainly at β grain boundary, and the grain boundary (GB) α precipitates show different morphologies, i.e., a globular allotriomorph and an elongated film like morphology, even at the same grain boundary. Such a morphological difference seems to be attributed to the grain boundary inclination, in other words, the grain boundary orientation.

The grain boundary fractograph (b) shows a rod like phase elongated in two directions on the grain boundary. The traces of these directions are shown in the figure. Considering the morphology of GB α shown in (a), the rod like phase is thought to be GB α precipitates. In addition, only one growth direction of the GB α precipitates on some grain boundaries implies

Fig.1 SEM back scattered image (a) and SEM grain boundary fractograph (b) of a specimen aged at 923 K for 400 ks. Morphology of grain boundary α phase depends on the grain boundary inclination (a), and the α phase elongate to two directions as indicated in (b).

that a certain restriction on the growth direction should be exist.

Crystallography

Figure 2 shows TEM micrograph of GB α precipitates formed at a grain boundary in the specimen aged at 923 K. Orientation relationship among GB α and matrix β grains and deviation from an exact Burgers OR, i.e., (0001)α // (110)β, [11$\bar{2}$0]α // [1$\bar{1}$1]β, are indicated in Table 1. The GB α precipitates, marked as α1 and α2, are related by a near-Burgers OR in different variants with respect to β1 grain, while no specific OR was observed with another β grain. Although an intragranular α precipitates and a β matrix grain are in very close to a Burgers OR in the present alloy, those α precipitates at the β grain boundary showed not a little deviation from the OR. In case of α1 precipitate, its OR is much closer to a Potters OR, i.e., (1$\bar{1}$01)α // (110)β, [11$\bar{2}$0]α // [1$\bar{1}$1]β, rather than a Burgers OR. Since these two ORs are related only by approximately 1.5 degree rotation around a common [11$\bar{2}$0]α (// [1$\bar{1}$1]β) axis, for the sake of convenience we would describe the deviated ORs including a Potters OR as a "near-Burgers OR", whose deviation angle less than 5 degree from the exact one.

Furthermore, angles between {110}β planes and the grain boundary plane are also indicated in Table 1. It is noteworthy that the (1$\bar{1}$0)β1 plane, in which (0$\bar{1}$11)α1 and (0$\bar{1}$11)α2 planes are almost parallel, has the minimum angle with the grain boundary plane.

Figure 3 shows another example of GB α in the specimen aged at 923 K for 400 ks. The α allotriomorph has a near-Burgers OR with respect to both β grains. Deviation angles between the planes are summarized in Table 2. As well as the GB α shown in Fig. 2, one of {1$\bar{1}$01}α planes is in a nearly parallel relationship with ($\bar{1}$01)β1 plane, which has the minimum angle with the grain boundary plane.

Such a crystallographical feature that one of {1$\bar{1}$01}α planes have a near-parallel relationship with the (110)β plane, which is the most close to parallel to the grain boundary plane, was confirmed in all GB α precipitates examined.

Figure 4 indicates the smallest deviation angles between a{1$\bar{1}$01}α plane and a {110}β plane and between a (0001)α plane and a {110}β plane of GB α precipitates formed at 18 different grain boundaries with respect to both β grains. In the graph GB α precipitates are fixed to have a near-Burgers OR to β1 grain, which is shown in left side. It is clear that the GB α precipitates are able to be classified into three types according to the deviation angles. Type A has a near-Burgers OR with one β grain and no specific OR to another β grain. Type B has a near-Burgers OR with one β

Fig. 2 TEM micrograph of a specimen aged at 923 K for 400 ks. Orientation relationship is summarized in Table 1.

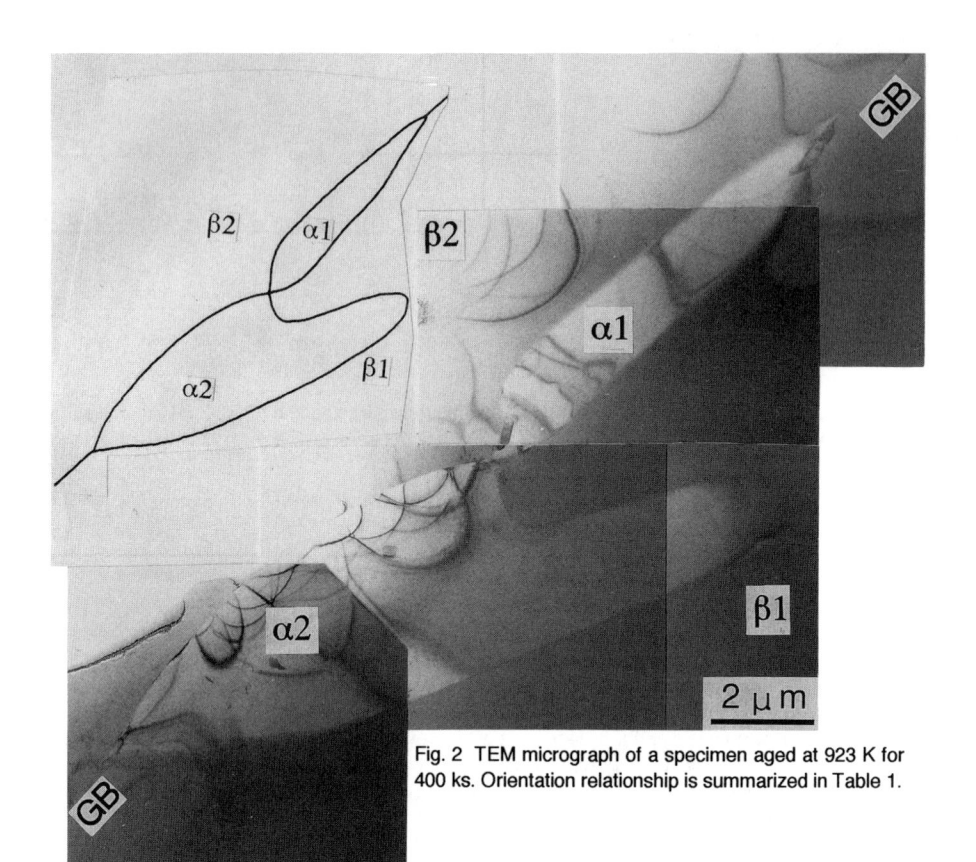

Table 1
Orientation relationship and angles between {110}β planes and the grain boundary plane of the grain boundary α and the adjacent β grains shown in Fig. 2.

misorientation between β grains		
[$\overline{0.1155}$ 0.2193 0.9688] β1 / 45.3 deg.		
	variant of GB α	deviation (deg.)
α1 / β1	$(0\bar{1}11)_{α1}$ / $(1\bar{1}0)_{β1}$	0.15
	$(0001)_{α1}$ / $(101)_{β1}$	1.68
	$[\bar{2}110]_{α1}$ / $[\bar{1}\bar{1}1]_{β1}$	0.73
α2 / β1	$(0\bar{1}11)_{α2}$ / $(1\bar{1}0)_{β1}$	1.19
	$(0001)_{α2}$ / $(0\bar{1}1)_{β1}$	0.46
	$[\bar{2}110]_{α2}$ / $[111]_{β1}$	0.36

angle between {110}β and grain boundary plane (deg.)			
$(1\bar{1}0)_{β1}$	**26.4**	$(110)_{β2}$	65.8
$(\bar{1}01)_{β2}$	29.0	$(101)_{β2}$	67.9
$(1\bar{1}0)_{β2}$	32.7	$(101)_{β1}$	70.2
$(\bar{1}01)_{β1}$	33.6	$(110)_{β1}$	74.0
$(011)_{β1}$	56.1	$(0\bar{1}1)_{β1}$	86.4
$(011)_{β2}$	62.3	$(0\bar{1}1)_{β2}$	88.1

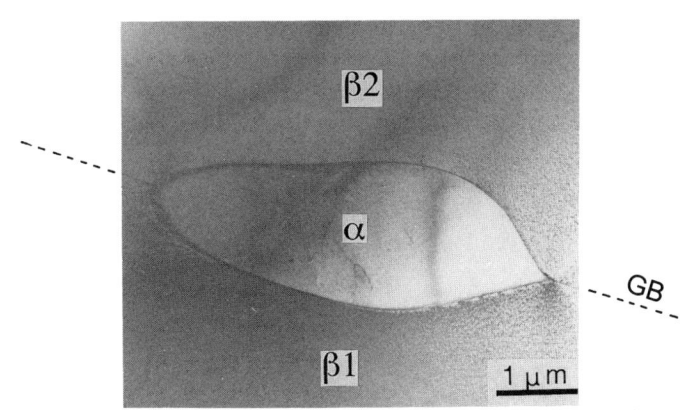

Fig. 3 TEM micrograph of a specimen aged at 923 K for 400 ks. Orientation relationship is summarized in Table 2.

Table 2 Orientation relationship and angles between {110}β planes and the grain boundary plane of the grain boundary α and the adjacent β grains shown in Fig. 3.

misorientation between β grains		
[$\overline{0.6953}$ 0.6040 0.3895] β1 / 12.5 deg.		
β grain	variant of GB α	deviation (deg.)
β1	$(1\bar{1}01)\alpha / (\bar{1}01)\beta1$	0.35
β1	$(0001)\alpha / (1\bar{1}0)\beta1$	1.91
β1	$[11\bar{2}0]\alpha / [111]\beta1$	0.59
β2	$(0\bar{1}11)\alpha / (1\bar{1}0)\beta2$	2.20
β2	$(0001)\alpha / (0\bar{1}1)\beta2$	4.16
β2	$[\bar{2}110]\alpha / [111]\beta2$	2.03
angle between {110}β and grain boundary plane (deg.)			
$(\bar{1}01)\beta1$	**9.1**	$(110)\beta1$	61.7
$(101)\beta2$	12.5	$(011)\beta2$	65.6
$(110)\beta2$	51.4	$(1\bar{1}0)\beta1$	68.1
$(0\bar{1}1)\beta1$	52.1	$(0\bar{1}1)\beta2$	69.4
$(1\bar{1}0)\beta2$	55.7	$(\bar{1}01)\beta2$	77.8
$(011)\beta1$	59.2	$(101)\beta1$	81.9

grain and a near-$\{1\bar{1}01\}\alpha // \{110\}\beta$ OR to another β grain. Type C has a near-$\{1\bar{1}01\}\alpha // \{110\}\beta$ OR to both of the β grains. The frequency of appearance of Types A, B and C are 44%, 39% and 17%, respectively. α1 and α2 GB α precipitates shown in Fig. 2 are Type A and the GB α shown in Fig. 3 is Type B. Since the smallest misorientation angle between $\{1\bar{1}01\}\alpha$ and $\{110\}\beta$ planes for an exact Burgers OR is less than 2 degree, a near-Burgers OR is a special case of a near-$\{1\bar{1}01\}\alpha //$ $\{110\}\beta$ OR. Hence, common crystallographical features in these three Types are (1) a near-$\{1\bar{1}01\}\alpha // \{110\}\beta$ relationship with respect to at least one β grain and (2) this $\{110\}\beta$ (// $\{1\bar{1}01\}\alpha$) plane is the most close to parallel to the grain boundary plane. In addition, a tendency of smaller deviation angle between a $\{1\bar{1}01\}\alpha$ plane and a $\{110\}\beta$ plane rather than a $(0001)\alpha$ plane and a $\{110\}\beta$ plane was observed in many cases. In other words, GB α precipitates tend to have a Potter OR rather than a Burgers OR. The reason for this is not clear at present, however, a $\{1\bar{1}01\}\alpha$ plane, as well as a (0001)α and a {1$\bar{1}$00}α planes, is predicted to have good coherency with a {110}β plane [14-16].

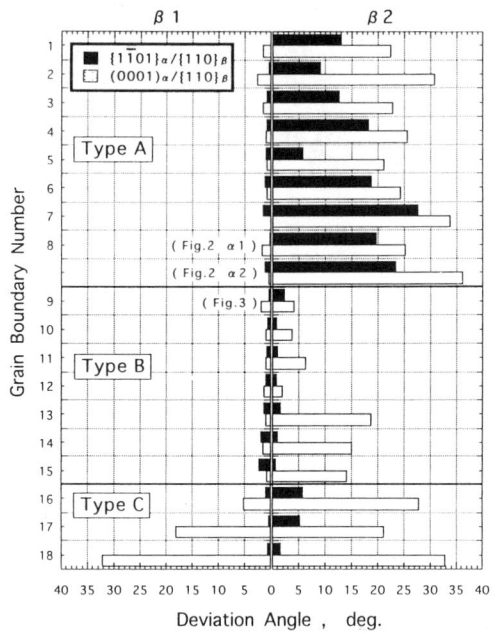

Fig. 4 The smallest deviation angles between a {1$\bar{1}$01}α plane and a {110}β plane and between a (0001)α plane and a {110}β plane of GB α precipitates formed at 18 different grain boundaries with respect to both β grains.

Variant selection

As described above, GB α precipitates have, at least, a near-{1$\bar{1}$01}α // {110}β OR with respect to one of the adjacent β grains. The important fact is that this {110}β plane is always the most close to parallel to the grain boundary plane. Therefore, the selection of only one {110}β plane from twelve {110}β planes in the adjacent β grains reduces 24 near-Burgers variants (12 variants are possible in each β grain) to 2.

Moreover, the coherency between GB α and adjoining β grains should be another effective factor to restrict the variant. Table 3 indicates the smallest deviation angles between (0001)α and {110}β planes and between {1$\bar{1}$01}α and {110}β planes with respect to both β grains, in case of the GB α shown in Fig. 3. Another Burgers variant, (0001)α // (011)β, [11$\bar{2}$0]α // [1$\bar{1}$1]β, is assumed by rotating the observed variant 70.53 degree around the ($\bar{1}$01)β plane normal axis and the calculated smallest deviation angles are indicated as well. As can be seen, the GB α with the actual variant as well as the assumed variant have good coherency with the β1 grain, however, the deviation angles against the β2 grain are much larger in the assumed variant than in the actual variant. It implies that coherency of GB α with the assumed variant is poor with respect to the opposite β grain, i.e., β2 grain, hence, we could not observe the assumed variant. The result strongly suggests that the coherency between GB α and a β grain, which is opposite to a near-Burgers one, is important for the variant selection as well as the grain boundary orientation.

The variant of GB α precipitate is, therefore, initially selected by the grain boundary orientation, followed by the coherency between the GB α and the adjacent β grains, as schematically illustrated in Figure 5. It can be surmised that this crystallographical rule for the variant selection leads to two variants in Type A GB α and a single variant in Types B and C GB α. Because the GB α precipitates with above two variants have two different close packed direction parallel relationship, i.e., $<11\bar{2}0>\alpha$ // $<111>\beta$, they will grow toward these two $<11\bar{2}0>\alpha$ directions as suggested by Furuhara [11]. The observed two growth direction in Fig. 1 (b) supports this prediction. According to this hypothesis, GB α will have a globular allotriomorph or an irregularly elongated film morphology by the impingement of different variant α phase, and a smoothly elongated film morphology by coalescence of a single variant α phase after long period of aging. Furthermore, if the grain boundary changes its orientation, the variant of GB α will change and show different morphology. Such an example is shown in Fig. 1 (a).

Table3 The smallest deviation angles between (0001)α and {110}β planes and between {1$\bar{1}$01}α and {110}β planes with respect to both β grains, in case of an observed and an assumed variant of the GB α shown in Fig. 3.

	β1	(deg.)	β2	(deg.)
Observed near-Burgers variant (0001)α / (1$\bar{1}$0)β1 [11$\bar{2}$0]α / [111]β1	(0001)α/($\bar{1}$10)β1	1.9	(0001)α/(0$\bar{1}$1)β2	4.2
	(1$\bar{1}$01)α/($\bar{1}$01)β1	0.4	(0$\bar{1}$11)α/($\bar{1}$10)β2	2.2
	($\bar{1}$101)α/(0$\bar{1}$1)β1	3.4	(01$\bar{1}$1)α/($\bar{1}$01)β2	5.4
	(0$\bar{1}$11)α/(011)β1	8.7	(1$\bar{1}$01)α/(101)β2	7.8
	(01$\bar{1}$1)α/(101)β1	10.6	($\bar{1}$101)α/(110)β2	11.7
	($\bar{1}$011)α/(110)β1	27.6	(10$\bar{1}$1)α/(011)β2	25.3
	(10$\bar{1}$1)α/(110)β1	30.4	($\bar{1}$011)α/(011)β2	32.7
	(1$\bar{1}$01)α/($\bar{1}$10)β1	59.7	(0$\bar{1}$11)α/(0$\bar{1}$1)β2	57.9
Assumed near-Burgers variant (0001)α / (011)β1 [11$\bar{2}$0]α / [1$\bar{1}$1]β1	(0001)α/(011)β1	1.6	(0001)α/($\bar{1}$10)β2	10.2
	(1$\bar{1}$01)α/($\bar{1}$01)β1	0.4	(1$\bar{1}$01)α/(101)β2	8.2
	($\bar{1}$101)α/(110)β1	3.1	(10$\bar{1}$1)α/(0$\bar{1}$1)β2	11.8
	(10$\bar{1}$1)α/($\bar{1}$10)β1	8.1	(1$\bar{1}$01)α/(011)β2	14.1
	($\bar{1}$011)α/(101)β1	9.6	($\bar{1}$011)α/($\bar{1}$01)β2	21.3
	(01$\bar{1}$1)α/(0$\bar{1}$1)β1	28.2	(01$\bar{1}$1)α/(110)β2	22.5
	(0$\bar{1}$11)α/(0$\bar{1}$1)β1	29.4	(0$\bar{1}$11)α/($\bar{1}$01)β2	32.0
	(1$\bar{1}$01)α/(011)β1	60.0	(1$\bar{1}$01)α/($\bar{1}$10)β2	52.9

a {110}β // Grain boundary plane

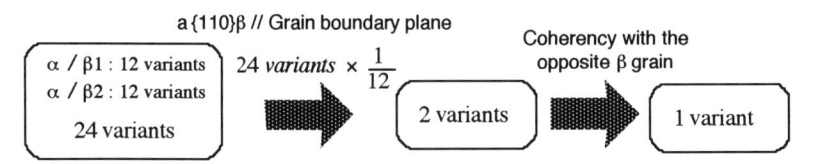

α / β1 : 12 variants
α / β2 : 12 variants
24 variants
→ 24 variants × $\frac{1}{12}$ → 2 variants → Coherency with the opposite β grain → 1 variant

Fig. 5 Schematic illustration of the variant selection.

Conclusions

The morphology and crystallography of α phase formed at β grain boundary in a Ti-22V-4Al alloy was investigated with TEM and SEM. The crystallography was analyzed by Kikuchi patterns among the grain boundary α phase and β grains. The conclusions are as follows:

1. Grain boundary α precipitates show globular allotriomorph or film morphology. A strong dependence is observed between their morphology and crystallography.

2. According to crystallographical feature, they are able to be classified in three types; Type A: a near-Burgers OR to one β grain and no specific OR to another one. Type B: a near-Burgers OR to one β grain and a near-$\{1\bar{1}01\}\alpha$ // $\{110\}\beta$ OR to another one. Type C: a near-$\{1\bar{1}01\}\alpha$ // $\{110\}\beta$ OR to both β grains. The near-$\{1\bar{1}01\}\alpha$ // $\{110\}\beta$ OR is observed in all types of the grain boundary α precipitates.
3. Variants selection in the grain boundary α precipitate concerned to the grain boundary orientation is observed. A $\{1\bar{1}01\}\alpha$ plane is confirmed to be almost parallel to one of the twelve $\{110\}\beta$ planes in the adjacent β grains, which has the minimum angle with the grain boundary plane. Hence, the variants are restricted to 2 at the most, and are reduced to one in the case of good coherency to the opposite β grain.
The morphology and crystallography of the grain boundary α phase are well predicted by this hypothesis.

References

1. M.Hillert, The Decomposition of Austenite by Diffusional Process (Interscience, New York, 1962), 197-248.
2. P.L.Ryder and W.Pitsch, "The crystallographic analysis of grain boundary precipitation," Acta Met., 14(1966), 1437-1448.
3. P.L.Ryder, W.Pitsch and R.F.Mehl, "Crystallography of the precipitation of ferrite on austenite grain boundaries in a Co-20% Fe alloy," Acta Met., 15(1967), 1431-1440.
4. D. Vaughan, "The precipitation of θ' at high angle boundaries in an Al-Cu alloy," Acta Met., 18(1970), 183-187.
5. A.D.King and T.Bell, "Crystallography of grain boundary proeutectoid ferrite," Metall. Trans., 6A(1975), 1419-1429.
6. K.Ameyama, T.Maki and I.Tamura, "Morphology and crystallography of the precipitation of austenite at ferrite grain boundaries in two-phase stainless steel," J.Japan Inst. Metals, 50(1986), 606-611.
7. K. Ameyama et al., "Morphology of proeutectoid ferrite at austenite grain boundaries in low carbon steels," Tetsu-to-Hagane, 74(1988), 1839-1845.
8. K.Ameyama and T.Maki, "Precipitation of austenite at deformation twin boundaries of ferrite in two phase stainless steel," Scripta Metall. Mater., 24(1990), 173-178.
9. K.Ameyama, G.C.Weatherly and K.T.Aust, "A study of grain boundary nucleated widmanstatten precipitates in a two-phase stainless steel," Acta Metall. Mater., 40(1992), 1835-1846.
10. T. Furuhara and H.I.Aaronson, "Crystallography and interfacial structure of proeutectoid α grain boundary allotriomorphs in a hypoeutectoid Ti-Cr alloy," Acta Metall. Mater., 39(1991), 2887-2899.
11. T.Furuhara et al., "Crystallography of grain boundary α precipitates in a β titanium alloy," Metall. Trans., 27A(1996), in press.
12. J.K.Lee and H.I.Aaronson, "Influence of faceting upon the equilibrium shape of nuclei at grain boundaries - I," Acta Met., 23(1975), 799-808.
13. J.K.Lee and H.I.Aaronson, "Influence of faceting upon the equilibrium shape of nuclei at grain boundaries - II," Acta Met., 23(1975), 809-820.
14. T. Furuhara and H.I.Aaronson, "Computer modeling of partially coherent B.C.C.:H.C.P. boundaries," Acta Metall. Mater., 39(1991), 2857-2872.
15. Y.Mou and H.I.Aaronson, "O-lattice modeling of ledged, partially coherent B.C.C.:H.C.P. boundaries," Acta Metall. Mater., 42(1994), 2133-2144.
16. Y.Mou and H.I.Aaronson, "The role of unrotated lines in O-lattice modeling, with applications to flat B.C.C.:H.C.P. boundaries," Acta Metall. Mater., 42(1994), 2145-2157.

TITANIUM-TANTALUM ALLOY DEVELOPMENT

*J.D. Cotton, J.F. Bingert, P.S. Dunn, D.P. Butt and +R.W. Margevicius

Materials Science and Technology Division
Los Alamos National Laboratory
Los Alamos, NM 87545

Abstract

Research has been underway at Los Alamos National Laboratory for several years to develop an alloy capable of containing toxic materials in the event of a fire involving a nuclear weapon. Due to their high melting point, good oxidation resistance, and low solubility in molten plutonium, alloys based on the Ti-Ta binary system have been developed for this purpose. The course of the alloy development, along with processing and property data, are presented in this overview.

This work was supported by the Department of Energy under contract #W-7405-ENG-36.

*Presently at Boeing Defense & Space Group, PO Box 3999, Seattle, WA 98124-2499
+Presently at Westinghouse Electric Corp., 310 Beulah Rd., Pittsburgh, PA 15235-5098

Introduction

Fire resistance is an important aspect of the safety of nuclear weapons during storage and handling. One method by which to enhance the fire resistance is to provide reliable containment to prevent the release of toxic materials in the event of a fire. The development of effective containment may be accomplished by an appropriate alloy which must display several attributes: liquid metal (typically Pu) corrosion resistance, oxidation resistance, sufficient strength and ductility, moderate density and processability. This paper describes how each of these attributes has been addressed within the context of an unusual alloy development program.

Service Design Goals

The initial design goals require survival at 1000 °C for 2 hours, in contact with molten Pu and air on opposing sides of the containment wall. In addition, the material must support its own weight at these conditions, and have a preferred density of 6 to 8 g/cc, Processability requirements necessitate the ability to cast, form and weld the containment material with a minimum of difficulty. These constraints were used to guide the alloy development.

Approach

Preliminary work indicated that the most challenging aspect of this program would be liquid metal corrosion resistance. In the absence of intermetallic compound formation, liquid metal corrosion is dominated by dissolution of the containment material into the liquid. This process proceeds at a rate proportional to the solubility of the containment material in the liquid phase at the temperature of interest [1]. Thus, it is logical to choose a containment material which has a low solubility in liquid Pu. A comparison of various metals with melting points above 1000 °C indicates that tungsten (W) and Ta are two of the best metals in this regard, having 1000 °C solubilities in liquid Pu of 0.05 and 0.5 at.%, respectively [2]. Based on this metric alone, W is clearly the best choice for containment, and historically, this has been practiced for liquid Pu refining hardware. However, its high density (19.3 g/cc) and difficult processability make Ta the better choice in most cases.

Another advantage of Ta is that its oxidation resistance can be improved by alloying with Ti [3]. In fact, the oxidation rate in air is reportedly a minimum at approximately 5 wt.% Ta, with a moderate increase to about 90 wt.% Ta and a marked increase above above this level. The addition of Ti to Ta also decreases the melting point, as shown by the phase diagram in Figure 1 [4, after 5]. This facilitates casting and welding, and has been shown to have no detrimental effect on ductility [6]. Furthermore, the Ti-Ta system is one of a class of β-isomorphous Ti alloys that are generally ductile and heat treatable, which would provide some measure of property control. Finally, the addition of Ti to Ta reduces the density in an efficient manner, due to the low density of Ti (4.5 g/cc). The density requirement essentially confines the Ta contents to between 30 to 60 wt.% Ta, using the constant volume assumption of:

$$\rho_{Ti-Ta} = \frac{100}{\frac{wt.\%Ti}{\rho_{Ti}} + \frac{wt.\%Ta}{\rho_{Ta}}}$$

However, Ti is extensively soluble in liquid Pu, displaying a solubility limit of 37.1 at.% at 1000 °C [2]. This behavior is counter to the oxidation resistance trend, which improves at Ti-rich compositions, in addition to the density. Thus, a need exists to optimize the Ta content for these properties, as well as to verify the processing characteristics of this alloy system. Additionally, microstructural characterization is required to understand and control the alloy properties adequately.

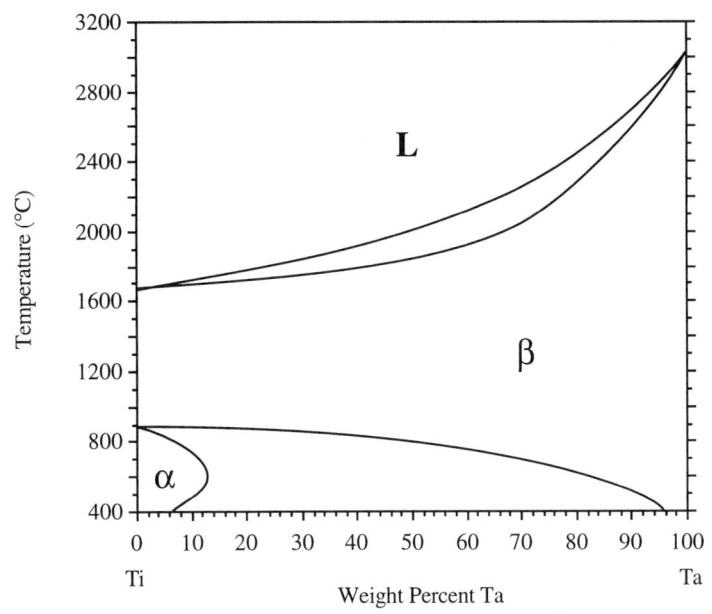

Figure 1. Ti-Ta Binary Phase Diagram [4,5].

Previous Work

Research of Ti-Ta alloys has been relatively limited. One study concerning liquid metal corrosion behavior of Ti-Ta alloys has been published in the open literature [7], and this is discussed below in the context of this program. The mechanical properties of Ti-Ta alloys have been reviewed in a previous paper [8]. Other work in this system is confined to phase equilibria [5,9-17] and oxidation behavior [3,18-20]. Some of the issues relevant to casting Ti-Ta alloys are described in a related paper [21].

Experimental Procedures

The experimental procedures are described in detail elsewhere [4,7,8,19,21]. In general, however, the initial castings were produced by arc melting 5N pure Ti plate with commercial purity Ta sheet in proportions to attain 5, 10, 20, 40, 60 and 80 wt.% Ta in the alloys. Extensive remelting was employed to promote homogeneity. To produce sufficient material for hot rolling to sheet, several buttons were melted together to produce slabs approximately 100 x 100 x 21 mm in size. Later, the plasma arc melting process was used to create billets up to 152 mm in dia. and 300 mm in length [21]. Sheet product was used for most experiments and was produced by hot rolling in air at 900 °C to appropriate reductions.

Liquid Pu corrosion resistance was usually measured by exposure experiments in vacuum or argon at 700 to 1100 °C for two hours, and also for various liquid volume/contact area (V/A) values. These were evaluated by measurement of time-to-perforation and metallography. Oxidation resistance was measured in both laboratory air and 20%O_2-80%Ar, at a variety of temperatures and times. Thermogravimetric analysis (TGA) was used to record weight changes. Metallography was also utilized to ascertain the amount of structural alloy lost to oxidation and α-case formation.

Mechanical properties were evaluated by tensile testing of sheet specimens according to ASTM E8 for a one-inch gauge length. In addition, tensile strength at 1000 °C was determined on a thermomechanical test system for selected alloys in vacuum. Density was measured by the Archimedes' method.

Results and Discussion

A portion of the TGA data are summarized in Figure 2 and compared with a commercial oxidation-resistant Ti alloy, β21S. It can be observed that below 60 wt.% Ta the alloys are approximately equivalent in their total weight gain, and competitive with β21S. However, the more pertinent method to evaluate material losses due to oxidation and α case formation is by metallography, which was combined with like data for liquid Pu corrosion and presented in Figure 3. Experimentally measured densities are given in Table I, and match closely with the theoretical values. To maintain a constant areal density, the compatibility test specimen thickness was varied with Ta content. Thus, the data are plotted as % thickness remaining in the test specimen. These results indicate that while the maximum material lost due to liquid Pu dissolution occurs at 60 wt.% Ta, oxidation material lost increases at an increasing rate with Ta content. The sum of these two processes results in the bold curve, which indicates a worst case composition at 80 wt.% Ta and increasing goodness with Ti content. For the results presented here, the best performance occurs at 40 wt.% Ta. This may improve further with increasing Ti content, although it is anticipated that at some point the curve will reach a maximum due to increasing solubility in the molten Pu.

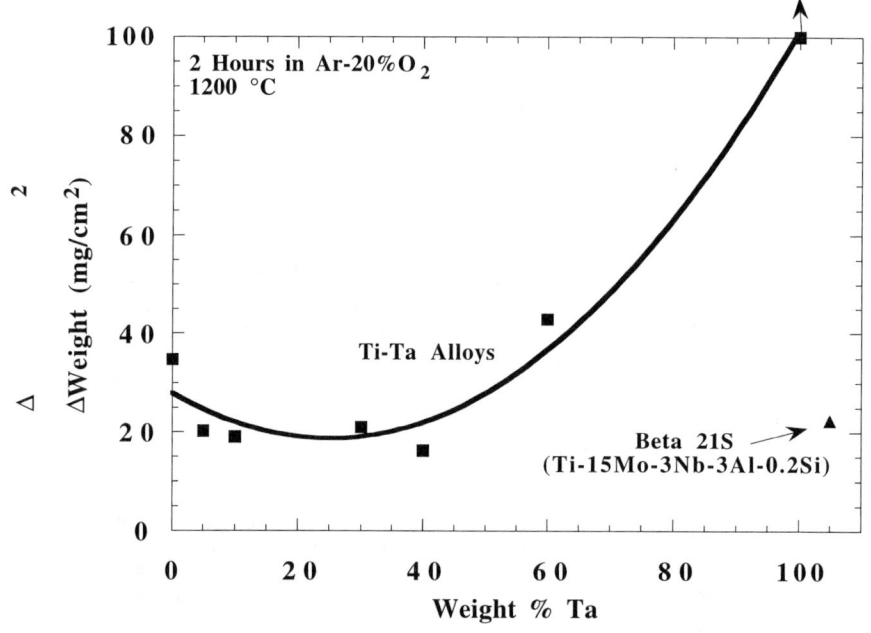

Figure 2. TGA Data for the Oxidation of Ti-Ta Alloys (and β21S) in 20% O_2-80% Ar.

The room-temperature tensile properties are given in Table II for β-annealed (900 °C/1 hour + water-quench) heat treatment conditions, and also the 1000 °C tests [8]. These mechanical property values were deemed adequate for the application.

In the course of this development, investigations of the weldability (electron beam) and machinability were undertaken, and no unusual behavior was noted, with the exception of a highly tenacious surface oxide which required bead blasting plus electrochemical means to remove. Both cold and hot forming of sheet into small cups was possible, the latter in air at

800-850 °C with minimal interstitial element pickup. In general, these alloys behaved similar to other β-Ti alloys in their response to processing and heat treatment.

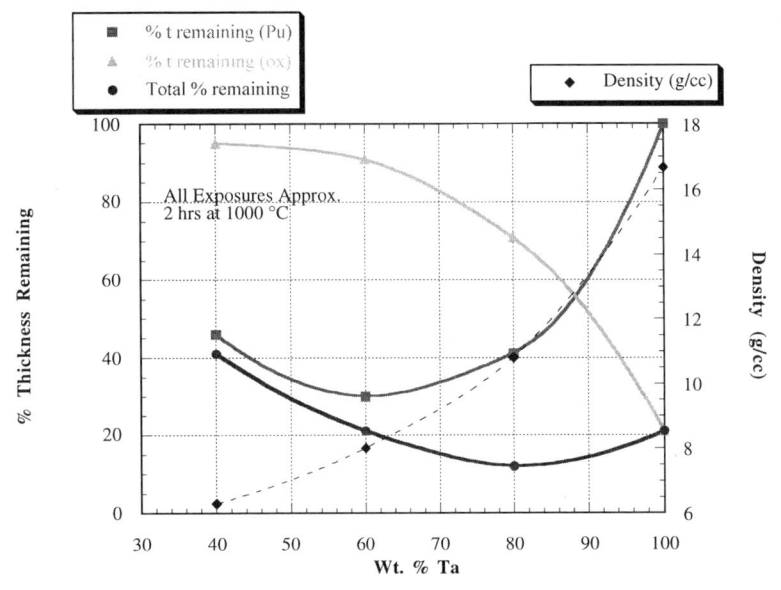

Figure 3. Combined Effects of Oxidation and Liquid Pu Corrosion on Material Thickness Losses for Ti-Ta Alloys.

Table I. Measured Densities of Ti-Ta Alloys Compared to Pure Ti and Ta.

Composition (wt.%)	Density (g/cc)
Ti	4.54
60Ti-40Ta	6.3
40Ti-60Ta	8.0
20Ti-80Ta	10.8
Ta	16.6

Table II. Mean Tensile Properties for β-Quenched Ti-Ta Alloys [8].

	Room Temperature			1000 °C
Composition (Wt.%)	TYS MPa (ksi)	UTS MPa (ksi)	Total Elong. (%)	UTS MPa (ksi)
60Ti-40Ta	310 (44.9)	639 (92.7)	23	20 (2.9)
40Ti-60Ta	421 (61.1)	643 (93.3)	24	43 (6.3)
20Ti-80Ta	606 (87.8)	639 (92.6)	8	159 (23.0)

Summary

A program was undertaken to develop an alloy resistant to liquid Pu corrosion, and to oxidation, at 1000 °C. This has been accomplished within the constraints of the program.

References

1. M.G. Fontana and N.D. Green, Corrosion Engineering, (New York, NY, USA: McGraw-Hill, 1978) 290-1.
2. F.H. Ellinger et al., "Constitution of Plutonium Alloys," Los Alamos National Laboratory, Report #LA-3870 (1968).
3. R.F. Voytovich and E.I. Golovko, "Oxidation of Ti-Ta and Ti-Nb Alloys," Russ. Metally. USSR, 1 (1970), 183-7.
4. J.D. Cotton et al., "Microstructure and Mechanical Properties of Ti-40 wt. % Ta (Ti-15 at. % Ta)," Metall. Trans. A 25A (1994) 461-72.
5. J.L. Murray, Phase Diagrams of Binary Titanium Alloys (Metals Park, OH, USA: ASM International, 1987), 302-6.
6. W.J. Tomlinson and R. Rushton, "The Fabricability and Hardness of Concentrated Binary Tantalum Alloys Containing Titanium, Hafnium, Vanadium, Holybdenum, Tungsten, Rhenium and Palladium." J. Less-Common Met. 115 (1986) L1-L4.
7. J.D. Cotton et al., , "Nature of Dissolution of Binary Tantalum-Titanium Alloys by Molten Plutonium," in Actinide Processing: Methods and Materials, ed. B. Mishra, (Warrendale, PA, USA: TMS, 1994) 45-56.
8. J.D. Cotton et al., "Tensile Deformation of Titanium-Tantalum Alloys," in High-Temperature, High Performance Materials for Rocket Engines and Space Applications, ed. K. Upadhya, (Warrendale, PA, USA: TMS, 1995) 49-59.
9. P. Duwez, "The Martensite Transformation Temperature in Titanium Binary Alloys," Trans. A.S.M. 45 (1953) 934-40.
10. D.J. Maykuth et al., "Titanium-Tungsten and Titanium-Tantalum Systems," JOM (Trans. AIME) Feb. (1953) 231-7.
11. P.B. Budberg and K.I. Shakova, "Phase Diagram of the Titanium-Tantalum System," Iz. Akad. Nauk SSSR, Neorg. Mater., 3 (1967) 656-60.
12. P.N. Nikitin and V.S. Mikheyev, "Solubility of Tantalum in α-Titanium" Fiz. Metal. Metalloved., 23 (1969) 1127-9.
13. D. Summers-Smith, "The Constitution of Tantalum-Titanium Alloys," J. Inst. Met., 81 (1952) 73-6.
14. K.A. Bywater and J.W. Christian, "Precipitation Reactions in Titanium-Tantalum Alloys," Phil. Mag. A, 25 (1972) 1275-89.
15. K.A. Bywater and J.W. Christian, "Martensitic Reactions in Titanium-Tantalum Alloys," Phil. Mag. A, 25 (1972) 1249-74.
16. S.G. Fedotov et al., "Phase Transformations During Heating of Metastable Alloys of the Ti-Ta System," Phys. Met. Metall., 62 (1986) 109-13.
17. T. Yamane and J. Ueda, "Transmission Electron Microscope Structure of a Ti-9 Wt.% Ta Alloy Quenched from β Region," Acta Metall., 14 (1966) 348-9.
18. Y.S. Chen and C.J. Rosa, "Oxidation Characteristics of Ti-4.37 wt.% Ta Alloy in the Temperature Range 1258-1473 K," Oxidation of Metals, 14 (1980) 167-85.
19. Hanrahan et al., "High Temperature Oxidation of Titanium-Tantalum Alloys," 19th DOE Compatibility, Aging and Service Life Conf., (Los Alamos, NM, USA: Los Alamos National Laboratory, 1994).
20. D.A. Prokoshkin et al, "Kinetics of Oxidation of Ta-Ti Alloys," Iz. Akad. Nauk. SSSR (Metally.), 5 (1984) 178-80.
21. P.S. Dunn et al, "Plasma Arc Melting of Titanium-Tantalum Alloys," Proc. 1994 Inter. Symp. Liquid Metal Processing and Casting, (Santa Fe, NM, USA: 1994) 85-97.

EFFECTS OF HEAT TREATMENT ON MICROSTRUCTURE, STRENGTH, AND TOUGHNESS OF Timetal-21S

R.K. Bird, T.A. Wallace, and W.D. Brewer

NASA Langley Research Center
Hampton, VA 23681

Abstract

The effects of heat treatment on the tensile properties and fracture toughness of 0.56-mm-thick Timetal-21S sheet were studied over the temperature range -54°C to +177°C, with the purpose of developing appropriate microstructures and phase distributions to yield high strengths with as high a toughness as possible. β and α/β anneals with various quenches and ages, as well as direct aging, were investigated. β-annealing, water-quenching and aging generally produced microstructures of fine alpha plates within the β matrix resulting in the best combinations of strength and toughness. α/β annealing and aging resulted in thick grain boundary alpha with an associated low toughness. Low temperature ages tended to produce ω phase and low toughness. The mechanical behavior of the Timetal-21S sheet was highly dependent upon temperature, but the properties over the temperature range studied seemed to follow the same general strength/toughness trends as the room temperature properties.

Introduction

β titanium alloys are of interest for a variety of aerospace applications, including subsonic and supersonic airframe structures, because of their versatility. β alloys generally have high strength to density ratios, deep hardenability, can be heat treated to a wide range of properties, and have potential for lower cost because of being strip producible. In particular, Timetal-21S has been given considerable attention because of the additional benefits of improved oxidation and corrosion resistance (ref. 1). In this study, the effects of heat treatment on the tensile properties and toughness of 0.56-mm-thick Timetal-21S sheet were studied over a range of temperatures, with the purpose of developing appropriate microstructures and phase distribution to maximize strength/toughness combinations. The heat treatments included β and α/β anneals with various quenches and ages, as well as direct aging. Optical microscopy, scanning and transmission electron microscopy, and x-ray analyses were used to define the microstructures. Toughness was measured by the J-integral method using compact tension specimens. Heat treatments, microstructure, tensile and fracture properties were correlated from -54°C to 177°C.

Materials and Specimens

The Timetal-21S (Ti-15Mo-2.7Nb-3Al-0.25Si, wt%) material used was in the form of 0.56-mm thick cold-rolled sheet. Subsized tensile specimens were machined in accordance with ASTM specification E8 (ref. 2). The specimens were oriented such that the specimen length was parallel to the sheet rolling direction. Compact tension specimens for fracture toughness measurements were machined from Timetal-21S sheet in accordance with ASTM specification E1152 (ref. 3). The specimens were oriented such that the direction normal to the crack plane was parallel to the sheet rolling direction and the crack growth direction was perpendicular to the sheet rolling direction.

Experimental Procedures

Heat treatment:
β-annealing was conducted at 899°C for 0.5 hour. The β transus for Timetal-21S is 816°C (ref. 4). Before annealing the specimens were sealed in evacuated bags constructed of commercially pure titanium foil. After annealing, the samples were cooled to room temperature by either water quenching or furnace cooling. The samples were removed from the bags before any subsequent aging treatments. α/β annealing was conducted near the β transus at 810°C for times from 0.5 hour to 2 hours followed by a furnace cool to the aging temperature. Aging was conducted at temperatures from 454°C to 621°C for times from 8 hours to 80 hours followed by a furnace cool. The α/β anneals and all aging treatments were conducted in a vacuum furnace with the samples wrapped in tantalum foil to prevent oxidation. Direct aging consisted of aging the cold-rolled Timetal-21S samples without a prior anneal.

Test procedures:
Tensile and toughness tests were conducted at -54°C, room temperature, and 177°C. Tensile tests were conducted in accordance with ASTM specification E8 (ref. 2) using a closed-loop hydraulic test machine with automated data acquisition. Strain was measured using back-to-back extensometers with 1.0-inch gage length.

Fracture toughness of Timetal-21S was measured using the J-R curve determination method described in ASTM specification E1152 (ref. 3). The specimens were tested using a closed-loop hydraulic test machine with automated test control and data acquisition. A clip gage with 0.25-inch gage length was attached to the "mouth" of the specimen to monitor crack length via the compliance method. The automated data acquisition system converted the raw data measurements to J versus Δa. J_{Ic} was calculated using the ASTM specification E813 (ref. 5). Fracture toughness, K_{JIc}, was calculated from J_{Ic} using the following equation from reference 6:

$$K = (J \cdot E)^{1/2} ; \qquad E = \text{modulus}$$

To conduct the cryogenic and elevated temperature tensile and toughness tests, the specimens and grips were enclosed within an environmental chamber that utilized liquid nitrogen for cooling and fan-circulated resistance-heated air for heating. All compact tension specimens were fatigue precracked at room temperature, regardless of the toughness test temperature. The specimens were cooled or heated to the desired temperature and allowed to soak for 5 minutes prior to testing. The desired temperature was maintained within ±2°C for the duration of the test.

Metallurgical analysis:
Specimen fracture surfaces were cleaned in acetone and examined using scanning electron microscopy (SEM). Microstructures were examined by SEM and transmission electron microscopy (TEM). SEM samples were prepared by cross-sectioning and mounting in epoxy, polishing, and lightly etching. TEM specimens were mechanically ground, dimpled, and ion milled at room temperature. Volume fractions of constituent phases within bulk specimens were estimated by x-ray diffraction analysis.

Results and Discussion

A preliminary heat treatment study was conducted to develop a wide range of tensile properties and microstructures for cold-rolled Timetal 21S sheet. The heat treatments were then refined in attempts to obtain microstructures resulting in high strength with as high a fracture toughness as possible. Table I summarizes the results, including the final heat treatments, yield strength, toughness, and microstructural phase information.

Table I. Effect of heat treatment on strength, toughness, and microstructure of Timetal-21S sheet.

Anneal conditions	Aging conditions	Code	Yield strength (MPa)	K_{IIc} (MPa√m)	Microstructural features	Vol. frac. α
Direct Age (no anneal)	521°C, 11 hr	A	1379	54.9	fine α	41%
	621°C, 8 hr	B	1034	78.5	coarse α	42%
β anneal @ 899°C, .5 hr; WQ to RT	449°C, 11 hr	C	1400	42.0	fine α, some ω	–
	521°C, 11 hr	D	1421	59.3	fine α	34%
	538°C, 11 hr	E	1317	55.6	intermediate-sized α	38%
	566°C, 11 hr	F	1324	54.4	intermediate-sized α	42%
	621°C, 8 hr	G	1076	82.5	coarse α	39%
β anneal @ 899°C, .5 hr; FC to RT	521°C, 11 hr	H	1310	53.3	fine α	39%
α/β anneal @ 810°C, 2 hr; FC to 454°C	454°C, 11 hr	I	1021	72.6	thick gb α, fine α, some ω	29%
	454°C, 80 hr	L	1338	51.8	thick gb α, fine α, some ω	40%
α/β anneal @ 810°C, .5 hr; FC to 454°C	454°C, 71 hr	M	1338	46.6	thick gb α, fine α, some ω	–

WQ = water quench; FC = furnace cool; gb = grain boundary

Toughness and yield strength values from table I are plotted in figure 1. The letters in the figure correspond to the heat treatment codes in table I. The dashed lines represent the upper and lower bounds of the data showing the overall strength-toughness trend. As expected the general trend was for the toughness to decrease as the strength increased. But, within each area of the data (higher strength or higher toughness) certain heat treatments had the effect of increasing both the strength and the toughness. With one exception, the data from material that was β-annealed, water-quenched, and aged (conditions C,D,E,F,G) fell within the upper portion of the data band. However, material in condition C exhibited very low toughness. The microstructures of the material in these conditions are shown in figure 2 and are generally characterized by α plate precipitates within a β matrix. The prior β grain size was approximately the same in all cases with a very thin layer of grain boundary α. The microstructure associated with aging at 621°C (condition G), figure 2a, contained coarse α plates and corresponded to a high toughness and low strength. Figure 2b shows that aging at 566°C (condition F) resulted in a smaller α plate size, higher strength, and lower toughness. Aging at 538°C (condition E) produced a microstructure and properties very similar to that produced at the 566°C aging temperature. However, aging at 521°C (condition D) resulted in a much finer precipitation of α plates (figure 2c), a higher strength, and a slightly higher toughness. Figure 2d shows the microstructure after aging at 449°C (condition C). This microstructure exhibited very fine α plate precipitates. TEM electron diffraction analysis indicated that the microstructure also contained ω phase, which has been reported to have an embrittling effect (ref. 7) and may have contributed to the very low toughness associated with this condition. In addition, the 449°C aging condition resulted in colonies of small α precipitates with a triangular prism morphology near the prior β grain boundaries (ref. 8), which may have also been a contributing factor to the low toughness.

A detailed microstructural study of β-annealed, water-quenched, and aged cold-rolled Timetal-21S sheet (ref. 8) indicated that the character of the α plate precipitates was dependent on the aging temperature and had a significant impact on properties. In general, type 1 α exhibits a Burgers relationship and is coherent with the parent β phase, while type 2 α exhibits a more complex relationship with the β phase and is incoherent. A complete description of the differences between these two types of α can be found in references 8 and 9. Aging at 521°C produced a mixture of type 1 and type 2 α which resulted in good strength and toughness. Aging at 566°C resulted in predominately type 2 α and a lower strength and toughness. The α phase after aging at 449°C was exclusively type 1, which corresponded to the lowest toughness. Thus a balanced combination of the two types of α may be critical to developing a good combination of strength and toughness.

Figure 1 shows that α/β anneals (conditions I,L,M) produced data that lie along the lower portion of the data band. Therefore, for each area of the data (higher strength or higher toughness), the materials with the α/β anneal seemed to have less desirable strength/toughness combinations. Figure 3 shows the microstructure after a 2-hour α/β anneal and aging at 454°C for 11 hours (condition I). α/β annealing resulted in a thick, nearly-continuous layer of primary α along the prior β grain boundaries, which contributed to the lower toughness values. In addition, the β grain size was an order of magnitude smaller than that associated with the β anneals. Very fine α plates precipitated within the β matrix. The total volume fraction of α (grain boundary and intragranular plates) was 29%. Increasing the aging time to 80 hours (condition L) resulted in a large increase in the amount of α plate precipitates within the grains and produced a large strength increase and toughness decrease. Decreasing the annealing time to 0.5 hour and aging at 454°C for 71 hours (condition M) resulted in the same strength level and a reduction in toughness (see table I).

Comparing conditions I and G illustrates the effect of the grain boundary α on the material performance. Both conditions produced similar strength levels, but the α/β anneal resulted in a lower toughness. Figure 4 shows the fracture surface of a compact tension specimen that was α/β annealed and aged at 454°C for 11 hours (condition I). The end of the fatigue precrack is on the left-hand side of the photo and the crack propagation direction is towards the right. The load axis is normal to the plane of the page. Although some ductile dimpled rupture was evident, a significant portion of the failure occurred by intergranular fracture, indicating that the crack propagated along the grain boundary α and resulted in a relatively low toughness. Figure 5, on the other hand, shows that β-annealing, water quenching, and aging at 621°C for 8 hours (condition G) produced a compact tension specimen fracture surface dominated by ductile dimpled rupture, resulting in a higher toughness.

Figure 1 and table I also show that direct aging the cold-rolled sheet without an anneal (conditions A,B) produced strength/toughness data that lie within the middle portion of the data band. Both direct aging heat treatments produced approximately the same volume fraction of α, but dramatically different α plate sizes and mechanical properties. The 621°C age (condition B) produced coarse α, relatively low strength, and high toughness. The 521°C age (condition A) produced much finer α plates, high strength, and lower toughness.

The effect of the pre-aging heat treatment on the properties can be demonstrated using the data in figure 1 and table I. Conditions A, D, and H involved aging at 521°C for 11 hours, but with different starting conditions. Of these three conditions, the β-anneal and water-quench (condition D) produced the highest strength and toughness. Direct aging (condition A) produced the highest volume fraction of α, probably because of the abundance of nucleation sites in the cold-rolled microstructure, but resulted in slightly lower strength and toughness. Furnace cooling after β-annealing (condition H) allowed some grain boundary α to form prior to aging, reducing the amount of strengthening α plate precipitates that formed during aging. This condition resulted in the lowest strength and toughness of the three.

β-annealing, water-quenching, and aging at 521°C for 11 hours (condition D) produced the best strength/toughness combination within the high-strength region of figure 1. This heat treatment was selected to determine if refinement of the aging condition could produce improvements in both strength and toughness. Specimens were β-annealed and water-quenched, then aged at 521°C for 1, 4, and 48 hours to determine the effects of under- and over-aging on the properties. Figure 6 shows the yield strength and toughness as a function of aging time at 521°C. The yield strength increased from 855 MPa to 1317 MPa during the first 4 hours of aging. The strength peaked at 1421 MPa for an age time of 11 hours, and slightly decreased to 1393 MPa after the 48-hour age. The toughness exhibited a large decrease from 66.2 MPa$\sqrt{m}$ to 54.2 MPa$\sqrt{m}$ during the first 4 hours of aging, and peaked at 59.3 MPa$\sqrt{m}$ for the 11-hour age. The toughness exhibited a small decrease to 55.7 MPa$\sqrt{m}$ after aging at 48 hours, in a manner similar to the yield strength. The reason for the low toughness associated with the 4-hour age is not clear, but TEM analysis (ref. 8) suggested that the morphology of the α phase may contribute to the lower toughness. This aging study confirmed that the 11-hour age resulted in the best strength and toughness for a high-strength condition.

Although the material that was β-annealed, water-quenched, and aged at 521°C for 11 hours produced the best strength/toughness combination for a high-strength condition, direct aging at 521°C for 11 hours may be the more desirable heat treatment. This direct aging treatment, while producing slightly lower properties than the β-anneal and water-quench, offers a much simpler and less expensive heat treatment. Thus, material that was direct aged at 521°C for 11 hours was selected to determine the effect of temperature on strength and toughness. Figure 7 shows K_{JIc} plotted against the yield strength at -54°C, room temperature, and 177°C. Also shown is the room temperature data range that was shown previously in figure 1. (Excluded from this data range was the ω-bearing condition and the α/β annealed conditions.) At -54°C, the toughness was the lowest of all conditions evaluated (33.5 MPa$\sqrt{m}$), but the accompanying yield strength was the highest (1572 MPa). As the temperature was increased from -54°C to 177°C, the strength decreased by 23% and toughness increased by 97%. The room temperature and 177°C data points lie within the range of strength/toughness data measured at room temperature for the other heat treatments. In fact, if the room temperature data range is extrapolated out to higher strengths, the cryogenic data point would fall within that range. Thus, the elevated and cryogenic properties followed the same strength/toughness trend as did the room temperature data.

Concluding Remarks

Timetal-21S sheet (0.56 mm thick) was processed using a variety of heat treatments and the tensile properties and toughness behavior was determined. Toughness was measured by calculating K_{JIc} from J-R tests using compact tension specimens.

A wide range of room temperature yield strength/toughness combinations were produced by the selected heat treatments. In general, the toughness decreased as the yield strength increased. Heat treatments involving a β-anneal, water-quench, and aging tended to produce the best set of strength/toughness combinations. For similar strength levels, α/β annealing and aging tended to result in lower toughness values than those produced by β-anneal/water-quench/age heat treatments. The α/β anneal resulted in a thick layer of grain boundary α which contributed to the lower toughness.

Material that was β-annealed, water-quenched, and aged at 521°C for 11 hours produced a high yield strength (1421 MPa) with a moderate toughness (59.3 MPa$\sqrt{m}$). This heat treatment produced a microstructure containing fine α plate precipitates within a β matrix. Material that was direct aged at 521°C for 11 hours produced yield strength and toughness values of 1379 MPa and 54.9

MPa√m, respectively. Although these are slightly lower values than those produced by the β-anneal/water-quench/age heat treatment, direct aging may be preferable because of its simplicity.

The strength and toughness of Timetal-21S that was direct aged at 521°C for 11 hours was strongly dependent on temperature. As the test temperature was varied from -54°C to 177°C, the yield strength decreased by 23% and the toughness increased by 97%. This behavior was consistent with the strength/toughness trend measured at room temperature using a variety of heat treatments.

References

1. Bania, P.J.: "Beta Titanium Alloys and Their Role in the Titanium Industry". Beta Titanium Alloys in the 1990's, D. Eylon, R.R. Boyer, and D.A. Koss, Eds. The Minerals, Metals, and Materials Society, Warrendale, PA, 1993, pp. 3-14.

2. ASTM E8-91, "Standard Test Methods of Tension Testing of Metallic Materials", 1992 Annual Book of ASTM Standards, vol. 03.01, American Society of Testing and Materials, 1992, pp. 130-149.

3. ASTM E1152-87, "Standard Test Method for Determining J-R Curves", 1992 Annual Book of ASTM Standards, vol. 03.01, American Society of Testing and Materials, 1992, pp. 847-857.

4. Bania, P.J. and Parris, W.M.: "Beta-21S: A High Temperature Metastable Beta Titanium Alloy", Titanium 1990: Products and Applications, Vol. II, Proceedings of the 1990 International Conference, Titanium Development Association, Dayton, Ohio, pp. 784-793.

5. ASTM E813-89, "Standard Test Method for J_{IC}, A Measure of Fracture Toughness", 1992 Annual Book of ASTM Standards, vol. 03.01, American Society of Testing and Materials, 1992, pp. 732-746.

6. Hertzberg, Richard W. Deformation and Fracture Mechanics of Engineering Materials, New York: John Wiley & Sons, 1983.

7. Donachie, Matthew J., Jr., ed. Titanium: A Technical Guide. ASM International, Metals Park, OH, 1988.

8. Li, Q. and Crooks, R.: "Isothermal Decomposition of Timetal-21S". Proceedings of the 125th TMS Annual Meeting, 1996.

9. Williams, J.C.: "Precipitation in Titanium Base Alloys", Titanium Technology: Present Status and Future Trends, F.H. Froes, D. Eylon, and H.B. Bomberger, eds. The Titanium Development Association, 1985, pp. 75-86.

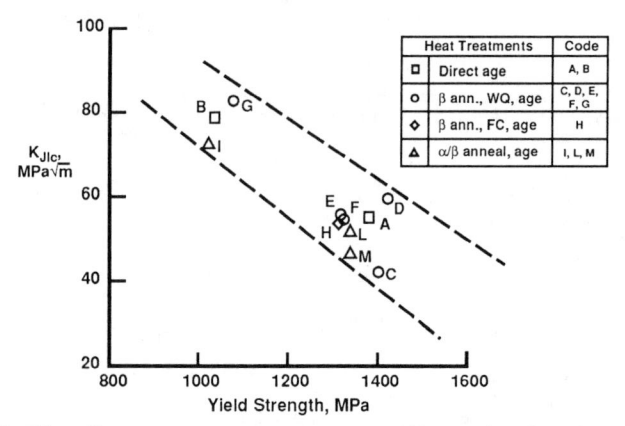

Figure 1. Effect of heat treatment on room temperature yield strength and toughness of cold-rolled Timetal-21S sheet.

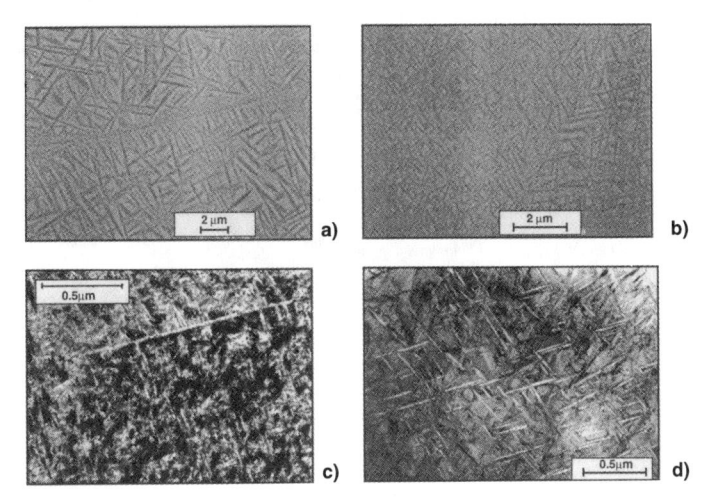

Figure 2. Microstructure of β-annealed and water-quenched cold-rolled Timetal-21S sheet aged at (a) 621°C for 8 hours, (b) 566°C for 11 hours, (c) 521°C for 11 hours, and (d) 454°C for 11 hours.

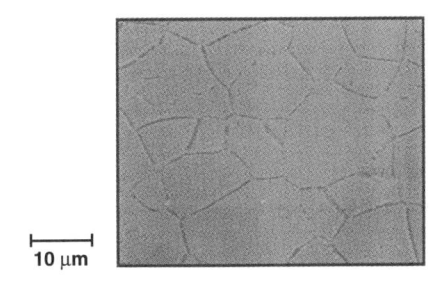

Figure 3. Micrographs of cold-rolled Timetal-21S sheet α/β-annealed, furnace cooled to 454°C, and aged at 454°C for 11 hours.

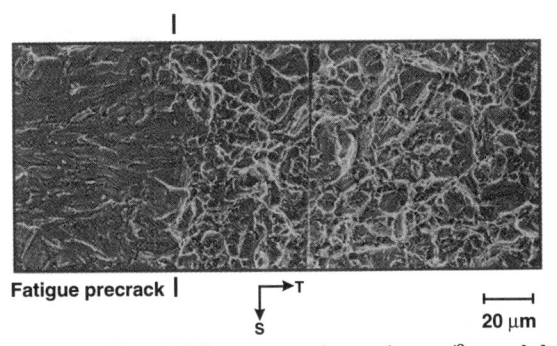

Figure 4. Fracture surface of Timetal-21S compact tension specimen. α/β-annealed, furnace cooled to 454°C, and aged at 454°C for 11 hours.

Fatigue precrack

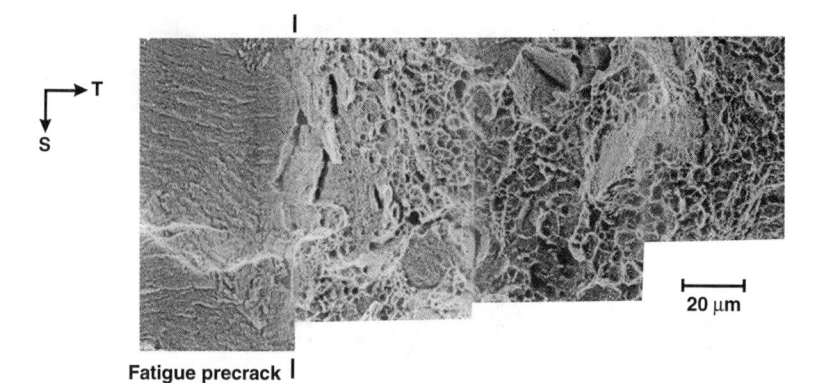

Figure 5. Fracture surface of Timetal-21S compact tension specimen. β-annealed, water-quenched, aged at 621°C for 8 hours.

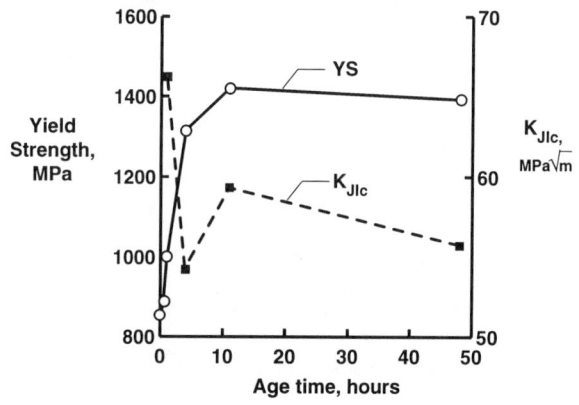

Figure 6. Effect of aging time on room temperature yield strength and toughness. β-annealed, water-quenched, aged at 521°C.

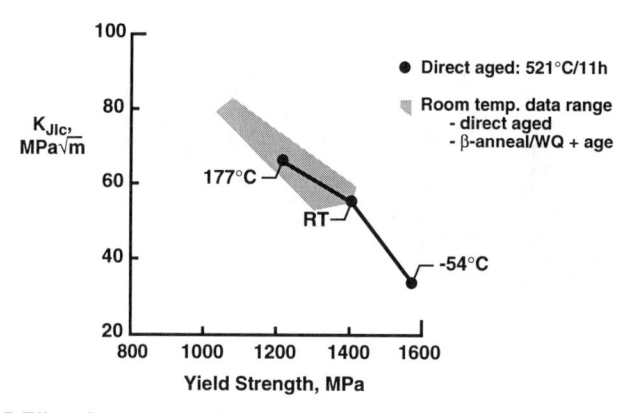

Figure 7. Effect of temperature on yield strength and toughness of cold-rolled Timetal-21S sheet. Direct aged at 521°C for 11 hours.

Isothermal Decomposition of Timetal 21s

Qiong Li and Roy Crooks

Analytical Services and Materials, Inc.,
Metallic Materials Branch, NASA Langley Research Center,
Hampton, Virginia

Abstract

The decomposition of the metastable β-phase titanium alloy Timetal-21s was studied by transmission electron microscopy after β solution, quenching and isothermal aging at temperatures ranging from 450°C - 565°C. These results were compared to available strength and toughness data. Diffraction patterns indicated the presence of athermal ω-phase in the as-quenched condition; with type I α, and then type II α developing after further aging. Several stages were observed in the precipitation and growth of α, and these were characterized by their morphologic and crystallographic changes. In early stages, some of the α were observed in triangular-cross-section, prism-shaped structures formed by the junction of three twin-related α variants. Further aging produced α plates joined in "V" shapes. Single variant plates of type I α were also seen in aged material. The formation of the tri-variant alpha differs from the single variant variety by the operation of a sympathetic nucleation mechanism, probably associated with the low energy of the α/α coherent twin interface. Material heat treated to produce triangular α and low volume fractions of type II α exhibited relatively low toughness.

This work was conducted under funding of NASA Contract NAS1-19708

Introduction

TIMETAL-21S (Ti-15Mo-3Nb-3Al-0.2Si w/o) is a metastable β alloy which exhibits good oxidation resistance and has therefore been of interest for high temperature applications [1]. Metastable β alloys can be solution treated at an elevated temperature within the single phase β–field, quenched to preserve a metastable single phase in the two phase region, and aged to precipitate second phases. These precipitation processes can lead to significant mechanical property changes [2]. Four types of precipitates develop during the quenching and subsequent aging of metastable β alloys: i) athermal omega (ω_{ath}) ii) isothermal omega (ω_{iso}), iii) Type I alpha (α_I), and iv) Type II alpha (α_{II}). The ω_{ath} phase forms during quenching from the single β phase region, while the ω_{iso} phase forms at relatively low isothermal aging temperatures. The hexagonal phases α_I and α_{II} differ in their orientation relationships with the bcc (β) matrix; a Burgers relationship, $<11\bar{2}0>_\alpha \parallel <111>_\beta$, $(0001)_\alpha \parallel (110)_\beta$ [3], is found for α_I, but not for α_{II}. A few studies have addressed the effects of heat treatment on the microstructure of Timetal-21s [4,5]. A better understanding of the relationship between heat treatments, phase transformations and property changes in this alloy should provide a basis for the development of optimum properties.

Experimental Procedure

Samples of cold-rolled, 0.56 mm thick sheet, were solution treated in the β phase region (900°C for 0.5 hour), water quenched, and then aged at various temperatures and times. Transmission electron microscopy (TEM) samples were compared from the as-quenched condition, an isochronal aging study (11 hours at 450°C, 522°C and 565°C) and an isothermal aging study (1, 4 and 11 hours at 522°C). A related study of the mechanical properties has been reported by Bird et al. [6]. TEM specimens were mechanically ground, dimpled and then ion milled with liquid nitrogen cooling. Samples of the as-quenched material were ion-milled without cooling, to avoid a low temperature martensitic reaction. The thin foils were examined in a Philips EM 420T operated at 120 kV. On-zone selected area diffraction (SAD) patterns were used to identify precipitate reflections.

Results and Discussion

Microstructural changes from the as-quenched condition were evaluated for samples from the isochronal and isothermal aging studies, and are presented with reference to their corresponding mechanical properties.

As-quenched alloy

TIMETAL-21S is not a single phase β alloy in the as-quenched condition, since athermal ω forms during quenching [5]. The ω_{ath} phase can be identified by diffuse streaks in selected area electron diffraction (SAD) pattern [7,8]. Figure 1 shows the $(011)_\beta$ SAD pattern from an as-quenched sample of TIMETAL 21S. The β-ω transformation is a displacement-controlled reaction; and short-range-correlated displacements resulting from the collapse of the $(111)_\beta$ planes give rise to the diffuse streaks. Attempts at direct observation of the ω phase in this condition were not successful, however, neither in dark field (DF) images from the diffuse streaks nor in the bright field (BF) image; which suggests that the ω phase transformation was not complete.

Isochronal aging study

Several heat treatments were used to assess mechanical property variations [6], and three of these were selected for a comparison of the effect of temperature on the precipitate microstructure at an exposure time of 11 hours. The aging treatments and resultant toughness (K_{JIc}) and yield strength are shown in Table 1.

Table 1

	11h 450°C	11h 522°C	11h 565°C
K_{JIc} (MPa √m)	42	60	54
YS (MPa)	1400	1421	1324

The range of mechanical properties shown above is due to microstructural changes brought about by precipitation processes, which can be well characterized by TEM. The TEM observations on the conditions described in Table 1 are given below.

11h 450°C (lowest toughness, medium yield strength): The predominant precipitate phase is α_I. Figure 2 shows a bright field image of α_I plates from an intergranular region. The $(111)_\beta$ SAD pattern (Figure 3) shows a Burgers relationship between the hexagonal α and the cubic β matrix, confirming the presence of α_I. There is also evidence of the athermal omega phase; Figure 4 shows the $(011)_\beta$ SAD pattern with diffuse streaks similar to those in Figure 1. Very small α_I precipitates were observed near some grain boundaries in the form of three-variant structures. These structures, which appeared as equilateral triangles when viewed near the $<111>_\beta$ direction (Figure 5), were about 50 nm across. A more detailed description of this structure is given later.

11h 522°C (high toughness, high yield strength): The principle phase in this condition is α_I (see Figure 14) with a volume fraction higher than that in the 11h 450°C condition. Some α_{II} phase was also observed. The SAD pattern in Figure 6 shows reflections due to α_I and a characteristic ring pattern due to α_{II} precipitates. A narrow (~30nm wide) band of α_I phase was observed along some grain boundaries, and sideplate α_I was present near some grain boundaries. Interfacial dislocations were not found at the α_I precipitate/ β matrix interface in this study although they have been reported for other β alloys such as Beta C and Ti 15-3-3-3 etc.[9]. This suggests that the α_I precipitate is coherent with the β matrix, which may be a factor in the strengthening mechanisms of this alloy.

11h 565°C (medium toughness, low yield strength): The precipitate microstructure is dominated by α_{II}. The characteristic ring patterns due to this precipitate are visible on the $<111>_\beta$ SAD pattern in Figure 7. α_{II} can form either in plate or blocky morphology in the matrix (see Figure 8). Only small amounts of α_I precipitation were also observed in the matrix, this is evident from the weak intensities of the α_I reflections in Figure 7. Grain boundary precipitates included both α_I and α_{II} in 100nm wide bands.

Isothermal aging study

Microstructural changes were evaluated as a function of time at a temperature of 522°C. Table 2 indicates the aging treatments used and the corresponding toughness and yield strength data.

Table 2

	1h 522°C	4h 522°C	11h 522°C
K_{JIc} (MPa √m)	65	53	60
YS (MPa)	993	1317	1421

1h 522°C (high toughness, low yield strength): The predominant phase is metastable β. Both the volume fraction and number density of precipitates are quite low (Figure 9). The presence of ω_{ath} in matrix is indicated by intense, diffuse streaks shown in the $(011)_\beta$ SAD pattern of Figure 10a. Some α_I precipitate plates are present; Figure 10b shows strong α_I reflections in a $(111)_\beta$ SAD pattern from an area containing precipitate plates, but α_{II} was not observed.

4h 522°C (low toughness, high yield strength): Both α_I and α_{II} are present (Figure 11). Some triangular α_I phase was also observed (Figure 12), but was located intragranularly in contrast to the near grain boundary location of the 11h 450°C condition; in addition, the triangular α_I phase here is larger and its morphology is an "open" triangle enclosing β phase. Corresponding to this microstructure, K_{JIc} is rather low (namely, 53 MPa$\sqrt{m}$) but yield strength is fairly high (1300 MPa). The $(111)_\beta$ SAD pattern in Figure 13 shows the coexistence of α_I and α_{II} precipitates. Both the size and volume fraction of α_I and α_{II} are larger than in the 1h 522°C condition.

11h 522°C (high toughness, high yield strength): The predominant phase is α_I. Figure 14 shows a high number density and volume fraction of α_I and α_{II} precipitates. The α_I precipitates in this condition are larger than in the 4h 522°C condition. No triangular α_I was observed, but large two variant plate structures were observed, producing a "V" shape.

Summary of aging comparisons.

Higher aging temperatures and times favored the development of α_{II}. The volume fraction and size of α_I and α_{II} showed an increase with increasing aging time at 522°C. The small solid triangular α_I phase present during earlier stages of aging gives way to large, open triangles and V-shaped, bi-variant α_I particles.

The two conditions with relatively low fracture toughness, namely 11h 450°C and 4h 522°C, were found to contain triangular α_I precipitates, and low volume fractions of α_{II}. The lower toughness of the 4h 522°C material is particularly surprising since further aging results in increases in both strength and toughness. The reason for the lower K_{JIc} of these microstructures has not been thoroughly investigated, but is likely to be due to either a positive effect of the α_{II} or a negative effect of the triangular α_I. Considerations of the effects of the triangular α_I would benefit from an understanding of the details of its development.

Triangular α_I

The TEM investigation was used to establish the crystallographic relationship between the triangular α_I variants and the β matrix phase, and this is shown, schematically, in Figure 15. This figure indicates that the triangular α_I not only conforms to the Burgers relationship: $<11\bar{2}0>_\alpha \parallel <111>_\beta$ and $(0001)_\alpha \parallel (011)_\beta$, but also satisfies a twin relationship between each α_I variant. The alpha twin planes $(01\bar{1}1)$ and $(0\bar{1}11)$ are coincident with the β (110) and $(01\bar{1})$, consistent with the Burgers relationship [10]. The SAD patterns for the twin and Burgers relationship along $<111>_\beta$ orientation can not be distinguished from each other. The twin relationship can, however, can be recognized from a high resolution image. Figure 16 is a high resolution image of the triangular α_I precipitate obtained by using only the α_I reflections. The (0001) planes of α_I phase with d-spacing of 0.47 nm can be seen. The twin relationship between the three variants of α_I can be recognized from the lattice fringes in the image.

In the present study, the triangular α_I has been shown to form in a manner different from that suggested by Inaba et al. [9]. They concluded that the triangular α_I was a hollow tetrahedron enclosing some β phase. In the present investigation, the triangular α_I phase was found to result from a triangular prism growth morphology. Figure 17 is α near B= $<111>_\beta$ BF image showing the single phase, three variant nature of the triangular precipitates. The SAD pattern of this region shows that the triangular phase is α_I. Figure 18 which is an off $<111>_\beta$ DF image showing an inclined view of the triangular prism; parallel edges of the prism which are along $<111>_\beta$ are clearly seen. A triangular prism can account for geometrical aspects of the α_I morphology, the Burgers orientation relationships of the three variants of α_I with the matrix, and the twin orientation relationships between the α_I variants. Based on these considerations, the tetrahedral shape described by Inaba et al. is not a viable morphology for the triangular α_I in this alloy.

Nucleation mechanism of α_I The multi-variant morphologies observed for α_I can be accounted for on the basis of a sympathetic nucleation mechanism [11] on a low energy α/α interface. The low interfacial energy usually associated with coherent twin boundaries [12] between the α_I variants is consistent with this mechanism for the formation of the multivariant structures.

Summary

The TEM characterization of TIMETAL 21S produced the following conclusions:
- Athermal ω is formed during quenching and gradually diminishes during aging.
- Aging at different temperatures featured the formation of several types of alpha: α_I at 450°C, a mixture of α_I and α_{II} at 522°C, and α_{II} at 565°C.
- The α_I in TIMETAL 21S is coherent with the β matrix.
- Early stages of α_I formed tri-variant prisms with triangular cross sections. The three variants of α_I were separated by coherent twin interfaces and form via a sympathetic nucleation mechanism.
- Low fracture toughness was associated with the presence of triangular α_I and a lack of α_{II}.
- The closed, tri-variant prisms were replaced by open triangles and bi-variant V-shaped α_I in later stages.
- Single (plate) and multi-variant α_I were observed simultaneously, and appear to develop by different mechanisms.

Acknowledgments

The authors are grateful to the Metallic Materials Branch at NASA Langley Research Center for their support this work, to R. K. Bird for providing the experimental data and to C. G. Rhodes, R.Shenoy, W. Brewer and T. Wallace for their invaluable discussion and suggestions.

References

[1] P.J. Bania, Iron Steel Inst. Jpn. Int., 31 (8) (1991), 45-47.
[2] J.C.Williams, Titanium Science and Technology, vol. 3, R.I. Jaffee and H.M. Burte, eds., (New York, NY:Plenum Press, 1973), 1433-94, .
[3] W.G. Burgers, Physica, (1) (1934), 561-86.
[4] M.W. Mahoney, P.L.Martin and D.A. Hardwick,"Microstructural stability of Beta-21S" Titanium '92 Science and Technology, Ed. by F.H. Froes and I. Caplan, TMS, (1993) 161-168.
[5] K. Chandhuri and J.H. Perepezko: "Microstructural Study of the Titanium Alloy Ti-15Mo-2.7Nb-3Al-0.2Si (TIMETAL 21S)", Metall. Trans. , (24A) (1994), 1109-1118, .
[6] R.K. Bird, "Effects of Heat Treatment on Microstructure, Strength, and Toughness of Timetal 21S Sheet", Advances in the Science and Technology of Titanium Alloy Processing, (Warrendale, PA: TMS, 1996) in press.
[7] D. de Fontaine, Acta. Metall. (18) (1970), 275-79.
[8] D. de Fontaine, N.E. Paton and J.C.Williams, Acta. Metall. (19) (1971), 1153-1162.
[9] T.Inaba, K.Ameyama and M.Tokizane,"Morphology of α Precipitate in Ti-15V-3Cr-3Sn-3Al b Titanium Alloy", J. Japan Inst. Metals, (56) (1992), 881-888.
[10] D.I. Potter, J. Less-Common Metals, (31) (1973), 299.
[11] E.Sarath Kumar Menon and H.I. Aaronson, "Morphology, Crystallography and Kinetics of Sympathetic Nucleation", Acta. Metall. (35) (1987), 549-563.
[12] P. Haasen, Physical Metallurgy, (Cambridge: Cambridge University Press, 1978).

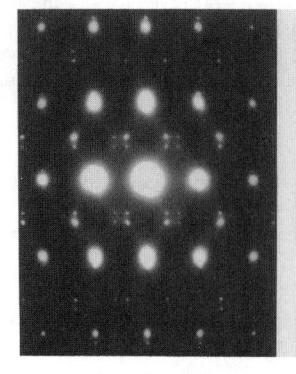

Figure 1. <011>$_\beta$ SAD pattern showing athermal ω reflections, as-quenched condition.

Figure 2. BF image of α_I precipitates in the 11h 450°C condition.

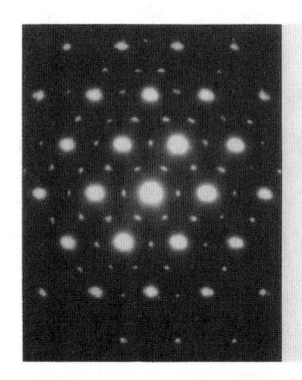

Figure 3. <111>$_\beta$ SAD pattern in the 11h 450°C condition.

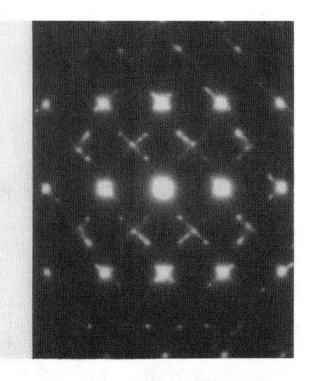

Figure 4. <011>$_\beta$ SAD pattern in the 11h 450°C condition.

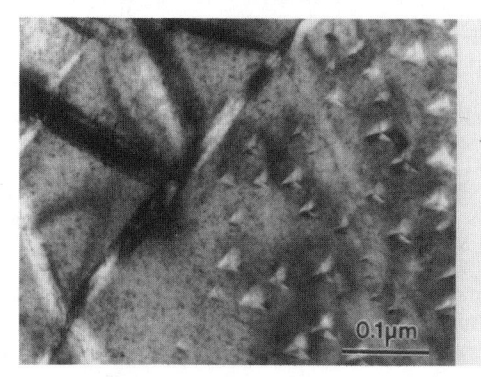

Figure 5. BF image of α_I near grain boundary in the 11h 450°C condition.

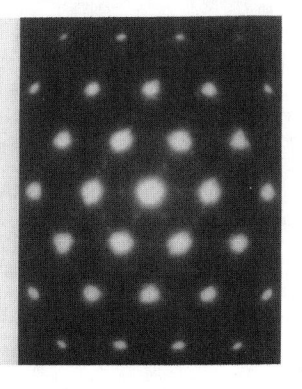

Figure 6. <111>$_\beta$ SAD pattern in the 11h 522°C condition.

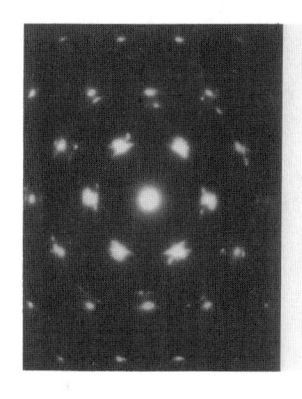

Figure 7. <111>$_\beta$ SAD pattern in 11h 565°C condition.

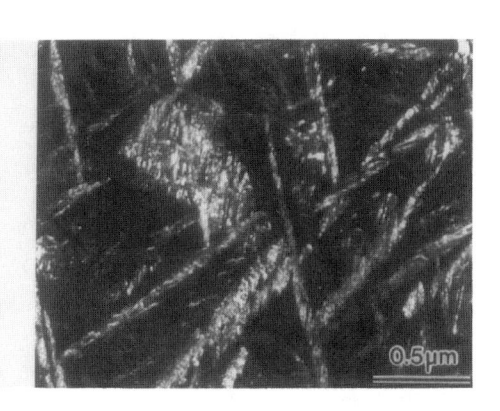

Figure 8. DF image of α_{II} precipitates in the 11h 565°C condition.

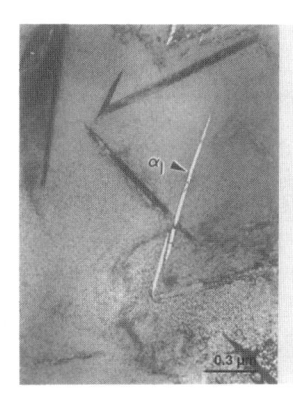

Figure 9. BF image of α_I in the 1h 522°C condition.

Figure 10. In the 1h 522°C condition a) <011>$_\beta$ pattern from β matrix, b) <111>$_\beta$ SAD pattern in the presence α_I.

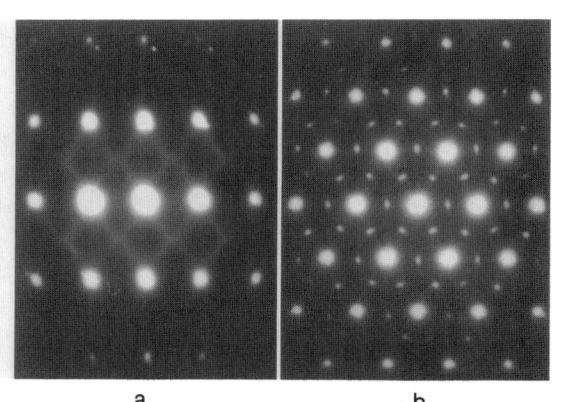

Figure 11. BF image of α_I in the 4h 522°C condition.

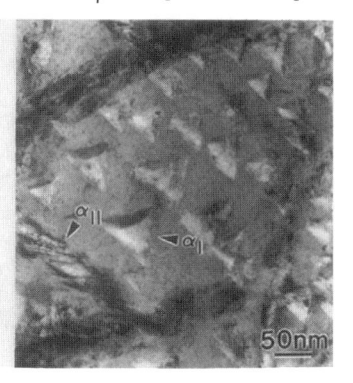

Figure 12. Triangular α_{II} in the 4h 522°C condition.

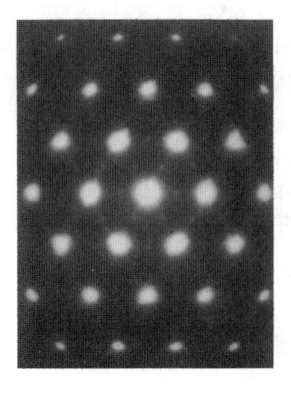

Figure 13. <111>$_\beta$ SAD pattern in the 4h 522°C condition.

Figure 14. α_I and α_{II} precipitates in the 11h 522°C condition.

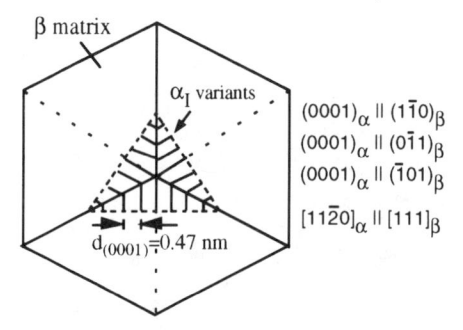

Figure 15. Schematic of triangular α_I in β matrix.

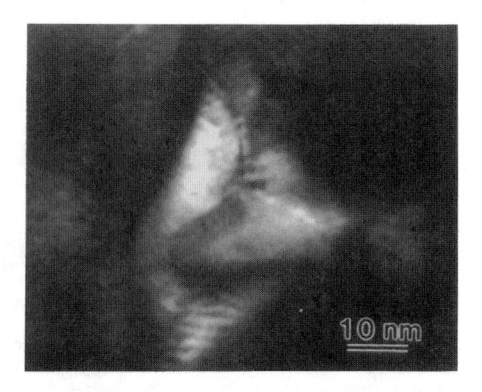

$(0001)_\alpha \parallel (1\bar{1}0)_\beta$
$(0001)_\alpha \parallel (0\bar{1}1)_\beta$
$(0001)_\alpha \parallel (\bar{1}01)_\beta$
$[11\bar{2}0]_\alpha \parallel [111]_\beta$

$d_{(0001)} = 0.47$ nm

Figure 16. High resolution image of the triangular α_I.

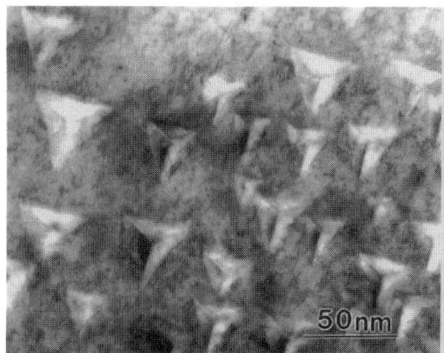

Figure 17. BF image of the triangular α_I near <111>$_\beta$ orientation.

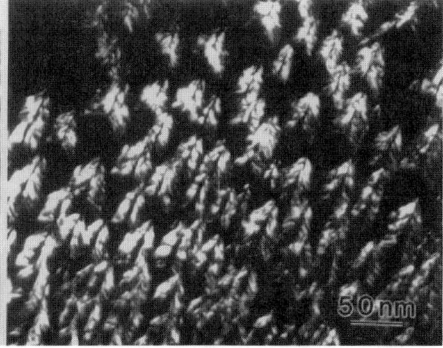

Figure 18. DF image of triangular α_I, off <111>$_\beta$ orientation.

EFFECTS OF MICROSTRUCTURE AND OXYGEN CONTENT ON THE FRACTURE BEHAVIOUR OF THE α + β TITANIUM ALLOY Ti-4Al-4Mo-2Sn-0.5Si wt.% (IMI 550).

L. J. Hunter and M. Strangwood.

School of Metallurgy and Materials / IRC in Materials for High Performance Applications, The University of Birmingham, Edgbaston, Birmingham. B15 2TT. UK. Fax 44 21 414 5232.

Abstract

The effect of microstructure and oxygen content on the room temperature tensile strength and fracture toughness of the α + β titanium alloy Ti-4Al-4Mo-2Sn-0.5Si wt.% (IMI 550), are assessed in this paper, studied by varying the solution treatment temperature and oxygen content. A detailed micromechanical study indicates that failure is controlled by interfacial regions between primary α and transformed β and at interfaces within the transformed β. A transmission electron microscopy characterisation of such interfacial regions has also been undertaken and has been related to variations in both microstructure and oxygen content.

Introduction

The fracture toughness and strength of α + β titanium alloys, used for aerospace applications, are important considerations in the design of components. Improvements in fracture toughness to existing commercial α + β titanium alloys, such as Ti-6Al-4V wt.%, have been made by both microstructural and compositional modifications (1). However, increases in fracture toughness levels are usually achieved at the expense of the material's strength.

Ti-4Al-4Mo-2Sn-0.5Si wt.% (IMI 550) is a medium strength α + β titanium alloy which is currently under development for extended use in aerospace applications. IMI 550 has the advantage of higher strength, increased creep resistance and lower susceptibility to stress corrosion cracking, compared with commercial Ti-6Al-4V wt.% (IMI 318) (1).

Oxygen strongly influences microstructural and mechanical properties. It is always present in significant concentrations in commercial products. IMI 550 is typically produced with oxygen levels up to 2500 ppm, with the majority of casts in the 1700 - 2100 ppm range (1). Oxygen is thus often used as an alloying element to achieve desired strength levels.

A standard heat treatment, recommended for engineering applications, consists of solution treatment at a temperature of 900°C for 1 hour per 25 mm of section thickness, followed by air cooling to ambient temperature and ageing for 24 hours at a temperature of 500°C (2). This results in a tensile strength of approximately 1100 MPa and a fracture toughness of 70 MPa.m$^{1/2}$.

A central theme of this study, along with previous work (1, 3), is to quantify the effects of microstructure (primary α content) and composition (oxygen content) on the fracture toughness and strength of Ti-4Al-4Mo-2Sn-0.5Si wt.%. This requires a detailed investigation to elucidate the micromechanisms of ductile failure in Ti-4Al-4Mo-2Sn-0.5Si wt.%.

Experimental Procedure

Ingots (MT995, BD39164 and RU4277) of Ti-4Al-4Mo-2Sn-0.5Si wt.% (IMI 550) were produced at IMI Titanium ltd, by vacuum arc melting. Three levels of oxygen content, in the range 400 - 2100 ppm were produced, and the exact chemical compositions are outlined in table I.

Table I: Chemical compositions.

	Al wt.%	Mo wt.%	Sn wt.%	Si wt.%	Fe wt.%	O ppm	N ppm	H ppm
MT995:	4.17	4.15	2.08	0.48	0.05	400	20	15
BD39164:	4.16	4.07	2.01	0.45	0.06	1450	35	10
RU4277:	4.10	4.14	2.10	0.48	*	2100	58	*

* Levels not determined.

The ingots were α + β forged, to produce rectangular bar. The β-transus approach curves, for each ingot, were established by IMI Titanium plc and the material was then subjected to the non-standard heat treatments outlined in table II, to obtain nominally 50 and 20% primary α phase constants for each oxygen level.

Table II: Heat treatment schedule and resulting microstructures.

	Solution treatment	Cooling rate	Age
MT995	3 hours at 870°C	Air cool	24 hours at 500°C + air cool
MT995	2 hours at 920°C	Air cool	24 hours at 500°C + air cool
BD39164	2 hours at 900°C	Air cool	24 hours at 500°C + air cool
BD39164	2 hours at 933°C	Air cool	24 hours at 500°C + air cool
RU4277	30 mins at 900°C	Air cool	24 hours at 500°C + air cool
RU4277	30 mins at 940°C	Air cool	24 hours at 500°C + air cool

Microscopy samples were ground and polished to a 0.05 μm finish and etched in Kroll's II reagent (2% HF, 10% HNO$_3$ and 88% H$_2$O vol.%). Small scale microstructural features were resolved using transmission electron microscopy (TEM). Foils were prepared from 3 mm diameter discs, which were spark machined from 0.5 mm thick slices. These were then mechanically ground to a thickness of 100 - 150 μm, dimpled to a depth of 40 μm and perforated by ion beam thinning, using a Gatan precision ion polishing system. The foils were examined in a Philips CM20 operating at 200 kV and equipped with a Link QX 2000 energy dispersive X-ray spectroscopy (EDS) system, using an ultra-thin window detector.

Crack opening displacement (COD) values have been evaluated in air at ambient temperature, in accordance with BS5762:1979 (4). Room temperature tensile tests, of the six heat treated conditions, were carried out by IMI Titanium ltd. Fracture surfaces were examined using an Hitachi S4000 field emission gun (FEG) scanning electron microscope (SEM).

In-situ crack path studies were undertaken to further investigate the micro-mechanisms of failure. Fatigue pre-cracked compact tension (CT) specimens (14.6 mm X 14 mm X 3.5 mm) were tested at a slow strain rate, in vacuum, in an Hitachi S4000 FEG SEM. The surface was mechanically ground, polished and etched, as outlined previously, prior to fatigue pre-cracking.

A further study of crack paths was undertaken by interrupting a COD test, before it attained maximum load. The specimen was then sectioned, mounted, polished and etched. Crack paths, as a function of microstructure, were then determined both near the surface and in the bulk of the specimen.

Results and Discussion
Effect of Heat Treatment on Microstructure

Scanning electron micrographs of $\alpha + \beta$ Ti-4Al-4Mo-2Sn-0.5Si wt.%, subjected to the heat treatments outlined in table I, are shown in figure 1.

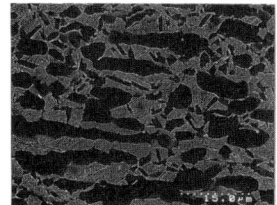

MT995: 3 hours 870°C air cool + age 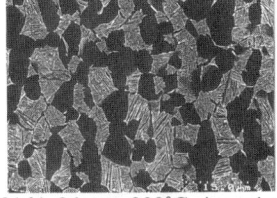   BD39164: 2 hours 900°C air cool + age

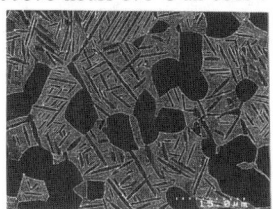

RU4277: 30 mins 900°C air cool + age 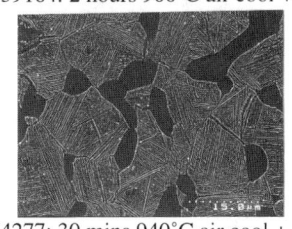   RU4277: 30 mins 940°C air cool + age

Figure 1. Micrographs of the various heat treated conditions of Ti-4Al-4Mo-2Sn-0.5Si wt.%.

It can be seen that for each heat treated condition, the microstructure consists of a mixture of primary α and transformed β phases. If an alloy has been $\alpha + \beta$ finish forged, air cooled and solution treated at a lower temperature than the finishing temperature, a low density of smaller, elongated α phase develops in addition to the equiaxed primary α, as seen in RU4277 (30 mins 900°C air cool + age). All of the microstructures below contain a very high density of α / β interfaces per unit volume. The 20% primary α air cooled and aged microstructures, for all three oxygen levels, are all closely similar, hence only one representative micrograph is included. Grain boundary α is observed along the prior β grain boundaries.

The use of TEM allows resolution of the transformed β structure, and reveals a fine interwoven secondary α lath structure in a mixture of α'' and small amounts of retained β and ω phase. A change in structure is also observed along the primary α / transformed β interfaces, as shown in figure 2.

The interfacial region is found to vary with changes in both the composition and solution treatment temperature. The low and medium oxygen contents (MT995 and BD39164) reveal a striated layer which exhibits a hexagonal crystallographic structure and is approximately 0.1 μm thick in the medium oxygen content material and much thinner in the low oxygen content material. The high oxygen content (RU4277), which is the standard oxygen content subjected to the standard heat treatment (outlined previously), reveals a thin, monolithic layer which is deduced to be retained β phase from the very high molybdenum levels in the interfacial region (table III).

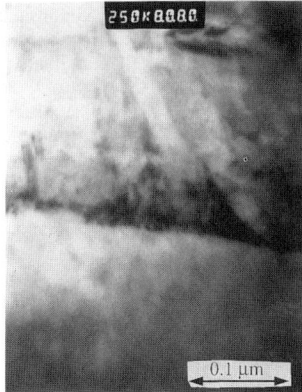

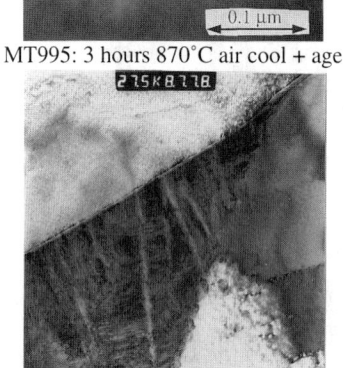

MT995: 3 hours 870°C air cool + age   BD39164: 2 hours 900°C air cool + age

RU4277: 30 mins 900°C air cool + age   BD39164: 2 hours 900°C air cool + age

Figure 2. Transmission electron micrographs showing the transformed β structure and interfacial region for various heat treated conditions of Ti-4Al-4Mo-2Sn-0.5Si wt.%.

Compositional analysis of the phases indicated a dependence on oxygen content and heat treatment, as shown in table III. Variations in the compositions of the different regions for the three casts are consistent with the role of oxygen, as an α-stabiliser, and the variation in solution treatment temperature. It can be seen that the molybdenum content is high in the transformed β phase, compared with the primary α phase, but the highest level is attained in the primary α / transformed β interfacial region, indicating the presence of a microchemical gradient in the concentration of molybdenum. The microchemical gradient in molybdenum would arise from non-equilibrium effects during air cooling as α grows into the remaining β. Incomplete redistribution of the rejected molybdenum results in enrichment of the interfacial region, forming retained β or the striated hexagonal close packed layer.

Table III: Effect of oxygen content and heat treatment on phase composition.
MT995: 3 hours 870°C air cool + 24 hours 500°C air cool.

Wt. %	Al	Si	Sn	Mo
Primary α	4.74	0.49	2.39	1.07
Interfacial region	2.67	0.43	2.55	**16.98**
Transformed β	2.97	0.42	2.86	12.62

BD39164: 2 hours 900°C air cool + 24 hours 500°C air cool.

Wt. %	Al	Si	Sn	Mo
Primary α	4.02	0.62	1.83	1.57
Interfacial region	2.76	0.61	2.63	**16.10**
Transformed β	2.33	0.82	4.01	13.19

RU4277: 30 minutes 900°C air cool + 24 hours 500°C air cool.

Wt. %	Al	Si	Sn	Mo
Primary α (equiaxed)	5.69	0.65	2.68	1.05
Primary α (lath)	4.27	1.76	2.45	1.48
Interfacial region	2.31	1.20	2.58	**22.28**
Transformed β	3.00	0.78	2.57	19.30

Effect of Heat Treatment on Yield Strength and Crack Opening Displacement (COD)
Figure 3 summarises the trends in COD and yield strength. It is apparent from figure 3 that there is a decrease in COD as yield strength increases. An increase in oxygen content results in a reduction in COD. Figure 3 indicates that as primary α content decreases, from 50 to 20%, the effect of oxygen content increases. It is noted that, for a nominally 20% primary α microstructure, there is a 61% increase in fracture toughness when the oxygen content is decreased from 1450 to 400 ppm. The significance of this improvement is further emphasised by the fact that it was achieved with only a 11% loss in yield stress. For a nominally 50% microstructure, there is a 27% increase in toughness when the oxygen content is decreased from 1450 to 400 ppm. This is accompanied by a 5% loss in yield stress.

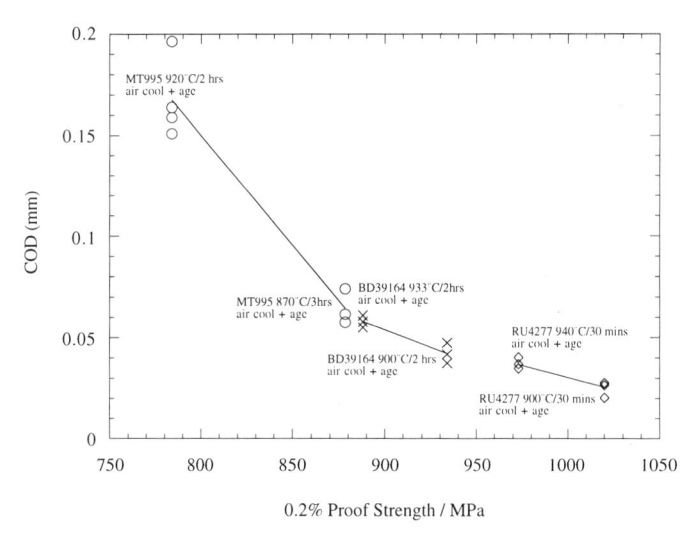

Figure 3. COD versus yield strength for various heat treatments and various oxygen contents of Ti-4Al-4Mo-2Sn-0.5Si wt.% (IMI 550).

Fractography
Fractography indicates that in all COD specimens, the fracture surface appearance, for all of the various heat treatments was similar, thus indicating that the failure mechanism was closely

similar in all of the microstructures that are studied here. A typical fractograph, figure 4, shows two matching halves, of a COD specimen (MT995 870°C 3 hours / air cool + age).

It is observed that the crack prefers to follow primary α / transformed β interfaces and decohesion at the interface is observed (arrow 1). Figure 4 also shows where a whole colony of transformed β has decohered and been removed (arrow 2). When the advancing crack traverses transformed β, interfacial decohesion is again observed, leaving behind well-defined microvoids (arrow 3). Voids are also observed occasionally to have nucleated from silicides of composition Ti_5Si_3, approximately 1 μm in diameter (arrow 4) but this is a local effect and is not considered to play a significant role in the fracture mechanism.

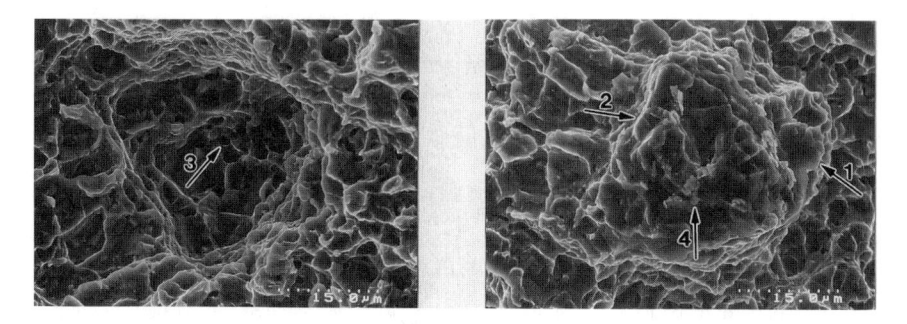

Figure 4. SEM micrograph of two matching halves of a COD specimen (MT995).

In-situ COD tests showed a sharp crack (initiated by fatigue prior to testing) propagating through a transformed β grain, as shown in figure 5a, and then upon reaching a transformed β / primary α interface, the crack arrested and blunted, as shown in figure 5b. Increasing the load further, resulted in microvoids initiating ahead of the main crack tip by decohesion of interfacial regions within the transformed β phase and between the primary α and transformed β. Upon continual loading, the microvoids increased in size, subsequently coalescing until they formed a continuous path to the main crack, resulting in crack extension.

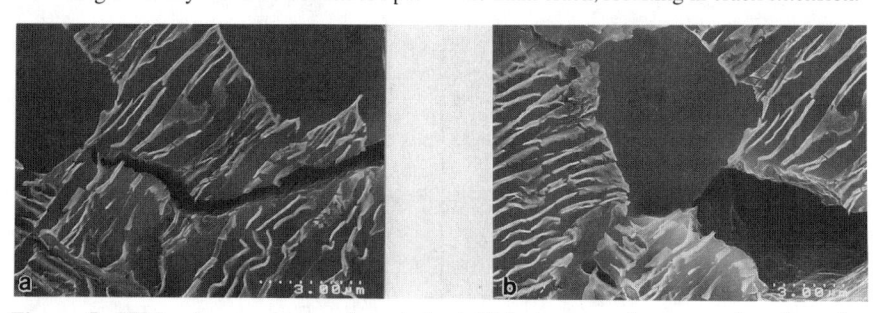

Figure 5. SEM micrograph showing **a)** the initial sharp crack propagating through a transformed β grain and **b)** blunted crack at the transformed β / primary α interface.

An examination of the crack paths in both *in-situ* and COD specimens, for the microstructures shown in figure 1, are shown in figure 6.

Figure 6a is taken in the bulk of a COD specimen whereas figure 6c is taken near the surface of a COD specimen and where there is less constraint present during crack growth. No change in mechanism is observed and hence the crack path observed in the *in-situ* test is considered to be representative of crack extension in the plane strain regions.

A preference for crack propagation through interfacial regions within the transformed β phase and between the primary α and transformed β, as indicated in figure 4, is confirmed, by these *in-situ* observations and thus emphasises the importance of these regions in controlling the fracture process.

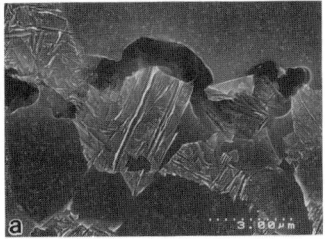

MT995: 3 hours 870°C air cool + age

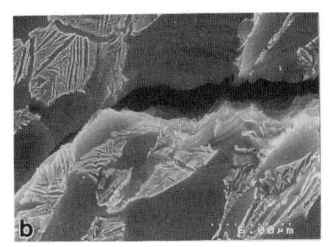

BD39164: 2 hours 900°C air cool + age

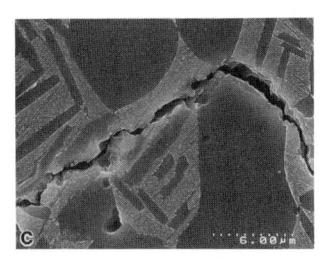

RU4277: 30 mins 900°C air cool + age

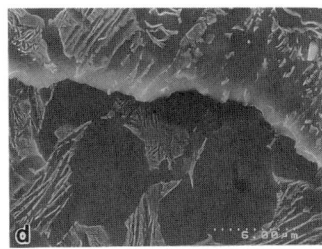

BD39164: 2 hours 933°C air cool + age

Figure 6. SEM micrographs showing the crack path in (**a**) + (**c**) COD specimens and (**b**) + (**d**) *In-situ* CT specimens.

The probable reason for the susceptibility of these interfaces to void initiation is that they are sites for strain localisation. Void initiation takes place preferentially in the interfacial region due to a change in structure and hence plausibly a breakdown in strain compatibility. The long interface around the primary α particles and along the grain boundary α phase provides a continuous path of easy void growth. When the void or crack reaches the end of a primary α grain, it will seek out an adjacent primary α grain, propagating along more tortuous secondary α interfaces, within the transformed β phase, where necessary. Hence, fracture toughness increases with a lower volume fraction of primary α, for any given oxygen content. However, the difference in fracture toughness levels could also be related to a compositional variation of the phases present. It is very difficult to separate the effects of phase composition and proportion and morphology of the β transformation product as it is impossible to change one without the other.

An increase in the alpha stabilising element, oxygen leads to an increase in yield strength of the α phase and micro variations in composition which cause changes in the nature of both the interface phase and the transformed β phase. Hence, there is plausibly an increase in strain localisation between the primary α phase and transformed β or interface phase and fracture toughness levels drop.

Conclusions

1. Optical microscopy reveals the presence of two phases α and transformed β. However, transmission electron microscopy indicates the presence of an interface phase along primary α / transformed β interfaces. The size and microchemistry of the interfacial region is found to vary with changes in oxygen content.

2. An increase in crack opening displacement levels can be achieved in Ti-4Al-4Mo-2Sn-0.5Si wt.% (IMI 550), by use of a higher solution treatment temperature, within the α + β phase field, to reduce the volume fraction of primary α regions. This is believed to be due to a reduced number of primary α / transformed β interfaces acting as preferential crack paths.

3. An increase in oxygen content results in a reduction in crack opening displacement.

4. The crack path is along interfacial regions predominantly and thus emphasises the importance of the interfacial regions between the primary α, transformed β and secondary α regions in controlling fracture toughness.

Acknowledgements

This work has been jointly funded by the Engineering and Physical Sciences Research Council (EPSRC) and The University of Birmingham. Thanks are extended to Prof. J.F. Knott and Prof. M.H. Loretto for the provision of laboratory facilities, and to IMI Titanium ltd for supplying material, compositional analyses and tensile test results.

References

1. R.F. Vaughan, P.A. Blenkinsop and J.A. Hall, "The Effect of Oxygen Content on the Mechanical Properties of an α + β Titanium Alloy Ti-4Al-4Mo-2Sn-0.5Si (IMI 550)," Titanium '80 Science and Technology, (1980), 1645-1651.

2. IMI Titanium plc, Company Brochure, Titanium Alloy IMI 550, (1984).

3. R.F. Vaughan, P.A. Blenkinsop and D.F. Neal, "The Effect of Variations in Heat Treatment on the Mechanical Properties of an α + β Titanium Alloy IMI 550 (Ti-4Al-4Mo-2Sn-0.5Si)," Proceedings of the Third International Conference on Titanium, (1976), 2047-2060.

4. British Standards Institution, "Methods for Crack Opening Displacement (COD) Testing," BS5762:1979.

PHASE STABILITY IN A Ti-45.5Al-2Nb-2Cr ALLOY

V. Seetharaman*, C. M. Lombard** and S. L. Semiatin**

*Materials and Processes Division, UES, Inc., 4401 Dayton-Xenia Road, Dayton, OH 45432-1894
**Materials Directorate, Wright Laboratory, WL/MLLN, Wright Patterson Air Force Base, OH 45433-7817

Abstract

High temperature phase equilibria in a Ti-45.5Al-2Nb-2Cr alloy were investigated via metallography, X-ray diffraction and electron microprobe analysis. The alloy was extensively hot worked through a sequence of isothermal forging and pack rolling operations coupled with intermediate, recrystallization heat treatments. Specimens from the rolled sheets were subjected to prolonged isothermal heat treatments in the temperature range 900°C-1400°C for durations ranging up to 100 hours, followed by water quenching as well as slow cooling. The results indicated that a three phase equilibrium involving γ, α / α_2, and β / β_2 phases exists in the temperature range 900°C-1250°C. While the volume fraction of β_2 appears to reach a maximum at ~1200°C, that of α increases rapidly with temperature in the range 1200°C-1300°C. Heat treatment in the single phase α field (1300°C-1340°C) led to pronounced grain growth, whereas annealing in the $\alpha + \beta$ field (T>1340°C) caused retardation in the growth of α grains. The influence of heat treatment temperature and cooling rate on microstructure evolution in this alloy is discussed in the light of available information on phase stability in Ti-Al-X systems, where X is a beta stabilizing element.

Introduction

Phase equilibria and solid state phase transformations occurring in Ti-Al alloys containing 40-50 at.% Al have been investigated in detail (1-3). These studies have shown that depending upon composition, processing, and heat treatment, a rich variety of microstructures can be generated in these alloys through transformations involving three major phases: α / α_2, β / β_2 and γ. For example, a Ti-48 at.% Al alloy may exhibit an equiaxed, lamellar, Widmanstatten, or massive gamma microstructure depending on the rate at which it is cooled from the single phase α field. These microstructures, in turn, exert specific influences on mechanical properties such as tensile strength and ductility, creep resistance, and fatigue life.

Addition of transition metals like niobium, chromium, manganese or vanadium in small quantities (≤ 4 at.% total) leads to significant improvement in strength, ductility and oxidation resistance of near gamma titanium aluminides. The influence of these elements (X) on the phase equilibria and microstructural evolution in Ti-Al-X alloys has been studied to a limited extent (4-7). Huang and Hall (4) have demonstrated that the principal effect of Cr addition is to lower the alpha transus temperature and thus decrease the volume fraction of α_2. Masahashi and Mizuhara (5) have compared the microstructures of Ti-47Al-3 X alloys where X = Cr, Mo or W and have found that W is the most potent stabilizer of β / β_2.

The objectives of the current work on the phase stability in a Ti-45.5Al-2Nb-2Cr alloy were: (i) to examine the long term stability of microstructures at temperatures below the alpha transus and (ii) follow microstructure development during heat treatment at temperatures close to the $\alpha / \alpha + \beta$ boundary with an ultimate goal of grain size control.

Experimental Procedures

The nominal composition (in atomic percent) of the titanium aluminide alloy used in the present work was Ti-45.5Al-2Nb-2Cr. The alloy was isothermally forged to a thickness of 12.5 mm, followed by multipass pack rolling to yield a final sheet thickness of approximately 2 mm (8,9). The interstitial contents of the rolled sheets were analyzed to be oxygen -850 ppm and nitrogen -65 ppm (in weight). Specimens for heat treatments were cut from ground sheets, wrapped in tantalum foils, and sealed in evacuated quartz tubes. For heat treatments at temperatures in the range 900-1200°C, the specimens were directly heated to the specified temperature, held isothermally for 100 hours, and then water quenched. For 1260°C, the heat treatment time was chosen as 20 hours. For temperatures above 1300°C, the specimens were subjected to an intermediate treatment at 1200°C for 30 minutes, transferred directly to the high temperature furnace at $1300 \leq T \leq 1400$°C, held for 20 minutes, followed by water quenching or controlled cooling at 10°C/min. to 900°C and air cooling at temperatures below 900°C.

Microstructural characterization of the heat treated specimens was carried out using polarized light microscopy and scanning electron microscopy in the back-scattered electron (BSE) imaging mode. Electron microprobe analysis (EMPA) using wavelength dispersive spectroscopy was performed to obtain chemical compositions of the different phases and microconstituents. X-ray diffractometry (XRD) of the heat treated specimens was carried out at room temperature using monochromated copper K α radiation. Quantitative metallographic techniques such as point counting and linear intercept methods were used to estimate volume fraction, grain size, and particle size of the different phases.

Results

The microstructure of the as-rolled Ti-45.5Al-2Nb-2Cr sheets consisted of equiaxed gamma grains together with unglobularized $\alpha_2 + \gamma$ lamellar grains. When the rolled sheet was heat treated at 1200°C for two hours, the remnant $\alpha_2 + \gamma$ lamellar grains underwent static globularization to produce fully equiaxed gamma grains (8). Figure 1 shows the microstructures of specimens aged at 900°C, 1000°C, 1100°C and 1200°C for 100 hours. While the specimen aged at 900°C reveals some amount of elongated lamellar grains, those exposed to higher temperatures are essentially free from lamellar structures. All the microstructures shown in Figure 1 contain mixtures of three phases γ, α_2 and β_2. While γ is the major (dark) phase and constitutes the continuous matrix, angular particles / plates of α_2 and β_2 are located mainly along the grain boundaries. It is often difficult to distinguish between α_2 and β_2 in BSE micrographs based on their contrast alone, particularly when the particles are very fine. An increase in the aging temperature from 900°C to 1200°C leads to gradual increases in the mean grain size of gamma, volume fraction of α_2 / β_2 and the size of the α_2 / β_2 particles (Table I). Compositions of the different phases obtained by EMPA of selected specimens are furnished in Table II. It is clear that β_2 is substantially enriched in Cr and depleted in Al, while γ has limited solubility for Cr. As the aging temperature increases, the Cr content of β_2 decreases steadily and the Al contents of both γ and α_2 increase. Figure 2a displays a portion of the X-ray diffraction pattern obtained from the specimen aged at 1200°C. While the peaks associated with γ and α_2 phases are quite intense, a relatively weak peak of $(110)_{\beta2}$ can also be seen. The intensities of this peak and of $(2021)_{\alpha2}$ normalized with the intensity of $(111)_\gamma$ increase with aging temperature (Figure 2b). To a first approximation, this suggests increases in the volume fractions of both α_2 and β_2 with temperature in the range 900°C to 1200°C. While the amount of β_2 increases steeply between 1100°C and 1200°C, that of α_2 tends to saturate above 1100°C.

Table I. Summary of the Microstructural Analyses

Heat Treatment	Grain Size (μm)	Precipitate Size (μm)	Morphology	Type	Volume Fraction γ	$\alpha_2 + \beta_2$
900°C/100 h	7 (γ)	2	Angular	$\alpha_2 + \beta_2$	0.90	0.10
		0.5	Plate	$\alpha_2 + \beta_2$		
1000°C/100 h	8 (γ)	3	Angular	$\alpha_2 + \beta_2$	0.90	0.10
		1	Plate	$\alpha_2 + \beta_2$		
1100°C/100 h	12 (γ)	3	Angular	$\alpha_2 + \beta_2$	0.85	0.15
		1	Plate	$\alpha_2 + \beta_2$		
1200°C/100 h	20 (γ)	5-15	Angular	$\alpha_2 + \beta_2$	0.80	0.20
		2	Plate	$\alpha_2 + \beta_2$		
1260°C/20 h	35 (α_2)	10-20	Equiaxed	γ	0.30	0.70
		1-2	Plates	γ		
1321°C/20 min.	250 (α_2)					1.0 *
1360°C/20 min.	60 ($\gamma_M / \alpha_2 + \gamma$)	1-3	Grain Boundary Layer	β_2	0.97 **	0.03

* 100% α_2 ** Massive gamma and lamellar structures

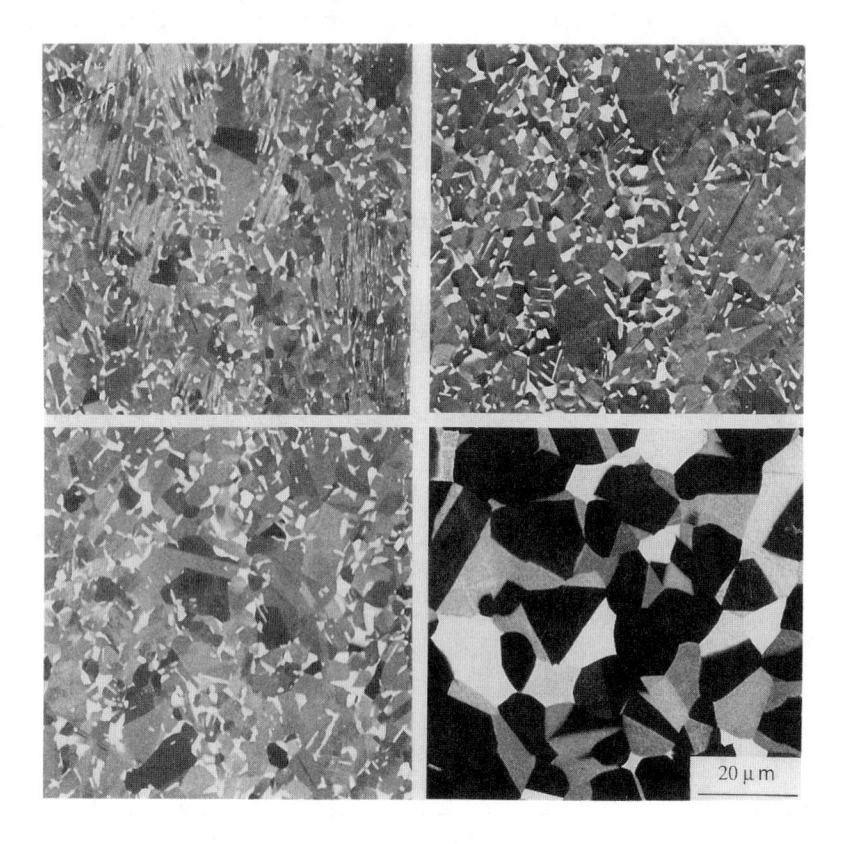

Figure 1: Scanning electron micrographs of specimens heat treated at (a) 900°C (b) 1000°C (c) 1100°C and 1200°C for 100 hours and water quenched.

Heat treatment at 1260°C followed by water quenching leads to microstructure consisting of ~70% α_2 and 30% γ (Figure 3a). No β / β_2 could be detected in the specimen. Comparison of this structure with that of the specimen heat treated at 1200°C (Figure 1d) suggests that the proportions of the α and γ phase change rapidly at temperatures above 1200°C. Water quenching of the specimen annealed at 1320°C produces a single phase microstructure with a mean grain size of ~250 µm (Figure 3b). XRD of this specimen confirmed this phase to be α_2. These observations are consistent with (a) an estimated alpha transus of ~1300°C for this alloy (8,10) and (b) reports that the ordering of $\alpha \rightarrow \alpha_2$ is very rapid (1,2).

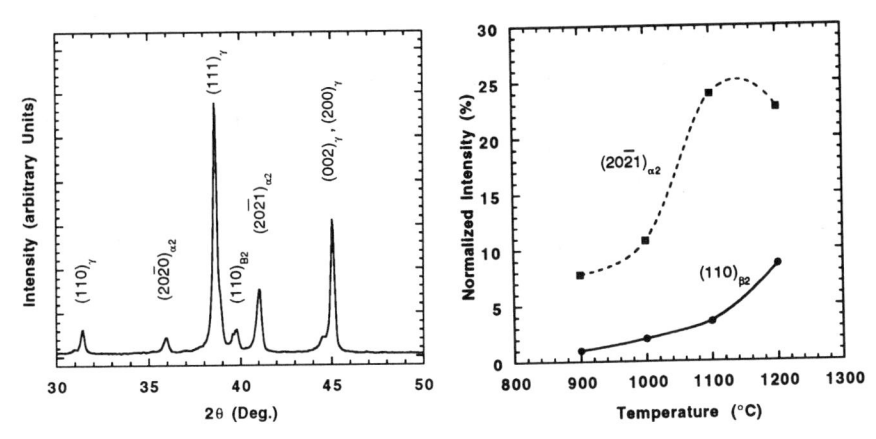

Figure 2: (a) X-ray diffractogram obtained after heat treatment at 1200°C for 100 hours + water quenched, and (b) Normalized intensities of $\{110\}_{\beta_2}$ and $(20\bar{2}1)_{\gamma_2}$ peaks as a function of the treatment temperature.

Table II. Electron Microprobe Analyses of the Heat Treated Specimens

Heat Treatment	Structural Feature	\multicolumn{4}{c}{Average Chemical Composition (at .%)}	Identity
		Al	Nb	Cr	Ti	
1000°C/100 h + W.Q.	Matrix Grains	45.9	1.9	1.6	50.6	γ
	Gray Particles	36.5	1.8	2.0	60.2	α_2
	Bright Particles	33.8	2.4	8.1	55.7	β_2
1100°C/100 h + W.Q.	Matrix Grains	46.2	1.9	1.6	50.3	γ
	Gray Particles	37.9	1.7	2.3	58.1	α_2
	Bright Particles	35.8	2.1	5.9	56.2	β_2
1200°C/100 h + W.Q.	Matrix Grains	46.8	1.9	1.4	49.9	γ
	Gray Particles	40.2	1.8	2.4	55.6	α_2
	Bright Particles	37.5	2.1	4.6	55.8	β_2
1260°C/20 h + W.Q.	Angular/Plate Shaped Particles	48.7	1.9	1.1	48.3	γ
	Matrix (light)	43.8	1.9	2.1	52.2	α_2
	Matrix (gray)	43.7	1.9	2.1	52.3	α_2
1360°C/20 min. + W.Q.	Fine Lamellar Structure	45.2	1.9	1.7	51.2	$\gamma + \alpha_2$
	Feathery Structure	45.2	1.9	1.7	51.2	γ_M
	Intergranular Particles / Layer	37.2	2.4	5.7	54.7	β_2

Annealing at 1360°C followed by water quenching produces a drastically different microstructure (Figures 4a and 4b). It consists of 60 μm grains containing ~40% feathery structure and the remaining volume made up of fine 'lamellae-like' structure. The average compositions of these two regions are identical and correspond approximately to the nominal composition of the alloy. A thin, discontinuous layer of β_2 along the grain boundaries is also discernible. The feathery structure results from the massive transformation of $\alpha \rightarrow \gamma_M$ (2).

The fine lamellae-like structure is most likely the consequence of a shear dominated, bainite type transformation of α. Clearly, additional work is needed to identify the mechanism(s) responsible for the formation of this structure on quenching. In view of the large differences in the compositions of the β layer and the matrix (Table II), one can anticipate a local increase in the Al content and depletion in the Cr content in the regions adjacent to the β layer. This Al enriched region is more likely to undergo massive transformation or lamellae-like decomposition than the regions far away from the β layer. Once nucleated, the γ lamellae can grow very rapidly across the entire grain. This argument would also explain as to why the specimen water quenched from 1320°C exhibits a single phase α_2 structure whereas the specimen quenched from 1360°C contains mixtures of massive transformation product and 'lamellae-like' structures, together with grain boundary β / β_2.

An independent set of experiments in which the specimens were annealed at intervals of 20°C in the range 1340°C to 1400°C, followed by controlled cooling at the rate of 10°C/min. indicated pronounced refinement of α grains. Figure 5(a) shows a typical microstructure containing well developed lamellae along with intergranular particles of $\beta_2 + \gamma$. Figure 5(b) reveals a broad minimum in the grain size vs. annealing temperature at ~1380°C for 10°C/min. cooling rate. This figure also includes grain size data for water quenched specimens. These results suggest that the presence of an optimum fraction of β in the form of a thin intergranular layer is required for retarding the growth of α grains.

Figure 3: Optical microstructure of the specimen aged at (a) 1260°C for 20 hours and (b) 1320°C for 20 min., followed by water quenching.

Discussion

The total volume fraction of α_2 and β_2 (Table I) is plotted against the heat treatment temperature in Figure 6. This figure also includes data computed using the binary Ti-Al phase diagram (3) and the pseudo-binary diagram determined by Semiatin, et.al. (10). It is clear that for T>1250°C, the experimental data and the computed data are in reasonable agreement.

However, the experimental measurements for T≤1200°C show significant deviations from the data computed using the binary diagram. This discrepancy can be ascribed to the existence of three phases $\gamma + \alpha_2 / \alpha + \beta_2$, and it becomes necessary to use ternary diagrams in order to analyze phase equilibria in this alloy at T≤1250°C. Figure 7 is a vertical section of a Ti-Al-Cr phase diagram at ~3.5 at.%Cr (11). If we assume that niobium is approximately half as effective as chromium, then the total chromium equivalent of the current alloy is 3 at.%. Since this value is close to the 3.5 at %.Cr used for construction of this diagram, the use of this diagram for analyzing phase equilibria in Ti-45.5Al-2Nb-2Cr alloy is justified. Chemical compositions of the different phases obtained by EMPA in the current work on Ti-45.5Al-2Nb-2Cr alloy are superimposed on this diagram. It is clear that the current experimental data on Ti-45.5Al-2Nb-2Cr system agree quite well with the positions of the phase boundaries found in the Ti-Al-Cr system. Accordingly, one should anticipate $\alpha + \beta$, α, $\alpha + \gamma$, and $\alpha + \gamma + \beta$ in the temperature ranges 1360°C-1430°C, 1300°C-1350°C, 1250°C-1300°C and 1200°C-1250°C, respectively. These expectations are in complete agreement with experimental findings (Table I). However, the diagram suggests only two phases $\gamma + \beta$ at T<1200°C, whereas $\gamma + \alpha_2 + \beta_2$ are found in specimens aged at 900°C-1100°C. These results suggest that the decomposition of the $\alpha \rightarrow \beta + \gamma$ at T<1200°C may be kinetically limited and that it may be retarded further by the ordering of $\alpha \rightarrow \alpha_2$ at T<1050°C.

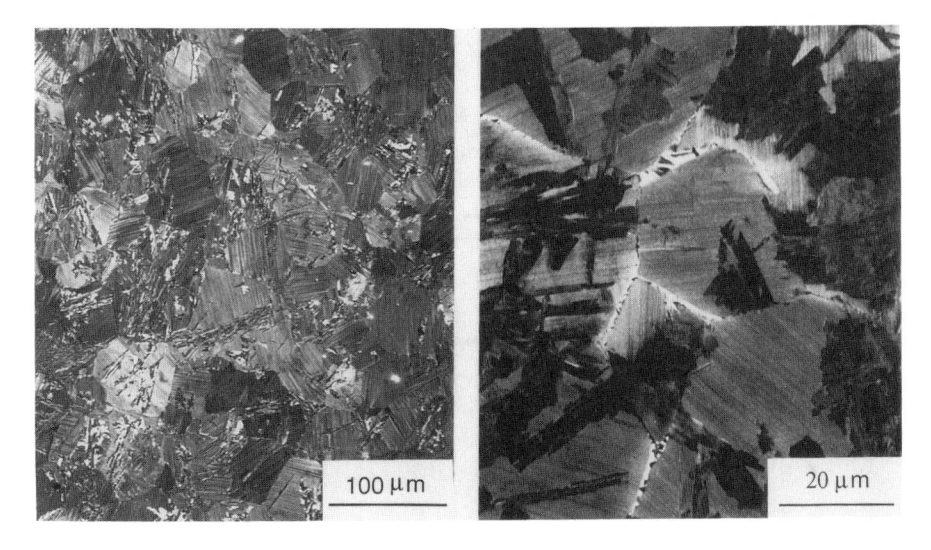

Figure 4: (a) Optical and (b) scanning electron micrographs of the specimen heat treated at 1360°C for 20 min. + W.Q.

Masahashi and Mizuhara (5) have shown that the Ti-47Al-3Cr alloy contains predominantly $\beta + \gamma$ after thermomechanical processing. However, when this alloy was subsequently heat treated at 1050°C for 288 h, the microstructure contained mainly a mixture of $\gamma + \alpha_2$. These results appear to contradict predictions based on the phase diagram shown in Figure 7. These discrepancies can be partly attributed to differences in the oxygen contents. Figure 7 was constructed using high purity alloys containing very low oxygen, whereas the alloy used in the present work contained ~850 wppm oxygen. Oxygen dissolves preferentially in α_2 and stabilizes it as the expense of β / β_2. With an increase in the oxygen content, the three phase region marked α + β + γ in Figure 7 is likely to expand or move to lower temperatures. The observation of γ + α$_2$ + β$_2$ at temperatures below 1200°C is in accordance with this argument.

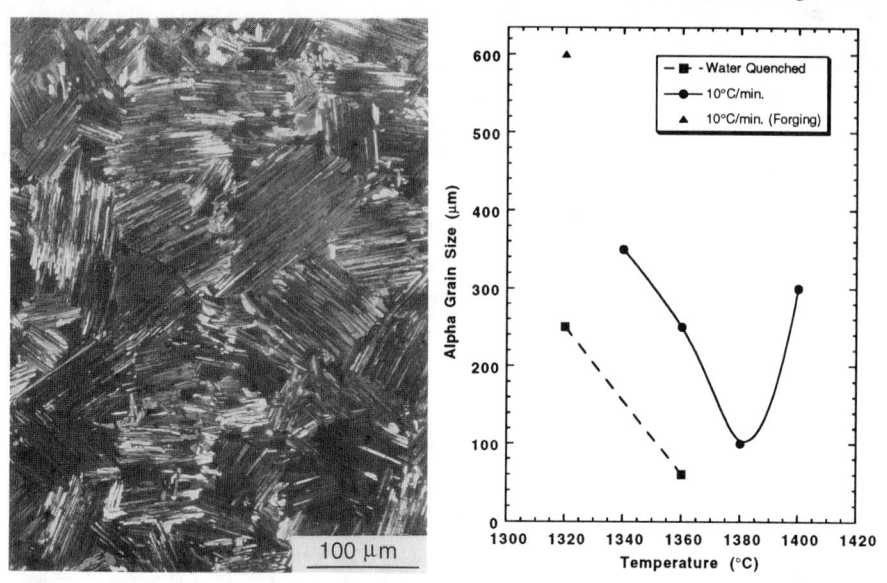

Figure 5: (a) Optical micrograph of the specimen annealed at 1380°C for 20 min. + cooled at 10°C/min. and (b) mean grain size of α vs. heat treatment temperature.

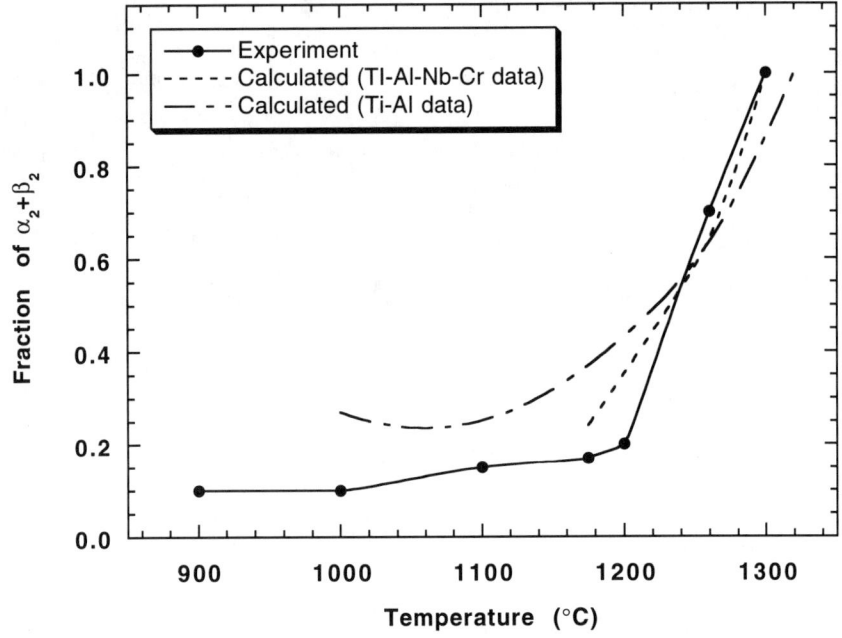

Figure 6: Volume fraction of α$_2$ + β$_2$ vs. temperature for Ti-45.5Al and Ti-45.5Al-2Cr-2Nb alloys.

Ahmed et. al., (12) have studied high temperature phase equilibria in the Ti-45.5Al-2Nb-2Cr alloy using differential thermal analysis and in situ high temperature diffractometry. They have concluded that the alloy does not reside in single phase α field at any temperature and that it transforms directly from $\gamma + \alpha / \alpha_2 + \beta$ to $\gamma + \beta$. In contrast, the current work shows that the alloy undergoes a series of transformations on heating: $\gamma + \alpha / \alpha_2 + \beta / \beta_2 \rightarrow \gamma + \alpha \rightarrow \alpha \rightarrow \alpha + \beta$ and that the temperature window for the single phase α is ~50°C. XRD patterns obtained by Ahmed et. al. at temperatures above 1100°C reveal fairly strong peaks of Al_2O_3. If we assume that the Al content of the alloy is reduced by 2-3 at.% within a thin region below the surface due to oxidation, then it is possible to explain the absence of single phase α in the results of Ahmed et. al. with the help of Figure 7.

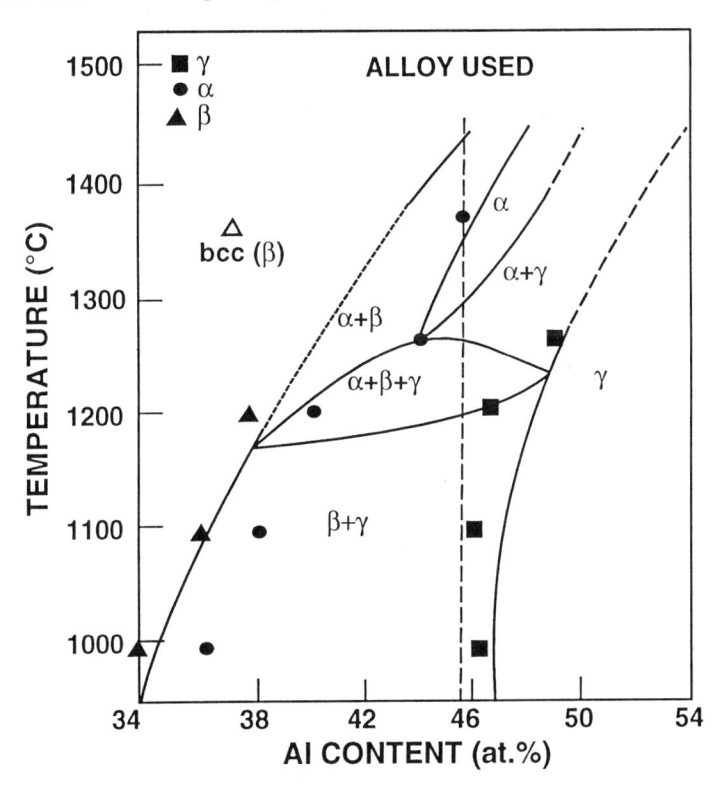

Figure 7: Isopleth of the Ti-Al-Cr system at Cr = 3.5 at.% (11). Experimental data for Ti-45.5Al-2Nb-2Cr are superimposed.

Conclusions

Microstructural and microchemical characterization of the Ti-45.5Al-2Nb-2Cr alloy sheets heat treated at temperatures in the range 900-1400°C led to the following conclusions: (i) equilibration at T<1250°C leads to mixtures of $\gamma + \alpha_2 + \beta_2$ whereas exposure in the range 1250°C < T < 1300°C results in the formation of $\gamma + \alpha_2$, (ii) the alloy remains in the single phase α field over the temperature range 1300°C-1340°C, leading to rapid grain growth, and (iii) annealing at temperatures in the $\alpha + \beta$ field (T≥1340°C) results in reduced growth of the α grains due to the presence of intergranular β layer.

Acknowledgments

This work was performed as part of the in-house research activities of the Processing Science Group, Materials Directorate, Wright Laboratory, Wright-Patterson AFB, OH. One of the authors (VS) was supported by Air Force Contract No. F33615-92-C-5900.

References

1. S.A. Jones and M.J. Kaufman, "Phase Equilibria and Transformations in Intermediate Titanium-Aluminum Alloys", Acta Metall. Mater., 41 (1993), 387-398.

2. Y. Yamabe, M. Takeyama, and M. Kikuchi, "Microstructure Evolution Through Solid-Solid Phase Transformations in Gamma Titanium Aluminides", in Y-W. Kim, et.al., eds., Gamma Titanium Aluminides (Warrendale, PA: TMS, 1995), 111-129.

3. C. McCullough, J.J. Valencia, C.G. Levi, and R. Mehrabian, "Phase Equilibria and Solidification in Ti-Al Alloys", Acta Metall., 37 (1989), 1321-1336.

4. S.C. Huang and E.L. Hall, "The Effects of Cr Additions to Binary TiAl-Base Alloys", Metall. Trans., 22A (1991), 2619-2627.

5. N. Masahashi and Y. Mizuhara, "Phase Stability of β Phase in TiAl Intermetallics", Y-W. Kim, et.al., Gamma Titanium Aluminides (Warrendale, PA: TMS, 1995), 165-172.

6. M.A. Morris, M. Leboeuf and E. Batawi, "Effect of Thermomechanical Treatments on Phase Distribution and Microstructure Evolution of a Ti-48Al-2Mn-2Nb Alloy", Scripta Metall. et Mater., 32 (1994), 71-76.

7. T. Ahmed and H.M. Flower, "Partial Isothermal Sections of Ti-Al-V Ternary Diagram", Mater. Sci. Tech., 10 (1994), 272-288.

8. S.L. Semiatin and V. Seetharaman, "Deformation and Microstructure Development During Hot-Pack Rolling of a Near Gamma Titanium Aluminide Alloy", Metall. Mater. Trans. A, 26A (1995), 371-381.

9. V. Seetharaman and S.L. Semiatin, "Microstructures and Mechanical Properties of Rolled Sheets of a Gamma Titanium Aluminide Alloy", in Y-W. Kim, et. al., Gamma Titanium Aluminides (Warrendale, PA: TMS, 1995), 753-760.

10. S.L. Semiatin, V. Seetharaman and V.K. Jain, "Microstructural Development during Conventional and Isothermal Forging of a Near Gamma Titanium Aluminide", Metall. Mater. Trans. A, 25A (1994), 2753-2768.

11. M. Takeyama, Tokyo Institute of Technology, Tokyo, Japan, unpublished research (1993).

12. T. Ahmed, S. Guillard, H.J. Rack, V. Seetharaman and J.C. Chestnutt, "High Temperature Phase Stability of Ti-(45.5-48)Al-2Nb-2Cr", in Y-W. Kim, et. al., Gamma Titanium Aluminides (Warrendale, PA: TMS, 1995), 203-210.

Microstructure Evolution in an Orthorhombic Titanium

Aluminide as a Function of Temperature-Strain-Time

R.W. Hayes
*C.G. Rhodes

Metals Technology Inc.
Northridge, Ca 91324

*Rockwell International Science Center
Thousand Oaks, Ca 91360

Abstract

Ti-22Al-23Nb that had been creep tested at 650°C and 704°C was examined using SEM and TEM to characterize microstructural features that develop at the primary alpha-2-matrix interfaces during creep deformation. In addition tensile specimens from Ti-22Al-23Nb and Ti-22Al-21Nb-2Ta were pre-strained at room temperature to 5% strain prior to a 650°C/100 hour heat treatment to investigate the effect of pre-strain on the development of the primary alpha-2-matrix interfacial microstructural features. Detailed examination indicates that the primary alpha-2-matrix interfacial microstructural features are alpha-2 phase which has nucleated at those interfaces and grows into the matrix. Neither creep strain nor room temperature pre-strain influences the nucleation of interfacial alpha-2 phase. However with increasing temperature the interfacial alpha-2 phase becomes more continuous indicating a strong temperature effect on its growth.

Introduction

Recent microstructural observations following high-temperature creep deformation of orthorhombic titanium aluminides such as Ti-22Al-23Nb (atomic %) have revealed the presence of microstructural features which at first appear to be cavities (1-3). These features are found at the primary alpha-2-matrix interfaces that are normal to the applied stress direction. It could be argued that high localized stresses build up at these interfaces during creep deformation which would lead eventually to the nucleation and growth of cavities. However, further examination of the deformed microstructure also reveals the presence of these cavity like features along primary alpha-2-matrix interfaces oriented parallel to the applied stress direction. Such observations are not in agreement with expected cavity nucleation sites during high-temperature creep. In addition, these cavity-like features are also found within the microstructure of the grip sections of creep specimens, calling into question the possibility that these microstructural features are indeed creep cavities. In this paper we present evidence which demonstrates that the microstructural features appearing at the primary alpha-2-matrix interfaces are a solid phase corresponding to the alpha-2 structure.

Advances in the Science and Technology of
Titanium Alloy Processing
Edited by I. Weiss, R. Srinivasan,
P.J. Bania, D. Eylon, and S.L. Semiatin
The Minerals, Metals & Materials Society, 1997

Material and Experimental

Sheet samples of two orthorhombic alloys were acquired from Mr. Paul Smith, Materials Lab, Wright-Patterson AFB. The nominal compositions of the two alloys in atomic percent are:

Ti-22Al-23Nb
Ti-22Al-21Nb-2Ta

The oxygen content of each of the alloys has also been analyzed and was found to be 0.12 wt, pct. for the Ti-22Al-23Nb and 0.085 wt. pct. for the Ti-22Al-21Nb-2Ta respectively. The 0.762 mm sheet was received in the rolled plus annealed condition (Fig. 1).

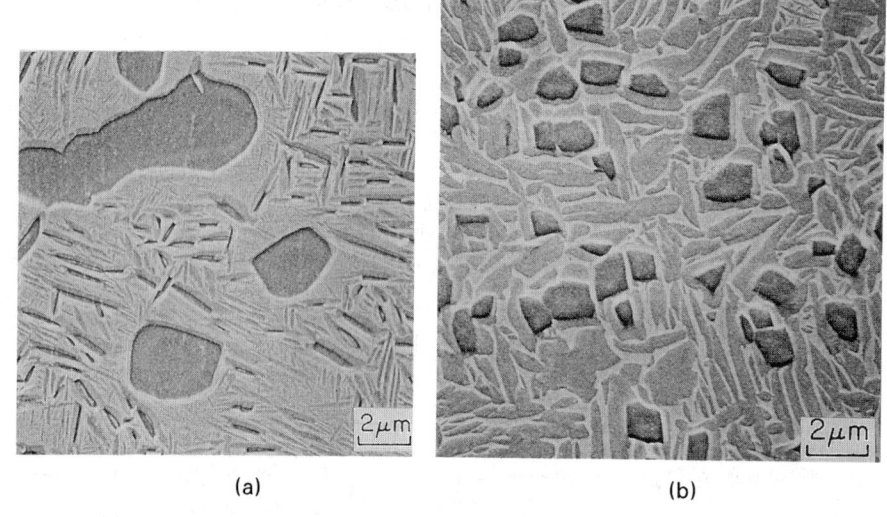

Fig. 1 Microstructures of starting sheet material (a) Ti-22Al-23Nb; (b) Ti-22Al-21Nb-2Ta.

These alloys consist of remote primary alpha-2 within a matrix of ordered b.c.c. B2 beta phase and fine orthorhombic laths precipitated within the beta phase. There is a rim of orthorhombic phase enclosing the primary alpha-2 particles being thin in the Ti-22Al-23Nb (Fig. 1a) and thick in the Ti-22Al-21Nb-2Ta (Fig. 1b). The orthorhombic phase corresponds to the Ti_2AlNb stoichiometric composition where the Nb atoms occupy specific unit cell sites (4). Given the ternary ordering within the orthorhombic phase, one might expect an enhancement of the high-temperature creep resistance of this phase. This has indeed recently been found (5).

Creep tests were performed at 650°C/172.35 MPa/600 hours and at 704°C/103.4 MPa/235 hours on the Ti-22-Al-23Nb alloy in a flowing argon atmosphere. The creep strain was measuring using an averaging type extensometer employing dual LVDT'S.

Thin fails for transmission electron microscopy were taken in the plane of the sheet by punching a 0.3175 cm disc. The disc was then dished prior to electrolytic thinning. The samples were examined in a Philips CM30 Stem.

Results

Samples were selected to examine (i) the effect of creep strain, (ii) the effect of pre-strain followed by isothermal exposure and (iii) the effect of temperature on the nature and characteristics of the microstructural features appearing at the primary alpha-2 matrix interfaces. Each of these variables will be treated separately.

Effect of Creep Strain

Ti-22Al-23Nb was creep tested at 704°C/103.4 MPa/235 hours and discontinued at a creep strain of 2.6%. A comparison of the grip and gage sections of this creep specimen are presented in Fig. 2.

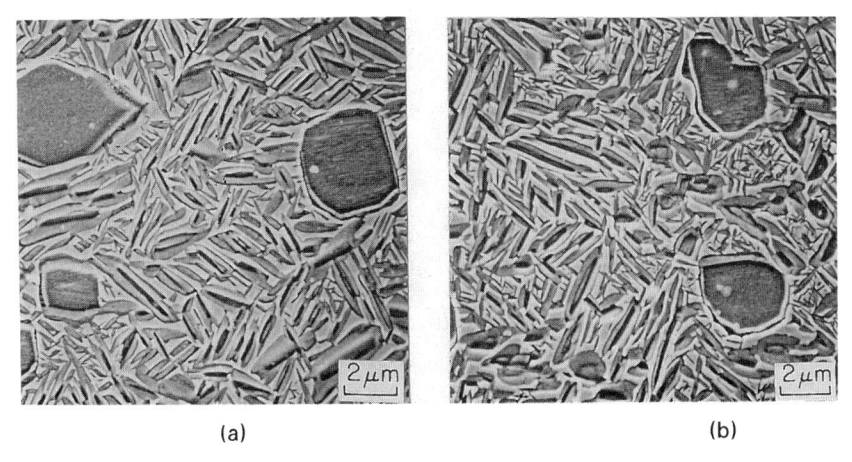

(a)   (b)

Fig. 2 Ti-22Al23Nb Creep tested at 704°C/103.4 MPa/235hours. (a) grip section; (b) gage section. Note dark phase at surface of primary alpha-two particles that is absent in Figure 1.

Virtually all of the primary alpha-2 particles contain the cavity like microstructural features around their peripheries in both the grip and gage section microstructures. There is also no obvious difference in the amount of the interfacial microstructural features when the grip and gage section microstructures are compared. From these results it is evident that creep strain does not play a significant role in the nucleation and growth of the interfacial microstructural features. This result also casts strong doubt on the possibility that these interfacial microstructural features are cavities brought about by creep deformation.

Effect of Pre-Strain

If the interfacial microstructural features are cavitation resulting from accumulated strain generated during sheet rolling, their growth may be exacerbated by inducing additional strain. Thus pre-straining the sheet material at room temperature followed by an isothermal heat treatment within the creep temperature regime (650°C) may cause more cavities to nucleate and grow along the primary alpha-2-matrix interfaces. This follows the results found from these types of experiments on a number of Ni-based superalloys [6,7]. Thus, if this were the case in the present materials, one would expect to see an increase in the density of cavity-like features along the primary alpha-2-matrix interfaces within the gage section as opposed to the grip section. When these comparisons were made we found no significant difference in the amount of cavity like features between the grip and gage sections of either Ti-22Al-23Nb or the Ti-22Al-21Nb-2Ta. These results indicate that pre-straining at room temperature does not lead to an increase in the density of the interfacial microstructural features in either alloy.

Effect of Temperature

To assess the role of temperature on the characteristics of the interfacial microstructural features, we have examined grip sections of Ti-22Al-23Nb creep tested at 650°C/172.35 MPa/600 hours and at 704°C/103.4 MPa/235 hours. The creep test at 650°C was terminated at a creep strain of 2.3% very close to the creep strain of the 704°C specimen. Thus, creep strain can be eliminated as a variable in this comparison.

A pair of SEM micrographs showing the grip section microstructures of the two creep specimens is presented in Fig. 3.

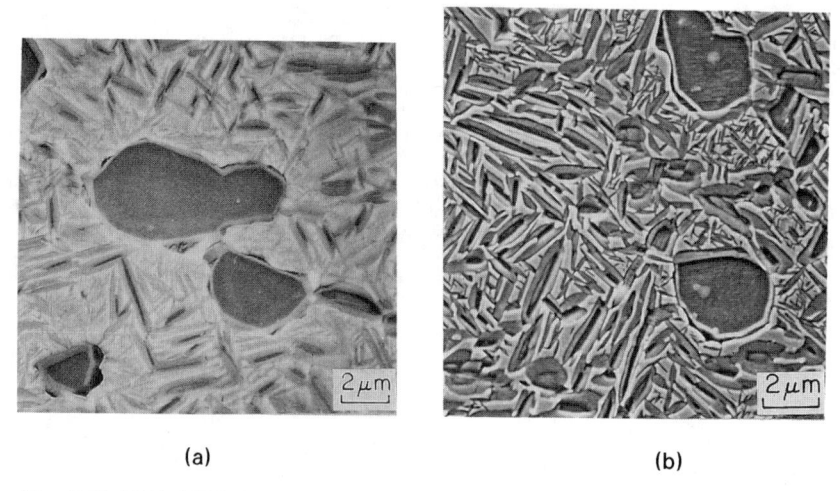

(a)   (b)

Fig. 3 Ti-22Al-23Nb isothermally treated. (a) 650°C/600 hrs; (b) 704°C/306 hrs. Dark phase at surface of primary alpha-two particles is nodular in (a), and continuous in (b).

The microstructure of the 650°C creep specimen (Fig. 3a) reveals a discontinuous distribution of the interfacial microstructural features. The microstructural features appear to grow as nodules on the outer surface of the primary alpha-2 particle thus giving them their cavity-like appearance. We also note that a rim of orthorhombic-phase is generally found on the primary alpha-2-particles (Fig. 3). In these cases, the cavity- like microstructural features actually nucleate at the orthorhombic rim-matrix interface. In all cases, the interfacial microstructural features grow into the matrix orthorhombic plus beta phase from the primary alpha-2-matrix interfaces. The microstructure of the 704°C creep specimen (Fig. 3b) reveals that the interfacial microstructural features have become more continuous and not nodular as in the 650°C specimen. This indicates a temperature dependence of the growth of the interfacial microstructural features.

Transmission Electron Microscopy

To gain a better understanding of the nature of the primary alpha-2-matrix interfacial microstructural features, transmission electron microscopy (TEM) analysis was performed on the grip sections of the Ti-22Al-23Nb 704°C creep specimen and the Ti-22Al-23Nb-2Ta pre-strained and heat treated specimen. The objective of the TEM investigation was first to determine if the interfacial features were cavities and second, to characterize these features if they were notcavities.

TEM examination of the grip section of Ti-22Al-21Nb-2Ta, pre-strained and annealed at 650°C, revealed that the alpha-2-matrix interface features were not cavities, (Fig. 4).

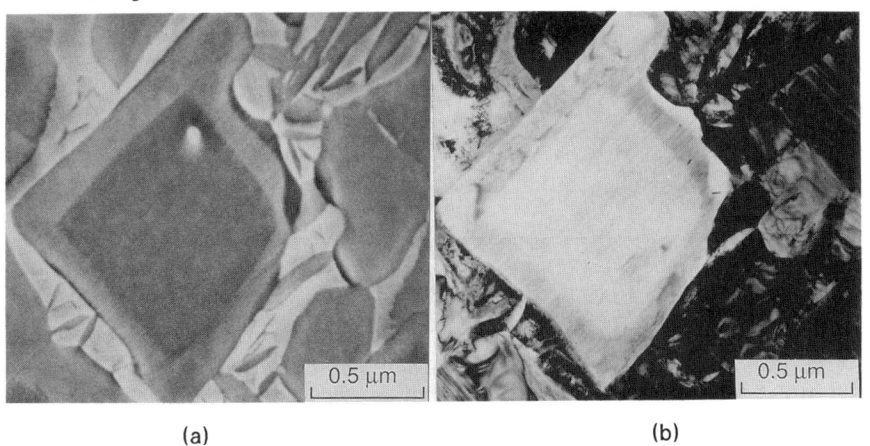

(a)   (b)

Fig. 4 Ti-22Al-21Nb-2Ta illustrating the appearance of dark phase in TEM. (a) SEM of primary alpha-two particle; (b) TEM of same particle. Dark phase at surface of primary alpha-two particle imaged in SEM appears light in TEM.

The interface features that appear dark in the SEM image, (Fig. 4a), can be seen as light regions in the TEM image, (Fig. 4b). Of course, holes or voids in TEM samples will appear light in the image, so a tilting experiment was conducted to demonstrate that the feature is indeed a solid phase, (Fig. 5).

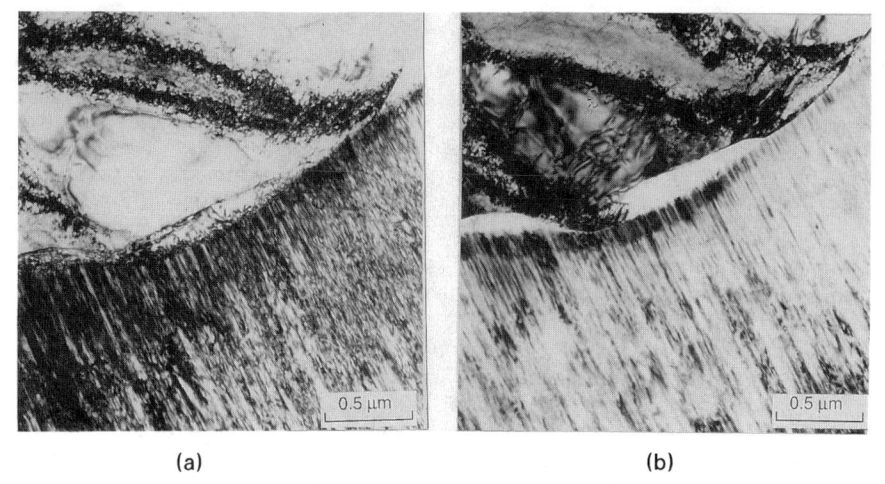

Fig. 5 Demonstrating that surface phase imaged in TEM is not a cavity. (a) tilted to greatest contrast, surface phase images bright; (b) same area as (a) after tilting to show structure in surface phase.

Identification of the alpha-23matrix interface phase was accomplished by means of electron energy dispersive spectroscopy (EDS) and electron diffraction (SAD and CBED). EDS analysis was performed on discreet spots at regular intervals from the center of a primary alpha-2 particle, through the rim orthorhombic phase, and into the interfacial feature. A summary of those results is given in Table 1. The interfacial feature has a composition similar to the alpha-2 phase and quite different from the orthorhombic phase at which it has nucleated.

Table 1. Chemical Analysis by EDS (at. pct.)

AREA	Ti	Al	Nb	Ta
Primary Alpha-Two	60.7	25.3	12.7	1.3
Rim Orthorhombic	56.3	24.8	17.2	1.8
Interface Feature	63.7	24.6	10.4	1.3

Electron diffraction patterns, either selected area or convergent beam, were taken in the same manner at the EDS analysis, moving across a primary alpha-two particle into the interfacial region. A typical result is shown in Fig. 6.

(a)

(b)

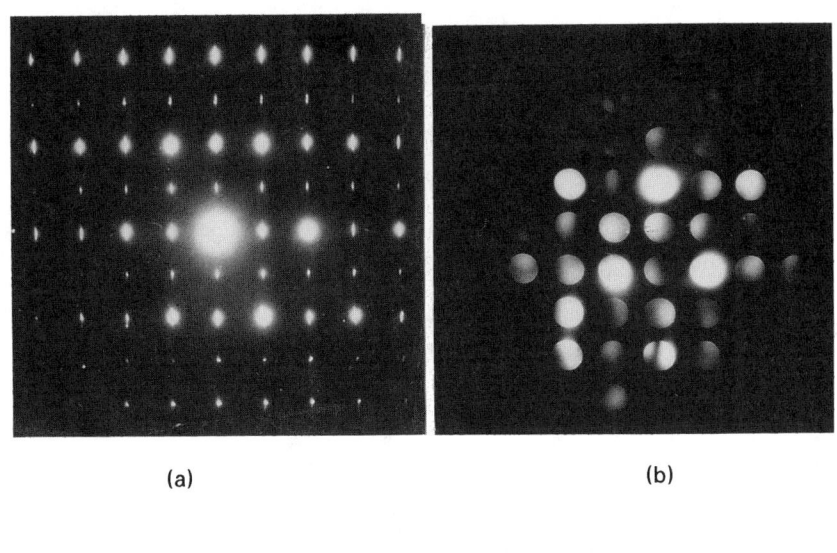

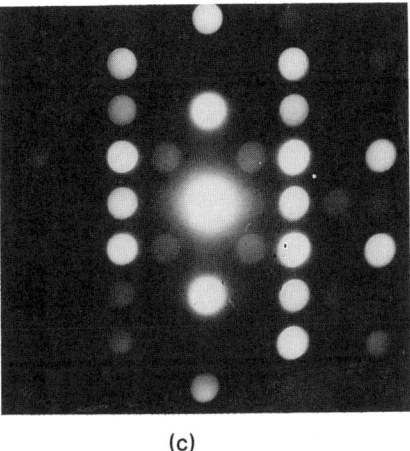

(c)

Fig. 6 Demonstrating that surface phase has same crystal structure as primary alpha-two particle. (a) SAD from primary alpha-two in [1120] orientation; (b) CBED from rim orthorhombic phase in [110] orientation; (c) CBED from surface phase in [1120] hcp orientation.

The primary alpha-two particle was tilted to a [1120] zone, (Fig. 6a), and the other patterns were taken without further tilting. The orthorhombic rim phase is in a [110] zone, which follows from the orientation relation between alpha-two and orthorhombic phases (4). It should be noted that the relationship [0001]a$_2$//[100]0 leads to the other two {1120}a$_2$//{110}0). Finally, the interfacial feature exhibits a diffraction pattern identical to that given by the underlying primary alpha-two, (Fig. 6c). This result, combined with the EDS data lead to the conclusion that the interfacial features are indeed alpha-two phase.

Discussions

At first glance, the microstructural features appearing at the primary alpha-2-matrix interfaces following creep deformation of the Ti-22Al-23Nb alloy indeed appear to be cavities. Creep cavities are expected to nucleate and grow at boundaries or interfaces between hard particle and soft matrix oriented normal to the applied stress direction. Recent work [8] has shown that the primary alpha-2 phase in the Ti-22Al-23Nb matrix is the hard phase within the softer b.c.c. B2-orthorhombic matrix. Thus, under creep conditions the nucleation and growth of cavities along the primary alpha-2-matrix interfaces oriented normal to the applied stress direction may be expected. However, it has also been found that these cavity-like features also appear along primary alpha-2-matrix interfaces oriented parallel to the applied stress direction of gage section microstructures. In addition, these interfacial microstructural features have also been found in the grip section microstructures of creep specimens. Clearly, creep cavities would not be expected in the grip section microstructure unless their formation was related to material processing.

It has been shown that room temperature pre-strain followed by an isothermal heat treatment within a temperature regime high enough to promote diffusion can result in grain boundary cavitation of Ni-base superalloys [6,7]. During room temperature pre-straining of these alloys, planar slip bands form and impinge upon grain boundary M_6C or $M_{23}C_6$ carbides resulting in the formation of small microcracks. During the isothermal heat treatment, vacancies feed into the microcracks causing them to open up into cavities. If the cold rolling of the orthorhombic alloys into sheet is thought of as the pre-strain operation which is followed by an annealing treatment (996°C in the case of the present Ti-22Al-23Nb alloy), one might be able to rationalize the presence of cavities along interfaces parallel to the applied creep stress direction as well as the presence of cavities along the primary alpha-2-matrix interfaces in the grip section microstructures. The above argument requires that very high stresses must be developed at the primary alpha-2-matrix interfaces during cold rolling which would initiate the microcracks. That this may occur has not been determined. However, if such a scenario were the case, one would expect that room temperature pre-straining followed by the 650°C/100 hour isothermal heat treatment should lead to an increase in the cavity density within the gage section microstructure relative to that of the grip section. Our experimental results indicated that this was not the case, suggesting that these interfacial microstructural features are not the result of processing history.

TEM and SAD analysis indicated that the primary alpha-2-matrix interfacial microstructural features were consistent with the ordered alpha-2 structure. These results were found for both orthorhombic alloys. The nucleation and growth of the interfacial alpha-2 phase was found to vary significantly with temperature. At 650°C nucleation appears to be more isolated with the alpha-2 phase growing as nodules from the primary alpha-2-matrix interface and into the matrix (Fig. 3a). At 704°C, the interfacial alpha-2 phase nucleation appears to be more uniform and it grows as a nearly continuous sheath (Fig. 3b).

The nucleation and growth of ordered alpha-2 phase at the primary alpha-2-matrix interfaces appears to be in contrast with recent phase diagram studies of the Ti-22Al-23Nb alloy [9]. This work suggests that the orthorhombic phase is in equilibrium with the ordered b.c.c. B2 beta phase at 650°C.

One possibility for the stabilization of the alpha-2 phase under the present creep conditions would be the diffusion of oxygen into the material during creep testing. If oxygen segregates to the primary alpha-2-matrix interfaces, it is possible that the alpha-2 phase may become the stable phase in these local regions. However, the isothermal heat treatments following the pre-straining experiments were conducted in quartz capsules which were evacuated, back filled with argon and sealed. Thus, the possibility of oxygen pick up during heat treatment of these specimens is unlikely, yet the interfacial primary alpha-2-matrix alpha-2 phase was prominent in these specimens. From these results it would appear that the explanation of the nucleation and growth of the primary alpha-2-matrix interfacial alpha-2 under the present experimental conditions must lie somewhere other than the ingress of oxygen into the material. Clearly, further detailed work is needed in order to understand the nucleation and growth of alpha-2 phase at the primary alpha-2-matrix interfaces during creep deformation of the Ti-22Al-23Nb alloy and other orthorhombic alloys of similar composition.

Conclusions

Microstructural features which develop at the interfaces between primary alpha-2 particles and the matrix of orthorhombic alloys such at Ti-22Al-23Nb have been observed following creep deformation of these alloys within the temperature range 650-760°C. At first these features appear to be creep cavities. Several observations such as (i) the presence of these interfacial microstructural features at interfaces parallel to the applied creep stress direction and (ii) the presence of these microstructural features within the grip section microstructures lead to the conclusion that these microstructural features are not creep cavities. Detailed TEM and SAD analysis performed in this study has shown that the primary alpha-2-matrix interfacial microstructural features are consistent with the ordered DO_{19} alpha-2 structure. We have suggested that oxygen ingress into the material during creep deformation may not explain the presence of the interfacial alpha-2 phase. Clearly, more detailed work is needed in order to understand the nucleation and growth of the primary alpha-2-matrix interface phase under creep deformation conditions.

References

[1] M.A. Foster, P.R. Smith and D.B. Miracle, Scripta Metall. et Mater., vol 33, pp 975-981 (1995).

[2] T. Pollock, Carnegie Mellon University, unpublished research, (1995).

[3] R. Zordan, Allison Engine Co., unpublished research, (1995).

[4] D. Banerjee, A.K. Gogia, T.K. Nandy and V.A. Joshi, Acta Metall., vol 36, pp 871-882 (1988).

[5] T.K. Nandy, R.S. Mishra and D. Banerjee, Scripta Metall et Mater., vol 28, pp 569-574 (1993).

[6] B.F. Dyson, M.S. Loveday and M.J. Rodgers, Proc. R. Soc. 349A, p. 245 (1976).

[7] M. Kikuchi, K. Shiozawa and J.R. Weertman, Acta Metall. vol 29, p. 1747 (1981).

[8] P.R. Smith, J.A. Graves and C.G. Rhodes, Metall. Trans A, vol 25A, pp 1267-1283 (1994).

[9] C.G. Rhodes, Rockwell International Science Center, unpublished research, (1996).

Hydrogen Technology of Semiproducts and Finished Goods Production from High-Strength Titanium Alloys

Alexander A.Ilyin, Vladimir K.Nosov, Michail Yu.Kollerov, Alexander A.Krastilevsky, Svetlana V.Scvortsova, Alexei V.Ovchinnikov

Moscow State University of Aviation Technology after K.Tsiolkovsky
Metals Science and Heat Treatment Technology Department
Petrovka Str.27, K-31, Moscow, Russia, 103767

Abstract. Influence of hydrogen alloying on structure and deformation mechanism of high-strength titanium alloys VT22I and Ti-10-2-3 at the room temperature was investigated. It was found that alloys containing 0,3-0,4% wt. of hydrogen are deformed by means of compression and rolling on more than 80% without intermediate annealing. Further heat treatment of deformed alloys in air atmosphere or vacuum may form different types of structure providing both high ductility and high strength properties. Hydrogenization alloys can be used for production sheets by cold rolling and cold stamping at the room temperature with further strengthening during hydrogen removing by vacuum annealing.

Introduction

Titanium alloys, with only a few exceptions, commercial titanium and high-alloyed β–alloys, have reduced ductility at cold deformation (rolling, drawing, stamping), what limited range of products and semiproducts from high-strength (α+β)–titanium alloys. Recently considerable attention has been focused on hydrogen plastification of titanium alloys during hot deformation, as a method of their workability improving [1].

This work is a continuation of investigations to substitute hot deformation for cold one, it will allow us to refuse of complicated and expensive equipment and thus significantly lowered cost of semiproducts and products from titanium alloys. Receiving new semiproducts (sheets, foils, wire) from high-strength alloys, such as Ti-10-2-3 and VT22I, can greatly widen the range of products, it also allows to obtain composite materials.

Objects and Methods of Investigations

Ti-10-2-3 and VT22I alloys are regarded to (α+β)-titanium alloys of transition class with temperature of martensite transus near room temperature. Chemical composition of those alloys is in Table I.

Table I. Chemical composition of investigated alloys[1]

Alloys	Ti	Al	Mo	V	Cr	Fe	C	O	N	H
		\multicolumn{5}{c	}{Alloying elements, %}	\multicolumn{4}{c}{Impurities, %}	
VT22I	base	2,9	4,8	4,9	0,85	0,9	0,10	0,15	0,05	0,004
Ti-10-2-3	base	3,2	-	10,0	-	2,2	0,12	0,14	0,04	0,004

The rolled rods 20 mm in diameter are used as initial billets. Hydrogen content in the rods did not exceed 0.004%. Hydrogenation up to the given concentrations was fulfilled on Sievert's equipment by thermodiffusion method in the atmosphere of high-purity hydrogen, produced during titanium hydride decomposition. After heating to the temperature 50°C higher Ac_3 transus of hydrogenated alloys for an hour these billets were quenched in water. The temperature Ac_3 was selected by experimental quenching method.

Tests on compression of samples with 15 mm in diameter were made on hydraulic press PM-125 with the force 1.25 MH at the room temperature with the rate of deformation $\dot{\varepsilon}=2.4*10^{-2}$ s^{-1}. Ductility of alloys was estimated by the limit of deformation degree (ε_{lm}), it corresponds to the formation of the first macroscopic crack, and resistance to deformation - by specific compression force (q_i). The yield strength (YS) and specific forces of compression at 20% deformation (q_{20}) were observed as characteristics of deformation resistance.

Structure of samples was studied by means of opticmicroscope NEOPHOT-30 and electronic transmission microscope EM-125K with thin foil method. X-ray structure analysis was made on diffractometer DRON-4 in filtered K_α copper radiation.

Experiment Results and Discussion

1.1 Hydrogen Influence on Phase Composition and Deformability.
Samples of investigated alloys with initial hydrogen content after quenching from β-field have structure of mechanically unstable β-phase and small quantity of α"-martensite (15-20%). With increasing of hydrogen content the volume fraction of martensite decreases and their structure is represented only β-phase in case of hydrogen content more than 0.1% for VT22I and 0.2% for Ti-10-2-3 (Table II).

The curve of mechanical properties (YS, q_{20}, ε_{lm}) of quenched VT22I depending on hydrogen content can be divided into two main stages (Figure 1). While heightening concentration up to 0.3%, the intensive increasing of yield strength is observed as well as deformation limit increasing. At the same time the specific compression forces at 20% deformation significantly lowers. At the second stage at hydrogen content higher 0.3% yield strength also increases but slower than at the first stage, but ε_{lm} reaches its upper level (about 80%) limited by power of deforming equipment, but not by alloy ductility. Specific forces of compression at 20% deformation remains nearly constant and at the hydrogen content over 0.5% their values are lower the yield strength of alloy.

The same curves of mechanical properties depending on hydrogen content were observed on Ti-10-2-3 alloy. The only distinction from the VT22I alloy consists of the following: the maximum values of ε_{lm} are achieved at the higher hydrogen concentrations (by 0.05-0.1%) (Table II).

---

[1] Here and further containing of components is in wt. %.

Table II. Phase composition of quenched Ti-10-2-3 and VT22I alloys samples before and after deformation by means compression to different degree.

Alloy	Hydrogen concentration %	Phase composition of samples after deformation to degree (ε)			Limit of deformation degree, %
		$\varepsilon=0$	$\varepsilon=8-10\%$	$\varepsilon=\varepsilon_{lm}$	
Ti-10-2-3	0,004	$\beta+\alpha"$	$\beta+\alpha"$	BCT $+\alpha"$	24
	0,1	$\beta+\alpha"$	$\beta+\alpha"$	BCT $+\alpha"$	28
	0,2	$\beta\,(+\alpha")$	BCT$+\alpha"$	BCT $(+\alpha")$	51
	0,3	β	$\beta\,(+\alpha")$	β	76
	0,5	β	β	β	80*
	0,7	β	β	β	80
VT22I	0,004	$\beta+\alpha"$	$\beta+\alpha"$	BCT $+\alpha"$	28
	0,1	$\beta\,(+\alpha")$	BCT $+\alpha"$	BCT $+\alpha"$	33
	0,2	β	BCT $+\alpha"$	BCT$(+\alpha")$	56
	0,3	β	$\beta\,(+\alpha")$	$\beta\,(+\alpha")$	80
	0,5	β	β	β	80
	0,7	β	β	β	80

* - This deformation degree was maximum for applied equipment.

It is well known, that in the alloys of transition class metastable β-phase can be unstable under strains and stresses [2]. So, in this work, attempts were made to find a connection between mechanical properties change during compression and structure transformations, developing during cold deformation in alloys. For this purpose the X-ray structure analysis of samples was made in the direction of compression axis after deformation 8-10% and after the upper achieved deformation degree. The first mentioned deformation was selected according to crystallographically deformation of $\beta \Rightarrow \alpha"$ transformation which takes place in mechanically unstable β-phase, when stresses needed for deformation development did not exceed the "real" yield strength of alloy, caused by dislocation slipping [3].

The investigations have shown, that in the VT22I samples at low degrees of deformation $\alpha"$-martensite is forming, its volume fraction lowers with hydrogen content increasing. The diffraction pattern differs greatly for samples with limit of deformation degree. Diffraction peaks of martensite in samples with hydrogen content up to 0.2% are diminishing and widening, and at the hydrogen content higher 0.2%H disappears completely. β-phase reflects widen and displace towards higher Braggs angles. For some samples, deformed with limited degree, the structure analysis in the perpendicular direction to compression axis was made. In general the X-ray patterns does not change, but β-reflects displace towards low Braggs angles in comparison with their place in samples without deformation. Such changes in diffraction patterns may be considered as tetragonal distortion of crystal lattice of β-phase, which has been observed earlier on other ($\alpha+\beta$) and β-titanium alloys [4].

The same results were received during X-ray analysis of deformed samples of Ti-10-2-3. While compressing of quenched samples of VT22I and Ti-10-2-3 some transformations can take place simultaneously or sequentially. At low degree of deformation $\beta \Rightarrow \alpha"$ transus is realized, and at higher degree the β-phase with bulk centered tetragonal (BCT) is formed.

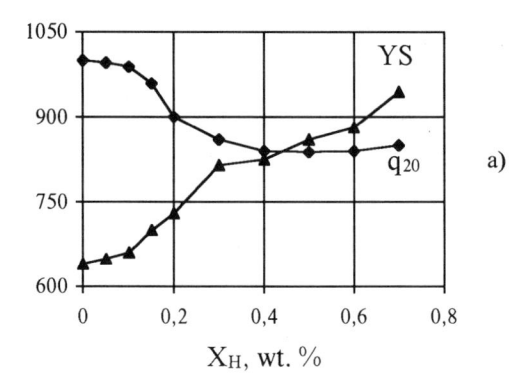

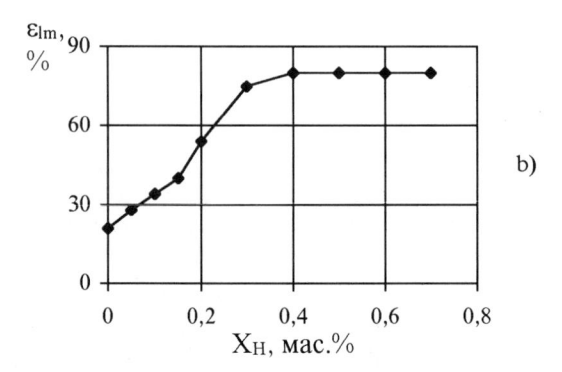

Figure 1 - Hydrogen influence on yield´strength (YS), specific forces of compression at 20% deformation (q_{20}) (a) and the limit of deformation degree (ε_{lm}) (b) of VT22I quenched samples.

Hydrogen strongly influences on completeness of the transformations increases β-phase stability towards β⇨α" transformation [2], and decreases volume fraction of martensite, appearing during deformation. As the result of this process YS of investigated alloys increases with hydrogen content heightening and reaches "real" yield strength. At the hydrogen content higher 0.2% (for VT22I) and 0.3% (for Ti-10-2-3) β-phase becomes mechanically stable and does not have any changes during deformation. Thereby in the alloys, having small content of hydrogen at the first stages of deformation, the phases with low symmetry of crystal structure (α", BCT) form. In this structure the processes of slip during further deformation are complicated in comparison with deformation of BCC-phase. At higher hydrogen concentrations any transformations during deformation does not take place, and shape changes of samples are due to slipping mechanism. It causes the forces lowering at higher degree of deformation and increases alloy ductility.

Thus hydrogen considerably influences on regularities of structure transformations at cold deformation of titanium alloys Ti-10-2-3 and VT22I. Leading to β⇨α" transformation suppression hydrogen favours lowering of deformation forces and increasing alloys ductility. Alloys VT22I and Ti-10-2-3 with hydrogen content higher 0.3% can be deformed by 80% under compression at room temperature without failure and may be used for sheets and foils production by means of cold rolling.

1.2 Formation of Structure and Properties of Hydrogenated alloys during rolling and heat treatment.

For investigation of structure formation and properties of alloys during cold rolling the billets 10 mm thick and sheets 2.2 mm were rolled at room temperature with deformation for each pass about 20% without intermediate annealing until the final thick of 0.1-0.2 mm, which is the limit for applied type equipment (until rollers joining). At some stages of cold rolling of sheets and foils, samples were cut for future investigation of structure and properties. According to the scheme an attempts of cold rolling of Ti-10-2-3 and VT22I with initial (0.004%) hydrogen content were made.

Fulfilled investigations have shown, samples from VT22I with 0.3% H and Ti-10-2-3 with 0.36% H can be rolled without cracking until the 0.1-0.12 mm thick from sheet 2.2 mm thick as well as from billet 10 mm thick. Small cracks 0.5-1 mm depth were observed near the edge of strip at the deformation degree higher 70%. In the samples with initial content of hydrogen (0.004%) cracks formed at the rolling with 50% degree, and at further deformation samples were failed.

In the samples alloyed by hydrogen the structure is represented equiaxed β-grains in the plane of rolling and with grains stretched in the rolling direction. In the samples with initial hydrogen content the two phase (α+β)-structure during cold rolling nearly does not changes, and cracks appear at higher degrees of deformation. Texture of cold deformed samples also depends on hydrogen concentration. If in the sheets with initial hydrogen content the main component of texture is {100}<110>, and in the hydrogenated samples this texture is {111}<110>.

During further heat treatment in the structure of sheets polygonization and recrystallization processes take place and at the same time β-phase decomposition can occur. Moreover, temperature-rate condition of β-decomposition will mainly depend on atmosphere in which heat treatment takes place. If heat treatment is in the air then the alloy composition does not change. When heat treatment is in the vacuum, at the temperature higher 400°C intensive hydrogen removing from the sample takes place, it leads to decrease of stability β-phase to decomposition and increase Ac_3 temperature of alloys. Results of X-ray analysis and electronic microscope research of phase composition and structure of sheets from Ti-10-2-3 after heat treatment in the air and in the vacuum is generalized on diagram at Fig. 2. During heat treatment in the air in β-phase polygonozation (400-450°C) and recrystallization (550-700°C) processes develop. Under the long exposure time (more 10 hours) in the 350-550°C temperature range α+β-structure forms in which β-phase may keep initial deformed state ($β_d$) or polygonized ($β_p$) or recrystallized ($β_r$) structure depending on determined temperature. During heat treatment in the vacuum β-phase decomposition takes place at the more wide temperature range and significantly smaller exposure time, Fig. 2. Precipitations of α-phase hinder β-phase recrystallization therefore $β_r$-structure formation is possible only at the 750°C or higher temperature.

Thus, by means of heat treatment of cold deformed sheets we may obtain different types of structures. Single recrystallized β-phase obtaining by annealing in the air will provide leads to high ductility and stampability at the room temperature. Fine (α+β)-structure forming by vacuum annealing leads to increase of strength.

For technological properties testing of cold rolled sheets with 1 mm thick samples were cut. One part of samples was annealed in the air at the temperature providing recrystallized structure formation (650°C, 0.5 h). Mechanical properties of samples are represented in Table III for Ti-10-2-3 alloy.

For samples alloyed by hydrogen after recrystallizing annealing the elongation exceeds 13% and reduction of area - 50%. The yield strength of alloy is lower 600 MPa, what allows to use annealed sheet for future cold stamping.

Table III. Mechanical properties of sheets from Ti-10-2-3.

Treatment	Hydrogen concentration after treatment, %	YS, MPa	UTS, MPa	EL, %	RA, %
Cold rolling	0.36	820	830	2.4	18.5
Annealing at 650°C, 0.5h	0.36	560	750	13.5	54.5
Vacuum annealing at 650°C, 3 h	0.005	1030	1090	11.2	49.4
Vacuum annealing at 780°C, 0.5 h and ageing 450°C, 8 h	0.004	1120	1170	6.5	36.2

Made experiments for sheets stampability from VT22I and Ti-10-2-3 alloys have shown, that in drawing (more 1,8), flanging (more 1,6) and bending (bending radius is smaller than 3 sheet thickness) hydrogenated high strength titanium alloys are as good as commercial titanium and high-alloyed β-alloys and have not tendency to forming festoons.

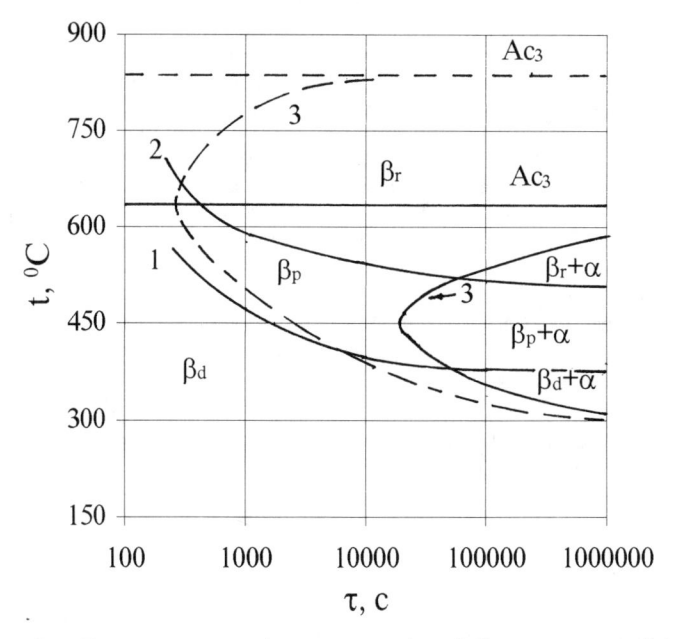

Figure 2 - Temperature and exposure time influence on conditions of start of polygonization (1), recrystallization (2) and β-phase decomposition of Ti-10-2-3 alloy with 0,36% hydrogen (—— heating in the air ; ······· in the vacuum).

For future application of products from hydrogenated semiproducts hydrogen concentration must be reduced until 0.008-0.004% by means of vacuum annealing. During vacuum annealing the formation of extremely fine α-phase takes place, which causes alloy strengthening. Varying of the vacuum annealing and ageing temperatures one may control the volume fraction and morphology of α-phase, and correspondingly, the level of mechanical properties of titanium products, see Table III.

Presented strength characteristics of alloys in this work are not maximum and searching of heat treatment optimum conditions (temperature of vacuum annealing, further cooling rates, ageing conditions) allows for VT22I and Ti-10-2-3 alloys to increase UTS till 1250 MPa with satisfied ductility characteristics. According to these characteristics the investigated alloys are as good as high-alloyed β-alloys and significantly exceed commercial titanium.

Conclusions

The results of present investigation have shown, hydrogen essentially influences on phase composition and mechanical properties of high strength titanium alloys. Hydrogen stabilizes β-phase and hinders the development of martensitic transformations during deformation. As a result of it the basic mechanism of hydrogenization alloys deformation at room temperature is slip, which in single β-phase takes place under significantly lower deformation forces comparing with to β+α"-structure, taking place during first stages of deformation for alloys with initial hydrogen concentration. Ductility increasing and deformation forces decreasing of high strength titanium hydroganized alloys VT22I and Ti-10-2-3 allow to obtain sheets and foils from these alloys by cold rolling without intermediate annealings. Further heat treatment in the air or in the vacuum allows to obtain different types of structures and mechanical properties of alloys. So, treatment in the air $650^{\circ}C$ forms recrystallized β-structure providing high characteristics of ductility and stampability at room temperature.

Heat treatment in the vacuum provides hydrogen removing till safe from the point of view hydrogen embrittlement concentrations (0,004-0,008%) and fine α-phase formation which strengthens alloys till UTS=1150 MPa and ductility is satisfied. Strength properties of sheets and foils from VT22I and Ti-10-2-3 after vacuum annealing considerably exceeds ones of the same semiproducts which have been produced from other titanium alloys by now.

References

1. V.K.Nosov and B.A.Kolachev, Hydrogen Plastification under Hot Deformation of Titanium Alloys (Moscow, Metallurgia Publishing Company, 1986), 118.
2. Alexander A.Ilyin, Phase and Structure Transformations in Titanium Alloys Mechanism and Kinetics (Moscow, Nauka Publishing Company, 1994), 304.
3. A.A.Ilyin, M.Y.Kollerov, S.V.Scvortsova "Titanium Alloys with Shape Memory Effect and Their Perspective Technological Application", (Titanium -1990. Products and Application Proceedings of the Technical. Program From the 7-th International Conference). Published by Titanium Development Association, v.II, (1990), 746-754.
4. N.V.Maltsev and N.I.Kashnikov, "Cold Plastic Deformation Degree Influence on Ageing Processes of Titanium Alloys VT16", FMM, 56 (1983), 1115-1169.

Acknowledgments

The research described in this publication was made possible in part by Grants №NAJ000 and NAJ300 from the International Science Foundation and Russian Government.

JOINING AND CASTING

A Study of Diffusion-Bonding of Dissimilar Titanium Alloys

By

Gopal Das and Carter Barone

Pratt & Whitney
P.O. Box 109600
West Palm Beach, Fl 33410-9600

Abstract

The microstructure and mechanical properties of diffusion-bonded XD gamma TiAl to Tiadyne 3515 ™ (Alloy C), a β-titanium alloy using TiCuNi 54 filler alloy as the transient liquid phase (TLP) were studied. The post-bond heat treatment led to the broadening of the TiCuNi/Alloy C interface while the XD gamma TiAl/TiCuNi interface remained virtually unaffected. This was coupled with coarsening of the microstructure of the bond as well as Alloy C matrix. Several intermetallic phases were identified in the bond region by X-ray, electron microprobe and TEM. The hardness peaked in the bond close to the XD gamma TiAl/TiCuNi interface and was significantly higher than that of either XD gamma TiAl or Tiadyne 3515 ™ (Alloy C). The peak hardness remained unaffected by post-bond heat treatment. The 3-point bend strength of the bond was improved by post-bond heat treatment; however, tensile properties were not affected. The formation of intermetallic phases as well as the bond width were reduced by using thinner filler interlayer. This improved the tensile properties of the bond. All failures occurred at the bond.

Introduction

Titanium alloys have been routinely used in the aerospace structural applications. Recently, ordered titanium aluminides, such as gamma TiAl have become of interest and are emerging as potential high temperature materials owing to their low density, high melting temperature, high modulus, and good creep, fatigue and oxidation resistance (1). However, use of gamma TiAl is limited because of its poor room temperature ductility. Although significant progress has been made in the development of these aluminides, only a limited effort has been directed to the joining of gamma TiAl to itself or to other classes of titanium alloys (2-5).

Since gamma TiAl is susceptible to solid state cracking due to its lack of ductility at room and intermediate temperatures (2), conventional fusion welding techniques may not be successful for joining gamma TiAl to other titanium alloys. To overcome the disadvantage of fusion welding, diffusion bonding of gamma TiAl to Tiadyne 3515 ™ (Alloy C) via a transient liquid phase is considered to be a promising joining method. Commercially available TiCuNi 54 is used as a TLP interlayer. The bonding is performed above the melting temperature of the interlayer. Post- bond heat treatments are designed to homogenize the microstructure of the bond by controlled diffusion with the aim of improving bond properties. However, post-bond heat treatments may influence the microstructure and properties of the base metals, especially Tiadyne 3515 ™ (Alloy C) by grain growth (6).

The present investigation examines the suitability of diffusion bonding TiAl to Tiadyne 3515 ™ (Alloy C) using commercially available TiCuNi 54 filler interlayer as a transient liquid phase. The effect of post-bond heat treatment on the microstructure of the bond as well as the mechanical properties were evaluated. Results are discussed in terms of the microstructure of the bond and its effects on the bond properties.

Materials and Procedures

The gamma TiAl alloy used for this investigation has the following composition: Ti-47 Al-2Mn-2Nb-0.8 vol% TiB_2 and was produced using the XD process as patented by the Martin Marietta Research Laboratories. Hereafter it is referred to as XD TiAl. TiB_2 particulate phase was introduced during the ingot processing for grain refinement. The investment cast ingot was hot isostatically pressed (HIP'd) at 2300°F/25 ksi/4 hr followed by heat treatment at 1850°F/50 hr. The resulting microstructure consisted of lamella colonies of Ti_3Al (DO19 crystal structure) and gamma TiAl (L1o crystal structure) and of equiaxed gamma TiAl. TiB_2 particulates were randomly distributed throughout the matrix.

Tiadyne 3515 ™ (Alloy C), a β- titanium alloy has the following chemical composition: Ti-35V-15Cr. It combines high elevated temperature strength properties and a markedly improved resistance to sustained combustion. It was obtained in the form of wrought rod. Typical microstructure consisted of beta grains, and very fine TiC precipitates.

TiCuNi 54 filler alloy, of the composition Ti-27.5 Cu-18.5 Ni, was obtained in the form of thin foils from Wesgo. A typical microstructure is shown in Figure 1. It consists of two outer layers of Ti and a central layer of CuNi alloy with an overall thickness of 70 μm. The composition of the central layer was determined by energy dispersive spectroscopy to be 60 wt.% Ni and 40 wt.% Cu. A few TiCuNi 54 foils were rolled down to a thickness of ~ 25 μm at room temperature, chemically cleaned, and vacuum annealed at 1200°F/2 hr. These thinner foils were used to determine the effect of foil thickness on the microstructure and mechanical properties of the bond.

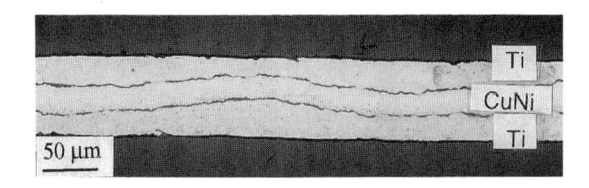

Figure 1 - Photomicrograph of polished and etched TiCuNi 54 filler alloy showing two outer Ti layers and a central layer of CuNi alloy.

Fabrication and Specimen Preparation

Both XD TiAl and Tiadyne 3515 ™ (Alloy C) were obtained in the form of rods. The rods were surface ground to 0.5" diameter and the faying faces were ground to 32 RMS finish. These were cleaned chemically and the interlayer was placed between the XD TiAl and Tiadyne 3515 ™ (Alloy C) rods as shown in Figure 2. This assembly was bonded in vacuum with a fixed load of 3.5 lbs using the following loading procedure: 1600°F/20 min ---> 1780°F/2 hr ---> cool to 1000°F @ 150°F/hr ---> rapid helium quench to room temperature. A few samples were also bonded in the same way as described above, except that the assembly was held at 1780°F/4 hr. Post-bond heat treatment was done in vacuum at 1780°F for 8 hr on specimens fabricated at 1780°F/2 hr. One specimen was bonded using 25 μm thick interlayer, and the assembly was held at 1780°F for 12 hr.

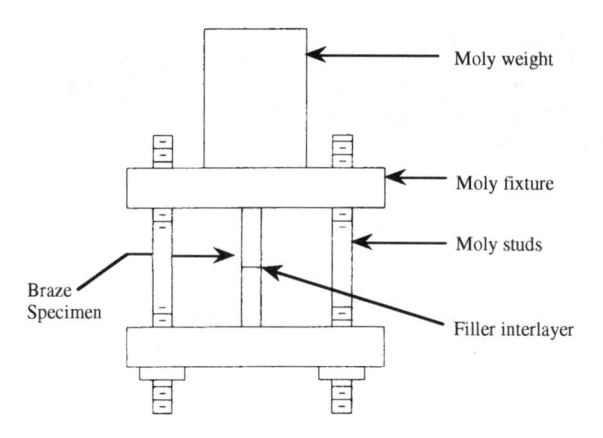

Figure 2 - Schematic diagram of the bond assembly.

3-point bend test specimens ~0.5" wide x 0.1" thick x 3" long were cut from the bonded specimen by electro-discharge machining. The surfaces of each specimen were polished successively with 400, 600, 1200 and 4000 grit SiC papers. The bend tests were conducted at room temperature using a spacing of 2.8". For tension tests, cyclindrical specimens were used. The gage section was surface ground, and polished successively with 400, 600, 1200 and 4000 grit SiC papers. Tensile tests were conducted at room temperature and at 800°F.

Materials Characterization

Microstructure of the bond was examined with optical and scanning electron microscopes (SEM). Microhardness indentations were introduced across the bond interfaces on polished specimens using a Vickers hardness tester with a 300-g load. Specimens for metallography were prepared using standard techniques and modified Kroll's reagent (5 vol.% HF - 12 vol.% HNO_3 - 83 vol % H_2O) was used as an etchant. Compositional variations across the bond interface were determined by microprobe as well as by energy dispersive spectroscopy (EDS) on metallographic cross-sections. X-ray diffraction patterns from bond areas were collected with a Siemens automatic diffractometer (D500) for phase identification. The microstructural characterization of selected specimens was done in a Philips EM400T-FEG transmission electron microscope with the goal to identify different phases present in the bond area. Discs 3 mm in diameter were cut from the bond-line of bonded specimens by electro-discharge machining. The thickness of each disc was then reduced to 100 μm by hand grinding on SiC paper. The centers of these discs were dimpled before ion-thinning to electron transparency in an argon ion mill. Fracture surfaces were examined in an AMRAY 1910 FE field emission SEM.

Results

Microstructure. Figure 3 shows cross-sectional microstructures of bonds in XD TiAl/TiCuNi/Alloy C for different fabrication and heat treatment conditions. In all cases, good bonds were achieved with the molten interlayer filling the gap along the mating surfaces and no imperfections such as pores were found in the bond. The microstructure of the bond consisted of Ti$_3$Al - phase adjacent to the XD TiAl matrix interspersed with numerous particles of different shapes and sizes and a nearly featureless interdiffusion zone toward the Alloy C matrix. The total thickness of the bond layer was ~110 μm and 140 μm for specimens that had been bonded with 70 μm thick TiCuNi 54 at 1780°F/2 hr and 1780°F/4 hr respectively (Figures 3a & 3b). When a 1780°F/2 hr bonded specimen was subjected to a post-bond heat treatment at 1780°F for 8 hr, the TiCuNi/Alloy C interface appeared to grow substantially compared to the TiCuNi/XD TiAl interface (Figure 3c). The microstructure of the bond coarsened, and a band of fine precipitates (~ 17 μm thick) delineated the transition from the bond to the Alloy C matrix. The microstructure of the matrix, especially of Alloy C, revealed grain coarsening. Additionally, numerous fine precipitates were observed in Alloy C matrix. These were identified as TiC. The thickness of the bond fabricated at 1780°F/12 hr using 25 μm thick TiCuNi 54 was 100 μm, and the microstructure of the bond contained precipitates of a much reduced density (Figure 3d). Similar to the observation in post-bond heat treated sample (Figure 3c), a string of fine precipitates was found to align parallel to the bond/matrix interface.

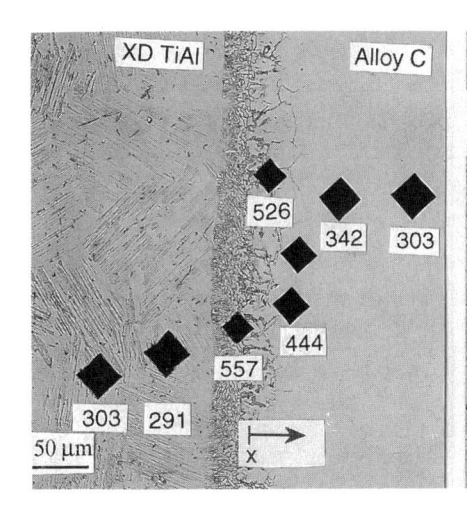

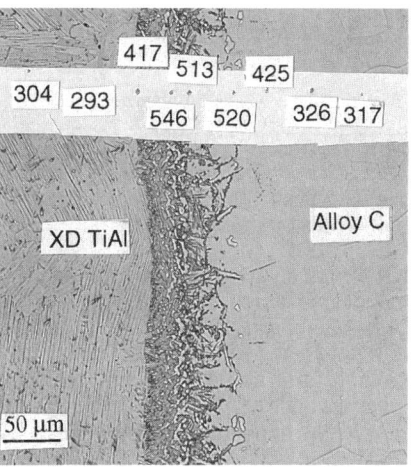

1780°F/2hr., 70μm thick TiCuNi54
(a)

1780°F/4hr., 70μm thick TiCuNi54
(b)

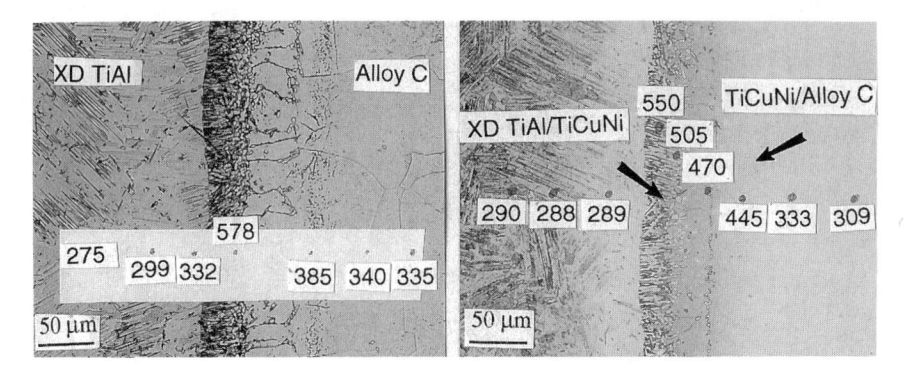

Post-bond heat treatment at 1780°F/12 hr. 1780°F/12hr., 25μm thick TiCuNi54
(c) (d)

Figure 3 - Light micrographs showing cross-sectional microstructures of bonds: The microhardness indentations as well as microhardness numbers are shown. The arow indicates the region along which concentrations of various elements were determined by microprobe.

X-ray diffraction patterns from the bond layers were generated on specimens subjected to different thermal conditions. The following phases were identified:

Material	Phases identified by X-ray diffraction
Alloy C/TiCuNi/ XD TiAl -Bonded at 1780°F/2 hr	Ti_3Al, Ti_2Ni, $Ti(Cu, Al)_2$, TiCu, $CuNi_2Ti$, β-Ti
Alloy C/TiCuNi/ XD TiAl -Bonded at 1780°F/2 hr + post bond heat treat at 1780°F/8 hr	Ti_3Al, TiAl, TiB, TiB_2, $Ti(Cu, Al)_2$, TiC, β-Ti

The X-ray diffraction results showed that as a result of the post-bond heat treatment at 1780°F/8 hr the nickel-rich phase had disappeared, and several other new phases developed.

Chemical compositions of various phases in the bond layer for different thermal conditions were determined by EDS. A typical example is shown in Figure 4 for Alloy C/TiCuNi/XD TiAl bonded at 1780°F/4 hr. Characterization by electron diffraction in TEM indicated the presence of Ti_3Al, Ti_2Ni, and TiNi in the bond layer of the sample fabricated at 1780°F/4 hr as shown in Figure 5. The compositions of # 1 and # 3 in Figure 4 could be matched with those of Ti_2Ni and TiNi respectively, assuming suitable partitioning of different elements in respective lattice sites.

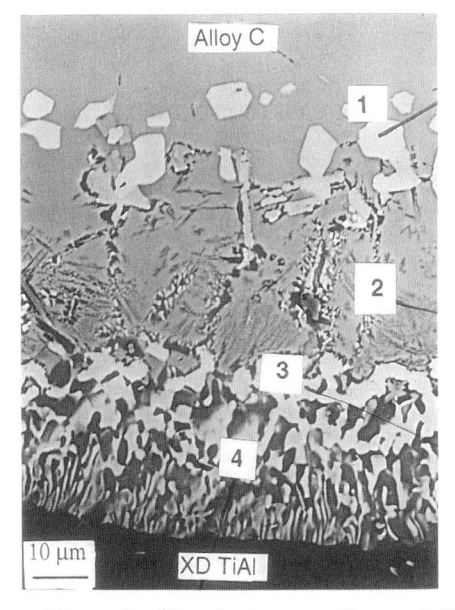

Element at.%	#1 Light Phase	#2 Gray Phase	#3 Light Phase	#4 Dark Phase
Al	10	22	25	32
Ti	58	58	43	64
V	1	7	3	0
Cr	1	2.5	1.8	0.1
Mn	0.5	1	1.8	0.6
Ni	23	5	17	0.6
Cu	6.5	4	8.7	0.7
Nb	0.1	1	0.5	2
	≈Ti$_2$Ni	Ti$_3$Al	≈TiNi	Ti$_3$Al

Figure 4 - Chemical compositions of different phases present in the bond layer of a specimen fabricated at 1780°F/4 hr.

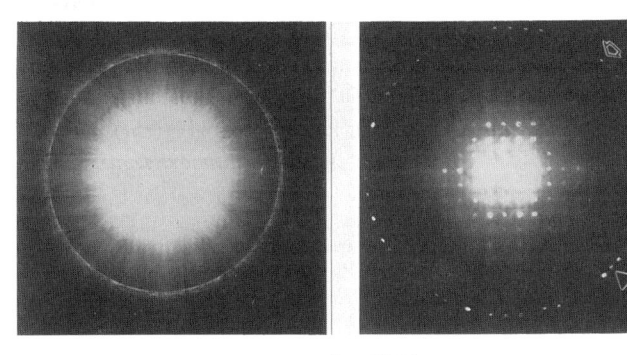

[0001]

(a) Ti$_3$Al

[001]

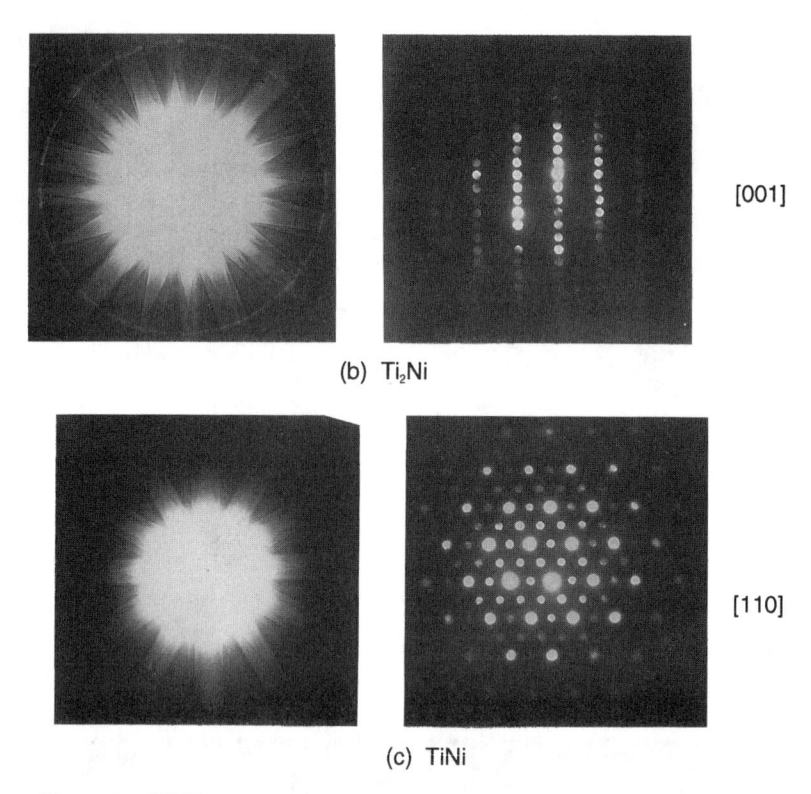

(b) Ti₂Ni

[110]

(c) TiNi

Figure 5 - CBED patterns from various locations in the bond layer of a specimen fabricated at 1780°F/4 hr showing the presence of (a) Ti₃Al, (b) Ti₂Ni, and (c) TiNi phases.

Typical concentration profiles of various elements across the bond layer from position x toward and into the Alloy C matrix (see Figure 3a) in specimens, bonded at 1780°F/2 hr, are shown in Figure 6. It is seen that Al from XD TiAl, Cu, and Ni from TlCuNi interlayer have diffused toward Alloy C while V and Cr from Alloy C have diffused toward the XD TiAl. On the XD TiAl/TiCuNi side, no diffusion of Ni and Cu from TiCuNi interlayer into the XD TiAl matrix was detected. Post-bond heat treatment at 1780°F/8 hr yielded similar concentration profiles with the exception that the diffusion zone had grown considerably larger and that individual elements had diffused to greater distances into the Alloy C matrix.

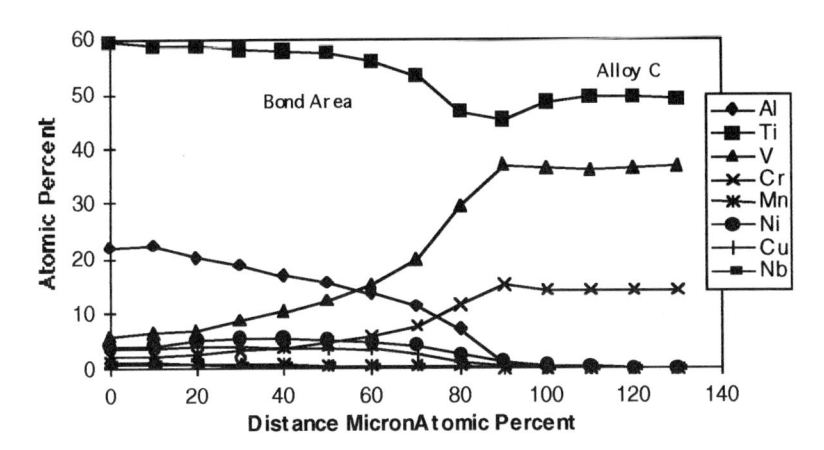

Figure 6 - Concentration profiles of various elements across the bond layer from position X to Alloy C in a specimen bonded at 1780°F/2 hr. (See Figure 3a)

Microhardness. Microhardness measurements were made across the bond layers for different heat treatment conditions. A typical example is shown in Figure 3 and the numbers in the micrographs represent the Vickers hardness values at those locations. The microhardness values are, in general, higher in the bond, reaching a maximum value in the bond close to the XD TiAl/TiCuNi interface. The peak hardness in the bond is considerably higher than either of the matrices and was not affected by post-bond thermal treatment. Within the bond the hardness goes down toward the Alloy C matrix. The specimen bonded with 25 μm thick TiCuNi interlayer displayed a similar behavior to that described above for 70 μm thick interlayer.

Mechanical Properties

Bend Properties. Table I summarizes the room temperature bend properties (3-point bending) of bonds made with 70 μm thick TiCuNi 54 filler interlayer at 1780°F/2 hr as well as bonds subjected to post-bond heat treatment at 1780°F/8 hr. The room temperature bend properties of as-bonded specimens are considerably lower than those of the parent material. However, post-bond heat treatment appeared to significantly enhance the bond strength. Still, the bend strength is about 60% of that of XD TiAl. The fracture took place along the bond /matrix interface. A typical fractograph is shown in Figure 7 which suggests that the failure occurred in a brittle manner.

Table I. Room Temperature 3-Point Bend Test Data.

Material	Flexural Strength (ksi)	Modulus (Msi)	Remarks
XD Gamma TiAl	127, 137	22, 23	
Alloy C	228	15.8	No Failure
XD Gamma TiAl/TiCuNi/Alloy C Bonded at 1780°F/2 hr	48, 55	17.5, 17.5	
XD Gamma TiAl/TiCuNi/Alloy C Bonded at 1780°F/2 hr + Heat Treat at 1780°F/8 hr	79	18	

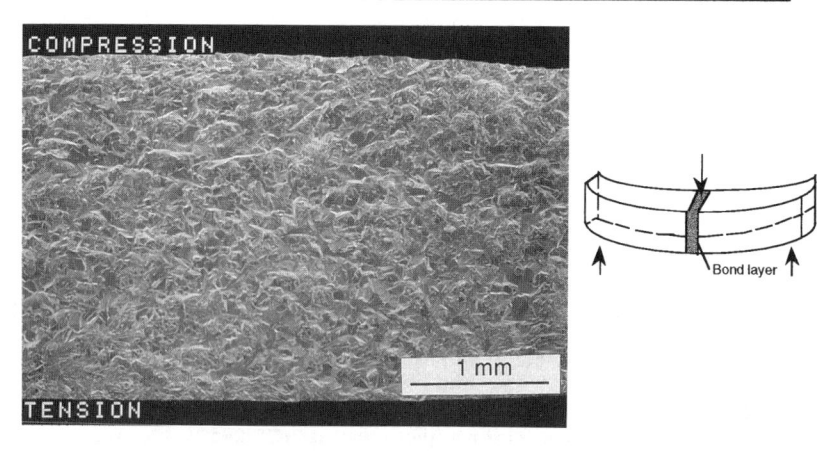

Figure 7 - Fractograph showing brittle failure in a diffusion-bonded specimen tested in 3-point bending at room temperature.

Tensile Properties. Table II illustrates the tensile properties of bonds made with 70 μm and 25 μm TiCuNi interlayers. It can be seen that 800°F tensile strength of the bond processed with 70 μm thick TiCuNi 54 at 1780°F/2hr was about 65% of that of the XD TiAl with a failure strain of only 0.3%. Post-bond heat treatment at 1780°F/8hr did not have any beneficial effect on the tensile properties of the bond. The fracture took place at the bond and appeared flat and was identical for both thermal conditions. A typical example is shown in Figure 8 a for a bond processed at 1780°F/2hr. However, the specimen bonded with a 25 μm interlayer at 1780°F/12hr displayed a substantially higher strength, about 81% of that of XD TiAl. Also, the strain to failure was a bit higher, i.e., 0.5%. The fracture took place at the bond and the fractograph appeared rugged (Figure 8b). Thinner interlayer coupled with a lower density of intermetallic phases and a reduced bond width has higher tensile strength than bonds processed with a thicker interlayer. Still, the strain to failure of the bonded specimen was very low, indicating that the bonds were brittle.

Table II. Tensile Properties of XD Gamma TiAl/TiCuNi/Alloy C.

Material	Ultimate Tensile Strength (ksi)	Modulus (Msi)	Failure Strain (%)	Remarks
XD Gamma TiAl	81, 82	24, 24	1.4, 1.4	RT Tension
Alloy C	145	16.8	20	RT Tension
As-Bonded XD Gamma TiAl/TiCuNi/Alloy C	52, 54	18, 20	0.3, 0.3	800°F Tension 70 μm thick TiCuNi Foil
Post-Bond Heat Treated XD Gamma TiAl/TiCuNi/Alloy C	53, 54	18, 20	0.3, 0.3	800°F Tension 70 μm thick TiCuNi Foil
As-Bonded XD Gamma TiAl/TiCuNi/Alloy C	66	23.5	0.5	RT Tension 25 μm thick TiCuNi Foil 1780°F/12 hr

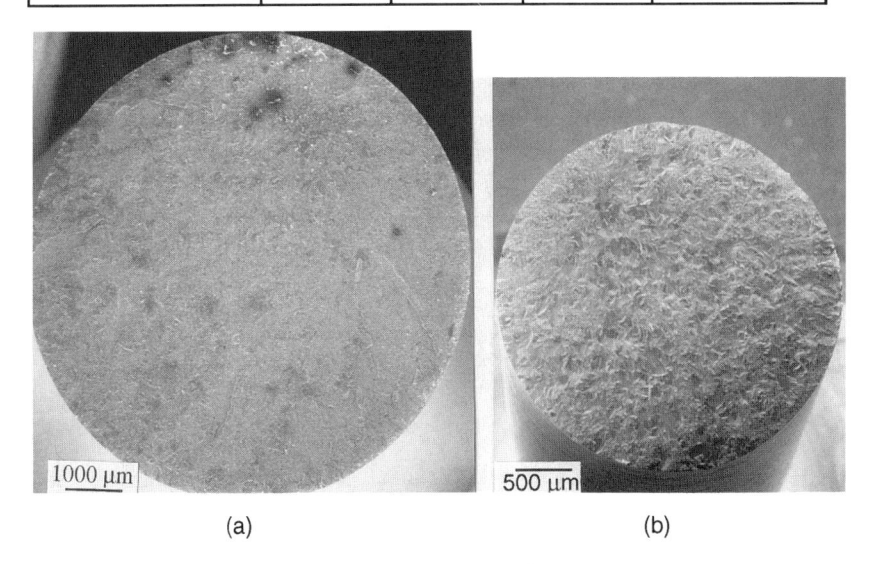

(a) (b)

Figure 8 - Fractographs showing brittle failure in a bonded specimen tested in tension at room temperature: (a) specimen bonded with 70 μm thick TiCuNi filler alloy, and (b) specimen bonded with 25 μm thick TiCuNi filler alloy.

Summary and Discussion

Defect - free diffusion-bonding of XD TiAl to Alloy C was achieved using two different thicknesses of TiCuNi 54 interlayer as the transition liquid phase. Several intermetallic phases, such as Ti_3Al, Ti_2Ni, $Ti(Cu, Al)_2$, $TiCu$, $CuNi_2Ti$,

TiAl were formed in the bond region. The density of intermetallic phases as well as the bond width were reduced for thinner TiCuNi interlayer but the formation of intermetallic phases could not be suppressed. It has been reported that presence of these intermetallic phases becomes an hindrance to homogenization of the microstructure of the bond during post-bond heat treatment (4).

The mechanical properties of the diffusion-bonded specimens are strongly dependent on the microstructure of the bond. The intermetallic phases formed in the bond are very brittle and responsible for poor mechanical properties of the bond as observed in the present study. Similar observations were reported earlier (4, 7). Incorporation of a thinner TiCuNi interlayer has led to an improvement of the tensile properties. This may be explained by the formation of a reduced level of intermetallic phases in the bond. These results appear to indicate that a complete suppression of brittle intermetallic phases is required to achieve a good bond strength. This could perhaps be accomplished by using thinner TiCuNi interlayers with reduced concentrations of Cu and Ni. Experiments in this area are in progress.

Acknowledgments

The authors thank Messrs. Randy Pennington, Bill Tress, Wilson Horne, Bob Surace of P&W and Dr. R. Ayer, STEM, Inc. for their assistance with specimen fabrication, and microstructural characterization by X-ray, SEM and TEM. The authors are grateful to Professor G. E. Welsch (CWRU, Cleveland) for reviewing the manuscript. This work was supported by Pratt & Whitney, FL.

References

1. H.A. Lipsitt, High Temperature Ordered Intermetallic Alloys, Mat. Res. Soc. Symp. Proc., Vol.39, 1985, p.351.
2. R.A. Patterson, P. Martin, P.L. Damkroger, and L. Christodoulou, Titanium Aluminide : Electron Beam Weldability, Welding Research Supplement (January 1990), 39-s-44s.
3. P. Yan, R.E. Somekh, and E.R. Wallach, "Solid-State Bonding of TiAl with Interlayers", Proc. of the 3rd National Conference on Trend in Welding Research, Gatlinburg, TN , 1992, p. 1063-1067.
4. P. Yan and E.R. Wallach, "Diffusion-bonding of TiAl", Intermetallics 1 (1993) 83-97.
5. P.L. Threadgill, "The prospects for joining titanium aluminides", Materials Science and Engineering A 192/193 (1995) 640-646.
6. E. Lugscheider, and L. Martinez, " Entwicklung neuer Titanbasislote zum Fugen von TiAl6V4. DVS - Berichte Bd 132, DVS- Verlag", s.36ff ,1990.
7. L.C. Den, S.F. Chang, J.S. Lee, P.W Kao, and K. -C. Hsieh, "Microstructure Study for Brazing of Ti-Al6 V4 with Ti-15Cu-15 Ni Filler Metal. Institute of Materials Science and Engineering, National Sun Yat-Sen University Kaohsiung, Taiwan, Republic of China, Unpublished paper.

Explosion Welds Between Titanium and Dissimilar Metals

John G. Banker
CLAD Metal Products, Inc.

Vonne D. Linse
Regal Technology Corporation

Abstract

Explosion welded transition joints provide a means for making fully welded structural connections between titanium and dissimilar metals such as aluminum, stainless steels and nickel alloys. The unique aspects of explosion bonding titanium are discussed. Mechanical test data is presented on several metal combinations. Frequently thin interlayers of other metals are employed in the bonding operation to enhance product features, such as strength, toughness or leak tightness. Data on performance of specific interlayer metals is presented.

Introduction

Titanium is a preferred material of construction for many equipment and process applications (Ref. 1) It exhibits unique performance features resulting from the combination of low density, high strength and stiffness, corrosion resistance, and reliable fabricability. As with many ideal solutions for a specific design, it is sometimes not the best solution for all aspects of the design. Optimized designs frequently require a combination of different alloys or materials. Bimetallic components comprised of titanium bonded to a dissimilar metal provide solution for optimizing the design. There are basically two types of bimetallic applications with highly different product feature requirements: clad products and bimetal couplings.

Clad: In clad products the titanium is primarily present for surface performance, such as corrosion resistance, while the base metal is the primary structural component. The bond between the two metals is essentially limited to shear loading under normal performance conditions. A common example is a steel pressure vessel with titanium cladding on the interior surface.

Couplings: Bimetal couplings typically require structural features for both the titanium and second metal, as well as the bond. Both the titanium and second metal are commonly selected for mechanical performance characteristics. The bond between the two may be subject to tensile, shear, impact, and fatigue conditions. Bimetallic, titanium-to-stainless steel tube couplings are a primary example.

Manufacture of Bimetallic Materials

Although bimetallic components between titanium and dissimilar metals are easy to imagine and draw on paper, they are not that simple to make. Titanium reacts readily with most other metals during fusion welding and other hot bonding processes. The reaction results in the formation of intermetallic compounds which typically exhibit very poor mechanical performance. Consequently, the designer is required to use either mechanical connections or turn to cold welding processes. Explosion welding is a reliable, proven solution for making structurally reliable welds between titanium and most other metals. Explosion bonded titanium-steel clad plate has provided reliable performance in chemical process applications for over thirty years. Explosion bonded transition joints have similarly provided structurally reliable bimetallic couplings between titanium and steels, nickel alloys, aluminum, and other metals. (Ref. 2)

Explosion bonded titanium clad plate will not be discussed in further detail in this paper. Another paper in this publication addresses this subject extensively (Ref. 3). However, much of the technical discussion of this paper is applicable when higher strength, or higher hardness titanium alloys are required in clad metal applications.

Explosion Welding Process

Explosion welding is a solid state metal-joining process that uses explosive force to create an electron-sharing metallurgical bond between two metal components. The explosive detonation above the plates creates a near instantaneous temperature spike, but there is minimal thermal energy and no time for heat transfer to the metals. Significant localized heating occurs in the future bondzone region due to the energy of collision; however, this thermal spike also lasts only microseconds. Consequently, there is no metallurgically significant bulk temperature increase in the metals. Because of the absence of bulk heating, explosion welded products do not exhibit many of the metallurgical problems typical of dissimilar metal joints. The microstructures, mechanical properties and corrosion properties of the wrought parent components are not significantly altered during the bonding process. There are no heat affected zones. Brittle intermetallic layers are not formed (Ref. 4-7).

Because of the unique safety and noise/vibration considerations explosion welding is typically performed in relatively isolated facilities by companies specializing in explosives operations. The dissimilar metal joining capabilities of explosion welding are made available for typical fabricator shop usage through the concept of welding transition joints. In transition joint manufacture the component metals are typically explosion welded in plate or cylinder form and are then provided as blocks, strips or tubular couplings to equipment fabricators. Transition joints are commonly produced for joining titanium to aluminum, copper, steel, stainless steel, and nickel alloys. In fabrication, the titanium components are welded to the titanium face of the transition joint and the other metal components welded to the other face. Figure 1 depicts the transition joint concept.

Explosion welding can be used to join most metal combinations. However, not all bondzones exhibit ideal properties of strength, toughness, ductility, leak tightness, etc. The bondzone properties that are required for a particular component are determined by the technical requirements of the specific application. The explosion bond is typically tailored to provide the optimum combination of required features. For many components, the optimum bondzone performance features are best achieved with the addition of one or more interlayer metals between the two primary metals. An interlayer metal is frequently used when higher strength titanium alloys are explosion welded to a dissimilar metal.

Explosion Welding of Titanium Alloys

Due to their strong ability to react metallurgically with iron to form iron-titanium intermetallic compounds, titanium alloys can present a challenge when being explosion welded to steels and other iron-based and nickel-based alloys. If the process energetics which cause the formation of these imbrittling intermetallic compounds, and the process parameters which control them, are well understood, then reliable high quality explosion welds can be made in titanium-steel. It does, however, require close control of the explosion welding process.

For explosion welding there are two process limitations which are critical to parameter selection (Ref. 4). These limit the range of manufacturability of reliable titanium-steel welds. These limitations are graphically presented in Figure 2. The first and probably most critical limitation is the upper collision energy boundary at which the energy of formation of the titanium-iron intermetallic is reached. At this point the intermetallic starts to form in sufficient quantities along the weld interface to cause the weld to be brittle. This boundary is relatively constant for most titanium alloys of interest for commercial applications. The second important limitation is the lower collision energy boundary at which the pressure caused by the collision becomes sufficient to cause bonding conditions. At this point dynamic yielding and flow of the faying surfaces of the metals results in jetting which allows welding conditions to occur. This boundary is primarily dictated by the dynamic yield strengths of the two materials being welded. As the dynamic yield strength increases, the lower energy boundary moves upward toward the upper energy boundary. Once these limits merge, structurally reliable explosion welds are not possible.

It is then the region between the upper and lower collision boundaries which dictates the latitude in process collision parameters which will result in high quality explosion weld between two metals. For the titanium-iron system, this allowable range of welding parameters is relatively narrow at best. It is sufficient to allow the lower strength, commercially pure (CP) titanium grades to be directly explosion clad to low to medium yield strength steels. As the titanium and/or steel are alloyed and their strength levels begin to increase, the lower limit for explosion welding to start moves upward. At higher strengths, the lower limit becomes sufficiently close to, or even reaches and exceeds, the upper limit so there is no latitude for the process to achieve a direct, high quality weld between the two metals. It then becomes necessary to employ a third, intermediate metal layer between the two which is compliant in terms of metallurgical compatibility and weldability. There are a number of compliant interlayer metals which can be used to successfully weld high strength titanium to high strength steel. They include, CP titanium, vanadium, tantalum, niobium, low carbon steel, low alloy aluminum and copper alloys. Selection of the specific interlayer material will depend on a number of factors primarily related to the weld or joint performance required by the application. Cost will also obviously vary depending on which interlayer metal is selected.

Interlayer Materials

When interlayers are needed for bonding purposes, the properties of the interlayer significantly affect overall component characteristics. The thickness of interlayers typically ranges from 0.01-inch to 0.25-inch, depending upon the required joint properties. When interlayers are very thin, typically 0.010-inch, the interlayer behaves mechanically as a thin film. The joint tensile properties will be much higher than are typical for standard tensile specimens of the interlayer metal. The bond shear properties are however representative of the interlayer metal. Thicker interlayers generally result in bond properties that are typical of the interlayer metal, both in tension and shear.

Tables I and II presents bondzone mechanical test data on transition joints between titanium alloys and both steel and nickel alloys. A direct bond between these metal combinations results in low strength and toughness at best. The data show the effect of various interlayer metals, and thicknesses, upon bond tensile and shear strengths and toughness. Table III presents bondzone test data of a variety of metals combinations.

Current Market Applications and Test Programs

Explosion welded titanium transitions are currently used in a variety of market applications. These include applications in the process industries and in shipboard installations for joining unalloyed titanium to steel, aluminum, and copper. Transition joints between alloy titanium and nickel alloys or stainless steels are more commonly used in aerospace and aircraft applications. Currently aerospace applications are primarily in tubular connections. Explosion bonded tubular transitions generally employ a cylindrical or conical (scarf) bondzone configuration. Tensile and impact loading of the dissimilar metal bond is minimized in these designs. Figure 3 shows a concentrically bonded tubular transition joints between Ti-6-4 and

AISI 4130 alloy steel. Helium leak tightness of these bondzones is typically less than 1×10^{-9} atm-cc/sec. Test programs are have been performed for a variety of structural transition joint applications in which alloy titanium is joined to high strength steel or nickel alloy components.

Explosion welded titanium transition joints have been in use in shipboard applications for over ten years. Titanium provides superior corrosion resistance in saltwater environments. Traditionally ship construction has been mainly either aluminum or steel. The installation of titanium components onto the existing fleet has resulted in dissimilar metal connections. Where mechanical connections are used, galvanic corrosion is enhanced in the crevice between the dissimilar metals. Transition joints provide a means for eliminating the crevice and controlling corrosion. (Ref. 8-10) Shipboard applications typically utilize CP titanium bonded to aluminum, carbon steel, stainless steel, or copper-nickel.

Summary

Structurally reliable welds between titanium and dissimilar metals can be made using explosion bonded welding transition joints. Interlayer metals are commonly used to enhance transition joint properties. A thin tantalum interlayer between alloy titanium and Inconel provides a transition joint similar in strength to the titanium component. Other interlayer metals which are effective in achieving good strength and toughness are vanadium, niobium, unalloyed titanium plus unalloyed nickel, copper-nickel, and iron. Titanium transition joints have been used extensively in aerospace, shipboard, and process industry applications.

References:

1. Donachie, M.J.Jr., Titanium, A Technical Guide, ASM International, Metals Park, OH, 1988.
2. Banker, J.G., "Explosion Welded Transition Joints for Structural Welds between Titanium and Dissimilar Metals", Proceedings of the Technical Program from the 1994 International Conference, International Titanium Association, Boulder, CO, pp 62-69.
3. Banker, J.G., & Forrest, A.L., "Titanium/Steel Explosion Clad for Autoclaves and Reactors", Presentation at 1996 TMS Annual Meeting, February 8, 1996, TMS, Warrendale, PA.
4. Linse, Vonne D., "Procedure Development and Process Considerations for Explosion Welding", ASM Handbook Volume 6, 1993.
5. Patterson, R. Alan, "Fundamentals of Explosion Welding", ASM Handbook Volume 6, 1993.
6. Banker, J.G. & Reineke, E.G., "Explosion Welding", ASM Handbook Volume 6, 1993.
7. Pocalyko, A., "Explosion Clad Metals", Encyclopedia of Chemical Technology, Volume 15, Third edition, John Wiley & Sons, 1981, pp 275-296.
8. McKinney, C.R., and Banker, J.G, "Explosion Bonded Metals for Marine Structural Applications", Society of Naval Architects and Marine Engineers, Nov. 1970, p 285-292.
9. US Patent 5,296,647, Assembly for Connecting an Electrical Box to a Plate with a Bimetallic Flange
10. US Patent 5,190,831, Bonded Titanium/Steel Components

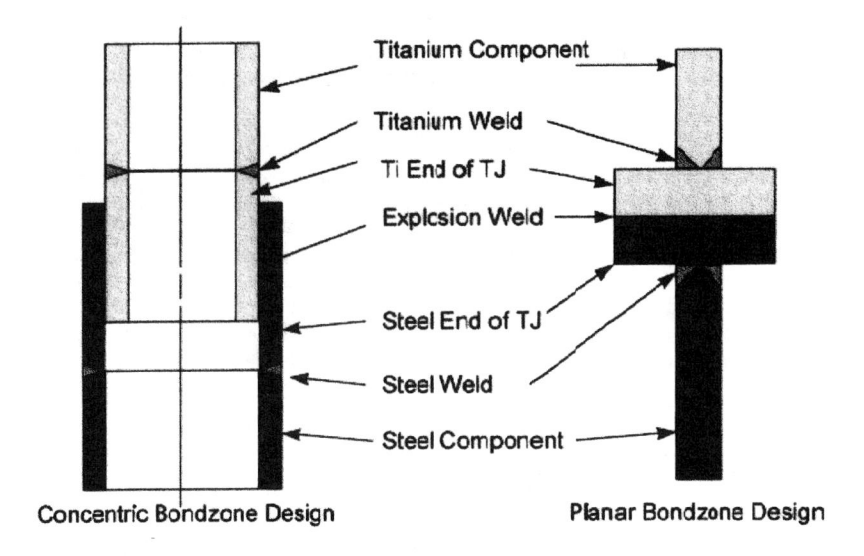

Figure 1: Concept for making welds between titanium and dissimilar metals using bimetallic explosion welded transition joint (TJ) components.

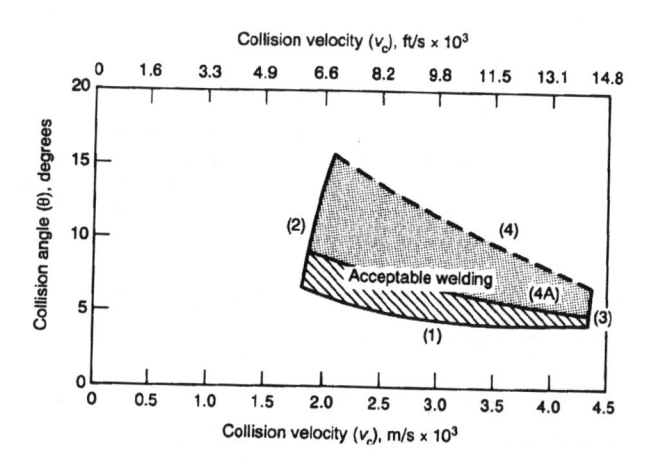

Figure 2: Plot of collision angle vs collision velocity obtain typical process parameter envelope. Lower collision energy limit is (1). Upper collision energy limit is (4A). (Ref. 4)

Figure 3: Concentrically bonded transition joint between AISI 4130 Steel and Titanium 6-4, with an interlayer of vanadium. As explosion bonded at top. After machining at bottom.

Table I

Typical Properties for Alloy Titanium Explosion Welded to Alloy Steels and Nickel Alloy

Titanium Alloy	Base Metal	Interlayer Metal	Shear (psi)	Tensile (psi)	Toughness (Note)
6-2-4-2	Inconel 718	Ta		124,400	VG
6-4	Inconel 718	Ta	34,300	120,200	VG
6-4	Inconel X750	Ta		119,000	VG
6-4	RA330	Ta		51,000	VG
6-4	304L S.St.	Ta	33,000	85,000	VG
6-4	17-4PH S.St.	Ta	48,000	79,000	VG
6-4	AerMet 100	Ta	120,000		
6-2-4-2	Inconel 718	Nb		80,500	G
6-2-4-2	Inconel 718	V	36,000	102,900	VG
6-4	Inconel 718	Ti**+Ni	60,000		M
6-4	Inconel 718	Ti**	29,000		P
6-4	Inconel 718	Ni	49,000		P

Note: Toughness VG=Very Good G=Good   M=Moderate P=Poor   **B265 Gr 1

Table II

Test Results for Inconel 718 Bonded to Titanium Alloys with Various Interlayers Showing Effects of Interlayer Thickness and Post Weld Heat Treatment (Project B)

Titanium Alloy	Interlayer Alloy	Thick (in)	Shear Strength (psi) As-Welded	800F HT	Tensile Strength (psi) As-Welded	800F HT
Ti-6-4	V	0.010	45,000	43,700	107,000	65,700
Ti-6-4	V	0.020	41,700	41,000	89,900	-
Ti-6-2-4-2	V	0.010	41,800	41,100	103,300	65,400
Ti-17*	V	0.010	42,900	43,000	101,200	-
Ti-6-4	Ti**	0.020	50,800	62,800	94,900	65,100

Notes: * Alloy Ti--5Al-4.5Cb-3.5Mo-2Sn
 ** Ti is ASTM B265 Grade 1

Table III

Typical Properties for CP Titanium Explosion Bonded to Other Metals With Various Interlayers

Metal w/Interlayer	Ultimate Strength (psi)	Toughness (Note)
Aluminum 1100	15,000	VG
Aluminum 5456	None	None
w/Tantalum	45,000	VG
w/Aluminum 1100	15,000	VG
Copper	20,000	G
Nickel	40,000	G
Copper-Nickel (70-30)	50,000	VG
Low Carbon Steel (C1008)	40,000	G
Structural Steel (A36 or Eq)	45,000	M
w/C1008	40,000	G
w/Copper-Nickel (70-30)	50,000	VG
Alloy Steel (4340 or Eq)	None	None
w/Tantalum	80,000	VG

Titanium/Steel Explosion Bonded Clad for Autoclaves and Reactors

John G. Banker
CLAD Metal Products, Inc.

Arthur L. Forrest
A.L.Forrest Consulting

Abstract

Titanium solves corrosion, maintenance and environmental problems in many reactor and autoclave applications. Titanium clad construction permits autoclave designers a great deal of flexibility combined with significant cost reduction. Titanium alloys can be selectively applied in specific regions of the autoclave to accommodate local environmental conditions. Resistance to general corrosion, crevice corrosion, abrasion, and ignition conditions can be accomplished selectively and economically. Explosion clad pressure vessel fabrication and performance have been proven through three decades of experience.

Introduction

Titanium provides superior corrosion resistance in a broad range of chemical environments (Ref. 1,2). However, the initial cost of titanium equipment is high when compared with that of many of the less corrosion resistant alloys. This is particularly true where the elevated design pressures and temperatures necessitate heavy wall thicknesses. The value of titanium is realized through the improved performance and extended service life of the equipment. The initial cost of titanium equipment can typically be significantly reduced by the use of clad. Explosion cladding of a relatively thin layer of titanium onto a thicker, lower cost structural base metal, such as steel, has proven to be a reliable and cost effective alternative. Over the past 30 years, properly constructed titanium explosion clad pressure vessels have demonstrated highly reliable performance. Figure 1 presents a cost comparison of titanium/steel clad with solid titanium at thicknesses commonly used in pressure vessels.

Advantages of Clad Construction for Hydrometallurgy Autoclaves

In addition to lower metal costs, titanium explosion clad offers many advantages for autoclaves.
1. Fabrication costs are lower than for solid titanium.
2. Since titanium is no longer the strength component, the titanium alloy can be chosen for other features such as corrosion resistance, erosion/abrasion resistance, ignition resistance. The designer is no longer limited to alloys contained in the Pressure Vessel Code table of allowable working stresses.
3. The titanium cladding alloy selection can be varied selectively within the autoclave to provide unique performance features in specific regions of the autoclave.
>Alloys containing platinum group metals, such as Grades 7 or 11, can be used where crevice corrosion is anticipated.
>Highly abrasion resistant alloys, such as Grade 5, can be used where severe abrasion and erosion are anticipated.
>Highly ignition resistant alloys, such as Ti-45Nb, can be applied in regions where oxygen impingement or rubbing surfaces pose a potential ignition threat.
>The lower cost unalloyed grades can be used in regions where there is simply a corrosion problem.
4. The titanium eliminates the need for thick refractory brick significantly reducing pressure vessel diameter and weight, shell thickness, welding and fabrication costs, and transportation costs. Figure 2 presents the approximate weight reduction which is somewhat proportionately related to total cost.
5. Components outside of the corrosion envelope, such as supports, stiffeners, agitator mounts and external jackets can be fabricated from low cost steel.

Titanium Clad Applications and Performance Experience

Chemical Process Applications: Since the late 1960's titanium explosion clad has been the standard material of construction in the chemical process industry when titanium corrosion resistance is required in combination with high pressures and/or high temperatures. Chemical and petrochemical process companies have installed hundreds of pressure vessels and thousands of heat exchangers fabricated from titanium explosion clad. A large proportion of these units has been installed in petrochemical plants producing terephthalic acid (TPA) and purified terephthalic acid (PTA.) TPA and PTA are building block chemicals for the manufacture of polyester. In the past two decades these plants have proliferated around the world. These agitated reactors, which operate at temperatures between 200°C (390°F) and 250°C (480°) and maximum pressures of 2.34MPa (340psi), have excellent service records. The trend has been to increasingly larger vessels and higher temperatures, resulting is a significant advantage to clad construction. Titanium clad PTA reactors of over 7.7m (300in) diameter and 75mm (3in) wall thickness have recently been placed in service. Figure 3 shows a mid-size PTA vessel during fabrication.

Hydrometallurgy Applications: Titanium has been shown to provide excellent corrosion resistance in many hydrometallurgical ore treatment environments (Ref. 3). Operators of sulfuric acid oxidative pressure leaching autoclaves agree "that titanium, if it can be used, is the best material... and most are changing over to titanium" (Ref. 4) for internals and slurry piping. A significant number of titanium and titanium clad pressure vessels have been fabricated for production autoclaving, Table I. Additionally, many others have employed titanium and titanium clad components in selected areas, Table II. Figure 4 shows a gold leaching autoclave with titanium clad agitator mounting covers and blind flanges.

Explosion Clad

Process Overview: Explosion cladding is a solid state metal-joining process that uses explosive force to create an electron-sharing metallurgical bond between two metal components. Although the explosive detonation generates considerable heat, there is no time for heat transfer to the component metals; therefore, there is no appreciable temperature increase in the metals. Due to the absence of heating, the microstructures, mechanical properties and corrosion properties of the wrought parent components are not significantly altered during explosion bonding. There are no heat affected zones, and brittle intermetallic layers are not formed. Figure 5 presents a typical photomicrograph of a titanium/steel explosion clad bondzone. The wavy morphology is characteristic of the explosion bonding process. (Ref. 5-8)

Because of the unique safety and noise/vibration considerations explosion cladding is performed in relatively isolated facilities by companies specializing in explosives operations. Product sizes are normally limited only by the size availability of the component metals and the explosive detonation limitations of the manufacturing facility. Titanium clad plates with widths of 4m (156in) and lengths of 10m (390in) are commonly produced. The titanium cladding thickness typically ranges between 2mm (0.08in) and 19mm (0.75in), dependent upon the application. The steel base metal typically ranges between 12mm (0.5in) and 500mm (20in), dependent upon pressures. Explosion clad materials are typically supplied to equipment manufacturers in the form of flat plates and discs, formed heads, and cylindrical shapes. Clad cylindrical components, such as pipes and nozzle necks, can be manufactured. Additional information is available through the International Explosion Metalworking Association (IEMA)[1], an organization of major manufacturers worldwide which is dedicated to the development of the industry.

Explosion Bonded Titanium/Steel Clad: At the present time there are no ASTM or ASME standard specifications for titanium clad. (Since the titanium is not part of the strength design allowance, ASME Code compliance is only required for the base steel.) IEMA member companies have established recommended industry standards for titanium clad which are presented in IEMA Guidenote 100.

Shear strength testing is the most common method for determining clad bondzone mechanical properties. The test design and method presented in ASTM A-264 is typically used; however, it is common practice to double the lug width to reduce variability. Guidenote 100 specifies a minimum value of 137mpa (20,000 psi). Bond shear strengths are typically in the 240 MPa (35,000 psi) to 308MPa (45,000 psi) range.

---

[1] IEMA, 6220 Old Brompton Road, Boulder, CO USA 80301

The bond integrity of clad metals is normally inspected using ultrasonic testing. For critical service reactors, Guidenote 100 requires inspection of 100% of the clad surface and a minimum sound bond of 98%. The actual frequency of occurrence of rejectable nonbonds is typically less than 0.005% on an area basis.

Titanium and Steel Alloy Options: All of the titanium alloys can be clad using the explosion bonding process. However, the optimum bond mechanical properties are developed when the yield strengths of both the cladding and base metal are below 344 MPa (50,000 psi). Consequently the optimum bond strength and toughness of titanium cladding results from a combination of Titanium Grade 1, or similar, clad to a moderate strength pressure vessel steel, such as ASME SA516 Grade 70. (Titanium Grades 11, 17, and 26 exhibit similar yield strength and similar bond performance to Grade 1.) Although higher strength titanium alloys such as Grades 2 and 12, can be directly bonded to steel, the bond strength is typically low and clad may disbond during subsequent fabrication. When cladding higher strength titanium grades, it is common practice to use an interlayer metal between the alloy titanium and steel. Grade 1 titanium is the most commonly used interlayer for clad pressure vessel applications (Ref 9). Alternately, other alloys can be applied to the Grade 1 base using processes such as weld overlay or strip cladding. For example, in regions requiring high erosion resistance, Grade 5 or 12 can be either weld overlay deposited onto Grade 1 cladding, or wear plates can be welded directly to the Grade 1 cladding. Table III lists many of the currently available titanium alloys and highlights their specific features (Ref. 10).

Heat Treatment : During the explosion cladding operation, the bondzone wave formation results is significant localized cold work, yielding a bond which exhibits high strength but low ductility. It is standard practice to stress relieve titanium/steel clad in the 538°-635°C (1000°-1175° F) range to improve bond toughness.

Forming: During vessel fabrication the clad plates are formed into cylinders and heads. Cylinder rolling is either performed cold or in the temperature range of 538° to 620°C (1000° to 1150° F), dependent upon forming equipment capability. Hot pressing is the preferred head forming technique. Forming temperatures should be maintained below 815°C) 1500°F and hold times should be minimized.

Titanium Clad Fabrication

Titanium and steel cannot be directly fusion welded to each other due to brittle intermetallic formation. Clad fabrication is typically accomplished using a batten strap technique as depicted in Figure 6 . The insert approach, as depicted in Figure 7 is also infrequently used. In both cases, the titanium cladding is removed from the area around all edges where steel welds are to be made, typically 12mm (0.5in) inward from the steel weld prep edge. The steel base metal is prepared and welded using conventional steel fabrication procedures. All stress relief and NDT of the steel pressure vessel, required for Code compliance, is then performed. The vessel is then cleaned up and prepared for fabrication of the titanium cladding. The batten strap technique is normally used when the titanium is 5mm or thinner. In the batten strap technique, a filler-metal strip is inserted into the space where the titanium has been removed. The choice of filler is dependent upon proprietary fabrication preferences; commonly used materials include copper, steel, and titanium. A wider strip of titanium, the batten strap, is then placed over the weld area. The batten strap is welded along the edges with fillet welds. Backside purging with inert gas is necessary for reliable welds. When titanium cladding is around 8mm or heavier, the insert design of Figure 5 can be used. This approach provides a flush clad surface. In critical service installations, a batten strap can be added on top of the welded insert to provide additional corrosion protection. When fillet welding batten strips on clad, the chilling, heat sink effect of the base steel must be taken into consideration, otherwise good fillet root penetration may not be achieved. Large diameter nozzles are frequently fabricated from clad plate using these same procedures as for vessel fabrication. Typically, small diameter nozzles are lined with solid titanium or bimetallic, titanium-steel, sleeves. Attachments between nozzles and the vessel body are made using procedures similar those employed on the vessel circumferential and longitudinal butt welds.

The fabrication of clad vessels requires special precautions since these are primarily steel fabrications which typically produce large amounts of dirt, dust and sparks. During steel fabrication titanium surfaces must be protected from grinding, welding and cutting sparks to avoid localized iron contamination of the clad surface. Common techniques are to cover the titanium with fire-resistant paper or coat it with anti-spatter compound. Forming tools must be thoroughly cleaned prior to cold forming. Surface protection with heavy paper is desirable to protect the titanium from impressed contaminants. Hot formed components such should be pickled or ground afterwards. It is best to transfer the vessel to a clean shop after steel fabrication is complete and prior to titanium welding. Alternatively, since vessels are normally clad on the inside, the interior of clean clad steel vessel can provide suitable conditions for titanium welding, once the openings are sealed off. Care must be taken to assure that contamination from the steel welding shop is not dragged in later.

Welding: Steel welding is performed using the same methods that would be used for a bare steel vessel. The heat from steel welds, and of standard post weld stress relief will not deleteriously affect the bond. However, as noted above, care must be taken to prevent titanium surface contamination.

Titanium welding is performed using procedures which are standard for solid titanium components (Ref. 11,12). These welds require considerably greater attention to cleanliness and atmospheric protection than steel welds. At room temperature, upon exposure to air titanium reacts instantly to form an impervious, self-healing oxide film, angstroms thick. This film gives titanium its corrosion resistance in oxidizing media, or passivated reducing acid environments, such as is found in oxidative sulfuric acid pressure leaching autoclaves. Diffusion of oxygen at temperatures up to 425°C (800°F). takes place at a negligible rate. At higher temperatures this reaction proceeds at an increasing rate, until, at welding temperatures, reaction is extremely rapid. Molten titanium will tear apart the water molecule to get at the oxygen, which diffuses instantly into liquid titanium. Even small amounts of oxygen (>0.8%) will totally embrittle titanium, rendering it unsuitable as a structural metal (Ref. 13). At temperatures found in the weld puddle and heat affected zone (H.A.Z.) titanium reacts with virtually anything, resulting in contamination and embrittlement, inclusions or porosity. Titanium will not react with argon or helium. Normally argon is used to protect the weld and H.A.Z. during welding and cooling to below 800°F. When making production welds, a trailing shield attached to the torch is usually necessary to permit reasonable welding speeds while maintaining protection while above 800°F. Additional care may be needed to assure that the backside of the weld is properly shielded. Inadequate shielding will result in visible discoloration of the weld. A silver color indicates adequate shielding. Discoloration ranging from straw to blue to gray and white is indicative of increasingly inadequate shielding and poor weld quality. Maintaining adequate shielding is mandatory; this rule is absolutely unforgiving and cannot be transgressed, even for an instant. If contamination occurs, the only remedy is removal of the contaminated metal, by cutting it out. Once this simple principle is understood, virtually all the techniques for welding titanium become apparent. Welding titanium then becomes a matter of cleaning and protection, and thinking, more than anything else.

Caution! When performing titanium welding inside any vessel safety precautions must be taken. While argon is not poisonous, its job is to displace air with an inert atmosphere during welding, and it will not support human life. For people it's like water, and a person can drown in it. Unlike water, it is invisible and your feet don't get wet when you're standing in it. Workers welding or purging inside closed vessels must be carefully monitored and should carry alarm badges to guard against drop in oxygen content. Safety procedures in this regard must be detailed and well enforced.

Conclusion

Titanium clad solves corrosion, maintenance and environmental problems in many reactor and autoclave applications. Titanium clad construction permits autoclave designers a great deal of flexibility combined with significant cost reduction. Titanium alloys can be selectively applied in specific regions of the autoclave to maximize performance under anticipated localized environmental conditions at minimal cost. Explosion clad pressure vessel fabrication and performance have been demonstrated through three decades of experience. We will see increased application of titanium explosion clad in hydrometallurgy autoclaves.

REFERENCES
1) Donachie, M.J.Jr., Titanium, A Technical Guide, ASM International, Metals Park, OH, 1988.
2) McMaster, J.A. "The Use of Titanium in Pressure Vessel and Piping Construction" First ASTM Symposium on Titanium and Zirconium for Chemical Process Industries - 1975.
3) Shutz, R.W., & Covington, L.C., "Hydrometallurgical Applications of Titanium" Industrial Applications of titanium and Zirconium, ASTM STP830, R.T.Webster & C.S.Young. American Society for Testing and Materials, 1984, pp29-47.
4) "Seminar on Perspectives on Pressure Oxidation" Transcript of Proceedings March 24, 1992. Editors: J.M. Maycock, K.D. Lunde, Fluor Daniel Wright.
5) Pocalyko, A., "Explosively Clad Metals," Encyclopedia of Chemical Technology, Vol. 15, Third Edition, John Wiley & Sons, 1981, pp 275-296.
6) Patterson, R.A., "Fundamentals of Explosion Welding", ASM Handbook, Vol.6, Welding, Brazing, and Soldering, 1993, pp 160-164.
7) Banker, J.G., Reineke, E.G., "Explosion Welding", ASM Handbook, Vol. 6, Welding, Brazing, and Soldering, 1993, pp 303-305.
8) Linse, V.D., "Procedure Development and Process Considerations for Explosion Welding, ASM Handbook, Vol. 6, Welding, Brazing, and Soldering, 1993, pp896-900.
9) Banker, J.G., "Explosion Welded Transition Joints for Structural Welds between Titanium and Dissimilar Metals," Proceedings of the Technical Program form the 1994 International Conference, Titanium Development Association, pp62-69.
10) Shutz, R.W., "Recent Titanium Alloy and Product Developments for Corrosive Industrial Service," National Association of Corrosion Engineers, Corrosion 95, Paper No. 244.
11) McCue, D. & Irving, R.,"Gas Tungsten Arc Welding: It's Built to Handle Titanium" Welding Journal, November 1991
12) Titanium Welding Handbook and Video, International Titanium Association, Boulder, CO, 1994
13) Ogden, H,\.R. & Jaffee, H.R., "Effects of Carbon, Oxygen, and Nitrogen on the Mechanical Properties of Titanium and Titanium Alloys", BML Report #20, Batelle Memorial Laboratory
14) Fraser, Kevin, Kilborn Engineering, "Survey of Elevated Pressure Metallurgical Operations-Free World", Private Communication.

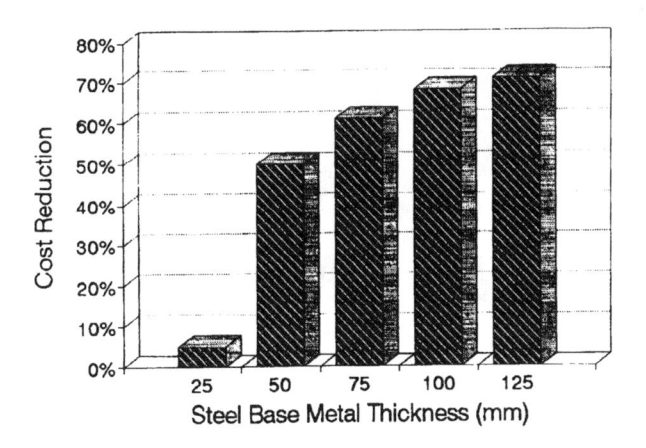

Figure 1: Cost comparison of explosion bonded titanium-steel clad, 3mm Titanium Grade 1 /ASME SA516 Grade 70, to equivalent thickness of Titanium Grade 2.

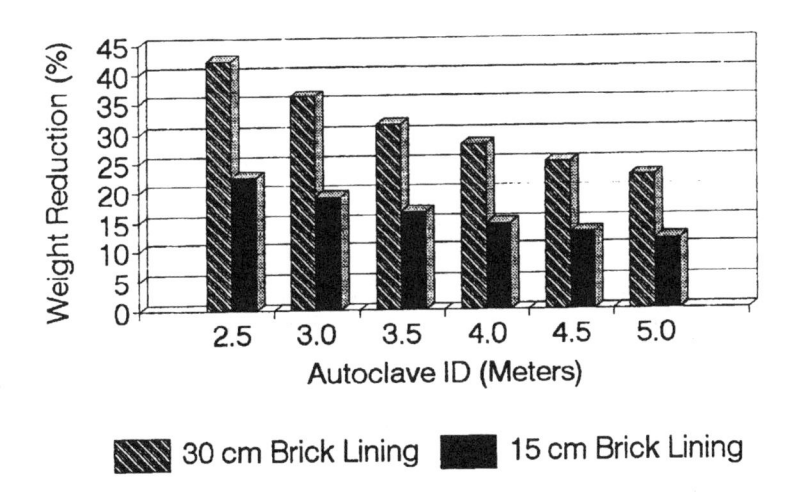

Figure 2: Approximate autoclave weight savings resulting from deletion of brick lining.

Figure 3: PTA process reactor, 4m diameter x 8m long, 7.5mm Titanium Grade 1 clad to 45mm thick steel, SA-516 Grade 70.

Figure 4: Gold processing autoclave with titanium explosion clad agitator mount covers and blind flanges.

Figure 5: Photomicrograph of titanium/steel explosion bonded clad bondzone, 50X

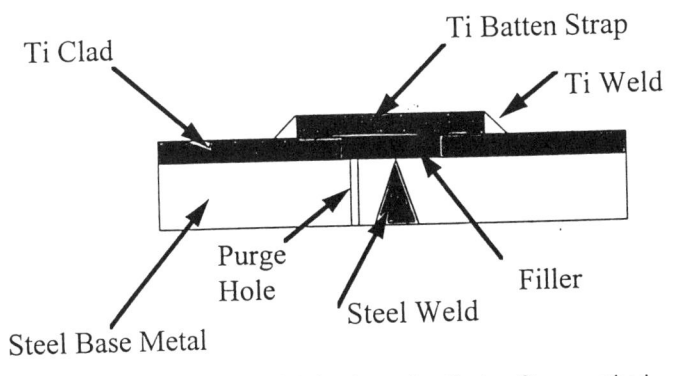

Figure 6: Schematic of titanium clad fabrication using Batten Strap method.

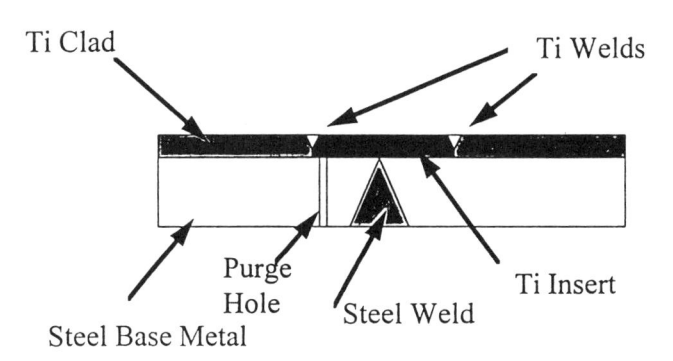

Figure 7: Schematic of titanium clad fabrication using Insert method.

Table I

TITANIUM AND TITANIUM-CLAD PRODUCTION AUTOCLAVES (Ref. 14)

Date Installed	Type	# Units	Design *	Size (OD x L) (m)	Temp (Deg.C)	Press. (Mpa)	Process (Note)
1970	Ti Clad	2	C/H/4	3.3 x 13.2	121	517	1
1970	Ti Clad	2	C/H/4	3.3 x 13.2	200	1900	2
1975	Ti Clad	1	C/H/3	1.8 x 5.4	140	1380	3
1975	Ti Clad	1	C/H/3	1.8 x 5.4	140	1380	3
1976	Ti	3	C/H/7	3.0 x 29.0	140	517	4
1976	Ti Clad	3	B/V/1	2.4 x 4.5	160	1034	3
1982	Ti	2	B/V/1	1.1 x 2.4	-	-	5
1984	Ti	1	C/H/4	0.75 x 3.6	-	-	6

* Design: C= Continuous   B=Batch   #=Number of Compartments
         H= Horizontal   V=Vertical

Process Notes:
1. Acid Zn Leach
2. Acid Fe Precipitation
3. Acid Ni-Cu-Co Leach
4. Acid Cl Leach of Cu
5. Acid Co-Cr-Ni Leach
6. Acid S/C Cu Leach

Table II

PRESSURE LEACHING PROJECTS WHICH HAVE USED TITANIUM and/or TITANIUM CLAD INTERNALS, AGITATORS, OR PIPING

Cominco	Trail, British Columbia, CANADA
Kidd Creek Mines	Timmins, Ontario, CANADA
Homestake-McLaughlin Mine	California, USA
Sao Bento Mineraco	Belo Horizonte, BRAZIL
First Miss Gold-Getchell Mine	Winnemucca, Nevada, USA
American Barrick-Goldstrike Mine	Elko, Nevada, USA
Placer Dome-Porgera Mine	Papua, NEW GUINEA
Placer Dome-Campbell Mine	Balmerton, Ontario, CANADA
Hudson's Bay Mining & Smelting	Flin Flon, Manitoba, CANADA
NERCO-Con Mine	Yellowknife, N.W.T., CANADA
Santa Fe Mining, Lone Tree	Valmy, Nevada, USA
Santa Fe Mining, Twin Creeks	Winnemucca, Nevada, USA
Niugini Gold	Lihir Island, Papua, NEW GUINEA

Table III
SELECTED TITANIUM ALLOYS AND PERFORMANCE FEATURES (Ref. 10)

Gr# (*)	Basic Alloy Components	Cladability to Steel(**)	Cost (***)	Features/Motivation for Alloy (****)
1	Ti (Chem. Pure)	Direct	1.1	Low Cost, Low Strength
2	Ti (less pure)	Interlayer	1.0	Low Cost, Moderate Strength
3	Ti (less pure)	Interlayer	1.0	Low Cost, Higher Strength
5	Ti+6AL+4V	Interlayer	1.2	High Strength and Erosion Resistance
7	Ti Gr2+0.15Pd	Interlayer	1.9	Crevice Corrosion Resistance
9	Ti-3Al-2.5V	Interlayer	1.3	High Strength and Erosion Resistance
11	Ti Gr1+ 0.15Pd	Direct	1.9	Crevice Corrosion Resistance
12	Ti+.3Mo+.8Ni	Interlayer	1.2	High Strength and Erosion Resistance
16	Ti Gr2 + .05Pd	Interlayer	1.4	Crevice Cor. Resist, Lower Cost
17	Ti Gr1 + .05Pd	Direct	1.4	Crevice Cor. Resist, Lower Cost
18	Ti Gr 9 + .05Pd	Interlayer	1.6	Crevice Corrosion Resistance
24	Ti Gr5 + .05Pd	Interlayer	1.6	Crevice. Corrosion Resist
26	Ti Gr1 + .1Ru	Direct	1.2	Crevice Cor. Resist, Lower Cost
NA	Ti-45Nb	Direct	4.0	Excellent Ignition Resistance

Legend:
* ASTM B265 Grade Designation
** Readily clad direct to steel, or interlayer recommended
*** Cost Ratio to Lowest Cost Alloy, Current Metal Prices at time of TMS Presentation
**** When Comp. shows "Gr.# + addition", alloy also exhibits features of the base Gr.

LASER WELDING OF Ti-6.8Mo-4.5Fe-1.5Al (TIMETAL® LCB) ALLOY

I. Weiss*, N. Stefansson*, W.A. Baeslack III**
J. Hurley***, P.G. Allen****, and P.J. Bania****

*Mechanical and Materials Engineering Department
Wright State University, Dayton, Ohio 45435

**College of Engineering
The Ohio State University, Columbus, Ohio 43210

***Edison Welding Institute
Columbus, Ohio 43201

****Henderson Technical Laboratory
Titanium Metals Corporation, Henderson, Nevada 89009

Abstract

Laser welding was utilized to obtain high-integrity welds in 1.5 mm thick LCB sheet. Welds were characterized by narrow fusion and heat-affected zones and a relatively fine beta grain size of about 100 μm. Postweld heat treatment at 593°C (1100°F) for 4 hours resulted in fine alpha precipitation in the base and weld metals thus producing material with high strength above 1275 MPa (187 KSI) and a low ductility of about 2%. Heat treatment at 649°C (1200°F) for 8 hours resulted in coarse alpha precipitation both along beta grain boundaries and in beta grain interiors. This material exhibited a lower tensile strength of about 1080 MPa (160 KSI) with higher ductility about 4%. An intermediate level of strength of about 1150 MPa (170 KSI) and significantly improved ductility exceeding 8% was achieved using a duplex heat treatment of 482°C (900°F) for 2 hours + 649°C (1200°F) for 3 hours. Fractography of the welded samples showed the fracture appearance of the fusion zone is mainly transgranular in nature for the as-welded material and intergranular in nature for the aged material. However, a mixture of dimpled type fracture and intergranular fracture mode was observed at the heat-affected zone and base metal for the aged samples.

Introduction

New lightweight materials for automotive applications are in great demand. In the past the utilization of titanium alloys was disregarded as a result of the high cost of the material. With development of newer and cheaper titanium alloys, titanium is now a viable choice for automotive components. Recently, a new low-cost beta titanium alloy has been developed. Known as TIMETAL® LCB (Ti-6.8Mo-4.5Fe-1.5Al in weight percent), the alloy uses the beta stabilizing element molybdenum, added in the form of a ferro-moly compound. In the solution-heat-treated and quenched condition, the alloy has a beta phase (BCC) structure at room temperature. It thus possesses excellent workability compared to alloys containing alpha phase. Aging the alloy at temperatures 155 to 185°C (280 to 330°F) below the transus temperature causes the alpha phase to precipitate. In this condition, the alloy has a yield strength greater than 1080 MPa (160 KSI) and a tensile elongation about 10% (1,2), which compares favorably with high strength steels.

As a structural alloy, processing LCB by fusion welding is an important factor in establishing potential use of the material. Metastable beta titanium alloys are weldable (3,4), however, the presence of large columnar beta grains in the fusion zone of conventional gas tungsten-arc welds reduces the fusion zone ductility relative to the base metal, especially in the aged condition. Laser welding produces narrow welds with a refined microstructure promoting weld zone ductility. The objective of this work was to evaluate the laser welding characteristics of TIMETAL® LCB sheet and determine the influence of postweld aging on the weld microstructure, mechanical properties and fracture characteristics.

Materials and Experimental Procedure

TIMETAL® LCB with nominal composition of Ti-6.8Mo-4.5Fe-1.5Al, in the form of 1.5 mm thick sheet, solution treated at 830°C (1525°F) for 15 min, AC was used ($T_\beta \approx 805°C$ (1480°F). The sheet was cut into 25 mm x 100 mm strips, with the sample length oriented either parallel (L samples) or perpendicular (T samples) to the sheet rolling direction. Prior to welding the samples were degreased in acetone. Full-penetration, bead-on-plate laser welds were produced in the strips using a GE 3000 CO_2 laser. A laser power of 3000 watts and beam traversing speed of 55 mm/sec was utilized (Fig.1a). Slower speed of 42 mm/sec (Fig.1b) and faster speeds of 74 mm/sec and 89 mm/sec (Figs 1c and 1d) were also tried, however, wide fusion zone, and lack of fusion material (undercut) were observed, respectively, as shown in the top (T) and bottom (B) of the laser welds of Figure 1. Laser welding was carried out in a helium-purged collapsible chamber in order to prevent contamination. Following welding, selected strips were heat treated. Three postweld heat treatments were selected: 593°C (1100°F) for 4 hours, 649°C (1200°F) for 8 hours, and a duplex heat treatment of 482° (900°F) for 2 hours+649°C (1200°F) for 3 hours. Characterization of the as-welded and postweld heat treated welds included the following: optical microscopy

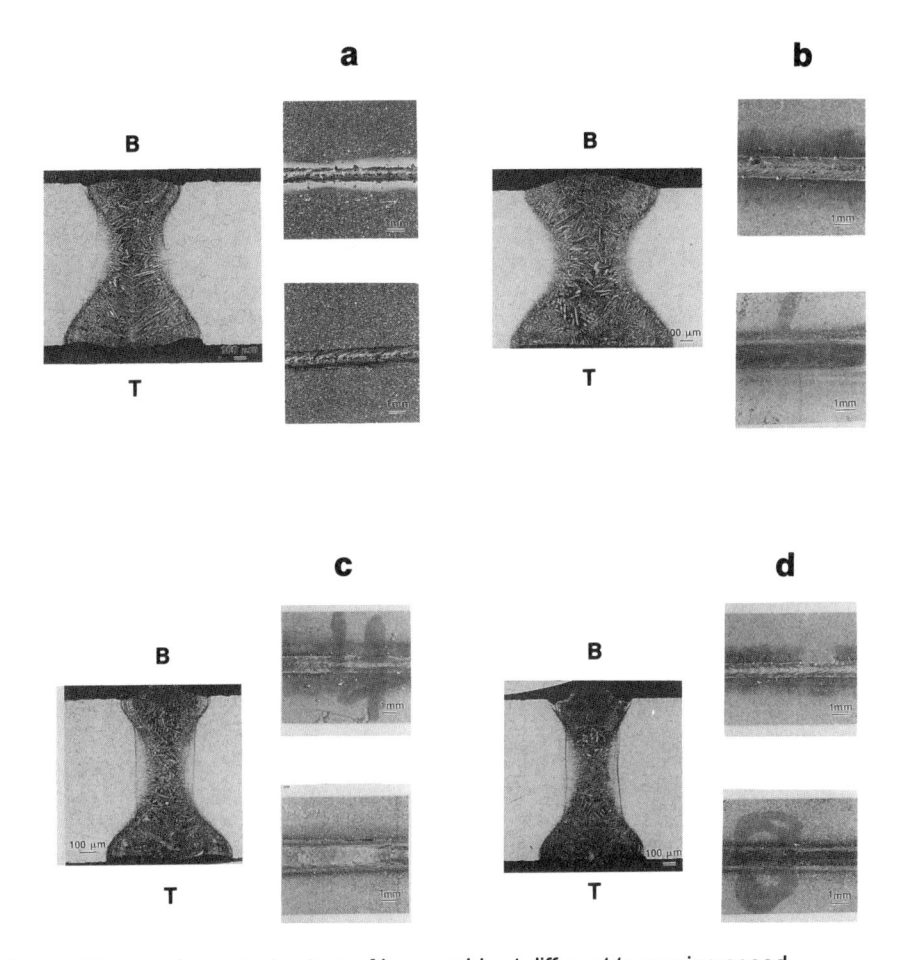

Fig. 1: Micro and macro structure of laser welds at different traversing speed
(a) 55 mm/sec (b) 42 mm/sec (c) 74 mm/sec and (d) 89 mm/sec.

analysis, microhardness testing (DPH with 1000 gr load), tensile testing of longitudinal and transverse-weld oriented specimens at room temperature and at a constant speed of 0.05 in/min and fractographic examination utilizing Scanning-Electron Microscopy (SEM).

Results and Discussion

Microstructure of as-welded and welded and aged material

Figure 2 shows the microstructure of a laser weld in LCB sheet. This material was solution heat treated at 830°C (1525°F) for 15 min, followed by air cooling. The base metal (BM) microstructure shows equiaxed beta grains of about 100μm in size (Fig. 2a). The weld fusion zone (FZ) is characterized by a "crown" which undergoes 3-D heat flow on the top surface (T), a narrow zone at the center of the sheet which experienced 2-D heat flow, and a smaller crown on the bottom surface (B) as shown in Figure 2b. In addition, a narrow heat affected zone (HAZ), and epitaxial nucleation and growth of columnar beta grains in the fusion zone directly from beta grains in the HAZ was observed (Fig. 2b). Figures 3,4 and 5 show the microstructure of laser welds in LCB sheet aged at 593°C (1100°F) for 4 hours (Fig.3), 649°C (1200°F) for 8 hours (Fig. 4) and 482°C (900°F) for 2 hours + 649°C (1200°F) for 3 hours (Fig. 5). A narrow HAZ with equiaxed grains of about 100 μm in size, columnar grains in the FZ and equiaxed grains in the BM were also observed in the aged material as shown in figures 3b, 4b, and 5b. The different grain contrast observed in the FZ, HAZ, and BM of the aged material was associated with the precipitation of the alpha phase and the subsequent etching of the material.

SEM observations of aged and etched material, revealed the morphology and size of the alpha phase. Acicular alpha phase (about 0.08 μm thick) aligned along three crystallographic orientations was observed in the FZ and HAZ of material aged at 593°C (1100°F) for 4 hours. In addition, coarse, continuous alpha phase was present at the beta grain boundaries as shown in figures 6a and 6b. Coarser alpha phase (about 0.2 μm thick) was observed in the BM of the same material (Fig. 6c). Aging at 649°C (1200°F) for 8 hours resulted in the precipitation of coarse alpha phase (about 0.5 μm thick) along beta grain boundaries and in beta grain interiors of the FZ, HAZ, and BM as shown in figure 7. Duplex heat treatment of 482°C (900°F), 2 hours + 649°C (1200°F), 3 hours, resulted in a fine alpha phase in the BM (Fig. 8c) and coarser alpha phase in the HAZ (Fig. 8b) and FZ (Fig. 8a). This variation in the alpha phase platelet size was opposite to that observed in material aged at 593°C (1100°F) for 4 hours, and can be rationalized in terms of alpha phase coarsening during the second part of the duplex heat treatment (aging at 649°C (1200°F) for 3 hours), especially for fine alpha platelets nucleating near the solidification subgrains typically found in the weld fusion zone (5,6).

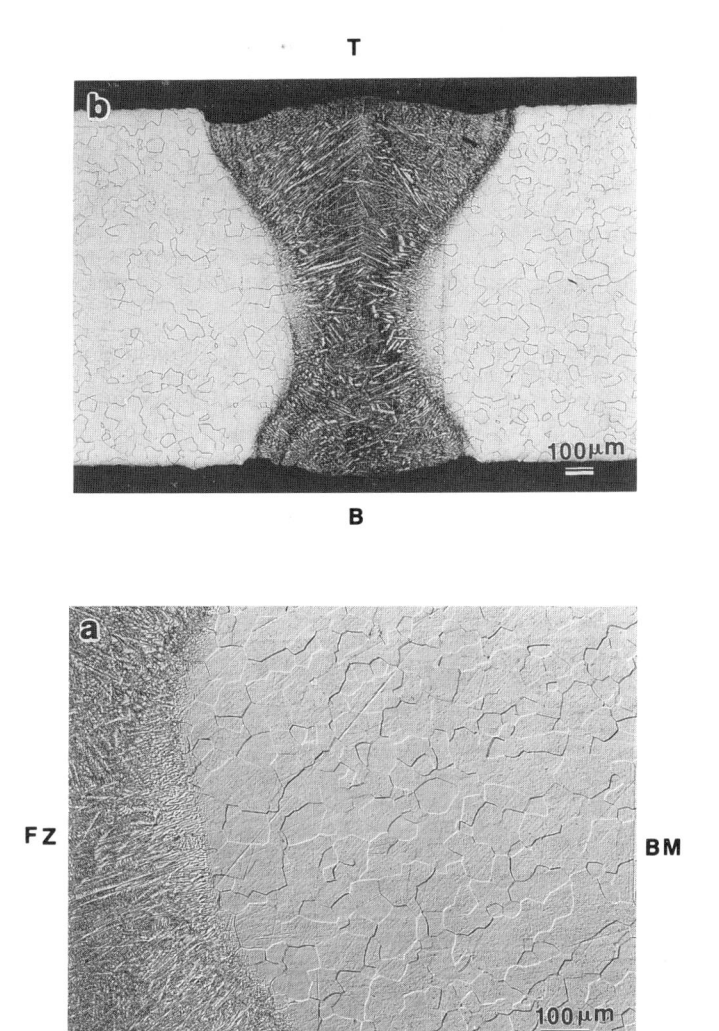

Fig. 2: Microstructure of laser weld in solution treated (829°C (1525°F), 15 min, AC) LCB sheet (a) microstructure of the fusion zone, the heat-affected zone and the base metal (b) low magnification.

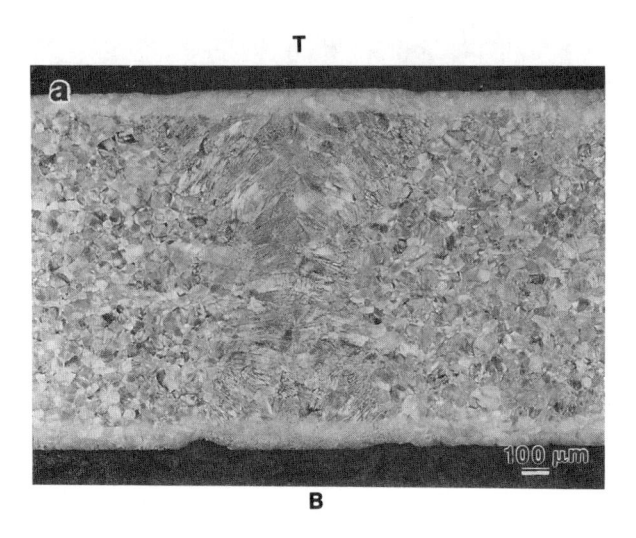

Fig. 3: Microstructure of laser weld heat treated at (593°C (1100°F), 4 hours, AC)
(a) low magnification (x50)   (b) higher magnification (x100).

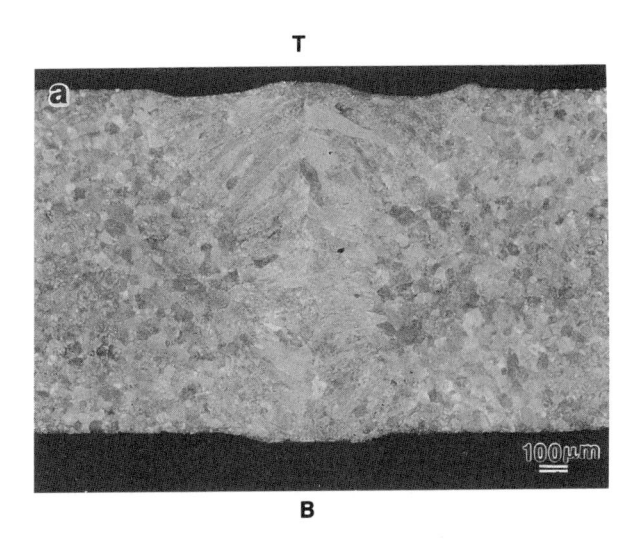

Fig. 4: Microstructure of laser weld heat treated at (649°C (1200°F), 8 hours, AC) (a) low magnification (x50) (b) higher magnification (x100).

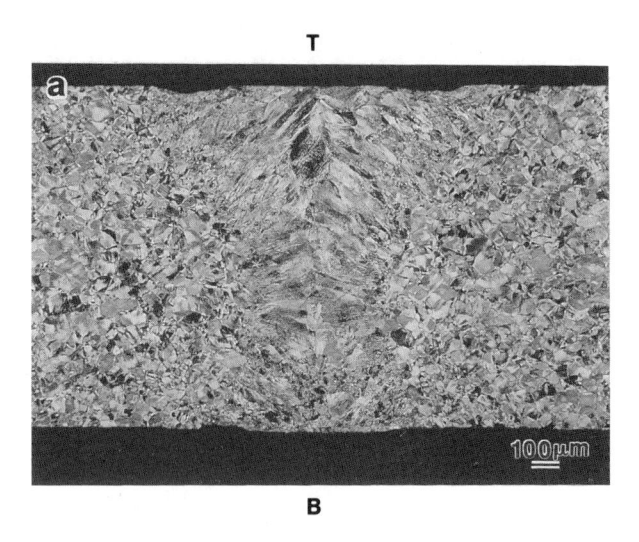

Fig. 5: Microstructure of laser weld duplex heat treated at (482°C (900°F), 2 hours + 649°C (1200°F), 3 hours, AC) (a) low magnification (x50) (b) higher magnification (x100).

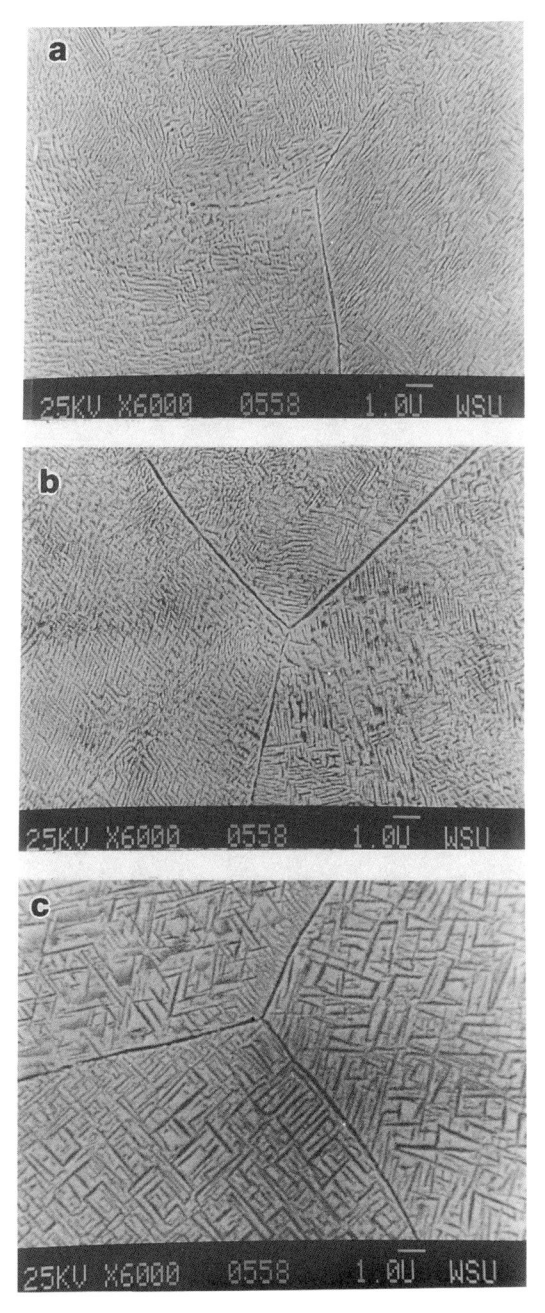

Fig. 6: SEM micrographs of alpha phase in laser weld heat treated at (593°C (1100°F), 4 hours, AC) (a) fusion zone   (b) heat-affeacted zone   (c) base metal.

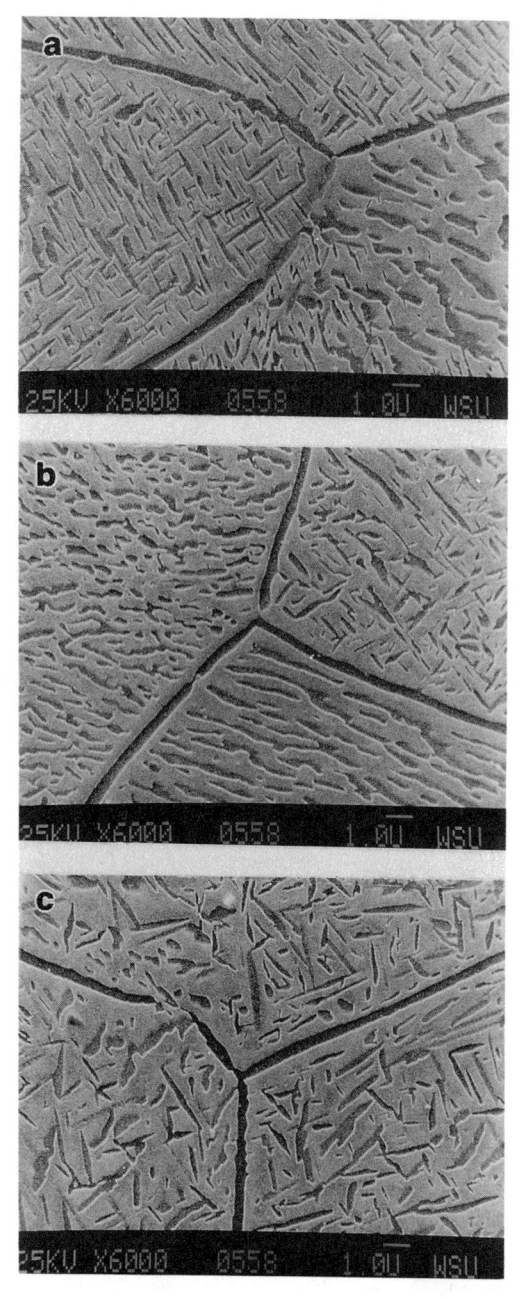

Fig. 7: SEM micrographs of alpha phase in laser weld heat treated at (649°C (1200°F), 8 hours, AC) (a) fusion zone   (b) heat-affected zone   (c) base metal.

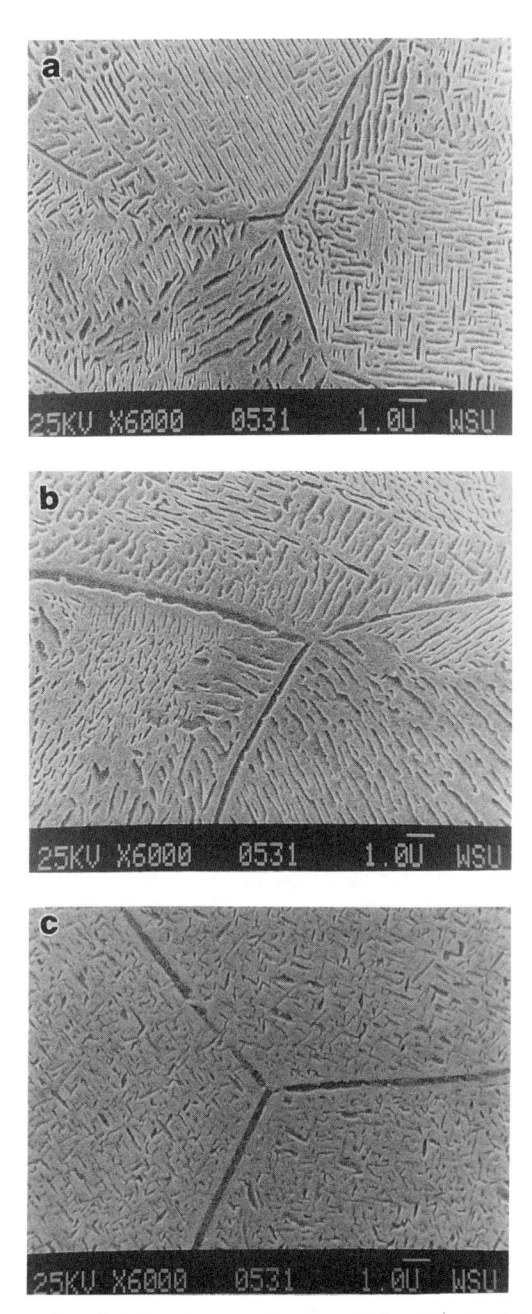

Fig. 8: SEM micrographs of alpha phase in laser weld heat treated at (482°C (900°F), 2 hours + 649°C (1200°F), 8 hours, AC) (a) fusion zone (b) heat-affected zone (c) base metal.

Mechanical Properties

Hardness traverses from the center of the laser weld to the base metal for the as-welded, and welded and aged samples are shown in figure 9. Hardness values were higher in the FZ and HAZ than in the BM for the as-welded samples and welded material aged at 593°C (1100°F) for 4 hours. This was consistent with the microstructural characteristics shown before where finer intragranular alpha phase was observed in the FZ and HAZ then in the BM. Coarse alpha phase was observed in the FZ, HAZ, and BM of material aged at 649°C (1200°F) for 8 hours and resulted in relatively constant hardness values across the weld zones. Lower hardness values in the FZ and HAZ, and higher hardness in the BM were observed following the duplex heat treatment of 482°C (900°F), 2 hours + 649°C (1200°F), 3 hours as shown in figure 9. This hardness trend was again consistent with the microstructural observations of finer alpha phase in the BM and coarser alpha phase in the FZ and HAZ.

Figure 10 shows the engineering stress-strain curves of longitudinal and transverse-weld oriented samples taken from longitudinal (Figs. 10a and 10c) and transverse (Fig. 10b) directions of the LCB sheet, and tested in tension at a constant speed of 0.05 in/min. The solution heat-treated specimens show a slight flow softening following yielding, whereas the aged samples deform at constant load after yielding as shown in figure 10.

Table 1 summarizes the results of the tensile tests for the as-welded and postweld aged specimens. Solution treated (895°C (1525°F), 15 min, AC) longitudinal and transverse-weld oriented samples (L and T samples) show tensile ductility in the range of 15 to 17% and yield stresses in the range of 1020 and 1050 MPa. Aged longitudinal-weld oriented samples (649°C (1200°F), 8 hours, AC, and 593°C (1100°F), 4 hours, AC) taken from longitudinal directions of the sheet (L samples) show higher yield stresses of 1060 MPa and 1250 MPa and lower ductilities of 4.2 and 2.4% respectively. Similar results were observed for longitudinal-weld oriented samples taken from the transverse direction of the sheet (T samples). The duplex heat treatment 482°C (900°F), 2 hours + 649°C (1200°F), 3 hours of longitudinal and transverse oriented-weld samples (L and T samples) resulted in intermediate yield stresses of 1155 Mpa, 1140 MPa and 1175 MPa and corresponding ductilities of 8, 7.6 and 4%. Some transverse-weld oriented samples taken from longitudinal direction of the sheet (L samples) show low ductility in the aged condition (649°C (1200°F), 4 hours) AC) the result of weld imperfection (undercut) present along the weld-base metal interface.

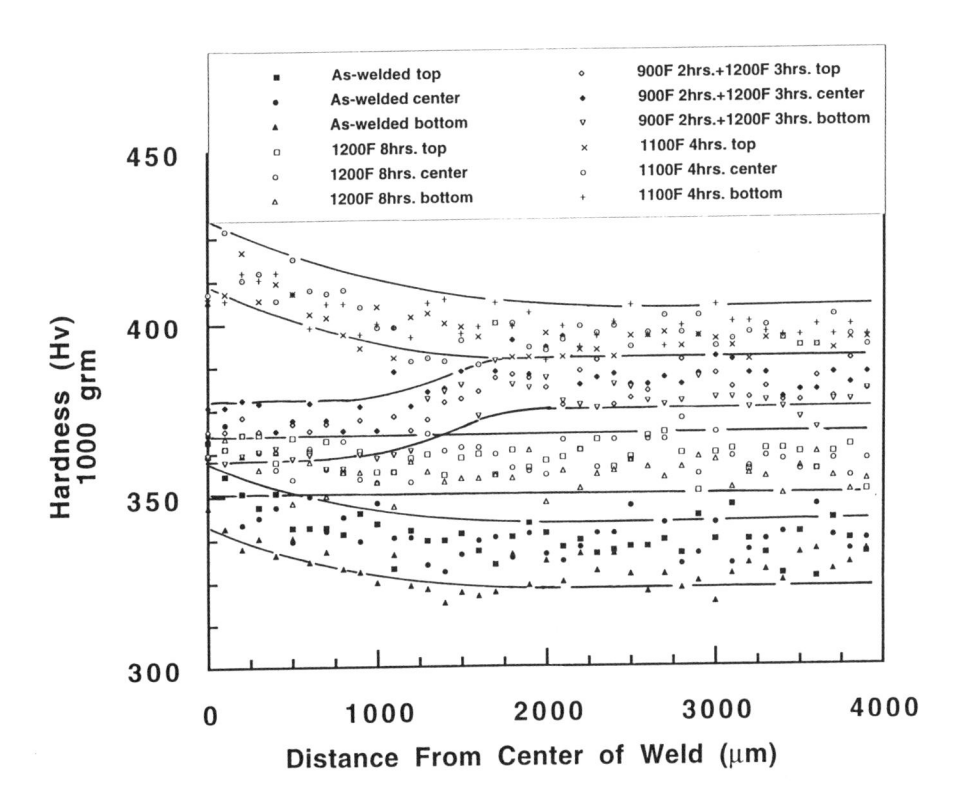

Fig. 9: Hardness traverses from the weld centerline to the base metal for the as welded, and postweld heat treated LCB sheet.

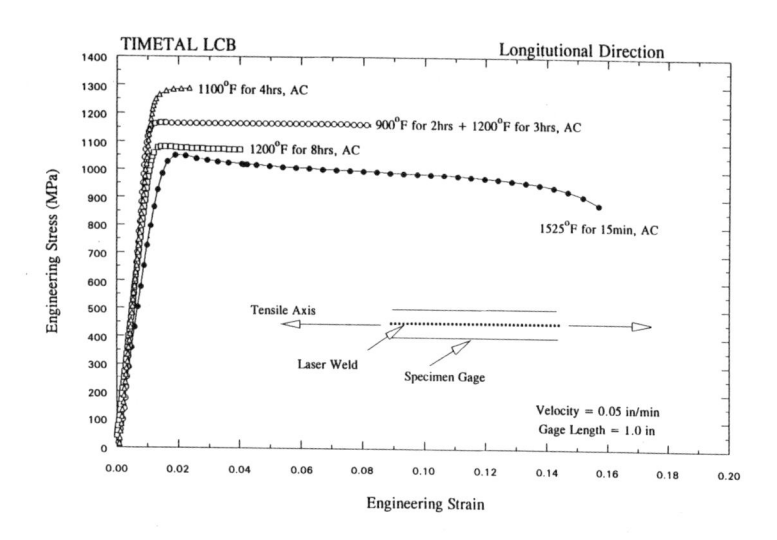

Fig. 10a: Engineering stress-strain curves of longitudinal-weld oriented sample taken from longitudinal direction of the sheet.

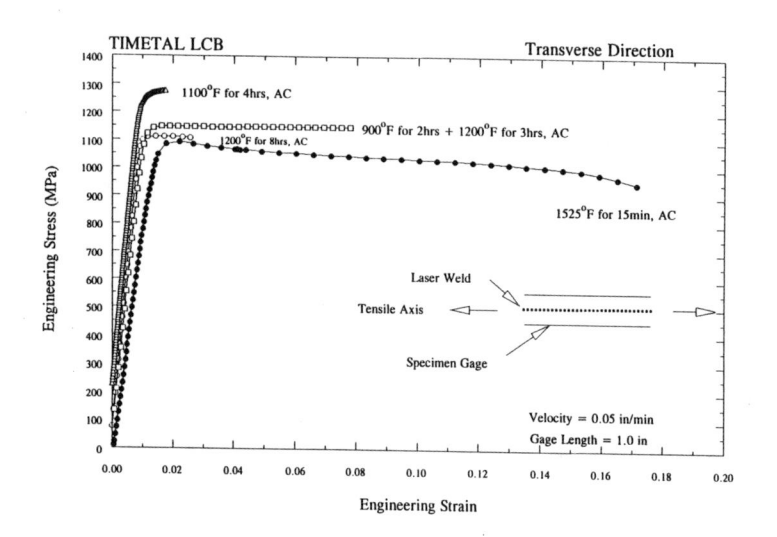

Fig. 10b: Engineering stress-strain curves of longitudinal-weld oriented sample taken from transverse direction of the sheet.

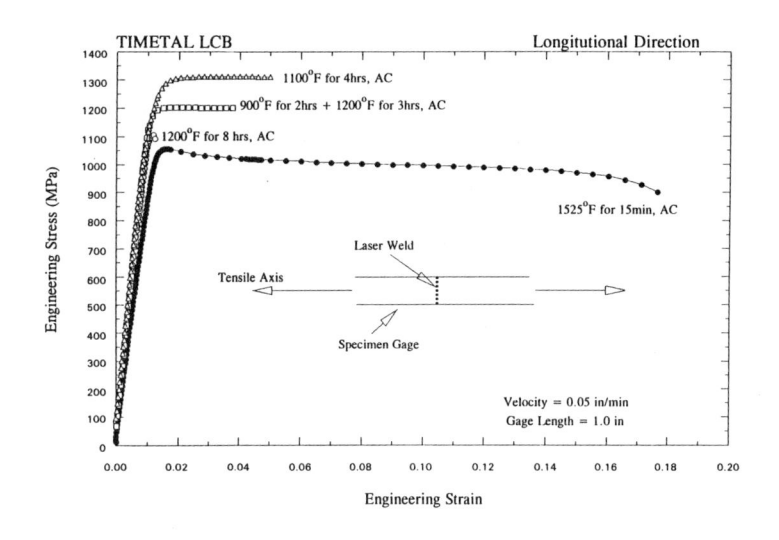

Fig. 10c: Engineering stress-strain curves of transverse-weld oriented sample taken from longitudinal direction of the sheet.

Condition: Longitudinal Direction, Longitudinal Weld	Yield Stress MPa/Ksi	UTS MPa/Ksi	e_f %
Solution treated at 1525°F, 15min, AC	1020/150	1060/156	15.7
Aged: 1100°F, 4hrs., AC	1250/184	1288/189	2.45
Aged: 1200°F, 8hrs, AC	1060/156	1079/160	4.20
Aged: 900°F, 2hrs.+1200°F, 3hrs., AC	1155/171	1174/173	8.22
Condition: Transverse Direction, Longitudinal Weld	Yield Stress MPa/Ksi	UTS MPa/Ksi	e_f %
Solution treated at 1525°F, 15min, AC	1050/154	1091/162	17.2
Aged: 1100°F, 4hrs., AC	1250/184	1276/189	1.82
Aged: 1200°F, 8hrs, AC	1085/161	1110/163	2.80
Aged: 900°F, 2hrs.+1200°F, 3hrs., AC	1140/167	1150/170	7.60
Condition: Longitudinal Direction, Transverse Weld	Yield Stress MPa/Ksi	UTS MPa/Ksi	e_f %
Solution treated at 1525°F, 15min, AC	1025/151	1055/155	15.5
Aged: 1100°F, 4hrs., AC	1265/186	1312/193	5.10
Aged: 1200°F, 8hrs., AC	1085/161	1105/162	1.40
Aged: 900°F, 2hrs.+1200°F, 3hrs., AC	1175/174	1202/177	3.90

Table 1: Mechanical properties of LCB laser welds.

Fracture Analysis

Figure 11 shows fracture surfaces of a longitudinal-weld oriented tensile specimen (L samples) in the as-welded condition. The fracture surfaces of the FZ (Figs. 11a and 11b) and the BM (Figs. 11c and 11d) are characterized by a ductile, transgranular appearance, formed by a void nucleation, growth and coalescence mechanism. Fracture surfaces of longitudinal-weld oriented tensile samples (L samples), heat treated at 593°C (1100°F) for 4 hours (Fig. 12), show a predominantly intergranular fracture with small areas showing cleavage and dimple type fracture appearance, especially in the FZ as shown in figures 12a and 12b. The fracture appearance of the HAZ (Figs. 12c and 13d) and the BM (Figs. 12e and 12f) indicate intergranular fracture with a large amount of dimpled surfaces. Aging at 649°C (1200°F) for 8 hours resulted in fracture surfaces shown in figure 13. Intergranular fracture with cracking along beta grain boundaries where coarse alpha phase is present (Fig. 7) was observed in the FZ as shown in figures 13a and 13b. The HAZ and the BM fracture surfaces show primarily ductile failure (Figs. 13d and 13e). The fracture morphology of longitudinal-weld oriented tensile samples (L samples) heat treated at 482°C (900°F), 2 hours + 649°C (1200°F), 3 hours (Fig. 14) was similar to the fracture characteristics of laser welds heat treated at 593°C (1100°F) for 4 hours, where intergranular fracture was observed in the FZ and substantial amount of dimpled product is present on the fracture surfaces of the HAZ and BM (Figs. 14d and 14f, respectively). Under this aging condition (duplex aging), the laser welds exhibited the highest ductility (Table 1). Fracture surfaces of transverse-weld oriented tensile samples (L samples) heat treated at 593°C (1100°F), 4 hours, (Figs. 15a,b), 649°C (1200°F), 8 hours (Figs. 15c, 15d) and 482°C (900°F), 2 hours + 649°C (1200°F), 3 hours (Figs. 15e and 15f) occurred intergranularly with large amount of cracking along beta grain boundaries with few areas containing dimpled type fracture as shown in figures 15b, 15d, and 15f. These tensile fractures initiated predominately on the BM side of the weld-base metal interface. Similar fracture morphologies has been observed in other beta titanium alloys (3,4,5).

Structure/Property/Fracture Relationships

Postweld heat treatment response of the TIMETAL® LCB alloy closely parallels that observed in previous heat treatment studies on metastable-beta titanium alloys (3,5), with an increase in heat treatment temperature promoting an increase in intragranular and grain boundary alpha coarseness, and correspondingly a decrease in hardness and strength across the weld zone. The generally finer alpha microstructure observed in the weld fusion zone versus the base metal, and correspondingly higher hardness, for the single-cycle heat treatments is also consistent with earlier studies, and results from the influence of a fine fusion zone solidification substructure in the fusion zone that may promote alpha nucleation.

The nucleation of fine alpha platelets during the first step in the duplex heat treatment, followed by microstructural coarsening during the second step of the duplex heat treatment, resulted in a grain boundary and intragranular alpha structure intermediate in coarseness relative to the two single-cycle heat treated specimens. Interestingly, this heat treatment was unique in promoting a fine alpha phase morphology in the base metal relative to the weld zone, which must result from differences in the response of

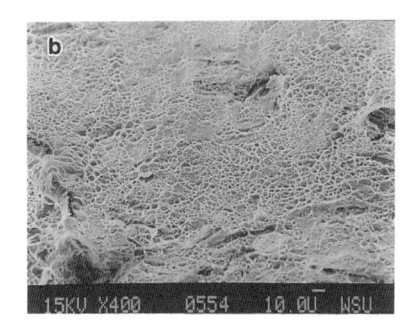

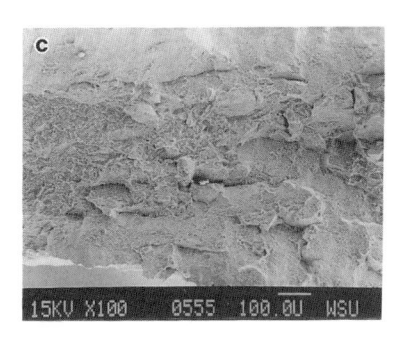

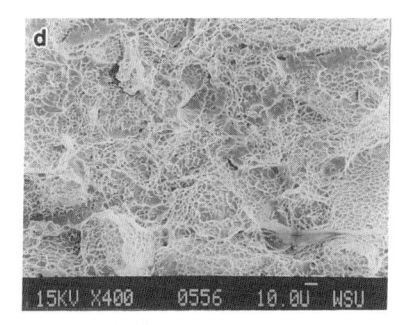

Fig. 11: SEM fractographs of longitudinal-weld oriented tensile specimen (L sample) in the as-welded condition) (a) and (b) fusion zone (c) and (d) base metal.

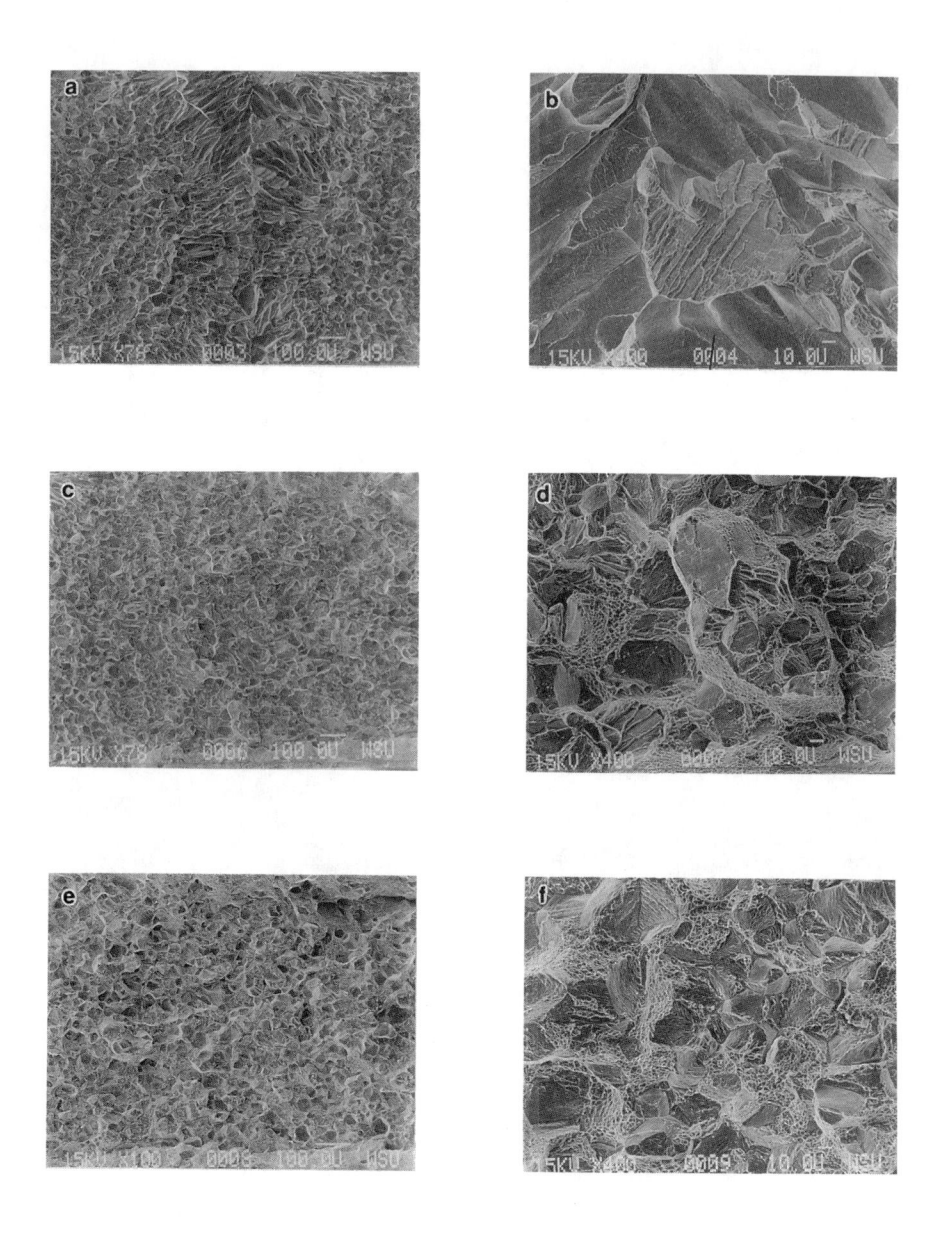

Fig. 12: SEM fractographs of longitudinal-weld oriented tensile specimen (L sample), heat treated at 593°C (1100°F) for 4 hours   (a) and (b) fusion zone   (c) and (d) heat-affected zone and   (e) and (f) base metal.

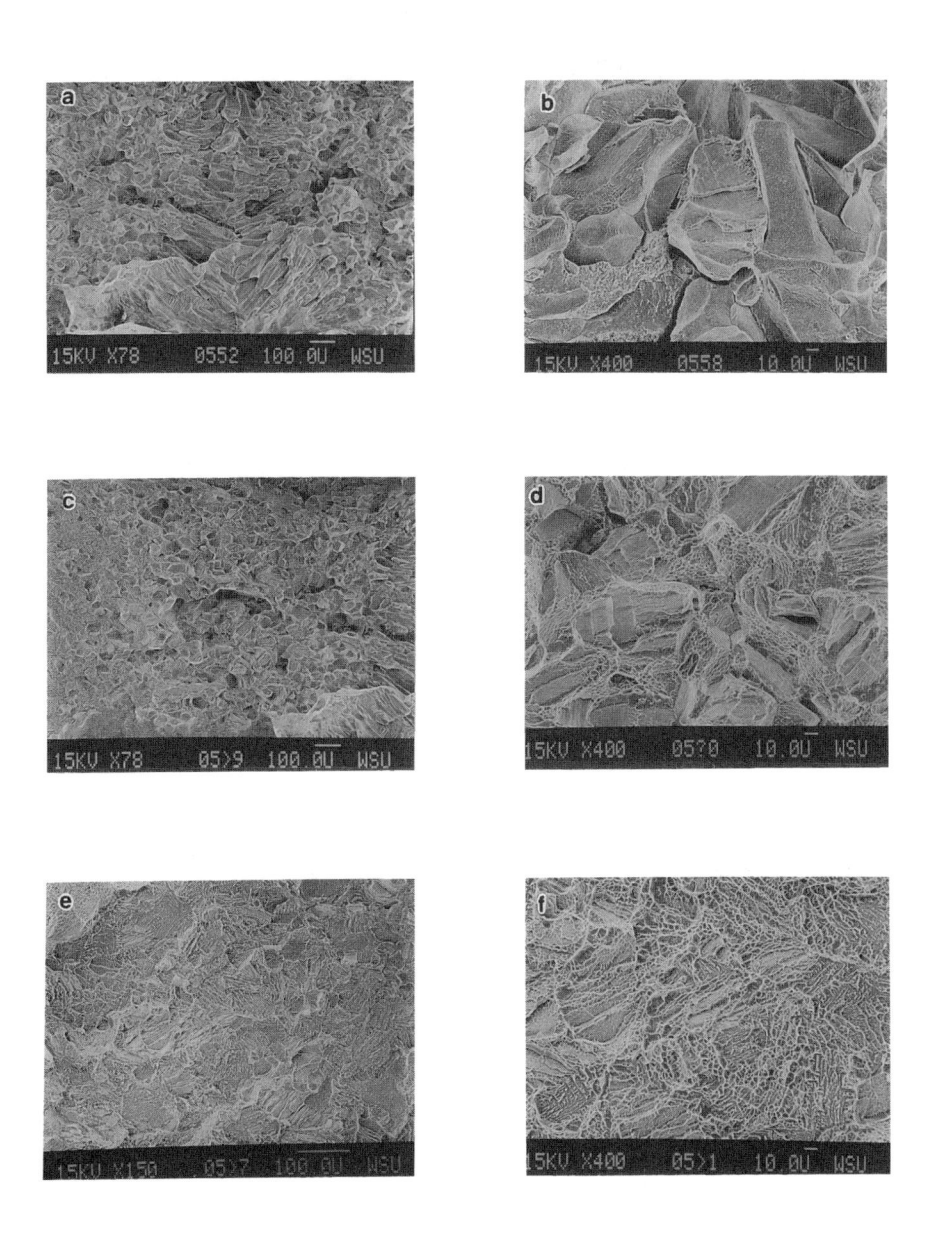

Fig. 13: SEM fractographs of longitudinal-weld oriented tensile specimen (L sample) heat treated at 649°C (1200°F) for 8 hours   (a) and (b) fusion zone   (c) and (d) heat-affected zone and   (e) and (f) base metal.

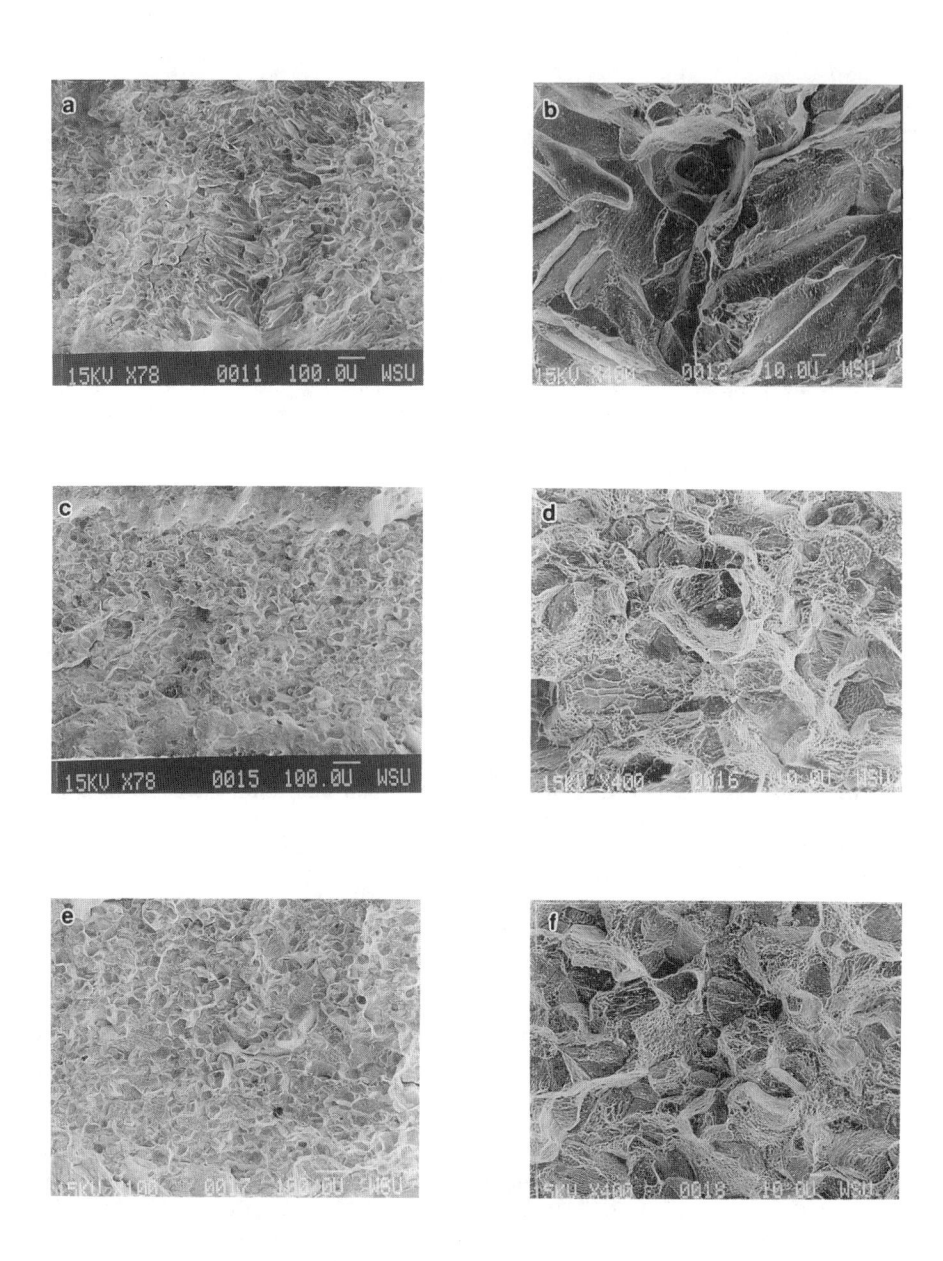

Fig. 14: SEM fractographs of longitudinal-weld oriented tensile specimen (L sample) heat treated at 482°C (900°F) for 2 hours + 649°C (1200°F) for 8 hours (a) and (b) fusion zone (c) and (d) heat-affected zone and (e) and (f) base metal.

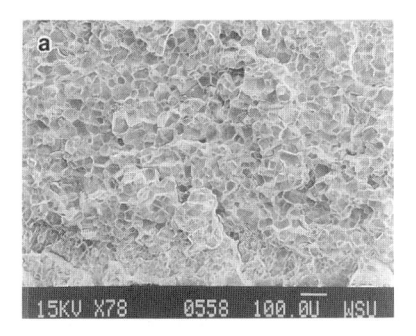

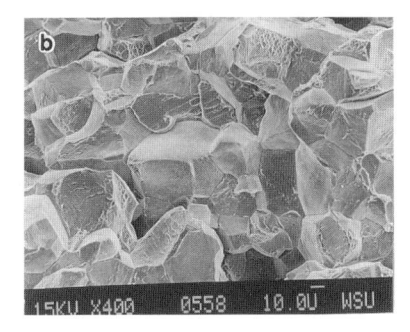

Fig. 15: SEM fractographs of transverse-weld oriented tensile specimen (L sample) heat treated at 593°C (1100°F) for 4 hours (a) low magnification (x78) (b) higher magnification (x400).

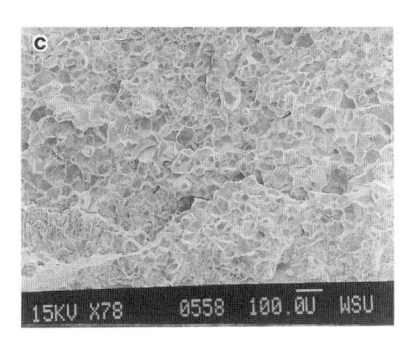

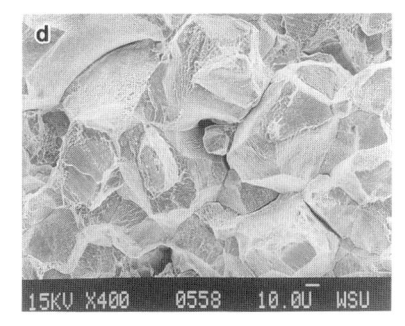

Fig.15(cont.) SEM fractographs of transverse-weld oriented tensile specimen (L sample) heat treated of 649°C (1200°F) for 8 hours (c) low magnification (x78) (d) higher magnification (x400).

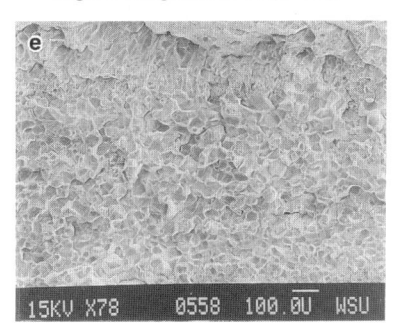

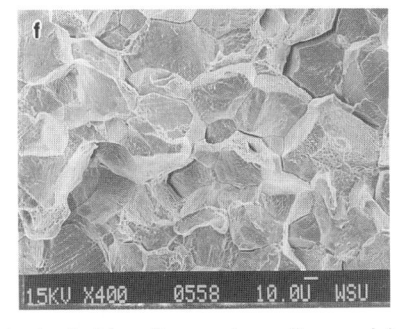

Fig.15(cont.) SEM fractographs of transverse-weld oriented tensile specimen (L sample) heat treated at 482°C (900°F) for 2 hours + 649°C (1200°F) for 8 hours (e) low magnification (x78) (f) higher magnification (x400).

these regions to nucleation at low temperatures. As shown in Table 1, the ductility of longitudinal-weld oriented specimens in the duplex heat treatment condition showed ductility levels remarkably higher than for the single-cycle heat treated specimens. A comparison of the fusion zone fracture surfaces for these different conditions showed predominantly intergranular fracture for all heat treatments, suggesting that the increase in ductility may be associated with the nature of crack initiation and propagation along prior-beta grain boundaries, versus a change in macroscopic crack path (e.g., from intergranular to transgranular). It has been well established that the ductility of transformed-beta microstructures observed in the weld fusion and near-HAZ's is a strong function of the thickness and continuity of the grain boundary alpha, and the strength and deformation behavior of the intragranular microstructure (as influenced by alpha coarseness and the beta or fine, transformed-beta matrix). Apparently, the duplex heat treatment promoted the evolution of a microstructure that optimized this combination and correspondingly ductility, with a lower intragranular strength versus the 593°C (1100°F), 4 hrs heat treatment, but a thinner, and less continuous grain boundary alpha phase versus the 649°C (1200°F), 8 hrs heat treatment.

Although not observed in the present study, it is important to note that the utilization of a duplex heat treatment can be particularly beneficial in the heat treatment of base metal and weldments in plate metastable-beta titanium alloys. In plate thicknesses, residual work in the solution heat-treated microstructure may be sufficiently low to preclude uniform alpha-phase precipitation. Nonuniform precipitation has also been observed in the weld HAZ in plates, where further solutionizing and grain growth in this region can lead to a very poor single-cycle aging response. The use of a duplex postweld heat treatment has been shown to be very successful in overcoming these difficulties, and would be expected to be effective in TIMETAL® LCB plate.

Conclusions

1. High-integrity welds with relatively fine beta grain structure were produced in LCB sheet using CO_2 laser welding process. The presence of beta phase in the fusion and heat affected zones during weld cooling promoted high ductility in the as-welded condition.

2. For longitudinal-weld oriented material; postweld aging at 593°C (1100°F) for 4 hours resulted in high base metal and weld tensile strength (1275 MPa (187 KSI) but with low ductility in the range of 1.8 to 2.4% in the weld zone. Coarser alpha phase produced during aging at 649°C (1200°F) for 8 hours resulted in a moderate weld tensile strength (1080 MPa (160 KSI) and a weld ductility in the range of 2.8 to 4.2%. The duplex postweld treatment of 482°C (900°F), 2 hours + 649°C (1200°F), 3 hours resulted in intermediate strength (1150 MPa (170 KSI) and higher ductility in the range of 7.6 to 8.2%.

3. LCB sheet in the as-welded condition failed in a ductile transgranular fracture mode. In the postweld heat treated condition, the material failed predominantly by intergranular fracture mode especially in the fusion zone. Intergranular fracture with dimpled fracture appearance on some of the fracture surfaces was observed in the heat-affected zone and base metal fractures.

References

1. P.G. Allen and A.J. Hutt, Proceedings of the 1994 International Titanium Conference, Titanium Development Association, Boulder, CO, p. 397, 1994.

2. "Low Cost Beta Titanium," Journal announcement, Mechanical Engineering, p. 64, July 1993.

3. D.W. Becker and W.A. Baeslack III, Welding Journal, Research Supplement, 59, (85S), 1980.

4. M.A. Greenfield and C.M. Pierce, Welding Journal, Research Supplement, 52, (524S), 1973.

5. P.S. Liu, K.H. Hou, W.A. Baeslack III and J. Hurley, Titanium '92 Science and Technology, Edited by F.H. Froes and I. Caplan, TMS, Warrendale, PA, Vol. 2, p. 1477, 1993.

6. P. G. Allen and P.J. Bania, Advances in the Science and Technology of Titanium Alloy Processing, edited by I. Weiss, R. Srinivasan, P.J. Bania, D. Eylon and L.S. Semiatin, TMS, Warrendale, PA., 1997.

Coating of Titanium with Chromium to Enable Porcelain-Titanium Bonding for Dental Restorations

R. R. Wang, and G. E. Welsch
School of Dentistry, and Department of Materials Science and Engineering,
Case Western Reserve University
Cleveland, OH 44106-4905

O. R. Monteiro, and I. G. Brown
Lawrence Berkeley National Laboratory
Berkeley, CA 94720

ABSTRACT

Failures that occur in titanium-ceramic restorations are of concern to clinicians. The application of porcelain to titanium for restorations requires heating to temperatures of $850^\circ C$ and higher. Porous, nonprotective oxide formation and an adherence problem of titanium at dental porcelain sintering temperatures are limiting factors for the fabrication of titanium-ceramic restorations. Chromium single layer and chromium/chromium oxide bi-layer coatings of up to 1 µm thickness were applied to titanium surfaces using the plasma immersion implantation and deposition method. Such coatings serve as an oxygen diffusion barrier on the titanium. Chromium must further develop its own chromia surface layer to which the porcelain overlayer is bondable. Initial experiments have shown this approach to be promising. Cross sections of Cr and Cr/Cr_2O_3 coated titanium were examined by various electron microscopy methods and by X-ray and electron diffraction analysis. Comparative adhesion strength measurements between the coatings and titanium were made before and after high temperature oxidation.

INTRODUCTION

The use of titanium (Ti) and its alloys in orthopedic and dental implants has dramatically increased in the past few years due to their excellent biocompatibility, desirable physical and chemical properties and low cost [1-4]. The thin native titanium oxide surface film plays a positive but still not fully understood role in the biocompatibility. The native Ti oxide allows bone cells to grow directly onto the Ti surface in a process named osseointegration [5]. Oxides of other metals usually lead to an inflammatory reaction. Recently Ti has been considered as a biomaterial for removable or fixed dental prostheses such as partial dentures or titanium-ceramic crowns [6-9].

The application of porcelain to titanium dental restorations requires firing the porcelain on titanium copings or frameworks at temperatures of $850^{\circ}C$ or higher. Heating of titanium at these temperatures in air leads to the formation of Ti oxide at the interface between the Ti and the porcelain. Even if in vacuum or in inert gas atmosphere the application of porcelain, which is essentially an alloyed SiO_2 glass, would form a titanium oxide interface layer because of titanium's high oxygen affinity. Thermally formed titanium oxide adheres poorly [10-11]. Spallation of the porcelain coating occurs in the TiO_2 interface layer due to stress generated by thermal expansion mismatch. It is the major cause for the failure of Ti-porcelain crowns.

In order to prevent the formation of a weak interface, a strategy to prevent Ti oxide formation at dental porcelain sintering temperatures ($850-950^{0}C$) and to generate an adherent oxide scale was devised. The procedure involved using intermediate metallic and/or ceramic layers, which were deposited on the Ti prior to the application of porcelain. Such layers must consist of biocompatible materials, must act as a barrier to the diffusion of oxygen, and must be strongly adherent to the Ti substrate. They must also be capable of establishing strong bonding to the porcelain. The use of metallic undercoats for enhancing adhesion of porcelain to metallic substrates for metal-ceramic crowns has been proposed previously for substrates of Ni-Cr and Co-Cr alloys [12] as well as gold [13]. Electroplating and flame spraying methods were used for depositing an intermediate layer on metal copings to enhance porcelain-metal bonding for dental prostheses. The effect of underlayers on titanium-porcelain prostheses, however, has not been studied and is the object of this investigation.

Chromium is a promising coating material for this particular application because it forms a dense Cr oxide layer and has low oxygen solubility. Cr is also inexpensive. It has been recognized that Cr is chemically compatible with dental porcelains when using Ni-Cr alloys for metal-ceramic restorations [14], and is capable of establishing strong bonding with dental porcelains.

Metal Plasma Immersion Ion Implantation and Deposition (MPIIID) method has been developed to combine physical vapor deposition from a metal plasma with pulsed implantation of the metal [15]. The advantage of this process is coatings with superior adhesion between the substrate and coating materials due to atomic mixing between the incoming ions and the substrate atoms. The amount of intermixing is controlled by the ion energy imparted during the implantation cycles. In addition to single phase coating, metal/metal oxide bi-layers can also be deposited by MPIIID in a controlled oxygen partial pressure.

EXPERIMENTAL

Deposition of Intermediate Layers (MPIIID)

Thin films of chromium were deposited on 4 cm by 1.5 cm by 0.1 cm titanium coupons. The Ti surface was ground and polished prior to the Cr deposition with silicon carbide paper and with 0.5µm alumina suspension to a mirror finish. Each sample was washed in an ultrasonic bath in methyl alcohol and acetone for 10 minutes before insertion in the vacuum chamber. The substrates were then mounted on a water-cooled substrate, with silicon wafer pieces mounted adjacently. The silicon wafers provided control substrates on which the film could be analyzed. Water-cooling the substrate prevented the sample temperature to increase due to the ion bombardment during the deposition. Although there were no actual measurements during deposition, specimen temperatures were expected not to exceed 50°C. Immediately prior to beginning the MPIIID process, the Ti coupons were sputter-cleaned in a DC glow discharge, with argon gas at 100 mT pressure. The DC voltage applied to the substrate holder was -500V with respect to ground. The deposited Cr coating thickness on Ti specimens ranged from 0.25µm to 1µm. In addition to single chromium films, bi-layers of Cr and Cr_2O_3 were also tried as diffusion barriers. The oxide layer was produced by MPIIID in a controlled oxygen partial pressure for bi-layer coated Ti samples.

Details on Metal Plasma Immersion Ion Implantation and Deposition (MPIIID) have been recently summarized [15, 16]. The technique consists basically of using a cathodic vacuum arc to generate the plasma, whose ions are then accelerated towards the substrate by pulse-biasing the latter. At the initial stages of the deposition, high voltages are used to bias the substrate in order to enhance adhesion. The accelerated ions are implanted in the substrate, resulting in a cohesive interface due to interatomic mixing. The deposition of metallic Cr films was conducted at a vacuum better than 5×10^{-6} Torr, in order to minimize the oxygen content of the film. When a Cr/Cr-oxide bi-layer was deposited, the Cr layer was deposited as described above. After the desired thickness of Cr was achieved, oxygen was admitted to the vacuum chamber, until a pre-determined pressure was reached between 20 and 100 mT. At this moment the plasma gun was re-started and the deposition continued. The deposited film consisted of amorphous chromium oxide of various ratios of Cr/CrO_x. The total thickness of bi-layer coatings was 1µm.

Oxidation treatment and porcelain deposition

After the undercoats of Cr or Cr/CrO_x films were applied, the Ti coupons were subjected to oxidation treatments that are similar to the conditions used during the porcelain firing. Commercial dental porcelains for metal-ceramic restorations have sintering temperatures between 850 and 950°C. Therefore, in evaluating the coating characteristics, oxidation treatments were conducted with and without porcelain application. Coupons were heated in air at 850 and 950°C at a rate of 50°C/min. The holding time at peak temperature was 15 minutes.

Application of the porcelain to Ti was performed according to commercially established recommendation. Four layers were consecutively applied. For the first layer the sample was heated to 850°C in vacuum at a rate of 50°C/min, and held at 850°C for 5 min prior to cool down. During the application of the second and third layers, the same procedure was used, but the time at 850°C was reduced to 1 minute. For the final layer, the procedure was the same as for the second and third layers, but it was conducted in air. Samples with a 1μm Cr layer and with a 0.5μm Cr/0.5μm Cr oxide bi-layer were used to test the adherence of porcelain coatings.

Characterization of the films.

The thicknesses of Cr and Cr/Cr-oxide coatings on Si substrate control samples were measured by scanning electron microscopy. Coated and cleaved silicon wafers were observed edge-on in a JEOL 8400 SEM at operating voltage of 30kV. This technique for measuring thickness was particularly advantageous in the case of the bi-layers because the Cr and chromium oxide have different fracture morphologies. The thicknesses of the individual layers could be directly measured.

SEM on fracture cross sections was also used to analyze the coated Ti coupons before and after oxidation to characterize the several layers that resulted from the oxidation treatments. The fractures were prepared by initially enveloping the entire sample in an epoxy layer and sawing it part way through, leaving an area of about 0.5 mm^2 of uncut material. The coupon was then immersed in liquid nitrogen and fractured with pliers. The liquid nitrogen temperature guaranteed that the fracture was brittle, therefore without distorting the layers. Energy dispersive X-ray spectroscopy was used to evaluate the elemental compositions of the layers.

X-ray diffraction was used to characterize the phases of the film deposits on the Ti coupons prior and after oxidation. X-ray diffraction spectra were obtained in a SIEMENS D-5000 diffractometer using Cu K$_\alpha$ radiation. Scans were performed at 40kV. Diffraction angles from 20° to 70° were measured with a step size of 0.02° and a ramping speed of $0.01°s^{-1}$.

Cross section samples of Cr/Cr-oxide bi-layer were prepared from the coated silicon control samples, and were studied in a JEOL200CX with a parallel EELS spectrometer. The Cr or Cr/CrO$_x$ coated Si samples were cleaved and glued together with epoxy. Slices were transversely cut from the sandwich specimens and were polished to about 200μm thickness. The thin slabs were then glued to a grid with an oval opening and thinned down by dimpling and ion-milling until the films achieved electron transparency. Electron diffraction studies were conducted on these samples.

Adhesion tests of the as deposited and oxidized films were carried out with a Sebastian-type puller built in-house. Because this is not a standardized test, the results were used only for comparison among the several tests. For the adhesion test an aluminum pin is glued to the film surface, and a load is applied pulling the pin away from the coupon. The load at which separation occurs is recorded and converted to a normal rupture stress. The upper limit of this test is limited by the strength of the glue, which in the present case corresponds to a stress of 70 MPa.

RESULTS

X-ray diffraction patterns of the bi-layer deposits are shown in Figure 1. The diffraction peaks were identified as those from Ti substrate and some weaker and broader ones from the Cr/Cr oxide bi-layer. Metallic films deposited by MPIIID at room temperature tend to be polycrystalline, whereas the oxide is expected to be amorphous. Figure 1 shows a small peak at 44.5°, which coincides with the strongest diffracting peak of Cr, and a broad maximum around 42° to 44°, which correspond to the strongest reflections from chromium oxide. The Cr peak is stronger in a typical Cr/Cr oxide sample because the very low glancing angle of the incident X-ray radiation makes the detection volume very shallow, and because of the thin layer of the Cr oxide.

The bi-layer deposited on the Ti coupons was also analyzed by transmission electron microscopy prior to the oxidation, and the results are presented in Figure 2. The electron diffraction ring pattern in Figure 2(a) shows that the chromium layer is polycrystalline, and with a very fine grain size, since the pattern was obtained from a selected area of a 89 nm in diameter. Electron diffraction from the chromium oxide film consisted of diffuse rings, characteristic of electron diffraction from an amorphous material. A bright field image of the composite bi-layer is shown in Figure 2(b). Figures 3(a) and 3(b) show the energy loss spectra of the microcrystalline Cr and the amorphous Cr-oxide films respectively. The stoichiometry of the oxide cannot be inferred from the spectrum. Only the chromium L 3-2 lines are seen from the crystalline layer in Figure 3(a). Figure 3(b) shows chromium L 3-2 and oxygen K lines in the spectrum obtained from the amorphous Cr oxide layer and no sign of titanium was detected in the MPIIID Cr.

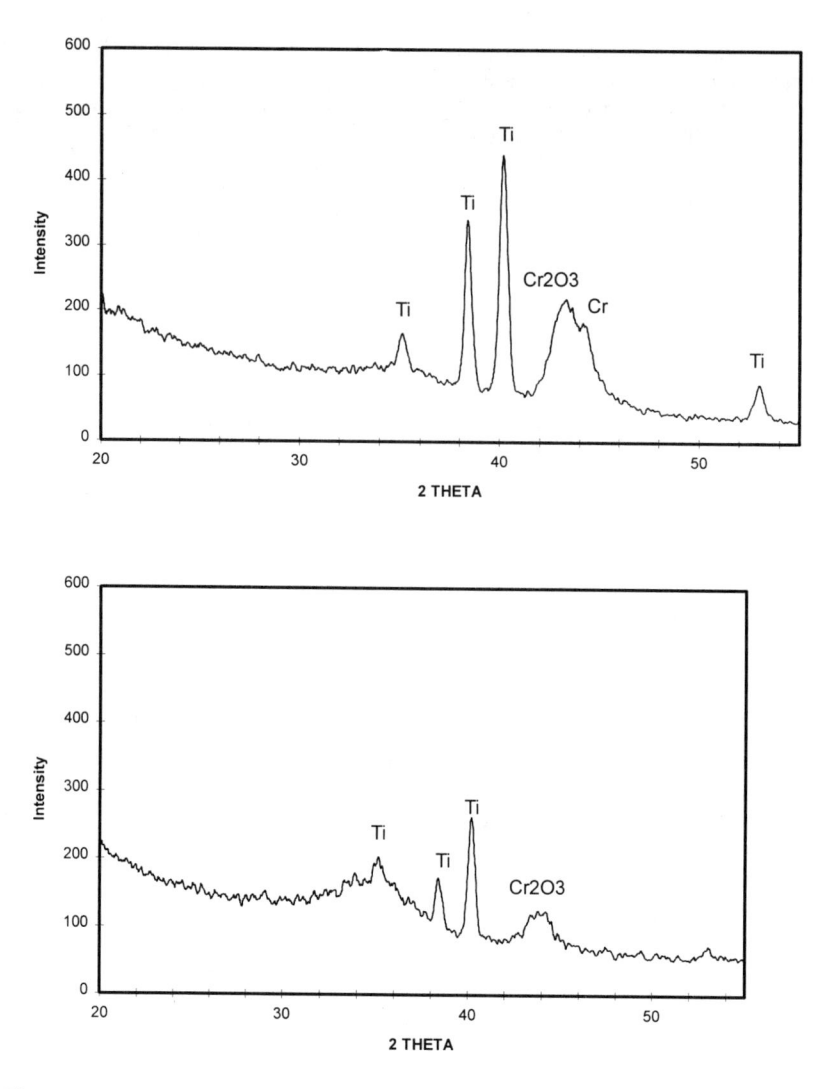

Figure 1: X-ray diffraction pattern form (a) a Ti coupon coated with 200 nm Cr and 200 nm of chromium oxide; and (b) a Ti coupon coated with 200 nm Cr and 700 nm of chromium oxide.

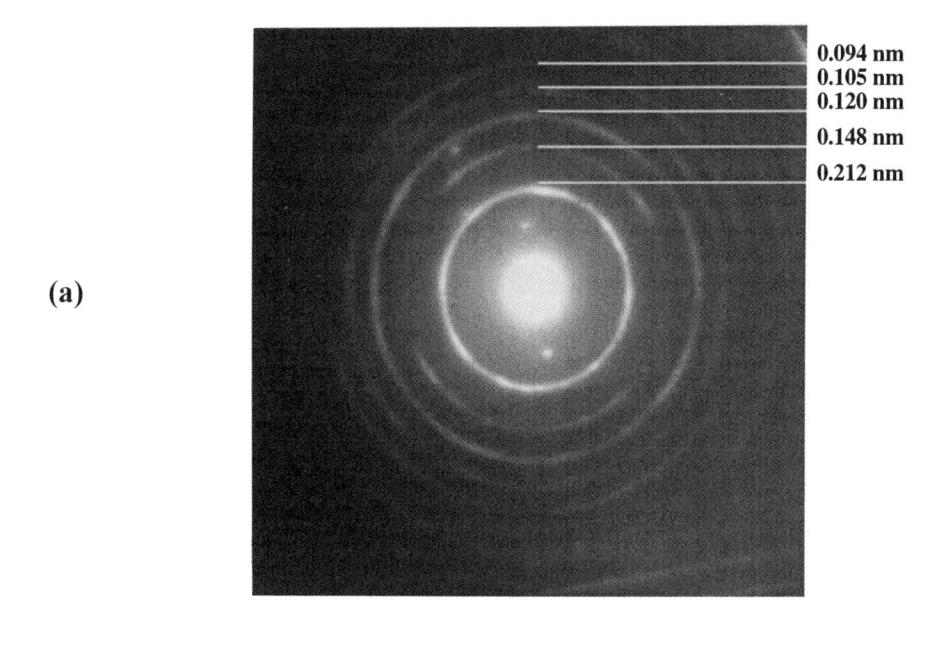

Figure 2: (a) Electron diffraction pattern of the Cr deposited film. (b) Bright field image of the Cr - Cr oxide bi-layer.

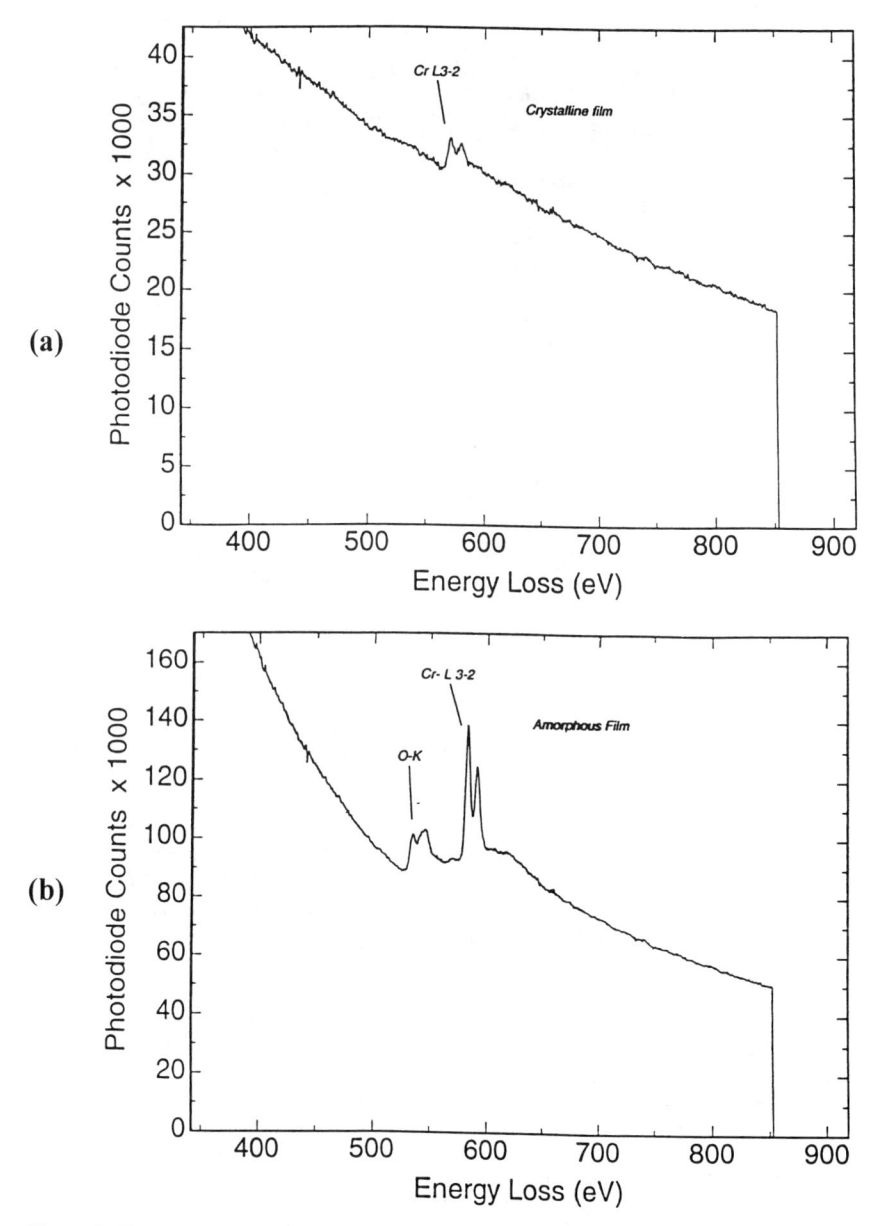

Figure 3: Electron energy loss spectra taken from TEM thin foils of Cr and Cr/Cr-oxide layers deposited by MPIIID. (a) Electron energy loss spectrum of Cr layer; (b) Electron energy loss spectrum of Cr/Cr-oxide layer.

Table 1 shows the results of adhesion tests for several films prior and after oxidation at 950°C for 15 minutes in air. The oxidation condition is similar to the process commonly used in the fabrication of dental crowns. Figure 4 is a schematic representation of the setup for measurement of adhesion. Before oxidation, all Cr and Cr/Cr-oxide coated samples had adhesion strength greater than 70 MPa. This means that the failure did not occur in the deposited Cr or Cr-oxide layer but at the aluminum pin-epoxy junction. After oxidation most samples had much lower adhesion strength values except samples Cr:Ti #08 and #09.

Samples Cr:Ti #08, #09, and #13B had undercoats with approximately same total thickness of 1μm. The behavior of the first two were similar within the experimental limitations of the adhesion test, and it was not possible to determine whether one performed better than the other. Sample #13B delaminated at a lower load than either #08 or #09. The results indicate that the Cr layer is more important as the diffusion barrier than the deposited oxide layer. The chromium oxide produced via MPIIID is not as impervious to oxygen as is metallic chromium.

Table 1: Adhesion tests performed at room temperature before and after 15 minute-oxidation at 950°C in air of coated Ti coupons, using a Sebastian-type puller.

Sample	MPIIID Cr (nm)	MPIIID CrO_x (nm)	Adhesion Before Oxidation (MPa)	After Oxidation (MPa)
Cr:Ti #02	250	0	> 70	< 5
Cr:Ti #05	300	0	> 70	< 5
Cr:Ti #07	500	0	> 70	< 10
Cr:Ti #08	1000	0	> 70	> 70
Cr:Ti #09	500	500	> 70	> 70
Cr:Ti #13B	200	700	> 70	17.8

In order to understand the failure mechanisms of the oxidized samples in Table 1, fracture cross sections were prepared and analyzed in the scanning electron microscope. Figure 5 shows the morphology of sample Cr:Ti #05, which delaminated at a very low stress after being oxidized. Energy dispersive X-ray analysis indicated that the powdery layer immediately adjacent to the Ti substrate consists of titanium oxide. The top layer is mostly Cr_2O_3 but it contains a substantial amount of Ti oxide as well.

Table 2 shows data for deposition thickness and oxide adhesion after two different temperatures, 850 and 950°C. At 850°C the Cr-layer held up as an oxygen diffusion barrier. The faster oxidation of 950°C consumed the 200 nm Cr layer completely, and having lost the oxygen diffusion barrier, titanium oxide formed. The loss in adhesion strength, Table 2, is a direct consequence.

Table 2: Adhesion tests of samples oxidized at 850°C and 950°C

Sample	Thickness of Cr-layer (nm)	Thickness of Cr_2O_3-layer (nm)	Oxidation Temperature (°C)	Rupture stress P/A after oxidation (MPa)
Cr:Ti #13A	200	500	850	42.8
Cr:Ti #13B	200	500	950	17.8

Figures 6a and 6b show x-ray diffraction spectra of samples Cr:Ti #13A and #13B after oxidation at 850°C and 950°C. It is clear from these figures that oxidation at the higher temperature consumed the Cr-layer and resulted in substantial formation of titanium oxide. At 850°C the predominant oxide phase is crystalline Cr_2O_3. The diffraction pattern of sample #13A is virtually that of crystalline chromium oxide, and perhaps some chromium, whereas the diffraction pattern of #13B shows just rutile. There is a direct correlation between the loss of adhesion and the formation of Ti oxide in those instances in which the Cr undercoat could not perform its role of an oxygen diffusion barrier.

Direct application of porcelain to Ti coupons without any Cr or Cr/Cr-oxide coatings resulted in poor adhesion strength. Porcelain layer often delaminated at temperature or during the cool-down. Large pieces of porcelain layer consistently detached from the titanium, with a grayish film, believed to be porous rutile, attached to its underside.

Porcelain was also applied to Ti coupons with a 1000 nm thick Cr coating and to coupons with a 500 nm Cr/500 nm Cr-oxide coating. Figure 7 is the SEM image of a fracture surface through the porcelain-coating (oxide)-titanium interface. Figure 8a is an EDX spectrum of porcelain remote from the porcelain-Cr coating interface. Figure 8b is the EDX spectrum of the region marked "porcelain" in Figure 7b of the fracture of the Ti/porcelain assembly along the Cr_2O_3/porcelain bond interface. This interface was sufficiently strong to pull some porcelain particles from the bulk of the porcelain. More import, the adhesion strengths at the Ti/Cr and Cr/Cr_2O_3 interfaces, were stronger as they did not fail.

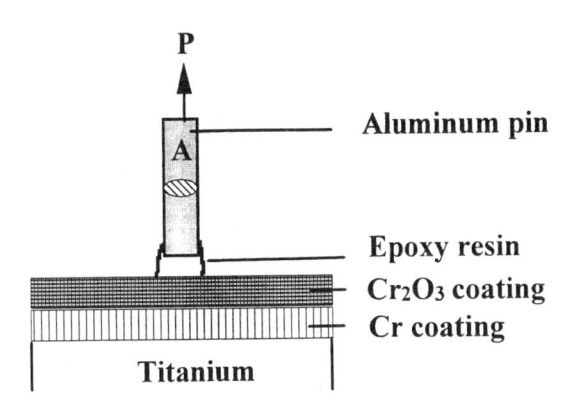

Figure 4 Schematic representation of the setup for adhesion strength test.

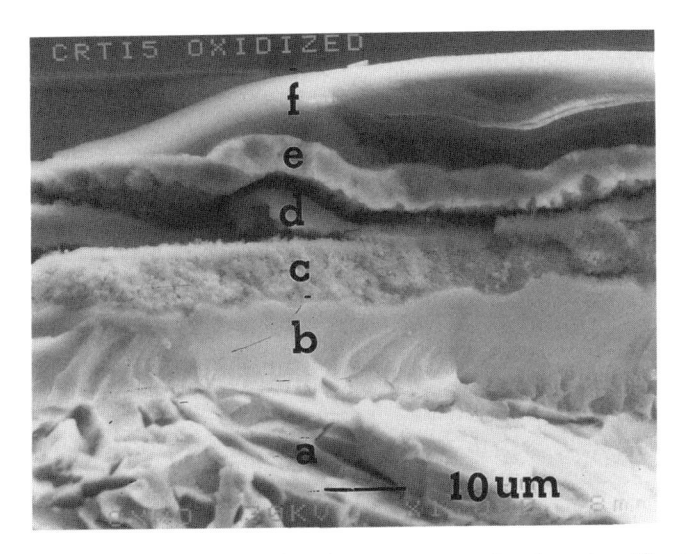

Figure 5 Scanning electron micrograph of fracture cross section through oxidized (15 min. 950°C, air) sample Cr:Ti #05. (a) Titanium base, no oxygen embrittlement. (b) Titanium surface layer, embrittled by oxygen. (c) Porous titanium oxide layer, believed to be mostly TiO_2. (d) Gap from spallation. (e) Mixed Cr and Ti oxide layers. (f) Glue enveloping specimen for mounting.

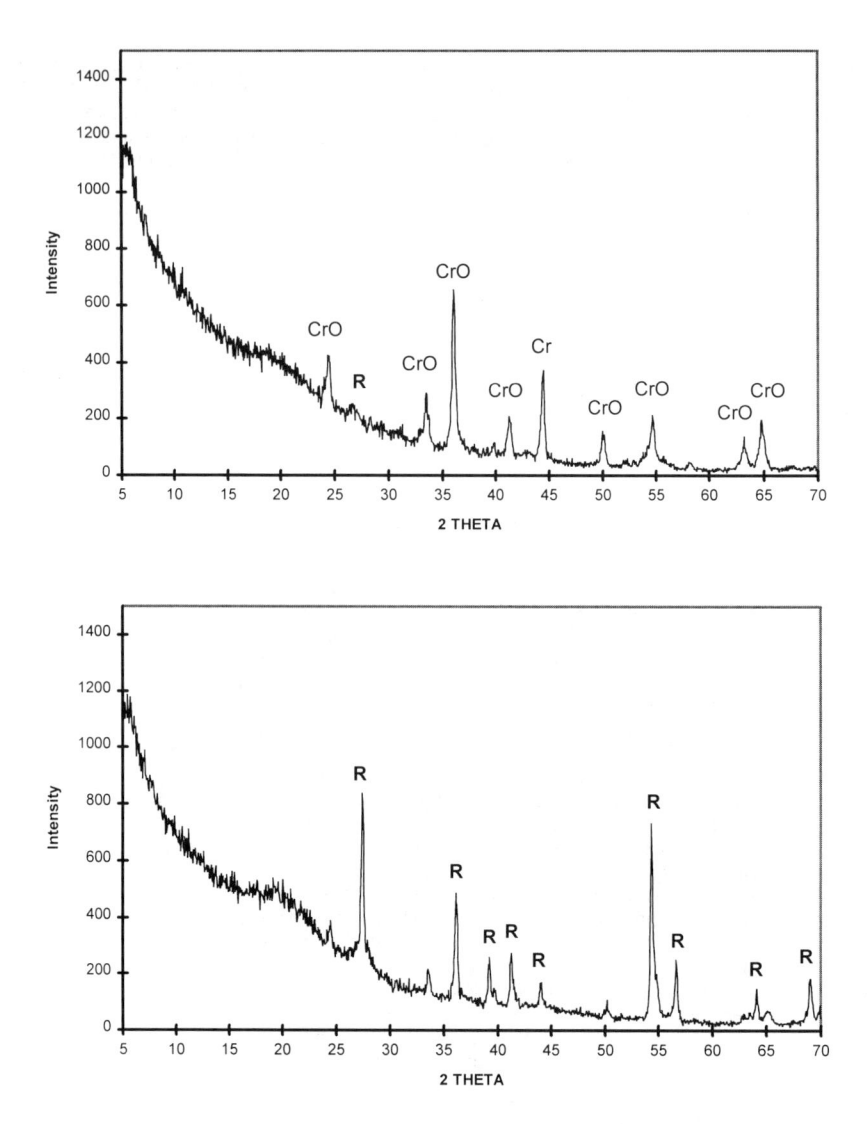

Figure 6 X-ray diffraction pattern from Ti specimen coated with Cr 200 nm/Cr$_2$O$_3$ 500 nm (a) after oxidation at 850°C for 15 minutes. (b) after oxidation at 950°C for 15 minutes. CrO stands for Cr$_2$O$_3$ and **R** for rutile.

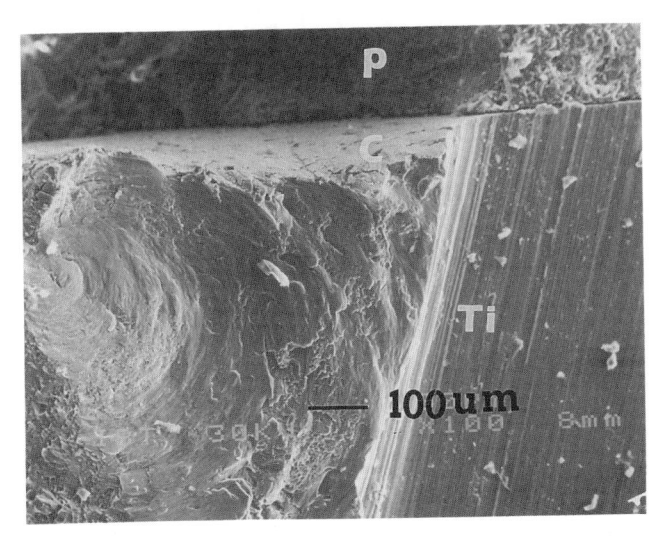

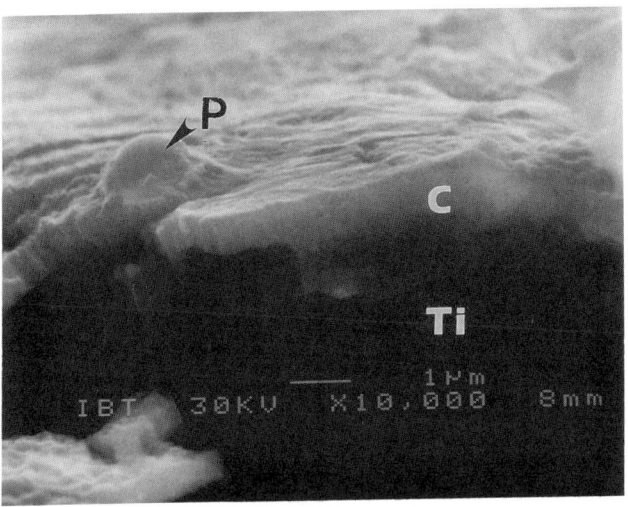

Figure 7 (a) SEM image of fracture surface through sample Cr:Ti #08 with a thick porcelain coating. P: porcelain; C: Cr layer; vertical striations on the right are the saw cut and further to the left is the fracture cross-section through the Ti substrate. (b) High magnification of the cross section showing adherent Cr/Cr_2O_3 film on titanium substrate. The partial adherence of porcelain (arrow) on the Cr oxide is evident from the porcelain particle that stuck to Cr-oxide and was pulled out from the porcelain body. P: porcelain; C: Cr/Cr_2O_3 layer.

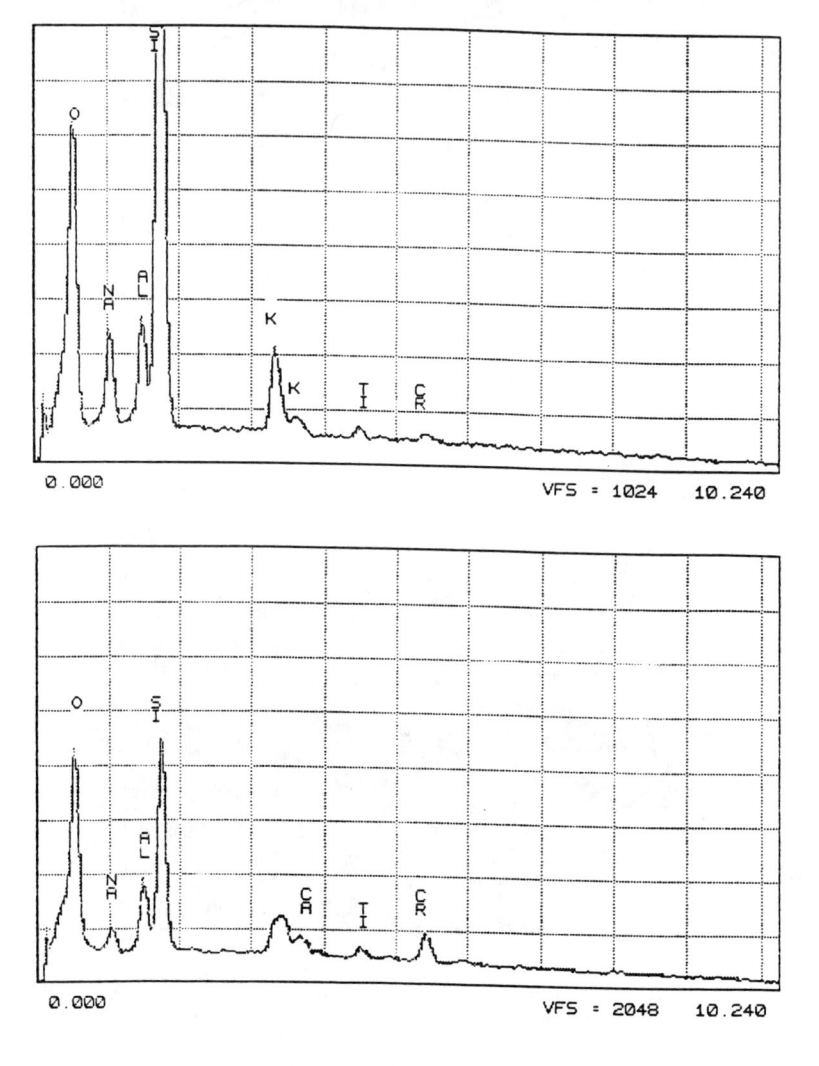

Figure 8 (a) EDX spectrum of porcelain in a region remote from the porcelain-Cr coating interface. (b) EDX spectrum of the interfacial film shown in Fig 7b. This spectrum is similar to that in 8a, except for an increased signal from Cr.

SUMMARY AND CONCLUSIONS

The effect of Cr and Cr/Cr_2O_3 surface films, deposited via MPIIID, on adhesion between titanium and porcelain was studied. The MPIIID deposition process conditions provided good adhesion between the as-deposited films and titanium substrate. High temperature oxidation experiments that simulate the conditions in porcelain firing were carried out in order to study the chromium film's effectiveness in preventing oxidation of the titanium and in enabling the bonding of porcelain to titanium.

Chromium films with thickness below 0.5μm were ineffective in preventing oxidation of titanium at 950°C for 15 minutes. Chromium films with thickness greater than or equal to 1 μm resulted in no measurable loss of adhesion upon oxidation at 950°C.

Loss of adhesion is basically caused by the formation of a porous titanium oxide layer. TiO_2 may form underneath oxidized chromium and eventually reach the surface. This layer is weak, and delamination of the chromium oxide layer occurs by fracture in this rutile layer. In fact in the case of thin Cr coatings, the phases after oxidation were a mixed chromium and titanium oxide layer on the outside, an intermediate TiO_2 layer, and the titanium substrate.

The use of bi-layers of chromium and chromium oxide did not result in an advantage over just a chromium layer. Porcelain was most successfully applied to chromium-coated titanium samples.

ACKNOWLEDGMENT

This work was supported by a research initiation grant from the National Institute of Health (NIH/NIDR # R03DE11032). Additional support was provided by the Director, Office of Basic Energy Sciences, Advanced Energy Project Division of the U.S. Department of Energy under Contract No. DE-AC03-76SF00098. The authors also would like to acknowledge the use of the facilities of the National Center of Electron Microscopy at LBNL.

REFERENCES

1. "Biodegradation of Restorative Metallic Systems" Effects and Side Effects of Dental Restorative Materials, ed: J Lemons, (Bethesda, MD: National Institute of Dental Research, 1992), 32-38.

2. David F. William, Biocompatibility of Clinical Implant Materials (Boca Raton, FL: CRC Press Inc., 1986), 44-49.

3. S. Mohammad, L. Wictorin, L. Ericson, and P. Thomsen, "Cast Titanium As Implant Material," Journal of Materials Science-Materials In Medicine, 6 (1995), 435-444.

4. H. G. Elcharkawi, "Residual Ridge Changes Under Titanium Plasma-Sprayed Screw Implant Systems," Journal of Prosthetic Dentistry, 62 (1989), 576-580.

5. T. Albrektsson, P. I. Branemark, and H.A. Hansson, "The Interface Zone of Inorganic Implants in Vivo: Titanium Implants in Bone," Ann. Biomed. Eng., 15 (1986), 1-27.

6. K.E. Boening, M. H. Walter, and P. O. Reppel, "Non-cast Titanium Restorations in Fixed Prosthodontics," J. Oral. Rehabil., 19 (1992), 281-287.

7. R. Blackman, N. Barghi, and C. Tran, "Dimensional Changes in Casting Titanium Removable Partial Denture Frameworks," J. Prosthet. Dent., 66 (1992), 309-315.

8. I. R. Harris, and J. L. Wicken, "A Comparison of the Fit of Spark-eroded Titanium Copings and Cast Gold Alloy Copings," Int. J. Prosthet., 9 (1994), 348-355.

9. P. K. Vallittu, and M. Kokonen, "Deflection Fatigue of Cobalt-Chromium, Titanium, And Gold Alloy Cast Denture Clasp," J. Prosthet. Dent., 74 (1995), 412-419.

10. G. E. Welsch, and A. I. Kahveci, "Oxidation Behavior of Titanium Alloys" in: T. L. Grobstein and J. Doychak eds. Oxidation of High Temperature Intermetallics (Metal Park, OH: ASM International, 1989), 207-218.

11. G. E. Welsch, A. I. Kahveci, and D. S. Friedman, "In situ TEM Investigation of Oxidation of Titanium Alumnide Alloys," In Bennett M. J. and Lorimer G. W. eds. Proceeding of the First International Conference on Microscopy of Oxidation, (Bookfield, VT: The Institute of Metals, 1991), 193-205.

12. P. Klimonda, O. Lingstuyl, B. Lavalle, and F. Dabosi, "The Use of a Flame-Sprayed Undercoat to Improve the Adherence of SiO_2-Al_2O_3 Dental Ceramics on Ni-Cr and Co-Cr Alloys," In: Surfaces and Interfaces in Ceramic and Ceramic-Metal Systems, ed: J. Pask and A. Evans, (New York, NY: Plenum Press, 1981), 477-486.

13. M. A. Salamah, and D. White, "Role of Nickel in Porcellain Enamelling", in: <u>Surfaces and Interfaces in Ceramic and Ceramic-Metal Systems</u> ed: J. Pask and A. Evans, (New York, NY: Plenum Press, 1981), 467-476.

14. William J. O'Brian and Robert G. Craig, ed., <u>Proceedings of Conference on Recent Developments in Dental Ceramics</u> (Columbus, OH: The American Ceramic Society, 1985), 66-83.

15. I. G. Brown et al., "New Developments in Metal Ion Implantation by Vacuum Arc Ion Sources and Metal Plasma Immersion" (Paper presented at the Materials Research Society Annual Meeting, Boston, MA, 27 November 1995), 5.

16. I. G. Brown et al., "Industrial Applications of Metal Plasma Immersion Implantation and Deposition" (Paper presented at the TMS Annual Meeting, Cleveland OH, 29 October 1995), 122.

THE CORRELATION OF PRIMARY CREEP RESISTANCE TO THE HEAT TREATED MICROSTRUCTURES IN INVESTMENT CAST TI-AL GAMMA ALLOYS

D.Y. Seo, T.R. Bieler, and D.E. Larsen*

Department of Materials Science and Mechanics
Michigan State University, East Lansing, MI 48824-1226, USA

* Howmet Corporation, Whitehall, MI, 49461, USA

Abstract

Cast gamma titanium aluminides are gaining acceptance as replacements for superalloy and steel components in aerospace, automotive and industrial applications. Components cast from these alloys can operate at temperatures up to 815 °C(1500 °F) at half the weight of the components they replace. Most applications of cast gamma components require good creep resisitance to meet component life requirements. Four heat treatments were developed and applied to investment cast Ti-45Al-2Nb-2Mn(at%)+0.8v%TiB$_2$ XD^{TM}, Ti-47Al-2Nb-2Mn(at%) + 0.8v%TiB$_2$ XD^{TM}, and Ti-47Al-2Nb-2Cr(at%) alloys in an effort to enhance creep properties with a decrease in heat treatment time compared to current practies. Results show that the creep resistance of Ti-47Al-2Nb-2Cr(at%) alloys can be significantly altered, up to 3X, through heat treatment. Some improvement was obtained in Ti-47Al-2Nb-2Mn(at%) + 0.8v%TiB$_2$ XD^{TM} and heat treatment had no effect on Ti-45Al-2Nb-2Mn(at%)+0.8v%TiB$_2$ XD^{TM} alloy. The variation, or lack of variation, in creep resistance with heat treatment can be explained by differences observed in the microstructures and textures produced by the various heat treatments.

Research funded by the Howmet Corp., Whitehall, MI. SEM images were obtained with the assistance of B. Simkin. The FEG-SEM used in this study was supported in part by the National Science Foundation through grant #DMR9302040.

Introduction

Gamma titanium aluminide can be heat treated to obtain four types of microstructures [1]. The best creep resistance has been observed in fully lamellar microstructures, but the duplex microstructure provides desirable room temperature ductility [2,3]. The effect of alloying elements such as Cr, Nb, V, Mn, W, Mo, and Si on properties in two phase Ti-Al alloys has been investigated [4-9]. Most of this research has focused on minimum creep rate conditions or stress-rupture properties. However, the primary creep resistance is important for practical applications. Gamma based TiAl exhibits primary creep strains that are typical for metals, where the minimum is reached after about 1% strain. In an effort to decrease the creep rate during primary creep deformation, additions of W, Mo, and Si have increased the time to reach 0.5% strain due to a dynamic precipitation process [10,11]. In related studies, observations of lamellar refinement were made that suggest that primary creep consists of a significant amount of mechanical twinning parallel to lamellar interfaces that occurs as an easy mode of deformation that hardens the microstructure at low strains [11~14]. In this paper we analyze the primary creep resistance (time to 0.5% creep strain) in three two-phase TiAl alloys that do not rely on the refractory elements for primary creep resistance. In these materials, the nature of primary creep strain is explored, as it is affected by changes in the microstructure that arise from different heat treatments. Two alloys containing TiB_2 particulate introduced with the XD^{TM} process and a third alloy without TiB_2 are compared.

Materials and Experimental Procedure

Three investment cast gamma TiAl alloys were used, Ti-45Al-2Nb-2Mn-0.8TiB$_2$ (0.8 TiB$_2$ refers to volume fraction) Ti-47Al-2Nb-2Mn-0.8TiB$_2$, and Ti-47Al-2Nb-2Cr. All materials were hot isostatically pressed (HIPed) at 1260°C for 4 hours at 172 MPa (2300°F, 25 ksi) in order to eliminate casting porosity. The heat treatments used were in the $\alpha_2+\gamma$ phase field at 900 or 1010°C (1650 or 1850°F) for 10 or 20 hours, and cooled in a static argon atmosphere (SAC). A two step heat treatment was also used consisting of 1177°C (in the $\alpha+\gamma$ phase field) for 5 hours and then in the $\alpha_2+\gamma$ field for 10 hours at 1010°C (2150 and 1850°F), respectively, and cooled in static argon. These heat treatments are illustrated on the phase diagram in Figure 1 [15]. The phase boundaries are not accurate for the quaternary alloys of this study. These specimens were then subjected to room temperature tensile tests and creep tests at 760°C, 138 MPa (1400°F, 20 ksi). The creep specimens were cooled under load once 0.5 % strain was reached.

For optical and SEM analysis, deformed samples were obtained from the gage section of the creep deformed specimen. Some undeformed samples were also cut from the grip section of tensile test specimens. The samples were mounted, mechanically polished, and etched in Kroll's reagent (2ml HF, 6 ml HNO$_3$, and 92 ml H$_2$O). For TEM investigation, 0.7 mm thick slices were cut such that the sample normal was parallel to the tensile axis. 3mm diameter discs were cut from the slices using an EDM machine and ground to about 0.1 mm thick. The disks were thinned using a double jet electropolishing system with a 10% sulfuric acid and methanol solution at -25°C and 20 volts. These foils were examined in a Hitachi H-800 transmission electron microscope operated at 200kV. For SEM analysis, a Cam Scan 44FE SEM was used at 25kV to obtain backscattered electron images that show phase contrast. Texture measurements were made on selected specimens and analysis was made using popLA software[16].

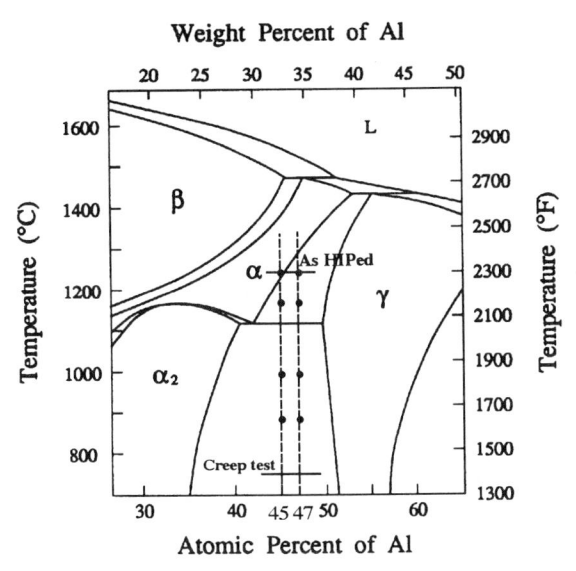

Figure 1- Phase diagram illustrating heat treatment and creep test condition[15].

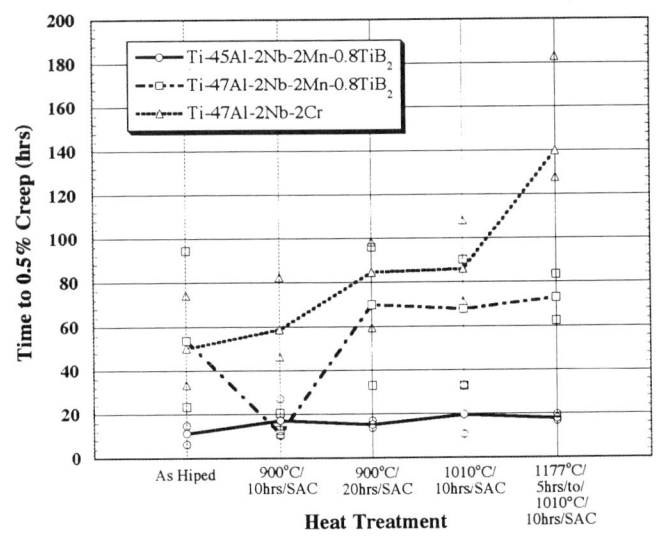

Figure 2- Time to 0.5% creep strain with respect to heat treatment. The lines go through the average of 3 values, the high and low values are indicated by additional symbols.

Results

Primary creep to 0.5% strain and Tensile tests

The time to 0.5% creep of the three titanium aluminide alloys varies with heat treatment as shown in Figure 2. The average value of three samples is plotted with the line, and high and low values are also plotted. From this plot, it is apparent that the 45 Al composition is insensitive to heat treatment. The 47 Al samples exhibit different responses to heat treatment. With TiB$_2$, the heat treatment for 10 hours at 900°C is damaging to primary creep resistance, but otherwise there is no significant improvement with increasing time at temperature in heat treatment, other than a smaller range of values with the two step heat treatment. For the alloy without TiB$_2$, the creep resistance is substantially improved with increasing amounts of time and temperature in heat treatment, but the variability in the samples is not significantly improved with heat treatment. In tensile tests elongation (0.8~1.5%) and strength was not strongly affected by heat treatment. Elongation was degraded slightly with heat treatment in the 47Al with TiB$_2$, and improved slightly in the 45Al with TiB$_2$.

Microstructures

The samples chosen for microstructural observation were the middle performance specimens. The microstructural variations in these specimens are summarized in Table I. Details of the microstructure for each combination of composition and heat treatment are described in more detail below.

Ti-45Al-2Nb-2Mn-0.8TiB$_2$

As shown in Figure 3.a, optical micrographs of the crept specimen show that many TiB$_2$ particles (indicated by arrows) are distributed randomly. There are several types of particles such as equiaxed, needle, and irregular shaped precipitates with a spacing of 30-50 μm. As annealing time increased no distinct change in the particles was observed, but in the case of the two-step heat treatment there appeared to be fewer particles. From micrographs of etched samples, a very fine duplex microstructure was observed in all specimens, regardless of heat treatment. There is a homogeneous distribution of 20-30 μm equiaxed and 30-100 μm lamellar grains. The volume fraction of lamellar grains increases slightly increased after heat treatment from as-HIPed to 900°C/20hrs, but no difference between the 900°C/20hrs and the two-step heat treatment was observed.

From analysis of backscattered images of undeformed and deformed specimens obtained in SEM, there are no distinguishable changes in the microstructure resulting from creep deformation (Figure 4.a,b). TiB$_2$ particles are distributed randomly, and they are located inside gamma grains, in grain boundaries between gamma (dark phase) and α_2 (bright phase) grains and in lamellar grains, but not inside α_2 grains, as shown in Figure 4. Single phase γ and α_2 grains represent about 40% of the volume fraction. The α_2 grains have large and angular shapes, but the γ grains tend to be more round and equiaxed. This indicates that equiaxed γ grains nucleate in several places in a given parent α grain, and that some of the parent α grains transform into lamellar grains, but others do not.

Very fine mechanical twins are observed in some α_2 grains in both deformed and undeformed specimens (see arrows in Figure 4). These finely twinned regions also exhibit contrast indicating formation of extremely fine lamellar microstructures consisting of both α_2 and γ lathes. Since they are much finer than the obvious lamellar grains, they are not likely to have

Table 1: Observations of the Effects of Heat Treatment on Three TiAl Alloys from optical microscopy

		As Hip'ed	900°C/10hrs/SAC	900°C/20hrs/SAC	1177°C/5hrs To 1010°C/10hrs/SAC
Ti-45Al-2Nb-2Mn-0.8TiB$_2$	lamellar % & grain size	25% / 30μm	/ 30~100μm	39% / 50~70μm	37% / 50~100μm
	equiaxed % & grain size	75% / 15μm	/ 20~30μm	61% / 20~30μm	63% / 20~30μm
	TiB$_2$	30~50μm spacing		gray phase observed	
	characteristics	Homogeneous	Lam patches	Lam patches	
Ti-47Al-2Nb-2Mn-0.8TiB$_2$	lamellar % & grain size	21% / 50~200μm	33% / 40~100μm	16% / 50~150μm	56% / 100~200μm
	equiaxed % & grain size	79% / 10~40μm	67% / 10~30μm	84% / 10~40μm	44% / 10~40μm
	TiB$_2$	30~50μm spacing	clustered	similar	10~40μm spacing
	characteristics	Heterogenous	Less Heterog	Homogeneous	Large lame. grains
Ti-47Al-2Nb-2Cr	lamellar % & grain size	51% / 50~400μm	46% / 50~150μm	43% / 40~150μm	18% / 50~100μm
	equiaxed % & grain size	49% / 5~150μm	54% / 5~150μm	57% / 5~150μm	82% / 5~150μm
	characteristics	Heterogenous	Heterogenous	Less Heterog	Homogeneous

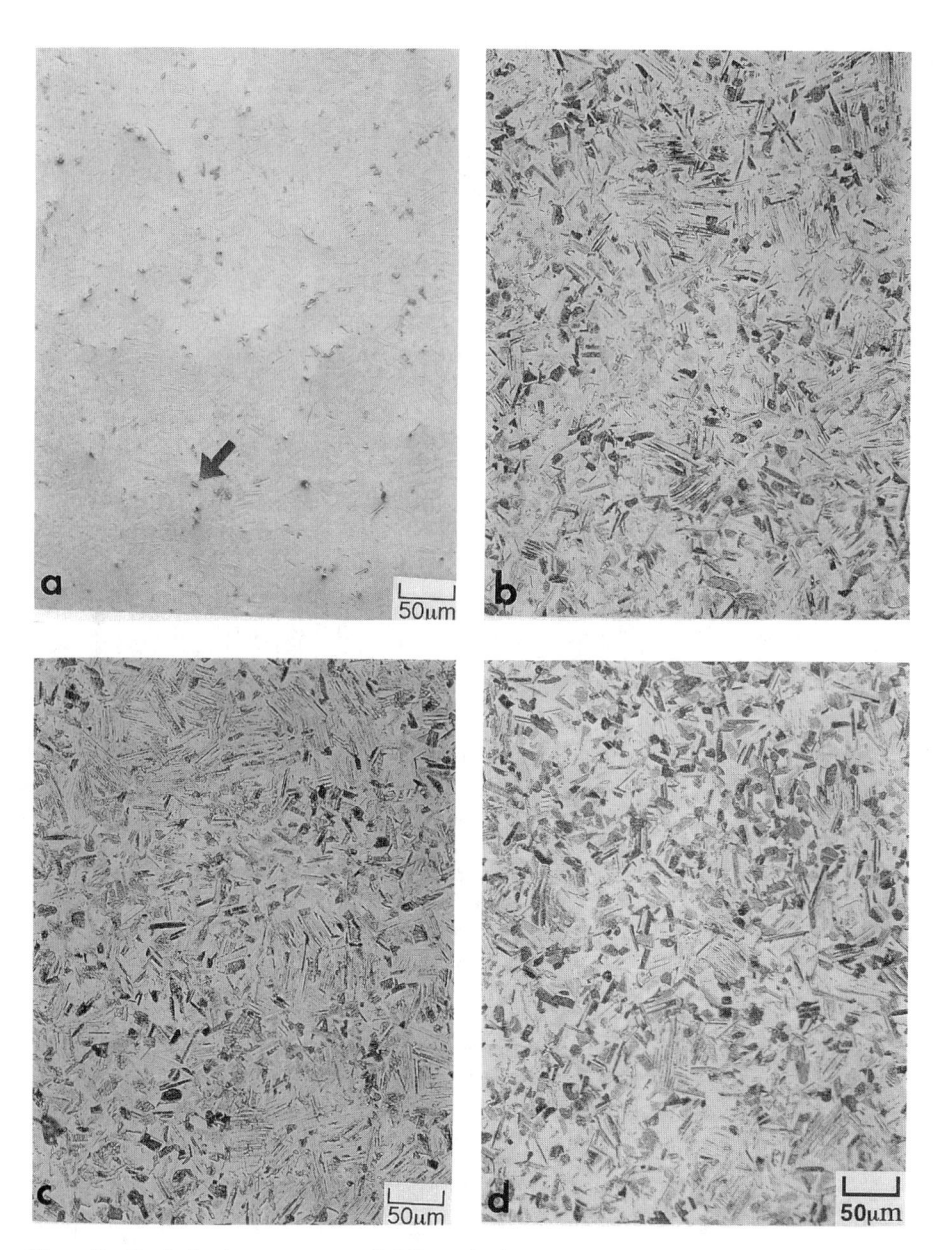

Figure 3 - Optical microstructures of deformed Ti-45Al-2Nb-2Mn-0.8TiB$_2$ a) as-HIPed, as polished, b) as-HIPed and etched, c) 900°C/20hrs heat treatment, and d) two-step heat treatment.

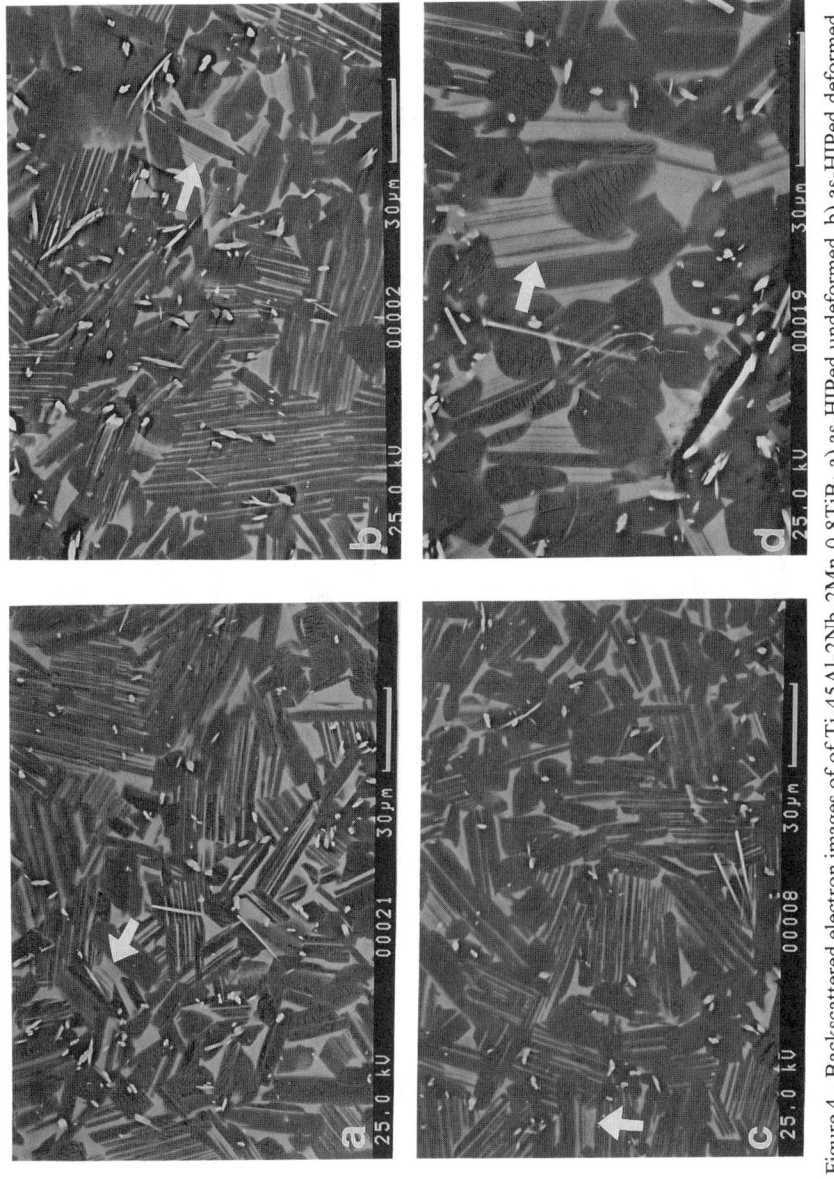

Figure 4 - Backscattered electron images of of Ti-45Al-2Nb-2Mn-0.8TiB$_2$ a) as-HIPed undeformed, b) as-HIPed deformed, c) 900°C/20hrs heat treatment, and d) two-step heat treatment.

resulted from thermally induced transformation, so they are more likely to have resulted from local stress concentrations [12,14,17]. These mechanical twins were observed in all heat treatments. Since the α_2 grains have an angular shape, stress concentrations are more likely to be focused at triple points in α_2 grains, where the fine mechanical twinning is most often observed.

The fine twinned regions are also observed in TEM images. They occur in the angular shaped α_2 grains, as illustrated in Figure 5.a. Regions of extremely fine mechanical twinning are also observed in some lathes of the lamellar grains (Figure 5.b). The width of the mechanically twinned region appears to be larger in regions with a higher stress concentration. Dislocations are also observed in the vicinity of TiB_2 particles, in Figure 5.c.

Ti-47Al-2Nb-2Mn-0.8TiB$_2$

The TiB_2 particles in the as-HIPed alloy in Figure 6.a appear larger (10-70 μm) than in the 45 Al specimen in Figure 3.a, but they have a similar spacing. Microstructures of as-HIPed specimens after creep deformation exhibit a heterogeneous microstructure illustrated in Figure 6.b. There are clusters of small equiaxed grains in some regions where there are fewer lamellar grains, and conversely, a few fine γ grains are observed in regions containing lamellar grains. After heat treatment in Figure 6.c and d the microstructures exhibit fewer small grain clusters. The volume fraction of equiaxed grains changes, and after the two step heat treatment, the lamellar volume fraction becomes largest. The maximum grain size of lamellar grains is about 200 μm. As time and temperature of heat treatment increase, microstructures became more homogeneous. The equiaxed grain size distribution becomes more homogeneous, and fewer clusters of finer γ grains were observed.

Ti-47Al-2Nb-2Cr

Although no TiB_2 is present in this alloy, there are second phases present (Figure 7.a). However, in contrast to the microstructures containing TiB_2, the as-HIPed microstructure exhibits very large 300-400 μm lamellar grains and about 200 μm pockets of equiaxed 10-20 μm grains (Figure 7.b). The microstructure is extremely heterogeneous, and it exhibits some of the same types of heterogeneities (clusters of small equiaxed grains) observed in Figure 6.b. With heat treatment, the volume fraction of the lamellar grains decreases with increasing time and temperature, and the microstructure becomes less heterogeneous. The pockets of equiaxed grains become fewer and there are fewer fine grains in the pockets.

In the backscattered electron SEM images of this alloy there is a smaller volume fraction of angular equiaxed α_2 grains between gamma grains and the α_2 laths are thinner in lamellar grains (Figure 8). The very fine mechanical twins seen in the 45 Al sample (Figure 4) were not observed in the smaller α_2 equiaxed grains in the 47 Al specimens. Very small bright phases can be seen either in α_2 equiaxed grains or boundaries of gamma grains (indicated by arrows) and these phases are β_2 phase which is produced by addition of Cr. They have an orientation relationship of $(0001)\alpha_2//\{111\}\gamma//\{110\} \beta_2$, and $<11\bar{2}0>\alpha_2//<1\bar{1}0>\gamma//<\bar{1}11>\beta_2$ [18]. Since Cr is known to strengthen the α_2 and γ phases, this may account for the lack of mechanical twinning observed in the α_2 phase in the Ti-47Al-2Nb-2Cr alloy [18]. There appears to be a decrease in the amount of β_2 phase with the increasing amounts of heat treatment (Figure 8.c,d). At lower magnifications, the distribution of large equiaxed grains correlates with higher aluminum content regions (which are darker in Figure 8.e). With deformation, some dynamic growth of α_2 regions drain the surrounding region of titanium, in the lamellar region illustrated in the center of Figure 8.f. Also there appears to be less segregation since there are fewer dark (alumium rich) regions after creep deformation (Figure 8.e~h).

TEM investigations indicate that a 170° angle relationship between lamellae is frequently

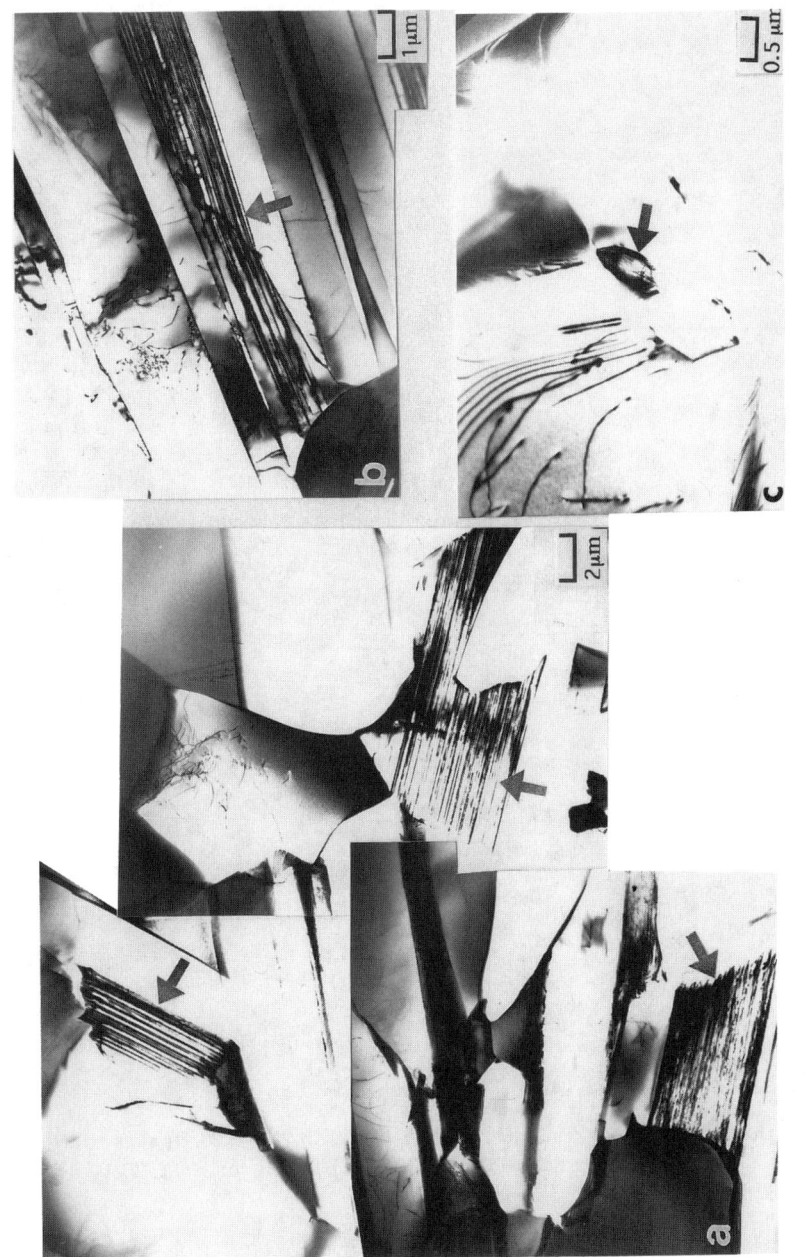

Figure 5 - Bright field images of as HIPed Ti-45Al-2Nb-2Mn-0.8TiB$_2$ after creep test a) Fine mechanical twins observed in α_2 grains b) Fine mechanical twins observed in lamellar regions, c) dislocations interactions with a TiB$_2$ particle.

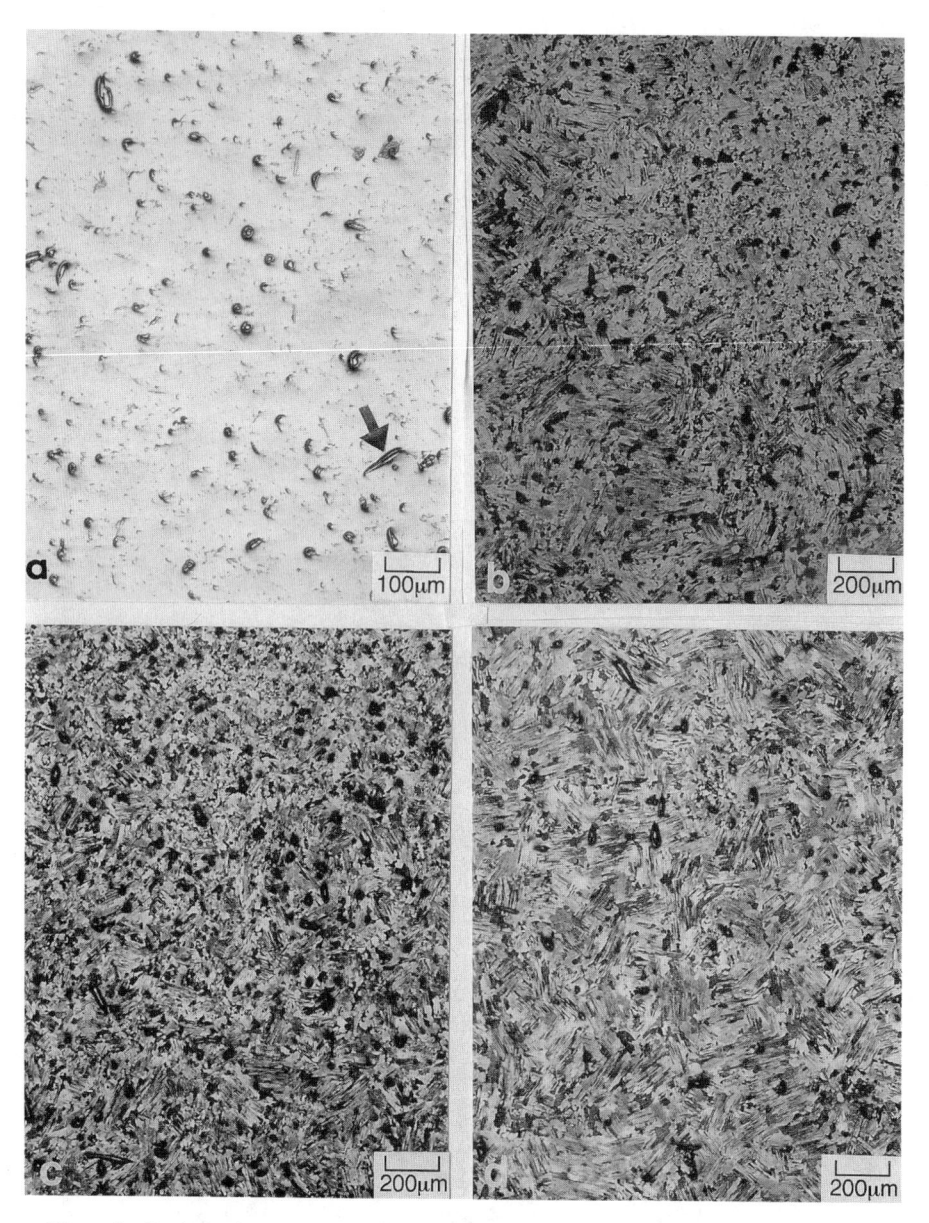

Figure 6 - Optical microstructures of deformed Ti-47Al-2Nb-2Mn-0.8TiB$_2$ a) as-HIPed, as polished, showing TiB$_2$ particles, b) as-HIPed and etched, c) 900°C/20hrs heat treatment, and d) two-step heat treatment.

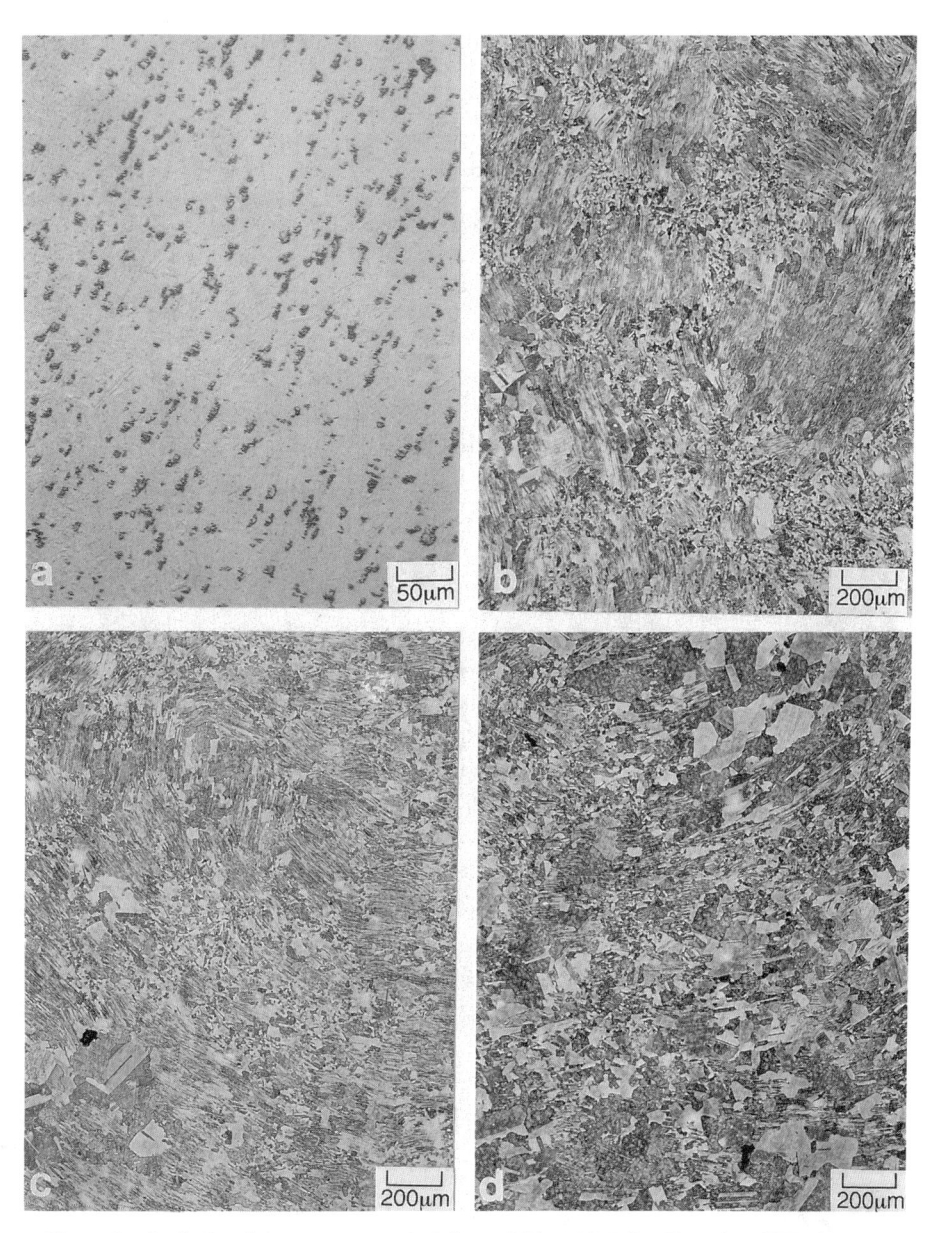

Figure 7 - Optical microstructures of deformed Ti-47Al-2Nb-2Cr a) as-HIPed, as polished, b) as-HIPed and etched, c) 900°C/20hrs heat treatment, and d) two-step heat treatment.

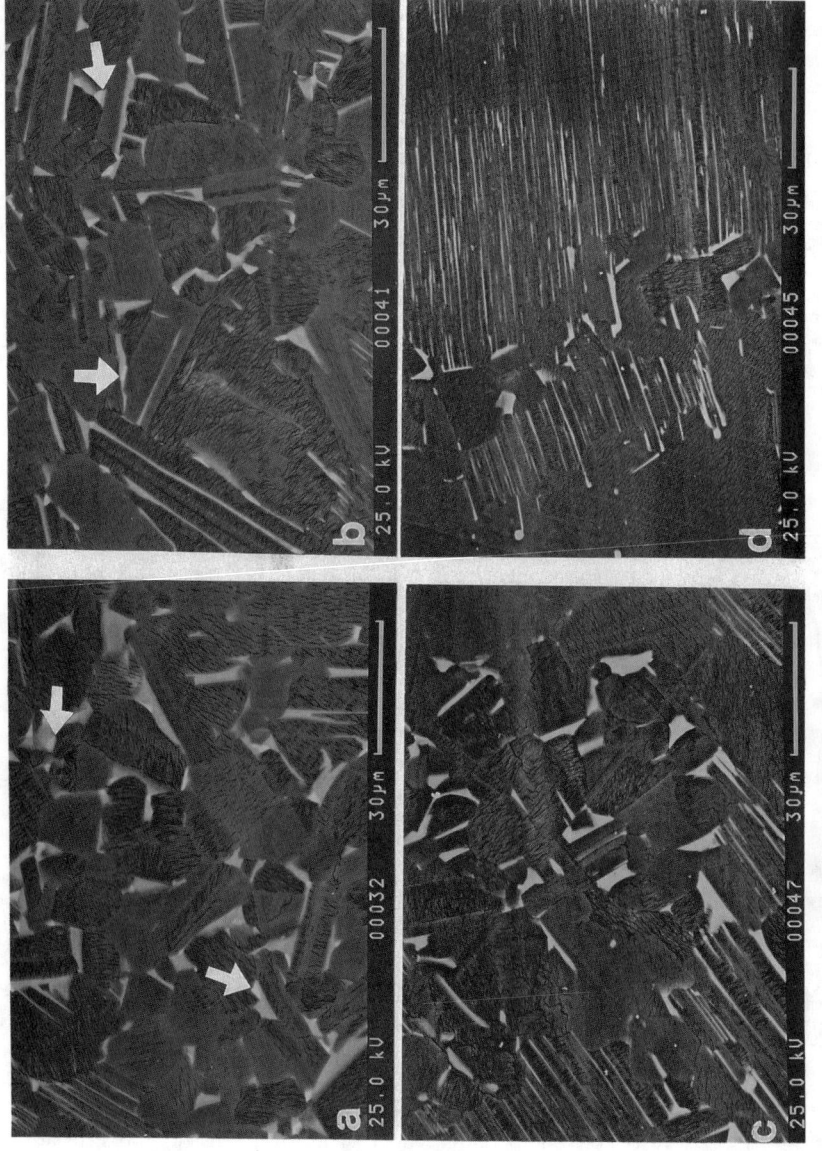

Figure 8 - Backscattered electron images of of deformed Ti-47Al-2Nb-2Cr a) as-HIPed, b) 900°C/20hrs heat treatment, and c) and d) two-step heat treatment.

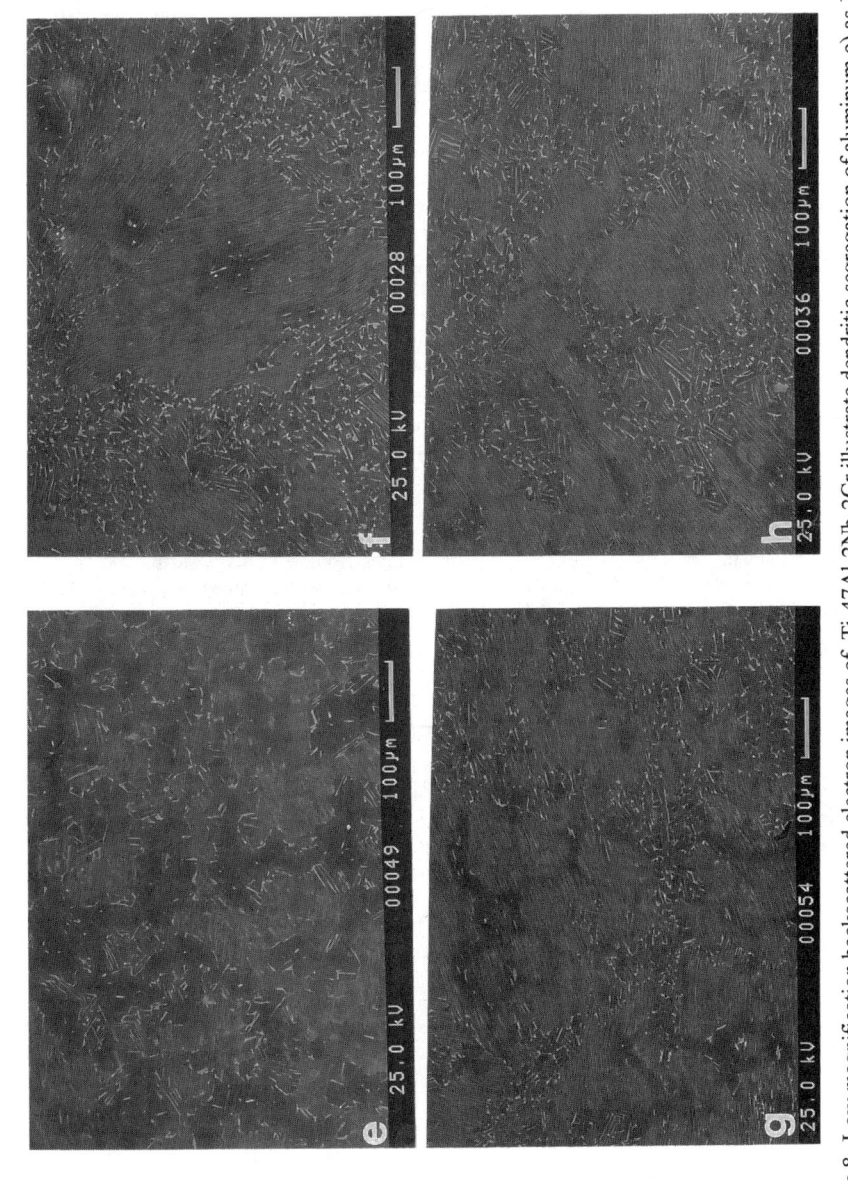

Figure 8- Low magnification backscattered electron images of Ti-47Al-2Nb-2Cr illustrate dendritic segregation of aluminum. e) as-HIPed undeformed, f) as-HIPed deformed, g) 900°C/20hrs heat treatment, undeformed, and h) 900°C/20hrs heat treatment, deformed.

observed in the as-HIPed and heat treated conditions. Typically, one of the two lamellar lathes has a finer spacing as shown in Figure 9.a. Subgrains developed in larger regions of lamellar grains during high temperature deformation in Figure 9.b. No subgrains had developed fully in equiaxed grains even though there were regions exhibiting a dislocation density greater than 2 X10^{12}/m^2 in Figure 9.e. From Figure 9.c, it appears that an equiaxed grain is consuming a lamellar region, but the α_2 lamellae resist consumption. After the two-step heat treatment, dynamic precipitation was observed in grain interiors and interfaces in Figure 9.e, and this may account for some the exceptional creep resistance [11].

By using the dark field method and tilting techniques, the spacing of γ laths, α_2/α_2 laths and the thickness of α_2 laths in several lamellar grains were measured. The data was obtained by measuring about 100-350 laths and taking the average as shown in Figure 10 and Table 2. The spacing generally decreases with more heat treatment. In the case of γ lamellar spacing, there was not much reduction of spacing from heat treatment, only about 8%. However, a 4% reduction of γ lamellar spacing after creep deformation was observed in the 900°C/20hr heat treated alloy. The refinement of the lamellar spacing could result from mechanical twinning parallel to lamellar boundaries, since this is an easy mode of deformation. Measured α_2/α_2 spacing shows that the spacing decreases from about 5% to 18% from the As HIPed condition to the one and two-step heat treatments. After deformation, in the one-step heat treatment there was a 22% reduction of the α_2/α_2 spacing, and a 10% reduction in the α_2 thickness. This may result from nucleation of new α_2 lamellae initiated by mechanical twinning, since a mechanical twin has the crystal structure of the α_2 phase for three layers in each interface [14]. The heat treatment and creep deformation cause the lamellar scale to become more refined.

Figure 11 shows inverse pole figures for 3 heat treatments. Figure 11.a shows that mechanical twining has a low schmid factor near 001, and increasing heat treatment tends to increase the volume fraction of crystal orientations that are hard to twin.

Discussion

Ti-45Al-2Nb-2Mn-0.8TiB$_2$

As shown in Figure 2, primary creep resistance does not improve much and only a slight increase occurs after heat treatment from the as-HIPed condition. There is also no appreciable change in microstructure. The poor primary creep resistance can be most clearly correlated with the large amount of fine mechanical twinning observed in the microstructure, particularly in the α_2 grains. Heat treatment did not affect the rate of generation of these fine mechanical twins. The fine microstructures can also account for poor creep resistance, since diffusional creep processes are accelerated with more interfacial area.

Ti-47Al-2Nb-2Mn-0.8TiB$_2$

The as-HIPed condition gives more creep resistance than the 45Al composition, but after heat treatment at 900°C for 10 hours the creep resistance decreases substantially. Heat treatment for 20 hours at the same temperature and the two step heat treatment provides slightly improved creep resistance. The microstructure of this alloy has a larger scale and the (TiB$_2$) particles are bigger and are more evenly distributed than in the 45Al alloy as shown in Figures 3 and 6. The particles were often observed with dislocation tangles nearby. Thus the primary creep resistance in all heat treatments are slightly improved except for the 900°C/10hrs case, where the only explanation is a less homogeneous microstructure and a higher lamellar volume fraction. With further heat treatment, the microstructure becomes more homogeneous and the

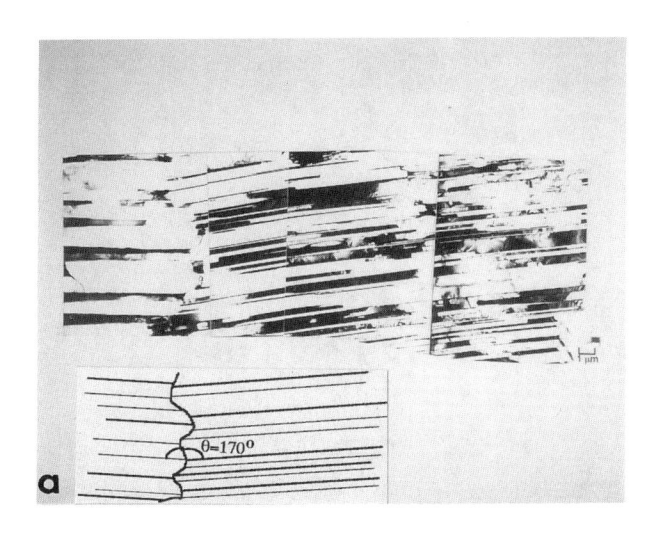

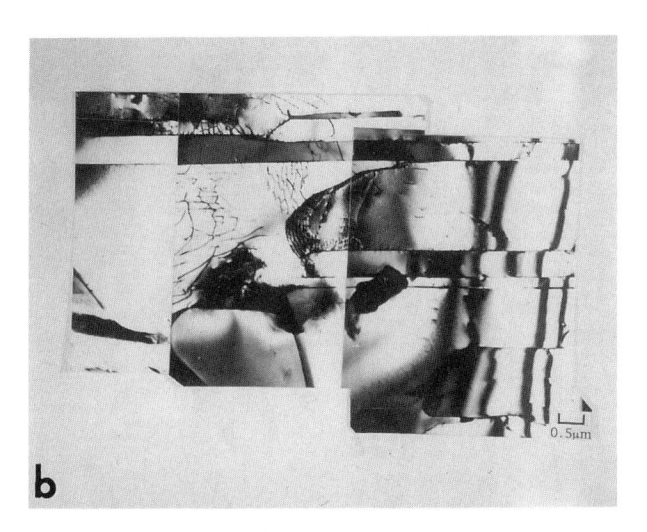

Figure 9 - Bright field images of as HIPed Ti-47Al-2Nb-2Cr after creep test a) commonly observed relationship between coarser and finer lamellar regions exhibit a 170° about a <110> axis relationship, b) subgrain boundaries develop in larger γ regions.

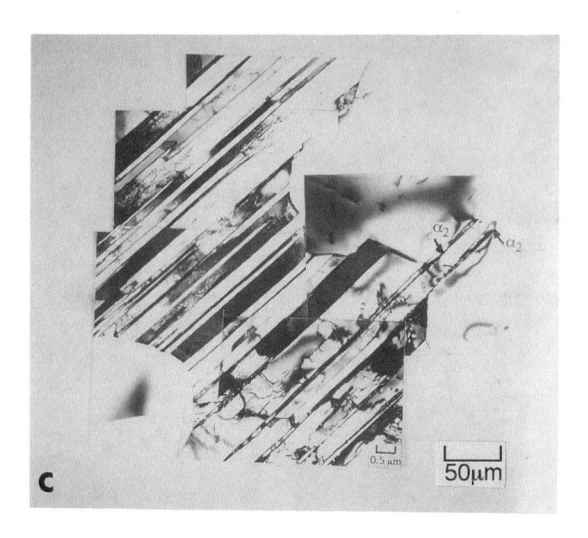

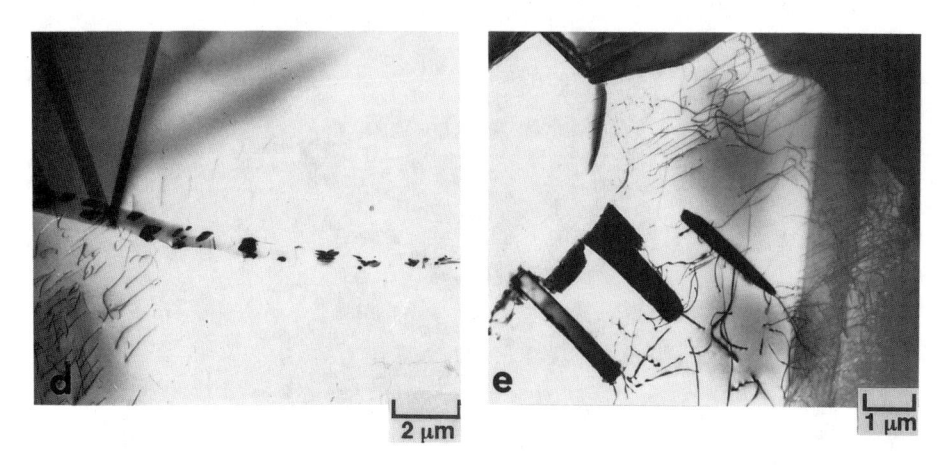

Figure 9 - Bright field images of Ti-47Al-2Nb-2Cr after creep test c) α_2 lathes resist absorption in a boundary where an equiaxed γ grain is consuming a lamellar grains(900°C/20hrs heat treatment), d) particles are located at grainboundary, and e) inside a γ grain after two-step heat treatment.

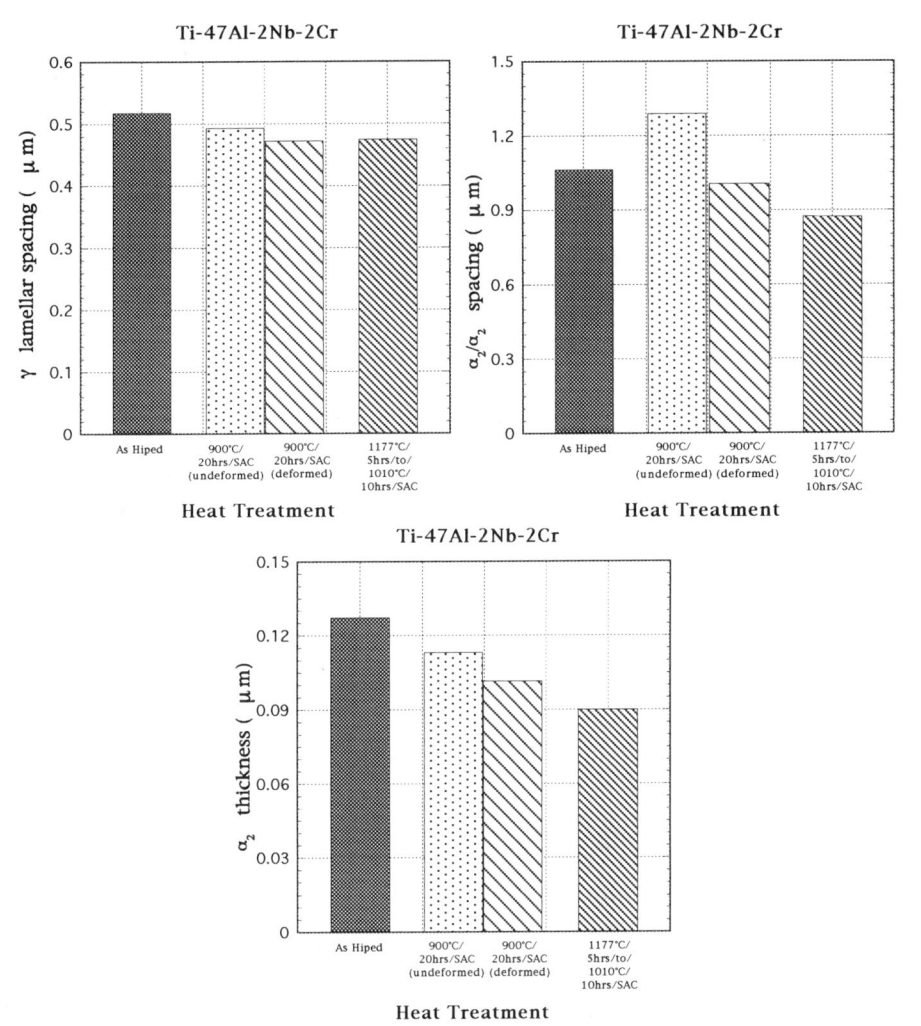

Figure 10 - Refinement of lamellar feature was observed with increasing heat treatment conditions, and resulting from deformation.

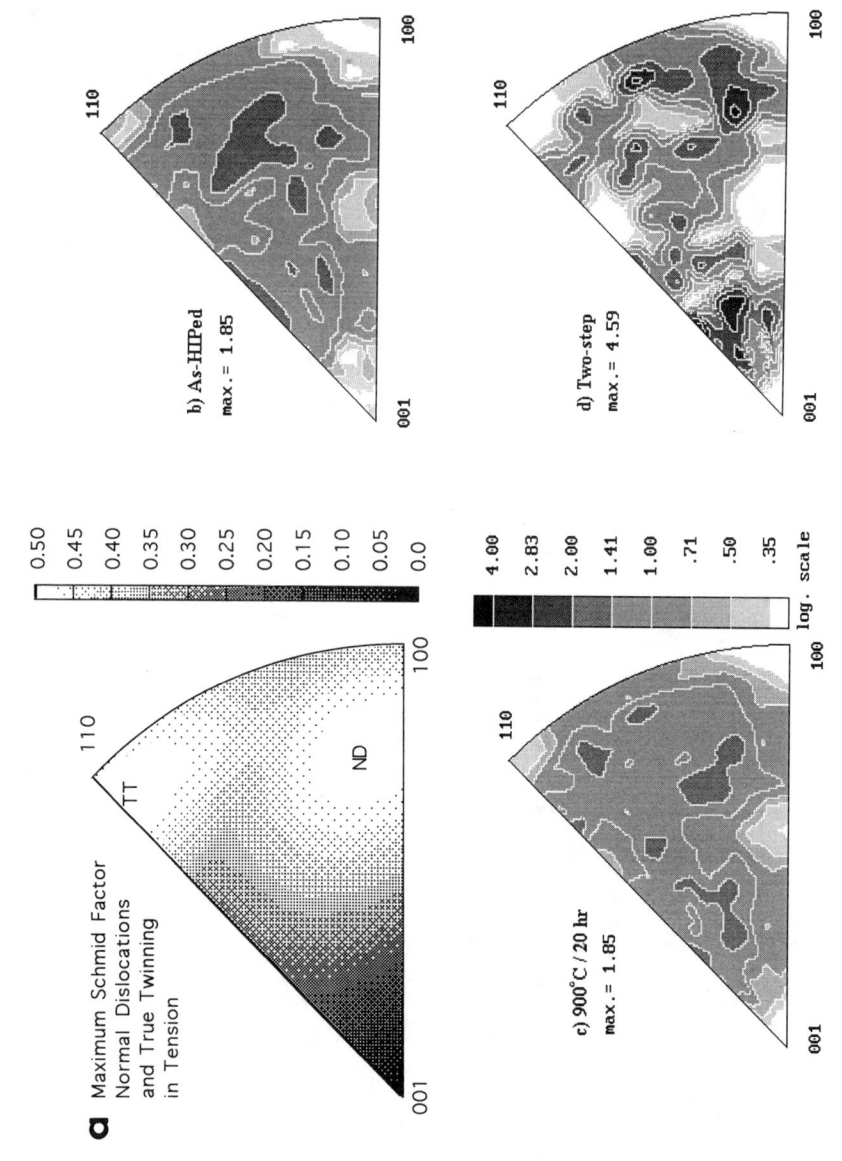

Figure 11- Inverse pole figures for 3 heat treatments a) Schimid factor, b) as-HIPed, c) 900°C/20hrs heat treatment, d) two step heat treatment after creep test.

number of small equiaxed grains decreases. However, it appears that the benefit gained by reducing the number of small grains was offset by the increasing the lamellar volume fraction after the two-step heat treatment.

Ti-47Al-2Nb-2Cr
This alloy shows the best primary creep resistance due to the fact that heat treatment provides substantial benefits (Figure 2). As time and temperature of heat treatment increase the creep resistance increases. The two-step heat treatment provides the best 0.5% creep resistance of about 140 hours. This microstructure has the smallest volume fraction (18%) of lamellar grains, a homogeneous microstructure, the least amount of pockets of small equiaxed grains and bigger equiaxed grains, which all correlate with better primary creep resistance.

In general, the lamellar microstructure gives a lower minimum creep rate [2]. However, this is not true for primary creep resistance, where initial deformation occurs easily in the lamellar microstructure. By comparing Figure 8 with 4, fine mechanical twins are found in the angular equiaxed α_2 grains, but not in the α_2 regions in the 47Al material. One difference is that the α_2 grains in the 45 Al material are larger and more complicated in shape, and this may lead to larger stress concentrations and a geometrical condition that permits mechanical twinning to accomodate the stress concentration. Figure 11 shows that resistance to mechanical twinning occurs when crystals are oriented near 001 and away from 441, and this desirable texture is strengthened with heat treatment. The increased volume fraction in a hard orientation for twinning may contribute to the excellent creep resistance resulting from the two-step heat treatment. Figure 10 shows that lamellar grains were refined after deformation. The correlation of refined lamellar spacing with better creep resistance indicates that the contribution of the lamellar microstructure to the strength can arise from a Hall-Petch relationship that reduces slip distances, and/or from a sub-grain strengthening effect, if a twin interface can be considered effectively similar to a low-angle subgrain boundary. Since the Hall-Petch effect is associated with a low temperature strengthening mechanism, and the large equiaxed grain size is consistent with better diffusion creep resistance, then the primary creep resistance at 760°C arises from a particular mixture of low and high temperature strengthening mechanisms. Consequently, deformation at a different temperature may result in a different balance of strain arising from low and high temperature mechanisms.

Conclusions

The effect of different heat treatments on primary creep resistance in three investment cast TiAl alloys were studied. Ti-45Al-2Nb-2Mn-0.8TiB$_2$ did not show much difference in primary creep resistance because of no change in the microstructures resulting from various heat treatments. The creep resistance was poor, due to very fine microstructures and very fine mechanical twining observed in α_2 grains. Some limited improvement in the Ti-47Al-2Nb-2Mn-0.8TiB$_2$ alloy was obtained with heat treatment. However, the heat treatments and especially the two step case significantly improved the creep resistance of the Ti-47Al-2Nb-2Cr alloy. This heat treatment led to the smallest volume fraction (18%) of lamellar grains, a homogeneous microstructure, the least amount of pockets of small equiaxed grains, and large equiaxed grains, which all correlate with better primary creep resistance. Texture analysis indicated that the desirable texture to minimize mechanical twining is to have crystals oriented near 001 and away from 441. This texture is strengthened with heat treatment and it may contribute to the excellent creep resistance obtained from the two-step heat treatment.

References

1. Y. W. Kim and D.M. Dimiduk, JOM, vol. 43, No.8, (1991), p.40.
2. S.C. Huang and Y.W. Kim, Scripta Metall., 25, (1991), p.1901.
3. Y. W. Kim, JOM, vol. 41, No.7, (1989), p.24.
4. M. Yamaguchi and H. Inui, Structural Intermetallics, eds. R. Darolia, et al., (TMS, Warrendale, PA, 1993), p.127
5. S. Tsuyama, S. Mitao and K. Minakawa, Mater. Sci. Eng., A153, 1992, p.427.
6. S.C. Huang, D.W. McKee, D.S. Shih and J.C. Chestnutt, Intermetallic Compounds-Structure and Mechanical Properties, ed. O. Izumi, (The Japan Institute of Metals, Sendai, 1991), p.363.
7. P.L. Martin and H.A. Lipsett, Proc. 4th Int. Conf. Creep & Fracture of Engineering Materials and Structures, (The Institute of Metals, London, 1990).
8. T. Maeda, M. Okada, Y. Shida and M. Nakanishi, Proc. 1989 Sapporo Meeting, (The Japan Institute of Metals, Sendai, 1989), p.238.
9. T. Maeda, M. Okada and Y. Shida, MRS Symp. Proc., vol 213, (MRS Pittsburgh, PA, 1991), p.555.
10. P.R. Bhowal, H.F. Merrick and D.E. Larsen, Mater. Sci. Eng., A192/193, 1995, p.685-690.
11. D.Y. Seo, S.U. An, T.R. Bieler, D.E. Larsen, P. Bhowal, H. Merrick, Gamma Titanium Aluminides, ed. Y-W. Kim, et al. (TMS, Las Vegas, NV 1995), p.745.
12. Z. Jin, S.W. Cheong and T.R. Bieler, Gamma Titanium Aluminides, ed. Y-W. Kim, et al. (TMS, Las Vegas, NV 1995), p.975.
13. T.R. Bieler, D.Y. Seo R.S. Beals, C.H. Wu, and S.L. Choi, Gamma Titanium Aluminides, ed. Y.W. Kim, et al. (TMS, Las Vegas, NV 1995), p.795.
14. Z. Jin and T.R. Bieler, Phil Mag., 70,(1995), p.819-836.
15. J. L. Murray, Binary Alloy Phase Diagrams, ed. T.B. Massalski, ASM, Metals Park, OH, 1986, p.173.
16. T.S. Kallend, U.F. Kocks, A.D. Rollett, and H.R. Wenk, Mater. Sci. Eng., A132, 1991, p.1-11.
17. Z. Jin, R. Beals and T.R. Bieler, Structural Intermetallics, eds. R. Darolia et al., (TMS, Warrendale, PA 1993), p.275.
18. Y. Zheng, L. Zhao and K. Tangri, Scripta Metall., 26, (1992), p.291.

Table 2 Average and range of dimensions describing lamellar microstructures

Ti-47Al-2Nb-2Cr	As HIPed	900°C/ 20hrs/SAC (undeformed)	900 °C/ 20hrs/SAC (deformed)	1177 °C/ 5hrs/to/1010°/ 10hrs/SAC
γ lamellar spacing	0.518 µm 0.05~2.43 µm	0.493 µm 0.07~2.0 µm	0.475 µm 0.083~1.65 µm	0.473 µm 0.125~1.98 µm
α_2/α_2 spacing	1.064 µm 0.55~1.68 µm	1.289 µm 0.1~5.25 µm	1.007 µm 0.15~2.41 µm	0.874 µm 0.25~2.42 µm
$\alpha 2$ thickness	0.127 µm 0.04~0.415 µm	0.113 µm 0.04~0.25 µm	0.102 µm 0.06~0.166 µm	0.090 µm 0.01~0.15 µm

Simulation of Multicomponent Evaporation in Electron Beam Melting and Refining[*]

A. Powell[**]   J. Van Den Avyle   B. Damkroger   J. Szekely[**]

Liquid Metal Processing Laboratory, MS 1134, Sandia National Laboratories, PO Box 5800, Albuquerque, NM 87185. Phone: (505) 845-3105, Fax: (505) 845-3430.

[**]Massachusetts Institute of Technology, 77 Massachusetts Ave. Rm. 8-135, Cambridge, Massachusetts 02139. Phone: (617) 253-3222/3236, Fax: (617) 253-8124.

Abstract

Experimental results and a mathematical model are presented to describe differential evaporation rates in electron beam melting of titanium alloys containing aluminum and vanadium. Experiments characterized the evaporation rate of commercially pure titanium, and vapor composition over titanium with up to 6% Al and 4.5% V content as a function of beam power, scan frequency and background pressure. The model is made up of a steady-state heat and mass transport model of a melting hearth and a model of transient thermal and flow behavior near the surface. Activity coefficients for aluminum and vanadium in titanium are roughly estimated by fitting model parameters to experimental results. Based on the ability to vary evaporation rate by 10-15% using scan frequency alone, we discuss the possibility of on-line composition control by means of intelligent manipulation of the electron beam.

---

[*]This work performed at Sandia National Laboratories is supported by the U. S. Department of Energy under contract number DE-ACO4-94AL85000.

Introduction

Although the electron beam cold hearth melting and refining process (shown schematically in figure 1) is extensively and increasingly used for the production of commercially pure (c.p.) titanium [1] [2], its use in producing titanium alloys has been limited by poor control of ingot composition which necessitates subsequent homogenization [3] [4]. This is due in part to irregularities in feed chemistry, which can be adjusted on-line, but also to frequent freezing and remelting of metal in the hearth and changes in throughput rate, which are very difficult to control.

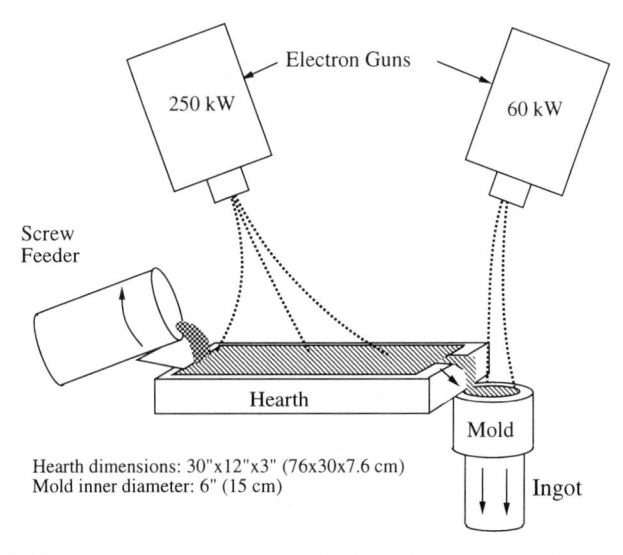

Figure 1: Electron beam melting facility at Sandia National Laboratories.

Because of this, control of composition downstream in the process could be very beneficial to electron beam melting of titanium alloys. The effect of beam scan frequency on evaporation rate, which has been extensively demonstrated experimentally [5] [6] [7], may prove to be a means of establishing such downstream control. The development of a process model is thus in progress, both to evaluate potential use of scan frequency in composition control, and if it appears promising, to aid in the development of a suitable on-line control system.

The work described here includes experiments necessary for development of such a model, resulting in estimates for the activity of aluminum and vanadium in titanium and characterization of evaporation rate, and also a surface layer model for the estimation of enhanced evaporation rate due to transient heating by the scanning beam.

Experiments

Two sets of experiments were run in the hearth of the Sandia National Laboratories Liquid Metal Processing Laboratory electron beam furnace. These examined the effect of process parameters on evaporation rate of c.p. titanium, as measured by vapor condensation rate a known distance above the melt surface, and on the relationship between melt and vapor chemistry in Ti-Al-V melting. The 250 kW gun was used alone at power levels from 150 to 265 kW, scan frequencies from 30 to 450 Hz, and beam spot size set from approximately 2 to 4 cm (though attempts to measure the actual spot size were unsuccessful), under background argon pressure between 0.13 and 40 Pa (10^{-5} to 3×10^{-3} torr).

Measurements of c.p. titanium evaporation rate were begun by holding process parameters constant for at least seven minutes in order to achieve steady-state pool geometry and skull heat transfer conditions. A water-cooled probe with vapor condensation substrate was then inserted to a known position and orientation in the furnace for approximately sixty seconds, and subsequently withdrawn. Thicknesses of films thus deposited were measured by optical microscopy of sectioned substrates.

In the alloy experiments, composition variation was achieved by starting the beam over a charge of as-received Ti-6%Al-4%V and taking melt and vapor condensate samples periodically as aluminum evaporated preferentially out of the melt, moving the composition down along the gray line shown in figure 2. After this, a steady state composition was established by melting alloy scrap and flowing it through the hearth to build an ingot at a constant rate, so that vapor samples at various beam and chamber conditions could be taken with constant melt composition. Finally, c.p. titanium was slowly added to the hearth in order to obtain melt and vapor samples at low vanadium concentrations. Electron probe microanalysis was used to determine the composition of all samples.

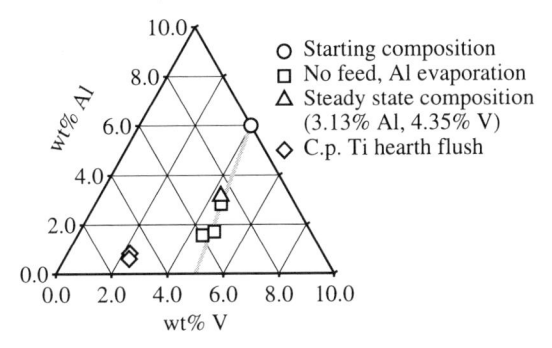

Figure 2: Hearth melt compositions for alloy experiments, shown in the Ti-Al-V system.

Experimental Results

Figure 3 shows the variation of pure titanium condensation rate at the substrate with process parameters. Inspection of the graph reveals the expected rise in evaporation rate with increasing beam power and decreasing spot size. Chamber pressure presents competing effects of gas focusing of the beam and interference with evaporant transport. However, the frequency correlation is the opposite of what was expected. Further analysis of the process showed that our beam deflection system could not track the imposed pattern at the highest frequencies, so the resulting pattern was narrower and considerably more intense, leading to an increase in maximum temperature and evaporation rate.

In order to separate out the effect of pattern size on evaporation rate, we calculated an expected evaporation rate by integrating Langmuir evaporation rates over temperature maps of the hearth surface taken at 30 and 450 Hz [8]. Based on this method, the temperature profile arising from the 450 Hz pattern is expected to lead to a 44% higher evaporation rate than at 30 Hz. The average observed rise of 25% at 0.13 Pa thus shows a net decrease of approximately 19% due to increased scan frequency.

Melt and vapor compositions were measured as described above, and are summarized in figure 4. The "evaporation ratio" ER shown in figure 4 is defined (for aluminum) as

$$ER_{Al} = \frac{\text{wt\%Al}_{\text{vapor}}/\text{wt\%Ti}_{\text{vapor}}}{\text{wt\%Al}_{\text{melt}}/\text{wt\%Ti}_{\text{melt}}}. \tag{1}$$

This is used because it can be shown [9] that this evaporation ratio is equivalent to

$$ER_{Al} = \gamma_{Al} \frac{\bar{p}_{vAl}}{\bar{p}_{vTi}} \sqrt{\frac{M_{Ti}}{M_{Al}}}, \tag{2}$$

where γ_i is the activity coefficient, $\bar{p}_{vi}$ is the vapor pressure of pure species i, and M_i its molecular weight, as long as aluminum activity follws Henry's law. Because these material properties do not vary with composition, the evaporation ratio is not expected to do so.

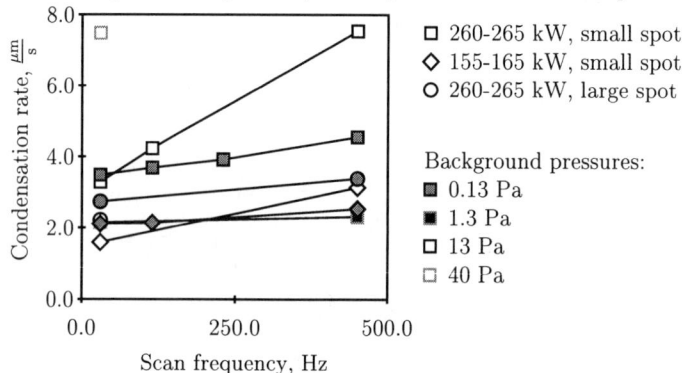

Figure 3: Effect of process parameters on measured condensation rate of pure titanium.

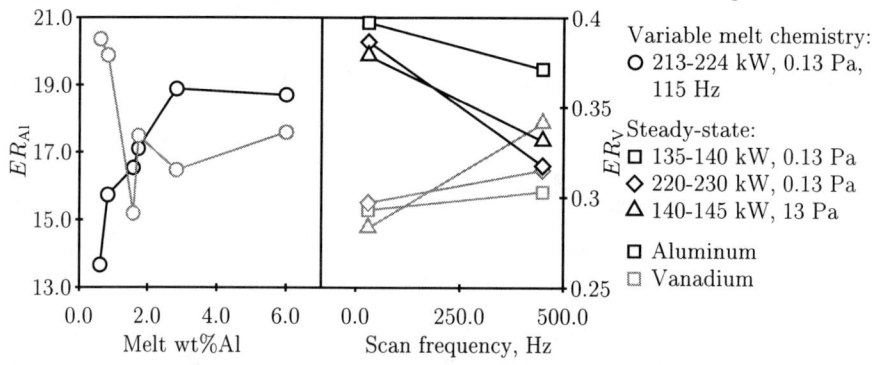

Figure 4: Effect of melt composition and process parameters on measured evaporation ratios of aluminum and vanadium in titanium.

Measured evaporation ratios fell in the range of 16-21 for aluminum and 0.28 to 0.34 for vanadium over the range of process parameters used in the steady-state experiment. The ratio of pure element vapor pressures varies considerably, from just over 800 at the melting point of titanium to around 140 at a 500°C superheat for aluminum, and from 0.16 to 0.37

over the same temperature range for solid vanadium (vapor pressure of undercooled liquid vanadium at these temperatures is not available).

Aluminum evaporation ratios were found to decrease with increasing scan frequency, increasing beam power, increasing background gas pressure, and decreasing Al and V content; Vanadium exhibited the opposite tendencies, presumably because of its lower evaporation rate than that of titanium. From these correlations, it would seem that the aluminum evaporation ratio is falling with increasing surface temperature, so the decreasing pure vapor pressure ratio controls that change. It is interesting that although the evaporation ratio is related to chemical parameters that are independent of composition, the biggest change in both evaporation ratios occurred when the vanadium concentration changed (figure 4; figure 2 indicates the lowest Al concentration coincides with very low V).

In order to calculate the activity coefficient γ_{Al}, we consider the quantity $\frac{\bar{p}_{vAl}}{\bar{p}_{vTi}}\sqrt{\frac{M_{Ti}}{M_{Al}}}$, which is equal to ER_{Al}/γ_{Al} and is a function of temperature alone. The average value of that quantity over a hearth surface temperature map is 317 at 30 Hz and 283 at 450 Hz. Based on average aluminum evaporation ratios of 20.3 at 30 Hz and 17.8 at 450 Hz, this gives us activity coefficients of 0.0640 and 0.0628 respectively. The same treatment for vanadium gives average ratios of 0.312 and 0.293, and dividing the average ER_V values of 0.291 and 0.320 by these yields activity coefficients for solid vanadium of 0.993 and 1.027, which indicates approximately ideal behavior of V in molten Ti.

Surface Layer Heat Transfer and Evaporation

When considering the beam's ability to control composition, one must consider both composition limits which the beam can produce under steady-state conditions, and the dynamic ability to vary composition between those limits in response to process changes. Dynamic control will be covered in a later paper, so we turn here to the steady-state case.

In a continuous solute removal reactor such as a melting hearth, the ratio of solute concentration at the exit to that at the inlet C_{out}/C_{in} is a function of flow rate Q, reaction area A, and the reaction rate constant k'' (equal to the solute flux divided by concentration). The nature of this function will depend on the flow patterns present in the hearth, with the extremes in behavior being a perfect mixing tank, which behaves as

$$\frac{C_{out}}{C_{in}} = \frac{1}{1 + \frac{k''A}{Q}}, \tag{3}$$

and a plug flow reactor, which gives

$$\frac{C_{out}}{C_{in}} = \exp\left(-\frac{k''A}{Q}\right). \tag{4}$$

If we neglect dead zones in the flow pattern, the relevant area is the molten area of the hearth, and flow rate is given by throughput divided by density. The reaction rate constant for species i in solvent s is calculated as the ratio of Langmuir evaporation rate to molar density in the melt, and since Henrian vapor pressure is given by $\gamma_i \bar{p}_{vi} X_i$ (using the same definitions as equations 1–2), this gives

$$k''_i = \frac{\text{molar flux}}{\text{molar density}} = \frac{\frac{\gamma_i \bar{p}_{vi} X_i}{\sqrt{2\pi M_i RT}}}{X_i \frac{\rho_s}{M_s}} = \frac{M_s}{\rho_s} \frac{\gamma_i \bar{p}_{vi}}{\sqrt{2\pi M_i RT}}. \tag{5}$$

For aluminum in titanium, the reaction rate constant k''_{Al} takes on values from 6.19 $\frac{\mu m}{s}$ at the melting point of titanium to 239 $\frac{\mu m}{s}$ at a 500°C superheat. Using top surface temperature maps of the hearth at Sandia [8], we calculate an average k''_{Al} of $98.3\frac{\mu m}{s}$ at 30 Hz, and combining this with a typical flow rate of 23 $\frac{cm^3}{s}$ and top surface area of 0.2 m^2 gives the ratio $\frac{k''A}{Q}$ as 0.86. This will lead to retention of 42% of the aluminum under plug flow conditions, and 54% under perfect mixing, the latter of which agrees well with 52% retention (3.13% ÷ 6%) observed in the steady state alloy chemistry experiment.

We now consider the effect of beam frequency on the time-averaged reaction rate constant. When the transient motion of the beam has an effect on the evaporation characteristics, modeling such characteristics would seem at first to require a time-stepped simulation of the beam's travels over the whole hearth, which would be extremely computationally intensive. However, if the beam moves through the pattern relatively quickly, the depth of the layer heated during one complete scan will be much smaller than the width of the beam spot, and as long as the Peclet number is small (indicating transient thermal convection can be neglected), vertical conduction will dominate heat transfer away from the surface.

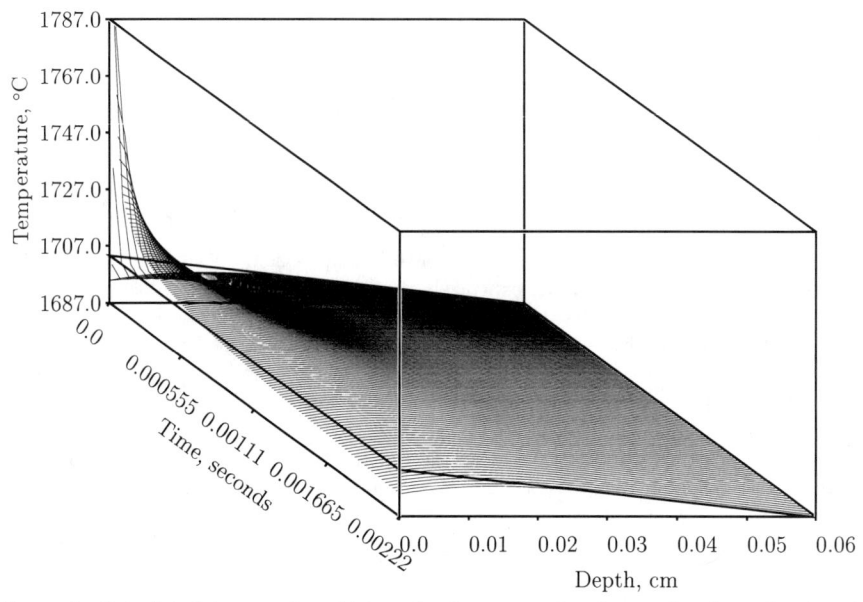

Figure 5: Calculated temperature history in the top surface layer of molten titanium hit by a 150 kW (net) beam at 450 Hz. For clarity, only every other timestep is shown here.

For this reason, temperature fluctuations near the surface can be considered locally one-dimensional, and the increase in losses for a given average surface temperature can be calculated by simply solving the heat conduction equation in one dimension. Unfortunately, the highly nonlinear nature of thermal losses at the surface makes the problem analytically intractable, but it is an easy problem to solve numerically, as has been done here.

The program written to solve this problem simulates transient heat transfer through the top layer of the melt, down to a depth $\delta_T = 4\sqrt{\alpha t}$ where α is the thermal diffusivity and t is the period of beam rastering (the inverse of the frequency). Below this depth, temperature does not change significantly. On the bottom of this surface layer temperature is held constant. On the top, heat flux is given by the difference between heat input from the beam and losses due to radiation and evaporation. The beam is modeled as the projection of a traveling Gaussian heat flux onto a point, which is thus a Gaussian distribution in time. Evaporation rate is assumed to follow ideal Langmuir evaporation into a vacuum.

This top surface layer simulation runs through several beam scan cycles until convergence is reached. A typical history for the last cycle of such a simulation is shown in figure 5. For accurate coupling with thermofluid simulations of the molten hearth, which linearize temperature distribution near the surface, a tangent line in the temperature-depth curve is drawn at the bottom of the simulated layer, and its intercept with the surface is considered to be the average surface temperature, as can be seen on figure 5. The time-averaged evaporative flux for a given species in solution is calculated from the surface temperature history, and plotted against the average surface temperature for several frequencies, as shown in figure 6.

For the simulations presented here, net beam power (less backscattering losses) was set to 150 kW, spot diameter to 2 cm, and overall scan pattern length to 2 m. Under these conditions, peak power density is about 24 kW/cm^2, and dwell time is one hundredth the beam scan period. Figure 6 shows calculated pure titanium evaporation rate ($\frac{kg}{m^2 \cdot sec}$) and aluminum reaction rate coefficient at three different frequencies and using the latter definition of average surface temperature. It is worth noting that evaporation rates at 450 Hz are not significantly different from those at constant temperature for titanium.

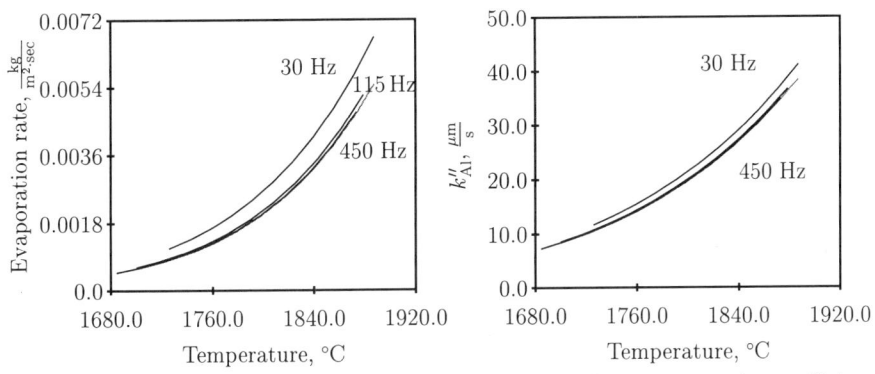

Figure 6: Calculated pure titanium evaporation rates and aluminum reaction coefficients due to a 150 kW (net) beam as a function of average surface temperature at 30, 115 and 450 Hz. The gray curves represent constant temperature evaporation rate.

Also, calculated increases in titanium evaporation at 30 Hz above constant temperature, which are about 40% near the melting point and 25% at a 200°C superheat, are of the same order of magnitude as the 19% net increase calculated from experimental data, and a 17% increase in titanium evaporation rate measured by Melde et al. [7, p. 79], when going from 20 ms to 300 ms dwell time. The higher values predicted by the model are due to the fact that the model deals with losses directly in the path of the beam, while most of the hearth is not in that path and will not exhibit such strong temperature fluctuations.

This frequency change has a somewhat smaller effect on the time-averaged aluminum reaction rate constant k''_{Al}, which is at most 11% higher at 30 Hz than at 450 Hz. Because aluminum concentration change varies with k''_{Al}, they will scale similarly; in the case of the Sandia furnace we can expect to be able to vary the concentration 5% (perfect mixing) to 10% (plug flow) by changing beam scan frequency alone.

Conclusions

Activity coefficients of aluminum and vanadium in titanium have been estimated at 0.063 and 1.0 based on melt and condensed vapor compositions and hearth temperature profiles. The only assumptions required for this estimation are Henrian behavior of the solutes and equal distribution of all three constituent elements throughout the vapor plume.

In addition, the mathematical model presented here gives a good estimate of the effect of beam scan frequency on pure titanium evaporation rate and aluminum activity. Based on this model, we conclude that because aluminum reactivity is not very sensitive to scan frequency, scan frequency alone probably can not produce the desired changes in composition, and it will be necessary to explore alternative pattern designs which produce large changes in evaporation rate without significantly affecting fluid flows and skull shape. The design of such patterns will require the coupling of this surface model with a more comprehensive model of fluid flow and heat transfer in the hearth or mold.

It is important to note that implementation of the type of process control described here will require the use of a beam deflection system capable of tracking patterns at very high frequencies, which is possible but somewhat expensive; and on-line measurement of melt chemistry, which is extremely difficult. These two considerations may hinder practical use of this technique for the foreseeable future.

References

[1] S.M. Tilmont and H. Harker, "Maximelt, an Update" (Paper presented at Electron Beam Melting and Refining State of the Art 1993), 214-225.

[2] DS. Lowe, "Electron Beam Cold Hearth Refining in Vallejo" (Paper presented at Electron Beam Melting and Refining State of the Art 1994), 69-77.

[3] J.C. Borofka, "Qualification of the 3.3 MW Maximelt EBCHR Furnace for Premium Quality Titanium Alloys" (Paper presented at Electron Beam Melting and Refining State of the Art 1992), 179-189.

[4] Dr. Entrekin, Axel Johnson Metals, private communication with author, TMS Annual Meeting, 14 February 1995.

[5] S. Schiller, A. von Ardenne and H. Förster. "Evaporation of Alloying Elements in EB-Melting–Possibility of Influence" (Paper presented at Electron Beam Melting and Refining State of the Art 1984), 49-69.

[6] M. Blum et al., "Results of Electron Beam Remelting of Superalloys and Titanium Alloys with a High-Frequency EB-Gun" (Paper presented at Electron Beam Melting and Refining State of the Art 1993), 102-115.

[7] C. Melde, M. Kramer, A. von Ardenne and M. Neumann, "The Super Deflection System – a Tool to Reduce the Evaporation Losses in EB Melting Process" (Paper presented at Electron Beam Melting and Refining State of the Art 1993), 69-80.

[8] M. Miszkiel, R. Davis, J. Van Den Avyle and A. Powell "Video Imaging and Thermal Mapping of the Molten Hearth in an Electron Beam Melting Furnace" (Paper presented at Electron Beam Melting and Refining State of the Art 1995).

[9] A. Powell et al., "Simulation of Multicomponent Losses in Electron Beam Melting and Refining at Varying Scan Frequencies" (Paper presented at Electron Beam Melting and Refining State of the Art 1995).

CASTING TECHNOLOGY FOR GAMMA TITANIUM ALUMINIDE VALVES

M. M. Keller, W. J. Porter, III, P. E. Jones, and D. Eylon
Graduate Materials Engineering
University of Dayton
Dayton, Ohio 45469-0240

Abstract

This paper reviews casting technology for gamma titanium aluminide automotive valves. It focuses on the permanent mold casting process examined by the EMTEC Automotive Valve Project. Over 800 Ti-47Al-2Nb-1.75Cr (at%) valves were produced using several variations of the permanent mold process: static, centrifugal, squeeze pin, and injection processes. Of the processes the injection process produced the finest as-cast microstructure. As-cast and as-HIP microstructures and tensile properties of the static and injection processes are presented as well as challenges to production scale up.

Introduction

Gamma titanium aluminide (γ-TiAl) alloys have been evaluated by both the aerospace and automotive industries for applications such as low pressure turbine blades and automotive exhaust valves. General Electric [1], General Motors [2], Volvo [3], and Ford Motor Company [4] have all reported successful engine tests of TiAl components. While γ-TiAl alloys show much potential, cost and processing are two serious impediments to the implementation of TiAl components. Investment casting offers a lower cost alternative to wrought processing, however the cost and cycle time associated with investment casting may not be suitable for the automotive industry which is concerned with producing high volume components at a very low cost. Permanent mold casting methods have been used to produce aluminum and steel components, but have not been widely used for titanium or TiAl alloys [5]. As a result there has been interest in exploring the feasibility of producing lightweight γ-TiAl intake and exhaust valves by permanent mold casting. The EMTEC Automotive Valve Project was established in response to this interest to investigate permanent mold casting of the General Electric alloy Ti-47Al-2Nb-1.75Cr [6]. During the course of this project over 800 valve blanks have been produced using several variations of the permanent mold process. Furthermore, TiAl valves from this project were road tested in two Chevrolet Corvettes. In both vehicles the γ-TiAl valves survived over 15,000 miles and exhibited no damage or defects upon completion of the road test (Figure 1). Other groups are also investigating permanent mold processing as a means to produce automotive exhaust valves. Daimler Benz also has reported successfully producing permanent mold cast TiAl valves for engine tests [7]. This paper will review and assess the variations of the permanent mold process which were investigated in the EMTEC project. Microstructural and mechanical properties will also be reported.

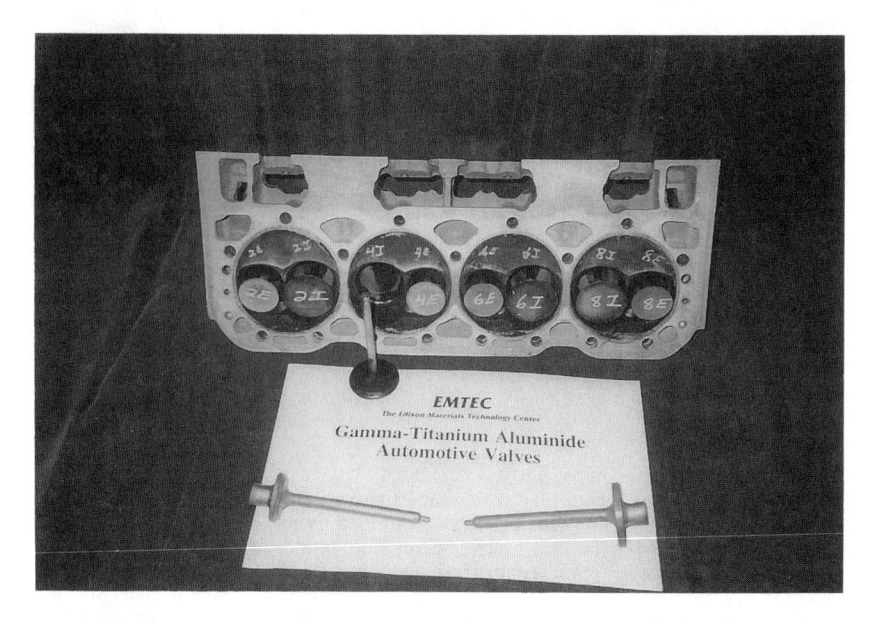

Figure 1: Cylinder head from road tested Chevrolet Corvette containing γ-TiAl alloy valves.

Material and Methods

The alloy selected by the EMTEC team was the General Electric alloy Ti-47Al-2Nb-1.75Cr (at%) [6]. This alloy was selected because it has an established record of good castability [8] and mechanical property data available for valve design. Figure 2 is a schematic of the sequence of melting, die casting, and finishing operations used to produce valves and Figure 3 is a schematic of the permanent mold injection casting process. All melting was done using a vacuum arc remelting (VAR) furnace at Howmet Corporation. The molten alloy was poured into a permanent steel mold. Figure 4 shows the castings and the dies.

Several variations of the permanent mold casting process were explored. The baseline process was the static permanent mold process. In this process the valves were cast using the gravity feed process [10]. Later, in an effort to decrease void content within the valves four methods of applying pressure during solidification were tried: gas boost, centrifugal casting, squeeze pins, and injection. Details of these processes have been described previously by Jones et al [10].

After the valve blanks were cast, they were radiographically and dimensionally inspected. Most of the castings were HIP'd at 1200°C/172MPa/3hr and were subsequently heat treated at 1176°C for 20 hours, furnace cooled to 260°C followed by 1010°C for 50 hours. A few of the valves were HIP'd at 1260°C/172MPa/4hr and did not undergo heat treatment. The valves were HIP'd and heat treated in the stem up position in order to minimize distortion resulting from porosity within the valves.

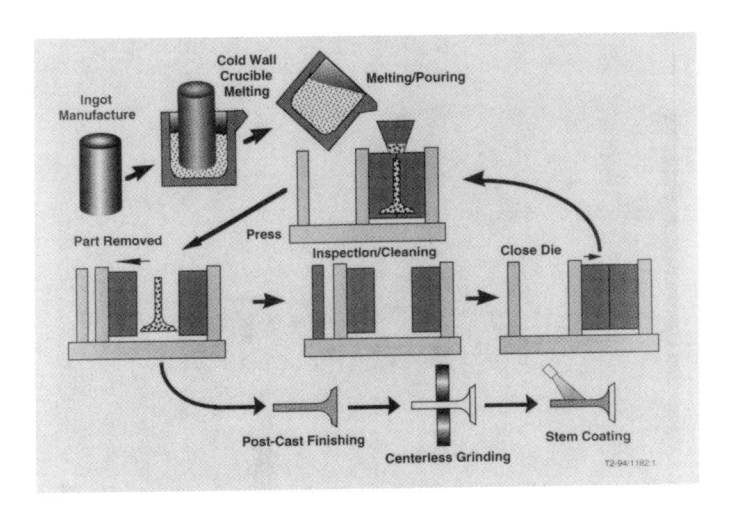

Figure 2: Schematic of the permanent mold process [9].

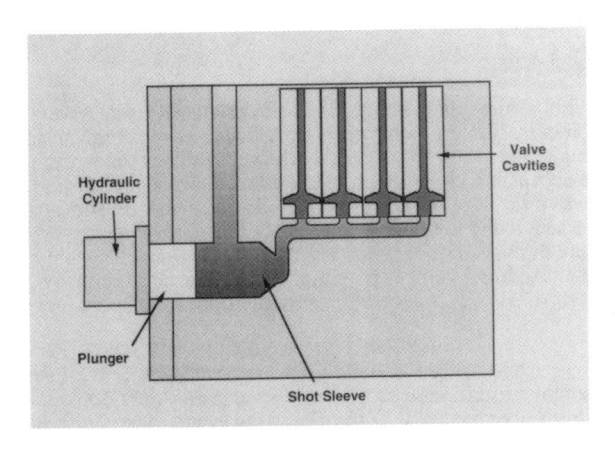

Figure 3: Schematic of the permanent mold injection casting process developed by Howmet for the EMTEC automotive valve project.

Figure 4: As-cast valves in the dies [9].

Results and Discussion

Casting Process Evaluations

In the course of this project over 800 valves were produced using several variations of the permanent mold process. Table 1 summarizes the advantages and disadvantages of each of these processes. Although the as-cast dimensions of the valve blanks produced using the static side gate process were acceptable, there was considerable dimensional distortion after HIP and heat treatment. This was a result of off-center porosity in the stems. In some cases the valve head were no longer perpendicular to the stem after HIP. To improve valve quality several methods of applying pressure during solidification were tried. It was found that applying pressure during solidification did show improvement by either reducing shrinkage or by moving the shrink closer to the centerline. However none of the processes produced pore free or low porosity as-cast valves.

Over 50 pours were made using the static permanent mold process. Inspection of the dies revealed no measurable distortion or erosion of the valve dies after 50 pours. In addition there was no increase in surface roughness of the dies. However surface finish of the as-cast valves was affected by the cleaning of the dies. A buildup of a metallic layer on the die surfaces resulted in a poor as-cast surface finish of the valves after several pours. It therefore became necessary to chemically clean the dies after every fifth pour. Although the durability of the steel dies through the 50 pours show great promise, more extensive trials of the dies would be required to determine whether the die life would be suitable for an automotive quantity production environment. In addition, the cleaning process used in this study would not be suitable for production casting, therefore another means of reducing the buildup on the dies would have to be developed.

Table 1: Casting Process Comparisons. ⊤ shows valve orientation and pressure application point during casting.

	PROCESS	PRESSURE METHOD	ADVANTAGES	DISADVANTAGES
1	STATIC, SIDE GATE INTO HEAD		Simple equipment	Must disassemble dies to remove castings, off center porosity
	A. No preheat	None	Baseline for all comparisons	Baseline for all comparisons
	B. No preheat	Inert gas	Improved surface finish	No significant improvement in fill
	C. 260°C preheat	None	Improved fill	Longer cycle time for preheat & cooling
2	STATIC, TOP GATE INTO HEAD	Hydrostatic	Better fill than Process 1	Lower alloy yield
	A. No preheat	Hydrostatic	Porosity closer to centerline than baseline	
	B. 260°C preheat	Hydrostatic	Porosity closer to centerline than baseline	No advantage over preheat, longer cycle time
3	CENTRIFUGAL, SIDE GATE INTO HEAD	Centrifugal	Best stem dimensions, void size reduction	Longer cycle time, more complex equipment, more difficult maintenance
	A. No preheat	Centrifugal	Less stem shrink than 1A	
	B. 260°C preheat	Centrifugal	Improved fill based on weight	Unacceptable cycle time
4	SQUEEZE PINS, TOP GATE INTO STEM	Hydraulically Actuated Pins	Potential to adapt mature die casting method	Narrow process window in which to apply pressure (tenths of seconds)
	A. No preheat			Equipment design & maintenance challenges.
5	INJECTION, BOTTOM GATE INTO HEAD	Hydraulically Actuated Shot Sleeve	Highest potential throughput, reduced void size	Equipment design challenges
	A. No preheat		Finest as-cast structure	

Microstructural Evaluation

Of the permanent mold processes, the injection process produced the finest as-cast microstructure. In Figure 5 the as-cast microstructure for the injection cast process is compared to that of the static cast process. The injection cast microstructure is highly segregated and significantly finer than that of the static cast structure. The finer as-cast microstructure for the injection cast process may be the result of metal being sprayed into the mold during the injection process, or of heat being lost to the shot sleeve.

Post HIP microstructures for static and injection processes are shown in Figure 6. The microstructural differences observed arose from the different HIP conditions which were used. The 1200°C/172MPa/3hr HIP cycle used for the static cast specimens is designed to produce a near-γ microstructure while the 1260°C/173MPa/4hr HIP cycle used for the injection cast specimens produces a duplex microstructure. However it should be noted that after the HIP

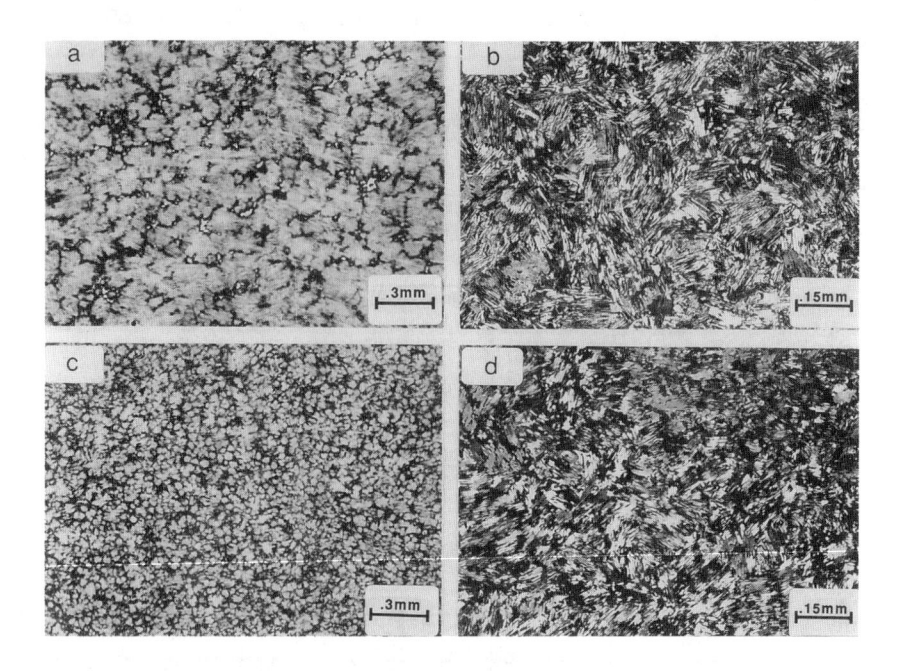

Figure 5: a) As-cast structure from the Static cast process (Etched, original magnification 50X); b) As-cast structure from Static cast process (Polarized light, original magnification 100X); c) As-cast structure from Injection cast process (Etched, original magnification 50X); d) As-cast structure from Injection cast process (Polarized light, original magnification 100X)

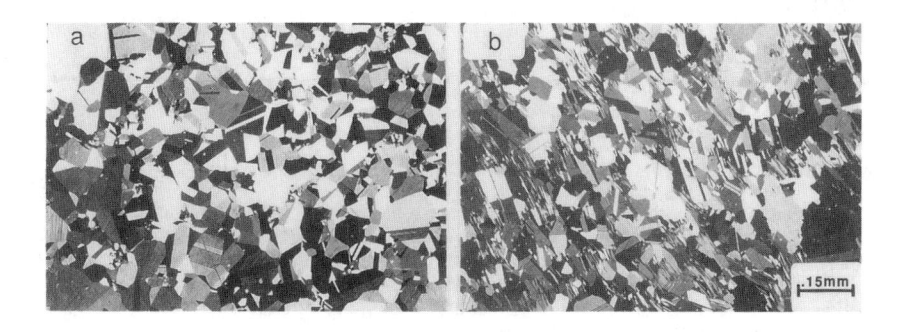

Figure 6: a) As-HIP structure from the Static cast process; b) As-HIP structure from the Injection cast process (Polarized light, original magnification 100X)

cycles similar equiaxed γ grain sizes are observed for both conditions. The similarity in the grain sizes of the as-HIP static and injection structures show that the finer features of the as-cast injection structure are not retained after the 1260°C HIP cycle.

Mechanical Properties

Table 2 compares the room temperature tensile properties for static and injection cast material HIP'd using the 1260°C/172MPa/4hr cycle. The static castings are stronger than the injection castings but have a lower ductility. The higher strength of the static castings may be explained by a lower aluminum level in the static cast material. Analysis of the first 50 heats showed aluminum variation of ± 0.5at% around the target level of 47at%. The tensile specimens for the static castings were taken from a heat with an aluminum level of 46.1at%, while the aim aluminum composition for the injection castings was 47at% Al. Previous work on this alloy showed that the room temperature yield strength decreases ~40MPa for every 1at% increase of aluminum level [8]. However aluminum level alone does not account for the difference in the yield strengths between the static and injection processes. Further study is required to try to account for the difference in the strengths between the two processes.

Table 2: Tensile properties at 23°C for static and injection cast processes.*

Process	0.2% YS (MPa)	UTS (MPa)	% Plastic Elongation
Static	409	466	1.5
Injection	302	411	2.2

*Average of 2 data points

Conclusions

With over 800 valves produced, permanent mold casting has been shown to be a viable method for producing TiAl automotive valves. The application of pressure during solidification does improve the casting fill; however none of the methods produced pore free or very low porosity as-cast valves. To maintain accurate dimensional control after HIP, it is essential to both minimize shrink and to move it to the centerline of the valve or to produce pore free valves that do not require HIP.

The permanent mold injection cast process show promise as a new method of producing fine grained TiAl components. The fine grain size and the highly segregated cast structure produced by the injection process offers a better starting point for microstructural refinement during HIP and heat treatment.

Acknowledgments

The support of the Edison Materials Technology Center, our EMTEC Valve Team, the National Science Foundation, and Wright Laboratory of Wright Patterson Airforce Base made this work possible. Their contributions are greatly appreciated.

REFERENCES

1. C. M. Austin and T. J. Kelly, "Development and Implementation Status of Cast Gamma Titanium Aluminide," *Structural Intermetallics*, R. Darolia et al. eds., TMS, 1993, pp. 143-150.

2. S. Hartfield-Wunsch, A. A. Sperling, R. S. Morrison, W. E. Dowling Jr., and J. E. Allison, "Titanium Aluminide Automotive Engine Valves," *Gamma Titanium Aluminides*, Y-W. Kim, R. Wagner, and M. Yamaguchi eds., TMS, Warrendale, PA, 1995, pp. 41-52.

3. D. E. Larsen Jr. and D. A. Wheeler, "Investment Cast Gamma and XD™ Gamma Titanium Aluminide Components for Aerospace and Automotive Applications," Paper No. 8, Investment Casting Institute: 42nd Annual Technical Meeting, 1994.

4. W. E. Dowling, Jr., J. E. Allison, L. R. Swank, and A. M. Sherman, "TiAl-based Alloys for Exhaust Valve Applications," SAE paper 930630, SAE, Warrendale, PA, 1993.

5. Y. Mae, Metallurgical Review of MMIJ, Vol. 8, No. 1, 1991, pp. 113-118.

6. S. C. Huang and M. F. X. Gigliotti, U.S. Patent No 4 879 092.

7. S. Hurta, H. Clemens, G. Frommeyer, H.-P. Nicolai, and H. Sibum, "Valves of Intermetallic γ-TiAl-Based Alloys: Processing and Properties," to be published in the Proceedings of the Eight World Conference on Titanium, Birmingham, UK, 1995.

8. D. Eylon, P. E. Jones et al., <u>EMTEC TR94-24 Part 1</u>, (EMTEC, Kettering, OH, 1994) pp. 69-137.

9. G. Colvin, "Permanent Mold Casting of Titanium Aluminide Automotive Valves," presented at the Titanium Development Association Meeting, San Diego, CA, Oct. 1994.

10. P. E. Jones,. W. J. Porter, III, D. Eylon, and G. Colvin, "Development of a Low Cost Permanent Mold Casting Process for TiAl Automotive Valves," *Gamma Titanium Aluminides*, Y-W. Kim, R. Wagner, and M. Yamaguchi eds., TMS, Warrendale, PA, 1995, pp. 53-62.

Thermohydrogen Treatment of Shape Casted Titanium Alloys

Alexander A.Ilyin, Andrey M.Mamonov, Yulia N.Kusakina

Moscow State University of Aviation Technology after K.Tsiolkovsky,
Metals Science and Treatment Technology Department
Petrovka St. 27, Moscow, K-31, Russia, 103767

Abstract. The processing schemes of castings from $(\alpha + \beta)$ - titanium alloys VT6 (Ti-6Al-4V) and VT23 (Ti-5.5Al-4V-2Mo-1- Cr-0.5Fe) are offered. They are based on HIP and THT (thermohydrogen treatment) combination. It is shown that the treatment leads to formation of fine intergrain $(\alpha +\beta)$-structure and removing porosity. This treatment increase strength of castings up to 11-21%, fatigue strength (10^7 cycles) for VT23 and number of cycles before failure (under stresses of 500 MPa) for VT6 in 2 times in comparison with cast condition. Impact strength and ductility are at satisfactory level.

Introduction

There are some advantages of shape casting technology for titanium alloy products production in comparison with application of deformed semiproducts. They are: 3-5 times increasing of material usage coefficient; elimination of low-productivity and expensive technology of casting processing into deformed semiproducts; considerable reduction of cutting processes; removing problems of stamps and cutting tools wearing; etc. [1].

However, despite of high economical effectiveness of shape casting technology, its application in industrial scale is limited by low level of mechanical and exploitation properties (especially fatigue properties) of casted material, which are determined by porosity, microliquation, and microstructure peculiarities such as coarse grains, thick plates inside the grains [2]. Elimination of porosity is achieved by means of high-temperature gasostatic treatment (HIP). However, HIP does not readily influence on casted metal structure at the zones, where porosity is absent [2]. Conventional heat treatment - annealing for lowering of residual casting stresses does not lead to the increasing alloy strength. Application of strengthening heat treatment, including rapid quenching, could not always be used for shape castings, it may cause internal dangerous stresses appearance.

There is a new effective method of structure formation for titanium alloys that is called thermohydrogen treatment (THT), it is based on reversible hydrogen alloying. This technology includes three main operations: hydrogen sorption till needed concentrations, heat treatment and exclusion of hydrogen by vacuum annealing [3]. Scientific base of THT has already been developed, and wide opportunities of this technology as a method of structure improving for different classes of alloys have also be shown by now [4,5]. Transformation of casted structure and increasing of mechanical properties, as a result of this treatment, is expected to be one of the most promising THT applications. Besides, THT does not need any rapid cooling and heating, and gives an opportunity to eliminate surface oxidation of castings. Method of mechanical properties improving of casted ($\alpha+\beta$) titanium alloys based on HIP and THT combination is described in this work.

Materials and Methods.

Casted VT6 (Ti-6Al-4V) and high-strength casted VT23 alloys were investigated. Cylindrical castings of this alloys (15 mm in diameter) were cut into samples for testaments after the removal of surface layers. THT of alloys was made by means of Sieverts' equipment. HIP was in argon atmosphere under 155 MPa pressure for 2 hours. Fatigue tests were made on smooth cylindrical samples by bending with rotating 100 Hz and index of asymmetry R=-1.

Results and discussion.

The aim of the treatment of casted VT6 and VT23 alloys is obtaining of fine intergrain ($\alpha+\beta$)-structure (without changing of primary β-grain dimensions), which provides high strength and fatigue strength. Development of THT regimes is based on analysis of interaction between alloys and hydrogen, phase equilibrium in alloy-hydrogen system, and influence of heat treatment of phase composition and structure of alloys, and studying of processes, taking place under vacuum processing etc. These investigations were made by authors earlier and published in works [6-8].

For example, it was established for ($\alpha + \beta$)-alloys, that maximum refining of α-plates, located in β-grain is achieved under 0,5-0,7% concentration of hydrogen. Optimum temperature of vacuum annealing was 750 ^{0}C.

Removing of porosity is one of the most serious problems of quality improving of shape castings and level of their properties. HIP increases density of casted material. However, initial casted structure of alloys after HIP does not practically change, because its formation is controlled only by gasostate processing temperature. As a result, positive HIP influence on mechanical properties is connected, mostly, with increasing of effective cross section of castings and removing of stress concentrators. At the same time, it is known, that alloying of titanium alloys by hydrogen leads to the realisation of hydrogen plastification effect, which is defined by lowering of yield strength alloys under deformation temperatures. From this point of view HIP of hydrogenated metal appears to be effective for both removing porosity and improving of final structure.

Thus two methods of thermohydrogen and gasostatic treatments combination are possible: 1) realisation of THT after HIP; 2) gasostatic treatment of preliminary hydrogenated material with further vacuum annealing for hydrogen removing (Figure 1).

For microliquation removing the casted samples before further treatment were subjected to homogenising annealing in vacuum at the temperature 1000^{0}C (VT6) and 970^{0}C (VT23) for 1 hour. Realisation of such schemes of VT6 alloy treatment gives the following results.

1. At the first scheme of treatment parameters of hot isostatic processing looked like ordinary. Here the goal is to eliminate porosity of material without changing its structure. On the base of literature data the following parameters of HIP were chosen: pressure in gasostate - 155 MPa, temperature - 950^{0}C.

The aim of THT is to transform thick plates in castings into the fine integrain structure. For achieving extremely fine structure hydrogen concentration was chosen as equal to 0.8%.

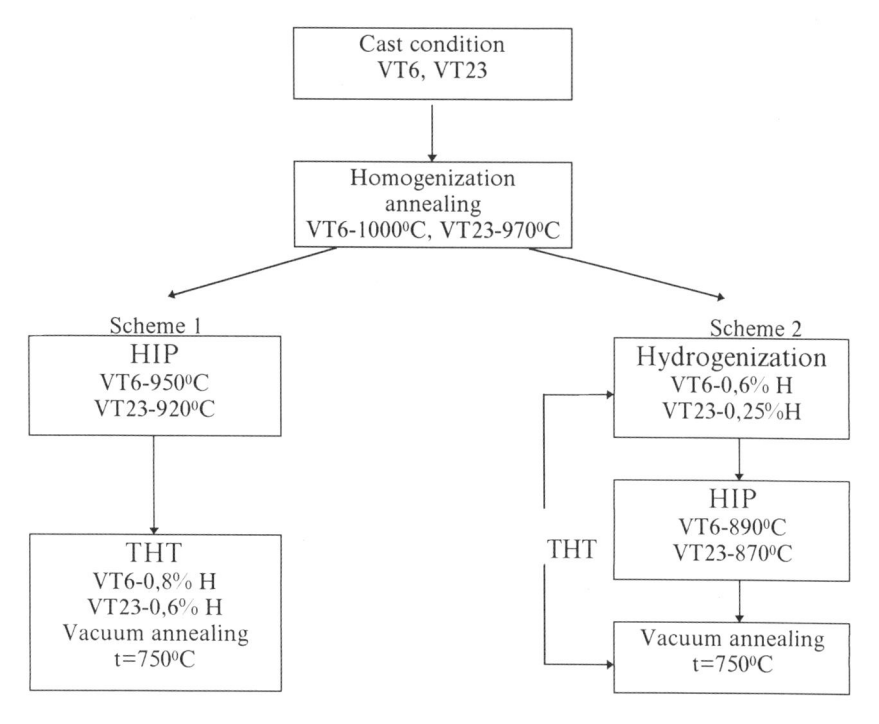

```
                    Cast condition
                       VT6, VT23
                            |
                            v
                    Homogenization
                       annealing
                 VT6-1000°C, VT23-970°C
                   /                  \
                  /                    \
          Scheme 1                       Scheme 2
            HIP                        Hydrogenization
          VT6-950°C                     VT6-0,6% H
          VT23-920°C                    VT23-0,25%H
                                             |
                                             v
                                            HIP
                                         VT6-890°C
                                         VT23-870°C
            THT                   THT        |
         VT6-0,8% H                           v
         VT23-0,6% H                   Vacuum annealing
       Vacuum annealing                    t=750°C
           t=750°C
```

Figure 1 - Technological schemes of treatment of castings of alloys VT6 and VT23.

2. The second scheme of treatment is based on HIP with hydrogen plastification (HP) effect using. In this case, while optimum hydrogen content is added, the strains, needed for plastic flow of metal, reduce, and it simplifies the process of pores closing during HIP. More over, the opportunity of temperature lowering during gasostating appears, what can diminish the power-consuming of the process, and simultaneously provide formation of more appropriate structure.

The optimum hydrogen concentrations for VT6 alloy, which gives us an opportunity for better HP effect development, is 0.25-0.35% at the temperature 800-900^{0}C. But such hydrogen concentration is insufficient for needed transformation of casted structure. So, the second scheme of HIP and THT combination was applied to samples with 0.6% of hydrogen. The HP effect in the process develops in not maximum degree, but it is strong enough for lowering of HIP temperature by 50-70 ^{0}C. Samples structure after treatment according to both schemes is similar, it is homogenous in section and is represented with fine α- phase in β- matrix and thin α- surroundings along the initial β-grain boundaries (Figure 2). The value of structure refining at 1 scheme of treatment somehow higher, then during scheme 2 realisation, what is caused by larger hydrogen concentration. At the same time duration of the vacuum annealing (at the temperature 750^{0}C) according to scheme 1 is longer for 0,5 hour, than during scheme 2. The effectiveness of both schemes of treatment was proved by mechanical tests on tension, impact strength (KCV) and low cycle fatigue (LCF) at the stresses level σ=500 MPa (Table I). The tests were made on samples in initial casted condition, after HIP, and also after treatment where HIP and THT were combined.

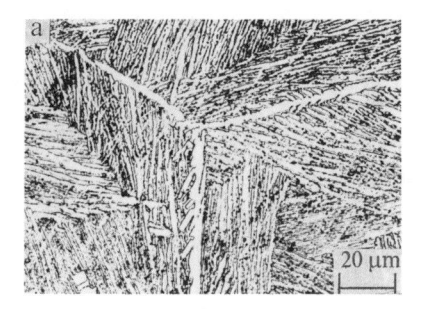

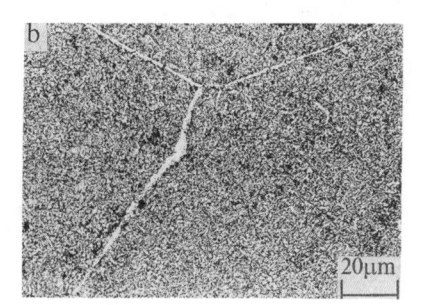

Figure 2 - Structure of samples of alloy VT6 in initial cast condition (a) and after treatment scheme 1 (b)

Analysis of changing in porosity was made by means of hydrostatic weighting, where density of samples was determined in initial state and after treatment. The results of density measuring are represented in Table I.

Table I. Mechanical properties and density of castings of alloy VT6 after different processing schemes.

Condition, processing scheme	Density, g/sm^3	UTS, MPa	YS, MPa	El, %	KCV MJ/m^2	N,cycles ·10^4 σ=500 MPa
Initial (cast)	4,421	920	830	8,2	0,44	49
After HIP	4,571	930	830	9,0	0,49	53
Scheme 1	4,576	1120	1080	10,3	0,48	97
Scheme 2	4,578	1100	1040	10,7	0,50	91
Rolled rod VT6*	4,580	–	–	–	–	–

*Note: For comparison, the density of hot rolled rod, where internal inhomogeneities are absent, and its chemical composition almost corresponds to cast condition (concentration range of Al and V is not higher 0,1%), are measured.

Analysis of given data shows that both given schemes of treatment practically totally eliminates porosity, more over HIP of preliminary hydrogenated alloy is more effective even at lower temperatures. It is connected with effect of hydrogen plastification. The main peculiarity of first scheme of treatment is the additional increase of castings density as the results of THT. It is probably connected with the fact, that phase transformations during the THT with increase of hydrogen concentration are accompanied with strong volume effects. Appearing here interface stresses cause a local plastic strains in the zone of macrodefects and partially removes them.

Strength properties increasing by 15-20% as well as fatigue increasing more than in 2 times are connected mainly with the cardinal changes in the structure. Appropriate level of ductility and impact strength is provided additionally by porosity absence and existence of more ductile α-surroundings around extremely strengthened grains.

During the first scheme of treatment in VT23 alloy the analogous structure changes takes place.

Microstructure of VT23 alloy in initial casted state is characterised by coarse initial β-grains with thin α-surroundings (Figure 3a). Intergrain structure is represented with α-plates, subdivided by β- layers. Gasostatic treatment (scheme 1) at 920°C for 2 hours leads to partial pore closing, what was proved by analysis of microsection series. The average size of pore remaining in the castings after HIP is about 0.05-0.1 mm according to metallographic analysis. Structure in the zone of dense material is represented with α-plates colonies. In the microvolumes, formed as the result of plastic flow of metal during pores closing, the distortion and fragmentation of α-plates were observed. During THT the fine and dispersed (α+β) - structure is formed, and initial β-grains are surrounded by thin α-surroundings (Figure 3b).

Effect of hydrogen plastification in VT23 alloy in higher degree is developing at hydrogen concentrations 0.25% at the temperature 850-900°C. During HIP of hydrogenated alloys and further vacuum annealing (scheme 2) the structure with more fine, while comparing with casted condition, α- lamellas form (Figure 3c). However the degree of dispersity of intergrain structure is significantly lower, than at using scheme 1.

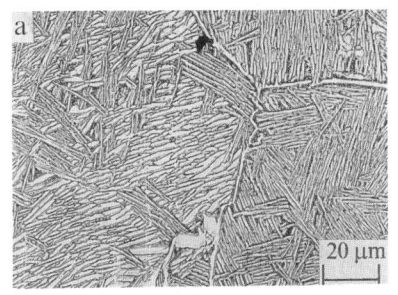

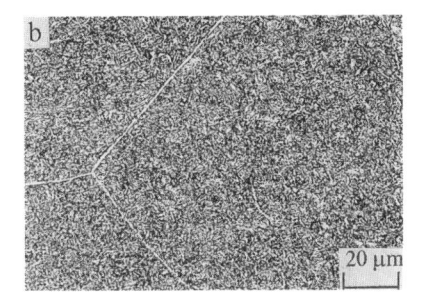

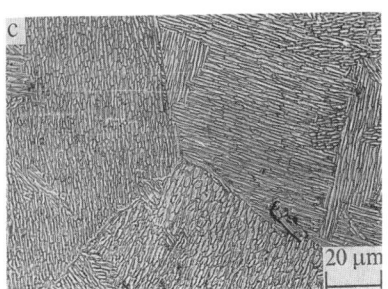

Figure 3 - Structure of samples of alloy VT23 in initial cast condition (a) and after treatment scheme 1 (b) and scheme 2 (c).

Analysis of density variations of samples from VT23 alloy have shown, that HIP of alloy with 0.25% of hydrogen (scheme 2) causes the higher increase of casting density, in comparison with HIP without hydrogen (Table II), in spite of the fact that HIP temperature is reduced by 50°C. It should be mentioned, that THT (scheme I) leads to additional samples density increasing, while removing pores with size smaller 0.1 mm. It is caused, as for VT6 alloy, by considerable interface stresses leading to local plastic deformation in the pore zones.

Mechanical tests were made on samples in initial casted condition, after HIP, and after treatment according to scheme 1 and 2 (Table II).

Results of analysis shows, that the treatment according to this suggested schemes leads to strengthening by 11-19% comparing with initial state and by 13-21% comparing with condition after HIP. Treatment according to the first scheme provides higher increase of strength properties due to more dispersity of intergrain (α+β)- structure. Characteristics of ductility and impact strength also increase by means of porosity elimination and structure transformations.

Table II. Mechanical properties of castings of VT23 alloy after treatment by different regimes.

Condition, processing scheme	Density, g/sm³	UTS, MPa	YS, MPa	El, %	RA, %	KCV, MJ/m²
Initial cast	4,418	1000	960	7,2	15,0	0,47
After HIP	4,556	980	940	9,4	18,9	0,59
Scheme 1	4,560	1190	1120	8,5	19,0	0,56
Scheme 2	4,563	1110	1060	8,9	19,6	0,56
Rolled rod	4,564	–	–	–	–	–

Note: RA - retardation area.

Fatigue properties were tested on samples in initial casted state after HIP, and also after treatment accordingly to the first scheme. The endurance limit σ^{-1} was tested on the base of 10^7 cycles. The results of tests are represented at Figure 4. Their analysis shows, that HIP increases endurance by 10-15% mainly by porosity of samples elimination. THT and HIP combination according to scheme 1 causes abrupt increase of endurance (in 2 times) in comparison with casted state. Obtained date of fatigue strength exceeds the properties of this alloy in deformed condition. It is caused by both additional increase of castings density and favourable structure formation.

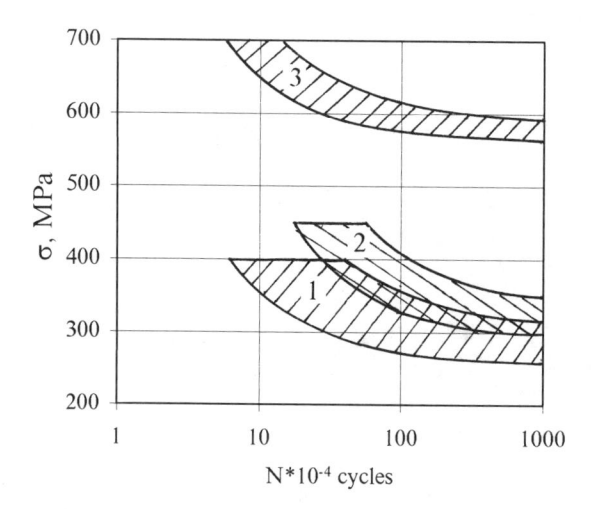

Figure 4 - Results of fatigue tests of samples of alloy VT23.
  1 - initial cast condition
  2 - condition after HIP
  3 - condition after treatment, scheme 1 (HIP+THT)

Obtained fine and dispersed intergrain structure has increased resistance to fatigue failure mainly by means of increase of fatigue crack nucleation work. It is testified, partially, by a big number of cycles before failure at the upper range of loading, equal to 0.6-0.8 of yield strength, is about 300000 cycles at $\sigma = 600$ MPa.

The suggested types of treatment also causes the stabilisation of alloy properties. Thus, for VT23 alloy in initial state at the loading 360 MPa, the upper number of cycles before failure is equal to 10^6, and the lower is about 10^5. For the same alloy after suggested treatment the upper and lower number of cycles before the failure weakly varies (75 thousands and 106 thousands correspondingly at σ = 700 MPa).

Analysis of failure nature of samples from VT23 after various types of treatment during impact strength and fatigue limit testes were done. It was learned out, that THT leads to general change in fracture surfaces on account of strong dispersion of intergrain structure and grain strengthening (Figure 5).

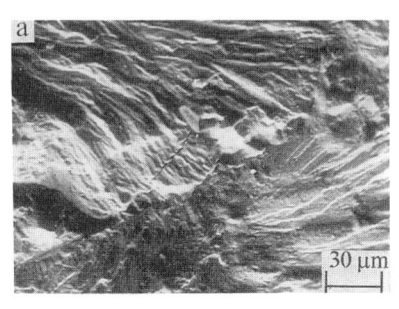

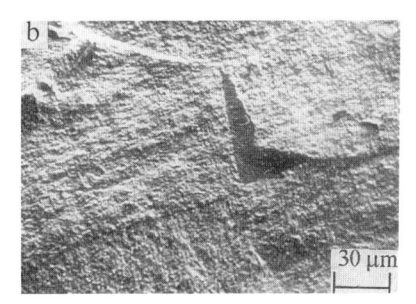

Figure 5 - Fractography of fatigue fractures of samples of alloy VT23:
    a - condition after HIP
    b - after treatment scheme 1 (HIP+THT)

In the initial casted condition and after HIP in fracture of impact samples fracture the square of faceted part is 10-12%. Chipping takes place along α-colonies, it results to several facets within the β-grain boundaries. After using schemes with THT the square of faceted part of fracture increases up to 40-55%.

While analysing of fatigue fractures, it was established that zone of fatigue crack in samples without THT had considerable relief, difference of height of it is up to 2 mm. Surface of fatigue crack forms as a result of many intergrain chipping within β- grain boundaries. Growth of fatigue crack is connected with internal structure of β-grains and is orientated relatively towards α-colonies. So its separate areas of fatigue crack does not lie in the plane of the main direction of its growth, but they are situated under certain angles to this direction.

Using of THT as a final operation strongly changes morphology of fatigue fracture of samples. Fatigue crack has smoothed surface without relief connected with internal structure of β-grains. Then each β-grain behaves as a monolith, and within it chipping forms almost coinciding to the plane of main direction of crack or situates under low angles to it. There are marks of microplastic deformation on the surface of crack.

Fatigue limit of samples with such type of fracture attains its height values.

Conclusions

Processing schemes, suggested in this work for cast (α+β) titanium alloys VT6 and VT23, are based on combination of HIP and THT, and ensure cardinal transformation of cast structure. Thick lamella intergrain structure transforms into fine α+β- structure. Initial size of β-grains preserves unchangeable, and thin continuous α-surrounding situates along its boundaries. Suggested schemes of treatment ensure removing of porosity in castings, and THT removes fine pores(up to 0.1 mm). The effect of hydrogen plastification in addition allows us to increase the density of castings even at the reduction of temperature of HIP by 50^0C-70^0C.

It was shown, that combination of HIP and THT of cast alloys VT6 and VT23 makes it possible to increase strength (by 15 - 20%) and especially fatigue characteristics of alloys:

fatigue limit of VT23 and number cycles before failure (at the stress $\sigma=500$ MPa) for samples of VT6 alloy increases in 2 times in comparison with cast condition, ductility and impact strength are at satisfactory level. It allows to forecast high efficiency and reliability of products, which were produced by shaped casting.

References

1. K.K.Yasinski."State and Perspectives of Development of Shaped Castings of Titanium". (Science. Production and Applications of Titanium in Conversion Conditions. Proc. 1st Int.Conf. on Titanium SNG-countries. Moscow, VILS, 1994), 234-243.
2. G.A.Bochvar, N.V.Yanovskaya, "Influence of High-Temperature HIP Treatment on the Process of Structure Formation and Mechanical Properties of Cast Titanium Alloys", Titanium , 1(1993), 21-23.
3. A.A.Ilyin. "Phase and Structure Transformations in Titanium Alloys, Alloyed by Hydrogen". Isvestiya Vusov. Cvetnaya metallurgia , 1 (1987), 96-101.
4. A.A.Ilyin, B.A.Kolachev, A.M.Mamonov. "Phase and Structure Transformations in Titanium Alloys under Thermohydrogenous Treatment". (Titanium' 92. Science and Technology Proc. of 7th World Conf. on Titanium. San Diego, California, USA, 1992), 941-946
5. A.A.Ilyin, A.M.Mamonov. "Thermohydrogen Treatment of Casted Titanium Alloys". Journal of Aeronautical Materials (China), v. 2, (1992) , 4-5.
6. A.A.Ilyin, A.M.Mamonov."Temperature-Concentrations Diagrams of Phase Composition Hydrogenated Multicomponent Alloys on Titanium Base". Metals RAN, 5 (1994), 71-78.
7. A.A.Ilyin, A.M.Mamonov, B.K.Nosov, V.M.Maystrov. "About the Influence of Hydrogen on Diffusion Mobility of Atoms in Metallic Sublattice of β-phase in Titanium Alloys". Metals RAN, 5 (1994), 99-103.
8. A.A.Ilyin, A.M.Mamonov, M.Y.Kollerov. "Scientific Base and Principles of Building of Processing Schemes of THT of Titanium Alloys". Metals RAN, 4(1994), 157-168.

COMPUTERIZED SPECIFICATION SYSTEM
FOR TITANIUM PROCESSING

Dan Z. Sokol

Renaissance Engineering, Inc.
First National Plaza ▪ Suite 1414
Dayton, Ohio USA 45402
513-224-1414 Fax: 513-224-1418

Abstract

In the materials industry, the product "design" is generally documented and communicated using hard copy *material and process specifications*. These voluminous documents describe everything about the desired end product, including the required processing methods, testing procedures, and material properties. Since specifications must be strictly followed in great detail, each specification (and revision) must be continually compared with processing and testing criteria. For example, MIL-T-9047 "Testing of Titanium and Titanium Alloy" covers both the acceptance limits and prescribed testing methods for chemical composition, mechanical properties, microstructure, melting methods, packaging, etc.

Based on support from the National Science Foundation and the National Institute of Standards and Technology, Renaissance developed a language and a computerized framework to automate many of the activities which depend upon specifications. Working with two companies in the titanium industry, a computerized tool has evolved which supports many of the facets of titanium processing, including testing, process planning, and certification.

Materials Industry Design Process

A major difference between material suppliers and discrete manufacturers is that most component and final product manufacturers document and communicate their product designs using drawings and bills of material. For example, a manufacturer that provides parts for aircraft engines will document his design with a dimensioned drawing, as well as a process list of manufacturing steps, description of tooling, etc. For a number of years, these manufacturers have used formalized systems for documenting and communicating the design. In most operations, these designs are stored electronically using the now familiar computer-aided design (CAD) systems. Design engineers work with manufacturing engineers to create all the necessary documentation for building and testing the product.

The electronic drawing acts as the foundation for the many related essential activities. For example, the drawing is frequently used to generate the bill of material required by manufacturing control systems. It is also often used as input to computer-aided manufacturing (CAM) to automatically create the numerical control codes needed for the machine tools. It also forms the basis of inspections in companies that use it for coordinate measuring machines (CMM). Concurrent engineering techniques have been successful in overlapping the lead times associated with the tasks allocated to these and other functions.

While techniques such as CAD can be of some assistance in defining the detail order specifications for dimensional characteristics, it is of little or no use in providing information regarding the product composition and quality specifications. Thus, in the material industries (i.e., suppliers of steel, titanium, composites, etc.), the product design is generally documented and communicated using hard copy material and process specifications. These voluminous textual specifications describe everything about the desired end "product" including the required processing methods to make the product, testing procedures, and material properties. Since the specifications are the "hub" of all engineering activity, the totally paper-based specifications have been a major obstacle to improving systems above the shop floor.

Unlike the discrete manufacturer, the material supplier performs extensive destructive testing in a laboratory and generates numerous documents that certify that the material meets the requirements described in the specifications. Whereas the majority of the industrial base has the availability of computerized tools like CAD for documenting and communicating product requirements, material suppliers are primarily paper-based. Material suppliers receive hundreds of new specifications or revisions every year and store multiple copies in various filing cabinets throughout the company (See Figure 1).

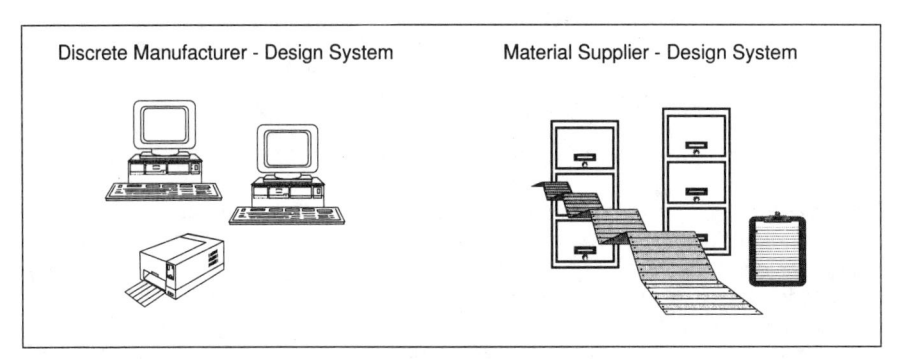

Figure 1 - Comparison of Design Systems

To appreciate the importance of material specifications, it is important to understand their format, content, and application. Figure 2 provides a simple illustration of the content in a typical 20-page material specification.

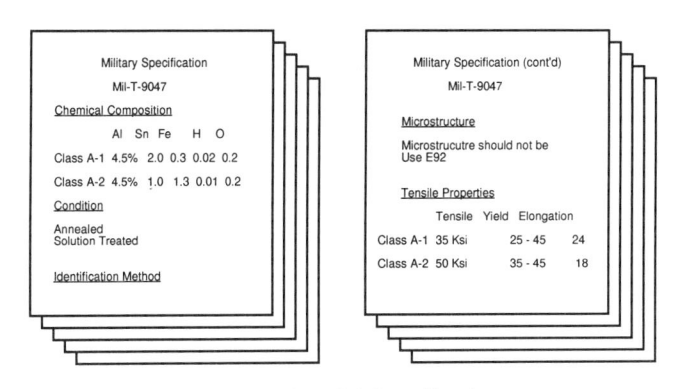

Figure 2 - Material Specifications

Since many of the applications for material are related to sea- or air-worthiness, the specifications are extremely comprehensive and voluminous covering hundreds of different material properties. Each specification must be strictly followed in great detail. As a result, each customer's specifications (and revisions) must be continually compared with processing and testing criteria. The maintenance of these paper-based specifications is, in itself, a major effort. Customer material specifications are received in the Documentation Departments (or similar groups generally within the Quality Control Department) where they are reviewed and marked-up in comparison to existing specification revisions on file. A response is drafted to the customer stating any comments, items open to negotiation, or exceptions to the specification.

As mentioned previously, the specifications are the hub of many potential concurrent engineering and processing functions in operation at material suppliers. Figure 3 shows the relationship with four other major functions at most material suppliers. For example, since the cost of material is often dependent upon the amount of testing and handling required, the specifications are used by Sales Order Entry personnel at the time of order-taking to aid in identifying the appropriate price for the order. Specifications are used by Production to define the appropriate operations to include in the production routing and data collection forms. Testing Laboratory personnel use the specifications to define the acceptance limits and methods required for a particular order. Certification uses data from the specification to ensure that the order has been completed and packed for shipment.

A customer typically places an order for material by defining the grade or type (e.g., certain alloys of titanium are used for structural applications while others are used for high-temperature areas), identifying the quantity (in pounds, feet, etc.), and referencing the specifications the order is required to meet. In many situations, the orders reference multiple specifications. Frequently, the multiple specifications are not mutually exclusive -- i.e., each specification may redundantly address the same properties. For example, two or more specifications referenced on the same order may have sections covering chemical composition. The result is that the various end users (material engineers, lab foremen, etc.) have to determine the appropriate methods and acceptance limits. For example, while Specification A may state that hydrogen content (a key parameter impacting the strength of titanium) cannot exceed 120 parts per million (ppm), Specification B may require material with a hydrogen content not exceeding 90 ppm. Someone in the operation is required to manually review each specification document to determine that the order should be produced to the most stringent requirements.

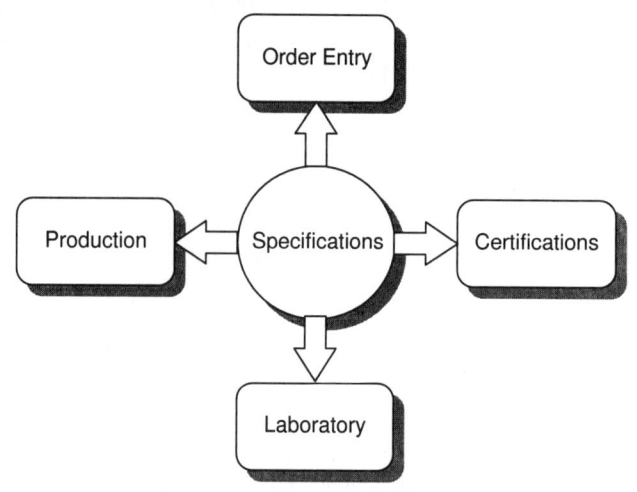

Figure 3 - Relationships between Functions

To further complicate matters, Specification A may require tensile tests using a heat treated sample. Specification B may also have a section that requires tensile testing using solution treated samples. Since a test sample used for a heat treated test cannot be used for a solution treated test, this situation requires a series of additional tests for each specification (which in turn, requires more material to be consumed for destructive testing). This issue must also be discerned via a thorough, manual review of each referenced specification.

On the basis of the specifications identified on the order, the Production Departments create the appropriate production routings, log sheets, and test requests. Again, the same issue concerning multiple specifications is pertinent to the Production Department. Since the orders are often specific to a particular revision (e.g., MIL-T-9047, Revision J), even when the Production Department receives a repeat order (same combination of grade, customer, and specifications), each specification and revision has to be reviewed again.

Once the production run is complete and all test and production data are compiled, the Certification Department reviews the data. Personnel in the Certification Department manually compare the processing, inspecting, and testing data versus the acceptance limits and methods defined in the specifications. At this point, the yields of the products are often impacted by incorrect paperwork. If one step was missed in reading a specification -- or as is more often the case, if the order for multiple specifications does not adhere to all the requirements -- the order can be rejected by the customer.

Magnitude of the Problem

The use of manual specifications is an extremely labor-intensive (and paper-intensive) process. Many material suppliers employ numerous persons in the Documentation and Certification Departments. These employees spend a majority of their time retrieving, sorting, and generating documentation related to specifications.

The manual use of specifications is not only labor-intensive and error-prone, but is also extremely time-consuming. Since many of the time-consuming review tasks are on the production "critical path," any delay in the cycle time delays shipment of the product.

A Method to the Madness

Potential resolutions to the problems have emphasized electronic word processing and graphics manipulation. However, these approaches with specifications fail to address the needs of organizing the quantitative specification data into a usable format and structure.

Many material suppliers have recognized the need to formalize the "order taking" process. To develop a solution, Renaissance Engineering worked with material suppliers to design and implement the Manufacturing Specification System (MASS™). The system provides the following capabilities:

Structured Specifications - The system provides a structured repository for the key elements of data buried in the specifications. Since it provides specific data items, it is more than a text retrieval system. The system has the ability to accommodate the broad range of specification formats used by material suppliers.

Intelligent Retrieval Methods - The system provides the ability for multiple modes of retrieval. It allows a user to extract the acceptance limits for a specific test from a particular specification. It allows the user to compare different specifications and highlight the differences between the specifications.

Linkage to Other Systems - The system is usable in other than just a stand-alone mode. That is, the data is structured to allow for use by many other systems. For example, the acceptance limits are available for use by lab information management systems.

Easy Method for Maintenance - The system allows for entry of data from new specifications and revisions. It dovetails with the work flow used in the specification review process. It also has the ability to adjust the structure for future material testing procedures.

Figure 4 summarizes the features and benefits from using the Manufacturing Specification System.

Features	Benefits
Immediate verification of multiple specifications - With the material specifications on-line for use throughout the company, personnel will be able to immediately determine requirements.	This will help to insure that there are no situations where there are conflicting specifications and help eliminate the conduct of unnecessary tests or processes. This will reduce the operating cost.
Complete checking of conformance - All Departments will have the ability to concurrently identify the complete set of production, testing, packing, shipping, etc., requirements for an order.	All requirements are defined up-front when engineering the solution. This will reduce cycle time as well as avoid rejects from incorrect paperwork.

Figure 4 - Summary of Features & Benefits

New Processes

The introduction of information technology in the form of the computerized Manufacturing Specification System has enabled material suppliers to define a more rigorous method for managing materials processing. For example, computerized specifications can be accessed by Order Entry personnel at the time of order-taking to aid in identifying the appropriate price for the material; Specifications can be used by Production personnel to define the appropriate operations to include in the production routing; Laboratory personnel can use the electronic specifications to define the acceptance limits and testing methods required; and Certification personnel can access the specifications to ensure that orders have been completed properly.

Bibliography

Brown, D. H. (1993) "PIM: A Tool for Concurrent Engineering," Computer-Aided Engineering, vol. 1, no. 1, Page 64.

Miller, DEL. (1993) Concurrent Engineering Design: Integrating Best Practices for Process Improvement, Detroit, Society of Manufacturing Engineers.

Niebel, B.W. (1982) "Designing for Manufacturing," Handbook of Industrial Engineering, New York, John Wiley & Sons.

Smith, G.A., (1993) "Next Step in Product Design," Enterprise, vol. 1, no. 1, pp. 6-7.

SUBJECT INDEX

A

Abrasive cutting, 301, 302, 303, 304
Adiabatic, 225, 228, 229, 232
Aging
    effect on properties, 352, 356-365
    Ti-10V-2Fe-3Al, 356, 357
    Ti-3Al-8V-6Cr-4Mo-4Zr, 358, 359, 361, 362, 365
Aging response, 233, 234, 236, 238, 240
Aging, direct
    effect on Timetal® LCB properties, 358, 360, 361
Aging, indirect
    effect on Timetal® LCB properties, 358, 360, 361
Alloys
    β-CEZ®, 379
Alpha alloys, 350, 352, 353, 355, 356, 361, 362, 364
Alpha phase, 481, 482, 483, 484
Alpha precipitation, 252, 253, 254, 255, 256, 257, 356, 358-362
Alpha Titanium-Hydrogen alloys, 109-115
    dynamic strain aging, 114
    flow stress, 112, 113
    hydrogen-induced softening, 109, 111, 112, 113, 114, 115
    mechanical anisotropy, 113, 114
Alpha-Beta alloys, 350-357, 361, 421, 422, 423, 426
    cavitation, 28
    extrusion defects, 34
    flow-through defects, 34
    globularization kinetics, 7
    globularization mechanisms, 10
    hot workability, 28
    ingot breakdown, 5
    ingot production, 4
    ingot structure, 5
    laps, 33
    secondary hot working, 13
    shear banding, 30
    strain-induced porosity, 28
    superplasticity, 20
    texture development, 16
    warm working, 25
    wedge cracking, 28
Alpha-Beta rolling, 452
Alpha-Two precipitation, 351, 352
Alpha-Two Titanium Aluminide alloys
    ingot breakdown, 38
    secondary hot working, 41
    sheet rolling, 41
    superplasticity, 41
    texture development, 42
Analysis
    Kikuchi pattern analysis, 459, 460
Annealing, 169, 171, 350-352, 362
Autoclaves, 549, 550

B

Beta alloys, 35, 233, 234, 240, 352, 356-364
Beta anneal, 353, 356
Beta rolling, 452
Beta titanium, 117-123
Bio material, 586
Biomedical implants, 283, 284
Bonding, 539-542, 585, 599

C

Casting, 631, 632, 635, 637, 639, 640, 641, 642
Cavitation, 28, 63, 163-166
Cavity growth rate, 165-166
Ceramic, 585, 586, 587
Clad, 539, 549, 550, 551, 552
Coating, 585, 586, 587, 588, 594
Cold extrusion, 271, 272, 273, 274, 276, 277, 279
Cold forming, 211, 212, 213, 214
Cold heading, 320, 325
Cold processing, 283, 284, 285, 288, 289
Cold roll plus age, 362, 363, 365
Cold roll plus solution treat plus age, 362, 363
Cold workability, 293, 294, 296, 299
Cold working, 241, 246, 249
Complex shapes, 369, 377
Composite, 293

Composites
  Ti-6%Al-4%V matrix, 185-192
  TiC reinforcement, 185-192
Compression tests, 201, 202, 204, 206, 241, 243
Computer aided engineering, 125, 126, 127, 131
Computer-Aided Design (CAD), 647, 648
Computer-Aided Manufacturing (CAM), 648
Computerized manufacturing system, 652
Computerized system, 647, 648, 652
Concurrent engineering, 125, 127, 128
Constitutive equations, 201, 203
Controlled-dwell extrusion, 54
Cooling curves, 369, 371
Cooling rate (from ST temperature) effect on properties, 353, 355, 356
Corona 5®, 451
Corrosion, 469, 549
Coupling, 539
CP Titanium, 201, 202
Crack closure, 383
Crack growth
  macrocracks, 379-386
  microcracks, 379-386
Creep cavities, 507, 508, 512
Creep resistance, 353, 355
Creep strain, 507, 508, 509
Creep test, 604, 605
Creep, 605, 606, 616, 621
Cryogenic machining, 309, 310, 315
Crystallography
  BCC, 459, 460
  crystallographical feature, 459, 460, 461, 465, 466
  crystallography of α precipitate, 459, 460, 461, 465, 466
  HCP, 459, 460
Cutting force, 339, 340, 341, 344

D

Deformation mechanics, 125, 128, 138
Deformation, 169, 171, 172 233, 240, 259, 260, 262, 331, 333, 517, 518, 519, 521
Dental restoration, 585, 586, 593
Diffusion bonding, 527, 528, 529

Dissimilar metals, 539, 540
Dissimilar titanium alloys, 527
Documents, 647
Duplex anneal, 353
  Ti-6Al-2Sn-4Zr-2Mo, 353
  Ti-6Al-4V, 353
Duplex microstructures, 608, 609, 612, 613, 614, 615
Dynamic recrystallization, 159

E

Electron beam, 623, 624
Equal channel angular extrusion, 56
Equiaxed microstructure, 169, 170, 173, 174
Evaporation, 623, 624, 625, 626, 627, 629
Explosion welding, 539-546
Explosive clad, 549, 550, 551, 552
Extrusion defects, 34

F

Failure, 585, 593
Fasteners, 319, 320, 325
Fatigue, 307, 639, 641, 644, 645
Fatigue crack growth
  arrest, 401-403
  interfacial reactions, 402, 403
  microstructure, 397, 401-403
Finite element, 309, 310, 311, 315
Flow behavior, 405, 411
Flow localization, 93, 94
  bulging, 94, 96
  cracking, 93, 96, 97
  shear bands, 93, 94, 97
Flow-through defects, 34
Foil processing, 293, 294, 296, 299
Foils, 219, 220, 224
Formability, 259, 260, 268
Fractography, 576-582
Fracture toughness, 457
Fracture, 225, 229, 230, 259, 264, 489, 493, 494

G

Gamma Titanium Aluminide, 219, 220, 224, 387, 389, 390, 498, 631, 637
  banding, 160
  B effects, 154
  homogenization, 155, 157
  wrought processing, 153-160
Gamma Titanium Aluminide alloys
  cavitation, 63
  controlled dwell extrusion, 54
  conventional hot extrusion, 54
  conventional hot forging, 53
  equal channel angular extrusion, 56
  globularization kinetics, 7, 48
  homogenization, 47
  hot workability, 60
  ingot breakdown, 48
  ingot microstructure, 45
  ingot production, 45
  isothermal forging, 48, 58
  recrystallization kinetics, 51
  shear banding, 64
  sheet rolling, 56
  superplastic sheet forming, 58
  texture development, 59
  wedge cracking, 60
Glass lubrication
  shape of glass pad, 87
  mathematical model, 90-91
Globularization mechanisms, 10, 48
Globular structure, 197-198
Grain boundary
  grain boundary orientation, 459-461, 465, 466
  grain boundary plane, 459-466
  grain boundary precipitate, 459-461, 463-466
Grain growth behavior, 387, 393, 394
Grain growth, 502

H

Hall-Petch relationship, 621
Hardness, 572, 573
Heat treatment, 331, 332, 335, 349-367, 369, 374, 375, 421-426, 517, 522, 561, 562, 564, 572, 576, 577, 582, 604, 605
  aging, 474-475
  α/β-anneal, 474-475
  β-anneal, 474-475
  β-transus, 397, 399, 403
  interfacial reactions, 397, 399-401, 403
  microstructure, 397, 399, 403
Heat treatments, 497
Heating cycles, 369, 371
Heating rate
  to aging temperature, 358
  to solution treat temperature, 362, 363
Holes, 301, 302, 306, 307
Homogenization, 47
Hot Deformation
  microstructure, 104-108
  m-values, 102-103
  stress-strain curve, 102-104
  tensile tests, 102-104
Hot extrusion, 54
  standard flat faced radiused die, 84
  metal flow
    plane strain FEM simulation of metal, 88-90
    flow, 88-90
  improvement in tolerances, 92
  flow stress curves of Ti-6Al-4V, 85
Hot isostatic pressing, 387, 389, 391, 392, 393
Hot rolling, 201, 202
Hot workability, 28, 60, 155
Hot working process design, 125, 126, 128, 139
Hot working, 35, 201, 202
Hydrogen processing, 517, 520
Hydrogen, 331, 332, 333, 334, 335, 339, 340, 341, 342
Hydrogen-induced hardening, 119-120, 123

I

IMI 550®, 489, 490
Induction heating of billets
  mathematical modeling, 85-86
  temperature profiles, 86
Ingot breakdown, 5, 35, 36, 38, 44, 48
Ingot production, 4, 45
Ingot structure, 5, 45
Ingots, 201, 202

Integrated product definition or design (IPD), 125, 126, 129, 131
Interfaces, 507, 514, 539-542, 544-546
Isothermal decomposition, 481, 482
Isothermal forging, 20, 48, 58

K
Knowledge intensive design, 125, 136, 138

L
Lamellar microstructure, 169, 170, 173, 174
Lamellar refinning, 617, 618, 619
Lamellar structure, 194-196
Laps, 33
Laser welding, 561, 562, 563, 564, 572, 576, 577, 582
Linked models, 125, 126, 128, 140
Liquid metal, 468
Liquid nitrogen cooling, 309, 310, 313, 314, 315
Localized shear, 225, 229

M
Machining, 309, 310, 339, 340, 344
Manufacturing, 211, 212, 213, 214, 319, 320, 325
Mechanical alloying, 387, 389, 392, 393
Mechanical twining, 606, 609, 610, 611, 616, 621
Mechanically alloyed, 405-411
Melting, 623, 624, 625
Metal matrix composite (MMC), 271, 272, 276
Metastable Beta alloys, 421, 422, 424, 425, 426
Microstructural characterization, 498
Microstructure, 422-426, 435, 436, 437, 438, 439, 489, 491, 492, 563-571
    α-phase, 475-477
    β-phase, 475-477
    bi-modal, 379
    necklace type, 379
Microstructure evolution, 162-165
Mill anneal, 350-352, 362
Modeling, 75-82, 205, 207, 369, 370, 371, 374, 375, 623, 624, 630

Ti-6242®, 76-81
    microstructure, 75-82
Morphology
    morphological feature, 459, 460, 465, 466
    morphology of α precipitate, 459, 460, 465, 466

N
Nanocrystalline, 405-411
Nanograins, 387, 388, 389, 390, 391, 392, 393
Near Beta Titanium alloys
    β-CEZ®, 379

O
Omega phase, 481, 482, 483
Omega precipitation, 358-361
Ordering
    Ti$_3$Al, 418-420
        diffraction pattern, 418
Orientation
    a near-$\{1\bar{1}01\}\alpha$ // $\{110\}\beta$ orientation relationship (OR), 459, 462, 463, 466
    Burgers OR, 459, 461, 463, 466
    grain boundary orientation, 459-466
    near-Burgers OR, 459, 462, 463, 466
    orientation relationship (OR), 459-466
Orthorhombic titanium aluminide alloys
    hot working, 43
    texture development, 44
Orthorhombic, 507, 508
Oxidation, 469, 470, 471
Oxygen effects on tensile toughness, 453, 456, 457
Oxygen, 435, 436, 437, 438, 439, 489, 490, 492

P
Peritectic solidification, 154
Permanent mold casting, 631, 632, 634, 635, 637
Phase
    α phase, 459-466
    β phase, 459-466
    grain boundary (GB) phase, 459-466
Phase stability, 497

Phase transformation, 435, 436, 438, 439
Plasma melt overflow, 219, 220, 224
Porcelain, 586, 587, 594, 599
Pressure vessel, 549, 550, 551
Processing, 331, 333, 334, 335, 647
    through transus, 379
Punching test, 225, 226, 231

Q
Quench delay
    effect on properties (Ti-6Al-4V and Sp-700), 356, 357

R
Reactors, 549
Recrystallization kinetics, 51
Recrystallization, 169, 171, 173, 174
Recrystallization anneal, 353
Refining, 623, 624
Residual work, 233, 240
Rolling, 219, 220, 224

S
Secondary hot working, 13, 37, 41
Segregation, 47, 154, 155, 156
Shear band, 241, 243
Shear bands, 30, 64, 225, 228, 229, 232
Sheet rolling, 41, 56
Sheet, 259, 260
Shot peen plus age, 362, 364
Simulation, 623, 624, 628, 629
Solution treat temperature
    effect on properties, 355, 356, 362, 363
Solution treat, 355, 356
Solution treated, 242, 259, 260
Specialized heat treatments, 361-365
Specification, 647, 648, 649, 650, 651
Springs, 211, 212, 213, 214
Starting microstructure, 93, 96, 97, 99, 100
    α grains, 93, 94, 99
    Widmanstätten, 93, 94, 97, 98, 99, 100
Steady-state flow, 120
    activation parameters, 120-123
        Gibbs energy, 120, 121
        activation volume, 122, 123
Strain hardening, 271, 273
Strain-induced porosity, 28

Stress corrosion
    Ti-6Al-4V, 415-418
        continuous cooling, 415, 416
        TTT curve, 417, 418
Stress relief, 350, 352
Stress-corrosion resistance
    Ti-6Al-4V, 351, 352
Strip cast, 219, 220, 221, 224
Structure formation
    processes
        globularization, 197, 198, 199
        polygonization, 197, 198
        recrystallization, 194, 195, 197-199
        spheroidization, 197, 198, 199
Structure regulation
    deformation modes, 198, 199
Subgrain boundary, 616, 617
Superplasticity, 20, 41, 58, 161, 405, 406, 408, 409

T
Tantalum, 467
Temperature distribution, 309, 310, 311, 314
Tensile properties, 457
    ductility, 166-168
Tensile property, 572, 574, 575
Tensile test, 606
Tests
    ductility, 379-386
    fracture toughness, 379-386
        J-integral, 475-477
    macrocrack propagation, 379-386
    microcrack propagation, 379-386
    S-N curves, 379-386
    tensile properties
        yield strength, 475-477
    yield strength, 379-386
Texture development, 16, 42, 44, 59
Texture, 616, 620, 621
Thermohydrogen treatment, 639, 640, 641, 642, 643, 645
Thermomechanical processing (TMP), 125, 126, 127-128, 135
    composites, 185-192
        flow stress, 185, 187, 189
        densification, 185, 190, 192
        microstructure, 185, 190-192

Ti-13Nb-13Zr, 283
Ti-45.5Al-2Cr-2Nb, 47, 48, 161
Ti-6Al-2Sn-4Zr-2Mo, 7, 169
Ti-6Al-4V
    effects of heat treatment, 351-354, 356, 357
TiCuNi, 529
Timetal® 21S, 259, 260, 264, 268, 435, 436, 439, 481, 482
Timetal® LCB, 241, 360, 561, 562, 582
Titanium, 93-100, 421-426, 467
    $\alpha + \beta$ alloys, 101
    Ti-6Al-4V, 101
Titanium alloys, 125, 126, 127, 134, 135, 271, 272, 273, 275, 319, 320, 325, 339, 340, 517, 519
Titanium Aluminide, 293, 295, 299, 300, 405-411
Titanium forgings
    design, 145-147
    manufacture, 147-149
    mechanical properties, 150-152
Titanium-Hydrogen alloys, 117-123
    transus temperatures, 118
Tool life, 309, 310, 339, 340
Torsion test, 169, 171
Transient Liquid Phase (TLP), 527, 528
TTT diagram, 417, 418, 436, 439
    Ti-6Al-4V, 351
    Timetal® LCB, 360

V

Valves, 631, 632, 634, 637
Variant
    variant selection, 459, 464, 465, 466
Vent screens, 301, 302

W

Warm working, 25, 242, 243, 246, 250
Waterjet, 301, 302, 303, 304
Wedge cracking, 28, 60
Wire drawing, 283, 285, 289
Work hardening rate, 110-111
Work hardening, 246

X

X-ray diffraction, 497
XD® Gamma TiAl, 527-537

AUTHOR INDEX

A
Allen, P.G., 233, 561
Ameyama, K., 459

B
Baeslack III, W.A., 561
Bania, P.J., 233, 561
Banker, J.G., 539, 549
Barone, C., 527
Bieler, T.R., 603
Bingert, J.F., 467
Bird, R.K., 473
Bowen, P., 397
Boyer, R.R., 349, 379
Brand, K., 387
Brewer, W.D., 473
Briggs, R., 413
Broadwell, R.G., 143
Broderick, T.F., 259
Brown, I.G., 585
Brun, M., 193
Bugle, C.M., 283
Butt, D.P., 467

C
Chaudhury, P.K., 93
Chen, X., 387
Cotton, J.D., 467
Crooks, R., 481

D
Daigle, K.P., 283
Damkroger, B., 623
Damodaran, D., 83
Das, G., 527
Davidson, J.A., 283
Ding, Y., 309
Drew, R.A.L., 185
Dunn, P.S., 467

E
Egorova, Y.B., 339
Esposito, T., 83
Eylon, D., 169, 631

F
Fanning, J.C., 451
Feng, C.R., 435
Forrest, A.L., 549
Froes, F.H., 169, 169, 405
Fujiwara, H., 459
Furrer, D., 75

G
Galyon, K., 83
Gartside, M.B., 101
Gaspar, T.A., 219
Ghosh, A.K., 161
Guillard, S., 93

H
Hardwick, D.A., 153
Harley, J., 561
Hawbolt, E.R., 201
Hayes, R.W., 507
Hebeisen, J., 387
Hong, S.Y., 309
Howson, T.E., 143
Hunter, L.J., 489

I
Ilyin, A.A., 331, 517, 639
Imam, I.M., 435

J
Jackson, A., 241
Jonas, J.J., 109, 117
Jones, P.E., 631

K
Kawakami, H., 459
Keller, M.M., 631
Kolachev, B.A., 331, 339
Kollerov, M.Y., 517
Koren, M., 379
Kosaka, Y., 225
Kosin, J.E., 225
Krastilevsky, A.A., 517
Kuhlman, G.W., 125
Kusakina, Y.N., 639

L
Larsen, D.E., 603
LeClair, S.R., 241
Li, Q., 481
Linse, V.D., 539
Lombard, C.M,. 161, 497
Lozhko, A.N., 369
Lütjering, G., 349, 379

M
Malkin, G.Z., 369
Mamonov, A.M., 639
Margevicius, R.W., 467
Martin, P.L., 153, 451
Mishra, R.S., 405
Mitchell, A., 201, 459
Monteiro, O.R., 585
Mukherjee, A.K., 405
Mukhopadhyay, D.K., 387, 405

N
Nosov, V.K., 331, 517

O
Ovchinnikov, A.V., 517
Overcoglu, M.L., 387

P
Pepka, C., 211
Peters, J.O., 379
Phelps, H.R., 301
Poole, W.J., 201
Porter III, W.J., 631
Powell, A., 623
Puschnik, H., 379

R
Ray, K., 201
Reshad, J., 259
Rhodes, C.G., 507
Robare, E.W., 283
Roessler, R., 293
Rollins, J., 75
Russo, P.A., 421

S
Saqib, M., 241
Scvortsova, S.V., 517

Seetharaman, V., 3, 497
Semiatin, S.L., 3, 161, 259, 497
Senkov, O.N., 109, 117
Seo, D.Y., 603
Shachanova, G., 193
Shen, G., 75
Shivpuri, R., 83
Sokol, D.Z., 647
Srinivasan, R., 241, 259
Srisukhumbowornchai, N., 387
Stefansson, N., 241, 561
Strangwood, M., 489
Sukonnik, I.M., 219
Suryanarayana, C., 387, 405
Sweby, S.V., 397
Szekely, J., 623

T
Talaev, V.D., 339
Thirukkonda, M., 93

V
Van Den Avyle, J., 623
Volodin, V.A., 319
Vorobiov, I.A., 319

W
Wagener, H.W., 271
Wallace, T.A., 473
Wang, R.R., 585
Wanjara, P., 185
Weiss, I., 3, 169, 241, 259, 561
Welsch, G.E., 169, 585
Wojcik, C.C., 293
Wolf, J., 271
Wood, J.R., 421

Y
Yue, S., 185

Z
Zordan, R., 293